AF291061

Probability and Its Applications

Published in association with the Applied Probability Trust
Editors: S. Asmussen, J. Gani, P. Jagers, T.G. Kurtz

Probability and Its Applications

The *Probability and Its Applications* series publishes research monographs, with the expository quality to make them useful and accessible to advanced students, in probability and stochastic processes, with a particular focus on:

- Foundations of probability including stochastic analysis and Markov and other stochastic processes
- Applications of probability in analysis
- Point processes, random sets, and other spatial models
- Branching processes and other models of population growth
- Genetics and other stochastic models in biology
- Information theory and signal processing
- Communication networks
- Stochastic models in operations research

For further volumes:
http://www.springer.com/series/1560

Radu Zaharopol

Invariant Probabilities of Transition Functions

 Springer

Radu Zaharopol
Ann Arbor
Michigan
USA

ISSN 1431-7028   Probability and Its Applications
ISBN 978-3-319-05722-4   ISBN 978-3-319-05723-1 (eBook)
DOI 10.1007/978-3-319-05723-1
Springer Cham Heidelberg New York Dordrecht London

Library of Congress Control Number: 2014942551

Mathematics Subject Classification (2010): 47A35, 37A30, 37A17, 60J25, 37A50, 37A10, 28D10

Printed on acid-free paper

Springer is part of Springer Science+Business Media (www.springer.com)

Acknowledgements

It is a great pleasure to acknowledge the various kinds of support that I received from the following people: Eduard Emel'yanov showed interest in both my monograph [143] and this book by making a series of useful comments including the improvement in the Lasota-Yorke lemma that appears in Theorem 1.2.3; Thomas Hempfling has shown interest in the monograph and has offered me expert guidance throughout the publication process; Marius Iosifescu has always been available for extremely useful advice; he also invited me to the $9^{\text{ème}}$ Colloque Franco-Roumain de Math. Appl., 28 Août-2 Septembre 2008, Braşov, Romania, to give a talk on some of the topics discussed here; Marina Reizakis and her team at Springer have expertly prepared the book for publication; Daniel W. Stroock brought to my attention information concerning ergodic measures that appear in his books [119] and [29]; Tomasz Szarek sent me several of his works that I have used here; Daniël Worm showed great interest in the results on the KBBY decomposition that I obtained in this book and elsewhere (as I point out several times, he has extended the KBBY decomposition to Polish spaces and has obtained various other very interesting results related to the decomposition); and lastly Marina Zaharopol has offered me very useful advice.

Contents

Introduction

There is something funny about the title of this book: transition functions appear in the study of continuous-time Markov processes and these transition functions are usually assumed to be part of the theory of such Markov processes.

However, if we browse through the main matter of the book (the chapters and the appendices), we notice that, except for transition functions, there is no mention of anything related to continuous-time Markov processes. Further browsing reveals that all the applications of the results obtained in the book deal with ergodic theory and dynamical systems. Hence, the reader must be wondering what is going on here.

The results discussed herein do apply to the transition functions that appear in the theory of continuous-time Markov processes (in the sense that, in theory, the results can be applied to most of these transition functions). As people who work in the theory of continuous-time Markov processes know well, it is usually impossible to explicitly find the transition functions; therefore, roughly speaking, when dealing with such Markov processes, the point is to obtain information about the processes without having information about the transition functions.

By contrast, when dealing with flows, semiflows, and one-parameter convolution semigroups of probability measures, it is very easy to explicitly obtain their transition functions and to use these transition functions to illustrate our results.

That is why, in spite of the title, the book looks more like a work in ergodic theory. However, it should be emphasized that, as it stands now, the situation is very likely to change because it seems that one of the most tractable current open problems in the area of research discussed in this book is how to obtain the results described herein in terms of resolvents and/or generators rather than transition functions. This problem is described (together with other open problems) in Section 4 of [148] (the problem itself appears in Sect. 4.2). Daniel Worm and I have already started to work in this direction in [134], where, as a by-product, we also obtain an extension and a strengthening of a theorem of Hunt discussed in Heyer's monograph [45].

The main reason for writing this book was to offer a complete (and, of course, rigorous) presentation of the ergodic decomposition for transition functions defined on locally compact separable metric spaces. For transition probabilities defined on such spaces, the ergodic decomposition was obtained and discussed in detail in [146].

In order to explain the importance of and the issues concerning the ergodic decomposition, we will first consider transition probabilities, and then we discuss the transition function case.

Let (X, d) be a locally compact separable metric space, let $\mathcal{B}(X)$ be the Borel σ-algebra on X, and let μ be a probability measure on $(X, \mathcal{B}(X))$.

A measurable function $\alpha : X \to X$ is said to be *measure preserving* if $\mu(A) = \mu(\alpha^{-1}(A))$ for every $A \in \mathcal{B}(X)$.

Such a map generates a unique transition probability, say P (see Sect. 1.1 for details), and the fact that α is measure preserving means precisely that μ is an invariant probability for P (see Sect. 1.2).

The measure μ is said to be *ergodic* if *there do not exist* two nonzero mutually singular measures ν_1 and ν_2 on $(X, \mathcal{B}(X))$ such that $\mu = \nu_1 + \nu_2$ and such that α is measure preserving with respect to both ν_1 and ν_2 (that is, such that $\nu_i(A) = \nu_i(\alpha^{-1}(A))$ for every $A \in \mathcal{B}(X)$ and every $i = 1, 2$).

Given the map α, we can consider a linear positive contraction $S : B_b(X) \to B_b(X)$ defined by $Sf = f \circ \alpha$ for every $f \in B_b(X)$, where $B_b(X)$ is the Banach space of all real-valued bounded Borel measurable functions on X.

In 1931, Birkhoff proved that for every $f \in B_b(X)$, the sequence of function averages $\left(\dfrac{1}{n} \displaystyle\sum_{k=0}^{n-1} f(\alpha^k(x)) \right)_{n \in \mathbb{N}}$ converges μ-a.e.; that is, the sequence $\left(\dfrac{1}{n} \displaystyle\sum_{k=0}^{n-1} S^k f \right)_{n \in \mathbb{N}}$ converges μ-a.e. If, in addition, μ is an ergodic invariant probability measure, then, for every $f \in B_b(X)$, the limit of $\left(\dfrac{1}{n} \displaystyle\sum_{k=0}^{n-1} f(\alpha^k(x)) \right)_{n \in \mathbb{N}}$ (or of $\left(\dfrac{1}{n} \displaystyle\sum_{k=0}^{n-1} S^k f \right)_{n \in \mathbb{N}}$) exists, and is constant μ-a.e. (For details on Birkhoff's theorem, see, for instance, Mañé's monograph [70].)

Now, let us look at Birkhoff's theorem from another point of view: let $\alpha : X \to X$ be a measurable function (in the sense, of course, that $\alpha^{-1}(A) \in \mathcal{B}(X)$ for every $A \in \mathcal{B}(X)$), but we do not consider any measure on the measurable space $(X, \mathcal{B}(X))$. Then, using α, we can still define a transition probability P and the operator $S : B_b(X) \to B_b(X)$ as we did before. It can be shown that if P has invariant probability measures, then P also has ergodic invariant probability measures.

Assume that P has ergodic invariant probability measures, and let $\{\mu_\theta \mid \theta \in \Theta\}$ be the collection of all distinct ergodic invariant probability measures of P. Then Birkhoff's theorem tells us that given $\theta \in \Theta$, for every $f \in B_b(X)$, the limit of $\left(\dfrac{1}{n} \displaystyle\sum_{k=0}^{n-1} f(\alpha^k(x)) \right)_{n \in \mathbb{N}}$ (or of $\left(\dfrac{1}{n} \displaystyle\sum_{k=0}^{n-1} S^k f \right)_{n \in \mathbb{N}}$) exists and is constant μ-a.e. (the constant depending on f, of course). This means that it is likely that there exists a vector subspace $\mathcal{A}$ of $B_b(X)$ sufficiently large and that there is a way to define a collection $\{A_\theta \mid \theta \in \Theta\}$ of nonempty mutually disjoint subsets of X such that for every $\theta \in \Theta$ and $f \in \mathcal{A}$, the limit of $\left(\dfrac{1}{n}\sum\limits_{k=0}^{n-1} f(\alpha^k(x))\right)_{n\in\mathbb{N}}$ does exist and depends on θ only, in the sense that for every $\theta \in \Theta$ and every $x \in A_\theta$, the limit of $\left(\dfrac{1}{n}\sum\limits_{k=0}^{n-1} f(\alpha^k(x))\right)_{n\in\mathbb{N}}$ exists, depends on θ and is independent of x.

Thus, as soon as Birkhoff's 1931 theorem became available, people working in ergodic theory tried to obtain a satisfactory approach to the following problem:

Let $\alpha : X \to X$ be a measurable map, let P and S be as above, assume that P has invariant ergodic probability measures, and let $\{\mu_\theta \mid \theta \in \Theta\}$ be the collection of all distinct invariant ergodic probability measures of α. Define in a natural manner the subspace $\mathcal{A}$ and a partition $(B_\rho)_{\rho\in\Lambda}$ of X such that for every $\theta \in \Theta$, there exists a $\rho \in \Lambda$ such that $A_\theta = B_\rho$.

Such a partition $(B_\rho)_{\rho\in\Lambda}$ is called the *ergodic decomposition of X* defined by α (or by P).

People realized that it is likely that even more general transition probabilities might define ergodic decompositions.

The first significant results in the direction of articulating ergodic decompositions defined by transition probabilities were obtained in the pioneering papers by Kryloff and Bogoliouboff [54], Beboutoff [9], and Yosida [135, 136] (see also Section 4 of Chapter 13 of Yosida's monograph [138]). In view of the above-mentioned works, we decided to call the ergodic decomposition discussed in [146] and in this book the KBBY decomposition.

Yosida's approach to the ergodic decomposition in his monograph [138] is the most general among the works [9, 54, 135, 136] and [138] mentioned in the previous paragraph. Still, Yosida's decomposition in [138] is valid only under significant restrictions (for instance, if a transition probability P is induced by a map $\alpha : X \to X$, in order for the results of Section 4 of Chapter 13 of [138] to be valid for P, it is necessary that α be continuous and that $\alpha^{-1}(K)$ be a compact subset of X whenever $K \subseteq X$ is compact).

In [146] I proved that any transition probability on a locally compact separable metric space (X, d) defines a KBBY ergodic decomposition of X, provided that $\mathcal{A} = C_0(X) =$ the Banach subspace of $B_b(X)$ of all continuous functions that vanish at infinity.

By the time the paper [146] was submitted for publication, I had already obtained the results announced in Section 7 of [146] concerning the ergodic decomposition defined by transition functions on locally compact separable metric spaces. These results are discussed in detail in Chaps. 5 and 6 of this monograph. As far as we know, even though people working in ergodic theory believed that the ergodic decomposition is probably valid for many flows and semiflows, no one had any idea how to articulate the decomposition for (the transition functions defined by) any of the various flows and semiflows of interest, and no one even thought about obtaining the decomposition for more general transition functions.

After the paper [146] appeared in print, D. T. H. Worm and S. C. Hille obtained in [131] the KBBY ergodic decomposition for transition probabilities defined on Polish spaces. Even though at this time certain minor aspects of the decomposition cannot be detailed in the more general case of Polish spaces, it is important to realize that the extension to these spaces is highly nontrivial because for many nonlocally compact Polish spaces (X, d) it is true that the Banach space $C_0(X)$ of all real-valued continuous functions on X that vanish at infinity is the zero space, so, for such a Polish space (X, d), our approach to the KBBY decomposition in [146] cannot be used.

Soon after they obtained the results of [131], D. T. H. Worm and S. C. Hille studied in [132] the KBBY ergodic decomposition defined by a very general type of transition function on Polish spaces. In order to obtain the decomposition defined by the transition functions, they succeeded in associating to each transition function a transition probability which has the property that it has the same ergodic decomposition as the transition function; therefore, the study of the KBBY decomposition of the transition function can be replaced by the study of the decomposition defined by the transition probability.

For other results on the KBBY ergodic decomposition in Polish spaces, see the thesis [130] of D. T. H. Worm, and the papers of D. T. H. Worm and S. C. Hille [133] and of T. Szarek and D. T. H. Worm [122].

A very interesting and complete study of the supports of ergodic invariant probability measures (employing results obtained in Chapter 2 of [143]) and of the KBBY decomposition (using results of [146], Worm and Hille [131], and Worm's thesis [130]) of certain transition probabilities appears in a recent paper of Alkurdi, Hille and van Gaans [4]. The transition probabilities under consideration in [4] stem from discrete-time Markov processes used to model impulsive random interventions in a dynamical system that is one-dimensional and deterministic.

There is another approach to the ergodic decomposition. This approach has limitations, but allows for a more direct use of Birkhoff's result. For measurable transformations, the decomposition appeared in the pioneering work of Rohlin [101]. The decomposition was extended nicely to transition probabilities on locally compact separable metric spaces to transition probabilities on locally compact separable metric spaces by Hernández-Lerma and Lasserre in [42] (see also Chap. 5 of their monograph [43]). In order to describe their approach, let $\alpha : X \to X$, S and P be as above. Let $\mathcal{M}(X)$ be the Banach space of all real-valued signed Borel measures on X. It can be shown that there exists a unique positive contraction $T : \mathcal{M}(X) \to \mathcal{M}(X)$ such that, for every $x \in X$, $T\delta_x$ is a probability measure defined by $T\delta_x(A) = \mathbf{1}_A(\alpha(x))$ for every Borel subset A of X, where δ_x is the Dirac measure concentrated at x. Assume that P has invariant probability measures, and let $\mu \in \mathcal{M}(X)$ be such an invariant probability (fixed throughout the discussion). Hernández-Lerma and Lasserre showed that there exists a Borel measurable subset Z of X such that $\mu(Z) = 1$ and for every $x \in Z$

there exists a weak* accumulation point ϕ_x of the sequence $\left(\dfrac{1}{n}\displaystyle\sum_{k=0}^{n-1} T^k \delta_x\right)_{n\in\mathbb{N}}$ such that ϕ_x is an invariant ergodic probability measure for P. Then they obtained an ergodic decomposition defined by these measures ϕ_x, $x \in Z$. The problem with this approach is that not all invariant ergodic probability measures of P are obtained in this manner. In the extreme case that the measure μ is itself ergodic, we obtain that $\phi_x = \mu$ for every $x \in Z$, even though P might have infinitely many distinct invariant ergodic probability measures.

In this book (in Chaps. 5 and 6) we show that under very general conditions, transition functions defined on locally compact separable metric spaces admit KBBY decompositions surprisingly similar to the KBBY decomposition for transition probabilities.

In Chap. 7 we discuss various results dealing with Feller transition functions, and, in particular, with Feller transition functions equicontinuous in the mean. Among the results that appear in the chapter are criteria for unique ergodicity and the continuous-time mean ergodic theorems announced in [147].

Chapters 5–7 contain essentially new results, except for Sect. 6.1, which contains preliminaries on invariant ergodic probability measures (that are known).

The focus of the book is on the new results presented in the last three chapters.

The organization of the volume is as follows: in the first chapter we survey various results on transition probabilities that we have obtained in the last 10 years and that are extended and/or used in this book.

Thus, this first chapter can be considered as an extended abstract of the entire book.

The next three chapters (Chaps. 2–4) contain preliminaries needed in the last three chapters. It should be mentioned here that some of the material discussed in Chaps. 2–4 are just surveys of well-known results from the literature, while others are new, and, in the author's opinion, sometimes quite interesting (such as Corollary 3.2.12, which is a consequence of a theorem of Dunford and Schwartz).

Chapter 5 and Sect. 6.2 contain the results on the KBBY decomposition announced in Section 7 of [146]. Section 6.3 (the last section of Chap. 6) discusses a connection between the KBBY decomposition for transition functions and the Choquet-Phelps theory of ergodic measures.

In the last chapter of the book (Chap. 7) we deal with Feller transition functions. Essentially, in this chapter we extend most of the results of [143] to Feller transition functions.

As explained at the beginning of the introduction, at this time, the only examples that can be used to illustrate the theory are essentially transition functions defined by flows, semiflows, and one-parameter convolution semigroups of probability measures. Since the flows, the semiflows, and the one-parameter convolution semigroups are usually quite sophisticated objects, we have written two appendices in which we try to introduce these objects as gently as possible; thus, for flows on spaces of cosets, we try to address a concern that has been stated in various publications by S. G. Dani and G. A. Margulis: namely, to write introductory materials that will allow a larger number of people to familiarize themselves with these beautiful areas.

In the two appendices we apply the principle used when teaching undergraduate calculus service courses (not for math majors): "Tell the truth, nothing but the truth, but not the whole truth." The difference between the application of the above principle to a service calculus course and the use of the principle in our appendices is that everything that we write is rigorous. But then, the question is: what kind of truth do you hide? Well, in this book, throughout the appendices and the chapters, we do not even mention that the natural setting for many of the examples (the flows, say) is in Lie groups.

The transition functions defined by the flows, the semiflows, and the one-parameter convolution semigroups of probability measures discussed in the appendices are defined in Chap. 2, and from then on, these transition functions are used to illustrate the theory.

The book is addressed to mathematicians who work in ergodic theory and/or dynamical systems and/or the theory of Markov processes. It is also addressed to people who are not mathematicians by title (they are often biologists, physicists, or economists who are interested in, say, interacting particle systems) but are mathematicians at heart (from a formal point of view, this means that they studied rigorous mathematics at some point in their life and that they are willing and able to read rigorous mathematics). As was the case with our 2005 Birkhäuser monograph, we have tried hard in this book to make the topics accessible to newcomers to the areas discussed.

There are at least two ways to systematically study the results discussed in this book. The first approach is to completely ignore the examples; in this case, the prerequisites for reading the book are almost the same as the prerequisites for reading our 2005 book (basic notions and results in functional analysis, general topology, and measure theory); the difference is that, in this book, we assume a willingness to "learn as you go" (or to actually know, of course) the Bochner integral (as discussed, for instance, in Appendix E, pp. 350–357 of Cohn's book [20] and the Dunford and Schwartz integral (that is, the extension of the Bochner integral discussed in the monograph [30] by Dunford and Schwartz).

The second approach consists of studying everything, including the examples. In this approach one has to know, or to be ready to "learn as you go" the topics discussed in the two appendices.

Let us conclude with a few words on the notation used in the book: as usual, the end of a proof is denoted by $\square$; to signify the end of an example, we use $\blacksquare$; and finally, $\blacktriangle$ is used to indicate the end of an observation or a group of observations.

Chapter 1
Preliminaries on Transition Probabilities

Our goal in this chapter is to briefly review several facts about transition probabilities. We will follow [146] and [143]. For additional information on transition probabilities, including many topics that we do not cover here, see, for instance, the monographs of Hernández-Lerma and Lasserre [43], Lasota and Mackey [57], Meyn and Tweedie [77], Revuz [97], and M. Rosenblatt [102].

There are several reasons for starting the book with a chapter on transition probabilities. As we will see in the next chapter in Sect. 2.1 (and as, I am sure, many readers know already), transition functions are families of transition probabilities, families which satisfy certain equalities known as the Chapman-Kolmogorov equations; thus, the transition probabilities that create a transition function can be thought of as building blocks that are glued together by the Chapman-Kolmogorov equations into the transition function, and, therefore, a good understanding of transition probabilities helps in understanding the transition functions. In particular, the fact that one or several transition probabilities that belong to a transition function has or have certain properties translates into relevant information about the transition function as a whole. Finally, the results on transition functions and their invariant probabilities that we discuss in this book have counterparts for transition probabilities that we obtained in [146] and [143] (but, of course, the proofs are very different, and, generally, more sophisticated when dealing with transition functions); hence, a review of these results for transition probabilities offers a good prelude for the topics that we discuss in this book.

In Sect. 1.1 we define the transition probabilities, the Markov pairs, the Feller transition probabilities, and the $C_0(X)$-equicontinuous transition probabilities. Also in this section we discuss several examples, which, like all the examples in this book, have only a didactic purpose in the sense that we use the examples to illustrate various results.

In Sect. 1.2 we define the invariant probabilities, the ergodic invariant probabilities, and the sets of maximal probability. Also in this section, we briefly review various operators that can be defined given a transition probability P and an invariant probability measure μ for P in order to recall a very useful consequence of

R. Zaharopol, *Invariant Probabilities of Transition Functions*, Probability and Its Applications 44, DOI 10.1007/978-3-319-05723-1_1,
© Springer International Publishing Switzerland 2014

the Hopf ergodic theorem. We end the section with a brief discussion of the Lasota-Yorke lemma, a result which is very useful when dealing with Feller transition probabilities or Feller transition functions.

In Sect. 1.3, we briefly review the Kryloff-Bogoliouboff-Beboutoff-Yosida (KBBY) decomposition for transition probabilities that we obtained in [146]. In Chaps. 5 and 6 in this book, we obtain the KBBY decomposition for a rather general type of transition function, and it is instructive to compare the results discussed in this section with the results obtained in Chaps. 5 and 6. It is of interest to point out that, as it stands today, due to the remarkable work by Worm and Hille [131], most of the results reviewed in Sect. 1.3 are valid in the more general setting of Polish spaces rather than the locally compact separable metric spaces that are under consideration in this book. Moreover, several results in Sect. 1.4.1 hold true in Polish spaces, as well.

In the last section of this chapter (Sect. 1.4) we briefly review various results dealing with supports of invariant probability measures for Feller transition probabilities, uniquely ergodic Feller transition probabilities, and equicontinuous (Feller) transition probabilities that appear in [143]. As in the case of the KBBY decomposition, in Chap. 7 of this book, we will compare the results obtained there with the corresponding results reviewed in Sect. 1.4. Several results in Sect. 1.4.2 have been obtained independently by Worm and Hille (see [133] and Section 7.3 of Worm [130]).

1.1 Basic Definitions and Results

Let $\mathbb{S}$ be a nonempty set and let $v : \mathbb{S} \to \mathbb{R}$ be a function.

As in the theory of vector lattices, we say that v is positive if $v(x) \geq 0$ for every $x \in \mathbb{S}$. We indicate that v is positive using the notation $v \geq 0$.

Note that the constant zero function is positive and that the set $\mathbb{R}$ of all real numbers can be thought of as the collection of all real-valued functions defined on a singleton (a *singleton* is a set that has exactly one element).

Let $a \in \mathbb{R}$. Since for us a is positive if $a = 0$ or $a > 0$, and since the words "positive number" are sometimes used to mean "strictly positive number", we will prefer the wording "a is a nonnegative number" instead of "a is a positive number". Of course, the above wording preference will also be used if a is, for example, a rational number or an integer.

The symbol $\mathbb{N}$ is reserved for the set of all natural numbers (that is, the set of all strictly positive integers).

Given a sequence $(a_n)_{n \in \mathbb{N}}$ of real numbers, we say that $(a_n)_{n \in \mathbb{N}}$ *diverges to* $+\infty$, or *to* $-\infty$, if $\lim_{n \to +\infty} a_n$ exists and is equal to $+\infty$, or to $-\infty$, respectively.

Let (X, d) be a locally compact separable metric space. (Throughout this book X stands for a locally compact separable metric space unless explicitly stated otherwise.) Also, let $\mathcal{B}(X)$ be the σ-algebra of all Borel subsets of X (that is, $\mathcal{B}(X)$ is the σ-algebra generated by the open subsets of X).

If $x \in X$ and $r \in \mathbb{R}$, $r > 0$, then we use the notation $B(x, r)$ for the open ball centered at x and of radius r in X; that is, $B(x, r) = \{y \in X \mid d(x, y) < r\}$.

For future reference, note that (X, d) is *second countable*; that is, it has a countable basis for the topology defined by the metric d (in other words, there exists an at most countable collection $\mathcal{U}$ of open subsets of X that has the following property: for every $x \in X$ and every neighborhood V of x, there exists a $U \in \mathcal{U}$ such that $x \in U \subseteq V$). For details, see Section D.32 of Appendix D, p. 348, of Cohn's book [20].

As usual, we say that a map $P : X \times \mathcal{B}(X) \to \mathbb{R}$ is a *transition probability* (on (X, d)) if the following two conditions are satisfied:

(*i*) For every $x \in X$, the map $\mu_x : \mathcal{B}(X) \to \mathbb{R}$ defined by $\mu_x(A) = P(x, A)$ for every $A \in \mathcal{B}(X)$ is a probability measure.

(*ii*) For every $A \in \mathcal{B}(X)$, the function $g_A : X \to \mathbb{R}$ defined by $g_A(x) = P(x, A)$ for every $x \in X$ is a Borel measurable function.

If Y is a nonempty set and $\mathcal{Y}$ is a σ-algebra of subsets of Y, we will denote by $B_b(Y, \mathcal{Y})$ the Banach space of all real-valued bounded measurable functions defined on the measurable space $(Y, \mathcal{Y})$, where the norm on $B_b(Y, \mathcal{Y})$ is the uniform (sup) norm: $\|f\| = \sup\limits_{y \in Y} |f(y)|$ for every $f \in B_b(Y, \mathcal{Y})$. If it is clear from the context which σ-algebra $\mathcal{Y}$ is under consideration, we will use the notation $B_b(Y)$ rather than $B_b(Y, \mathcal{Y})$. Thus, for instance, if Y is the (locally compact separable) metric space X that has been under consideration so far, then $B_b(X)$ stands for $B_b(X, \mathcal{B}(X))$ and is the Banach space of all real-valued bounded Borel measurable functions on X. We denote by $\mathcal{M}(Y, \mathcal{Y})$ the Banach space of all real-valued signed measures on $(Y, \mathcal{Y})$, where the norm on $\mathcal{M}(Y, \mathcal{Y})$ is the total variation norm. As in the case of $B_b(Y, \mathcal{Y})$, we use the notation $\mathcal{M}(Y)$ rather than $\mathcal{M}(Y, \mathcal{Y})$ whenever there is no doubt about which σ-algebra $\mathcal{Y}$ is under consideration. In particular, $\mathcal{M}(X)$ is the Banach space of all real-valued signed measures defined on $(X, \mathcal{B}(X))$.

If $\mu \in \mathcal{M}(Y, \mathcal{Y})$ then, as usual, we will denote by μ^+ and μ^- the positive and the negative part of μ, respectively. Thus, μ^+ and μ^- are measures (positive elements of $\mathcal{M}(Y, \mathcal{Y})$), and $\mu = \mu^+ - \mu^-$ by the Jordan decomposition theorem. The *variation* of μ is denoted by $|\mu|$; thus, $|\mu| = \mu^+ + \mu^-$. Of course the *total variation norm* $\|\mu\|$ of μ is defined by $\|\mu\| = |\mu|(Y)$. For additional details, see, for instance, Section 4.1 of Cohn [20].

If $f \in B_b(Y, \mathcal{Y})$ and $\mu \in \mathcal{M}(Y, \mathcal{Y})$, we will often use the notation $\langle f, \mu \rangle$ for $\int_Y f(x)\, d\mu(x)$.

Now, let $P : X \times \mathcal{B}(X) \to \mathbb{R}$ be a transition probability on (X, d).

We can use P to define two operators $S : B_b(X) \to B_b(X)$ and $T : \mathcal{M}(X) \to \mathcal{M}(X)$ as follows:

$$Sf(x) = \int f(y)\, d\mu_x(y) \quad \text{or} \quad Sf(x) = \int f(y) P(x, dy) \tag{1.1.1}$$

for every $f \in B_b(X)$ and $x \in X$, where μ_x, $x \in X$, are the probability measures that appear in condition (i) in the definition of a transition probability and $P(x, \mathrm{d}y)$ stands for $\mathrm{d}\mu_x(y)$, and

$$T\mu(A) = \int P(x, A)\,\mathrm{d}\mu(x) \tag{1.1.2}$$

for every $\mu \in \mathcal{M}(X)$ and $A \in \mathcal{B}(X)$.

Note that the maps S and T are well-defined in the sense that Sf does indeed belong to $B_b(X)$ for every $f \in B_b(X)$, and that $T\mu$ is a real-valued signed Borel measure for every $\mu \in \mathcal{M}(X)$.

The operators S and T have the following easy to prove properties:

- Both S and T are linear bounded operators; moreover, these operators are positive in the sense that $Sf \geq 0$ for every $f \in B_b(X)$, $f \geq 0$, and $T\mu \geq 0$ whenever $\mu \in \mathcal{M}(X)$, $\mu \geq 0$;
- The operator T is a contraction (that is, $\|T\| \leq 1$), and is even a Markov operator (i.e. $\|T\mu\| = \|\mu\|$ for every $\mu \in \mathcal{M}(X)$, $\mu \geq 0$);
- Like T, the operator S is a contraction, as well;
- The constant one function $\mathbf{1}_X : X \to \mathbb{R}$ (defined by $\mathbf{1}_X(x) = 1$ for every $x \in X$) has the property that $S\mathbf{1}_X = \mathbf{1}_X$;
- The operators S and T satisfy the equality

$$\langle Sf, \mu \rangle = \langle f, T\mu \rangle \tag{1.1.3}$$

for every $f \in B_b(X)$ and $\mu \in \mathcal{M}(X)$.

We refer to the ordered pair (S, T) as the *Markov pair defined* (or *generated*) *by* P.

If $A \in \mathcal{B}(X)$, we will denote by $\mathbf{1}_A$ the *characteristic function* (also known as the *indicator function*) *of A;* that is, $\mathbf{1}_A$ is the real-valued function on X defined by

$$\mathbf{1}_A(x) = \begin{cases} 1 & \text{if } x \in A \\ 0 & \text{if } x \in X \setminus A. \end{cases}$$

As usual a function is said to be *simple* if it has only finitely many values.

If $x \in X$, we will use the notation δ_x for the *Dirac (probability) measure concentrated at x;* that is, δ_x is the probability measure that belongs to $\mathcal{M}(X)$ and has the property that $\delta_x(\{x\}) = 1$.

In the next lemma we collect a few formulas that will be used often in the book.

Lemma 1.1.1. *Let (S, T) be the Markov pair defined by P. Also, let $x \in X$, $A \in \mathcal{B}(X)$, and $f \in B_b(X)$. Then:*

(i) $S\mathbf{1}_A(x) = P(x, A);$
(ii) $T\delta_x(A) = P(x, A);$
(iii) $Sf(x) = \langle Sf, \delta_x \rangle = \langle f, T\delta_x \rangle.$

Proof. Using the equalities (1.1.1) and (1.1.2) of this section, we obtain (*i*) and (*ii*), respectively.

In order to obtain (*iii*), we use first the definition of δ_x, and then the equality (1.1.3) of this section. $\square$

Note that using (*i*) or (*ii*) of the above lemma, we obtain that if (S, T) is a Markov pair defined by a transition probability P, then P is the unique transition probability that generates (S, T); that is, if P and P' are two transition probabilities that define the same Markov pair (S, T), then $P = P'$.

Given the transition probability P, one can recursively define a sequence $(P_n)_{n\in\mathbb{N}}$ of maps, $P_n : X \times \mathcal{B}(X) \to \mathbb{R}$ for every $n \in \mathbb{N}$, as follows: set $P_1 = P$, and, assuming that P_n has been defined, set $P_{n+1}(x, A) = \int_X P_n(y, A)P(x, dy)$ for every $n \in \mathbb{N}$, $x \in A$, and $A \in \mathcal{B}(X)$.

It is easy to see that P_n is a transition probability for every $n \in \mathbb{N}$. Following M. Rosenblatt [102], we will call P_n the nth step transition probability defined by P, $n \in \mathbb{N}$.

Let (S_n, T_n) be the Markov pair defined by P_n, $n \in \mathbb{N}$. Clearly, $(S_1, T_1) = (S, T)$, where (S, T) is the Markov pair defined by P.

The next proposition deals with two basic properties of the sequences $((S_n, T_n))_{n\in\mathbb{N}}$ and $(P_n)_{n\in\mathbb{N}}$.

Proposition 1.1.2. (*a*) $S_n = S^n$ and $T_n = T^n$ for every $n \in \mathbb{N}$.
(*b*) $P_{m+n}(x, A) = \int_X P_m(y, A)P_n(x, dy)$ for every $m \in \mathbb{N}$, $n \in \mathbb{N}$, $x \in X$ and $A \in \mathcal{B}(X)$.

Proof. (*a*) We first prove that $S_n = S^n$ for every $n \in \mathbb{N}$. The proof is by induction on n.

Clearly, the assertion is true for $n = 1$.

Now, assume that the assertion is true for $n \in \mathbb{N}$. We have to prove that $S_{n+1} = S^{n+1}$.

Let $f = \mathbf{1}_A$ for some $A \in \mathcal{B}(X)$. Using (*i*) of Lemma 1.1.1, the definition of P_{n+1}, then again (*i*) of Lemma 1.1.1, the induction hypothesis, and the definition of S in this order, we obtain that

$$S_{n+1}\mathbf{1}_A(x) = P_{n+1}(x, A) = \int_X P_n(y, A)P(x, dy) = \int_X S_n\mathbf{1}_A(y)P(x, dy)$$

$$= \int_X S^n\mathbf{1}_A(y)P(x, dy) = S(S^n\mathbf{1}_A)(x) = S^{n+1}\mathbf{1}_A(x)$$

for every $x \in X$.

Since $S_{n+1}\mathbf{1}_A = S^{n+1}\mathbf{1}_A$ for every $A \in \mathcal{B}(X)$, and since both S_{n+1} and S^{n+1} are linear operators, it obviously follows that $S_{n+1}f = S^{n+1}f$ whenever $f \in B_b(X)$ and f is a simple function.

Taking into consideration that the set of all real-valued simple measurable functions on X is dense in $B_b(X)$ (with respect to the uniform (sup) norm, of course), and that both operators S_{n+1} and S^{n+1} are continuous, we obtain that $S_{n+1} = S^{n+1}$.

We now prove that $T_n = T^n$ for every $n \in \mathbb{N}$.

Using the equality (1.1.3) of this section, the fact that $S_n = S^n$ that we have just proved, and using again (1.1.3) n times, we obtain that

$$T_n \mu(A) = \langle \mathbf{1}_A, T_n \mu \rangle = \langle S_n \mathbf{1}_A, \mu \rangle = \langle S^n \mathbf{1}_A, \mu \rangle = \langle \mathbf{1}_A, T^n \mu \rangle = T^n \mu(A)$$

for every $\mu \in \mathcal{M}(X)$, $A \in \mathcal{B}(X)$, and $n \in \mathbb{N}$. Thus, $T_n \mu = T^n \mu$ for every $\mu \in \mathcal{M}(X)$, so $T_n = T^n$ for every $n \in \mathbb{N}$.

(b) Let $m \in \mathbb{N}$, $n \in \mathbb{N}$, $x \in X$, and $A \in \mathcal{B}(X)$. Then, using (i) of Lemma 1.1.1, the fact that $S_k = S^k$ for all $k \in \mathbb{N}$ that we proved in (a) of this proposition, the equality (1.1.1) of this section, and using again (i) of Lemma 1.1.1, we obtain that

$$P_{m+n}(x, A) = S_{m+n}\mathbf{1}_A(x) = S^{m+n}\mathbf{1}_A(x) = S^n(S^m\mathbf{1}_A(x)) = S_n(S_m\mathbf{1}_A(x))$$

$$= \int S_m \mathbf{1}_A(y) P_n(x, \mathrm{d}y) = \int P_m(y, A) P_n(x, \mathrm{d}y).$$

$\square$

The equality that appears in (b) in the above proposition is known as the *Chapman-Kolmogorov equation*. As we will see in the next chapter a "continuous-time" version of the equation is at the core of the definition of the transition function.

As is well-known, the discrete-time time-homogeneous Markov processes are intimately related to transition probabilities in the sense that, to every such Markov process, one can associate a transition probability, and, conversely, for every transition probability P one can construct a discrete-time time-homogeneous Markov process whose associated transition probability is P (for details, see Section 1.2 of Revuz's monograph [97]).

As usual we denote by $\mathrm{supp}\,\mu$ the support of μ, where $\mu \in \mathcal{M}(X)$. Note that since we assume that X is a locally compact separable metric space, the supports of the elements of $\mathcal{M}(X)$ are well defined (see p. 226 of Cohn's book [20]).

Given $x \in X$, the set $\bigcup_{k=0}^{\infty} \mathrm{supp}(T^k \delta_x)$ is called the *orbit of x under the action of T* (or *of P*, or *of (S, T)*) and is denoted by $\mathcal{O}_T^{(\mathrm{TP})}(x)$ or $\mathcal{O}_P^{(\mathrm{TP})}(x)$. If T or P are clearly understood from the context, we will drop the subscript, and the orbit will be denoted simply by $\mathcal{O}^{(\mathrm{TP})}(x)$. The closure $\overline{\mathcal{O}^{(\mathrm{TP})}(x)}$ of $\mathcal{O}^{(\mathrm{TP})}(x)$ is called the *orbit-closure of x* (*under the action of T* (or *of P*, or *of (S, T)*)). As we will discuss soon, if P and (S, T) are defined by a measurable transformation (see the comments preceding Example 1.1.6), then the orbits and orbit-closures defined here are precisely the usual forward orbits and forward orbit-closures as defined in ergodic theory, dynamical systems, and topological dynamics (see, for example, the monographs by Furstenberg [38], Gottschalk and Hedlund [40], Mañé [70], and Rudolph [104]). We say that T (or (S, T) or P) is *minimal* if every orbit under the action of T is dense in X (that is, if $\mathcal{O}_T^{(\mathrm{TP})}(x)$ is dense in X for every $x \in X$). Finally, T (or (S, T), or P) is said to be *trivially minimal* if $\mathrm{supp}(T\delta_x) = X$ for every $x \in X$.

We will denote by $C_b(X)$ the Banach subspace of $B_b(X)$ of all real-valued continuous bounded functions defined on X. Thus, the norm on $C_b(X)$ is the uniform (sup) norm.

Given a transition probability P and the Markov pair (S, T) defined by P, we say that P is a *Feller transition probability* if $Sf \in C_b(X)$ whenever $f \in C_b(X)$. In this case, (S, T) is said to be a *Markov-Feller pair.*

The definition of a Markov-Feller pair given here is the same as on p. 53 of [146]. Note that, as pointed out there, even though the present definition is different from the definition given in Section 1.1 of [143], any result that is valid for the Markov-Feller pairs as defined here is valid for the Markov-Feller pairs as defined in Section 1.1 of [143] and vice versa.

As expected, the Feller transition probabilities and the corresponding Markov-Feller pairs have various interesting properties that are not true in the non-Feller case, in general. Let us point out now one such property that will be used later. The result is due to Lasota and Myjak (see Proposition 3.1 of [59]):

Proposition 1.1.3 (Lasota and Myjak). *Let P be a Feller transition probability on (X, d), and let (S, T) be the Markov-Feller pair defined by P. If $\mu \in \mathcal{M}(X)$, $\mu \geq 0$, and $\nu \in \mathcal{M}(X)$, $\nu \geq 0$, are such that $\mathrm{supp}\,\mu \subseteq \mathrm{supp}\,\nu$, then $\mathrm{supp}(T\mu) \subseteq \mathrm{supp}(T\nu)$.*

The above proposition has the following consequence (see Corollary 3.1 of Lasota and Myjak [59]):

Corollary 1.1.4. *Let P and (S, T) be as in Proposition 1.1.3, and let $\mu \in \mathcal{M}(X)$ and $\nu \in \mathcal{M}(X)$ be such that $\mu \geq 0$, $\nu \geq 0$, and $\mathrm{supp}\,\mu = \mathrm{supp}\,\nu$. Then $\mathrm{supp}(T\mu) = \mathrm{supp}(T\nu)$.*

We will denote by $C_0(X)$ the Banach subspace of $B_b(X)$ of all real-valued continuous functions on X that vanish at infinity. Thus, $C_0(X)$ as a Banach space in its own right is endowed with the uniform (sup) norm. Since every real-valued continuous function on X that vanishes at infinity is necessarily bounded, it follows that $C_0(X)$ is also a Banach subspace of $C_b(X)$.

It is well-known (see, for instance, Theorem 7.3.5, pp. 220–223 of Cohn's book [20]) that there exists an isometric isomorphism Θ from the topological dual $(C_0(X))^*$ of $C_0(X)$ (the *topological dual of a Banach space E* is the Banach space of all real-valued linear bounded functionals on E, and is usually denoted by E^*) onto $\mathcal{M}(X)$ which acts as follows: $\int_X f \, \mathrm{d}\Theta(\varphi) = \varphi(f)$ for every $\varphi \in (C_0(X))^*$ and $f \in C_0(X)$. In view of this fact, we will identify $(C_0(X))^*$ with $\mathcal{M}(X)$, and we will think of the elements of $(C_0(X))^*$ as elements of $\mathcal{M}(X)$ and vice versa, via the map Θ, but without mentioning Θ explicitly.

Let Y and Z be two real vector spaces, and let ψ be a real-valued function defined on $Y \times Z$.

We say that the ordered pair (Y, Z) is a *dual system with respect to* ψ if the following conditions are satisfied:

1. $\psi(ay_1 + by_2, z) = a\psi(y_1, z) + b\psi(y_2, z)$ for every $y_1 \in Y$, $y_2 \in Y$, $z \in Z$, $a \in \mathbb{R}$, and $b \in \mathbb{R}$.
2. $\psi(y, az_1 + bz_2) = a\psi(y, z_1) + b\psi(y, z_2)$ for every $y \in Y$, $z_1 \in Z$, $z_2 \in Z$, $a \in \mathbb{R}$, and $b \in \mathbb{R}$.
3. If $\psi(y, z) = 0$ for every $z \in Z$, then $y = 0$.
4. If $\psi(y, z) = 0$ for every $y \in Y$, then $z = 0$.

It is the custom (and we will adhere to it) to use the notation $\langle \cdot, \cdot \rangle$ for ψ, and $\langle y, z \rangle$ for $\psi(y, z)$, $y \in Y$, $z \in Z$.

Let (Y, Z) be a dual system with respect to $\langle \cdot, \cdot \rangle$.

Let $\sigma(Y, Z)$ be the collection of all the subsets U of Y which have the following property: for every $y \in U$ there exist $\varepsilon \in \mathbb{R}$, $\varepsilon > 0$, $n \in \mathbb{N}$, and n elements $z_1, z_2, \ldots, z_n$ of Z such that

$$\{y' \in Y \mid |\langle y', z_i \rangle - \langle y, z_i \rangle| < \varepsilon \text{ for every } i = 1, 2, \ldots, n\} \subseteq U.$$

It is easy to see that $\sigma(Y, Z)$ is a topology on Y. We call $\sigma(Y, Z)$ the *Z-weak topology of Y*. Note that a sequence $(y_n)_{n \in \mathbb{N}}$ of elements of Y is convergent in the Z-weak topology of Y if and only if there exists a $y \in Y$ such that the sequence $(\langle y_n, z \rangle)_{n \in \mathbb{N}}$ converges to $\langle y, z \rangle$ for every $z \in Z$. If $(y_n)_{n \in \mathbb{N}}$ converges to y in the Z-weak topology of Y, we say that $(y_n)_{n \in \mathbb{N}}$ is *Z-weakly convergent* (or *converges Z-weakly*); the limit y is called the *Z-weak limit of* $(y_n)_{n \in \mathbb{N}}$ and is denoted by $Z\text{-w-}\lim\limits_{n \to +\infty} y_n$.

Similarly, let $\sigma^*(Y, Z)$ be the collection of all the subsets V of Z with the following property: for every $z \in V$ there exist $\varepsilon \in \mathbb{R}$, $\varepsilon > 0$, $n \in \mathbb{N}$, and n elements $y_1, y_2, \ldots, y_n$ of Y such that

$$\{z' \in Z \mid |\langle y_i, z' \rangle - \langle y_i, z \rangle| < \varepsilon \text{ for every } i = 1, 2, \ldots, n\} \subseteq V.$$

As in the case of $\sigma(Y, Z)$, it is not difficult to see that $\sigma^*(Y, Z)$ is a topology on Z. We call $\sigma^*(Y, Z)$ the *Y-weak* topology of Z*. A sequence $(z_n)_{n \in \mathbb{N}}$ of elements of Z converges in the Y-weak* topology of Z if and only if there exists a $z \in Z$ such that the sequence $(\langle y, z_n \rangle)_{n \in \mathbb{N}}$ converges to $\langle y, z \rangle$ for every $y \in Y$. Whenever $(z_n)_{n \in \mathbb{N}}$ converges to z in the Y-weak* topology of Z, we say that $(z_n)_{n \in \mathbb{N}}$ is *Y-weak* convergent* (or *converges Y-weak**); the limit z is called the *Y-weak* limit of* $(z_n)_{n \in \mathbb{N}}$ and we use the notation $Y\text{-w}^*\text{-}\lim\limits_{n \to +\infty} z_n$ for z.

If Y is a Banach space and Z is the topological dual of Y, then (Y, Z) is a dual system with respect to the map $(y, z) \mapsto z(y)$ for every $y \in Y$ and $z \in Z$. In this case, we call $\sigma(Y, Z)$ and $\sigma^*(Y, Z)$ simply the *weak* and the *weak** topology on Y and on Z, respectively, and they are the usual weak and weak* topologies discussed in introductory courses in functional analysis. For additional details on dual systems and on various topologies that can be defined using these systems, see Aliprantis and Burkinshaw's monograph [3].

We now return to our discussion of the space $\mathcal{M}(X)$, where (X, d) is a locally compact separable metric space.

Since we may and do think of $\mathcal{M}(X)$ as the topological dual of $C_0(X)$, it follows that a sequence $(\mu_n)_{n\in\mathbb{N}}$ of elements of $\mathcal{M}(X)$ *converges in the weak* topology of* $\mathcal{M}(X)$ if there exists a $\mu \in \mathcal{M}(X)$ such that, for every $f \in C_0(X)$, the sequence $(\langle f, \mu_n\rangle)_{n\in\mathbb{N}}$ (that is, the sequence $\left(\int_X f \, d\mu_n\right)_{n\in\mathbb{N}}$) converges to $\langle f, \mu\rangle = \int_X f \, d\mu$. In this case, we say that μ is the *weak* limit of* $(\mu_n)_{n\in\mathbb{N}}$ (or that $(\mu_n)_{n\in\mathbb{N}}$ *converges to* μ *in the weak* topology of* $\mathcal{M}(X)$, or that $(\mu_n)_{n\in\mathbb{N}}$ *weak* converges to* μ) and we use the notation w*- $\lim_{n\to+\infty} \mu_n$ for μ.

Now consider the pair $(C_b(X), \mathcal{M}(X))$ and note that it is a dual system with respect to the map $(f, \mu) \mapsto \langle f, \mu\rangle = \int_X f \, d\mu$ for every $f \in C_b(X)$ and $\mu \in \mathcal{M}(X)$. In view of this fact, we say that a sequence $(\mu_n)_{n\in\mathbb{N}}$ of elements of $\mathcal{M}(X)$ *converges in the* $C_b(X)$*-weak* topology of* $\mathcal{M}(X)$ (or that $(\mu_n)_{n\in\mathbb{N}}$ is $C_b(X)$*-weak* convergent*, or that $(\mu_n)_{n\in\mathbb{N}}$ *converges* $C_b(X)$*-weak**) if there exists a $\mu \in \mathcal{M}(X)$ such that the sequence $\left(\int_X f(x) \, d\mu_n(x)\right)_{n\in\mathbb{N}}$ converges to $\int_X f(x) \, d\mu(x)$ for every $f \in C_b(X)$. In this case, we call μ the $C_b(X)$*-weak* limit of* $(\mu_n)_{n\in\mathbb{N}}$ and we denote μ by $C_b(X)$-w*- $\lim_{n\to+\infty} \mu_n$.

Observe that, given a sequence $(\mu_n)_{n\in\mathbb{N}}$ of elements of $\mathcal{M}(X)$, the $C_b(X)$-weak* convergence of $(\mu_n)_{n\in\mathbb{N}}$ is known as weak convergence in probability theory, and the weak* convergence of $(\mu_n)_{n\in\mathbb{N}}$ is called vague convergence in probability theory.

Note that both the weak* limit and the $C_b(X)$-weak* limit are unique whenever they exist because the corresponding topologies on $\mathcal{M}(X)$ are Hausdorff.

It is obvious that $C_b(X)$-weak* convergence implies weak* convergence in $\mathcal{M}(X)$. In general, weak* convergence does not imply $C_b(X)$-weak* convergence. However, if $(\mu_n)_{n\in\mathbb{N}}$ is a sequence of probability measures, $\mu_n \in \mathcal{M}(X)$ for every $n \in \mathbb{N}$, and if $\mu \in \mathcal{M}(X)$ is a probability measure, then the weak* convergence of $(\mu_n)_{n\in\mathbb{N}}$ to μ implies the $C_b(X)$-weak* convergence of $(\mu_n)_{n\in\mathbb{N}}$ to μ; this is a well-known result (see, for instance, p. 71 of Högnäs and Mukherjea [48]) and is not difficult to prove (use the fact that the elements of $\mathcal{M}(X)$ are *regular* signed real-valued Borel measures).

In the next proposition, we discuss several useful criteria for deciding if a transition probability is Feller or not.

Proposition 1.1.5. *Let P be a transition probability on (X, d), and let (S, T) be the Markov pair defined by P. The following assertions are equivalent:*

(a) *$Sf \in C_b(X)$ for every $f \in C_0(X)$.*
(b) *For every convergent sequence $(x_n)_{n\in\mathbb{N}}$ of elements of X, the sequence $(T\delta_{x_n})_{n\in\mathbb{N}}$ converges in the weak* topology of $\mathcal{M}(X)$ to $T\delta_x$, where $x = \lim_{n\to+\infty} x_n$.*

(c) For every convergent sequence $(x_n)_{n\in\mathbb{N}}$ of elements of X, the sequence $(T\delta_{x_n})_{n\in\mathbb{N}}$ converges $C_b(X)$-weak to $T\delta_x$, where $x = \lim\limits_{n\to\infty} x_n$.*

(d) P is a Feller transition probability.

Proof. $(a) \Rightarrow (b)$ The proof of the implication is obvious if we take into consideration the fact that $Sf(y) = \langle Sf, \delta_y\rangle = \langle f, T\delta_y\rangle$ for every $f \in C_0(X)$ and $y \in X$.

$(b) \Rightarrow (c)$ Let $(x_n)_{n\in\mathbb{N}}$ be a convergent sequence of elements of X, and let $x = \lim\limits_{n\to+\infty} x_n$. Since we assume that (b) holds true, the sequence $(T\delta_{x_n})_{n\in\mathbb{N}}$ weak* converges to $T\delta_x$. Since $T\delta_x$ and $T\delta_{x_n}$, $n \in \mathbb{N}$, are probability measures, using a comment made before this proposition after defining $C_b(X)$-weak* convergence, we obtain that $(T\delta_{x_n})_{n\in\mathbb{N}}$ converges $C_b(X)$-weak* to $T\delta_x$.

$(c) \Rightarrow (d)$ The proof is similar to the proof of the implication $(a) \Rightarrow (b)$, but here we use the fact that $Sf(y) = \langle f, T\delta_y\rangle$ for every $f \in C_b(X)$ and $y \in X$.

$(d) \Rightarrow (a)$ is obvious. $\square$

Let $\mathcal{A}$ be a nonempty set of real-valued functions defined on X. We say that $\mathcal{A}$ is *equicontinuous* if the following condition is satisfied:

(EQ)   For every convergent sequence $(x_n)_{n\in\mathbb{N}}$ of elements of X and for every $\varepsilon \in \mathbb{R}$, $\varepsilon > 0$, there exists an $n_\varepsilon \in \mathbb{N}$ such that $|f(x_n) - f(x)| < \varepsilon$ for every $n \in \mathbb{N}, n \geq n_\varepsilon$, and every $f \in \mathcal{A}$, where $x = \lim\limits_{n\to+\infty} x_n$.

Note that if $\mathcal{A}$ is equicontinuous, then each real-valued function that belongs to $\mathcal{A}$ is continuous.

If $\mathbf{J}$ is a nonempty set and $\gamma : \mathbf{J} \to B_b(X)$ is a function, then we say that γ is *equicontinuous* if the range $\{\gamma(j) \mid j \in \mathbf{J}\}$ of γ is equicontinuous. In particular, a sequence $(f_n)_{n\in\mathbb{N}}$ of elements of $B_b(X)$ is said to be *equicontinuous* if the range $\{f_n \mid n \in \mathbb{N}\}$ of $(f_n)_{n\in\mathbb{N}}$ is equicontinuous.

Now, let P be a transition probability defined on (X, d) and let (S, T) be the Markov pair defined by P. We say that P (or (S, T), or S) is $C_0(X)$-*equicontinuous* (or *equicontinuous with respect to* $C_0(X)$) if the sequence $(S^n f)_{n\in\mathbb{N}\cup\{0\}}$ is equicontinuous whenever $f \in C_0(X)$.

Note that if S is $C_0(X)$-equicontinuous, then Sf is a continuous bounded function whenever $f \in C_0(X)$. Thus, using Proposition 1.1.5, we obtain that (S, T) is a Markov-Feller pair.

For additional information on equicontinuity of sets of real-valued functions defined on a locally compact separable metric space, see the subsection *Equicontinuity* in Section 1.3 of [143]. For a more detailed discussion of equicontinuity of Markov-Feller pairs, including various kinds of equicontinuity that can be defined for these pairs, see the beginning of Section 4.1 (pp. 76–80) of [143]. Note that $C_0(X)$-equicontinuity is the most general type of equicontinuity among the kinds of equicontinuity discussed in [143]. However, for the results discussed in [143] and here, we do not need a more restrictive notion of equicontinuity. Since $C_0(X)$-equicontinuity is the only kind of equicontinuity used in this book, we will often use, simply, the term *equicontinuous* for a $C_0(X)$-equicontinuous transition probability P, or Markov-Feller pair (S, T), or first operator S of a Markov-Feller pair.

We will now conclude this section with examples of transition probabilities. Basically, we will discuss only two large families of transition probabilities which will be useful throughout the book when dealing with examples of transition functions. The examples that follow are by no means the only important examples of transition probabilities. For instance, we do not discuss here transition probabilities defined by iterated function systems (i.f.s.) with probabilities, which are special cases of *OMIGT processes* (the OMIGT (*Onicescu-Mihoc-Iosifescu-Grigorescu-Theodorescu*) processes, known also as *random systems with complete connections* have their origin in the pioneering 1935 paper of Onicescu and Mihoc [84], and have been made available to a large segment of the mathematical community in the monographs of Iosifescu and Grigorescu [49], and Iosifescu and Theodorescu [50]; see also Norman's book [83] and the paper of Stenflo [116]); for details on i.f.s. with probabilities, see, for example, the monograph by Lasota and Mackey [57], and the papers of Barnsley [5], Barnsley and Demko [6], Barnsley, Demko, Elton and Geronimo [7], Centore and Vrskay [17], Edalat [32], Lasota and Myjak [58–61], and [62], Lasota and Yorke [64], Myjak and Szarek [80], Nicol, Sidorov and Broomhead [82], Stenflo [114–117], and [118], Szarek [120], Vrscay [125], and our papers [140, 141], and [142]. Note that we do not discuss here the many examples of transition probabilities that appear in the study of discrete-time time-homogeneous Markov processes in the monograph by Meyn and Tweedie [77].

The transition probabilities that will appear in our examples below will be of two kinds: transition probabilities defined by measurable functions and transition probabilities defined by convolutions.

We first turn our attention to transition probabilities defined by measurable functions.

As usual, let (X, d) be a locally compact separable metric space.

A function $w : X \to X$ is said to be *Borel measurable* (or simply *measurable* if there is no danger of confusion) if $w^{-1}(A) \in \mathcal{B}(X)$ where $A \in \mathcal{B}(X)$. Note that any continuous function $u : X \to X$ is Borel measurable (for a proof see, for instance, Proposition II-1-1 on p. 31 of Neveu's monograph [81]).

Given a measurable function $w : X \to X$, we can define a map $P_w : X \times \mathcal{B}(X) \to \mathbb{R}$ as follows: $P_w(x, A) = \delta_{w(x)}(A) = \mathbf{1}_A(w(x))$ for every $x \in X$ and $A \in \mathcal{B}(X)$. It is easy to see that P_w is a transition probability. We call P_w the *transition probability induced* (or *generated*, or *defined*) *by* w.

Let (S_w, T_w) be the Markov pair defined by P_w. Then, using the equalities (1.1.1) and (1.1.2) of this section, we obtain that

$$S_w f(x) = \int f(y)\, d\delta_{w(x)}(y) = f(w(x)) \tag{1.1.4}$$

for every $f \in B_b(X)$ and $x \in X$, and that

$$T_w \mu(A) = \int \mathbf{1}_A(w(x)) \, d\mu(x) = \int \mathbf{1}_{w^{-1}(A)}(x) \, d\mu(x) = \mu(w^{-1}(A)) \qquad (1.1.5)$$

for every $\mu \in \mathcal{M}(X)$ and $A \in \mathcal{B}(X)$.

As in the case of P_w, we will often say that (S_w, T_w) is the *Markov pair induced* (or *generated*, or *defined*) *by* w.

Observe that T_w maps Dirac measures to Dirac measures since $T_w(\delta_x) = \delta_{w(x)}$ for every $x \in X$. Because of the manner in which T_w acts on Dirac measures, a discrete-time time-homogeneous Markov process whose transition probability is P_w, is said to be *deterministic*.

Note that the orbit, as defined after Proposition 1.1.2, of $x \in X$ under the action of T_w is $\mathcal{O}_{T_w}^{(\mathrm{TP})}(x) = \{w^n(x) \mid n \in \mathbb{N} \cup \{0\}\}$ and is the same as the orbit, as defined in Sect. A.3, of x under the action $(w_n)_{n \in \mathbb{N} \cup \{0\}}$ (of the additive semigroup $\mathbb{N} \cup \{0\}$) on X defined by $w_k(y) = w^k(y)$ for every $k \in \mathbb{N} \cup \{0\}$ and $y \in X$. The definition of orbits and orbit-closures given in Sect. A.3 are the traditional definitions that appear in ergodic theory, dynamical systems, and topological dynamics.

Note also that the definition of minimality given after Proposition 1.1.2 agrees with the usual one that appears in ergodic theory, dynamical systems, and topological dynamics and is given after Proposition A.3.3, in the sense that T_w is minimal according to the definition given after Proposition 1.1.2 if and only if the action $(w_n)_{n \in \mathbb{N} \cup \{0\}}$ is minimal in the spirit of the definition given after Proposition A.3.3.

It is easy to see that P_w is a Feller transition probability if and only if w is a continuous function.

Our first example of a transition probability appears in Example 5.2.5, pp. 66–67 of Hernández-Lerma and Lasserre's monograph [43] (see also Example 3.2 of [146]), and is very useful in illustrating how non-Feller transition probabilities fail to have certain properties that Feller transition probabilities do have.

Example 1.1.6. Let $X = [0, 1]$, and let d be the usual distance on $[0, 1]$ defined in terms of the absolute value: $d(x, y) = |x - y|$ for every $x \in [0, 1]$ and $y \in [0, 1]$. Now let $w : [0, 1] \to [0, 1]$ be defined by

$$w(x) = \begin{cases} \dfrac{x}{2} & \text{if } x \in (0, 1] \\ 1 & \text{if } x = 0. \end{cases}$$

Clearly, (X, d) is a compact metric space (hence, locally compact and separable), and w is a measurable map. Therefore, it makes sense to consider the transition probability P_w and the Markov pair (S_w, T_w) defined by w. Note that P_w is not a Feller transition probability (hence (S_w, T_w) is not a Markov-Feller pair) because w fails to be continuous (at zero). ∎

Note that in the above example X is a compact metric space. In such a case, any real-valued continuous function on X is bounded, and vanishes at infinity because the function has compact support; therefore $C_0(X) = C_b(X)$ and, as usual, we will use the notation $C(X)$ for any of the spaces $C_0(X)$ or $C_b(X)$.

Example 1.1.7 (Rotations of the Unit Circle). Let $\mathbb{R}/\mathbb{Z}$ be the compact metric group discussed in Example A.2.8 (the unit circle), let $a + \mathbb{Z} = \hat{a}$ be an element of $\mathbb{R}/\mathbb{Z}$, and consider the function $v_{a+\mathbb{Z}} : \mathbb{R}/\mathbb{Z} \to \mathbb{R}/\mathbb{Z}$ defined by $v_{a+\mathbb{Z}}(x + \mathbb{Z}) = a + x + \mathbb{Z}$ for every $x + \mathbb{Z} \in \mathbb{R}/\mathbb{Z}$; in the "hat" notation, we consider the map $v_{\hat{a}} : \mathbb{R}/\mathbb{Z} \to \mathbb{R}/\mathbb{Z}$ defined by $v_{\hat{a}}(\hat{x}) = \widehat{a + x}$ for every $\hat{x} \in \mathbb{R}/\mathbb{Z}$. The function $v_{\hat{a}}$ is known as the *rotation of the unit circle by a* (or *by the angle $2\pi a$*). Since $v_{\hat{a}}$ is a homeomorphism of $\mathbb{R}/\mathbb{Z}$, we can use $v_{\hat{a}}$ to define a transition probability $P_{\hat{a}} = P_{v_{\hat{a}}}$ as follows: $P_{\hat{a}} : \mathbb{R}/\mathbb{Z} \times \mathcal{B}(\mathbb{R}/\mathbb{Z}) \to \mathbb{R}$, $P_{\hat{a}}(\hat{x}, A) = \delta_{v_{\hat{a}}(\hat{x})}(A) = \delta_{\widehat{a+x}}(A) = \mathbf{1}_A(\widehat{a + x})$ for every $\hat{x} \in \mathbb{R}/\mathbb{Z}$ and $A \in \mathcal{B}(\mathbb{R}/\mathbb{Z})$. The Markov pair $(S_{v_{\hat{a}}}, T_{v_{\hat{a}}})$ defined by $v_{\hat{a}}$ will be denoted by $(S_{\hat{a}}, T_{\hat{a}})$ and is actually a Markov-Feller pair because $v_{\hat{a}}$ is a continuous function. Moreover, $(S_{\hat{a}}, T_{\hat{a}})$ is also equicontinuous.

Note that if b is a coset representative of $\hat{a}$, then $w_b = v_{\hat{a}}$, where w_b is the map of the flow $(w_t)_{t \in \mathbb{R}}$ of the rotations of the unit circle defined in Example A.3.4 that corresponds to $t = b$.

Note also that if $a \in \mathbb{Q}$, then the orbit of any $\hat{x} \in \mathbb{R}/\mathbb{Z}$ under the action of $T_{\hat{a}}$ is a finite subset of $\mathbb{R}/\mathbb{Z}$. If $a \notin \mathbb{Q}$, it can be shown (see, for example, Theorem 8.3, p. 53 of Robinson's book [99]) that every orbit under the action of $T_{\hat{a}}$ is dense in $\mathbb{R}/\mathbb{Z}$; that is, $T_{\hat{a}}$ is minimal. $\blacksquare$

Example 1.1.8 (Translations of the Torus). Let $n \in \mathbb{N}$, $n \geq 2$, and let $\mathbb{R}^n/\mathbb{Z}^n$ be the n-dimensional torus (see Example A.2.9).

Let $\mathbf{v} \in \mathbb{R}^n$, $\mathbf{v} = (v_1, v_2, \ldots, v_n)$, and let $u_{\mathbf{v}} : \mathbb{R}^n/\mathbb{Z}^n \to \mathbb{R}^n/\mathbb{Z}^n$ be defined by $u_{\mathbf{v}}(\mathbf{x} + \mathbb{Z}^n) = \mathbf{v} + \mathbf{x} + \mathbb{Z}^n$ for every $\mathbf{x} + \mathbb{Z}^n \in \mathbb{R}^n/\mathbb{Z}^n$, or, in "hat" notation, $u_{\mathbf{v}}(\hat{\mathbf{x}}) = \widehat{\mathbf{v} + \mathbf{x}}$ for every $\hat{\mathbf{x}} \in \mathbb{R}^n/\mathbb{Z}^n$.

The map $u_{\mathbf{v}}$ is a translation of the n-dimensional torus by $\hat{\mathbf{v}}$, so $u_{\mathbf{v}}$ is a homeomorphism of $\mathbb{R}^n/\mathbb{Z}^n$. Using $u_{\mathbf{v}}$, we can define a transition probability $P_{\mathbf{v}} = P_{u_{\mathbf{v}}}$ on $\mathbb{R}^n/\mathbb{Z}^n$ as follows: $P_{\mathbf{v}}(\hat{\mathbf{x}}, A) = \delta_{\widehat{\mathbf{v}+\mathbf{x}}}(A) = \mathbf{1}_A(\widehat{\mathbf{v} + \mathbf{x}})$ for every $\hat{\mathbf{x}} \in \mathbb{R}^n/\mathbb{Z}^n$ and every Borel subset A of $\mathbb{R}^n/\mathbb{Z}^n$. Since $u_{\mathbf{v}}$ is continuous, $P_{\mathbf{v}}$ is a Feller transition probability. The Markov-Feller pair defined by $P_{\mathbf{v}}$ is denoted by $(S_{\mathbf{v}}, T_{\mathbf{v}})$ and is defined as follows:

$$S_{\mathbf{v}} : B_b(\mathbb{R}^n/\mathbb{Z}^n) \to B_b(\mathbb{R}^n/\mathbb{Z}^n), \quad S_{\mathbf{v}}f(\hat{\mathbf{x}}) = f(\widehat{\mathbf{v} + \mathbf{x}})$$

for every $f \in B_b(\mathbb{R}^n/\mathbb{Z}^n)$ and $\hat{x} \in \mathbb{R}^n/\mathbb{Z}^n$, and

$$T_{\mathbf{v}} : \mathcal{M}(\mathbb{R}^n/\mathbb{Z}^n) \to \mathcal{M}(\mathbb{R}^n/\mathbb{Z}^n), \quad T_{\mathbf{v}}\mu(A) = \mu(u_{\mathbf{v}}^{-1}(A)) = \mu(-\hat{\mathbf{v}} + A)$$

for every $\mu \in \mathcal{M}(\mathbb{R}^n/\mathbb{Z}^n)$ and $A \in \mathcal{B}(\mathbb{R}^n/\mathbb{Z}^n)$.

It is easy to see that $(S_{\mathbf{v}}, T_{\mathbf{v}})$ is equicontinuous.

In general, the orbits under the action of $T_{\mathbf{v}}$ are not dense in $\mathbb{R}^n/\mathbb{Z}^n$. However, if the numbers $1, v_1, v_2, \cdots, v_n$ are rationally independent (the rational independence of a finite set of numbers is defined before Example A.3.5), it can be shown (see

Lemma 1, p. 66, Section 3.1 of Cornfeld, Fomin and Sinai's monograph [22]) that all the orbits under the action of $T_{\mathbf{v}}$ are dense in $\mathbb{R}^n/\mathbb{Z}^n$; that is, $T_{\mathbf{v}}$ is minimal.

Note that if $(w_t)_{t \in \mathbb{R}}$ is the rectilinear flow on $\mathbb{R}^n/\mathbb{Z}^n$ with velocity $\mathbf{v}$ discussed in Example A.3.5, then $u_{\mathbf{v}} = w_1$. ∎

In our next example we will consider Feller transition probabilities and Markov-Feller pairs defined by continuous functions on $\mathrm{PSL}(2, \mathbb{R})$. Recall that, as pointed out in Example A.2.10 in which $\mathrm{PSL}(2, \mathbb{R})$ was defined, $\mathrm{PSL}(2, \mathbb{R})$ is a locally compact separable metric group, so it makes sense to consider transition probabilities and Markov pairs on $\mathrm{PSL}(2, \mathbb{R})$.

Example 1.1.9 (Maps on $\mathrm{PSL}(2, \mathbb{R})$*).* Let $h \in \mathrm{SL}(2, \mathbb{R})$, and consider the function $u_h : \mathrm{PSL}(2, \mathbb{R}) \to \mathrm{PSL}(2, \mathbb{R})$ defined by $u_h(\hat{g}) = \hat{g}\hat{h}$ for every $\hat{g} \in \mathrm{PSL}(2, \mathbb{R})$, where $\hat{h}$ is the equivalence class $hL = \{-h, h\}$.

Clearly, u_h is a homeomorphism because $\mathrm{PSL}(2, \mathbb{R})$ is a topological group. The Feller transition probability $P_h = P_{u_h}$ generated by u_h is defined as follows: $P_h(\hat{g}, A) = \delta_{\hat{g}\hat{h}}(A) = \mathbf{1}_A(\hat{g}\hat{h})$ for every $\hat{g} \in \mathrm{PSL}(2, \mathbb{R})$ and every Borel subset A of $\mathrm{PSL}(2, \mathbb{R})$. The Markov-Feller pair induced by u_h is denoted (S_h, T_h) and the operators S_h and T_h are defined by

$$S_h f(\hat{g}) = f(\widehat{gh}) \text{ for every } f \in B_b(\mathrm{PSL}(2, \mathbb{R})) \text{ and } \hat{g} \in \mathrm{PSL}(2, \mathbb{R})$$

and

$$T_h \mu(A) = \mu(A\hat{h}^{-1}) \text{ for every } \mu \in \mathcal{M}(\mathrm{PSL}(2, \mathbb{R})) \text{ and } A \in \mathcal{B}(\mathrm{PSL}(2, \mathbb{R})).$$

Note that if $h = \begin{pmatrix} e^{\frac{s}{2}} & 0 \\ 0 & e^{-\frac{s}{2}} \end{pmatrix}$ for some $s \in \mathbb{R}$, then $u_h = w_s$, where $(w_t)_{t \in \mathbb{R}}$ is the geodesic flow on $\mathrm{PSL}(2, \mathbb{R})$ (for the definition of the flow, see Example A.3.6); note also that if $h = \begin{pmatrix} 1 & s \\ 0 & 1 \end{pmatrix}$ or $h = \begin{pmatrix} 1 & 0 \\ s & 1 \end{pmatrix}$ for some $s \in \mathbb{R}$, then $u_h = v_s^{(1)}$ or $u_h = v_s^{(2)}$, respectively, where $(v_t^{(1)})_{t \in \mathbb{R}}$ and $(v_t^{(2)})_{t \in \mathbb{R}}$ are the horocycle flows discussed in Example A.3.7. ∎

Example 1.1.10 (Maps on Spaces of Cosets of $\mathrm{PSL}(2, \mathbb{R})$*).* Let Γ be a lattice in $\mathrm{PSL}(2, \mathbb{R})$ (for the definition of lattices see the discussion preceding Proposition B.1.4) and let $h \in \mathrm{SL}(2, \mathbb{R})$.

Using h we can define two maps $\mathbf{u}_h^{(L)} : (\mathrm{PSL}(2, \mathbb{R})/\Gamma)_L \to (\mathrm{PSL}(2, \mathbb{R})/\Gamma)_L$ and $\mathbf{u}_h^{(R)} : (\mathrm{PSL}(2, \mathbb{R})/\Gamma)_R \to (\mathrm{PSL}(2, \mathbb{R})/\Gamma)_R$ as follows: $\mathbf{u}_h^{(L)}(\hat{x}\Gamma) = \hat{h}\hat{x}\Gamma$ for every $\hat{x}\Gamma \in (\mathrm{PSL}(2, \mathbb{R})/\Gamma)_L$ and $\mathbf{u}_h^{(R)}(\Gamma\hat{x}) = \Gamma\hat{x}\hat{h}$ for every $\Gamma\hat{x} \in (\mathrm{PSL}(2, \mathbb{R})/\Gamma)_R$. Note that, as in Example 1.1.9, we use the "hat" notation for the elements of $\mathrm{PSL}(2, \mathbb{R})$ because these elements are cosets. Thus, $\hat{h}$ is the coset $\{-h, h\}$ defined by h.

Since Γ is a discrete subgroup of $\mathrm{PSL}(2,\mathbb{R})$, we can use Proposition A.2.5 in order to infer that $(\mathrm{PSL}(2,\mathbb{R})/\Gamma)_{\mathrm{L}}$ and $(\mathrm{PSL}(2,\mathbb{R})/\Gamma)_{\mathrm{R}}$ are locally compact separable metrizable topological spaces. Using (b) of Proposition A.3.3 applied to the actions of $\mathrm{SL}(2,\mathbb{R})$ on $(\mathrm{PSL}(2,\mathbb{R})/\Gamma)_{\mathrm{L}}$ and on $(\mathrm{PSL}(2,\mathbb{R})/\Gamma)_{\mathrm{R}}$ defined by $(g,\hat{x}\Gamma)\mapsto \hat{g}\hat{x}\Gamma$ and $(g,\Gamma\hat{x})\mapsto \Gamma\hat{x}\hat{g}$, respectively, for every $g\in \mathrm{SL}(2,\mathbb{R})$, $\hat{x}\Gamma\in(\mathrm{PSL}(2,\mathbb{R})/\Gamma)_{\mathrm{L}}$, and $\Gamma\hat{x}\in(\mathrm{PSL}(2,\mathbb{R})/\Gamma)_{\mathrm{R}}$ (note that we can use (b) of Proposition A.3.3 because the standard surjection from $\mathrm{SL}(2,\mathbb{R})$ onto $\mathrm{PSL}(2,\mathbb{R})$ is a continuous group homomorphism), we obtain that the functions $\mathbf{u}_h^{(\mathrm{L})}$ and $\mathbf{u}_h^{(\mathrm{R})}$ are continuous. Accordingly, the functions $\mathbf{u}_h^{(\mathrm{L})}$ and $\mathbf{u}_h^{(\mathrm{R})}$ define Feller transition probabilities.

Let $P_h^{(\mathrm{L})}=P_{\mathbf{u}_h^{(\mathrm{L})}}$ and $P_h^{(\mathrm{R})}=P_{\mathbf{u}_h^{(\mathrm{R})}}$ be the transition probabilities defined by $\mathbf{u}_h^{(\mathrm{L})}$ and $\mathbf{u}_h^{(\mathrm{R})}$, respectively. Then $P_h^{(\mathrm{L})}(\hat{x}\Gamma, A) = \delta_{\hat{h}\hat{x}\Gamma}(A) = \mathbf{1}_A(\hat{h}\hat{x}\Gamma)$ for every $\hat{x}\Gamma \in (\mathrm{PSL}(2,\mathbb{R})/\Gamma)_{\mathrm{L}}$ and every Borel subset A of $(\mathrm{PSL}(2,\mathbb{R})/\Gamma)_{\mathrm{L}}$, and $P_h^{(\mathrm{R})}(\Gamma\hat{x}, A) = \delta_{\Gamma\hat{x}\hat{h}}(A) = \mathbf{1}_A(\Gamma\hat{x}\hat{h})$ for every $\Gamma\hat{x} \in (\mathrm{PSL}(2,\mathbb{R})/\Gamma)_{\mathrm{R}}$ and every $A \in \mathcal{B}((\mathrm{PSL}(2,\mathbb{R})/\Gamma)_{\mathrm{R}})$.

The transition probabilities $P_h^{(\mathrm{L})}$ and $P_h^{(\mathrm{R})}$ generate two Markov-Feller pairs: $(S_h^{(\mathrm{L})}, T_h^{(\mathrm{L})}) = (S_{u_h^{(\mathrm{L})}}, T_{u_h^{(\mathrm{L})}})$ and $(S_h^{(\mathrm{R})}, T_h^{(\mathrm{R})}) = (S_{u_h^{(\mathrm{R})}}, T_{u_h^{(\mathrm{R})}})$, respectively. The operators

$$S_h^{(\mathrm{L})} : B_b((\mathrm{PSL}(2,\mathbb{R})/\Gamma)_{\mathrm{L}}) \to B_b((\mathrm{PSL}(2,\mathbb{R})/\Gamma)_{\mathrm{L}})$$

and

$$S_h^{(\mathrm{R})} : B_b((\mathrm{PSL}(2,\mathbb{R})/\Gamma)_{\mathrm{R}}) \to B_b((\mathrm{PSL}(2,\mathbb{R})/\Gamma)_{\mathrm{R}})$$

are defined by $S_h^{(\mathrm{L})} f(\hat{x}\Gamma) = f(\hat{h}\hat{x}\Gamma)$ for every real-valued bounded Borel measurable function f defined on $(\mathrm{PSL}(2,\mathbb{R})/\Gamma)_{\mathrm{L}}$ and every $\hat{x}\Gamma \in (\mathrm{PSL}(2,\mathbb{R})/\Gamma)_{\mathrm{L}}$, and by $S_h^{(\mathrm{R})} f(\Gamma\hat{x}) = f(\Gamma\hat{x}\hat{h})$ for every $f \in B_b((\mathrm{PSL}(2,\mathbb{R})/\Gamma)_{\mathrm{R}})$ and every $\Gamma\hat{x} \in (\mathrm{PSL}(2,\mathbb{R})/\Gamma)_{\mathrm{R}}$. The operators

$$T_h^{(\mathrm{L})} : \mathcal{M}((\mathrm{PSL}(2,\mathbb{R})/\Gamma)_{\mathrm{L}}) \to \mathcal{M}((\mathrm{PSL}(2,\mathbb{R})/\Gamma)_{\mathrm{L}})$$

and

$$T_h^{(\mathrm{R})} : \mathcal{M}((\mathrm{PSL}(2,\mathbb{R})/\Gamma)_{\mathrm{R}}) \to \mathcal{M}((\mathrm{PSL}(2,\mathbb{R})/\Gamma)_{\mathrm{R}})$$

are defined by $T_h^{(\mathrm{L})}\mu(A) = \mu(\widehat{h^{-1}}A)$ for every $\mu \in \mathcal{M}((\mathrm{PSL}(2,\mathbb{R})/\Gamma)_{\mathrm{L}})$ and every $A \in \mathcal{B}((\mathrm{PSL}(2,\mathbb{R})/\Gamma)_{\mathrm{L}})$, and by $T_h^{(\mathrm{R})}\nu(B) = \nu(B\widehat{h^{-1}})$ for every $\nu \in \mathcal{M}((\mathrm{PSL}(2,\mathbb{R})/\Gamma)_{\mathrm{R}})$ and every $B \in \mathcal{B}((\mathrm{PSL}(2,\mathbb{R})/\Gamma)_{\mathrm{R}})$, where

$$\widehat{h^{-1}}A = \{\widehat{h^{-1}}\hat{x}\Gamma \mid \hat{x}\Gamma \in A\}$$

and

$$\widehat{Bh^{-1}} = \{\Gamma \hat{x} \widehat{h^{-1}} \mid \Gamma \hat{x} \in B\}.$$

Note that if $h = \begin{pmatrix} e^{-\frac{s}{2}} & 0 \\ 0 & e^{\frac{s}{2}} \end{pmatrix}$ for some $s \in \mathbb{R}$, then $\mathbf{u}_h^{(R)} = w_s^{(\Gamma)}$, where $(w_t^{(\Gamma)})_{t \in \mathbb{R}}$

is the geodesic flow on $(\mathrm{PSL}(2, \mathbb{R})/\Gamma)_R$ discussed in Example B.1.8. If $h = \begin{pmatrix} 1 & s \\ 0 & 1 \end{pmatrix}$

or $h = \begin{pmatrix} 1 & 0 \\ s & 1 \end{pmatrix}$ for some $s \in \mathbb{R}$, then $\mathbf{u}_h^{(R)} = \bar{v}_s^{(1\Gamma)}$ or $\bar{v}_s^{(2\Gamma)}$, respectively, where

$(\bar{v}_t^{(1\Gamma)})_{t \in \mathbb{R}}$ and $(\bar{v}_t^{(2\Gamma)})_{t \in \mathbb{R}}$ are the horocycle flows on $(\mathrm{PSL}(2, \mathbb{R})/\Gamma)_R$ discussed in
(a) of Example B.1.9.   ■

Example 1.1.11 (Maps on Spaces of Cosets of $\mathrm{SL}(n, \mathbb{R})$, $n \geq 2$). Let $n \in \mathbb{N}, n \geq 2$,
and let Γ be a lattice in $\mathrm{SL}(n, \mathbb{R})$ (note that $\mathrm{SL}(n, \mathbb{R})$ does have lattices because, as
pointed out in Example B.1.5, $\mathrm{SL}(n, \mathbb{Z})$ is a lattice in $\mathrm{SL}(n, \mathbb{R})$).

Given $h \in \mathrm{SL}(n, \mathbb{R})$ we can define two maps

$$u_h^{(L)} : (\mathrm{SL}(n, \mathbb{R})/\Gamma)_L \to (\mathrm{SL}(n, \mathbb{R})/\Gamma)_L$$

and

$$u_h^{(R)} : (\mathrm{SL}(n, \mathbb{R})/\Gamma)_R \to (\mathrm{SL}(n, \mathbb{R})/\Gamma)_R$$

as follows: $u_h^{(L)}(x\Gamma) = hx\Gamma$ for every $x\Gamma \in (\mathrm{SL}(n, \mathbb{R})/\Gamma)_L$, and $u_h^{(R)}(\Gamma x) = \Gamma x h$
for every $\Gamma x \in (\mathrm{SL}(n, \mathbb{R})/\Gamma)_R$.

Using Proposition A.2.5 as in Example 1.1.10, we obtain that $(\mathrm{SL}(n, \mathbb{R})/\Gamma)_L$
and $(\mathrm{SL}(n, \mathbb{R})/\Gamma)_R$ are locally compact separable metrizable topological spaces.
Since the actions of $\mathrm{SL}(n, \mathbb{R})$ on $(\mathrm{SL}(n, \mathbb{R})/\Gamma)_L$ and on $(\mathrm{SL}(n, \mathbb{R})/\Gamma)_R$ defined by
$(g, x\Gamma) \mapsto gx\Gamma$ and by $(g, \Gamma x) \mapsto \Gamma xg$, respectively, for every $g \in \mathrm{SL}(n, \mathbb{R})$,
$x\Gamma \in (\mathrm{SL}(n, \mathbb{R})/\Gamma)_L$, and $\Gamma x \in (\mathrm{SL}(n, \mathbb{R})/\Gamma)_R$, are continuous by (b) of
Proposition A.3.3, it follows that the maps $u_h^{(L)}$ and $u_h^{(R)}$ are continuous, as well,
and, in fact, $u_h^{(L)}$ and $u_h^{(R)}$ are homeomorphisms. Consequently, $u_h^{(L)}$ and $u_h^{(R)}$ define
Feller transition probabilities, which, in turn generate Markov-Feller pairs.

We will use the notations $P_{(h)}^{(L)} = P_{u_h^{(L)}}$, $(S_{(h)}^{(L)}, T_{(h)}^{(L)}) = (S_{u_h^{(L)}}, T_{u_h^{(L)}})$, $P_{(h)}^{(R)} =$

$P_{u_h^{(R)}}$, and $(S_{(h)}^{(R)}, T_{(h)}^{(R)}) = (S_{u_h^{(R)}}, T_{u_h^{(R)}})$.

The transition probabilities $P_{(h)}^{(L)}$ and $P_{(h)}^{(R)}$, and the Markov-Feller pairs
$(S_{(h)}^{(L)}, T_{(h)}^{(L)})$ and $(S_{(h)}^{(R)}, T_{(h)}^{(R)})$ generated by $u_h^{(L)}$ and $u_h^{(R)}$, respectively, are defined in
a manner similar to the corresponding transition probabilities and Markov-Feller
pairs discussed in Example 1.1.9. For instance, the transition probability

$$P_{(h)}^{(R)} : (\mathrm{SL}(n, \mathbb{R})/\Gamma)_R \times \mathcal{B}((\mathrm{SL}(n, \mathbb{R})/\Gamma)_R) \to \mathbb{R}$$

is defined by $P_{(h)}^{(\mathrm{R})}(\Gamma x, A) = \delta_{\Gamma x h}(A) = \mathbf{1}_A(\Gamma x h)$ for every $\Gamma x \in (\mathrm{SL}(n,\mathbb{R})/\Gamma)_{\mathrm{R}}$ and $A \in \mathcal{B}((SL(n,\mathbb{R})/\Gamma)_{\mathrm{R}})$, and the operators

$$S_{(h)}^{(\mathrm{R})} : B_b((\mathrm{SL}(n,\mathbb{R})/\Gamma)_{\mathrm{R}}) \to B_b((\mathrm{SL}(n,\mathbb{R})/\Gamma)_{\mathrm{R}})$$

and

$$T_{(h)}^{(\mathrm{R})} : \mathcal{M}((\mathrm{SL}(n,\mathbb{R})/\Gamma)_{\mathrm{R}}) \to \mathcal{M}((\mathrm{SL}(n,\mathbb{R})/\Gamma)_{\mathrm{R}})$$

are defined by $S_{(h)}^{(\mathrm{R})} f(\Gamma x) = f(\Gamma x h)$ for every $f \in B_b((\mathrm{SL}(n,\mathbb{R})/\Gamma)_{\mathrm{R}})$ and $\Gamma x \in (\mathrm{SL}(n,\mathbb{R})/\Gamma)_{\mathrm{R}}$ and by $T\mu(A) = \mu(Ah^{-1})$ for every $\mu \in \mathcal{M}((\mathrm{SL}(n,\mathbb{R})/\Gamma)_{\mathrm{R}})$ and $A \in \mathcal{B}((\mathrm{SL}(n,\mathbb{R})/\Gamma)_{\mathrm{R}})$.

Note that if $n = 2$ and $h = \begin{pmatrix} 1 & s \\ 0 & 1 \end{pmatrix}$ or $h = \begin{pmatrix} 1 & 0 \\ s & 1 \end{pmatrix}$ for some $s \in \mathbb{R}$, then $u_h^{(\mathrm{R})} = v_s^{(1\Gamma)}$ or $v_s^{(2\Gamma)}$, respectively, where $(v_t^{(j\Gamma)})_{t \in \mathbb{R}}$, $j = 1$ or 2, are the two horocycle flows discussed in (b) of Example B.1.9. Note also that the two 2×2 matrices $\begin{pmatrix} 1 & s \\ 0 & 1 \end{pmatrix}$ and $\begin{pmatrix} 1 & 0 \\ s & 1 \end{pmatrix}$ are unipotent elements of $\mathrm{SL}(2,\mathbb{R})$ (for the definition of a unipotent element, see the discussion preceding Theorem B.4.11). In general, if $\mathbf{Y}$ is a unipotent element in $\mathrm{SL}(n,\mathbb{R})$, $n \geq 2$, and if we take $h = \mathbf{Y}$ in the above discussion, then $u_{\mathbf{Y}}^{(\mathrm{R})} = w_1^{(\mathbf{Y})}$, where $(w_k^{(\mathbf{Y})})_{k \in \mathbb{Z}}$ is the standard action of $\mathbb{Z}$ on $(\mathrm{SL}(n,\mathbb{R})/\Gamma)_{\mathrm{R}}$ defined by $\mathbf{Y}$; the matrix $\mathbf{Y}$ does not have to be unipotent for the discussion here; however, the case when $\mathbf{Y}$ is unipotent is quite relevant because we can then use Ratner's Theorems B.4.11 and B.4.12. ∎

Example 1.1.12 (Maps on $\mathbb{S}_n$). Let $n \in \mathbb{N}$, $n \geq 2$.

Recall (see the beginning of Sect. B.4.1 and Proposition B.4.1) that $\mathbb{S}_n$ is the compact (separable) metric multiplicative semigroup of all $n \times n$ column stochastic matrices endowed with the metric defined by the restriction of the norm $\|\cdot\|_1$ on $\mathbf{M}(n,\mathbb{R})$ to $\mathbb{S}_n$, which was discussed before Proposition B.4.3.

Given $\mathbf{B} \in \mathbb{S}_n$, we can define a map $u_{\mathbf{B}} : \mathbb{S}_n \to \mathbb{S}_n$, $u_{\mathbf{B}}(\mathbf{X}) = \mathbf{B}$ for every $\mathbf{X} \in \mathbb{S}_n$. Since matrix multiplication is a well-defined algebraic operation (that defines the semigroup structure) on $\mathbb{S}_n$, it follows that $u_{\mathbf{B}}$ is well-defined in the sense that $u_{\mathbf{B}}(\mathbf{X}) \in \mathbb{S}_n$ for every $\mathbf{X} \in \mathbb{S}_n$. Also, it is easy to see that $u_{\mathbf{B}}$ is a continuous map. Therefore, $u_{\mathbf{B}}$ defines a Feller transition probability $P_{\mathbf{B}} = P_{u_{\mathbf{B}}}$, $P_{\mathbf{B}}(\mathbf{X}, A) = \delta_{\mathbf{BX}}(A) = \mathbf{1}_A(\mathbf{BX})$ for every $\mathbf{X} \in \mathbb{S}_n$ and every Borel subset A of $\mathbb{S}_n$. If $(S_{\mathbf{B}}, T_{\mathbf{B}}) = (S_{u_{\mathbf{B}}}, T_{u_{\mathbf{B}}})$ is the Markov-Feller pair defined by $u_{\mathbf{B}}$, then the operators $S_{\mathbf{B}} : B_b(\mathbb{S}_n) \to B_b(\mathbb{S}_n)$ and $T_{\mathbf{B}} : \mathcal{M}(\mathbb{S}_n) \to \mathcal{M}(\mathbb{S}_n)$ are defined by $S_{\mathbf{B}} f(\mathbf{X}) = f(\mathbf{BX})$ for every $f \in B_b(\mathbb{S}_n)$ and $\mathbf{X} \in \mathbb{S}_n$, and $T_{\mathbf{B}}\mu(A) = \mu(\mathbf{B}^{-1} A)$ for every $\mu \in \mathcal{M}(\mathbb{S}_n)$ and $A \in \mathcal{B}(\mathbb{S}_n)$ (for the definition of the set $\mathbf{B}^{-1} A$, see the equality (A.1.4)), respectively.

It is easy to see that $(S_{\mathbf{B}}, T_{\mathbf{B}})$ is equicontinuous.

Note that if $\mathbf{A} \in \mathbb{S}_n$ and we let $\mathbf{B} = \exp_s(s\mathbf{A})$ for some $s \in [0, +\infty)$, then $u_\mathbf{B} = \pi_s^{(\mathbf{A})}$ where $(\pi_t^{(\mathbf{A})})_{t \in [0,+\infty)}$ is the semiflow defined in Sect. B.4.1.   ■

Example 1.1.13 (Maps on $\mathbb{P}_n$). Let $n \in \mathbb{N}$, $n \geq 2$, and let $\mathbb{P}_n$ be the compact metric space of all n-dimensional stochastic column vectors in $\mathbb{R}^n$, where the metric on $\mathbb{P}_n$ is the metric defined by the restriction of the norm $\|\cdot\|_1$ to $\mathbb{P}_n$.

Let $\mathbf{B} \in \mathbb{S}_n$ and let $v_\mathbf{B} : \mathbb{P}_n \to \mathbb{P}_n$ be defined by $v_\mathbf{B}\mathbf{x} = \mathbf{Bx}$ for every $\mathbf{x} \in \mathbb{P}_n$. Note that $v_\mathbf{B}$ is well-defined in the sense that $\mathbf{Bx}$ is indeed an element of $\mathbb{P}_n$. Since $v_\mathbf{B}$ is a continuous function, it makes sense to consider the Feller transition probability $P_{v_\mathbf{B}}$ defined by $v_\mathbf{B}$. We will use the notation $P_{(\mathbf{B})}$ for $P_{v_\mathbf{B}}$. Thus, $P_{(\mathbf{B})} : \mathbb{P}_n \times \mathcal{B}(\mathbb{P}_n) \to \mathbb{R}$ is defined by $P_{(\mathbf{B})}(\mathbf{x}, A) = \delta_{\mathbf{Bx}}(A) = \mathbf{1}_A(\mathbf{Bx})$ for every $\mathbf{x} \in \mathbb{P}_n$ and every Borel subset A of $\mathbb{P}_n$. The Markov-Feller pair defined by $v_\mathbf{B}$ is denoted by $(S_{(\mathbf{B})}, T_{(\mathbf{B})})$ and the operators $S_{(\mathbf{B})} : B_b(\mathbb{P}_n) \to B_b(\mathbb{P}_n)$ and $T_{(\mathbf{B})} : \mathcal{M}(\mathbb{P}_n) \to \mathcal{M}(\mathbb{P}_n)$ are defined by $S_{(\mathbf{B})}f(\mathbf{x}) = f(\mathbf{Bx})$ for every $f \in B_b(\mathbb{P}_n)$ and $\mathbf{x} \in \mathbb{P}_n$, and $T_{(\mathbf{B})}\mu(A) = \mu(\mathbf{B}^{-1}A)$ for every $\mu \in \mathcal{M}(\mathbb{P}_n)$ and $A \in \mathcal{B}(\mathbb{P}_n)$, where $\mathbf{B}^{-1}A = v_\mathbf{B}^{-1}(A) = \{\mathbf{y} \in \mathbb{P}_n \,|\, \mathbf{y} = \mathbf{Bx}$ for some $\mathbf{x} \in A\}$.

We observe that $(S_{(\mathbf{B})}, T_{(\mathbf{B})})$ is equicontinuous.

Note that if $\mathbf{A} \in \mathbb{S}_n$ and $\mathbf{B} = \exp_s(s\mathbf{A})$ for some $s \in [0, +\infty)$, then $v_\mathbf{B} = \varphi_s^{(\mathbf{A})}$, where $(\varphi_t^{(\mathbf{A})})_{t \in [0,+\infty)}$ is the semiflow defined in Sect. B.4.1.   ■

Our last example of transition probabilities in this section deals with transition probabilities defined by convolutions of measures. These transition probabilities are quite different from the transition probabilities defined by measurable functions and are closely related to random walks (for details on the relationship between convolutions of measures and random walks see, for instance, the monographs by Heyer [46], Högnäs and Mukherjea [48], Meyn and Tweedie [77], Revuz [97], and M. Rosenblatt [102]).

In order to discuss the example, we need some preparation.

Let (X, d) be a locally compact separable metric space.

As usual, given a collection $\mathcal{C}$ of real-valued functions defined on X, we say that $\mathcal{C}$ is *uniformly bounded* if there exists an $M \in \mathbb{R}$, $M \geq 0$, such that $|f(x)| \leq M$ for every $f \in \mathcal{C}$ and $x \in X$. A sequence $(f_n)_{n \in \mathbb{N}}$ of real-valued functions on X is said to be *uniformly bounded* if the range $\{f_n \,|\, n \in \mathbb{N}\}$ of $(f_n)_{n \in \mathbb{N}}$ is uniformly bounded.

Proposition 1.1.14. *Let $\mathcal{F}$ be a vector subspace of $B_b(X)$ and assume that:*

(i) $C_0(X) \subseteq \mathcal{F}$.
(ii) (Invariance to Pointwise Convergence) If $(g_n)_{n \in \mathbb{N}}$ is a uniformly bounded sequence of elements of $\mathcal{F}$ such that $(g_n)_{n \in \mathbb{N}}$ converges pointwise on X, and if g is the pointwise limit of $(g_n)_{n \in \mathbb{N}}$ on X (that is, if $g : X \to \mathbb{R}$ is defined by $g(x) = \lim_{n \to +\infty} g_n(x)$ for every $x \in X$), then $g \in \mathcal{F}$.

Then $\mathcal{F} = B_b(X)$.

We will now discuss an application of Proposition 1.1.14 to the operation of convolution of signed measures (for the definition of the operation of convolution, see the discussion preceding Proposition B.2.1).

Proposition 1.1.15. *Let $(H, \cdot, d)$ be a locally compact separable metric semigroup (for the definition of a locally compact separable metric semigroup, see Sect. A.2), and let $\mu \in \mathcal{M}(H)$ and $\nu \in \mathcal{M}(H)$. Then $\int_H f(z)\, d(\mu * \nu)(z) = \int_H \int_H f(xy)\, d\mu(x)\, d\nu(y)$ for every $f \in B_b(H)$.*

Proof. Taking into consideration the manner in which the integrals with respect to signed measures are defined and using the fact that $\mu * \nu = \mu^+ * \nu^+ - \mu^- * \nu^+ - \mu^+ * \nu^- + \mu^- * \nu^-$, where $\mu = \mu^+ - \mu^-$ and $\nu = \nu^+ - \nu^-$ are the Jordan decompositions of μ and ν, respectively, we obtain that it is enough to prove the proposition under the assumption that $\mu \geq 0$ and $\nu \geq 0$.

Thus, assume that $\mu \geq 0$ and $\nu \geq 0$, and set

$$\mathcal{F} = \{ f \in B_b(H) \mid \int_H f(z)\, d(\mu * \nu)(z) = \int_H \int_H f(xy)\, d\mu(x)\, d\nu(y) \}.$$

In view of the definition of the operation of convolution we obtain that $C_0(X) \subseteq \mathcal{F}$. Clearly, $\mathcal{F}$ is a vector subspace of $B_b(X)$, and using the Lebesgue dominated convergence theorem, we obtain that $\mathcal{F}$ satisfies condition *(ii)* of Proposition 1.1.14. Thus, $\mathcal{F}$ satisfies all the conditions of Proposition 1.1.14, so, using the proposition we obtain that $\mathcal{F} = B_b(H)$. $\square$

We are now ready to discuss the transition probabilities defined by convolutions of measures.

Example 1.1.16 (Convolutions of Measures and Convolution Operators). Let $(H, \cdot, d)$ be a locally compact separable metric semigroup.

Let $\mu \in \mathcal{M}(H)$ be a probability measure.

Let $P_\mu : H \times \mathcal{B}(H) \to \mathbb{R}$ be defined by $P_\mu(x, A) = \mu * \delta_x(A)$ for every $x \in H$ and $A \in \mathcal{B}(H)$, where $*$ stands for the operation of convolution.

It is easy to see that P_μ is a transition probability. Indeed, if $A \in \mathcal{B}(H)$, then, by Proposition 1.1.14, $P_\mu(x, A) = \int_H 1_A(ux)\, d\mu(u)$ for every $x \in H$, so, using Proposition B.2.1, we obtain that the map $x \mapsto P_\mu(x, A)$, $x \in H$, is measurable; if $x \in H$, then $\mu * \delta_x$ is a probability measure by Proposition B.3.1, so the map $A \mapsto P_\mu(x, A)$, $A \in \mathcal{B}(H)$ is a probability measure. We will often refer to P_μ as the *transition probability defined by* μ.

Let (S_μ, T_μ) be the Markov pair defined by P_μ.

Using Proposition 1.1.15, we obtain that

$$S_\mu f(x) = \int_H f(y)\, d(\mu * \delta_x)(y) = \int_H f(zx)\, d\mu(z)$$

for every $f \in B_b(H)$ and $x \in H$. Note that, by Proposition B.2.1, $S_\mu f \in C_b(H)$ whenever $f \in C_b(H)$, so P_μ is a Feller transition probability and (S_μ, T_μ) is a Markov-Feller pair.

Again using Proposition 1.1.14, we also obtain that

$$
\begin{aligned}
T_\mu \nu(A) &= \int_H P_\mu(x, A)\, d\nu(x) = \int_H (\mu * \delta_x)(A)\, d\nu(x) \\
&= \int_H \left(\int_H \int_H 1_A(yz)\, d\mu(y)\, d\delta_x(z) \right) d\nu(x) \\
&= \int_H \left(\int_H 1_A(yx)\, d\mu(y) \right) d\nu(x) = (\mu * \nu)(A)
\end{aligned}
$$

for every $\nu \in \mathcal{M}(H)$ and $A \in \mathcal{B}(H)$; that is,

$$
T_\mu \nu = \mu * \nu \tag{1.1.6}
$$

for every $\nu \in \mathcal{M}(H)$. Since T_μ satisfies the above equality (1.1.6) for every $\nu \in \mathcal{M}(H)$, we say that T_μ is a (the) *convolution operator (defined by μ)*. Also, we refer to (S_μ, T_μ) as the *Markov pair defined by μ*.

Note that if $\mu \in \mathcal{M}(H)$ is a probability measure, and, for every $u \in [0, +\infty)$, we let T_u be the convolution operator defined by $\mu_u = \exp_s(u\mu)$, then the family $(T_u)_{u \in [0,+\infty)}$ is a one-parameter semigroup of convolution operators in the sense that $T_{u+v} = T_u T_v$ for every $u \in [0, +\infty)$ and $v \in [0, +\infty)$. This is so, because, by Proposition B.3.2, $(\mu_u)_{u \in [0,+\infty)}$ is a one-parameter convolution semigroup of probability measures. Observe that the set of all convolution operators defined by probability measures is a semigroup under the operation of composition of operators because $T_{\nu_1} T_{\nu_2} = T_{\nu_1 * \nu_2}$ for all probability measures $\nu_1 \in \mathcal{M}(H)$ and $\nu_2 \in \mathcal{M}(H)$, and because, by Proposition B.3.1, $\nu_1 * \nu_2$ is a probability measure. ∎

1.2 Invariant Probabilities

In Sect. 1.1 we defined the transition probabilities and their associated Markov pairs, and we singled out two important kinds of transition probabilities: the Feller transition probabilities and, among these Feller transition probabilities, we distinguished the equicontinuous ones. Our goal in this section is to discuss basic facts about invariant probabilities for transition probabilities. The definitions, the results, and the constructions discussed in the previous section and in this one are basic in the sense that they have appeared in various forms in many places in the literature, and will be used in the remaining sections of this chapter and throughout the book.

As in Sect. 1.1, let (X, d) be a locally compact separable metric space.

Also, let P be a transition probability on (X, d) and let (S, T) be the Markov pair defined by P.

An element μ of $\mathcal{M}(X)$ is said to be an *invariant element for T* (or *for P*, or *for (S,T)*) if $T\mu = \mu$. An invariant element for T will also be called *T-invariant* or *P-invariant*.

Note that if P_w is the transition probability defined by a measurable function $w : X \to X$, then $\mu \in \mathcal{M}(X)$ is an invariant element for P_w if and only if $\mu(w^{-1}(A)) = \mu(A)$ for every $A \in \mathcal{B}(X)$; note also that if $(H, \cdot, d)$ is a locally compact separable metric semigroup, if $\mu \in \mathcal{M}(H)$ is a probability measure, and if T_μ is the convolution operator defined by μ (see Example 1.1.16), then an element ν of $\mathcal{M}(H)$ is T_μ-invariant if and only if $\mu * \nu = \nu$.

Since the zero element (the zero measure) in $\mathcal{M}(X)$ is always invariant, we are interested in the case in which T has nonzero invariant measures.

Note that if T has nonzero invariant elements, then T has invariant probabilities because if $\mu \in \mathcal{M}(X)$, $\mu \neq 0$, is an invariant element for T, then $\dfrac{|\mu|}{\|\mu\|}$ is an invariant probability measure of T; thus, T has nonzero invariant elements if and only if T has invariant probability measures. Using the above observations and the fact that any nonzero T-invariant positive measure is a scalar multiple of a T-invariant probability measure, we obtain that for most purposes, in order to understand the structure of the set of all T-invariant elements, it is enough to understand the structure of the set of all probability measures that are invariant for T.

We pause for a moment to point out a few examples of invariant probability measures: the Haar-Lebesgue measure on $\mathbb{R}/\mathbb{Z}$ (discussed in Example B.1.6) is an invariant probability measure for the Markov-Feller pair $(S_{\hat{a}}, T_{\hat{a}})$ defined by the rotation of the unit circle by a, $a \in \mathbb{R}$ (see Example 1.1.7); the Haar-Lebesgue measure on $\mathbb{R}^n/\mathbb{Z}^n$ that appears in Example B.1.7 is an invariant probability measure for the Markov-Feller pair $(S_\mathbf{v}, T_\mathbf{v})$ defined by the translation $u_\mathbf{v}$ of the n-dimensional torus in Example 1.1.8; if Γ is a lattice in $\mathrm{PSL}(2, \mathbb{R})$, the standard $\mathrm{PSL}(2, \mathbb{R})$-invariant probability measures $\nu_{(\mathrm{PSL}(2,\mathbb{R})/\Gamma)_\mathrm{L}}$ and $\nu_{(\mathrm{PSL}(2,\mathbb{R})/\Gamma)_\mathrm{R}}$ (see Sect. B.1 for terminology and notation) are invariant probability measures for the Markov-Feller pairs $(S_h^{(\mathrm{L})}, T_h^{(\mathrm{L})})$ and $(S_h^{(\mathrm{R})}, T_h^{(\mathrm{R})})$ of Example 1.1.10, respectively; finally, the standard $\mathrm{SL}(n, \mathbb{R})$-invariant probability measures $\nu_{(\mathrm{SL}(n,\mathbb{R})/\Gamma)_\mathrm{L}}$ and $\nu_{(\mathrm{SL}(n,\mathbb{R})/\Gamma)_\mathrm{R}}$ are invariant probability measures for the Markov-Feller pairs $(S_{(h)}^{(\mathrm{L})}, T_{(h)}^{(\mathrm{L})})$ and $(S_{(h)}^{(\mathrm{R})}, T_{(h)}^{(\mathrm{R})})$ of Example 1.1.11, respectively.

There are various kinds of invariant probabilities for Markov pairs, and we will discuss some of them in this section. For us here, the most important invariant probability measures will be the ergodic ones. A T-invariant probability measure μ is said to be *ergodic* if the following assertion *does not* hold true for μ:

(E) There exists a Borel measurable subset A of X such that $\mu(A) > 0$ and $\mu(X \setminus A) > 0$, and such that the measures $\mu_1 : \mathcal{B}(X) \to \mathbb{R}$ and $\mu_2 : \mathcal{B}(X) \to \mathbb{R}$ defined by $\mu_1(B) = \mu(B \cap A)$ and $\mu_2(B) = \mu(B \cap (X \setminus A))$ for every $B \in \mathcal{B}(X)$ are both invariant for T.

The fact that a T-invariant probability measure μ is ergodic means that we cannot write μ as a sum of two nonzero mutually singular T-invariant measures. We can compare the role played by the ergodic T-invariant probability measures for T-invariance with the role played by molecules for a substance. A molecule is the smallest part of a substance that preserves all the properties of that substance; similarly, an ergodic T-invariant probability measure is the "smallest" T-invariant probability in the sense of assertion (E) that preserves the T-invariance.

The transition probability P (or the Markov pair (S, T) defined by P, or the operator T in the Markov pair) is said to be *uniquely ergodic* if P has exactly one invariant probability measure. The reason for the terminology stems from the fact that if P is uniquely ergodic, then the unique invariant probability measure is ergodic, and from the fact that if P has only one ergodic invariant probability measure, then P cannot have any other invariant probability measures. We will return to these facts in the next section.

We call P (or (S, T), or T) *strictly ergodic* if P is uniquely ergodic and if the support of the (unique) invariant probability measure is the entire space X.

A probability measure $\mu^* \in \mathcal{M}(X)$ is said to be *attractive for P* (or *for (S, T)*, or *for T*) if the sequence $(T^n \mu)_{n \in \mathbb{N} \cup \{0\}}$ converges to μ^* in the weak* topology of $\mathcal{M}(X)$ for every probability measure $\mu \in \mathcal{M}(X)$. It is easy to see that if P is a Feller transition probability and if P has an attractive probability measure μ^*, then μ^* is invariant for P (use the fact, pointed out before Proposition 1.1.5, that if a sequence of probability measures weak* converges to a probability measure ν, then the sequence also converges $C_b(X)$-weak* to ν) and P is uniquely ergodic.

Let A be a Borel subset of X. We say that A is a *set of maximal probability for P* (or *for (S, T)*, or *for T*) if either P does not have invariant probability measures, or else $\mu(A) = 1$ for every invariant probability measure μ for P. Thus, if P has invariant probabilities, a subset A of X, $A \in \mathcal{B}(X)$, is a set of maximal probability if all the invariant probability measures for P are concentrated on A.

Our goal now is to briefly recall several facts about almost everywhere (a.e.) convergence of sequences of functions and the Hopf ergodic theorem in order to discuss a certain behavior of the first operator S of the Markov pair (S, T) defined by a transition probability P, behavior that is generated by the presence of an invariant probability measure for P.

Let $(Y, \mathcal{Y}, \nu)$ be a measure space, and let $\mathbb{M}(Y, \mathcal{Y}, \nu)$ be the vector space of all equivalence classes of real-valued measurable functions on Y, where two functions f_1 and f_2 belong to the same class if and only if $f_1 = f_2$ ν-almost everywhere (ν-a.e.). If f is a real-valued measurable function on Y, we denote by $\bar{f}$ the element of $\mathbb{M}(Y, \mathcal{Y}, \nu)$ defined by f. As usual, we say that a sequence $(\bar{f}_n)_{n \in \mathbb{N}}$ of elements of $\mathbb{M}(Y, \mathcal{Y}, \nu)$ *converges ν-a.e.* if there exist $\bar{f} \in \mathbb{M}(Y, \mathcal{Y}, \nu)$, a real-valued measurable function g in the class $\bar{f}$ and a sequence $(g_n)_{n \in \mathbb{N}}$ of real-valued measurable functions on Y such that g_n belongs to $\bar{f}_n$ for every $n \in \mathbb{N}$, and such that $(g_n)_{n \in \mathbb{N}}$ converges to g ν-a.e. In this case, $\bar{f}$ is called the *ν-a.e. limit of $(\bar{f}_n)_{n \in \mathbb{N}}$*, or we say that $(\bar{f}_n)_{n \in \mathbb{N}}$ *converges to $\bar{f}$ ν-a.e.* It is easy to see that $(\bar{f}_n)_{n \in \mathbb{N}}$ converges ν-a.e. if and only if there exists an $\bar{f} \in \mathbb{M}(Y, \mathcal{Y}, \nu)$ such that for every sequence $(h_n)_{n \in \mathbb{N}}$ of real-valued measurable functions defined on Y such that h_n is in the class $\bar{f}_n$ for every $n \in \mathbb{N}$ and for every function h in the class $\bar{f}$, the sequence

$(h_n)_{n\in\mathbb{N}}$ converges to h ν-a.e. It is also easy to see that if $(\bar{f}_n)_{n\in\mathbb{N}}$ converges ν-a.e., then the ν-a.e. limit is unique.

Throughout the book, we will consider various vector subspaces of $\mathbb{M}(Y,\mathcal{Y},\nu)$, the most important of them being the usual spaces $L^p(Y,\mathcal{Y},\nu)$, $1 \le p < +\infty$, and the space $L^\infty(Y,\mathcal{Y},\nu)$ whenever ν is a σ-finite measure. As is generally the case, when dealing with $L^p(Y,\mathcal{Y},\nu)$, we will think of it as a Banach space endowed with its standard norm $\|\cdot\|_p$, $1 \le p \le +\infty$.

Let E be a Banach space. A linear bounded operator $Q : E \to E$ is said to be a *contraction (of E)* if $\|Q\| \le 1$.

Now, let $p = +\infty$, or $p \in \mathbb{R}$, $p \ge 1$, and let $L^p(Y,\mathcal{Y},\nu)$ be the usual Banach space. If $\bar{f} \in L^p(Y,\mathcal{Y},\nu)$, then we say that $\bar{f}$ is a *positive element of* $L^p(Y,\mathcal{Y},\nu)$ if there exists a real-valued measurable function g on Y such that g belongs to the class $\bar{f}$ and $g \ge 0$ ν-a.e.; naturally, we will use the notation $\bar{f} \ge 0$ in order to indicate that $\bar{f}$ is a positive element. A linear operator $Q : L^p(Y,\mathcal{Y},\nu) \to L^p(Y,\mathcal{Y},\nu)$ is called a *positive operator* if $Q\bar{f} \ge 0$ whenever $\bar{f}$ is a positive element of $L^p(Y,\mathcal{Y},\nu)$.

Theorem 1.2.1 (The Hopf Ergodic Theorem). *Assume that* $(Y,\mathcal{Y},\nu)$ *is a probability space, and let* $Q : L^1(Y,\mathcal{Y},\nu) \to L^1(Y,\mathcal{Y},\nu)$ *be a positive contraction such that* $Q\bar{1}_Y = \bar{1}_Y$. *Then:*

(a) *The sequence* $(\frac{1}{n}\sum_{k=0}^{n-1} Q^k \bar{f})_{n\in\mathbb{N}}$ *converges* ν-*a.e. to an element of* $L^1(Y,\mathcal{Y},\nu)$
   for every $\bar{f} \in L^1(Y,\mathcal{Y},\nu)$.
(b) *If* $\bar{f} \in L^1(Y,\mathcal{Y},\nu)$ *and* $\bar{g}$ *is the* ν-*a.e. limit of* $(\frac{1}{n}\sum_{k=0}^{n-1} Q^k \bar{f})_{n\in\mathbb{N}}$, *then*
   $\int \bar{f}\,d\nu = \int \bar{g}\,d\nu$.

For a proof of the Hopf ergodic theorem and additional details, see Theorem 3.5, pp. 128–129, and Section 3.3 of Krengel's monograph [53]. Note that our formulation of Hopf's theorem is less general than that of the above mentioned theorem in Krengel [53]; we prefer our formulation here because it is easier to use in our setting while preserving the full flavor of Hopf's theorem.

Let us return to our usual setting in this section. Thus, we assume given a locally compact separable metric space (X,d), a transition probability P defined on (X,d), and the Markov pair (S,T) defined by P.

Assume that T has invariant probabilities and let μ be such an invariant probability.

Let B_μ be the vector subspace of $\mathcal{M}(X)$ of all elements of $\mathcal{M}(X)$ that are absolutely continuous with respect to μ. It is well-known that if we endow B_μ with the total variation norm (that is, the norm inherited from $\mathcal{M}(X)$), then B_μ is a Banach subspace of $\mathcal{M}(X)$ and is Banach space isomorphic and isometric to $L^1(X,\mathcal{B}(X),\mu)$. There is a standard Banach space isomorphism and isometry ψ from B_μ onto $L^1(X,\mathcal{B}(X),\mu)$ defined as follows: $\psi(\nu) = \bar{f}_\mu$ for every $\nu \in B_\mu$, where f_μ is a Radon-Nikodým derivative of ν with respect to μ, and $\bar{f}_\mu$ is the element of $L^1(X,\mathcal{B}(X),\mu)$ (the equivalence class) defined by f_μ. Note that ψ is well-defined in the sense that a Radon-Nikodým derivative of $\nu \in B_\mu$ with respect to μ is μ-integrable, so it defines an element of $L^1(X, \mathcal{B}(X), \mu)$, and two Radon-Nikodým derivatives of ν with respect to μ define the same element of $L^1(X, \mathcal{B}(X), \mu)$. It is easy to see that ψ is a Banach space isomorphism and an isometry onto.

Using Lemma 5.1 of Lin [67], or Proposition 1.1 in Chapter 4 of Revuz [97], we obtain that $T\nu \in B_\mu$ for every $\nu \in B_\mu$, so we can use T to define another operator $U^{(\mu)} : L^1(X, \mathcal{B}(X), \mu) \to L^1(X, \mathcal{B}(X), \mu)$ as follows: $U^{(\mu)}(\bar{f})$ is the equivalence class of a Radon-Nikodým derivative of $T(\bar{f}\mu)$ with respect to μ. It is easy to see that $U^{(\mu)}$ is a positive contraction of $L^1(X, \mathcal{B}(X), \mu)$.

Let $U^{(\mu)'} : L^\infty(X, \mathcal{B}(X), \mu) \to L^\infty(X, \mathcal{B}(X), \mu)$ be the dual of $U^{(\mu)}$. Using Proposition 1.4 in Chapter 4 of Revuz's monograph [97], we obtain that $U^{(\mu)'}\bar{f} = \overline{Sf}$ for every $f \in B_b(X)$. Thus, using the fact that μ is a T-invariant probability measure, we obtain that

$$\int U^{(\mu)'} \bar{1}_A \, d\mu = \int S1_A \, d\mu = \langle 1_A, T\mu \rangle = \int 1_A \, d\mu = \mu(A) \qquad (1.2.1)$$

for every $A \in \mathcal{B}(X)$.

Since the above equalities (1.2.1) hold true for every $A \in \mathcal{B}(X)$, it follows that $U^{(\mu)'}$ has a unique extension to a linear operator $V^{(\mu)} : L^1(X, \mathcal{B}(X), \mu) \to L^1(X, \mathcal{B}(X), \mu)$ in the sense that there exists a unique linear operator $V^{(\mu)}$ from $L^1(X, \mathcal{B}(X), \mu)$ to $L^1(X, \mathcal{B}(X), \mu)$ such that $V^{(\mu)}\bar{f} = U^{(\mu)'}\bar{f}$ for every $\bar{f} \in L^\infty(X, \mathcal{B}(X), \mu)$.

It is not difficult to see that $V^{(\mu)}$ is a positive contraction, and, since $V^{(\mu)}\bar{1}_X = U^{(\mu)'}\bar{1}_X = \overline{S1_X} = \bar{1}_X$, we obtain that $V^{(\mu)}$ satisfies all the conditions of Theorem 1.2.1. This means, in particular, that if $f \in C_0(X)$, then the sequence $(\frac{1}{n}\sum_{k=0}^{n-1} V^{(\mu)^k}\bar{f})_{n\in\mathbb{N}}$ converges μ-a.e., and $\int \bar{f} \, d\mu = \int \bar{g} \, d\mu$, where $\bar{g}$ is the μ-a.e. limit of $(\frac{1}{n}\sum_{k=0}^{n-1} V^{(\mu)^k}\bar{f})_{n\in\mathbb{N}}$. In view of our discussion so far, we obtain the following theorem:

Theorem 1.2.2. *Let P be a transition probability defined on the locally compact separable metric space (X, d), let (S, T) be the Markov pair defined by P, assume that T has nonzero invariant elements, and let μ be a T-invariant probability measure. If $f \in C_0(X)$, then the sequence $(\frac{1}{n}\sum_{k=0}^{n-1} S^k f)_{n\in\mathbb{N}}$ converges μ-a.e. If g is a μ-a.e. limit of $(\frac{1}{n}\sum_{k=0}^{n-1} S^k f)_{n\in\mathbb{N}}$, then g is μ-integrable and $\int g \, d\mu = \int f \, d\mu$.*

We will now conclude this section by briefly reviewing a very useful result for dealing with invariant measures of Feller transition probabilities, namely the Lasota-Yorke lemma. The lemma was obtained by A. Lasota and J.A. Yorke in [64], and was used there to study iterated function systems with probabilities. We will follow here [33] and the subsection *The Lasota-Yorke Lemma* in Section 1.2 of [143].

Let P be a Feller transition probability and let (S, T) be the Markov-Feller pair defined by P.

Let $\mu \in \mathcal{M}(X)$. Then we can extend μ to a map $\tilde{\mu} : C_b(X) \to \mathbb{R}$ defined as follows: $\tilde{\mu}(f) = \int f(x)\,d\mu(x)$ for every $f \in C_b(X)$. It is easy to see that $\tilde{\mu}$ is a bounded linear functional on $C_b(X)$. We call $\tilde{\mu}$ the *standard extension of μ to $C_b(X)$*. If $\tilde{\mu}$ has the property that $\tilde{\mu}(Sf) = \tilde{\mu}(f)$ for every $f \in C_0(X)$, then $T\mu = \mu$, because in this case $\langle f, T\mu \rangle = \langle Sf, \mu \rangle = \tilde{\mu}(Sf) = \tilde{\mu}(f) = \langle f, \mu \rangle$ for every $f \in C_0(X)$. The Lasota-Yorke lemma is an extension of the above observation to arbitrary positive linear functionals on $C_b(X)$.

Theorem 1.2.3. *Let $\phi : C_b(X) \to \mathbb{R}$ be a positive linear functional such that $\phi(Sf) = \phi(f)$ for every $f \in C_0(X)$. Then the restriction μ_ϕ of ϕ to $C_0(X)$ is T-invariant (that is, $T\mu_\phi = \mu_\phi$) provided that we think of μ_ϕ as an element of $\mathcal{M}(X)$.*

For a proof of the theorem see Theorem 1.2.4 of [143]. Note that in [143] I also assumed that $\phi(1_X) = 1$, but, as pointed out by Emel'yanov, I did not use the assumption in the proof of Theorem 1.2.4 of [143]. The more elegant statement of the Lasota-Yorke lemma given here appears in Proposition 1.4 of [33].

1.3   The Ergodic Decomposition of Kryloff, Bogoliouboff, Beboutoff and Yosida

Our goal in this section is to briefly review an ergodic decomposition for transition probabilities that we have obtained in [146]. Since the decomposition stems from the pioneering works of Kryloff and Bogoliouboff [54], Beboutoff [9], and Yosida [135] and [136] (see also Section 4 in Chapter 13 of [138]), we call it the *Kryloff-Bogoliouboff-Beboutoff-Yosida* (or *KBBY*) *decomposition* (for a brief outline of the emergence of the decomposition in [9, 54, 135], and [136], see the subsection *The KBBY Decomposition* in Section 1.2 of [143]). The KBBY decomposition is valid for any transition probability defined on a locally compact separable metric space, and, as we will see in Chaps. 5 and 6 in this book, the decomposition is also valid for a rather large class of transition functions.

For a more general approach to most of the results in this section, see the paper by Worm and Hille [131].

Let (X, d) be a locally compact separable metric space, let P be a transition probability defined on (X, d), and let (S, T) be the Markov pair defined by P. The KBBY decomposition defined by P (or by (S, T)) is a splitting of the space X in terms of the convergence properties of the sequences $(\frac{1}{n}\sum_{k=0}^{n-1} S^k f(x))_{n\in\mathbb{N}}$, $f \in C_0(X)$, $x \in X$.

The decomposition has various features, the most important being the following:

– It allows us to associate in a natural manner a Borel measurable subset of X to each ergodic T-invariant probability measure such that the measure is concentrated on the set, and for every two distinct ergodic T-invariant probabilities, the corresponding subsets are disjoint; thus, the decomposition offers a "system of reference" for the ergodic invariant probability measures;

- The decomposition allows us to express each T-invariant probability measure as a convex combination, in integral form, of ergodic T-invariant probabilities;
- It allows us to obtain implicitly various criteria for the existence of T-invariant probability measures.

We will now describe the decomposition in detail.
Set

$$\Omega^{(\mathrm{TP})}(P) = \left\{ x \in X \;\middle|\; \begin{array}{l} L((S^n f(x))_{n \in \mathbb{N} \cup \{0\}}) = 0 \text{ for} \\ \text{every } f \in C_0(X) \text{ and every} \\ \text{Banach limit } L \end{array} \right\}$$

and $\Gamma^{(\mathrm{TP})}(P) = X \setminus \Omega^{(\mathrm{TP})}(P)$ (for details on Banach limits (used in the definition of $\Omega^{(\mathrm{TP})}(P)$ and, implicitly, in the definition of $\Gamma^{(\mathrm{TP})}(P)$, as well) see Royden [103], which is a widely used real analysis textbook, or the subsection *Banach Limits* of Section 1.3 of [143], and the references therein).

The superscript $^{(\mathrm{TP})}$, which stands for transition probability, will always be used when dealing with the sets of the KBBY decomposition defined by a transition probability, and we will use no superscript when we consider the corresponding sets of the decomposition defined by a transition function because the transition functions are our main object of study in this book. Also, if it will be clear from the context which transition probability is under consideration, we will use the notation $\Omega^{(\mathrm{TP})}$ and $\Gamma^{(\mathrm{TP})}$ rather than $\Omega^{(\mathrm{TP})}(P)$ and $\Gamma^{(\mathrm{TP})}(P)$, respectively, and the same convention applies to all the other sets of the KBBY decomposition defined by a transition probability.

Let $x \in X$, let L be a Banach limit, and let $\varepsilon_x^{(L)} : C_0(X) \to \mathbb{R}$ be a function defined by $\varepsilon_x^{(L)}(f) = L((S^n f(x))_{n \in \mathbb{N} \cup \{0\}})$ for every $f \in C_0(X)$. Using the fact that $\varepsilon_x^{(L)}$ is a linear functional, and since $\varepsilon_x^{(L)}$ is also positive (the positivity of $\varepsilon_x^{(L)}$ means that $\varepsilon_x^{(L)}(f) \geq 0$ for every $f \in C_0(X)$, $f \geq 0$), we obtain that $\varepsilon_x^{(L)}$ is continuous. Since $\varepsilon_x^{(L)}$ belongs to the topological dual of $C_0(X)$, we will think of $\varepsilon_x^{(L)}$ as an element of $\mathcal{M}(X)$. If $\varepsilon_x^{(L)} \neq 0$, then we call $\varepsilon_x^{(L)}$ an *elementary measure* (*for* P, or *for* T, or *for* (S, T)).

For every $f \in B_b(X)$ and $n \in \mathbb{N}$, let $\mathbf{A}_n^{(S)} f : X \to \mathbb{R}$ be defined by

$$\mathbf{A}_n^{(S)} f(x) = \frac{1}{n} \sum_{k=0}^{n-1} S^k f(x) \text{ for every } x \in X.$$

Set

$$\mathcal{D}^{(\mathrm{TP})} = \left\{ x \in X \;\middle|\; \begin{array}{l} \text{the sequence } \left(\mathbf{A}_n^{(S)} f(x)\right)_{n \in \mathbb{N}} \\ \text{converges to zero for} \\ \text{every } f \in C_0(X) \end{array} \right\}$$

and $\Gamma_0^{(\mathrm{TP})} = X \setminus \mathcal{D}^{(\mathrm{TP})}$.

We call $\mathcal{D}^{(\mathrm{TP})}$ the *dissipative part of X defined* or *generated by P* (or *defined* or *generated by (S, T)*). We say that P (or (S, T)) is *dissipative* if $\mathcal{D}^{(\mathrm{TP})} = X$.

Note that $\Gamma_0^{(\mathrm{TP})} \subseteq \Gamma^{(\mathrm{TP})}$.

Note also that, using Theorem 1.2.2, we obtain that a dissipative transition probability cannot have invariant probability measures; hence, such a transition probability cannot have nonzero invariant elements of $\mathcal{M}(X)$.

It is easy to find dissipative transition probabilities. For instance, if $u : \mathbb{R} \to \mathbb{R}$ is defined by $u(x) = x + 1$ for every $x \in \mathbb{R}$, then the transition probability P_u defined by u on $\mathbb{R}$ is dissipative. A more sophisticated example (but still easy to handle) is the transition probability P_h defined on $\mathrm{PSL}(2, \mathbb{R})$ in Example 1.1.9, where $\begin{pmatrix} e^{\frac{1}{2}} & 0 \\ 0 & e^{-\frac{1}{2}} \end{pmatrix}$.

Note that if (X, d) is a compact metric space, then the dissipative part of any transition probability P on (X, d) is empty because if (S, T) is the Markov pair defined by P, then the sequence $(\frac{1}{n} \sum_{k=0}^{n-1} S^k \mathbf{1}_X(x))_{n \in \mathbb{N}}$ converges to 1 for every $x \in X$, and $\mathbf{1}_X$ is a continuous function that vanishes at infinity if X is compact. Thus, for instance, the dissipative parts of the transition probabilities defined in Examples 1.1.6–1.1.8, 1.1.12 and 1.1.13 are all empty.

A major concern when dealing with the sets that appear in the KBBY decomposition is whether or not these sets are measurable.

In the next proposition we deal with the measurability of $\mathcal{D}^{(\mathrm{TP})}$ and $\Gamma_0^{(\mathrm{TP})}$.

Proposition 1.3.1. *The sets $\mathcal{D}^{(\mathrm{TP})}$ and $\Gamma_0^{(\mathrm{TP})}$ are Borel subsets of X.*

Now set

$$\Gamma_{\mathrm{c}}^{(\mathrm{TP})} = \left\{ x \in \Gamma_0^{(\mathrm{TP})} \;\middle|\; \begin{array}{l} \text{for every } f \in C_0(X), \text{ the sequence} \\ \left(\mathbf{A}_n^{(S)} f(x) \right)_{n \in \mathbb{N}} \text{ is convergent} \end{array} \right\}.$$

The definition of $\Gamma_{\mathrm{c}}^{(\mathrm{TP})}$ "strongly suggests" that we define, for every $x \in \Gamma_{\mathrm{c}}^{(\mathrm{TP})}$, a map $\varepsilon_x^{(\mathrm{TP})} : C_0(X) \to \mathbb{R}$ as follows: $\varepsilon_x^{(\mathrm{TP})}(f) = \lim_{n \to \infty} \frac{1}{n} \sum_{k=0}^{n-1} S^k f(x)$ for every $f \in C_0(X)$.

Clearly, the functions $\varepsilon_x^{(\mathrm{TP})}$, $x \in \Gamma_{\mathrm{c}}^{(\mathrm{TP})}$, are well-defined in the sense that $\lim_{n \to \infty} \frac{1}{n} \sum_{k=0}^{n-1} S^k f(x)$ exists for every $f \in C_0(X)$ and $x \in \Gamma_{\mathrm{c}}^{(\mathrm{TP})}$. It is also easy to see that $\varepsilon_x^{(\mathrm{TP})}$, $x \in \Gamma_{\mathrm{c}}^{(\mathrm{TP})}$, are linear functionals, are nonzero (because $\Gamma_{\mathrm{c}}^{(\mathrm{TP})} \subseteq \Gamma_0^{(\mathrm{TP})}$), and are positive, in the sense that $\varepsilon_x^{(\mathrm{TP})}(f) \geq 0$ for every $f \in C_0(X)$, $f \geq 0$, $x \in \Gamma_{\mathrm{c}}^{(\mathrm{TP})}$. Thus, $\varepsilon_x^{(\mathrm{TP})}$, $x \in \Gamma_{\mathrm{c}}^{(\mathrm{TP})}$, are also continuous, so we may and do think of $\varepsilon_x^{(\mathrm{TP})}$, $x \in \Gamma_{\mathrm{c}}^{(\mathrm{TP})}$, as positive nonzero elements of $\mathcal{M}(X)$. Using the same arguments as in the case in which P is a Feller transition probability, we obtain that $\varepsilon_x^{(\mathrm{TP})}$, $x \in \Gamma_{\mathrm{c}}^{(\mathrm{TP})}$, are elementary measures. The measures $\varepsilon_x^{(\mathrm{TP})}$, $x \in \Gamma_{\mathrm{c}}^{(\mathrm{TP})}$, are called *standard elementary measures*.

It seems of interest to us that one can obtain rather nice characterizations of the supports of the standard elementary measures (that is, one can obtain what we like to call "formulas" for these supports). Since we will discuss such "formulas" for supports of standard elementary measures defined by transition functions, let us briefly outline here the "formulas" for the supports when these measures are defined by transition probabilities.

Let $(\nu_n)_{n\in\mathbb{N}}$ be a sequence of probability measures, $\nu_n \in \mathcal{M}(X)$ for every $n \in \mathbb{N}$. We will use the notation $\tilde{\nu}_n$ for the average of $\nu_1, \nu_2, \ldots, \nu_n$ (so, $\tilde{\nu}_n = \dfrac{1}{n}\displaystyle\sum_{k=1}^{n}\nu_k$) for every $n \in \mathbb{N}$.

Set

$$
G^{\sim}\left((\nu_n)_{n\in\mathbb{N}}\right) = \left\{ x \in X \;\middle|\; \begin{array}{l} \text{there exists an open neighborhood } U \text{ of } x \\ \text{such that the sequence } (\tilde{\nu}_n(U))_{n\in\mathbb{N}} \\ \text{converges to zero} \end{array} \right\}
$$

and

$$
G_{\sim}\left((\nu_n)_{n\in\mathbb{N}}\right) = \left\{ x \in X \;\middle|\; \begin{array}{l} \text{there exists an open neighborhood } U \text{ of } x \\ \text{and a subsequence } \left(\tilde{\nu}_{n_l}\right)_{l\in\mathbb{N}} \text{ of } (\tilde{\nu}_n)_{n\in\mathbb{N}} \\ \text{such that the sequence } \left(\tilde{\nu}_{n_l}\right)_{l\in\mathbb{N}} \\ \text{converges to zero} \end{array} \right\}.
$$

Also set

$$
\overline{\mathrm{Lsupp}}_{n\to\infty}\,\nu_n = X \setminus G^{\sim}\left((\nu_n)_{n\in\mathbb{N}}\right)
$$

and

$$
\underline{\mathrm{Lsupp}}_{n\to\infty}\,\nu_n = X \setminus G_{\sim}\left((\nu_n)_{n\in\mathbb{N}}\right).
$$

Since $G^{\sim}\left((\nu_n)_{n\in\mathbb{N}}\right)$ and $G_{\sim}\left((\nu_n)_{n\in\mathbb{N}}\right)$ are open subsets of X, it follows that $\overline{\mathrm{Lsupp}}_{n\to\infty}\,\nu_n$ and $\underline{\mathrm{Lsupp}}_{n\to\infty}\,\nu_n$ are closed. We call the sets $\overline{\mathrm{Lsupp}}_{n\to\infty}\,\nu_n$ and $\underline{\mathrm{Lsupp}}_{n\to\infty}\,\nu_n$ the *mean upper limit support* (*m.u.l.s.* and) the *mean lower limit support* (*m.l.l.s.* of) the sequence $(\nu_n)_{n\in\mathbb{N}}$, respectively. The reason for the terminology stems from the fact that we think of $\overline{\mathrm{Lsupp}}_{n\to\infty}\,\nu_n$ and $\underline{\mathrm{Lsupp}}_{n\to\infty}\,\nu_n$ as a kind of upper and lower limit of the supports of the averages $\dfrac{1}{n}\displaystyle\sum_{k=1}^{n}\nu_k$, $n \in \mathbb{N}$, respectively. Note that $\underline{\mathrm{Lsupp}}_{n\to\infty}\,\nu_n \subseteq \overline{\mathrm{Lsupp}}_{n\to\infty}\,\nu_n$ for every sequence $(\nu_n)_{n\in\mathbb{N}}$ of probability measures, $\nu_n \in \mathcal{M}(X)$ for every $n \in \mathbb{N}$. If a sequence $(\nu_n)_{n\in\mathbb{N}}$, $\nu_n \in \mathcal{M}(X)$, $\nu_n \geq 0$, $\|\nu_n\| = 1$ for every $n \in \mathbb{N}$, has the property that $\overline{\mathrm{Lsupp}}_{n\to\infty}\,\nu_n = \underline{\mathrm{Lsupp}}_{n\to\infty}\,\nu_n$ then we say that $(\nu_n)_{n\in\mathbb{N}}$ has a *limit support in the mean* (*l.s.m.*), we denote by $\mathrm{Lsupp}_{n\to\infty}\,\nu_n$ any of the sets $\overline{\mathrm{Lsupp}}_{n\to\infty}\,\nu_n$ or $\underline{\mathrm{Lsupp}}_{n\to\infty}\,\nu_n$, and we call $\mathrm{Lsupp}_{n\to\infty}\,\nu_n$ the limit support in the mean (*l.s.m.*) of $(\nu_n)_{n\in\mathbb{N}}$.

We need the following proposition:

Proposition 1.3.2. *Let $(\nu_n)_{n\in\mathbb{N}}$ be a sequence of probability measures on $(X, \mathcal{B}(X))$. If the sequence of averages $(\tilde{\nu}_n)_{n\in\mathbb{N}}$ converges in the weak* topology of $\mathcal{M}(X)$, then $(\nu_n)_{n\in\mathbb{N}}$ has an l.s.m., and $\mathrm{Lsupp}_{n\to\infty}\nu_n = \mathrm{supp}\,\nu$, where ν is the weak* limit of $(\tilde{\nu}_n)_{n\in\mathbb{N}}$.*

If $(\nu_n)_{n\in\mathbb{N}}$ is a sequence of probability measures, $\nu_n \in \mathcal{M}(X)$ for every $n \in \mathbb{N}$, and the sequence $(\tilde{\nu}_n)_{n\in\mathbb{N}}$ of averages does not converge in the weak* topology of $\mathcal{M}(X)$, then $(\nu_n)_{n\in\mathbb{N}}$ may or may not have an l.s.m.

Using the l.s.m. we can obtain "formulas" for the supports of the standard elementary measures. We do this in the next theorem.

Theorem 1.3.3. *Let $x \in \Gamma_c^{(\mathrm{TP})}$. Then the sequence $(T^n\delta_x)_{n\in\mathbb{N}\cup\{0\}}$ has an l.s.m. and*
$$\mathrm{supp}\,\varepsilon_x^{(\mathrm{TP})} = \mathrm{Lsupp}_{n\to\infty}(T^n\delta_x).$$

Proof. Use the definition of $\Gamma_c^{(\mathrm{TP})}$ and Proposition 1.3.2. $\square$

Theorem 1.3.3 has the following consequence:

Corollary 1.3.4. *Let $x \in \Gamma_c^{(\mathrm{TP})}$. Then $\mathrm{supp}\,\varepsilon_x^{(\mathrm{TP})} \subseteq \overline{\mathcal{O}^{(\mathrm{TP})}(x)}$, where $\overline{\mathcal{O}^{(\mathrm{TP})}(x)}$ is the orbit-closure of x under the action of T.*

Proof. Let $x \in \Gamma_c^{(\mathrm{TP})}$, and assume that there exists a $y \in \mathrm{supp}\,\varepsilon_x^{(\mathrm{TP})}$ such that $y \notin \overline{\mathcal{O}^{(\mathrm{TP})}(x)}$. Then there exists an open neighborhood U of y such that $U \cap \overline{\mathcal{O}^{(\mathrm{TP})}(x)} = \emptyset$. Thus, $T^n\delta_x(U) = 0$ for every $n \in \mathbb{N} \cup \{0\}$, so $\limsup_{n\to\infty} \frac{1}{n}\sum_{k=0}^{n-1} T^k\delta_x(U) = 0$. Accordingly, $y \in G^{\sim}((T^n\delta_x)_{n\in\mathbb{N}\cup\{0\}})$; that is, $y \notin \overline{\mathrm{Lsupp}_{n\to\infty}(T^n\delta_x)} = \mathrm{Lsupp}_{n\to\infty}(T^n\delta_x)$. We have obtained a contradiction because, by Theorem 1.3.3, $\mathrm{Lsupp}_{n\to\infty}(T^n\delta_x) = \mathrm{supp}\,\varepsilon_x^{(\mathrm{TP})}$. Since the contradiction stems from our assumption that $y \in (\mathrm{supp}\,\varepsilon_x^{(\mathrm{TP})}) \setminus \overline{\mathcal{O}^{(\mathrm{TP})}(x)}$, we obtain that $\mathrm{supp}\,\varepsilon_x^{(\mathrm{TP})}$ is indeed a subset of $\overline{\mathcal{O}^{(\mathrm{TP})}(x)}$. $\square$

Note that assertion (a) of Theorem 2.2.1 of [143] holds true for any elementary measure of any transition probability (not necessarily Feller) because the proof given in [143] is also valid for elementary measures of transition probabilities that fail to be Feller. Thus, Corollary 1.3.4 offers another proof of (a) of Theorem 2.2.1 of [143] in the case in which the transition probability is not necessarily Feller, but the elementary measures are standard elementary measures.

We now return to the KBBY decomposition defined by the transition probability P.

Proposition 1.3.5. *The set $\Gamma_c^{(\mathrm{TP})}$ belongs to $\mathcal{B}(X)$.*

Now set $\Gamma_{\mathrm{cp}}^{(\mathrm{TP})} = \left\{x \in \Gamma_c^{(\mathrm{TP})} \,\big|\, \left\|\varepsilon_x^{(\mathrm{TP})}\right\| = 1\right\} = \{x \in \Gamma_c^{(\mathrm{TP})} | \varepsilon_x^{(\mathrm{TP})} \in \mathbb{P}(X)\}$, where $\mathbb{P}(X)$ is the set of all probability measures in $\mathcal{M}(X)$; that is, $\mathbb{P}(X) = \{\mu \in \mathcal{M}(X) | \mu \geq 0, \|\mu\| = 1\}$. Thus, $\Gamma_{\mathrm{cp}}^{(\mathrm{TP})}$ is the set of all $x \in \Gamma_c^{(\mathrm{TP})}$ such that $\varepsilon_x^{(\mathrm{TP})}$

is a probability measure. Note that the total variation norm of an elementary (not necessarily standard) measure cannot exceed 1.

Proposition 1.3.6. *The set $\Gamma_{\mathrm{cp}}^{\mathrm{(TP)}}$ is $\mathcal{B}(X)$-measurable.*

Note that every function $f \in B_b(X)$ is integrable with respect to $\varepsilon_x^{\mathrm{(TP)}}$ whenever $x \in \Gamma_{\mathrm{c}}^{\mathrm{(TP)}}$. Thus, for every $f \in B_b(X)$, the function

$$f^{\square}(x) = \begin{cases} \int\limits_{X} f(y)\,\mathrm{d}\varepsilon_x^{\mathrm{(TP)}}(y) & \text{if } x \in \Gamma_{\mathrm{c}}^{\mathrm{(TP)}} \\ 0 & \text{if } x \notin \Gamma_{\mathrm{c}}^{\mathrm{(TP)}} \end{cases} \tag{1.3.1}$$

is well-defined.

Proposition 1.3.7. *For every $f \in B_b(X)$, the function $f^{\square}$ is measurable.*

The functions $f^{\square}$, $f \in B_b(X)$, play a significant role in the study of the KBBY decomposition. For instance, the functions $f^{\square}$, $f \in C_0(X)$, can be used in the proof of Proposition 1.3.6.

A natural question at this point is whether all the elementary measures are T-invariant or not. It is tempting to believe that the elementary measures are T-invariant because if P is a Feller transition probability, then every elementary measure is T-invariant. However, if P is not Feller, it is no longer true that every elementary measure is T-invariant. There exist transition probabilities that have even standard elementary probability measures that are not invariant. A case in point is the transition probability P_w that appears in Example 1.1.6. Indeed, let (S_w, T_w) be the Markov pair defined by P_w and note that the sequence of averages $(\frac{1}{n} \sum_{k=0}^{n-1} S_w^k f(x))_{n \in \mathbb{N}}$ converges to $f(0)$ (because $S_w^l f(x) = f(w^l(x))$ for every $l \in \mathbb{N}$, so the sequence $(S_w^m f(x))_{m \in \mathbb{N} \cup \{0\}}$ converges to $f(0)$) for every $f \in C([0,1])$ and $x \in [0,1]$; we therefore obtain that $\Gamma_{\mathrm{cp}}^{\mathrm{(TP)}} = [0,1]$ and $\varepsilon_x^{\mathrm{(TP)}} = \delta_0$ for every $x \in [0,1]$, where δ_0 is the Dirac measure concentrated at 0; however, δ_0 is not a T_w-invariant measure because $T_w \delta_0 = \delta_1$.

The fact that the standard elementary probability measures are not necessarily T-invariant suggests to us that we should study the T-invariant standard elementary probability measures separately.

Thus, set $\Gamma_{\mathrm{cpi}}^{\mathrm{(TP)}} = \{x \in \Gamma_{\mathrm{cp}}^{\mathrm{(TP)}} \mid T\varepsilon_x^{\mathrm{(TP)}} = \varepsilon_x^{\mathrm{(TP)}}\}$.

Using Proposition 1.3.7, we obtain that the following proposition holds true.

Proposition 1.3.8. *The set $\Gamma_{\mathrm{cpi}}^{\mathrm{(TP)}}$ belongs to $\mathcal{B}(X)$.*

Now, in order for the set $\Gamma_{\mathrm{cpi}}^{\mathrm{(TP)}}$ to be of interest in the study of the invariant probabilities of P, we would like $\Gamma_{\mathrm{cpi}}^{\mathrm{(TP)}}$ to be a set "where the action is"; that is, we would like $\Gamma_{\mathrm{cpi}}^{\mathrm{(TP)}}$ to be a set of maximal probability. It turns out that $\Gamma_{\mathrm{cpi}}^{\mathrm{(TP)}}$ is indeed a set of maximal probability.

We will now outline the results needed to prove that $\Gamma_{\mathrm{cpi}}^{\mathrm{(TP)}}$ is a set of maximal probability, results which are also of interest in their own right.

First we need the following proposition:

Proposition 1.3.9. *The set $\Gamma_{\mathrm{c}}^{(\mathrm{TP})}$ is of maximal probability.*

For every $f \in B_b(X)$, set

$$A_f = \left\{ x \in X \;\middle|\; \text{the sequence } \left(\mathbf{A}_n^{(S)} f(x) \right)_{n \in \mathbb{N}} \text{ converges} \right\}$$

and let $f^* : X \to \mathbb{R}$ be defined as follows:

$$f^*(x) = \begin{cases} \lim\limits_{n \to \infty} \mathbf{A}_n^{(S)} f(x) & \text{if } x \in A_f \cap \Gamma_{\mathrm{c}}^{(\mathrm{TP})} \\ 0 & \text{if } x \notin A_f \cap \Gamma_{\mathrm{c}}^{(\mathrm{TP})}. \end{cases}$$

Note that using Proposition 1.3.5 and the fact that $A_f \in \mathcal{B}(X)$, we obtain that $f^* \in B_b(X)$ for every $f \in B_b(X)$.

Using Theorem 1.2.2 and the functions f^*, $f \in C_0(X)$, we obtain the following proposition:

Proposition 1.3.10. *The set $\Gamma_{\mathrm{cp}}^{(\mathrm{TP})}$ is of maximal probability.*

In the next proposition, we exhibit a significant relationship between the functions f^* and $f^{\square}$, $f \in B_b(X)$.

Proposition 1.3.11. *Assume that P has invariant probabilities, and let μ be such an invariant probability. Then $f^* = f^{\square}$ μ-a.e. for every $f \in B_b(X)$.*

Using Propositions 1.3.9–1.3.11, we obtain that $\Gamma_{\mathrm{cpi}}^{(\mathrm{TP})}$ is a set of maximal probability. For future reference we state this result in the next theorem.

Theorem 1.3.12. *The set $\Gamma_{\mathrm{cpi}}^{(\mathrm{TP})}$ is of maximal probability.*

Note that Theorem 1.3.12 has the following corollary, whose proof is obvious:

Corollary 1.3.13. *The following assertions are equivalent:*

(a) P has at least one invariant probability.
(b) The set $\Gamma_{\mathrm{cpi}}^{(\mathrm{TP})}$ is nonempty.

When we look at the probability measures $\varepsilon_x^{(\mathrm{TP})}$, $x \in \Gamma_{\mathrm{cpi}}^{(\mathrm{TP})}$, we notice that they have a certain minimality property discussed in Corollary 1.3.4: for every $x \in \Gamma_{\mathrm{cpi}}^{(\mathrm{TP})}$, the support of $\varepsilon_x^{(\mathrm{TP})}$ is included in the orbit-closure of x under the action of P. On the other hand, the probability measures that are invariant for P and ergodic possess a minimality property, as well, in the sense that if μ is such an invariant ergodic probability measure, then μ cannot be the sum of two nonzero mutually singular measures that are invariant for P. Thus, it makes sense to wonder whether the probability measures $\varepsilon_x^{(\mathrm{TP})}$, $x \in \Gamma_{\mathrm{cpi}}^{(\mathrm{TP})}$, are ergodic or not. It turns out that these measures are not necessarily ergodic even if the transition probability is Feller; for a remarkably simple example of a standard elementary (invariant) probability measure that fails to be ergodic, see Example 2.2.4 of [143], which was suggested by one of the anonymous referees of [143].

In view of the above discussion, a reasonable approach to studying the structure of the set of ergodic invariant probability measures for P is to try to find out if the set $\{\varepsilon_x^{(TP)} | x \in \Gamma_{cpi}^{(TP)}\}$ contains the ergodic probability measures that are invariant for P.

To this end, set

$$\Gamma_{cpie}^{(TP)} = \{x \in \Gamma_{cpi}^{(TP)} \mid \varepsilon_x^{(TP)} \text{ is ergodic}\}.$$

A subset A of X, $A \in \mathcal{B}(X)$, is called a *P-invariant set* (or an *invariant set for P*) if $P(x, A) = 1$ for every $x \in A$.

We will need the following lemma:

Lemma 1.3.14. *Let μ be an invariant probability measure for P. The following assertions are equivalent:*

(a) μ is ergodic.
(b) $\mu(A) = 0$ or 1 for every P-invariant set A.
(c) There exists a Borel subset B of X such that $\mu(B) = 1$ and such that the sequence $(\frac{1}{n} \sum_{k=0}^{n-1} S^k f(x))_{n \in \mathbb{N}}$ converges to $\int f(y) \, d\mu(y)$ for every $f \in C_0(X)$ and $x \in B$.

Proof. A proof of the equivalence of (b) and (c) appears in Lemma 3.3.1 of [143]. Note that even though we are dealing with Markov-Feller pairs in the above-mentioned lemma, the proof also remains valid for Markov pairs that are not necessarily Markov-Feller pairs. Note also that a proof of $(a) \Rightarrow (b)$ is incorporated in the proof of the second implication in Lemma 3.3.1 of [143]. Thus, we have to prove only the implication $(b) \Rightarrow (a)$.

To this end, assume that μ is not ergodic. Then, there exist two nonzero mutually singular measures μ_1 and μ_2 that are invariant for P such that $\mu = \mu_1 + \mu_2$.

Since μ_1 and μ_2 are mutually singular, it follows that there exists a subset A_1 of X, $A_1 \in \mathcal{B}(X)$ such that $\mu_1(X \setminus A_1) = 0$ and $\mu_2(A_1) = 0$.

Since $\mu = \mu_1 + \mu_2$ and since both μ_1 and μ_2 are nonzero measures, it follows that $0 < \mu(A_1) < 1$.

Taking into consideration that μ_1 is an invariant measure for P, we obtain that $0 = \mu_1(X \setminus A_1) = \int P(x, X \setminus A_1) \, d\mu_1(x)$; therefore, $P(x, X \setminus A_1) = 0$ for μ_1-a.e. $x \in X$. Since P is a transition probability, it follows that $P(x, A_1) = 1$ for μ_1-a.e. $x \in X$. Thus, there exists a subset A_2 of A_1, $A_2 \in \mathcal{B}(X)$ such that $\mu_1(A_2) = \mu_1(A_1)$ and such that $P(x, A_1) = 1$ for every $x \in A_2$.

Since μ_1 is invariant for P, it follows that $\mu_1(A_1) = \int P(x, A_1) \, d\mu_1(x)$ and $\mu_1(A_2) = \int P(x, A_2) \, d\mu_1(x)$. Taking into consideration that $P(x, A_2) \leq P(x, A_1)$ for every $x \in X$ (and that $\mu_1(A_1) = \mu_1(A_2)$), we obtain that $P(x, A_2) = P(x, A_1)$ for μ_1-a.e. $x \in X$, so $P(x, A_2) = 1$ for μ_1-a.e. $x \in X$. Therefore, there exists a subset A_3 of A_2, $A_3 \in \mathcal{B}(X)$ such that $\mu_1(A_3) = \mu_1(A_2)$ and such that $P(x, A_2) = 1$ for every $x \in A_3$.

In general, assume that we have constructed the sets $A_1, A_2, \ldots, A_k$ such that $A_l \in \mathcal{B}(X)$ for every $l = 1, 2, \ldots, k$, $A_1 \supseteq A_2 \supseteq \cdots \supseteq A_k$, $\mu_1(A_k) = \mu_1(A_1)$, and such that $P(x, A_l) = 1$ for every $x \in A_{l+1}$, $l = 1, 2, \ldots, k - 1$. Then using the fact that μ_1 is invariant for P, we obtain that $\mu_1(A_1) = \int P(x, A_1) \, d\mu_1(x)$ and $\mu_1(A_k) = \int P(x, A_k) \, d\mu_1(x)$. Since $P(x, A_k) \leq P(x, A_1)$ for every $x \in X$ and since $\mu_1(A_1) = \mu_1(A_k)$, it follows that $P(x, A_k) = P(x, A_1)$ for μ_1-a.e. $x \in X$, so $P(x, A_k) = 1$ for μ_1-a.e. $x \in X$. Therefore, there exists a subset A_{k+1} of A_k such that $A_{k+1} \in \mathcal{B}(X)$, $\mu_1(A_{k+1}) = \mu_1(A_k) = \mu_1(A_1)$, and such that $P(x, A_k) = 1$ for every $x \in A_{k+1}$.

Continuing in this way, we obtain a sequence $(A_k)_{k \in \mathbb{N}}$ of elements of $\mathcal{B}(X)$ such that $P(x, A_k) = 1$ for every $x \in A_{k+1}$, $A_{k+1} \subseteq A_k$, and $\mu_1(A_k) = \mu_1(A_{k+1})$ for every $k \in \mathbb{N}$.

Now set $A = \bigcap_{k \in \mathbb{N}} A_k$. Then $A \in \mathcal{B}(X)$, $0 < \mu(A) < 1$ (because $\mu_1(A) = \mu_1(A_1) = \mu(A_1)$ and $A \subseteq A_1$), and if $x \in A$, then $P(x, A) = 1$ because $P(x, A_k) = 1$ for every $k \in \mathbb{N}$.

We have obtained a contradiction because A is a P-invariant set and $0 < \mu(A) < 1$. The contradiction stems from our assumption that μ is not ergodic. $\qquad\square$

Set

$$
\Theta = \left\{ x \in \Gamma_{\mathrm{cpi}}^{(\mathrm{TP})} \;\middle|\; \int_{\Gamma_{\mathrm{cpi}}^{(\mathrm{TP})}} (f^*(y) - f^*(x))^2 \, d\varepsilon_x^{(\mathrm{TP})}(y) = 0 \text{ for every } f \in C_0(X) \right\}.
$$

Roughly speaking, the motivation for defining the set Θ can be described as follows: if $x \in \Theta$, then for every $f \in C_0(X)$ there exists a Borel measurable subset $A_{x,f}$ of X such that $\varepsilon_x^{(\mathrm{TP})}(A_{x,f}) = 1$ and such that f^* is constant on $A_{x,f}$; next, in view of the fact that $C_0(X)$ is a separable Banach space, we can hope that the set $A_{x,f}$, which depends on x and f, can be chosen independent of f, and, in this case, we can use (c) of Lemma 1.3.14 to conclude that $\varepsilon_x^{(\mathrm{TP})}$ is an ergodic invariant probability measure. Furthermore, we might be even able to prove that any ergodic invariant probability measure is of the form $\varepsilon_x^{(\mathrm{TP})}$ for some $x \in \Theta$.

We will now make all these ideas precise.

Proposition 1.3.15. *The set Θ is Borel measurable.*

In view of Proposition 1.3.15, we may now inquire if Θ is a set of maximal probability. The answer is given in the next theorem.

Theorem 1.3.16. *The set Θ is of maximal probability.*

The above theorem is proved using the following lemma:

Lemma 1.3.17. *Assume that P has at least one invariant probability, and let $\mu \in \mathcal{M}(X)$ be such an invariant probability. Also, let $f \in C_0(X)$. Then:*

(a) The function $g_f : \Gamma_{\mathrm{cpi}}^{(\mathrm{TP})} \to \mathbb{R}$ defined by

$$
g_f(x) = \int_{\Gamma_{\mathrm{cpi}}^{(\mathrm{TP})}} (f^*(y) - f^*(x))^2 \, d\varepsilon_x^{(\mathrm{TP})}(y)
$$

for every $x \in \Gamma_{\mathrm{cpi}}^{(\mathrm{TP})}$ *is measurable and the integral* $\int_{\Gamma_{\mathrm{cpi}}^{(\mathrm{TP})}} g_f(x)\,\mathrm{d}\mu(x)$ *is well-defined.*

(b) $\int_{\Gamma_{\mathrm{cpi}}^{(\mathrm{TP})}} g_f(x)\,\mathrm{d}\mu(x) = 0.$

Note that because we assume that P has at least one invariant probability, $\Gamma_{\mathrm{cpi}}^{(\mathrm{TP})}$ is nonempty by Corollary 1.3.13, so the assertions of the above lemma are meaningful.

Now let $\sim$ be a binary relation on Θ defined as follows: $x \sim y$ if and only if $\lim_{n\to\infty} \frac{1}{n} \sum_{k=0}^{n-1} S^k f(x) = \lim_{n\to\infty} \frac{1}{n} \sum_{k=0}^{n-1} S^k f(y)$ for every $f \in C_0(X)$, where $x \in \Theta$ and $y \in \Theta$. Note that the definition of $\sim$ makes sense because $\Theta \subseteq \Gamma_{\mathrm{cpi}}^{(\mathrm{TP})}$, so the limits used in defining $\sim$ do exist. Note also that $x \sim y$ if and only if $f^*(x) = f^*(y)$ for every $f \in C_0(X)$, where $x \in \Theta$ and $y \in \Theta$.

Clearly, $\sim$ is an equivalence relation. We will denote by $[x]$ the equivalence class of $x \in \Theta$ defined by $\sim$.

Proposition 1.3.18. *For every* $x \in \Theta$, *the set* $[x]$ *is a Borel subset of* X.

The results of the next theorem can be thought of as reaching the climactic stage in the development of the KBBY decomposition for a transition probability.

Theorem 1.3.19. *(a) For every* $x \in \Theta$, *the measure* $\varepsilon_x^{(\mathrm{TP})}$ *is an invariant ergodic probability measure and* $\varepsilon_x^{(\mathrm{TP})}([x]) = 1$ *(note that, by Proposition 1.3.18 the set* $[x]$ *is Borel measurable, so it makes sense to consider* $\varepsilon_x^{(\mathrm{TP})}([x]))$.

(b) Conversely, if μ *is an ergodic invariant probability measure for* P, *then* $\mu = \varepsilon_x^{(\mathrm{TP})}$ *for some* $x \in \Theta$. *Thus,* $\Theta = \Gamma_{\mathrm{cpie}}^{(\mathrm{TP})}$ *and the set* $\{\varepsilon_x^{(\mathrm{TP})} | x \in \Gamma_{\mathrm{cpie}}^{(\mathrm{TP})}\}$ *is precisely the set of all ergodic invariant probability measures for* P.

We will now discuss two consequences of Theorems 1.3.16 and 1.3.19.

Corollary 1.3.20. *The following assertions are equivalent:*

(a) There exists an invariant nonzero measure for P.
(b) $\Gamma_{\mathrm{cpie}}^{(\mathrm{TP})} \neq \emptyset$.
(c) There exists an ergodic invariant probability measure for P.

Corollary 1.3.21. *The following assertions are equivalent:*

(a) The transition probability P *is uniquely ergodic.*
(b) P *has exactly one ergodic invariant probability measure.*

Proof. $(a) \Rightarrow (b)$.   For a proof of this implication, see Corollary 6.7 of [146].
$(b) \Rightarrow (a)$.   Assume that P has exactly one ergodic invariant probability measure, and that P is not uniquely ergodic. Then, it follows that P has at least two distinct invariant probability measures, say μ and ν.

Since $\mu \neq \nu$, there exists an $f \in C_0(X)$ such that $\int_X f\,\mathrm{d}\mu \neq \int_X f\,\mathrm{d}\nu$.

Taking into consideration that $f \in C_0(X)$, we obtain that $\Gamma_c^{(TP)} \subseteq A_f$, where A_f is the subset of X corresponding to f defined before Proposition 1.3.10; hence $\Gamma_{cpie}^{(TP)} \subseteq A_f$.

Using Theorem 1.2.2 and the fact that, by Theorems 1.3.16 and 1.3.19, the set $\Gamma_{cpie}^{(TP)}$ is a set of maximal probability, we obtain that $\int_X f \, d\mu = \int_{\Gamma_{cpie}^{(TP)}} f^* \, d\mu$ and $\int_X f \, d\nu = \int_{\Gamma_{cpie}^{(TP)}} f^* \, d\nu$, where f^* is the function defined before Proposition 1.3.10. Accordingly, $\int_{\Gamma_{cpie}^{(TP)}} f^* \, d\mu \neq \int_{\Gamma_{cpie}^{(TP)}} f^* \, d\nu$.

Since $\int_{\Gamma_{cpie}^{(TP)}} f^* \, d\mu$ and $\int_{\Gamma_{cpie}^{(TP)}} f^* \, d\nu$ are not equal, it follows that the restriction of f^* to $\Gamma_{cpie}^{(TP)}$ cannot be a constant function.

We have obtained a contradiction because under the assumption that P has exactly one ergodic invariant probability measure, f^* has to be constant on $\Gamma_{cpie}^{(TP)}$. The contradiction stems from our assumption that P is not uniquely ergodic. $\square$

Note that if P has invariant probabilities, then $\Gamma_{cpie}^{(TP)} \neq \emptyset$, and if μ is an invariant probability measure for P, then using Theorem 1.2.1 (The Hopf Ergodic Theorem) and various results of this section, we obtain that

$$\int_X f(x) \, d\mu(x) = \int_{\Gamma_{cpie}^{(TP)}} \left(\int_X f(y) \, d\varepsilon_x^{(TP)}(y) \right) d\mu(x) \tag{1.3.2}$$

for every $f \in C_0(X)$. If we use the notation $\left\langle f, \varepsilon_x^{(TP)} \right\rangle = \int_X f(y) \, d\varepsilon_x^{(TP)}(y)$, $x \in \Gamma_{cpie}^{(TP)}$, then the above equality (1.3.2) becomes

$$\int_X f(x) \, d\mu(x) = \int_{\Gamma_{cpie}^{(TP)}} \left\langle f, \varepsilon_x^{(TP)} \right\rangle d\mu(x) \tag{1.3.3}$$

for every $f \in C_0(X)$. Since any element of $\mathcal{M}(X)$ can be thought of as a linear functional on $C_0(X)$, we can interpret the above equalities (1.3.2) and (1.3.3) as stating that any invariant probability measure for P can be expressed as a convex combination in integral form of ergodic measures, and this is a reason why we consider the ergodic invariant probability measures the "building blocks" for all the invariant probabilities.

We will conclude this section with an example. For the notation and the terminology used in it, see Example 1.1.11 (Maps on Spaces of Cosets of $SL(n, \mathbb{R})$, $n \geq 2$) and the topics discussed after Theorem B.4.10.

Example 1.3.22. Let $n \in \mathbb{N}$, $n \geq 2$, let Γ be a lattice in $SL(n, \mathbb{R})$, and let h be a unipotent element of $SL(n, \mathbb{R})$. Thus h is an $n \times n$ matrix in $SL(n, \mathbb{R})$ such that $(h - \mathbf{I}_n)^l = 0$ for some $l \in \mathbb{N}$.

Now let $u_h^{(R)} : (SL(n, \mathbb{R})/\Gamma)_R \to (SL(n, \mathbb{R})/\Gamma)_R$ be defined (as in Example 1.1.11) by $u_h^{(R)}(\Gamma x) = \Gamma x h$ for every $\Gamma x \in (SL(n, \mathbb{R})/\Gamma)_R$, and let $P_{(h)}^{(R)}$ be the (Feller) transition probability defined by $u_h^{(R)}$ on $(SL(n, \mathbb{R})/\Gamma)_R$.

Using one of Ratner's results (Theorem B.4.11), we obtain that $\Gamma_{\text{cpie}}^{(TP)}$ in this case is the entire space (that is, $\Gamma_{\text{cpie}}^{(TP)} = (SL(n, \mathbb{R})/\Gamma)_R$). Moreover, in Theorem B.4.12, Ratner obtains an impressively nice characterization of the ergodic invariant probability measures $\varepsilon_{\Gamma x}^{(TP)}$, $\Gamma x \in (SL(n, \mathbb{R})/\Gamma)_R$, in terms of the algebraic and topological setting in which the transition probability is defined. ∎

1.4 Feller Transition Probabilities

Our goal here is to briefly review several results of [143] that will be extended to Feller transition functions in Chap. 7 of this book.

The present section has two subsections. In the first subsection we review results of Chapters 2 and 3 of [143], while in the second one we discuss results of Chapter 4 of [143] on $C_0(X)$-equicontinuous (necessarily Feller) transition probabilities.

1.4.1 Supports of Elementary and Ergodic Invariant Measures, Minimality, Unique Ergodicity, and Generic Points

1.4.1.1 Elementary and Ergodic Invariant Measures

Let (X, d) be a locally compact separable metric space, let P be a Feller transition probability, and let (S, T) be the Markov-Feller pair defined by P (for the definitions of the Feller transition probabilities and their associated Markov-Feller pairs, see the discussion preceding Proposition 1.1.3). Unless explicitly stated otherwise, the transition probabilities under consideration in this section are assumed to be Feller (and, of course, the Markov pairs defined by these transition probabilities are Markov-Feller pairs).

In view of the results discussed in Sect. 1.3, we know that P defines a KBBY decomposition of the space X, so let $\Omega^{(TP)}$, $\Gamma^{(TP)}$, $\mathcal{D}^{(TP)}$, $\Gamma_0^{(TP)}$, $\Gamma_c^{(TP)}$, $\Gamma_{\text{cp}}^{(TP)}$, $\Gamma_{\text{cpi}}^{(TP)}$ and $\Gamma_{\text{cpie}}^{(TP)}$ be the sets of the decomposition defined by P.

As expected, the fact that the transition probability P is Feller has to have certain consequences concerning the decomposition, and probably the most significant such consequence is described in the next theorem.

Theorem 1.4.1. *Every elementary measure of P is T-invariant. In particular, the standard elementary measures are T-invariant, so $\Gamma_{\text{cp}}^{(TP)} = \Gamma_{\text{cpi}}^{(TP)}$.*

Using Theorem 1.4.1 we obtain several conditions which are both necessary and sufficient for the existence of invariant probabilities for P. For future reference, we state these conditions in the next theorem.

Theorem 1.4.2. *The following assertions are equivalent:*

(a) P has at least one invariant probability.

(b) $\Gamma_0^{(\mathrm{TP})} \neq \emptyset$.

(c) $\Gamma^{(\mathrm{TP})} \neq \emptyset$.

(d) There exist $x_0 \in X$ and compact $K \subseteq X$ such that $\limsup_{n \to +\infty} \mathbf{A}_n^{(S)} \mathbf{1}_K(x_0) > 0$.

(e) There exists $K \subseteq X$, K compact, and a probability measure $\mu_0 \in \mathcal{M}(X)$ such that $\limsup_{n \to \infty} \int_X \mathbf{A}_n^{(S)} \mathbf{1}_K \, d\mu_0 > 0$.

We have already pointed out in the comments that follow Corollary 1.3.4 that if $\varepsilon_x^{(L)(\mathrm{TP})}$ is an elementary measure (standard or not) for some $x \in \Gamma^{(\mathrm{TP})}$ and some Banach limit L, then $\operatorname{supp} \varepsilon_x^{(L)(\mathrm{TP})}$ is included in the orbit-closure $\overline{\mathcal{O}(x)}$ of x under the action of P even if P is not Feller. If P is a Feller transition probability and if $\varepsilon_x^{(L)(\mathrm{TP})}$ is an elementary measure such that $x \in \operatorname{supp} \varepsilon_x^{(L)(\mathrm{TP})}$, then, using a result of Lasota and Myjak (see Proposition 1.1.3), we can actually show that $\operatorname{supp} \varepsilon_x^{(L)(\mathrm{TP})} = \overline{\mathcal{O}(x)}$; for future reference, we state this fact in the next theorem.

Theorem 1.4.3. *Assume that the Feller transition probability P has elementary measures, and let $x \in X$ and L be a Banach limit such that $\varepsilon_x^{(L)(\mathrm{TP})}$ is an elementary measure for P and such that $x \in \operatorname{supp} \varepsilon_x^{(L)(\mathrm{TP})}$. Then $\operatorname{supp} \varepsilon_x^{(L)(\mathrm{TP})} = \overline{\mathcal{O}(x)}$.*

Note that if P is a Feller transition probability that has elementary measures, then we can always find $x \in X$ and a Banach limit L such that $\varepsilon_x^{(L)(\mathrm{TP})}$ is an elementary measure and $x \in \operatorname{supp} \varepsilon_x^{(L)(\mathrm{TP})}$. Indeed, by Theorem 1.4.1, P has nonzero invariant measures because P is a Feller transition probability and we assume that P has elementary measures; therefore, P has invariant probabilities. By Corollary 1.3.20, the set $\Gamma_{\mathrm{cpie}}^{(\mathrm{TP})}$ defined by P is nonempty. If $y \in \Gamma_{\mathrm{cpie}}^{(\mathrm{TP})}$, then $[y] \cap (\operatorname{supp} \varepsilon_y^{(\mathrm{TP})}) \neq \emptyset$ because, by (a) of Theorem 1.3.19, $\varepsilon_y^{(\mathrm{TP})}([y]) = 1$; thus, there exists an $x \in \operatorname{supp} \varepsilon_y^{(\mathrm{TP})}$ such that $x \in \Gamma_{\mathrm{cpie}}^{(\mathrm{TP})}$ and $\varepsilon_x^{(\mathrm{TP})} = \varepsilon_y^{(\mathrm{TP})}$. Finally, $\varepsilon_x^{(\mathrm{TP})} = \varepsilon_x^{(L)(\mathrm{TP})}$ for some Banach limit L because ergodic invariant probability measures are standard elementary measures, and, as pointed out when we defined the standard elementary measures, such a Banach limit L does indeed exist.

In view of the above discussion, we arrive at the following natural question: can we find "formulas" for the supports of ergodic invariant probability measures as we obtained for the supports of standard elementary measures? It turns out that we can, and we will now discuss such "formulas".

Theorem 1.4.4. *Assume that the set $\Gamma_{\mathrm{cpie}}^{(\mathrm{TP})}$ that appears in the KBBY decomposition defined by the Feller transition probability P is nonempty. If $x \in \Gamma_{\mathrm{cpie}}^{(\mathrm{TP})}$, then $\operatorname{supp} \varepsilon_x^{(\mathrm{TP})} = \bigcap_{y \in [x]} \overline{\mathcal{O}^{(\mathrm{TP})}(y)}$.*

Note that using the comments preceding Theorem 1.4.3, we obtain that $\operatorname{supp} \varepsilon_x^{(\mathrm{TP})} \subseteq \bigcap_{y \in [x]} \overline{\mathcal{O}^{(\mathrm{TP})}(y)}$ for every $x \in \Gamma_{\mathrm{cpie}}^{(\mathrm{TP})}$ even if the transition probability P is not Feller. The assumption that P is Feller is used in the proof of Theorem 1.4.4

only to prove that

$$\operatorname{supp} \varepsilon_x^{(\mathrm{TP})} \supseteq \bigcap_{y \in [x]} \overline{\mathcal{O}^{(\mathrm{TP})}(y)} \tag{1.4.1}$$

whenever $x \in \Gamma_{\mathrm{cpie}}^{(\mathrm{TP})}$ because, in order to prove the above inclusion (1.4.1), we use the Lasota and Myjak result Proposition 1.1.3, and the proposition is not true in general if P is not Feller.

The proof of Theorem 1.4.4 yields the following consequence:

Corollary 1.4.5. *Assume as in Theorem 1.4.4 that $\Gamma_{\mathrm{cpie}}^{(\mathrm{TP})}$ is nonempty. Then, for every $x \in \Gamma_{\mathrm{cpie}}^{(\mathrm{TP})}$, the set $(\operatorname{supp} \varepsilon_x^{(\mathrm{TP})}) \cap [x]$ is nonempty and $\operatorname{supp} \varepsilon_x^{(\mathrm{TP})} = \overline{\mathcal{O}^{(\mathrm{TP})}(y)}$ for every $y \in (\operatorname{supp} \varepsilon_x^{(\mathrm{TP})}) \cap [x]$.*

Note that a proof of the corollary can also be obtained by using Theorem 1.4.3 and the comments made after the theorem.

Both Theorem 1.4.4 and Corollary 1.4.5 offer descriptions of the supports of ergodic invariant probability measures. However, the corollary also tells us that the support of each ergodic invariant probability measure is the orbit-closure of some suitably chosen element in X. This kind of information is often useful when studying dynamical systems.

1.4.1.2 Minimality and Unique Ergodicity

We will now briefly discuss several facts about Feller transition probabilities and minimality.

As before, we assume given a Feller transition probability defined on a locally compact separable metric space (X, d).

We start with a characterization of minimality for Feller transition probabilities.

Proposition 1.4.6. *The following assertions are equivalent:*

(a) The transition probability P is minimal.

(b) $\sum_{n=1}^{\infty} P_n(x, U) > 0$ whenever $x \in X$ and U is a nonempty open subset of X, where P_n, $n \in \mathbb{N}$, are the (Feller) transition probabilities defined after Lemma 1.1.1 starting with $P_1 = P$.

Note that we use the fact that P is a Feller transition probability only in showing that the implication $(a) \Rightarrow (b)$ is true (when proving that $(a) \Rightarrow (b)$ we need the assumption that P is Feller because we use Proposition 1.1.3, which fails to be true if P is not Feller); that is, the implication $(b) \Rightarrow (a)$ is true even if the transition probability P fails to be Feller.

A transition probability which satisfies condition (b) of Proposition 1.4.6 is said to be *topologically connected* (see Skorokhod [108]).

It is easy to find Feller transition probabilities that are, or fail to be, minimal. For instance, if we consider the transition probability $P_{\hat{a}}$ on $\mathbb{R}/\mathbb{Z}$ defined by a rotation of the unit circle by a (see Example 1.1.7), then, as pointed out in Example 1.1.7, $P_{\hat{a}}$ is minimal if $a \notin \mathbb{Q}$ and fails to be minimal if $a \in \mathbb{Q}$. Another family of examples is obtained if we consider translations of the torus (see Example 1.1.8). Let $n \in \mathbb{N}$, $n \geq 2$, let $\mathbf{v} \in \mathbb{R}^n$, $\mathbf{v} = (v_1, v_2, \cdots, v_n)$, and let $P_{\mathbf{v}}$ be the transition probability on $\mathbb{R}^n/\mathbb{Z}^n$ defined in Example 1.1.8. Then, as pointed out in Example 1.1.8, $P_{\mathbf{v}}$ is minimal if the numbers $1, v_1, v_2, \cdots, v_n$ are rationally independent. Finally, another family of minimal transition probabilities that we mention here is obtained using the operation of convolution of measures. Let $(H, \cdot, d)$ be a locally compact separable metric group, let $\mu \in \mathcal{M}(H)$ be a probability measure such that $\overline{\cup_{n=1}^{\infty}(\mathrm{supp}\,(\mu^n))} = H$, and let P_{μ} be the transition probability defined in Example 1.1.16. Using (b) of Lemma B.2.2, we obtain that P_{μ} is a minimal transition probability. Note that, since H is a locally compact separable metric group, there are many probability measures μ in $\mathcal{M}(H)$ such that $\mathrm{supp}\,\mu = H$ (indeed, since H is separable, there exist sequences $(d_n)_{n\in\mathbb{N}}$ of elements of H such that the range $\{d_n \,|\, n \in \mathbb{N}\}$ of $(d_n)_{n\in\mathbb{N}}$ is dense in H; if $(d_n)_{n\in\mathbb{N}}$ is such a sequence, then $\mu = \sum_{k=1}^{\infty} \frac{1}{2^n}\delta_{d_n}$ is a probability measure that belongs to $\mathcal{M}(H)$ and that has the property that $\mathrm{supp}\,\mu = H$); if μ is such a probability measure, then the convolution operator T_{μ} defined in Example 1.1.16 is actually trivially minimal, rather than just minimal.

Even though we might be tempted to believe that if a minimal Feller transition probability has nonzero invariant elements, then the transition probability is uniquely ergodic, we emphasize that minimality does not imply unique ergodicity. On the other hand, it is a correct "feeling" that if a minimal Feller transition probability has invariant probabilities, then the support of each such invariant probability is the entire space. For future reference, we state this fact in the next proposition.

Proposition 1.4.7. *Assume that the Feller transition probability P is minimal and has invariant probabilities. If μ is such an invariant probability for P, then* $\mathrm{supp}\,\mu = X$.

If the set $\Omega^{(\mathrm{TP})}$ that appears in the KBBY decomposition defined by P is empty, then Proposition 1.4.7 has the following converse:

Proposition 1.4.8. *Assume that the Feller transition probability P has the property that $\Gamma^{(\mathrm{TP})} = X$ and that $\mathrm{supp}\,\mu = X$ for every invariant probability measure μ for P. Then P is a minimal transition probability.*

If we combine Propositions 1.4.7 and 1.4.8, we obtain the following theorem:

Theorem 1.4.9. *Assume that P has the property that $\Gamma^{(\mathrm{TP})} = X$. Then the following assertions are equivalent:*

(a) The transition probability P is minimal.
(b) $\mathrm{supp}\,\mu = X$ for every invariant probability μ for P, $\mu \in \mathcal{M}(X)$.

If (X, d) is compact, then $\Gamma^{(TP)} = X$ because $\Gamma_0^{(TP)} = X$ in this case, and, as pointed out after defining the set $\Gamma_0^{(TP)}$ in Sect. 1.3, $\Gamma_0^{(TP)} \subseteq \Gamma^{(TP)}$. Thus, in this case, Theorem 1.4.9 becomes:

Corollary 1.4.10. *If (X, d) is a compact metric space, then the following assertions are equivalent:*

(a) P is a minimal transition probability.
(b) If $\mu \in \mathcal{M}(X)$ is an invariant probability for P, then $\operatorname{supp} \mu = X$.

Corollary 1.4.10 complements results of Skorokhod [108] and is an extension of a known result in ergodic theory (the corollary is known if P is induced by a homeomorphism; for details, see Theorem 6.17 of the book by Walters [126]).

Proposition 1.4.8, Theorem 1.4.9 and Corollary 1.4.10 can be used to study the minimality of transition probabilities as shown in the next example.

Example 1.4.11. Let Γ be a lattice of $\mathrm{SL}(2, \mathbb{R})$ such that $(\mathrm{SL}(2, \mathbb{R})/\Gamma)_{\mathrm{L}}$ is a compact space and let $h = \begin{pmatrix} 1 & s \\ 0 & 1 \end{pmatrix}$ for some $s \in \mathbb{R}$, $s \neq 0$. Now consider the map $u_h^{(\mathrm{L})}$ and the transition probability $P_{(h)}^{(\mathrm{L})}$ defined in Example 1.1.11 (note that $P_{(h)}^{(\mathrm{L})}$ is the transition probability induced by $u_h^{(\mathrm{L})}$ and, as pointed out in Example 1.1.11, $P_{(h)}^{(\mathrm{L})}$ is a Feller transition probability).

Using a result of Hedlund (see Dani and Smillie's paper [28] for details concerning this result), we obtain that the horocycle flow $(v_t^{(1\Gamma \mathrm{L})})_{t \in \mathbb{R}}$ defined at (b) of Example B.1.9 is minimal.

Since $(v_t^{(1\Gamma \mathrm{L})})_{t \in \mathbb{R}}$ is a minimal flow, we can conclude from Theorem 2 of Dani and Smillie [28] that $\Gamma_{\mathrm{cpie}}^{(TP)} = (\mathrm{SL}(2, \mathbb{R})/\Gamma)_{\mathrm{L}}$ for the transition probability $P_{(h)}^{(\mathrm{L})}$, that $P_{(h)}^{(\mathrm{L})}$ is strictly ergodic and the unique invariant probability of $P_{(h)}^{(\mathrm{L})}$ is the standard $\mathrm{SL}(2, \mathbb{R})$-invariant probability measure $v_{(\mathrm{SL}(2,\mathbb{R})/\Gamma)_{\mathrm{L}}}$. Thus, using Corollary 1.4.10, we obtain that $P_{(h)}^{(\mathrm{L})}$ is a minimal transition probability. ∎

We have seen in Proposition 1.4.7 that if a Feller transition probability is minimal and has invariant probabilities, then the support of each of these invariant probabilities is the entire space on which the transition probability is defined. In particular, if the transition probability is uniquely ergodic, then the support of the unique invariant probability of the transition probability is the entire space. However, there exist uniquely ergodic Feller transition probabilities for which the supports of their invariant probabilities fail to be the entire space. Thus, a natural question that comes to mind is: can we find "formulas" for the supports of the invariant probabilities of uniquely ergodic Feller transition probabilities (that is, can we find characterizations of the supports of these invariant probabilities)?

We will now briefly discuss such "formulas".

As before, let P be a Feller transition probability defined on (X, d).

Set

$$\gamma^{(\mathrm{TP})} = \bigcap_{x \in \Gamma^{(\mathrm{TP})}} \overline{\mathcal{O}^{(\mathrm{TP})}(x)},$$

$$\gamma_0^{(\mathrm{TP})} = \bigcap_{x \in \Gamma_0^{(\mathrm{TP})}} \overline{\mathcal{O}^{(\mathrm{TP})}(x)},$$

$$\gamma_{\mathrm{c}}^{(\mathrm{TP})} = \bigcap_{x \in \Gamma_{\mathrm{c}}^{(\mathrm{TP})}} \overline{\mathcal{O}^{(\mathrm{TP})}(x)},$$

$$\gamma_{\mathrm{cp}}^{(\mathrm{TP})} = \bigcap_{x \in \Gamma_{\mathrm{cp}}^{(\mathrm{TP})}} \overline{\mathcal{O}^{(\mathrm{TP})}(x)},$$

where, of course, $\Gamma^{(\mathrm{TP})}$, $\Gamma_0^{(\mathrm{TP})}$, $\Gamma_{\mathrm{c}}^{(\mathrm{TP})}$ and $\Gamma_{\mathrm{cp}}^{(\mathrm{TP})}$ are the sets that appear in the KBBY decomposition of X defined by P.

The next theorem gives us the "formulas" we are interested in.

Theorem 1.4.12. *Assume that P is uniquely ergodic, and let μ be the invariant probability for P. Then* $\operatorname{supp}\mu = \gamma_{\mathrm{cp}}^{(\mathrm{TP})} = \gamma_{\mathrm{c}}^{(\mathrm{TP})} = \gamma_0^{(\mathrm{TP})} = \gamma^{(\mathrm{TP})}$.

Note that if P is uniquely ergodic (and Feller), then $\Gamma_{\mathrm{cp}}^{(\mathrm{TP})} = \Gamma_{\mathrm{cpie}}^{(\mathrm{TP})} = [x]$ for every $x \in \Gamma_{\mathrm{cpie}}^{(\mathrm{TP})}$; therefore, the equality $\operatorname{supp}\mu = \gamma_{\mathrm{cp}}^{(\mathrm{TP})}$ in the above theorem is a consequence of Theorem 1.4.4. Although $\operatorname{supp}\mu = \gamma_{\mathrm{cp}}^{(\mathrm{TP})}$ is already a "formula" for the support of μ, it is often more difficult to use than the other equalities in Theorem 1.4.12 because, in general, it is hard to check if an element x of X belongs to $\Gamma_{\mathrm{cp}}^{(\mathrm{TP})}$. For instance, if X is compact, the equality $\operatorname{supp}\mu = \gamma_0^{(\mathrm{TP})}$ allows us to obtain the following simple expression for $\operatorname{supp}\mu$:

Corollary 1.4.13. *Assume that (X,d) is a compact metric space, P is a uniquely ergodic Feller transition probability, and that μ is the invariant probability for P. Then* $\operatorname{supp}\mu = \bigcap_{x \in X} \overline{\mathcal{O}^{(\mathrm{TP})}(x)}$.

In June 1999 during a discussion with Furstenberg, he brought to my attention the following situation that occurs from time to time: We are given a Feller transition probability P defined on (X,d) which has invariant probabilities, is not uniquely ergodic, and has the property that there exists a closed subset F of X such that $\operatorname{supp}\mu = F$ for every invariant probability μ for P. Furstenberg then asked me if we could obtain a "formula" for F similar in spirit to the "formula" for the support of an attractive probability measure obtained in [142]. Even though the question was formulated for a compact metric space, in the next theorem we discuss such "formulas" in the more general setting of a locally compact separable metric space. Note that the same "formulas" that appear in Theorem 1.4.12 are valid for F, as well.

Theorem 1.4.14. *Assume that the Feller transition probability P defined on the locally compact separable metric space (X, d) has invariant probability measures, and that there exists a closed subset F of X such that $\operatorname{supp} \mu = F$ for every invariant probability measure μ for P. Then $F = \gamma_{\mathrm{cp}}^{(\mathrm{TP})} = \gamma_{\mathrm{c}}^{(\mathrm{TP})} = \gamma_{0}^{(\mathrm{TP})} = \gamma^{(\mathrm{TP})}$.*

Theorem 1.4.12, Corollary 1.4.13 and Theorem 1.4.14 can be restated using a terminology that is rather appealing to intuition, stems from the theory of Markov chains (discrete-time Markov processes defined on finite state spaces), and, as used here, can be found in Högnäs and Mukherjea's monograph [48] on p. 180. Following [48], we say that x *leads* to y if $y \in \overline{\mathcal{O}(x)}$, $x \in X$, $y \in X$.

Let A be a nonempty subset of X and let $y \in X$. We say that y is a *universal element with respect to A* if $y \in \bigcap_{x \in A} \overline{\mathcal{O}(x)}$ (that is, if every $x \in A$ leads to y). We say that y is a *universal element* if y is universal with respect to the entire space X. The term "universal element" was suggested to us by Furstenberg.

Using the above terminology, we can rephrase Theorem 1.4.12 as follows:

Theorem 1.4.15. *Let P be uniquely ergodic and let μ be the invariant probability measure for P. Then $\operatorname{supp} \mu$ is the set of all universal elements with respect to A whenever A is any of the following four subsets of X: $\Gamma_{\mathrm{cp}}^{(\mathrm{TP})}$, $\Gamma_{\mathrm{c}}^{(\mathrm{TP})}$, $\Gamma_{0}^{(\mathrm{TP})}$, or $\Gamma^{(\mathrm{TP})}$.*

Note that if P is a uniquely ergodic Feller transition probability and there exist universal elements (with respect to X), then these elements belong to the support of the invariant probability for P.

Corollary 1.4.13 becomes:

Corollary 1.4.16. *Assume that (X, d) is a compact metric space and that the Feller transition probability P is uniquely ergodic. Then the support of the invariant probability for P is the set of all universal elements (generated by P).*

Finally, Theorem 1.4.14 can be restated as follows:

Theorem 1.4.17. *Assume that the Feller transition probability P defined on the locally compact separable metric space (X, d) has invariant probabilities, and there exists a closed subset F of X such that $F = \operatorname{supp} \mu$ for every invariant probability μ for P. Then F is equal to the set of all universal elements with respect to B where B is any of the following four subsets of X: $\Gamma_{\mathrm{cp}}^{(\mathrm{TP})}$, $\Gamma_{\mathrm{c}}^{(\mathrm{TP})}$, $\Gamma_{0}^{(\mathrm{TP})}$, or $\Gamma^{(\mathrm{TP})}$.*

As in the case of Theorem 1.4.15, if there exist universal elements (with respect to X) generated by P, then these elements belong to F.

1.4.1.3 Generic Points

As before, we assume given a Feller transition probability P defined on a locally compact separable metric space (X, d), and we let (S, T) be the Markov-Feller pair defined by P.

An element x of X is called a *generic point* (see Furstenberg [38]) if the sequence $(\frac{1}{n} \sum_{k=0}^{n-1} S^k f(x))_{n \in \mathbb{N}}$ converges for every $f \in C_0(X)$. Thus, $x \in X$ is a generic point if and only if $x \in \mathcal{D}^{(\mathrm{TP})} \cup \Gamma_{\mathrm{c}}^{(\mathrm{TP})}$. The generic points are also known as *quasi-regular points* (see Oxtoby [85] and Krengel's book [53]).

Our goal here is to briefly discuss the use of generic points in the study of the unique ergodicity of Feller transition functions and in the study of the ergodic invariant probability measures of a Feller transition probability.

A generic point $x \in X$ is called *nonsingular* if $x \in \Gamma_{\mathrm{c}}^{(\mathrm{TP})}$. We say that $x \in X$ is *singular* if $x \in \mathcal{D}^{(\mathrm{TP})}$. Thus, a generic point $x \in X$ is nonsingular if and only if there exists an $f \in C_0(X)$ such that $\lim_{n \to \infty} \frac{1}{n} \sum_{k=0}^{n-1} S^k f(x) \neq 0$.

An element x of X is called a *dominant generic point for P* if x is a nonsingular generic point and if the following condition (called the *DGPTP condition*) is satisfied: $\lim_{n \to \infty} \frac{1}{n} \sum_{k=0}^{n-1} S^k f(y) \leq \lim_{n \to \infty} \frac{1}{n} \sum_{k=0}^{n-1} S^k f(x)$ whenever $y \in X$ and $f \in C_0(X)$ are such that the sequence $(\frac{1}{n} \sum_{k=0}^{n-1} S^k f(y))_{n \in \mathbb{N}}$ converges and $f \geq 0$.

The standard elementary measure defined by a dominant generic point is necessarily a probability measure. We state this fact in the next proposition:

Proposition 1.4.18. *Assume that the Feller transition probability P has dominant generic points, and let $x \in X$ be a dominant generic point for P. Then $x \in \Gamma_{\mathrm{cp}}^{(\mathrm{TP})}$ $(= \Gamma_{\mathrm{cpi}}^{(\mathrm{TP})})$.*

The next theorem shows the role played by the dominant generic points in the study of the unique ergodicity of Feller transition probabilities.

Theorem 1.4.19. *A Feller transition probability is uniquely ergodic if and only if it has at least one dominant generic point.*

Note that the generic points and the dominant generic points can be defined for any transition probability P, even if P is not Feller. However, Theorem 1.4.19 is not true, in general, for transition probabilities that are not Feller. For example, let P_w be the transition probability defined by the map $w : [0, 1] \to [0, 1]$ discussed in Example 1.1.6. Then any $x \in [0, 1]$ is a dominant generic point but P_w is not uniquely ergodic because there are no invariant probabilities for P.

Theorem 1.4.19 is relevant only in the noncompact case because if (X, d) is a compact metric space, the theorem can easily be deduced from known results.

If a Feller transition probability has a dominant generic point, then the point is not unique in general. By Theorems 1.4.1 and 1.4.19, $\Gamma_{\mathrm{cp}}^{(\mathrm{TP})} = \Gamma_{\mathrm{cpi}}^{(\mathrm{TP})} = \Gamma_{\mathrm{cpie}}^{(\mathrm{TP})}$ for such a transition probability, and any element of $\Gamma_{\mathrm{cpie}}^{(\mathrm{TP})}$ is a dominant generic point.

A transition probability defined on (the locally compact separable metric space) (X, d) is said to be *weak* mean ergodic* if $X = \mathcal{D}^{(\mathrm{TP})} \cup \Gamma_{\mathrm{c}}^{(\mathrm{TP})}$. Note that the definition also makes sense for transition probabilities that fail to be Feller. For instance, the transition probability P_w defined in Example 1.1.6 is weak* mean ergodic but is not Feller. The reason for our terminology lies in the fact that a transition probability P (or its associated Markov pair (S, T)) is weak* mean ergodic if and only if, for every $x \in X$, the sequence of averages $\left(\dfrac{1}{n} \sum_{k=0}^{n-1} T^k \delta_x \right)_{n \in \mathbb{N}}$ converges in the weak* topology of $\mathcal{M}(X)$.

Observe that a transition probability P is weak* mean ergodic if and only if each $x \in X$ is a generic point. Thus, if (S, T) is the Markov pair defined by P, and if P is weak* mean ergodic, then, for every $f \in C_0(X)$, the function $f^* : X \to \mathbb{R}$,

$$f^*(x) = \lim_{n \to +\infty} \frac{1}{n} \sum_{k=0}^{n-1} S^k f(x) \text{ for every } x \in X, \text{ is well-defined because the limits}$$

$$\lim_{n \to +\infty} \frac{1}{n} \sum_{k=0}^{n-1} S^k f(x), \ f \in C_0(X), \ x \in X, \text{ do exist.}$$

Given a nonempty set $\mathcal{A}$ of real-valued functions defined on X, we say that $\mathcal{A}$ has a *common (absolute) maximum* at $x_0 \in X$ if $f(x) \leq f(x_0)$ for every $f \in \mathcal{A}$ and $x \in X$. Thus $\mathcal{A}$ has a common maximum at $x_0 \in X$ if each $f \in \mathcal{A}$ attains an absolute maximum value at x_0.

In view of our discussion so far, we can restate Theorem 1.4.19 for weak* mean ergodic Feller transition probabilities as follows:

Corollary 1.4.20. *Assume that P is a weak* mean ergodic Feller transition probability, and set $\mathcal{A} = \{f^* \mid f \in C_0(X)\}$. Then P is uniquely ergodic if and only if there exists an $x_0 \in X$ such that $\mathcal{A}$ has a common maximum at x_0.*

The dominant generic points can also be defined "locally" and the new notion can be used to study the ergodicity of invariant probability measures of Feller transition probabilities. We will now discuss a result in this direction.

As always here, we assume given a Feller transition probability P defined on a locally compact separable metric space (X, d) and the Markov-Feller pair (S, T) defined by P.

Let A be a nonempty Borel subset of X and let $x_0 \in A$. In a similar manner as in the case of (global) dominant generic points, we say that x_0 is a *dominant generic point for A* if x_0 is a nonsingular generic point and if the following condition (called the *DGPTP-A condition*) is satisfied: $\lim_{n \to \infty} \frac{1}{n} \sum_{k=0}^{n-1} S^k f(x) \leq \lim_{n \to \infty} \frac{1}{n} \sum_{k=0}^{n-1} S^k f(x_0)$ for every $x \in A$ and $f \in C_0(X)$, $f \geq 0$, such that the sequence $(\frac{1}{n} \sum_{k=0}^{n-1} S^k f(x))_{n \in \mathbb{N}}$ is convergent.

Note that the notion of dominant generic point for a Borel subset of X makes sense even if the transition probability P is not Feller. Using this notion of dominant generic points for measurable subsets of X we obtain the following characterization of the invariant ergodic probability measures for P:

Theorem 1.4.21. *Assume that P has invariant probabilities, and let μ be such an invariant probability. Then, the measure μ is ergodic if and only if there exist $A \in \mathcal{B}(X)$ and $x_0 \in A$ such that $\mu(A) = 1$ and such that x_0 is a dominant generic point for A.*

Note that even though we assume in Theorem 3.3.2 of [143] that P is a Feller transition probability (that is, we deal with a Markov-Feller pair), the theorem and its proof are valid even if the Markov pair under consideration is not a Markov-Feller pair. Note also that in the proof of Theorem 1.4.21 we do not use the KBBY decomposition and, clearly, Corollary 1.3.21 is also a consequence of Theorem 1.4.21; that is, we obtain a new proof of the fact that if P is uniquely ergodic, then the unique invariant probability measure of P is an ergodic measure.

The equivalence relation $\sim$ that we defined on Θ after Lemma 1.3.17 turned out, by (b) of Theorem 1.3.19, to be an equivalence relation on $\Gamma_{\mathrm{cpie}}^{(\mathrm{TP})}$. However, the relation $\sim$ can be extended to $\Gamma_{\mathrm{c}}^{(\mathrm{TP})}$ in a natural manner as follows: if $x \in \Gamma_{\mathrm{c}}^{(\mathrm{TP})}$ and $y \in \Gamma_{\mathrm{c}}^{(\mathrm{TP})}$, then $x \sim y$ if and only if (by definition) $\lim_{n \to \infty} \frac{1}{n} \sum_{k=0}^{n-1} S^k f(x) = \lim_{n \to \infty} \frac{1}{n} \sum_{k=0}^{n-1} S^k f(y)$ for every $f \in C_0(X)$. As in Sect. 1.3, we denote by $[x]$ the equivalence class of $x \in \Gamma_{\mathrm{c}}^{(\mathrm{TP})}$.

It is easy to see that $[x]$, as a subset of X, belongs to $\mathcal{B}(X)$ whenever $x \in \Gamma_{\mathrm{c}}^{(\mathrm{TP})}$.

Note that the above discussion regarding $\sim$ is valid for any transition probability not necessarily Feller. In the Feller case we can use the extension of $\sim$ to $\Gamma_{\mathrm{c}}^{(\mathrm{TP})}$ in order to obtain the following useful criteria for the ergodicity of an invariant element of $\mathcal{M}(X)$:

Theorem 1.4.22. *Assume that μ is an invariant element of $\mathcal{M}(X)$ for the Feller transition probability P. The following assertions are equivalent:*

(a) μ is a probability measure and is ergodic.
(b) $\mu = \varepsilon_x^{(\mathrm{TP})}$ for some $x \in \Gamma_{\mathrm{c}}^{(\mathrm{TP})}$, and in this case $\varepsilon_x^{(\mathrm{TP})}([x]) = 1$.
(c) $\mu = \varepsilon_x^{(\mathrm{TP})}$ for some $x \in \Gamma_{\mathrm{cp}}^{(\mathrm{TP})}$, and in this case $\varepsilon_x^{(\mathrm{TP})}([x]) = 1$.

Observation. Note that if P and μ are as in Theorem 1.4.22, and $x \in \Gamma_{\mathrm{c}}^{(\mathrm{TP})}$ satisfies (b) of the theorem, then $x \in \Gamma_{\mathrm{cp}}^{(\mathrm{TP})}$ $(= \Gamma_{\mathrm{cpi}}^{(\mathrm{TP})}$ because P is a Feller transition probability) and x satisfies (c) of the above theorem; hence, if $x \in X$ satisfies (b) or (c) of Theorem 1.4.22, then $\int_{\Gamma_{\mathrm{cpi}}^{(\mathrm{TP})}} (f^*(y) - f^*(x))^2 \, \mathrm{d}\varepsilon_x^{(\mathrm{TP})}(y) = 0$ for every $f \in C_0(X)$, so $x \in \Gamma_{\mathrm{cpie}}^{(\mathrm{TP})}$. ▲

1.4.2 Equicontinuity

As mentioned at the beginning of this section our goal here is to briefly discuss several results dealing with equicontinuous transition probabilities. These transition probabilities are necessarily Feller as pointed out when we defined them after Proposition 1.1.5.

Two topics will be reviewed in this subsection: the first one is a set of criteria for the unique ergodicity of equicontinuous transition probabilities; the second topic consists of two strongly related mean ergodic theorems that are valid for equicontinuous transition probabilities.

We saw in Theorem 1.4.12 that if a Feller transition probability P is uniquely ergodic, then the sets $\gamma^{(\mathrm{TP})}$, $\gamma_0^{(\mathrm{TP})}$, $\gamma_c^{(\mathrm{TP})}$, $\gamma_{\mathrm{cp}}^{(\mathrm{TP})}$ are all nonempty, and each of them is equal to the support of the unique invariant probability measure for P. Thus, a natural question in this context is if the nonemptiness of any of the sets $\gamma^{(\mathrm{TP})}$, $\gamma_0^{(\mathrm{TP})}$, $\gamma_c^{(\mathrm{TP})}$ or $\gamma_{\mathrm{cp}}^{(\mathrm{TP})}$ implies the unique ergodicity of P. The existence of minimal Feller transition probabilities that have invariant probabilities and are not uniquely ergodic shows that even if the sets $\gamma^{(\mathrm{TP})}$, $\gamma_0^{(\mathrm{TP})}$, $\gamma_c^{(\mathrm{TP})}$ and $\gamma_{\mathrm{cp}}^{(\mathrm{TP})}$ are all equal to the entire space, the corresponding transition probability might fail to be uniquely ergodic. So, we are led to the next natural question: can we single out a large enough class of Feller transition probabilities which has the property that if a transition probability is in this class and at least one of the sets $\gamma^{(\mathrm{TP})}$, $\gamma_0^{(\mathrm{TP})}$, $\gamma_c^{(\mathrm{TP})}$ or $\gamma_{\mathrm{cp}}^{(\mathrm{TP})}$ is nonempty, then the transition probability is uniquely ergodic? It turns out that the equicontinuous transition probabilities have this property. More precisely, we have the following theorem:

Theorem 1.4.23. *Assume that P is an equicontinuous transition probability and has invariant probabilities. Then P is uniquely ergodic if and only if at least one of the sets $\gamma_{\mathrm{cp}}^{(\mathrm{TP})}$, $\gamma_c^{(\mathrm{TP})}$, $\gamma_0^{(\mathrm{TP})}$ or $\gamma^{(\mathrm{TP})}$ is nonempty. If any of these sets is nonempty, then all of them are nonempty and equal to the support of the unique invariant probability measure for P.*

We will now discuss several results, which can be used to prove Theorem 1.4.23, and which are of independent interest.

Proposition 1.4.24. *Let P be an equicontinuous transition probability, and assume that P has ergodic invariant probability measures (that is, assume that $\Gamma_{\mathrm{cpie}}^{(\mathrm{TP})}$ is nonempty). If $x \in \Gamma_{\mathrm{cpie}}^{(\mathrm{TP})}$, then $[x]$, as a subset of X, is a closed set.*

Proof. Let (S, T) be the Markov-Feller pair defined by P.

Let $x \in \Gamma_{\mathrm{cpie}}^{(\mathrm{TP})}$.

In order to prove that $[x]$ is a closed subset of X, we have to prove that for every convergent sequence $(y_l)_{l \in \mathbb{N}}$ of elements of $[x]$, the limit of $(y_l)_{l \in \mathbb{N}}$ is an element of $[x]$, as well.

To this end, let $(y_l)_{l \in \mathbb{N}}$ be a convergent sequence of elements of $[x]$, and set $y = \lim_{l \to \infty} y_l$. Also, let $f \in C_0(X)$ and $\varepsilon \in \mathbb{R}$, $\varepsilon > 0$.

Since P is equicontinuous, there exists an $l_\varepsilon \in \mathbb{N}$ such that $\left| S^k f(y_l) - S^k f(y) \right| < \frac{\varepsilon}{2}$ for every $l \in \mathbb{N}$, $l \geq l_\varepsilon$ and every $k \in \mathbb{N} \cup \{0\}$.

Since $y_{l_\varepsilon} \in [x]$, there exists an $n_\varepsilon \in \mathbb{N}$ such that

$$\left| \frac{1}{n} \sum_{k=0}^{n-1} S^k f(y_{l_\varepsilon}) - \left\langle f, \varepsilon_x^{(\mathrm{TP})} \right\rangle \right| < \frac{\varepsilon}{2}$$

for every $n \in \mathbb{N}$, $n \geq n_\varepsilon$, where $\varepsilon_x^{(\mathrm{TP})}$ is the ergodic invariant probability measure defined by x.

We obtain that

$$\left| \frac{1}{n} \sum_{k=0}^{n-1} S^k f(y) - \langle f, \varepsilon_x^{(TP)} \rangle \right| \leq \left| \frac{1}{n} \sum_{k=0}^{n-1} S^k f(y) - \frac{1}{n} \sum_{k=0}^{n-1} S^k f(y_{l_\varepsilon}) \right|$$

$$+ \left| \frac{1}{n} \sum_{k=0}^{n-1} S^k f(y_{l_\varepsilon}) - \langle f, \varepsilon_x^{(TP)} \rangle \right| < \frac{1}{n} \sum_{k=0}^{n-1} \left| S^k f(y) - S^k f(y_{l_\varepsilon}) \right| + \frac{\varepsilon}{2} < \frac{\varepsilon}{2} + \frac{\varepsilon}{2} = \varepsilon$$

for every $n \in \mathbb{N}$, $n \geq n_\varepsilon$.

We have therefore proved that the sequence $(\frac{1}{n} \sum_{k=0}^{n-1} S^k f(y))_{n \in \mathbb{N}}$ converges to $\langle f, \varepsilon_x^{(TP)} \rangle$ for every $f \in C_0(X)$. Using the observation made after Theorem 1.4.22, we obtain that $y \in \Gamma_{\mathrm{cpie}}^{(TP)}$ and $y \sim x$. $\square$

Proposition 1.4.24 has the following consequence:

Corollary 1.4.25. *Let P be an equicontinuous transition probability. Then* $\operatorname{supp} \varepsilon_x^{(TP)} \subseteq [x]$ *for every* $x \in \Gamma_{\mathrm{cpie}}^{(TP)}$.

Proof. Let $x \in \Gamma_{\mathrm{cpie}}^{(TP)}$. Using (a) of Theorem 1.3.19, we obtain that $\varepsilon_x^{(TP)}([x]) = 1$. Since $[x]$ is a closed subset of X by Proposition 1.4.24, it follows that $\operatorname{supp} \varepsilon_x^{(TP)} \subseteq [x]$. $\square$

Let P be a not necessarily Feller transition probability and let (S, T) be the Markov pair defined by P. We say that the ergodic measures of P (or of (S, T), or of T) are *disjointly supported* (in short, we say that P, or (S, T), or T has the *e.m.d.s. property*) if $(\operatorname{supp} \mu) \cap (\operatorname{supp} \nu) = \emptyset$ for every distinct pair of ergodic invariant probability measures μ and ν. Note that if P does not have invariant probabilities (so, naturally, P does not have ergodic invariant probability measures), or if P is uniquely ergodic (so, by Corollary 1.3.21, P has exactly one ergodic invariant probability measure), then P has the e.m.d.s. property.

Using Corollary 1.4.25 we obtain the following theorem:

Theorem 1.4.26. *If P is equicontinuous, then P has the e.m.d.s. property.*

Proof. The proof is obvious in view of Corollary 1.4.25 because if $\varepsilon_x^{(TP)}$ and $\varepsilon_y^{(TP)}$ are two distinct ergodic invariant probability measures for P for some $x \in \Gamma_{\mathrm{cpie}}^{(TP)}$ and $y \in \Gamma_{\mathrm{cpie}}^{(TP)}$, then $[x] \cap [y] = \emptyset$, so, by Corollary 1.4.25, $(\operatorname{supp} \varepsilon_x^{(TP)}) \cap (\operatorname{supp} \varepsilon_y^{(TP)}) = \emptyset$. $\square$

Note that Theorem 1.4.26 is Theorem 4.1.7 of [143]. The proof obtained here seems to be shorter and more informative than the proof of Theorem 4.1.7 in [143].

We will now illustrate the results discussed so far in this subsection on several examples.

Example 1.4.27. Let $\mathbb{R}/\mathbb{Z}$ be the compact metric group discussed in Example A.2.8 (the unit circle), let $a \notin \mathbb{Q}$, let $v_{\hat{a}}$ be the rotation of the unit circle by a defined in Example 1.1.7, and let $P_{\hat{a}}$ be the transition probability defined by $v_{\hat{a}}$.

We first note that $P_{\hat{a}}$ has invariant probabilities because, as pointed out at the beginning of Sect. 1.2 before defining the ergodic invariant probability measures, the Haar-Lebesgue measure on $\mathbb{R}/\mathbb{Z}$ (discussed in Example B.1.6) is an invariant probability measure for $P_{\hat{a}}$. Since, as discussed in Example 1.1.7, $P_{\hat{a}}$ is minimal because $a \notin \mathbb{Q}$, it follows that the sets $\gamma_{\mathrm{cp}}^{\mathrm{(TP)}}$, $\gamma_{\mathrm{c}}^{\mathrm{(TP)}}$, $\gamma_{0}^{\mathrm{(TP)}}$ and $\gamma^{\mathrm{(TP)}}$ are all equal to the entire space $\mathbb{R}/\mathbb{Z}$. Finally, since, as pointed out in Example 1.1.7, $P_{\hat{a}}$ is an equicontinuous transition probability, it follows that $P_{\hat{a}}$ is uniquely ergodic by Theorem 1.4.23. By Corollary 1.4.25, $\mathbb{R}/\mathbb{Z} = \Gamma_{\mathrm{cpie}}^{\mathrm{(TP)}} = [\hat{x}]$ and $\varepsilon_{\hat{x}}^{\mathrm{(TP)}} = $ the Haar-Lebesgue measure on $\mathbb{R}/\mathbb{Z}$ for every $\hat{x} \in \mathbb{R}/\mathbb{Z}$. Note that we also obtain that the sequence $(\frac{1}{n}\sum_{k=0}^{n-1} S_{\hat{a}}^{k} f(\hat{x}))_{n \in \mathbb{N}}$ converges to $\int_{\mathbb{R}/\mathbb{Z}} f(\hat{y})\, \mathrm{d}\lambda_{\mathbb{R}/\mathbb{Z}}(\hat{y})$ for every $f \in C(\mathbb{R}/\mathbb{Z})$ and $\hat{x} \in \mathbb{R}/\mathbb{Z}$, where $\lambda_{\mathbb{R}/\mathbb{Z}}$ is the Haar-Lebesgue measure on $\mathbb{R}/\mathbb{Z}$ and $S_{\hat{a}}$ is the first operator in the Markov-Feller pair defined by $P_{\hat{a}}$. ∎

Example 1.4.28. Let $n \in \mathbb{N}$, $n \geq 2$, let $\mathbf{v} \in \mathbb{R}^{n}$, $\mathbf{v} = (v_1, v_2, \cdots, v_n)$, be such that $1, v_1, v_2, \cdots, v_n$ are rationally independent (the definition of the rational independence of a finite set of numbers is given before Example A.3.5), let $u_{\mathbf{v}}$ be the translation of the torus $\mathbb{R}^{n}/\mathbb{Z}^{n}$ by $\hat{\mathbf{v}}$ defined in Example 1.1.8, and let $P_{\mathbf{v}}$ be the transition probability defined by $u_{\mathbf{v}}$ (see Example 1.1.8 again). Since (as pointed out at the beginning of Sect. 1.2 before defining the ergodic invariant probability measures) the Haar-Lebesgue measure on $\mathbb{R}^{n}/\mathbb{Z}^{n}$ is an invariant probability for $P_{\mathbf{v}}$ (so, $P_{\mathbf{v}}$ has invariant probabilities), and since $P_{\mathbf{v}}$ is minimal (because $1, v_1, v_2, \cdots, v_n$ are rationally independent; see Example 1.1.8 for details), it follows that the sets $\gamma_{\mathrm{cp}}^{\mathrm{(TP)}}$, $\gamma_{\mathrm{c}}^{\mathrm{(TP)}}$, $\gamma_{0}^{\mathrm{(TP)}}$ and $\gamma^{\mathrm{(TP)}}$ are all equal to $\mathbb{R}^{n}/\mathbb{Z}^{n}$. Thus, taking into consideration that $P_{\mathbf{v}}$ is equicontinuous and using Theorem 1.4.23, we obtain that $P_{\mathbf{v}}$ is uniquely ergodic, so the Haar-Lebesgue measure $\lambda_{\mathbb{R}^{n}/\mathbb{Z}^{n}}$ on $\mathbb{R}^{n}/\mathbb{Z}^{n}$ is the unique invariant measure for $P_{\mathbf{v}}$. Using Corollary 1.4.25, we obtain that $\mathbb{R}^{n}/\mathbb{Z}^{n} = \Gamma_{\mathrm{cpie}}^{\mathrm{(TP)}} = [\hat{\mathbf{x}}]$ and $\varepsilon_{\hat{\mathbf{x}}}^{\mathrm{(TP)}} = \lambda_{\mathbb{R}^{n}/\mathbb{Z}^{n}}$ for every $\hat{\mathbf{x}} \in \mathbb{R}^{n}/\mathbb{Z}^{n}$. In a similar manner as in Example 1.4.27, we obtain that the sequence $(\frac{1}{n}\sum_{k=0}^{n-1} S_{\mathbf{v}}^{k} f(\hat{\mathbf{x}}))_{n \in \mathbb{N}}$ converges to $\int_{\mathbb{R}^{n}/\mathbb{Z}^{n}} f(\hat{\mathbf{y}})\, \mathrm{d}\lambda_{\mathbb{R}^{n}/\mathbb{Z}^{n}}(\hat{\mathbf{y}})$ for every $f \in C(\mathbb{R}^{n}/\mathbb{Z}^{n})$ and $\hat{\mathbf{x}} \in \mathbb{R}^{n}/\mathbb{Z}^{n}$, where $S_{\mathbf{v}}$ is the first operator in the Markov-Feller pair defined by $P_{\mathbf{v}}$. ∎

For our next example, in which we will discuss convolution operators, we need some preparation.

Let $\mathcal{A}$ be a nonempty set of probability measures, $\mathcal{A} \subseteq \mathcal{M}(X)$, where, as usual, (X, d) is a locally compact separable metric space. We say that $\mathcal{A}$ is a *tight set of probability measures* if for every $\varepsilon \in \mathbb{R}$, $\varepsilon > 0$, there exists a compact subset K_{ε} of X such that $\mu(K_{\varepsilon}) > 1 - \varepsilon$ for every $\mu \in \mathcal{A}$. We say that a sequence $(\mu_n)_{n \in \mathbb{N}}$ of probability measures in $\mathcal{M}(X)$ is *tight* if the range $\{\mu_n \mid n \in \mathbb{N}\}$ of $(\mu_n)_{n \in \mathbb{N}}$ is a tight set of probability measures.

Now let $(H, \cdot, d)$ be a locally compact separable metric semigroup, let $\mu \in \mathcal{M}(H)$ be a probability measure, and let P_{μ} be the transition probability defined in

Example 1.1.16. Note that, as pointed out in Example 1.1.16, P_μ is a Feller transition probability.

We say that the probability measure μ is *equicontinuous* if P_μ is an equicontinuous transition probability. Note that the collection of all equicontinuous probability measures is fairly large. Indeed, if a probability measure ν has the property that the sequence $(\nu^n)_{n \in \mathbb{N}}$ is tight, then, by Proposition 4.2 of [144], ν is equicontinuous; also, if H is *discrete* (that is, H is a semigroup that has a finite number of, or countably many elements, and the metric d defines the discrete topology on H), then any probability measure in $\mathcal{M}(H)$ is equicontinuous because any transition probability defined on such a discrete locally compact separable metric semigroup is equicontinuous.

Example 1.4.29. Let $(H, \cdot, d)$ be a locally compact separable metric semigroup, let $\mu \in \mathcal{M}(H)$ be a probability measure, and set $A = \overline{\cup_{n=1}^{\infty} \mathrm{supp}(\mu^n)}$.

Now, let P_μ be the Feller transition probability defined in Example 1.1.16, and let (S_μ, T_μ) be the Markov-Feller pair defined by P_μ. Note that using the definition of T_μ (see the equality (1.1.6)), we obtain that a probability measure $\nu \in \mathcal{M}(H)$ is invariant for P_μ if and only if

$$\mu * \nu = \nu. \tag{1.4.2}$$

The above equality (1.4.2) is known as the *Choquet-Deny equation* (for details and references, see Section 1 of [144]).

It can be shown (see the proof of Theorem 4.4 of [144]) that Ax is a subset of the orbit-closure $\overline{\mathcal{O}^{(\mathrm{TP})}(x)}$ of x under the action of T_μ for every $x \in H$.

Now assume that μ is equicontinuous, and that P_μ has at least one invariant probability. If $\cap_{x \in H} Ax \neq \emptyset$, then $\cap_{x \in H} \overline{\mathcal{O}^{(\mathrm{TP})}(x)} \neq \emptyset$; therefore, using Theorem 1.4.23 we obtain that P_μ is uniquely ergodic. This means that there exists a unique probability measure $\nu \in \mathcal{M}(H)$ that satisfies the Choquet-Deny equation (the above equality (1.4.1)).

Note that if, in particular, the probability measure μ has the additional property that $\overline{\cup_{n=1}^{\infty} \mathrm{supp}(\mu^n)} = H$ (that is, $A = H$) and H has left zeroids, then using again Theorem 1.4.23, we obtain that P_μ is uniquely ergodic, so there exists a unique invariant probability ν that satisfies the Choquet-Deny equation; this is so, because $\cap_{x \in H} Hx$ is precisely the set of all left zeroids of H. ∎

Theorem 1.4.23 can be restated in terms of the universal elements defined before Theorem 1.4.15. For this purpose, given a transition probability P defined, as usual, on a locally compact separable metric space (X, d), let $\mathcal{U}_{\mathrm{cp}}^{(\mathrm{TP})}$, $\mathcal{U}_{\mathrm{c}}^{(\mathrm{TP})}$, $\mathcal{U}_{0}^{(\mathrm{TP})}$ and $\mathcal{U}^{(\mathrm{TP})}$ be the sets of all universal elements with respect to $\Gamma_{\mathrm{cp}}^{(\mathrm{TP})}$, $\Gamma_{\mathrm{c}}^{(\mathrm{TP})}$, $\Gamma_{0}^{(\mathrm{TP})}$ and $\Gamma^{(\mathrm{TP})}$, respectively, defined by P. In terms of the sets $\mathcal{U}_{\mathrm{cp}}^{(\mathrm{TP})}$, $\mathcal{U}_{\mathrm{c}}^{(\mathrm{TP})}$, $\mathcal{U}_{0}^{(\mathrm{TP})}$ and $\mathcal{U}^{(\mathrm{TP})}$, Theorem 1.4.23 can be restated as follows:

Theorem 1.4.30. *Let P be an equicontinuous transition probability, and assume that P has at least one invariant probability measure. Then P is uniquely ergodic* *if and only if at least one of the sets $\mathcal{U}_{\mathrm{cp}}^{(\mathrm{TP})}$, $\mathcal{U}_{\mathrm{c}}^{(\mathrm{TP})}$, $\mathcal{U}_0^{(\mathrm{TP})}$ or $\mathcal{U}^{(\mathrm{TP})}$ is nonempty. If any of these sets is nonempty, then all of them are nonempty and equal to the support of the unique invariant probability measure for P.*

The proof of the above theorem is obvious since $\mathcal{U}_{\mathrm{cp}}^{(\mathrm{TP})} = \gamma_{\mathrm{cp}}^{(\mathrm{TP})}$, $\mathcal{U}_{\mathrm{c}}^{(\mathrm{TP})} = \gamma_{\mathrm{c}}^{(\mathrm{TP})}$, $\mathcal{U}_0^{(\mathrm{TP})} = \gamma_0^{(\mathrm{TP})}$ and $\mathcal{U}^{(\mathrm{TP})} = \gamma^{(\mathrm{TP})}$.

Theorem 1.4.30 has the following consequence:

Corollary 1.4.31. *Let P be an equicontinuous transition probability.*

(a) If P has universal elements (with respect to X) and nonzero invariant elements of $\mathcal{M}(X)$, then P is uniquely ergodic. In this case, the set of all universal elements is included in the support of the unique invariant probability measure for P.

(b) If X is compact, then P is uniquely ergodic if and only if P has universal elements. If P is uniquely ergodic, then the support of the unique invariant probability measure for P is equal to the set of all universal elements generated by P.

Sometimes, we cannot use Theorem 1.4.23 (or Theorem 1.4.30), or Corollary 1.4.31 because it is hard to check the nonemptiness of the sets $\gamma_{\mathrm{cp}}^{(\mathrm{TP})}$, $\gamma_{\mathrm{c}}^{(\mathrm{TP})}$, $\gamma_0^{(\mathrm{TP})}$ and $\gamma^{(\mathrm{TP})}$ in Theorem 1.4.23 (or $\mathcal{U}_{\mathrm{cp}}^{(\mathrm{TP})}$, $\mathcal{U}_{\mathrm{c}}^{(\mathrm{TP})}$, $\mathcal{U}_0^{(\mathrm{TP})}$, and $\mathcal{U}^{(\mathrm{TP})}$ in Theorem 1.4.30), and it is difficult to find out if there exist universal elements in order to be able to consider Corollary 1.4.31. In such situations, the following result is often useful (for an application of the result, see Theorem 3.6 of [145]):

Theorem 1.4.32. *Let P be an equicontinuous transition probability and assume that P has nonzero invariant elements of $\mathcal{M}(X)$. If the orbit-closures of any pair of points in X under the action of P fail to be disjoint (that is, if the intersection of the orbit-closures of any pair of points of X under the action of P is nonempty), then P is uniquely ergodic.*

A remarkable feature of the equicontinuous transition probabilities is that the operators that appear in the Markov-Feller pairs defined by these transition probabilities satisfy rather general mean ergodic theorems. We will now briefly discuss these theorems and several related topics.

To this end, let P be a Feller transition probability defined on a locally compact separable metric space (X, d) and let (S, T) be the Markov-Feller pair defined by P.

Theorem 1.4.33 (Pointwise Mean Ergodic Theorem for Equicontinuous Transition Probabilities). *Assume that P is equicontinuous. Then the sequence $\left(\frac{1}{n}\sum_{k=0}^{n-1} S^k f(x)\right)_{n\in\mathbb{N}}$ converges for every $f \in C_0(X)$ and every $x \in X$; that is, the sequence of functions $\left(\frac{1}{n}\sum_{k=0}^{n-1} S^k f(x)\right)_{n\in\mathbb{N}}$ converges pointwise on X whenever $f \in C_0(X)$.*

Note that Theorem 1.4.33 is a mean ergodic theorem for the first operator S of the Markov-Feller pair (S, T) under the assumption that P is equicontinuous. The corresponding mean ergodic theorem for the operator T is:

Theorem 1.4.34 (Weak* Mean Ergodic Theorem for Equicontinuous Transition Probabilities). *Assume that P is equicontinuous and let $\mu \in \mathcal{M}(X)$. Then the sequence $(\frac{1}{n} \sum_{k=0}^{n-1} T^k \mu)_{n \in \mathbb{N}}$ converges in the weak* topology of $\mathcal{M}(X)$ and the weak* limit of the sequence is an invariant element for T; also w^*-$\lim_{n \to \infty} \frac{1}{n} \sum_{k=0}^{n-1} T^k \mu \geq 0$ whenever $\mu \geq 0$.*

It is easy to see that if Theorem 1.4.34 is true, then Theorem 1.4.33 holds true, as well, and vice versa. The approach taken in [143] was to prove that Theorem 1.4.34 is true, and then to use the result in order to prove Theorem 1.4.33.

In order to prove Theorem 1.4.34, we applied the following proposition to the sequences $(\frac{1}{n} \sum_{k=0}^{n-1} T^k \mu)_{n \in \mathbb{N}}$, $\mu \in \mathcal{M}(X)$:

Proposition 1.4.35. *Let $(\mu_n)_{n \in \mathbb{N}}$ be a sequence of elements of $\mathcal{M}(X)$, and let $\mu^* \in \mathcal{M}(X)$. The following assertions are equivalent:*

(a) The sequence $(\mu_n)_{n \in \mathbb{N}}$ converges in the weak topology of $\mathcal{M}(X)$ to μ^*.*

(b) For every subsequence $(\mu_{n_k})_{k \in \mathbb{N}}$ of $(\mu_n)_{n \in \mathbb{N}}$ there exists a further subsequence $\left(\mu_{n_{k_l}}\right)_{l \in \mathbb{N}}$ of $(\mu_{n_k})_{k \in \mathbb{N}}$ such that the sequence (of real numbers) $\left(\left\langle f, \mu_{n_{k_l}} \right\rangle\right)_{l \in \mathbb{N}}$ converges to $\langle f, \mu^* \rangle$ for every $f \in C_0(X)$.

(c) For every subsequence $(\mu_{n_k})_{k \in \mathbb{N}}$ of $(\mu_n)_{n \in \mathbb{N}}$ and every $g \in C_0(X)$, there exists a subsequence $\left(\mu^{(g)}{}_{n_{k_l}}\right)_{l \in \mathbb{N}}$ of $(\mu_{n_k})_{k \in \mathbb{N}}$ such that the real number sequence $\left(\left\langle g, \mu^{(g)}{}_{n_{k_l}} \right\rangle\right)_{l \in \mathbb{N}}$ converges to $\langle g, \mu^* \rangle$ (note that the subsequence $\left(\mu^{(g)}{}_{n_{k_l}}\right)_{l \in \mathbb{N}}$ depends on g).

In order to be able to use Proposition 1.4.35 in the proof of Theorem 1.4.34, we needed the following result:

Proposition 1.4.36. *Assume that P is an equicontinuous transition probability and let (S, T) be the Markov-Feller pair defined by P. Then for every subsequence $(\frac{1}{n_k} \sum_{j=0}^{n_k-1} S^j)_{k \in \mathbb{N}}$ of $(\frac{1}{n} \sum_{j=0}^{n-1} S^j)_{n \in \mathbb{N}}$ there exists a further subsequence $(\frac{1}{n_{k_l}} \sum_{j=0}^{n_{k_l}-1} S^j)_{l \in \mathbb{N}}$ of $(\frac{1}{n_k} \sum_{j=0}^{n_k-1} S^j)_{k \in \mathbb{N}}$ such that the sequence $(\frac{1}{n_{k_l}} \sum_{j=0}^{n_{k_l}-1} S^j f(x))_{l \in \mathbb{N}}$ converges for every $f \in C_0(X)$ and $x \in X$.*

Theorem 1.4.33 can be restated in terms of the KBBY decomposition as follows:

Theorem 1.4.37. *If P is an equicontinuous transition probability defined on (X, d), then $X = \mathcal{D}^{(\mathrm{TP})} \cup \Gamma_c^{(\mathrm{TP})}$.*

Note that the conclusion of Theorem 1.4.33 (or Theorem 1.4.37) also holds true for some transition probabilities that fail to be equicontinuous. For instance, $X = \Gamma_{\mathrm{cp}}^{(\mathrm{TP})}$ for the transition probability P_w of Example 1.1.6, in spite of the fact that P_w is not even Feller.

We will now illustrate the use of the pointwise mean ergodic theorem for equicontinuous transition probabilities Theorem 1.4.33 and the weak* mean ergodic theorem for equicontinuous transition probabilities Theorem 1.4.34 on several examples.

Example 1.4.38. Let $n \in \mathbb{N}$, $n \geq 2$, let $\mathbb{S}_n$ be the compact (separable) metric multiplicative semigroup of all $n \times n$ column stochastic matrices endowed with the metric defined by the restriction of $\|| \cdot \||_1$ on $\mathbf{M}(n, \mathbb{R})$ to $\mathbb{S}_n$ (see the beginning of Sect. B.4.1, Proposition B.4.1, the definition of $\|| \cdot \||_1$ given before Proposition B.4.3, and Example 1.1.12), let $\mathbf{B} \in \mathbb{S}_n$, let $P_{\mathbf{B}}$ be the Feller transition probability defined in Example 1.1.12, and let $(S_{\mathbf{B}}, T_{\mathbf{B}})$ be the Markov-Feller pair defined by $P_{\mathbf{B}}$.

As pointed out in Example 1.1.12, $P_{\mathbf{B}}$ is an equicontinuous transition probability, so using Theorem 1.4.33 we obtain that the sequence of averages $(\frac{1}{n} \sum_{k=0}^{n-1} f(\mathbf{B}^k \mathbf{X}))_{n \in \mathbb{N}}$ converges for every $f \in C(\mathbb{S}_n)$ and $\mathbf{X} \in \mathbb{S}_n$. In particular, if we define for every $i \in \{1, 2, \cdots, n\}$ and $j \in \{1, 2, \cdots, n\}$, $f_{ij} : \mathbb{S}_n \to \mathbb{R}$, $f_{ij}(\mathbf{X}) = x_{ij}$ for every $\mathbf{X} \in \mathbb{S}_n$, $\mathbf{X} = (x_{kl})_{\substack{k \in \{1,2,\cdots,n\} \\ l \in \{1,2,\cdots,n\}}}$, then $f_{ij} \in C(\mathbb{S}_n)$, so the sequence $(\frac{1}{n} \sum_{k=0}^{n-1} \mathbf{B}^k)_{n \in \mathbb{N}}$ converges in the standard (Euclidean) topology of $\mathbf{M}(n, \mathbb{R})$ because the sequence $(\frac{1}{n} \sum_{k=0}^{n-1} f_{ij}(\mathbf{B}^k \mathbf{I}_n))_{n \in \mathbb{N}}$ converges (here, as usual in this book, $\mathbf{I}_n$ is the $n \times n$ identity matrix) and $\frac{1}{n} \sum_{k=0}^{n-1} f_{ij}(\mathbf{B}^k \mathbf{I}_n)$ is the (i, j) entry of the matrix $\frac{1}{n} \sum_{k=0}^{n-1} \mathbf{B}^k$, $n \in \mathbb{N}$, $i \in \{1, 2, \cdots, n\}$, $j \in \{1, 2, \cdots, n\}$.

Note that the convergence of the sequence $(\frac{1}{n} \sum_{k=0}^{n-1} \mathbf{B}^k)_{n \in \mathbb{N}}$ in the standard topology of $\mathbf{M}(n, \mathbb{R})$ can also be deduced from Theorem 2.1, p. 73 of Krengel's monograph [53] (the mean ergodic theorem for power bounded linear operators in reflexive Banach spaces). Indeed, if we endow the vector space $\mathbb{R}^n$ of all column vectors with n components with the usual Euclidean norm, and we think of $\mathbf{B}$ as a linear operator on $\mathbb{R}^n$, then $\mathbf{B}$ is power bounded; since $\mathbb{R}^n$ with the Euclidean norm is a reflexive Banach space (even a Hilbert space), we can apply Theorem 2.1, p. 73 of Krengel [53], and we obtain that the sequence $(\frac{1}{n} \sum_{k=0}^{n-1} \mathbf{B}^k \mathbf{x})_{n \in \mathbb{N}}$ converges in the Euclidean norm of $\mathbb{R}^n$ for every $\mathbf{x} \in \mathbb{R}^n$, so, clearly, $(\frac{1}{n} \sum_{k=0}^{n-1} \mathbf{B}^k)_{n \in \mathbb{N}}$ converges in the standard topology of $\mathbf{M}(n, \mathbb{R})$.

From our discussion so far, we obtain that, in particular, the sequence of column vectors $(\frac{1}{n} \sum_{k=0}^{n-1} \mathbf{B}^k \mathbf{x})_{n \in \mathbb{N}}$ converges in the metric topology of $\mathbb{P}_n$ whenever $\mathbf{x} \in \mathbb{P}_n$ where $\mathbb{P}_n$ is the compact metric space of all n-dimensional stochastic column vectors (see Example 1.1.13 (Maps on $\mathbb{P}_n$)). The convergence of the sequences $(\frac{1}{n} \sum_{k=0}^{n-1} \mathbf{B}^k \mathbf{x})_{n \in \mathbb{N}}$, $\mathbf{x} \in \mathbb{P}_n$, in the metric topology of $\mathbb{P}_n$ can also be obtained directly by applying the pointwise mean ergodic theorem for equicontinuous transition probabilities Theorem 1.4.33 to the transition probability $P_{(\mathbf{B})}$ defined in Example 1.1.13 (note that, as pointed out in Example 1.1.13, $P_{(\mathbf{B})}$ is equicontinuous) and then using the maps $g_j : \mathbb{P}_n \to \mathbb{R}$, $g_j(\mathbf{x}) = x_j$ for every $\mathbf{x} \in \mathbb{P}_n$, $\mathbf{x} = \begin{pmatrix} x_1 \\ x_2 \\ \cdots \\ x_n \end{pmatrix}$, $j = 1, 2, \cdots, n$, in a manner similar to the manner in which we used the maps f_{ij}, $i = 1, 2, \cdots, n$, $j = 1, 2, \cdots, n$, defined on $\mathbb{S}_n$. $\blacksquare$

Example 1.4.39. Let $(H, \cdot, d)$ be a locally compact separable metric semigroup, assume that H has a neutral element, and let e be the neutral element of H. Let $\mu \in \mathcal{M}(H)$ be a probability measure, let P_μ be the transition probability defined by μ (see Example 1.1.16), and assume that μ is equicontinuous (for the definition of equicontinuous probability measures, see the discussion preceding Example 1.4.29).

Since μ is equicontinuous, it follows that P_μ is an equicontinuous transition probability, so we can apply the weak* mean ergodic theorem for equicontinuous transition probabilities Theorem 1.4.34 in order to infer that the sequence $(\frac{1}{n} \sum_{k=0}^{n-1} \mu^k * \nu)_{n \in \mathbb{N}}$ converges in the weak* topology of $\mathcal{M}(H)$ for every $\nu \in \mathcal{M}(H)$, where $\mu^0 = \delta_e$. In particular, for $\nu = \mu$, we obtain that the sequence $(\frac{1}{n} \sum_{k=1}^{n} \mu^k)_{n \in \mathbb{N}}$, and consequently $(\frac{1}{n} \sum_{k=0}^{n-1} \mu^k)_{n \in \mathbb{N}}$ both converge in the weak* topology of $\mathcal{M}(H)$. $\blacksquare$

The weak* mean ergodic theorem for equicontinuous transition probabilities (Theorem 1.4.34) tells us that if P is an equicontinuous transition probability defined on a locally compact separable metric space (X, d) and if (S, T) is the Markov-Feller pair defined by P, then the sequence $\left(\left\langle f, \frac{1}{n} \sum_{k=0}^{n-1} T^k \mu \right\rangle \right)_{n \in \mathbb{N}}$ converges whenever $\mu \in \mathcal{M}(X)$ and $f \in C_0(X)$. Thus, a natural question is if the sequences $\left(\left\langle f, \frac{1}{n} \sum_{k=0}^{n-1} T^k \mu \right\rangle \right)_{n \in \mathbb{N}}$, $\mu \in \mathcal{M}(X)$, converge even in the more general case when $f \in C_b(X)$. It turns out that there exist equicontinuous Markov-Feller pairs (S, T), $f \in C_b(X)$, and $\mu \in \mathcal{M}(X)$ such that the sequence $\left(\left\langle f, \frac{1}{n} \sum_{k=0}^{n-1} T^k \mu \right\rangle \right)_{n \in \mathbb{N}}$ fails to converge. Since the Markov pairs (S, T) which have the property that the sequences $\left(\frac{1}{n} \sum_{k=0}^{n-1} T^k \mu \right)_{n \in \mathbb{N}}$, $\mu \in \mathcal{M}(X)$, converge $C_b(X)$-weak* are of significant practical importance, it is of interest to find conditions for a Markov pair (S, T) that will guarantee that the sequences $\left(\left\langle f, \frac{1}{n} \sum_{k=0}^{n-1} T^k \mu \right\rangle \right)_{n \in \mathbb{N}}$, $\mu \in \mathcal{M}(X)$, $f \in C_b(X)$, do converge. In the next theorem we discuss such conditions.

Theorem 1.4.40. *Let P be an equicontinuous transition probability defined on a locally compact separable metric space (X, d) and let (S, T) be the Markov-Feller pair defined by P.*

(a) *Let $\mu \in \mathcal{M}(X)$ be a probability measure, let μ^* be the weak* limit of $(\frac{1}{n} \sum_{k=0}^{n-1} T^k \mu)_{n \in \mathbb{N}}$ (the existence of μ^* is assured by Theorem 1.4.34), and assume that the sequence $(\frac{1}{n} \sum_{k=0}^{n-1} T^k \mu)_{n \in \mathbb{N}}$ is tight. Then $(\frac{1}{n} \sum_{k=0}^{n-1} T^k \mu)_{n \in \mathbb{N}}$ converges $C_b(X)$-weak* to μ^*.*

(b) Assume that, for every probability measure $\mu \in \mathcal{M}(X)$, the sequence $(\frac{1}{n} \sum_{k=0}^{n-1} T^k \mu)_{n \in \mathbb{N}}$ is tight. Then, for every $\nu \in \mathcal{M}(X)$, the sequence $(\frac{1}{n} \sum_{k=0}^{n-1} T^k \nu)_{n \in \mathbb{N}}$ converges $C_b(X)$-weak to the weak* limit of $(\frac{1}{n} \sum_{k=0}^{n-1} T^k \nu)_{n \in \mathbb{N}}$, whose existence is guaranteed by Theorem 1.4.34.*

Example 1.4.41. As in Example 1.4.39, let $(H, \cdot, d)$ be a locally compact separable metric semigroup which has a neutral element, and let $\mu \in \mathcal{M}(H)$ be a probability measure. Now, assume that the sequence $(\mu^n)_{n \in \mathbb{N}}$ is tight (note that, by Proposition 4.2 of [144], the tightness of $(\mu^n)_{n \in \mathbb{N}}$ implies the equicontinuity of μ assumed in Example 1.4.39), let P_μ be the transition probability defined by μ, and let (S_μ, T_μ) be the Markov-Feller pair defined by P_μ.

Since we assume that $(\mu^n)_{n \in \mathbb{N}}$ is tight, it follows that the sequence of averages $(\frac{1}{n} \sum_{k=1}^{n} \mu^k)_{n \in \mathbb{N}}$ is also tight; that is, the sequence $(\frac{1}{n} \sum_{k=0}^{n-1} T_\mu^k \mu)_{n \in \mathbb{N}}$ is tight. Using (a) of Theorem 1.4.40, we obtain that $(\frac{1}{n} \sum_{k=0}^{n-1} T_\mu^k \mu)_{n \in \mathbb{N}}$ is $C_b(X)$-weak* convergent. Thus, $(\frac{1}{n} \sum_{k=1}^{n} \mu^k)_{n \in \mathbb{N}}$ and, therefore, $(\frac{1}{n} \sum_{k=0}^{n-1} \mu^k)_{n \in \mathbb{N}}$ are $C_b(X)$-weak* convergent sequences (note that the $C_b(X)$-weak* convergence of $(\frac{1}{n} \sum_{k=1}^{n} \mu^k)_{n \in \mathbb{N}}$ is one of the assertions of Theorem 2.13-(i) of Högnäs and Mukherjea [48]; thus, we have obtained here an alternative proof of the assertion).

Using Proposition B.3.1 and Proposition 2.3 of Högnäs and Mukherjea [48], we obtain that the set $\mathbb{P}(H)$ of all probability measures in $\mathcal{M}(H)$ becomes a topological semigroup when endowed with the operation of convolution of measures and the restriction to $\mathbb{P}(H)$ of $C_b(X)$-weak* topology of $\mathcal{M}(H)$. On the other hand, a careful reading of the proof of Proposition 4.3.6 of [143] reveals that the $C_b(X)$-weak* limit of $(\frac{1}{n} \sum_{k=1}^{n} \mu^k)_{n \in \mathbb{N}}$ (and of $(\frac{1}{n} \sum_{k=0}^{n-1} \mu^k)_{n \in \mathbb{N}}$, as well) is an element of $\mathbb{P}(H)$. Accordingly, the sequence $(\frac{1}{n} \sum_{k=0}^{n-1} \mu^k * \nu)_{n \in \mathbb{N}}$ converges in the $C_b(X)$-weak* topology of $\mathcal{M}(H)$ for every $\nu \in \mathbb{P}(H)$. Therefore, the sequence $(\frac{1}{n} \sum_{k=0}^{n-1} \mu^k * \nu)_{n \in \mathbb{N}}$ converges in the $C_b(X)$-weak* topology of $\mathcal{M}(H)$ for every $\nu \in \mathcal{M}(H)$. ∎

Given a transition probability P defined on a locally compact separable metric space (X, d) and given the Markov pair (S, T) defined by P we say that P (or (S, T), or S) is $C_0(X)$-*equicontinuous in the mean* (or, simply, equicontinuous in the mean) if the sequence of averages $(\frac{1}{n} \sum_{k=0}^{n-1} S^k f)_{n \in \mathbb{N}}$ is equicontinuous for every $f \in C_0(X)$. Clearly, if P is equicontinuous, then P is equicontinuous in the mean. It is also easy to see that if P is equicontinuous in the mean, then P is a Feller transition probability; indeed, if P is continuous in the mean, then $Sf \in C_b(X)$ (because $\frac{f + Sf}{2} \in C_b(X)$ as a result of the equicontinuity in the mean of P) for every $f \in C_0(X)$, so, by Proposition 1.1.5, P is Feller. All the results about equicontinuous transition probabilities discussed in this subsection are valid for transition probabilities that are equicontinuous in the mean with minor modification in the proofs.

Let P be a transition probability defined on a locally compact separable metric space (X, d) and let (S, T) be the Markov pair defined by P.

Assume that P is uniquely ergodic and let μ^* be the unique invariant probability of P. It is often a major concern to find μ^*. Since, in general, μ^* has a complicated structure, finding μ^* means approximating μ^* using the iterates of T. For instance, if μ^* is an attractive probability measure, then starting with any probability measure $\mu \in \mathcal{M}(X)$, if we calculate the iterates $T\mu, T^2\mu, \cdots, T^n\mu$, for very large values of n we obtain very good approximations of μ^*. However, it is not very often that μ^* is an attractive probability. A situation that appears much more often is that P is weak* uniquely mean ergodic (we say that P (or (S, T), or T) is *weak* uniquely mean ergodic* if the sequence $(\frac{1}{n}\sum_{k=0}^{n-1} T^k \mu)_{n\in\mathbb{N}}$ converges to μ^* in the weak* topology of $\mathcal{M}(X)$ whenever μ is a probability measure in $\mathcal{M}(X)$). For instance, using Propositions 1.2 and 1.3, both on p. 178 of Krengel [53], we obtain that if P is Feller and X is compact, then P is weak* uniquely mean ergodic. However, if X fails to be compact, even if P is an equicontinuous transition probability, the unique ergodicity of P does not imply its weak* unique mean ergodicity (for an example, see p. 96 of [143]). Thus, it is of interest to have conditions under which unique ergodicity implies weak* unique mean ergodicity. We conclude the section and the chapter with a result in this direction.

Theorem 1.4.42. *Assume that P is a Feller transition probability and that $X = \Gamma_{\mathrm{cp}}^{(\mathrm{TP})}$. The following assertions are equivalent:*

(a) P is uniquely ergodic.
(b) P is weak uniquely mean ergodic.*

Chapter 2
Preliminaries on Transition Functions and Their Invariant Probabilities

Our goal in this chapter is to introduce the transition functions and to discuss various related notions and known basic results that will be used throughout the book. Also in this chapter, we discuss several examples of transition functions which have a didactic purpose in the sense that we present these examples only to use them to illustrate the results of the book.

In Sect. 2.1, after going over the definition of a transition function, we define the family of Markov pairs generated by a transition function and discuss several basic properties of this family. We also present various types of transition functions. In a similar manner as in the case of transition probabilities (see Sect. 1.1) we define the orbits and the orbit-closures under the action of a transition function and study basic properties of these orbits and orbit-closures.

In Sect. 2.2, we discuss and list various examples of transition functions. The transition functions that we will consider here are of two kinds: transition functions defined by one-parameter semigroups or one-parameter groups of measurable functions, and transition functions defined by one-parameter convolution semigroups of probability measures. We should keep in mind that the transition functions discussed in this section are by no means the only important ones; the reason for our choice of examples is that they are better suited to illustrate the results of the book.

In the last section (Sect. 2.3), we discuss basic facts about invariant probabilities of transition functions.

2.1 Transition Functions

As pointed out in the abstract of this chapter, in this section we introduce and review basic properties of the transition functions and of certain concepts closely related to transition functions.

The transition functions under consideration in this book are often called homogeneous transition functions in probability theory and have their origin in the

R. Zaharopol, *Invariant Probabilities of Transition Functions*, Probability and Its Applications 44, DOI 10.1007/978-3-319-05723-1_2,
© Springer International Publishing Switzerland 2014

theory of continuous-time time-homogeneous Markov processes. These transition functions are discussed in virtually every book that deals with continuous-time Markov processes (see, for instance, Sections 36 and 42 of Bauer [8], Section A2 of Appendix in Beznea and Boboc [11], Chapter 1 of Blumenthal and Getoor [14], Section 18 of Borovkov [15], Section 1.2 of Chung and Walsh [18], Section 4.2 of Deuschel and Stroock [29], Chapters 2 and 3 of Vol. 1 of Dynkin [31], Chapter 4 of Ethier and Kurtz [35], the last three sections of Chapter 1 of Fukushima, Oshima and Takeda [36], Chapters 1 and 2 of Gihman and Skorohod [39], Section 2.4 of Hida [47], Section 1.4 of Mandl [69], Chapter 2 of Marcus and Rosen [71], Sections 9.3 and 10.3 of Meyer [76], Section 6.4 of Rao [88], Chapter 3 of Revuz and Yor [98], Section 3.1 of Rogers and Williams [100], Exercise 4.3.55 and Section 7.4 of Stroock [119], and Section 3.2 of Taira [124]). Later, it was noticed that, under very general conditions, one can associate a (homogeneous) transition function to any flow or semiflow. Thus, most results about transition functions are relevant to both the theory of continuous-time time-homogeneous Markov processes on one hand, and to ergodic theory and dynamical systems on the other (this fact is discussed in Section 8.4 of Dunford and Schwartz [30] and in Chapter 13 of Yosida [138]).

Throughout this section and the entire chapter we will use the notations established in Chap. 1.

Let (X, d) be a locally compact separable metric space, and let $\mathbb{T}$ stand for either the additive metric group $\mathbb{R}$ of all real numbers, where the metric $d_{\mathbb{R}}$ on $\mathbb{R}$ is the usual one defined in terms of the absolute value ($d_{\mathbb{R}}(s, t) = |s - t|$ for every $s \in \mathbb{R}$ and $t \in \mathbb{R}$), or else the additive metric semigroup $[0, +\infty)$, where the distance on $[0, +\infty)$ is the restriction of $d_{\mathbb{R}}$ to $[0, +\infty) \times [0, +\infty)$.

A family $(P_t)_{t \in \mathbb{T}}$ of transition probabilities defined on (X, d) is called a *transition function on* (X, d) if it has the property that

$$P_{s+t}(x, A) = \int_X P_s(y, A) P_t(x, \mathrm{d}y) \tag{2.1.1}$$

for every $s \in \mathbb{T}$, $t \in \mathbb{T}$, $x \in X$, and $A \in \mathcal{B}(X)$.

The reader has no doubt recognized the similarity between the above equality (2.1.1) and the equality that appears in (b) in Proposition 1.1.2. The equality (2.1.1) above is called the *Chapman-Kolmogorov equation for transition functions*, or the *continuous-time Chapman-Kolmogorov equation*, or, if there is no danger of confusion, simply, the *Chapman-Kolmogorov equation*.

The transition functions are also known as *Markov transition families in continuous time*.

Let $(P_t)_{t \in \mathbb{T}}$ be a transition function.

If $\mathbb{T} = \mathbb{R}$, and we want to emphasize that $\mathbb{T} = \mathbb{R}$, we will call $(P_t)_{t \in \mathbb{R}}$ an $\mathbb{R}$-transition function. Similarly, if $\mathbb{T} = [0, +\infty)$, we will sometimes call $(P_t)_{t \in [0,+\infty)}$ a $[0, +\infty)$-transition function.

Given an $\mathbb{R}$-transition function $(P_t)_{t \in \mathbb{R}}$, it is obvious that $(P_t)_{t \in [0,+\infty)}$ is also a transition function. We call $(P_t)_{t \in [0,+\infty)}$ the *restriction of* $(P_t)_{t \in \mathbb{R}}$ *to* $[0, +\infty)$.

The reader familiar only with the theory of continuous-time Markov processes may wonder why we define $\mathbb{R}$-transition functions when continuous-time Markov processes generate only $[0, +\infty)$-transition functions. The reason for allowing the time $\mathbb{T}$ to be the entire real line is that, as we will see in the next section, where we discuss examples of transition functions, the flows that appear in the study of dynamical systems generate in a very natural manner $\mathbb{R}$-transition functions, and we want the results of this book to be of use in both the theory of continuous time Markov processes and the theory of dynamical systems.

As before, let $(P_t)_{t \in \mathbb{T}}$ be a transition function.

As discussed in Sect. 1.1, for every $t \in \mathbb{T}$, the transition probability P_t generates a Markov pair (S_t, T_t) as follows: $S_t : B_b(X) \to B_b(X)$ is defined by

$$S_t f(x) = \int_X f(y) P_t(x, dy)$$

for every $f \in B_b(X)$ and $x \in X$, and $T_t : \mathcal{M}(X) \to \mathcal{M}(X)$ is defined by

$$T_t \mu(A) = \int_X P_t(x, A) d\mu(x)$$

for every $\mu \in \mathcal{M}(X)$ and $A \in \mathcal{B}(X)$. We say that $((S_t, T_t))_{t \in \mathbb{T}}$ is the *family of Markov pairs defined* (or *generated*) *by* $(P_t)_{t \in \mathbb{T}}$.

As in the case of transition functions, given a family $((S_t, T_t))_{t \in \mathbb{R}}$ of Markov pairs defined by a transition function $(P_t)_{t \in \mathbb{R}}$, we call $((S_t, T_t))_{t \in [0,+\infty)}$ the *restriction of* $((S_t, T_t))_{t \in \mathbb{R}}$ *to* $[0, +\infty)$. Clearly, the restriction $((S_t, T_t))_{t \in [0,+\infty)}$ of $((S_t, T_t))_{t \in \mathbb{R}}$ to $[0, +\infty)$ is the family of Markov pairs defined by the transition function $(P_t)_{t \in [0,+\infty)}$, which is the restriction of $(P_t)_{t \in \mathbb{R}}$ to $[0, +\infty)$.

Since the transition probability that defines a Markov pair is unique (see the comment made after Lemma 1.1.1), it follows that if the transition function $(P_t)_{t \in \mathbb{T}}$ defines a family $((S_t, T_t))_{t \in \mathbb{T}}$ of Markov pairs, then $(P_t)_{t \in \mathbb{T}}$ is the unique transition function with this property; that is, if $(P_t')_{t \in \mathbb{T}}$ is another transition function that defines $((S_t, T_t))_{t \in \mathbb{T}}$, then $P_t = P_t'$ for every $t \in \mathbb{T}$.

In the next proposition we discuss an important property of the operators S_t, $t \in \mathbb{T}$, and T_t, $t \in \mathbb{T}$, that appear in the family of Markov pairs $((S_t, T_t))_{t \in \mathbb{T}}$ defined by a transition function.

Proposition 2.1.1. *Let $(P_t)_{t \in \mathbb{T}}$ be a transition function and let $((S_t, T_t))_{t \in \mathbb{T}}$ be the family of Markov pairs defined by $(P_t)_{t \in \mathbb{T}}$. Then $S_{r+t} = S_r S_t$ and $T_{r+t} = T_r T_t$ for every $r \in \mathbb{T}$ and $t \in \mathbb{T}$; that is, the families $(S_t)_{t \in \mathbb{T}}$ and $(T_t)_{t \in \mathbb{T}}$ are one-parameter semigroups of operators if $\mathbb{T} = [0, +\infty)$, and one-parameter groups of operators if $\mathbb{T} = \mathbb{R}$.*

Proof. We first prove that $S_{r+t} = S_r S_t$ for every $r \in \mathbb{T}$ and $t \in \mathbb{T}$.

To this end, let $r \in \mathbb{T}$ and $t \in \mathbb{T}$.

Using (i) of Lemma 1.1.1 and the Chapman-Kolmogorov equation we obtain that

$$S_{r+t}\mathbf{1}_A(x) = P_{r+t}(x, A) = \int_X P_t(y, A)P_r(x, \mathrm{d}y)$$

$$= \int_X S_t\mathbf{1}_A(y)P_r(x, \mathrm{d}y) = S_r S_t\mathbf{1}_A(x)$$

for every $A \in \mathcal{B}(X)$ and $x \in X$. Accordingly, it follows that $S_{r+t}f = S_r S_t f$ for every real-valued simple measurable function f on X.

Now, if $f \in B_b(X)$ is not necessarily a simple function, then for every $\varepsilon \in \mathbb{R}$, $\varepsilon > 0$, there exists a simple function $g \in B_b(X)$ such that $\|f - g\| < \dfrac{\varepsilon}{2}$. Taking into consideration that $S_{r+t}g = S_r S_t g$ and that S_u, $u \in \mathbb{T}$ are positive contractions of $B_b(X)$, we obtain that

$$\|S_{r+t}f - S_r S_t f\| \leq \|S_{r+t}f - S_{r+t}g\| + \|S_r S_t g - S_r S_t f\|$$

$$= \|S_{r+t}(f - g)\| + \|S_r S_t(f - g)\|$$

$$\leq \|f - g\| + \|f - g\| < \frac{\varepsilon}{2} + \frac{\varepsilon}{2} = \varepsilon.$$

We have therefore proved that $\|S_{r+t}f - S_r S_t f\| < \varepsilon$ for every $f \in B_b(X)$ and $\varepsilon \in \mathbb{R}$, $\varepsilon > 0$. Thus, $S_{r+t} = S_r S_t$.

We now prove that $T_{r+t} = T_r T_t$ for every $r \in \mathbb{T}$ and $t \in \mathbb{T}$.

Thus, let $r \in \mathbb{T}$ and $t \in \mathbb{T}$.

Using the equality (1.1.3) and the fact that $S_{r+t} = S_r S_t$, that we have just proved, we obtain that

$$T_{r+t}\mu(A) = \langle \mathbf{1}_A, T_{r+t}\mu \rangle = \langle S_{r+t}\mathbf{1}_A, \mu \rangle = \langle S_r S_t\mathbf{1}_A, \mu \rangle$$

$$= \langle \mathbf{1}_A, T_r T_t\mu \rangle = T_r T_t\mu(A)$$

for every $\mu \in \mathcal{M}(X)$ and $A \in \mathcal{B}(X)$. $\qquad\square$

Note that according to the definitions of one-parameter semigroups and one-parameter groups (see Sect. A.1) these one-parameter semigroups and one-parameter groups are semigroup and group homomorphisms from $[0, +\infty)$ and $\mathbb{R}$ to a semigroup H, respectively. However, in Proposition 2.1.1 we did not specify the codomain H for $(S_t)_{t\in\mathbb{T}}$ and $(T_t)_{t\in\mathbb{T}}$. We did so because any semigroup of linear operators from $B_b(X)$ to $B_b(X)$, where the algebraic operation that defines the semigroup structure is the composition of operators, can play the role of H for $(S_t)_{t\in\mathbb{T}}$ provided that each S_t belongs to the semigroup, $t \in \mathbb{T}$, and, similarly, any semigroup of linear operators from $\mathcal{M}(X)$ to $\mathcal{M}(X)$ which is a semigroup with respect to the composition of operators can be used as H if T_t belongs to the semigroup for every $t \in \mathbb{T}$.

So far, after defining the transition functions, we have considered the families of Markov pairs generated by these transition functions and we have discussed some of their properties. A natural question at this point is: given two families $(S_t)_{t\in\mathbb{T}}$ and $(T_t)_{t\in\mathbb{T}}$ of positive linear contractions of $B_b(X)$ and $\mathcal{M}(X)$, respectively, under what conditions does there exist a transition function $(P_t)_{t\in\mathbb{T}}$ such that (S_t, T_t) is the Markov pair defined by P_t for every $t \in \mathbb{T}$? In the next proposition we discuss such conditions. In the proposition and, unless stated explicitly otherwise, throughout the book, the one-parameter semigroups or groups of operators are semigroup or group homomorphisms from $[0, +\infty)$ or $\mathbb{R}$, respectively, to a semigroup H of operators, where the semigroup structure of H is defined by the composition of operators.

Proposition 2.1.2. *Assume that $(S_t)_{t\in\mathbb{T}}$, $S_t : B_b(X) \to B_b(X)$ for every $t \in \mathbb{T}$, is a one-parameter semigroup or group of positive contractions of $B_b(X)$ if $\mathbb{T} = [0, +\infty)$ or $\mathbb{T} = \mathbb{R}$, respectively, and assume that $S_t \mathbf{1}_X = \mathbf{1}_X$ for every $t \in \mathbb{T}$. Also, let $(T_t)_{t\in\mathbb{T}}$, $T_t : \mathcal{M}(X) \to \mathcal{M}(X)$ for every $t \in \mathbb{T}$, be a one-parameter semigroup or group of positive contractions if $\mathbb{T} = [0, +\infty)$ or $\mathbb{T} = \mathbb{R}$, respectively, and assume that $\langle S_t f, \mu\rangle = \langle f, T_t \mu\rangle$ for every $f \in B_b(X)$, $\mu \in \mathcal{M}(X)$, and $t \in \mathbb{T}$. Then there exists a unique transition function $(P_t)_{t\in\mathbb{T}}$ such that $((S_t, T_t))_{t\in\mathbb{T}}$ is the family of Markov pairs defined by $(P_t)_{t\in\mathbb{T}}$.*

Proof. First note that, under the conditions of the proposition, for every $t \in \mathbb{T}$, the positive contraction T_t is a Markov operator because $T_t \mu(X) = \langle \mathbf{1}_X, T_t \mu\rangle = \langle S_t \mathbf{1}_X, \mu\rangle = \mu(X)$ for every $\mu \in \mathcal{M}(X)$, so $\|T_t \mu\| = \|\mu\|$ for every $\mu \in \mathcal{M}(X)$, $\mu \geq 0$.

For every $t \in \mathbb{T}$, let $P_t : X \times \mathcal{B}(X) \to \mathbb{R}$ be defined by $P_t(x, A) = S_t \mathbf{1}_A(x)$ for every $x \in X$ and $A \in \mathcal{B}(X)$.

Taking into consideration that the unicity of a transition function which has the property that $((S_t, T_t))_{t\in\mathbb{T}}$ is the family of Markov pairs defined by the transition function follows from the discussion that precedes Proposition 2.1.1, we obtain that in order to prove the proposition, we have to prove that the following three assertions are true:

(a) P_t is a transition probability for every $t \in \mathbb{T}$.
(b) (S_t, T_t) is the Markov pair defined by P_t for every $t \in \mathbb{T}$.
(c) The Chapman-Kolmogorov equation for transition functions holds true for $(P_t)_{t\in\mathbb{T}}$.

Proof of (a). Let $t \in \mathbb{T}$.

For every $A \in \mathcal{B}(X)$, it follows that $P_t(x, A) = S_t \mathbf{1}_A(x)$ for every $x \in X$; since $S_t \mathbf{1}_A$ belongs to $B_b(X)$, we obtain that the map $x \mapsto P_t(x, A)$ from X to $\mathbb{R}$ is measurable.

Let $x \in X$. Since T_t is a Markov operator, it follows that $T_t \delta_x$ is a probability measure. Since $P_t(x, A) = \langle S_t \mathbf{1}_A, \delta_x\rangle = \langle \mathbf{1}_A, T_t \delta_x\rangle = T_t \delta_x(A)$ for every $A \in \mathcal{B}(X)$, we obtain that the map $A \mapsto P_t(x, A)$ from $\mathcal{B}(X)$ to $\mathbb{R}$ is a probability measure.

Thus, P_t is a transition probability.

Proof of (b). Let $t \in \mathbb{T}$.

Since $S_t \mathbf{1}_A(x) = \int_X \mathbf{1}_A(y) P_t(x, \mathrm{d}y)$ for every $A \in \mathcal{B}(X)$ and $x \in X$, it follows that $S_t f(x) = \int_X f(y) P_t(x, \mathrm{d}y)$ for every simple measurable real-valued function f and $x \in X$. Using the fact that S_t is a linear contraction of $B_b(X)$, taking into consideration that the set of all measurable real-valued simple functions defined on X is dense in $B_b(X)$, and using the Lebesgue dominated convergence theorem, we obtain that $S_t f(x) = \int_X f(y) P_t(x, \mathrm{d}y)$ for every $f \in B_b(X)$ and $x \in X$. Finally, since

$$T_t \mu(A) = \langle S_t \mathbf{1}_A, \mu \rangle = \int_X S_t \mathbf{1}_A(x) \mathrm{d}\mu(x) = \int_X P_t(x, A) \mathrm{d}\mu(x)$$

for every $\mu \in \mathcal{M}(X)$ and $A \in \mathcal{B}(X)$, it follows that (S_t, T_t) is the Markov pair defined by P_t.

Proof of (c). Taking into consideration that for every $x \in X$ and $t \in \mathbb{T}$ the probability measures $T_t \delta_x$ and $\mu_x^{(t)}$ are equal, where $\mu_x^{(t)}$ is defined by $\mu_x^{(t)}(A) = P_t(x, A)$ for every $A \in \mathcal{B}(X)$, we obtain that

$$P_{r+t} \delta_x(A) = T_{r+t} \delta_x(A) = \langle \mathbf{1}_A, T_r T_t \delta_x \rangle$$

$$= \langle S_r \mathbf{1}_A, T_t \delta_x \rangle = \int_X S_r \mathbf{1}_A(y) \mathrm{d}T_t \delta_x(y)$$

$$= \int_X P_r(y, A) P_t(x, \mathrm{d}y)$$

for every $r \in \mathbb{T}$, $t \in \mathbb{T}$, $x \in X$, and $A \in \mathcal{B}(X)$.

Thus, $(P_t)_{t \in \mathbb{T}}$ is a transition function. $\qquad\square$

Let $(P_t)_{t \in \mathbb{T}}$ be a transition function defined on (X, d).

We say that $(P_t)_{t \in \mathbb{T}}$ satisfies the *standard measurability assumption (s.m.a.)* if, for every $A \in \mathcal{B}(X)$, the map $(t, x) \mapsto P_t(x, A)$, $(t, x) \in \mathbb{T} \times X$, is jointly measurable with respect to t and x; that is, the map is measurable with respect to the Borel σ-algebra on $\mathbb{R}$ and the product σ-algebra $\mathcal{L}(\mathbb{T}) \otimes \mathcal{B}(X)$, where $\mathcal{L}(\mathbb{T})$ is the σ-algebra of all Lebesgue measurable subsets of $\mathbb{T}$.

The s.m.a. is a rather common assumption, so common that it is sometimes incorporated in the definition of a transition function (see, for instance, p. 156 of Ethier and Kurtz [35]). We will use it frequently.

Our goal now is to discuss several useful reformulations of the s.m.a. To this end, we need the following two lemmas:

Lemma 2.1.3. *Let $\mathcal{A}$ be a collection of subsets of X, and assume that $\mathcal{A}$ satisfies the following three conditions:*

(a) $\mathcal{A}$ is closed under the formation of finite disjoint unions; that is, for every $n \in \mathbb{N}$ and n disjoint subsets $A_1, A_2, \cdots, A_n$ of X such that $A_i \in \mathcal{A}$ for every $i = 1, 2, \cdots, n$, it follows that $\cup_{j=1}^{n} A_j \in \mathcal{A}$.

(b) $\mathcal{A}$ is closed under the formation of proper differences; that is, if $A \in \mathcal{A}$ and $B \in \mathcal{A}$, and $A \subseteq B$, then $B \setminus A \in \mathcal{A}$.

(c) $\mathcal{A}$ is a monotone class; that is, if $(A_n)_{n\in\mathbb{N}}$ is a sequence of elements of $\mathcal{A}$ and if $(A_n)_{n\in\mathbb{N}}$ is monotone (the fact that $(A_n)_{n\in\mathbb{N}}$ is monotone means that $(A_n)_{n\in\mathbb{N}}$ is either increasing ($A_n \subseteq A_{n+1}$ for every $n \in \mathbb{N}$), or else decreasing ($A_n \supseteq A_{n+1}$ for every $n \in \mathbb{N}$)), then $\lim_{n\to+\infty} A_n \in \mathcal{A}$, where $\lim_{n\to+\infty} A_n = \cup_{n=1}^{\infty} A_n$ if $(A_n)_{n\in\mathbb{N}}$ is increasing and $\lim_{n\to+\infty} A_n = \cap_{n=1}^{\infty} A_n$ if $(A_n)_{n\in\mathbb{N}}$ is decreasing.

If the compact subsets of X belong to $\mathcal{A}$, then $\mathcal{B}(X) \subseteq \mathcal{A}$.

For a proof of the lemma, see Lemma 2.1 of [146].

Lemma 2.1.4. *Let G be an open subset of X. Then there exists an increasing sequence $(f_n)_{n\in\mathbb{N}}$ of elements of $C_0(X)$ such that $f_n \geq 0$ for every $n \in \mathbb{N}$ and such that the sequence of functions $(f_n)_{n\in\mathbb{N}}$ converges pointwise to $\mathbf{1}_G$ (that is, for every $x \in X$, the sequence of real numbers $(f_n(x))_{n\in\mathbb{N}}$ converges to 1 if $x \in G$, and to 0 if $x \notin G$).*

Proof. Let G be an open subset of X. Since the assertion of the lemma is obviously true if G is the empty set, we may and do assume that G is nonempty. The restriction d_G of the metric d to $G \times G$ is a metric on G, and (G, d_G) is a locally compact separable metric space in its own right. Thus, using Proposition 1.1.3 of [143] we obtain that (G, d_G) is σ-compact, so there exists an increasing sequence $(K_n)_{n\in\mathbb{N}}$ of compact subsets of G (the sets K_n, $n \in \mathbb{N}$, are compact in both the topology defined by the metric d_G on G and the topology defined by d on X) such that $G = \cup_{n=1}^{\infty} K_n$.

Since K_n, $n \in \mathbb{N}$, are compact subsets of X in the topology defined by d on X, we can use Proposition 7.1.8, p. 199, of Cohn's book [20] in order to obtain that there exists a $g_n \in C_0(X)$ (actually, the function g_n can be chosen with compact support) such that $\mathbf{1}_{K_n} \leq g_n \leq \mathbf{1}_G$ for every $n \in \mathbb{N}$.

Set $f_1 = g_1$, and $f_n = \sup_{1 \leq i \leq n} g_i$ for every $n \in \mathbb{N}$, $n \geq 2$. Then $f_n \in C_0(X)$ for every $n \in \mathbb{N}$, and the sequence $(f_n)_{n\in\mathbb{N}}$ is increasing. Moreover, $(f_n)_{n\in\mathbb{N}}$ converges pointwise to $\mathbf{1}_G$ because if $x \in X \setminus G$, then $f_n(x) = 0$ for every $n \in \mathbb{N}$, and if $x \in G$, then there exists an $n_x \in \mathbb{N}$ such that $x \in K_n$ for every $n \in \mathbb{N}$, $n \geq n_x$, so $f_n(x) = 1$ for every $n \in \mathbb{N}$, $n \geq n_x$. $\square$

In the next proposition we discuss the reformulations of the s.m.a. that we mentioned before Lemma 2.1.3.

Proposition 2.1.5. *Let $(P_t)_{t\in\mathbb{T}}$ be a transition function defined on (X, d) and let $((S_t, T_t))_{t\in\mathbb{T}}$ be the family of Markov pairs defined by $(P_t)_{t\in\mathbb{T}}$. The following assertions are equivalent:*

(a) For every $f \in C_0(X)$, the real-valued map $(t, x) \mapsto S_t f(x)$ for every $(t, x) \in \mathbb{T} \times X$ is jointly measurable with respect to t and x.

(b) $(P_t)_{t\in\mathbb{T}}$ satisfies the s.m.a.

(c) For every $f \in B_b(X)$, the real-valued map $(t, x) \mapsto S_t f(x)$ for every $(t, x) \in \mathbb{T} \times X$ is jointly measurable with respect to t and x.

(*d*) *For every $f \in C_b(X)$, the real-valued map $(t, x) \mapsto S_t f(x)$ for every $(t, x) \in \mathbb{T} \times X$ is jointly measurable with respect to t and x.*

Proof. $(a) \Rightarrow (b)$.   Set

$$
\mathcal{A} = \left\{ A \in \mathcal{B}(X) \;\middle|\;
\begin{array}{l}
\text{the real-valued map } (t, x) \mapsto P_t(x, A) \text{ for} \\
\text{every } (t, x) \in \mathbb{T} \times X \text{ is measurable with respect} \\
\text{to the Borel } \sigma\text{-algebra on } \mathbb{R} \text{ and } \mathcal{L}(\mathbb{T}) \otimes \mathcal{B}(X)
\end{array}
\right\}.
$$

The proof of the implication will be completed if we show that $\mathcal{A} = \mathcal{B}(X)$. To this end, we will use Lemma 2.1.3.

Note that $\mathcal{A}$ satisfies conditions (a), (b) and (c) of Lemma 2.1.3.

We now show that the compact subsets of X belong to $\mathcal{A}$.

For every $f \in B_b(X)$, let $\psi_f : \mathbb{T} \times X \to \mathbb{R}$ be defined by $\psi_f(t, x) = S_t f(x)$ for every $(t, x) \in \mathbb{T} \times X$.

Since $\psi_{\mathbf{1}_A}(t, x) = S_t \mathbf{1}_A(x) = P_t(x, A)$ and $\psi_{\mathbf{1}_{X \setminus A}}(t, x) = 1 - P_t(x, A)$ for every $A \in \mathcal{B}(X)$ and every $(t, x) \in \mathbb{T} \times X$, it follows that if $A \in \mathcal{A}$, then $X \setminus A$ belongs to $\mathcal{A}$, as well. Therefore, taking into consideration that every compact subset of X is closed, we infer that in order to prove that the compact subsets of X belong to $\mathcal{A}$, it is enough to show that the open subsets of X belong to $\mathcal{A}$.

Thus, let G be an open subset of X. Since the empty set $\emptyset$ belongs to $\mathcal{A}$ (because $P_t(x, \emptyset) = 0$ for every $(t, x) \in \mathbb{T} \times X$), we may and do assume that $G \neq \emptyset$.

By Lemma 2.1.4 there exists an increasing sequence $(f_n)_{n \in \mathbb{N}}$ of elements of $C_0(X)$ that converges pointwise to $\mathbf{1}_G$ and such that $f_n \geq 0$ for every $n \in \mathbb{N}$.

Since we assume that (a) is true, it follows that ψ_{f_n} is jointly measurable with respect to t and x for every $n \in \mathbb{N}$.

Since $\psi_{f_n}(t, x) = S_t f_n(x) = \langle S_t f_n, \delta_x \rangle = \langle f_n, T_t \delta_x \rangle = \int_X f_n(y)\, \mathrm{d}T_t \delta_x(y)$ for every $n \in \mathbb{N}$, and since, by the monotone convergence theorem, the sequence $(\int_X f_n(y)\, \mathrm{d}T_t \delta_x(y))_{n \in \mathbb{N}}$ converges to $\int_X \mathbf{1}_G(y)\, \mathrm{d}T_t \delta_x(y)$, it follows that $(\psi_{f_n}(t, x))_{n \in \mathbb{N}}$ converges to $\psi_{\mathbf{1}_G}(t, x)$ for every $(t, x) \in \mathbb{T} \times X$.

Thus, $\psi_{\mathbf{1}_G}$ is jointly measurable with respect to t and x. Since $\psi_{\mathbf{1}_G}(t, x) = S_t \mathbf{1}_G(x) = P_t(x, G)$ for every $(t, x) \in \mathbb{T} \times X$, it follows that $G \in \mathcal{A}$.

Since $\mathcal{A}$ satisfies all the conditions of Lemma 2.1.3, using the lemma we obtain that (b) holds true.

$(b) \Rightarrow (c)$.   Assume that (b) is true.

In terms of the functions ψ_f, $f \in B_b(X)$, the fact that $(P_t)_{t \in \mathbb{T}}$ satisfies the s.m.a. means that the function $\psi_{\mathbf{1}_A}$ is jointly measurable with respect to t and x for every $A \in \mathcal{B}(X)$. It is easy to see that, in this case ψ_f is jointly measurable with respect to t and x whenever f is a simple measurable real-valued function on X.

Now, if $f \in B_b(X)$ is not necessarily simple, then there exists a sequence $(f_n)_{n \in \mathbb{N}}$ of simple functions, $f_n \in B_b(X)$ for every $n \in \mathbb{N}$, such that $(f_n)_{n \in \mathbb{N}}$ converges uniformly to f.

Since S_t, $t \in \mathbb{T}$, are linear contractions of $B_b(X)$, and since $\psi_g(t, x) = S_t g(x)$ for every $g \in B_b(X)$ and every $(t, x) \in \mathbb{T} \times X$, it follows that $(\psi_{f_n})_{n \in \mathbb{N}}$ converges pointwise (on $\mathbb{T} \times X$) to ψ_f. Since, for every $n \in \mathbb{N}$, ψ_{f_n} is jointly measurable with respect to t and x, we obtain that ψ_f is jointly measurable with respect to t and x, as well.

$(c) \Rightarrow (a)$ is obvious.

$(d) \Leftrightarrow (a)$.   The equivalence of (a) and (d) is obtained using the fact that (a) and (c) are equivalent, which we just proved, and the inclusions $C_0(X) \subseteq C_b(X) \subseteq B_b(X)$, which are obvious. $\square$

Proposition 2.1.5 has the following consequence:

Corollary 2.1.6. *Let $(P_t)_{t \in \mathbb{T}}$ be a transition function defined on (X, d), assume that $(P_t)_{t \in \mathbb{T}}$ satisfies the s.m.a., and let $((S_t, T_t))_{t \in \mathbb{T}}$ be the family of Markov pairs defined by $(P_t)_{t \in \mathbb{T}}$. Then, for every $x \in X$, and $f \in B_b(X)$, the function $\psi_f^{(x)}$: $\mathbb{T} \to \mathbb{R}$ defined by $\psi_f^{(x)}(t) = S_t f(x)$ for every $t \in \mathbb{T}$ is measurable with respect to the Borel σ-algebra on $\mathbb{R}$ and the σ-algebra $\mathcal{L}(\mathbb{T})$ of all Lebesgue measurable subsets of $\mathbb{T}$.*

Proof. Since $(P_t)_{t \in \mathbb{T}}$ satisfies the s.m.a. we can use $(b) \Rightarrow (c)$ of Proposition 2.1.5 in order to conclude that the functions ψ_f, $f \in B_b(X)$, are jointly measurable with respect to t and x. Taking into consideration that $\psi_f^{(x)}(t) = \psi_f(t, x)$ for every $(t, x) \in \mathbb{T} \times X$ and $f \in B_b(X)$ and using a well-known fact from measure theory (see, for instance, Lemma 5.1.1, p. 155, of Cohn [20]) we obtain that the assertion of the corollary is true. $\square$

Let $(P_t)_{t \in \mathbb{T}}$ be a transition function defined on (X, d), and let $((S_t, T_t))_{t \in \mathbb{T}}$ be the family of Markov pairs defined by $(P_t)_{t \in \mathbb{T}}$.

We say that $(P_t)_{t \in \mathbb{T}}$ (or $((S_t, T_t))_{t \in \mathbb{T}}$, or $(S_t)_{t \in \mathbb{T}}$) is $C_0(X)$-*pointwise continuous* (or, simply, *pointwise continuous*) if, for every $f \in C_0(X)$ and $x \in X$, the real-valued function $t \mapsto S_t f(x)$, $t \in \mathbb{T}$, is continuous. Naturally, if X is compact, we might use also the term $C(X)$-*pointwise continuous*.

Although somewhat unusual, pointwise continuity is a surprisingly weak condition. It is satisfied by most transition functions associated to continuous-time Markov processes (for instance, all the transition functions associated with the Markov processes defined by the interacting particle systems discussed in Liggett's monographs [65] and [66] are pointwise continuous), by most transition functions defined by flows and semiflows, and by the transition functions defined by the one-parameter convolution semigroups of probability measures discussed in this book. We will have to impose this condition in order to obtain various results in the last three chapters in this book.

We say that $(P_t)_{t \in \mathbb{T}}$ is a *Feller transition function* if P_t is a Feller transition probability for every $t \in \mathbb{T}$. Thus, $(P_t)_{t \in \mathbb{T}}$ is a Feller transition function if $S_t f \in C_b(X)$ for every $f \in C_b(X)$ and $t \in \mathbb{T}$. Naturally, if $(P_t)_{t \in \mathbb{T}}$ is a Feller transition function, then $((S_t, T_t))_{t \in \mathbb{T}}$ is said to be the *family of Markov-Feller pairs defined by* $(P_t)_{t \in \mathbb{T}}$. The Feller transition functions have nice properties and we will discuss these transition functions in-depth in Chap. 6.

An interesting family of Feller transition functions are the $C_0(X)$-equicontinuous (or, simply, equicontinuous) transition functions. The transition function $(P_t)_{t\in\mathbb{T}}$ (or $((S_t, T_t))_{t\in\mathbb{T}}$ or $(S_t)_{t\in\mathbb{T}}$) is said to be $C_0(X)$-*equicontinuous* (or *equicontinuous*) if the set of functions $\{S_t f \,|\, t \in \mathbb{T}\}$ is equicontinuous for every $f \in C_0(X)$; thus, $(P_t)_{t\in\mathbb{T}}$ is equicontinuous if and only if for every $f \in C_0(X)$, for every convergent sequence $(x_n)_{n\in\mathbb{N}}$ of elements of X, and for every $\varepsilon \in \mathbb{R}$, $\varepsilon > 0$, there exists an $n_\varepsilon \in \mathbb{N}$ such that $|S_t f(x_n) - S_t f(x)| < \varepsilon$ for every $n \in \mathbb{N}$, $n \geq n_\varepsilon$, and every $t \in \mathbb{T}$, where $x = \lim_{n\to\infty} x_n$. If the transition probability $(P_t)_{t\in\mathbb{T}}$ is equicontinuous, then, for every $f \in C_0(X)$ and $t \in \mathbb{T}$, the function $S_t f$ is continuous and bounded; therefore, using Proposition 1.1.5, we obtain that $(P_t)_{t\in\mathbb{T}}$ is a Feller transition function. We will discuss various properties of equicontinuous transition functions in the last two sections of Chap. 6.

Let $x \in X$. Set $\mathcal{O}(x) = \bigcup_{t\in\mathbb{T}} \operatorname{supp}(T_t \delta_x)$ and $\mathcal{O}^{(F)}(x) = \bigcup_{\substack{t\in\mathbb{T} \\ t\geq 0}} \operatorname{supp}(T_t \delta_x)$. The sets $\mathcal{O}(x)$ and $\mathcal{O}^{(F)}(x)$ are called the *orbit* and the *forward orbit of x under the action of* $(T_t)_{t\in\mathbb{T}}$ (or $(P_t)_{t\in\mathbb{T}}$ or $((S_t, T_t))_{t\in\mathbb{T}}$), respectively. The closures $\overline{\mathcal{O}(x)}$ and $\overline{\mathcal{O}^{(F)}(x)}$ of $\mathcal{O}(x)$ and $\mathcal{O}^{(F)}(x)$ in the topology defined by the metric d on X are called the *orbit-closure* and the *forward orbit-closure of x under the action of* $(T_t)_{t\in\mathbb{T}}$ (or $(P_t)_{t\in\mathbb{T}}$ or $((S_t, T_t))_{t\in\mathbb{T}}$), respectively. For a subset A of X, set $\mathcal{O}(A) = \bigcup_{x\in A} \mathcal{O}(x)$ and $\mathcal{O}^{(F)}(A) = \bigcup_{x\in A} \mathcal{O}^{(F)}(x)$. The sets $\mathcal{O}(A)$, $\overline{\mathcal{O}(A)}$, $\mathcal{O}^{(F)}(A)$, and $\overline{\mathcal{O}^{(F)}(A)}$ are called the *orbit*, the *orbit-closure*, the *forward orbit*, and the *forward orbit-closure of A under the action of* $(T_t)_{t\in\mathbb{T}}$ (or $(P_t)_{t\in\mathbb{T}}$, or $((S_t, T_t))_{t\in\mathbb{T}}$), respectively. Note that, if $\mathbb{T} = [0, +\infty)$, then the orbits and orbit-closures are forward orbits and forward orbit-closures, respectively, and vice versa. Thus, the study of forward orbits and forward orbit-closures is of interest only when $\mathbb{T} = \mathbb{R}$.

The transition function $(P_t)_{t\in\mathbb{T}}$ (or $((S_t, T_t))_{t\in\mathbb{T}}$ or $(T_t)_{t\in\mathbb{T}}$) is said to be *minimal* or *forward minimal* if, for every $x \in X$, the orbit or the forward orbit of x under the action of $(P_t)_{t\in\mathbb{T}}$ is dense in X, respectively. Clearly, the study of the forward minimality of $(P_t)_{t\in\mathbb{T}}$ is of interest only when $\mathbb{T} = \mathbb{R}$. Also obvious is the fact that the forward minimality of $(P_t)_{t\in\mathbb{T}}$ always implies the minimality of $(P_t)_{t\in\mathbb{T}}$; however, (if $\mathbb{T} = \mathbb{R}$) the minimality of a transition function does not imply its forward minimality in general as we will show using a very simple example in the next section.

It is often of interest to know if the orbit of an element x of X under the action of the transition function $(P_t)_{t\in\mathbb{T}}$ is dense in X. For instance, if we know that the orbit of every $x \in X$ is dense in X, then we know that $(P_t)_{t\in\mathbb{T}}$ is a minimal transition function. In the next proposition we discuss a necessary and sufficient condition for an orbit to be dense in X.

In order to state the proposition, recall that if $(P_t)_{t\in\mathbb{T}}$ satisfies the s.m.a., then using Corollary 2.1.6, we obtain that, given $f \in B_b(X)$ and $x \in X$, the map $t \mapsto S_t f(x)$, $t \in \mathbb{T}$, is measurable (with respect to $\mathcal{L}(\mathbb{T})$ and the Borel σ-algebra on $\mathbb{R}$); if, in addition, $f \geq 0$, then the map is also positive, so the Lebesgue integral $\int_{\mathbb{T}} S_t f(x) \, dt$ exists (note that the integral could be equal to $+\infty$); in particular, if $f = \mathbf{1}_A$ for some $A \in \mathcal{B}(X)$, the integral $\int_{\mathbb{T}} P_t(x, A) \, dt$ exists.

Proposition 2.1.7. *Assume that the transition function $(P_t)_{t \in \mathbb{T}}$ satisfies the s.m.a. and is pointwise continuous, and let $x \in X$. The following assertions are equivalent:*

(a) The orbit $\mathcal{O}(x)$ of x under the action of $(P_t)_{t \in \mathbb{T}}$ is dense in X.
(b) $\int_{\mathbb{T}} P_t(x, U) \, dt > 0$ for every nonempty subset U of X.

Proof. $(a) \Rightarrow (b)$ Assume that $\overline{\mathcal{O}(x)} = X$, and let U be a nonempty open subset of X. Then $U \cap \mathcal{O}(x) \neq \emptyset$, so there exists a $t_0 \in \mathbb{T}$ such that $U \cap (\mathrm{supp}\,(T_{t_0}\delta_x)) \neq \emptyset$. Let $y \in U \cap (\mathrm{supp}\,(T_{t_0}\delta_x))$.

Using Proposition 7.1.8 of Cohn's book [20], applied to the compact set $\{y\}$ and to U, we obtain that there exists an $f \in C_0(X)$ such that $f(y) = 1$, $0 \leq f \leq \mathbf{1}_U$, and such that the support supp f of f is included in U.

Set $U_{\frac{1}{2}} = \left\{ z \in X \mid f(z) > \dfrac{1}{2} \right\}$. Then $U_{\frac{1}{2}}$ is an open subset of X. Also, $U_{\frac{1}{2}} \cap (\mathrm{supp}\,(T_{t_0}\delta_x)) \neq \emptyset$ because $y \in U_{\frac{1}{2}} \cap (\mathrm{supp}\,(T_{t_0}\delta_x))$. Thus, $T_{t_0}\delta_x(U_{\frac{1}{2}}) > 0$.

Taking into consideration that $\frac{1}{2}\mathbf{1}_{U_{\frac{1}{2}}} \leq f$, we obtain that

$$0 < \frac{1}{2}(T_{t_0}\delta_x)(U_{\frac{1}{2}}) = \left\langle \frac{1}{2}\mathbf{1}_{U_{\frac{1}{2}}}, T_{t_0}\delta_x \right\rangle \leq \langle f, T_{t_0}\delta_x \rangle$$

$$= \langle S_{t_0} f, \delta_x \rangle = S_{t_0} f(x).$$

Since $(P_t)_{t \in \mathbb{T}}$ is pointwise continuous and $S_{t_0} f(x) > 0$, it follows that there exists an $\varepsilon \in \mathbb{R}$, $\varepsilon > 0$, such that $[t_0, t_0 + \varepsilon) \subseteq \mathbb{T}$ and $S_t f(x) > 0$ for every $t \in [t_0, t_0 + \varepsilon)$.

Taking into consideration that $0 \leq f \leq \mathbf{1}_U$ and that S_t, $t \in \mathbb{T}$, are positive operators, we obtain that

$$0 < \int_{t_0}^{t_0+\varepsilon} S_t f(x) \, dt \leq \int_{t_0}^{t_0+\varepsilon} S_t \mathbf{1}_U(x) \, dt \leq \int_{\mathbb{T}} S_t \mathbf{1}_U(x) \, dt \leq \int_{\mathbb{T}} P_t(x, U) \, dt.$$

$(b) \Rightarrow (a)$ Assume that (b) is true, but $\mathcal{O}(x)$ is not dense in X.

Set $U = X \setminus \overline{\mathcal{O}(x)}$. Clearly, U is a nonempty open subset of X. Since $U \cap \mathcal{O}(x) = \emptyset$, it follows that $U \cap (\mathrm{supp}\, T_t\delta_x) = \emptyset$ for every $t \in \mathbb{T}$. Therefore, $P_t(x, U) = T_t\delta_x(U) = 0$ for every $t \in \mathbb{T}$, so $\int_{\mathbb{T}} P_t(x, U) \, dt = 0$.

We have obtained a contradiction which stems from our assumption that $\overline{\mathcal{O}(x)} \neq X$. $\qquad\square$

The above proposition has the following obvious consequence:

Corollary 2.1.8. *Assume that the transition function $(P_t)_{t \in \mathbb{T}}$ satisfies the s.m.a. and is pointwise continuous. The following assertions are equivalent:*

(a) $(P_t)_{t \in \mathbb{T}}$ is a minimal transition function.
(b) $\int_{\mathbb{T}} P_t(x, U) \, dt > 0$ for every $x \in X$ and every nonempty open subset U of X.

It is of interest to point out that a proposition similar to Proposition 2.1.7 can be stated for transition probabilities. We state it next.

Proposition 2.1.9. *Let P be a transition probability on (X, d), let (S, T) be the Markov pair defined by P, and let $x \in X$. The following assertions are equivalent:*

(a) The orbit $\mathcal{O}^{(\mathrm{TP})}(x)$ of x under the action of P is dense in X.
(b) $\sum_{n=0}^{\infty} P_n(x, U) > 0$ for every nonempty open subset U of X, where P_n, $n \in \mathbb{N}$, are the (not necessarily Feller) transition probabilities defined after Lemma 1.1.1 starting with $P_1 = P$, and P_0 is the transition probability defined by $P_0(y, A) = \mathbf{1}_A(y)$ for every $y \in X$ and $A \in \mathcal{B}(X)$ (note that the Markov pair defined by P_0 consists of the identity operators on $B_b(X)$ and $\mathcal{M}(X)$).

The proof of Proposition 2.1.9 follows along the lines of the proof of Proposition 2.1.7, but is significantly simpler.

Like Proposition 2.1.7, Proposition 2.1.9 has a consequence similar to Corollary 2.1.8 and can be combined with Proposition 1.4.6 as follows:

Corollary 2.1.10. *Let P be a transition probability on (X, d), and let (S, T) be the Markov pair defined by P. The following two assertions are equivalent:*

(a) P is a minimal transition probability.
(b) $\sum_{n=0}^{\infty} P_n(x, U) > 0$ for every $x \in X$ and every nonempty open subset U of X, where P_n, $n \in \mathbb{N} \cup \{0\}$, are the transition probabilities that appear in Proposition 2.1.9.

If P is a Feller transition probability, then each of the assertions (a) and (b) is also equivalent to:

(c) $\sum_{n=1}^{\infty} P_n(x, U) > 0$ for every $x \in X$ and every nonempty open subset U of X, where P_n, $n \in \mathbb{N}$, are the transition probabilities that appear in assertion (b) above.

The proof of the corollary is obtained easily using Propositions 2.1.9 and 1.4.6.

2.2 Examples

As mentioned in the abstract of this chapter, in this section we are going to present examples of transition functions. The transition functions under consideration are of two types: transition functions defined by one-parameter semigroups and one-parameter groups of measurable functions (actually, except for one example, all the transition functions of this type will be transition functions defined by semi-flows and flows), and transition functions defined by one-parameter convolution semigroups of probability measures. The reader might wonder at this point why we don't also discuss examples of transition functions associated with continuous-time time-homogeneous Markov processes. The reason is that we want to use the examples in order to illustrate various aspects of the theory developed in the work, and, at the time of writing, we are able to state and prove our results only by using transition functions and their associated families of Markov pairs; even though it can be shown that many results in this volume are valid for various transition functions associated to Markov processes, the problem is that for most Markov processes of interest, their transition functions cannot be given explicitly.

The section is organized into three subsections. In the first subsection we show how to associate a transition function and a family of Markov pairs to a one-parameter semigroup or a one-parameter group of Borel measurable functions defined on a locally compact separable metric space, and we study various properties of these transition functions and families of Markov pairs. In the second subsection we list the transition functions and the corresponding families of Markov pairs of various one-parameter semigroups and one-parameter groups of Borel measurable functions that we will use to illustrate various concepts and results throughout the book. Finally, in the last subsection we discuss the transition functions and the corresponding families of Markov pairs defined by one-parameter convolution semigroups of probability measures.

2.2.1 Transition Functions Defined by One-Parameter Semigroups or Groups of Measurable Functions: General Considerations

As usual, in this book, let (X, d) be a locally compact separable metric space and let $\mathbb{T}$ stand for the additive metric group $\mathbb{R}$ or the additive metric semigroup $[0, \infty)$.

Let $\mathbf{w} = (w_t)_{t \in \mathbb{T}}$ be a one-parameter semigroup or a one-parameter group of elements of $\mathbf{B}(X)$, where $\mathbf{B}(X)$ is the semigroup of all measurable functions $f : X \to X$, where the algebraic operation that defines the semigroup structure on $\mathbf{B}(X)$ is the operation of composition of functions.

The range $\mathbf{w}(\mathbb{T}) = \{w_t \,|\, t \in \mathbb{T}\}$ of $\mathbf{w}$ is a subsemigroup or a subgroup of $\mathbf{B}(X)$ if $\mathbb{T} = [0, \infty)$ or $\mathbb{T} = \mathbb{R}$, respectively, and in both cases w_0 is a neutral element for $\mathbf{w}(\mathbb{T})$, thought of as a semigroup (if $\mathbb{T} = [0, \infty)$) or as a group (if $\mathbb{T} = \mathbb{R}$) in its own right. However, since $\mathbf{B}(X)$ is only a semigroup rather than a group, w_0 may or may not be the neutral element of $\mathbf{B}(X)$ (that is, the identity map Id_X of X). The neutral element w_0 is equal to Id_X whenever $\mathbf{w}$ is a semiflow (in the case $\mathbb{T} = [0, \infty)$) or a flow (when $\mathbb{T} = \mathbb{R}$), where we think of semiflows and flows as families of maps rather than functions defined on $\mathbb{T} \times X$. Although, most of the time, we will deal with semiflows and flows rather than the more general one-parameter semigroups and one-parameter groups of elements of $\mathbf{B}(X)$, we will however encounter an example of a one-parameter semigroup of elements of $\mathbf{B}(X)$ which is not a semiflow.

Since w_t, $t \in \mathbb{T}$, are measurable functions, we can use them to define transition probabilities as we did in Sect. 1.1 before Example 1.1.6; that is, for every $t \in \mathbb{T}$, let $P_t^{(\mathbf{w})} : X \times \mathcal{B}(X) \to \mathbb{R}$ be defined by

$$P_t^{(\mathbf{w})}(x, A) = \delta_{w_t(x)}(A) = \mathbf{1}_A(w_t(x)) \tag{2.2.1}$$

for every $x \in X$ and $A \in \mathcal{B}(X)$. Then, as pointed out in Sect. 1.1, $P_t^{(\mathbf{w})}$ is a transition probability called the transition probability defined by w_t, $t \in \mathbb{T}$.

Proposition 2.2.1. *The family $(P_t^{(\mathbf{w})})_{t \in \mathbb{T}}$ of transition probabilities is a transition function.*

Proof. We have to prove that $(P_t^{(\mathbf{w})})_{t \in \mathbb{T}}$ satisfies the Chapman-Kolmogorov equation (the equality (2.1.1)).

To this end, let $s \in \mathbb{T}$, $t \in \mathbb{T}$, $x \in X$, and $A \in \mathcal{B}(X)$. Then, using the above equality (2.2.1), we obtain that

$$
\begin{aligned}
P_{s+t}^{(\mathbf{w})}(x, A) &= \mathbf{1}_A(w_{s+t}(x)) = \mathbf{1}_A(w_s(w_t(x))) \\
&= P_s^{(\mathbf{w})}(w_t(x), A) = \int_X P_s^{(\mathbf{w})}(y, A)\, \mathrm{d}\delta_{w_t(x)}(y) \\
&= \int_X P_s^{(\mathbf{w})}(y, A) P_t^{(\mathbf{w})}(x, \mathrm{d}y),
\end{aligned}
$$

where the last equality holds true because $\delta_{w_t(x)}$ is the probability measure that corresponds to $P_t^{(\mathbf{w})}$ and to $x \in X$ that appears at (i) in the definition of a transition probability (see Sect. 1.1). $\qquad\square$

We call $(P_t^{(\mathbf{w})})_{t \in \mathbb{T}}$ the *transition function defined* (or *generated*) *by* $\mathbf{w}$.

For every $t \in \mathbb{T}$, let $(S_t^{(\mathbf{w})}, T_t^{(\mathbf{w})})$ be the Markov pair defined by $P_t^{(\mathbf{w})}$. Thus, using the equalities (1.1.4) and (1.1.5) discussed in Sect. 1.1, we obtain that $S_t^{(\mathbf{w})} : B_b(X) \to B_b(X)$ is defined by

$$S_t^{(\mathbf{w})} f(x) = f(w_t(x)) \tag{2.2.2}$$

for every $f \in B_b(X)$ and $x \in X$, and $T_t^{(\mathbf{w})} : \mathcal{M}(X) \to \mathcal{M}(X)$ is defined by

$$T_t^{(\mathbf{w})} \mu(A) = \mu(w_t^{-1}(A)) \tag{2.2.3}$$

for every $\mu \in \mathcal{M}(X)$ and $A \in \mathcal{B}(X)$, whenever $t \in \mathbb{T}$.

We will also refer to the family $((S_t^{(\mathbf{w})}, T_t^{(\mathbf{w})}))_{t \in \mathbb{T}}$ of Markov pairs defined by the transition function $(P_t^{(\mathbf{w})})_{t \in \mathbb{T}}$ as the *family of Markov pairs defined by* $\mathbf{w}$.

Note that, by Proposition 2.1.1, the families $(S_t^{(\mathbf{w})})_{t \in \mathbb{T}}$ and $(T_t^{(\mathbf{w})})_{t \in \mathbb{T}}$ are one-parameter semigroups of operators if $\mathbb{T} = [0, \infty)$, and one-parameter groups of operators if $\mathbb{T} = \mathbb{R}$.

Also note that, when endowed with the operation of composition of operators, the sets $\{S_t^{(\mathbf{w})} | t \in \mathbb{T}\}$ and $\{T_t^{(\mathbf{w})} | t \in \mathbb{T}\}$ are semigroups with neutral elements $S_0^{(\mathbf{w})}$ and $T_0^{(\mathbf{w})}$ whenever $\mathbb{T} = [0, \infty)$, and are groups with neutral elements $S_0^{(\mathbf{w})}$ and $T_0^{(\mathbf{w})}$ whenever $\mathbb{T} = \mathbb{R}$, respectively. The operators $S_0^{(\mathbf{w})}$ and $T_0^{(\mathbf{w})}$ are the identity operators on $B_b(X)$ and $\mathcal{M}(X)$, respectively, if and only if w_0 is the identity map Id_X of X.

We say that $\mathbf{w} = (w_t)_{t \in \mathbb{T}}$ is a *measurable one-parameter semigroup* (if $\mathbb{T} = [0, \infty)$) or a *measurable one-parameter group* (if $\mathbb{T} = \mathbb{R}$) *of elements of* $\mathbf{B}(X)$ if the mapping $(t, x) \mapsto w_t(x)$ for every $(t, x) \in \mathbb{T} \times X$, is jointly measurable with respect to t and x in the sense that the mapping is measurable with respect to the σ-algebra $\mathcal{B}(X)$ on X and the product σ-algebra $\mathcal{L}(\mathbb{T}) \otimes \mathcal{B}(X)$ on $\mathbb{T} \times X$, where $\mathcal{L}(\mathbb{T})$ is the σ-algebra of all Lebesgue measurable subsets of $\mathbb{T}$. Note that if $\mathbf{w}$ is a semiflow or a flow, then the measurability of $\mathbf{w}$ as defined here is the same as the measurability of $\mathbf{w}$ as a semiflow or a flow as defined in Sect. A.3, before Example A.3.4.

Using formula (2.2.1) in this section, we obtain that if $\mathbf{w}$ is a measurable one-parameter semigroup or a measurable one-parameter group of elements of $\mathbf{B}(X)$, then the transition function $(P_t^{(\mathbf{w})})_{t \in \mathbb{T}}$ satisfies the s.m.a.

As expected, we say that $\mathbf{w} = (w_t)_{t \in \mathbb{T}}$ is *continuous in t* (or *continuous with respect to t*) if, for every $x \in X$, the mapping $t \mapsto w_t(x)$, $t \in \mathbb{T}$, is continuous with respect to the standard topology on $\mathbb{T}$ and the topology defined by the metric d on X.

In view of the equality (2.2.2) in this section, we obtain that if $\mathbf{w} = (w_t)_{t \in \mathbb{T}}$ is continuous in t, then $(P_t^{(\mathbf{w})})_{t \in \mathbb{T}}$ is $C_0(X)$-pointwise continuous.

We say that $\mathbf{w} = (w_t)_{t \in \mathbb{T}}$ is a *continuous one-parameter semigroup* (if $\mathbb{T} = [0, \infty)$) or a *continuous one-parameter group* (if $\mathbb{T} = \mathbb{R}$) *of elements of* $\mathbf{B}(X)$ if the mapping $(t, x) \mapsto w_t(x)$, $(t, x) \in \mathbb{T} \times X$, is jointly continuous with respect to t and x (that is, if the mapping is continuous with respect to the topology defined by the metric d on X, and the product topology on $\mathbb{T} \times X$ defined by the standard topology on $\mathbb{T}$ and the metric topology on X). Note that the notions of continuous one-parameter semigroup and continuous one-parameter group of elements of $\mathbf{B}(X)$ are natural extensions of the notions of continuous semiflow and continuous flow defined before Example A.3.4, in the sense that if $\mathbf{w}$ is a semiflow or a flow, then the continuity of $\mathbf{w}$ as defined here is the same as the continuity of $\mathbf{w}$ as a semiflow or a flow as defined in Sect. A.3.

Taking into consideration formula (2.2.2) in this section, we obtain that if $\mathbf{w}$ is a continuous one-parameter semigroup or group of elements of $\mathbf{B}(X)$, then the transition function $(P_t^{(\mathbf{w})})_{t \in \mathbb{T}}$ defined by $\mathbf{w}$ is a Feller transition function. Note that, since the continuity of $\mathbf{w}$ implies the measurability of $\mathbf{w}$, if $\mathbf{w} = (w_t)_{t \in \mathbb{T}}$

is continuous, then $(P_t^{(\mathbf{w})})_{t \in \mathbb{T}}$ satisfies the s.m.a. Also, since the continuity of $\mathbf{w} = (w_t)_{t \in \mathbb{T}}$ implies that $\mathbf{w}$ is continuous in t, we obtain that if $\mathbf{w}$ is continuous, then $(P_t^{(\mathbf{w})})_{t \in \mathbb{T}}$ is $C_0(X)$-pointwise continuous.

We say that the one-parameter semigroup or group $(w_t)_{t \in \mathbb{T}}$ of elements of $\mathbf{B}(X)$ is *equicontinuous with respect to* $t \in \mathbb{T}$ if for every convergent sequence $(x_n)_{n \in \mathbb{N}}$ of elements of X and for every $\varepsilon \in \mathbb{R}, \varepsilon > 0$, there exists an $n_\varepsilon \in \mathbb{N}$ such that $d(w_t(x_n), w_t(x)) < \varepsilon$ for every $n \in \mathbb{N}, n \geq n_\varepsilon$, and every $t \in \mathbb{T}$, where $x = \lim_{n \to +\infty} x_n$. Note that the definition of the equicontinuity with respect to $t \in \mathbb{T}$ of the one-parameter semigroup or group $(w_t)_{t \in \mathbb{T}}$ of elements of $\mathbf{B}(X)$ is a natural extension of the definition of the equicontinuity with respect to $t \in \mathbb{T}$ for semiflows and flows discussed before Proposition B.4.6.

Using the fact that every $f \in C_0(X)$ is uniformly continuous and the equality (2.2.2) of this section, we obtain that if $\mathbf{w} = (w_t)_{t \in \mathbb{T}}$ is a one-parameter semigroup or group of elements of $\mathbf{B}(X)$ that is equicontinuous with respect to $t \in \mathbb{T}$, then the transition function $(P_t^{(\mathbf{w})})_{t \in \mathbb{T}}$ defined by $\mathbf{w}$ is $C_0(X)$-equicontinuous.

As in the case of semiflows and flows (see the discussion following Proposition A.3.3, in the slightly more general case of one-parameter semigroups and one-parameter groups of elements of $\mathbf{B}(X)$, given such a one-parameter semigroup or group $\mathbf{w} = (w_t)_{t \in \mathbb{T}}$ and $x \in X$ we call the sets $\mathcal{O}(x) = \{w_t(x) \mid t \in \mathbb{T}\}$, $\overline{\mathcal{O}(x)}$, $\mathcal{O}^{(\mathrm{F})}(x) = \{w_t(x) \mid t \in \mathbb{T}, t \geq 0\}$ and $\overline{\mathcal{O}^{(\mathrm{F})}(x)}$ the *orbit, orbit-closure, forward orbit* and *forward orbit-closure* of x under the action of $\mathbf{w}$, respectively. If the action of $\mathbf{w}$ has to be emphasized we use the notations $\mathcal{O}_{\mathbf{w}}(x)$, $\overline{\mathcal{O}_{\mathbf{w}}(x)}$, $\mathcal{O}_{\mathbf{w}}^{(\mathrm{F})}(x)$ and $\overline{\mathcal{O}_{\mathbf{w}}^{(\mathrm{F})}(x)}$ instead of $\mathcal{O}(x)$, $\overline{\mathcal{O}(x)}$, $\mathcal{O}^{(\mathrm{F})}(x)$ and $\overline{\mathcal{O}^{(\mathrm{F})}(x)}$, respectively. If we consider the transition function $(P_t^{(\mathbf{w})})_{t \in \mathbb{T}}$ defined by $\mathbf{w}$, then the orbit, orbit-closure, forward orbit and forward orbit-closure of x under the action of $(P_t^{(\mathbf{w})})_{t \in \mathbb{T}}$ as defined in the previous section (Sect. 2.1) are precisely the orbit, orbit-closure, forward orbit and forward orbit-closure of x under the action of $\mathbf{w}$, respectively, as defined in this section. Accordingly, the notions of orbit, orbit-closure, forward orbit and forward orbit-closure under the action of a transition function as defined in Sect. 2.1 are natural and fairly significant extensions of the corresponding notions for semiflows and flows that have appeared in ergodic theory and dynamical systems and that we discussed after Proposition A.3.3.

2.2.2   *Transition Functions Defined by Specific One-Parameter Semigroups or Groups of Measurable Functions*

As pointed out at the beginning of the section, in this subsection we discuss in detail several transition functions and families of Markov pairs defined by one-parameter semigroups or groups of elements of $\mathbf{B}(X)$ for various locally compact separable metric spaces (X, d). In order to warm up, we start with a flow which, in all likelihood, is the simplest example of a flow.

Example 2.2.2. Let $X = \mathbb{R}$, where $\mathbb{R}$ is endowed with its usual metric d defined by $d(x, y) = |x - y|$ for every $x \in \mathbb{R}$ and $y \in \mathbb{R}$.

For every $t \in \mathbb{R}$, let $w_t : \mathbb{R} \to \mathbb{R}$ be defined by $w_t(x) = t + x$ for every $x \in \mathbb{R}$. Clearly, $\mathbf{w} = (w_t)_{t \in \mathbb{R}}$ is a flow defined on $(\mathbb{R}, d)$.

Let $(P_t^{(\mathbf{w})})_{t \in \mathbb{R}}$ be the transition function defined by $\mathbf{w}$. Then using the equality (2.2.1) of this section, we obtain that

$$P_t^{(\mathbf{w})}(x, A) = \delta_{t+x}(A) = \mathbf{1}_A(t + x) = \mathbf{1}_{A-t}(x)$$

for every $t \in \mathbb{R}$, $x \in \mathbb{R}$, and $A \in \mathcal{B}(\mathbb{R})$.

Let $((S_t^{(\mathbf{w})}, T_t^{(\mathbf{w})}))_{t \in \mathbb{R}}$ be the family of Markov pairs defined by $\mathbf{w}$. Then, for every $t \in \mathbb{R}$, the operator $S_t^{(\mathbf{w})} : B_b(\mathbb{R}) \to B_b(\mathbb{R})$ is defined by $S_t^{(\mathbf{w})} f(x) = f(t + x)$ for every $f \in B_b(\mathbb{R})$ and $x \in \mathbb{R}$, and the operator $T_t^{(\mathbf{w})} : \mathcal{M}(\mathbb{R}) \to \mathcal{M}(\mathbb{R})$ is defined by $T_t^{(\mathbf{w})} \mu(A) = \mu(A - t)$ for every $\mu \in \mathcal{M}(\mathbb{R})$ and every Borel subset A of $\mathbb{R}$.

Note that since the flow $\mathbf{w}$ is continuous, it follows that $(P_t^{(\mathbf{w})})_{t \in \mathbb{R}}$ is a Feller transition probability, and $(S_t^{(\mathbf{w})}, T_t^{(\mathbf{w})})$ is a Markov-Feller pair for every $t \in \mathbb{R}$.

Since $\mathbf{w}$ is continuous, it follows that $\mathbf{w}$ is also measurable, so $(P_t^{(\mathbf{w})})_{t \in \mathbb{R}}$ satisfies the s.m.a. Again using the continuity of $\mathbf{w}$, we obtain that $(P_t^{(\mathbf{w})})_{t \in \mathbb{R}}$ is $C_0(X)$-pointwise continuous.

Note that $\mathcal{O}(x) = \overline{\mathcal{O}(x)} = \mathbb{R}$ and $\mathcal{O}^{(\mathrm{F})}(x) = \overline{\mathcal{O}^{(\mathrm{F})}(x)} = [x, +\infty)$ for every $x \in \mathbb{R}$, so we see that the forward orbit of $x \in \mathbb{R}$ is a proper subset of the orbit of x. Note also that $(P_t^{(\mathbf{w})})_{t \in \mathbb{R}}$ is an example of a minimal transition function that fails to be forward minimal. ∎

We now discuss an example of a one-parameter semigroup of elements of $\mathbf{B}(X)$ which is not a semiflow and which has the property that the transition function defined by the one-parameter semigroup fails to be Feller.

Example 2.2.3. Let $X = [0, 1]$ and consider on X the usual metric d defined by the absolute value as follows: $d(x, y) = |x - y|$ for every $x \in [0, 1]$ and $y \in [0, 1]$. Clearly, (X, d) is a compact metric space.

Now let $\mathbf{w} = (w_t)_{t \in [0, +\infty)}$ be the family of mappings defined as follows: $w_t : X \to X$,

$$w_t(x) = \begin{cases} \frac{x}{2^t} & \text{if } 0 < x < 1 \\ 1 & \text{if } x = 0 \text{ or } x = 1 \end{cases}$$

for every $t \in [0, +\infty)$.

Using the definition of $\mathbf{w}$ and by studying the cases $x \in (0, 1)$ and $x \in \{0, 1\}$ separately we obtain that $w_{s+t}(x) = w_s(w_t(x))$ for every $s \in [0, +\infty)$, $t \in [0, +\infty)$, and $x \in [0, 1]$. Thus, $\mathbf{w} = (w_t)_{t \in [0, +\infty)}$ is a one-parameter semigroup of elements of $\mathbf{B}([0, 1])$. However, note that $\mathbf{w}$ is not a semiflow because w_0 is not the identity map $Id_{[0, 1]}$ of $[0, 1]$.

The reader has probably already noticed that the example discussed here is a "continuous-time" version of Example 1.1.6.

Let $(P_t^{(\mathbf{w})})_{t \in [0,+\infty)}$ be the transition function defined by $\mathbf{w}$.

We will now prove that $(P_t^{(\mathbf{w})})_{t \in [0,+\infty)}$ satisfies the s.m.a. As pointed out earlier in this section, in order to prove that $(P_t^{(\mathbf{w})})_{t \in [0,+\infty)}$ satisfies the s.m.a., it is enough to prove that the following assertion holds true:

Assertion. The one-parameter semigroup $\mathbf{w}$ is measurable.

Proof of Assertion. We have to prove that the mapping $\varphi : [0, +\infty) \times [0, 1] \to [0, 1]$ defined by $\varphi(t, x) = w_t(x)$ for every $(t, x) \in [0, +\infty) \times [0, 1]$ is measurable with respect to the Borel σ-algebra $\mathcal{B}([0, 1])$ and the product σ-algebra $\mathcal{L}([0, +\infty)) \otimes \mathcal{B}([0, 1])$, where $\mathcal{L}([0, +\infty))$ is the σ-algebra of all Lebesgue measurable subsets of $[0, +\infty)$.

To this end, set $E = [0, +\infty) \times (0, 1)$ and $F = [0, +\infty) \times \{0, 1\}$. Note that E and F are measurable subsets of $[0, +\infty) \times [0, 1]$ (in the sense that both E and F belong to $\mathcal{L}([0 + \infty)) \otimes \mathcal{B}([0, 1]))$. Note also that the restriction $\varphi_{|E}$ of φ to E is a continuous function (with respect to the standard topology on $[0, 1]$ (the topology defined by the metric d on $[0, 1]$) and the topology induced on E by the standard topology on $\mathbb{R}^2$), so $\varphi_{|E}^{-1}(B)$ belongs to $\mathcal{L}([0, +\infty)) \otimes \mathcal{B}([0, 1])$ whenever B is a Borel subset of $[0, 1]$.

Now let B be a Borel subset of $[0, 1]$. If $1 \in B$, then $\varphi^{-1}(B) = \varphi_{|E}^{-1}(B) \cup F$, so $\varphi^{-1}(B)$ belongs to $\mathcal{L}([0, +\infty)) \otimes \mathcal{B}([0, 1])$; if $1 \notin B$, then $\varphi^{-1}(B)$ belongs to $\mathcal{L}([0, +\infty)) \otimes \mathcal{B}([0, 1])$, as well, because $\varphi^{-1}(B) = \varphi_{|E}^{-1}(B)$.

Since $\varphi^{-1}(B) \in \mathcal{L}([0, +\infty)) \otimes \mathcal{B}([0, 1])$ for every $B \in \mathcal{B}([0, 1])$, it follows that $\mathbf{w}$ is measurable. $\square$

Since the above assertion holds true, we obtain that $(P_t^{(\mathbf{w})})_{t \in [0,+\infty)}$ satisfies the s.m.a.

Now, let $((S_t^{(\mathbf{w})}, T_t^{(\mathbf{w})}))_{t \in [0,+\infty)}$ be the family of Markov pairs defined by $\mathbf{w}$.

Using the equality (2.2.2) of this section and the fact that the map $t \mapsto w_t(x)$, $t \in [0, +\infty)$, is continuous for every $x \in [0, 1]$, we obtain that the map $t \mapsto S_t f(x)$, $t \in [0, +\infty)$, is continuous for every continuous function f and every $x \in [0, 1]$. Thus, $(P_t^{(\mathbf{w})})_{t \in [0,+\infty)}$ is pointwise continuous.

Again using the equality (2.2.2) of this section and the fact that w_t fails to be continuous at $x = 0$ for every $t \in [0, +\infty)$, we obtain that $(P_t^{(\mathbf{w})})_{t \in [0,+\infty)}$ is not a Feller transition function. $\blacksquare$

Example 2.2.4 (The Transition Function of the Flow of the Rotations of the Unit Circle). Let $X = \mathbb{R}/\mathbb{Z}$, where $\mathbb{R}/\mathbb{Z}$ is the commutative compact metric group known as the unit circle discussed in Example A.2.8. Also, let $\mathbf{w} = (w_t)_{t \in \mathbb{R}}$ be the flow of the rotations of the unit circle defined in Example A.3.4 and let $(P_t^{(\mathbf{w})})_{t \in \mathbb{R}}$ be the transition function defined by $\mathbf{w}$.

Using the equality (2.2.1) of this section, we obtain that

$$P_t^{(\mathbf{w})}(\hat{x}, A) = \delta_{\widehat{t+x}}(A) = \mathbf{1}_A(\widehat{t+x}) = \mathbf{1}_{A-\hat{t}}(\hat{x})$$

for every $t \in \mathbb{R}$, $\hat{x} \in \mathbb{R}/\mathbb{Z}$, and every Borel subset A of $\mathbb{R}/\mathbb{Z}$. Note that we use the "hat" notation here because it is significantly more convenient.

Let $((S_t^{(\mathbf{w})}, T_t^{(\mathbf{w})}))_{t \in \mathbb{R}}$ be the family of Markov pairs defined by $\mathbf{w}$. Using the equalities (2.2.2) and (2.2.3) of this section, we obtain that $S_t^{(\mathbf{w})} f(\hat{x}) = f(\widehat{t+x})$ for every $t \in \mathbb{R}$, $f \in B_b(\mathbb{R}/\mathbb{Z})$, and $\hat{x} \in \mathbb{R}/\mathbb{Z}$, and that $T_t^{(\mathbf{w})} \mu(A) = \mu(A - \hat{t})$ for every $t \in \mathbb{R}$, $\mu \in \mathcal{M}(\mathbb{R}/\mathbb{Z})$, and $A \in \mathcal{B}(\mathbb{R}/\mathbb{Z})$.

Note that for every $a \in \mathbb{R}$, $P_a^{(\mathbf{w})} = P_{\hat{a}}$, $S_a^{(\mathbf{w})} = S_{\hat{a}}$, and $T_a^{(\mathbf{w})} = T_{\hat{a}}$, where $P_{\hat{a}}$ and $(S_{\hat{a}}, T_{\hat{a}})$ are the transition probability and the Markov pair, respectively, defined and discussed in Example 1.1.7.

Since $\mathbf{w}$ is a continuous flow, it follows that $(P_t^{(\mathbf{w})})_{t \in \mathbb{R}}$ satisfies the s.m.a. (because the continuity of $\mathbf{w}$ implies that $\mathbf{w}$ is measurable), and that $(P_t^{(\mathbf{w})})_{t \in \mathbb{R}}$ is a pointwise continuous Feller transition function.

It is easy to see that the flow $\mathbf{w}$ is equicontinuous with respect to $t \in \mathbb{R}$; therefore, the transition function $(P_t^{(\mathbf{w})})_{t \in \mathbb{R}}$ is equicontinuous.

Finally, note that since $\mathbf{w}$ is both minimal and forward minimal, it follows that $(P_t^{(\mathbf{w})})_{t \in \mathbb{R}}$ is minimal and forward minimal, as well. ∎

Example 2.2.5 (The Rectilinear Flow on the Torus and its Transition Function). Let $n \in \mathbb{N}$, $n \geq 2$, let $\mathbf{v} \in \mathbb{R}^n$, $\mathbf{v} = (v_1, v_2, \cdots, v_n)$, let $\mathbb{R}^n/\mathbb{Z}^n$ be the n-dimensional torus defined in Example A.2.9, let $\mathbf{w} = (w_t)_{t \in \mathbb{R}}$ be the rectilinear flow on $\mathbb{R}^n/\mathbb{Z}^n$ with velocity $\mathbf{v}$ defined in Example A.3.5, and let $(P_t^{(\mathbf{w})})_{t \in \mathbb{R}}$ be the transition function defined by $\mathbf{w}$.

In view of the equality (2.2.1) of this section and using the "hat" notation for the elements of $\mathbb{R}^n/\mathbb{Z}^n$, we obtain that

$$P_t^{(\mathbf{w})}(\hat{\mathbf{x}}, A) = \delta_{\widehat{t\mathbf{v}+\mathbf{x}}}(A) = \mathbf{1}_A(\widehat{t\mathbf{v}+\mathbf{x}}) = \mathbf{1}_{A-\widehat{t\mathbf{v}}}(\hat{\mathbf{x}})$$

for every $t \in \mathbb{R}$, $\hat{\mathbf{x}} \in \mathbb{R}^n/\mathbb{Z}^n$, and every Borel subset A of $\mathbb{R}^n/\mathbb{Z}^n$.

If $((S_t^{(\mathbf{w})}, T_t^{(\mathbf{w})}))_{t \in \mathbb{R}}$ is the family of Markov pairs defined by $\mathbf{w}$, then using the equalities (2.2.2) and (2.2.3) of this section, we obtain that $S_t^{(\mathbf{w})} f(\hat{\mathbf{x}}) = f(\widehat{t\mathbf{v}+\mathbf{x}})$ for every $t \in \mathbb{R}$, $f \in B_b(\mathbb{R}^n/\mathbb{Z}^n)$ and $\hat{\mathbf{x}} \in \mathbb{R}^n/\mathbb{Z}^n$, and that $T_t^{(\mathbf{w})} \mu(A) = \mu(A - \widehat{t\mathbf{v}})$ for every $t \in \mathbb{R}$, $\mu \in \mathcal{M}(\mathbb{R}^n/\mathbb{Z}^n)$ and $A \in \mathcal{B}(\mathbb{R}^n/\mathbb{Z}^n)$.

Note that (for $t = 1$) $P_1^{(\mathbf{w})} = P_{\mathbf{v}}$, $S_1^{(\mathbf{w})} = S_{\mathbf{v}}$, and $T_1^{(\mathbf{w})} = T_{\mathbf{v}}$, where $P_{\mathbf{v}}$ and $(S_{\mathbf{v}}, T_{\mathbf{v}})$ are the transition probability and the Markov pair discussed in Example 1.1.8, respectively.

Since the rectilinear flow on the torus is continuous and equicontinuous with respect to $t \in \mathbb{R}$, it follows that $(P_t^{(\mathbf{w})})_{t \in \mathbb{R}}$ satisfies the s.m.a. and is a pointwise continuous equicontinuous Feller transition function.

Note that in general $(P_t^{(\mathbf{w})})_{t\in\mathbb{R}}$ is neither forward minimal nor minimal. However, if the entries $v_1, v_2, \cdots, v_n$ are rationally independent, then $\mathbf{w}$ is minimal (see Example A.3.5 for details), so $(P_t^{(\mathbf{w})})_{t\in\mathbb{R}}$ is minimal, as well, in this case. ∎

Example 2.2.6 (Transition Functions of Geodesic Flows). Our goal here is to discuss the transition functions of the geodesic flow on $\mathrm{PSL}(2,\mathbb{R})$, which is defined in Example A.3.6, and the geodesic flows on certain spaces of cosets of $\mathrm{PSL}(2,\mathbb{R})$, which are defined in Example B.1.8.

(a) (The Geodesic Flow on $\mathrm{PSL}(2,\mathbb{R})$ and its Transition Function). Let $\mathrm{PSL}(2,\mathbb{R})$ be the locally compact separable metrizable group defined in Example A.2.10. Since the elements of $\mathrm{PSL}(2,\mathbb{R})$ are cosets (of $L = \{\mathbf{I}_2, -\mathbf{I}_2\}$ in $\mathrm{SL}(2,\mathbb{R})$), we will use the "hat" notation when dealing with these elements.

Let $\mathbf{w} = (w_t)_{t\in\mathbb{R}}$ be the geodesic flow on $\mathrm{PSL}(2,\mathbb{R})$. Thus, for every $t \in \mathbb{R}$, $w_t :$ $\mathrm{PSL}(2,\mathbb{R}) \to \mathrm{PSL}(2,\mathbb{R})$ is the function defined by $w_t(\hat{g}) = \hat{g}\hat{g}_t$ for every $\hat{g} \in$ $\mathrm{PSL}(2,\mathbb{R})$, where $g_t = \begin{pmatrix} e^{\frac{t}{2}} & 0 \\ 0 & e^{-\frac{t}{2}} \end{pmatrix}$.

Let $(P_t^{(\mathbf{w})})_{t\in\mathbb{R}}$ be the transition function defined by $\mathbf{w}$. Then, using the equality (2.2.1) of this section, we obtain that

$$P_t^{(\mathbf{w})}(\hat{g}, A) = \delta_{\hat{g}\hat{g}_t}(A) = \mathbf{1}_A(\hat{g}\hat{g}_t) = \mathbf{1}_{A(\hat{g}_t)^{-1}}(\hat{g})$$

for every $t \in \mathbb{R}$, $\hat{g} \in \mathrm{PSL}(2,\mathbb{R})$ and every Borel subset A of $\mathrm{PSL}(2,\mathbb{R})$.

Now, let $((S_t^{(\mathbf{w})}, T_t^{(\mathbf{w})}))_{t\in\mathbb{R}}$ be the family of Markov pairs defined by $\mathbf{w}$. Then, using formula (2.2.2) of this section, we obtain that $S_t^{(\mathbf{w})} f(\hat{g}) = f(\hat{g}\hat{g}_t)$ for every $t \in \mathbb{R}$, $f \in B_b(\mathrm{PSL}(2,\mathbb{R}))$, and $\hat{g} \in \mathrm{PSL}(2,\mathbb{R})$; also, in view of the equality (2.2.3) of this section, we obtain that $T_t^{(\mathbf{w})}\mu(A) = \mu(A(\hat{g}_t)^{-1})$ for every $t \in \mathbb{R}$, $\mu \in$ $\mathcal{M}(\mathrm{PSL}(2,\mathbb{R}))$, and $A \in \mathcal{B}(\mathrm{PSL}(2,\mathbb{R}))$.

Note that if we set $h = g_s$ for some $s \in \mathbb{R}$, then $P_s^{(\mathbf{w})} = P_h$, $S_s^{(\mathbf{w})} = S_h$, and $T_s^{(\mathbf{w})} = T_h$, where P_h and (S_h, T_h) are the transition probability and the Markov pair discussed in Example 1.1.9.

Since, as shown in Example A.3.6, $\mathbf{w}$ is a continuous flow, it follows that $(P_t^{(\mathbf{w})})_{t\in\mathbb{R}}$ satisfies the s.m.a. and is a $C_0(\mathrm{PSL}(2,\mathbb{R}))$-pointwise continuous Feller transition function.

(b) The Transition Functions of the Geodesic Flows on Certain Spaces of Cosets of $\mathrm{PSL}(2,\mathbb{R})$. Let Γ be a lattice in $\mathrm{PSL}(2,\mathbb{R})$, and let $\mathbf{w}^{(\Gamma)} = (w_t^{(\Gamma)})_{t\in\mathbb{R}}$ be the geodesic flow on $(\mathrm{PSL}(2,\mathbb{R})/\Gamma)_{\mathrm{R}}$ defined in Example B.1.8. Thus, for every $t \in \mathbb{R}$, $w_t^{(\Gamma)} : (\mathrm{PSL}(2,\mathbb{R})/\Gamma)_{\mathrm{R}} \to (\mathrm{PSL}(2,\mathbb{R})/\Gamma)_{\mathrm{R}}$ is defined by $w_t^{(\Gamma)}(\Gamma\hat{x}) = \Gamma\hat{x}\hat{g}_t$ for every $\Gamma\hat{x} \in (\mathrm{PSL}(2,\mathbb{R})/\Gamma)_{\mathrm{R}}$, where $\hat{g}_t$ is the element of $\mathrm{PSL}(2,\mathbb{R})$ defined in (a) of this example and we continue to use the "hat" notation for the elements of $\mathrm{PSL}(2,\mathbb{R})$.

Let $(P_t^{(\mathbf{w}^{(\Gamma)})})_{t\in\mathbb{R}}$ be the transition function defined by $\mathbf{w}^{(\Gamma)}$. Using formula (2.2.1) of this section, we obtain that

$$P_t^{(\mathbf{w}^{(\Gamma)})}(\Gamma\hat{x}, A) = \delta_{\Gamma\hat{x}\hat{g}_t}(A) = \mathbf{1}_A(\Gamma\hat{x}\hat{g}_t) = \mathbf{1}_{A(\hat{g}_t)^{-1}}(\Gamma\hat{x})$$

for every $t \in \mathbb{R}$, $\Gamma\hat{x} \in (\mathrm{PSL}(2,\mathbb{R})/\Gamma)_R$, and every Borel subset A of $(\mathrm{PSL}(2,\mathbb{R})/\Gamma)_R$.

If $(S_t^{(\mathbf{w}^{(\Gamma)})}, T_t^{(\mathbf{w}^{(\Gamma)})})_{t\in\mathbb{R}}$ is the family of Markov pairs defined by $\mathbf{w}^{(\Gamma)}$, then using formula (2.2.2) of this section, we obtain that $S_t^{(\mathbf{w}^{(\Gamma)})}f(\Gamma\hat{x}) = f(\Gamma\hat{x}\hat{g}_t)$ for every $t \in \mathbb{R}$, $f \in B_b((\mathrm{PSL}(2,\mathbb{R})/\Gamma)_R)$, and $\Gamma\hat{x} \in (\mathrm{PSL}(2,\mathbb{R})/\Gamma)_R$, and using formula (2.2.3) also of this section, we obtain that $T_t^{(\mathbf{w}^{(\Gamma)})}\mu(A) = \mu(A(\hat{g}_t)^{-1})$ for every $t \in \mathbb{R}$, $\mu \in \mathcal{M}((\mathrm{PSL}(2,\mathbb{R})/\Gamma)_R)$, and $A \in \mathcal{B}((\mathrm{PSL}(2,\mathbb{R})/\Gamma)_R)$.

Note that if we set $h = g_s$ for some $s \in \mathbb{R}$, then $P_s^{(\mathbf{w}^{(\Gamma)})} = P_h^{(R)}$, $S_s^{(\mathbf{w}^{(\Gamma)})} = S_h^{(R)}$, and $T_s^{(\mathbf{w}^{(\Gamma)})} = T_h^{(R)}$ where $P_h^{(R)}$ and $(S_h^{(R)}, T_h^{(R)})$ are the transition probability and the Markov pair discussed in Example 1.1.10.

Since, as pointed out in Example B.1.8, $\mathbf{w}^{(\Gamma)}$ is a continuous flow, it follows that $(P_t^{(\mathbf{w}^{(\Gamma)})})_{t\in\mathbb{R}}$ satisfies the s.m.a. and is a $C_0((\mathrm{PSL}(2,\mathbb{R})/\Gamma)_R)$-pointwise continuous Feller transition function. ∎

Example 2.2.7 (Transition Functions of Horocycle Flows). We will now discuss the transition functions of the horocycle flows defined in Appendices A and B. Thus, we will consider the horocycle flows on $\mathrm{PSL}(2,\mathbb{R})$ discussed in Example A.3.7 and the horocycle flows on spaces of cosets of $\mathrm{PSL}(2,\mathbb{R})$ and $\mathrm{SL}(2,\mathbb{R})$ that are described in Example B.1.9.

(*a*) The Horocycle Flows on $\mathrm{PSL}(2,\mathbb{R})$. As usual in this book, we will use the "hat" notation when dealing with elements of $\mathrm{PSL}(2,\mathbb{R})$.

Let $\mathbf{v}^{(1)} = (v_t^{(1)})_{t\in\mathbb{R}}$ and $\mathbf{v}^{(2)} = (v_t^{(2)})_{t\in\mathbb{R}}$ be the two horocycle flows on $\mathrm{PSL}(2,\mathbb{R})$ defined in Example A.3.7. Thus, for every $i = 1, 2$ and $t \in \mathbb{R}$, the function $v_t^{(i)}$: $\mathrm{PSL}(2,\mathbb{R}) \to \mathrm{PSL}(2,\mathbb{R})$ is defined by $v_t^{(i)}(\hat{h}) = \hat{h}\hat{h}_t^{(i)}$ for every $\hat{h} \in \mathrm{PSL}(2,\mathbb{R})$, where $h_t^{(1)} = \begin{pmatrix} 1 & t \\ 0 & 1 \end{pmatrix}$ and $h_t^{(2)} = \begin{pmatrix} 1 & 0 \\ t & 1 \end{pmatrix}$.

Let $(P_t^{(\mathbf{v}^{(i)})})_{t\in\mathbb{R}}$ be the transition function defined by $\mathbf{v}^{(i)}$, $i = 1, 2$. Using formula (2.2.1) of this section, we obtain that

$$P_t^{(\mathbf{v}^{(i)})}(\hat{h}, A) = \delta_{\hat{h}\hat{h}_t^{(i)}}(A) = \mathbf{1}_A(\hat{h}\hat{h}_t^{(i)}) = \mathbf{1}_{A(\hat{h}_t^{(i)})^{-1}}(\hat{h})$$

for every $i = 1, 2$, $t \in \mathbb{R}$, $\hat{h} \in \mathrm{PSL}(2,\mathbb{R})$, and every Borel subset A of $\mathrm{PSL}(2,\mathbb{R})$.

If $i \in \{1, 2\}$ and $((S_t^{(\mathbf{v}^{(i)})}, T_t^{(\mathbf{v}^{(i)})}))_{t\in\mathbb{R}}$ is the family of Markov pairs defined by $\mathbf{v}^{(i)}$, then using formula (2.2.2) of this section, we obtain that $S_t^{(\mathbf{v}^{(i)})}f(\hat{h}) = f(\hat{h}\hat{h}_t^{(i)})$ for every $t \in \mathbb{R}$, $f \in B_b(\mathrm{PSL}(2,\mathbb{R}))$, and $\hat{h} \in \mathrm{PSL}(2,\mathbb{R})$, and using formula (2.2.3) also of this section, we obtain that $T_t^{(\mathbf{v}^{(i)})}\mu(A) = \mu(A(\hat{h}_t^{(i)})^{-1})$ for every $t \in \mathbb{R}$, $\mu \in \mathcal{M}(\mathrm{PSL}(2,\mathbb{R}))$, and $A \in \mathcal{B}(\mathrm{PSL}(2,\mathbb{R}))$.

Note that $P_s^{(\mathbf{v}^{(i)})} = P_{h_s^{(i)}}$, $S_s^{(\mathbf{v}^{(i)})} = S_{h_s^{(i)}}$, and $T_s^{(\mathbf{v}^{(i)})} = T_{h_s^{(i)}}$, where $P_{h_s^{(i)}}$ and $(S_{h_s^{(i)}}, T_{h_s^{(i)}})$ are the transition probability and the Markov pair defined by $u_{h_s^{(i)}}$ in Example 1.1.9, respectively, whenever $s \in \mathbb{R}$ and $i = 1$ or 2.

Since, as mentioned in Example A.3.7, the flows $\mathbf{v}^{(1)}$ and $\mathbf{v}^{(2)}$ are continuous, if follows that $(P_t^{(\mathbf{v}^{(1)})})_{t\in\mathbb{R}}$ and $(P_t^{(\mathbf{v}^{(2)})})_{t\in\mathbb{R}}$ satisfy the s.m.a. and are $C_0(\mathrm{PSL}(2,\mathbb{R}))$-pointwise continuous Feller transition functions.

(b) Horocycle Flows on Spaces of Cosets of $\mathrm{PSL}(2,\mathbb{R})$. Let Γ be a lattice in $\mathrm{PSL}(2,\mathbb{R})$, and let $\bar{\mathbf{v}}^{(1\Gamma)} = (\bar{v}_t^{(1\Gamma)})_{t\in\mathbb{R}}$ and $\bar{\mathbf{v}}^{(2\Gamma)} = (\bar{v}_t^{(2\Gamma)})_{t\in\mathbb{R}}$ be the two horocycle flows defined in (a) of Example B.1.9. Thus, for every $j \in \{1,2\}$ and $t \in \mathbb{R}$, the mapping $\bar{v}_t^{(j\Gamma)} : (\mathrm{PSL}(2,\mathbb{R})/\Gamma)_\mathrm{R} \to (\mathrm{PSL}(2,\mathbb{R})/\Gamma)_\mathrm{R}$ is defined by $\bar{v}_t^{(j\Gamma)}(\Gamma\hat{x}) = \Gamma\hat{x}\hat{h}_t^{(j)}$ for every element $\Gamma\hat{x}$ of $(\mathrm{PSL}(2,\mathbb{R})/\Gamma)_\mathrm{R}$, where $\hat{h}_t^{(j)}$ is the element of $\mathrm{PSL}(2,\mathbb{R})$ that appears in the definition of the horocycle flows on $\mathrm{PSL}(2,\mathbb{R})$ in (a) of this example.

Let $(P_t^{(\bar{\mathbf{v}}^{(j\Gamma)})})_{t\in\mathbb{R}}$ be the transition function defined by the flow $\bar{\mathbf{v}}^{(j\Gamma)}$, $j = 1$ or 2. In view of the formula (2.2.1) of this section, we obtain that

$$P_t^{(\bar{\mathbf{v}}^{(j\Gamma)})}(\Gamma\hat{x}, A) = \delta_{\Gamma\hat{x}\hat{h}_t^{(j)}}(A) = \mathbf{1}_A(\Gamma\hat{x}\hat{h}_t^{(j)}) = \mathbf{1}_{A(\hat{h}_t^{(j)})^{-1}}(\Gamma\hat{x})$$

for every $j \in \{1,2\}$, $t \in \mathbb{R}$, $\Gamma\hat{x} \in (\mathrm{PSL}(2,\mathbb{R})/\Gamma)_\mathrm{R}$ and every Borel subset A of $(\mathrm{PSL}(2,\mathbb{R})/\Gamma)_\mathrm{R}$.

If $((S_t^{(\bar{\mathbf{v}}^{(j\Gamma)})}, T_t^{(\bar{\mathbf{v}}^{(j\Gamma)})}))_{t\in\mathbb{R}}$ is the family of Markov pairs defined by $\bar{\mathbf{v}}^{(j\Gamma)}$, $j = 1,2$, then using formula (2.2.2) of this section, we obtain that $S_t^{(\bar{\mathbf{v}}^{(j\Gamma)})}f(\Gamma\hat{x}) = f(\Gamma\hat{x}\hat{h}_t^{(j)})$ for every $j = 1,2$, $t \in \mathbb{R}$, $f \in B_b((\mathrm{PSL}(2,\mathbb{R})/\Gamma)_\mathrm{R})$ and $\Gamma\hat{x} \in (\mathrm{PSL}(2,\mathbb{R})/\Gamma)_\mathrm{R}$, and by formula (2.2.3) also of this section, we obtain that $T_t^{(\bar{\mathbf{v}}^{(j\Gamma)})}\mu(A) = \mu(A(\hat{h}_t^{(j)})^{-1})$ for every $j \in \{1,2\}$, $t \in \mathbb{R}$, $\mu \in \mathcal{M}((\mathrm{PSL}(2,\mathbb{R})/\Gamma)_\mathrm{R})$, and $A \in \mathcal{B}((\mathrm{PSL}(2,\mathbb{R})/\Gamma)_\mathrm{R})$.

Observe that $P_t^{(\bar{\mathbf{v}}^{(j\Gamma)})} = P_{h_t^{(j)}}^{(\mathrm{R})}$, $S_t^{(\bar{\mathbf{v}}^{(j\Gamma)})} = S_{h_t^{(j)}}^{(\mathrm{R})}$, and $T_t^{(\bar{\mathbf{v}}^{(j\Gamma)})} = T_{h_t^{(j)}}^{(\mathrm{R})}$, where $P_{h_t^{(j)}}^{(\mathrm{R})}$ and $(S_{h_t^{(j)}}^{(\mathrm{R})}, T_{h_t^{(j)}}^{(\mathrm{R})})$ are the transition probability and the Markov pair defined by $u_{h_t^{(j)}}^{(\mathrm{R})}$ in Example 1.1.10, respectively, whenever $t \in \mathbb{R}$ and $j = 1$ or 2.

Since, as pointed out in (a) of Example B.1.9, the two flows $\bar{\mathbf{v}}^{(j\Gamma)}$, $j = 1$ or 2, are continuous, it follows that both $(P_t^{(\bar{\mathbf{v}}^{(1\Gamma)})})_{t\in\mathbb{R}}$ and $(P_t^{(\bar{\mathbf{v}}^{(2\Gamma)})})_{t\in\mathbb{R}}$ satisfy the s.m.a. and are $C_0((\mathrm{PSL}(2,\mathbb{R})/\Gamma)_\mathrm{R})$-pointwise continuous Feller transition functions.

Using the result of Hedlund [41] that we mentioned in (a) of Example B.1.9 (see also Theorem 1.9 in Chapter 4 of Bachir Bekka and Mayer [10]), we obtain that for every $\Gamma\hat{x} \in (\mathrm{PSL}(2,\mathbb{R})/\Gamma)_\mathrm{R}$ and $j \in \{1,2\}$, either $\Gamma\hat{x}$ is a periodic point for the flow $\bar{\mathbf{v}}^{(j\Gamma)}$, or else the orbit of $\Gamma\hat{x}$ under the action of the transition function

$(P_t^{(\bar{\mathbf{v}}^{(j\Gamma)})})_{t\in\mathbb{R}}$ is dense in $(\mathrm{PSL}(2,\mathbb{R})/\Gamma)_{\mathrm{R}}$. If Γ is cocompact, the two transition functions defined by the horocycle flows $\bar{\mathbf{v}}^{(j\Gamma)}$, $j = 1, 2$, are minimal.

(*c*) Horocycle Flows on Spaces of Cosets of $\mathrm{SL}(2,\mathbb{R})$. Let Γ be a lattice in $\mathrm{SL}(2,\mathbb{R})$, and let $\mathbf{v}^{(j\Gamma\mathrm{L})} = (v_t^{(j\Gamma\mathrm{L})})_{t\in\mathbb{R}}$ and $\mathbf{v}^{(j\Gamma\mathrm{R})} = (v_t^{(j\Gamma\mathrm{R})})_{t\in\mathbb{R}}$, $j = 1$ or 2, be the four horocycle flows defined in (*b*) of Example B.1.9.

Thus, for every $j = 1$ or 2, and every $t \in \mathbb{R}$, the function

$$v_t^{(j\Gamma\mathrm{L})} : (\mathrm{SL}(2,\mathbb{R})/\Gamma)_{\mathrm{L}} \to (\mathrm{SL}(2,\mathbb{R})/\Gamma)_{\mathrm{L}}$$

is defined by $v_t^{(j\Gamma\mathrm{L})}(x\Gamma) = h_t^{(j)}x\Gamma$ for every $x\Gamma \in (\mathrm{SL}(2,\mathbb{R})/\Gamma)_{\mathrm{L}}$, and the function

$$v_t^{(j\Gamma\mathrm{R})} : (\mathrm{SL}(2,\mathbb{R})/\Gamma)_{\mathrm{R}} \to (\mathrm{SL}(2,\mathbb{R})/\Gamma)_{\mathrm{R}}$$

is defined by $v_t^{(j\Gamma\mathrm{R})}(\Gamma x) = \Gamma x h_t^{(j)}$ for every $\Gamma x \in (\mathrm{SL}(2,\mathbb{R})/\Gamma)_{\mathrm{R}}$, where $h_t^{(j)}$ is the element of $\mathrm{SL}(2,\mathbb{R})$ defined in (*a*) of this example.

Let $(P_t^{(\mathbf{v}^{(j\Gamma\mathrm{L})})})_{t\in\mathbb{R}}$ and $((S_t^{(\mathbf{v}^{(j\Gamma\mathrm{L})})}, T_t^{(\mathbf{v}^{(j\Gamma\mathrm{L})})}))_{t\in\mathbb{R}}$ be the transition function and the family of Markov pairs defined by the flow $\mathbf{v}^{(j\Gamma\mathrm{L})}$, respectively, for every $j \in \{1,2\}$. Then using the formulas (2.2.1)–(2.2.3) of this section, we obtain that

$$P_t^{(\mathbf{v}^{(j\Gamma\mathrm{L})})}(x\Gamma, A) = \delta_{h_t^{(j)}x\Gamma}(A) = \mathbf{1}_A(h_t^{(j)}x\Gamma) = \mathbf{1}_{(h_t^{(j)})^{-1}A}(x\Gamma)$$

for every $j \in \{1,2\}$, $t \in \mathbb{R}$, $x\Gamma \in (\mathrm{SL}(2,\mathbb{R})/\Gamma)_{\mathrm{L}}$, and every Borel subset A of $(\mathrm{SL}(2,\mathbb{R})/\Gamma)_{\mathrm{L}}$; $S_t^{(\mathbf{v}^{(j\Gamma\mathrm{L})})} f(x\Gamma) = f(h_t^{(j)}x\Gamma)$ for every $j \in \{1,2\}$, $t \in \mathbb{R}$, $f \in B_b((\mathrm{SL}(2,\mathbb{R})/\Gamma)_{\mathrm{L}})$, and $x\Gamma \in (\mathrm{SL}(2,\mathbb{R})/\Gamma)_{\mathrm{L}}$; $T_t^{(\mathbf{v}^{(j\Gamma\mathrm{L})})}\mu(A) = \mu((h_t^{(j)})^{-1}A)$ for every $j \in \{1,2\}$, $t \in \mathbb{R}$, $\mu \in \mathcal{M}((\mathrm{SL}(2,\mathbb{R})/\Gamma)_{\mathrm{L}})$, and $A \in \mathcal{B}((\mathrm{SL}(2,\mathbb{R})/\Gamma)_{\mathrm{L}})$.

Similarly, let $(P_t^{(\mathbf{v}^{(j\Gamma\mathrm{R})})})_{t\in\mathbb{R}}$ and $((S_t^{(\mathbf{v}^{(j\Gamma\mathrm{R})})}, T_t^{(\mathbf{v}^{(j\Gamma\mathrm{R})})}))_{t\in\mathbb{R}}$ be the transition function and the family of Markov pairs defined by $\mathbf{v}^{(j\Gamma\mathrm{R})}$, respectively, for every $j \in \{1,2\}$. Then

$$P_t^{(\mathbf{v}^{(j\Gamma\mathrm{R})})}(\Gamma x, A) = \delta_{\Gamma x h_t^{(j)}}(A) = \mathbf{1}_A(\Gamma x h_t^{(j)}) = \mathbf{1}_{A(h_t^{(j)})^{-1}}(\Gamma x)$$

for every $j \in \{1,2\}$, $t \in \mathbb{R}$, $\Gamma x \in (\mathrm{SL}(2,\mathbb{R})/\Gamma)_{\mathrm{R}}$ and $A \in \mathcal{B}((\mathrm{SL}(2,\mathbb{R})/\Gamma)_{\mathrm{R}})$; $S_t^{(\mathbf{v}^{(j\Gamma\mathrm{R})})} f(\Gamma x) = f(\Gamma x h_t^{(j)})$ for every $j \in \{1,2\}$, $t \in \mathbb{R}$, $f \in B_b((\mathrm{SL}(2,\mathbb{R})/\Gamma)_{\mathrm{R}})$, and $\Gamma x \in (\mathrm{SL}(2,\mathbb{R})/\Gamma)_{\mathrm{R}}$; $T_t^{(\mathbf{v}^{(j\Gamma\mathrm{R})})}\mu(A) = \mu(A(h_t^{(j)})^{-1})$ for every $j \in \{1,2\}$, $t \in \mathbb{R}$, $\mu \in \mathcal{M}((\mathrm{SL}(2,\mathbb{R})/\Gamma)_{\mathrm{R}})$, and $A \in \mathcal{B}((\mathrm{SL}(2,\mathbb{R})/\Gamma)_{\mathrm{R}})$.

Note that $P_t^{(\mathbf{v}^{(j\Gamma\mathrm{L})})} = P_{(h_t^{(j)})}^{(\mathrm{L})}$, $S_t^{(\mathbf{v}^{(j\Gamma\mathrm{L})})} = S_{(h_t^{(j)})}^{(\mathrm{L})}$, $T_t^{(\mathbf{v}^{(j\Gamma\mathrm{L})})} = T_{(h_t^{(j)})}^{(\mathrm{L})}$, $P_t^{(\mathbf{v}^{(j\Gamma\mathrm{R})})} = P_{(h_t^{(j)})}^{(\mathrm{R})}$, $S_t^{(\mathbf{v}^{(j\Gamma\mathrm{R})})} = S_{(h_t^{(j)})}^{(\mathrm{R})}$ and $T_t^{(\mathbf{v}^{(j\Gamma\mathrm{R})})} = T_{(h_t^{(j)})}^{(\mathrm{R})}$, where $P_{(h_t^{(j)})}^{(\mathrm{L})}$ and $(S_{(h_t^{(j)})}^{(\mathrm{L})}, T_{(h_t^{(j)})}^{(\mathrm{L})})$ are the transition probability and the Markov pair defined by the mapping $u_{h_t^{(j)}}^{(\mathrm{L})} : (\mathrm{SL}(2,\mathbb{R})/\Gamma)_{\mathrm{L}} \to (\mathrm{SL}(2,\mathbb{R})/\Gamma)_{\mathrm{L}}$ discussed in Example 1.1.11, and, similarly,

$P^{(R)}_{(h_t^{(j)})}$ and $(S^{(R)}_{(h_t^{(j)})}, T^{(R)}_{(h_t^{(j)})})$ are the transition probability and the Markov pair defined by the function $u^{(R)}_{h_t^{(j)}} : (\mathrm{SL}(2,\mathbb{R})/\Gamma)_R \to (\mathrm{SL}(2,\mathbb{R})/\Gamma)_R$ also defined in Example 1.1.11, respectively, whenever $t \in \mathbb{R}$ and $j = 1$ or 2.

Since, as mentioned in (b) of Example B.1.9, the four horocycle flows under consideration here are continuous, it follows that the transition functions defined by these flows satisfy the s.m.a. and are pointwise continuous Feller transition functions. ∎

Example 2.2.8 (Transition Functions of Exponential Flows on Spaces of Cosets of $\mathrm{SL}(n,\mathbb{R})$). Let $n \in \mathbb{N}$, $n \geq 2$, let M be a closed subgroup of $\mathrm{SL}(n,\mathbb{R})$, let $\mathbf{A}$ be a trace zero $n \times n$ matrix, and let $\mathbf{u} = (u_t)_{t \in \mathbb{R}}$ and $\mathbf{v} = (v_t)_{t \in \mathbb{R}}$ be the exponential flows on $(\mathrm{SL}(n,\mathbb{R})/M)_L$ and $(\mathrm{SL}(n,\mathbb{R})/M)_R$, respectively, defined by $\mathbf{A}$. (See Sect. B.4.2 for details on these flows and some of the terminology used in this example.)

Thus, for every $t \in \mathbb{R}$,

$$u_t : (\mathrm{SL}(n,\mathbb{R})/M)_L \to (\mathrm{SL}(n,\mathbb{R})/M)_L$$

is defined by $u_t(\mathbf{X}M) = \exp(t\mathbf{A})\mathbf{X}M$ for every $\mathbf{X}M \in (\mathrm{SL}(n,\mathbb{R})/M)_L$ and

$$v_t : (\mathrm{SL}(n,\mathbb{R})/M)_R \to (\mathrm{SL}(n,\mathbb{R})/M)_R$$

is defined by $v_t(M\mathbf{X}) = M\mathbf{X}\exp(t\mathbf{A})$ for every $M\mathbf{X} \in (\mathrm{SL}(n,\mathbb{R})/M)_R$.

Let $(P_t^{(\mathbf{u})})_{t \in \mathbb{R}}$ and $((S_t^{(\mathbf{u})}, T_t^{(\mathbf{u})}))_{t \in \mathbb{R}}$ be the transition function and the family of Markov pairs defined by $\mathbf{u}$, respectively.

Then,

$$P_t^{(\mathbf{u})}(\mathbf{X}M, E) = \delta_{\exp(t\mathbf{A})\mathbf{X}M}(E) = \mathbf{1}_E(\exp(t\mathbf{A})\mathbf{X}M) = \mathbf{1}_{\exp(-t\mathbf{A})E}(\mathbf{X}M)$$

for every $t \in \mathbb{R}$, $\mathbf{X}M \in (\mathrm{SL}(n,\mathbb{R})/M)_L$, and every Borel subset E of $(\mathrm{SL}(n,\mathbb{R})/M)_L$; $S_t^{(\mathbf{u})}f(\mathbf{X}M) = f(\exp(t\mathbf{A})\mathbf{X}M)$ for every $t \in \mathbb{R}$, $f \in B_b((\mathrm{SL}(n,\mathbb{R})/M)_L)$, and $\mathbf{X}M \in (\mathrm{SL}(n,\mathbb{R})/M)_L$; $T_t^{(\mathbf{u})}\mu(E) = \mu(\exp(-t\mathbf{A})E)$ for every $t \in \mathbb{R}$, $\mu \in \mathcal{M}((\mathrm{SL}(n,\mathbb{R})/M)_L)$, and $E \in \mathcal{B}((\mathrm{SL}(n,\mathbb{R})/M)_L)$.

Similarly, let $(P_t^{(\mathbf{v})})_{t \in \mathbb{R}}$ and $(S_t^{(\mathbf{v})}, T_t^{(\mathbf{v})})_{t \in \mathbb{R}}$ be the transition function and the family of Markov pairs defined by $\mathbf{v}$, respectively.

Then,

$$P_t^{(\mathbf{v})}(M\mathbf{X}, E) = \delta_{M\mathbf{X}\exp(t\mathbf{A})}(E) = \mathbf{1}_E(M\mathbf{X}\exp(t\mathbf{A})) = \mathbf{1}_{E\exp(-t\mathbf{A})}(M\mathbf{X})$$

for every $t \in \mathbb{R}$, $M\mathbf{X} \in (\mathrm{SL}(n,\mathbb{R})/M)_R$, and $E \in \mathcal{B}((\mathrm{SL}(n,\mathbb{R})/M)_R)$; $S_t^{(\mathbf{v})}f(M\mathbf{X}) = f(M\mathbf{X}\exp(t\mathbf{A}))$ for every $t \in \mathbb{R}$, $f \in B_b((\mathrm{SL}(n,\mathbb{R})/M)_R)$, and $M\mathbf{X} \in (\mathrm{SL}(n,\mathbb{R})/M)_R$; $T_t^{(\mathbf{v})}\mu(E) = \mu(E\exp(-t\mathbf{A}))$ for every $t \in \mathbb{R}$, $\mu \in \mathcal{M}((\mathrm{SL}(n,\mathbb{R})/M)_R)$, and $E \in \mathcal{B}((\mathrm{SL}(n,\mathbb{R})/M)_R)$.

Observe that if M is a lattice in $\mathrm{SL}(n, \mathbb{R})$, and if we set $\Gamma = M$ and $h = \exp(A)$, then $P_1^{(\mathbf{u})} = P_{(h)}^{(\mathrm{L})}$, $(S_1^{(\mathbf{u})}, T_1^{(\mathbf{u})}) = (S_{(h)}^{(\mathrm{L})}, T_{(h)}^{(\mathrm{L})})$, $P_1^{(\mathbf{v})} = P_{(h)}^{(\mathrm{R})}$, and $(S_1^{(\mathbf{v})}, T_1^{(\mathbf{v})}) = (S_{(h)}^{(\mathrm{R})}, T_{(h)}^{(\mathrm{R})})$, where the transition probabilities $P_{(h)}^{(\mathrm{L})}$ and $P_{(h)}^{(\mathrm{R})}$, and the Markov pairs $(S_{(h)}^{(\mathrm{L})}, T_{(h)}^{(\mathrm{L})})$ and $(S_{(h)}^{(\mathrm{R})}, T_{(h)}^{(\mathrm{R})})$ are defined by the maps $u_h^{(\mathrm{L})}$ and $u_h^{(\mathrm{R})}$, respectively, and are discussed in Example 1.1.11.

Since, as pointed out in Sect. B.4.2, the flows $\mathbf{u}$ and $\mathbf{v}$ are continuous, it follows that $(P_t^{(\mathbf{u})})_{t \in \mathbb{R}}$ and $(P_t^{(\mathbf{v})})_{t \in \mathbb{R}}$ satisfy the s.m.a. and are pointwise continuous Feller transition functions.

Note that if $\mathbf{v}$ is a unipotent flow, we obtain from one of Ratner's theorems stated in Sect. B.4.2 (Theorem B.4.10) that if M is a lattice in $\mathrm{SL}(n, \mathbb{R})$, then, for every $M\mathbf{X} \in (\mathrm{SL}(n, \mathbb{R})/M)_{\mathrm{R}}$, the orbit-closure of $M\mathbf{X}$ under the action of $(P_t^{(\mathbf{v})})_{t \in \mathbb{R}}$ (which is the same as the orbit-closure of $M\mathbf{X}$ under the action of $\mathbf{v}$) is homogeneous (see Sect. B.4.2 for the definition of a homogeneous orbit-closure under the action of a flow). ∎

Example 2.2.9 (Semiflows on $\mathbb{S}_n$ and $\mathbb{P}_n$ and their Transition Functions). Our goal here is to discuss the transition functions and the families of Markov pairs generated by the two kinds of semiflows defined in Sect. B.4.1.

(a) Transition Functions of Semiflows on $\mathbb{S}_n$. Let $n \in \mathbb{N}$, $n \geq 2$, let $\mathbf{A} \in \mathbb{S}_n$, and let $\pi^{(\mathbf{A})} = (\pi_t^{(\mathbf{A})})_{t \in [0, +\infty)}$ be the semiflow defined in Sect. B.4.1. Thus, $\pi_t^{(\mathbf{A})}(\mathbf{X}) = (\exp_s(t\mathbf{A}))\mathbf{X} = e^{-t}(\exp(t\mathbf{A}))\mathbf{X}$ for every $t \in [0, +\infty)$ and $\mathbf{X} \in \mathbb{S}_n$.

Let $(P_t^{(\pi^{(\mathbf{A})})})_{t \in [0, +\infty)}$ and $((S_t^{(\pi^{(\mathbf{A})})}, T_t^{(\pi^{(\mathbf{A})})}))_{t \in [0, +\infty)}$ be the transition function and the family of Markov pairs defined by $\pi^{(\mathbf{A})}$, respectively.
Then

$$P_t^{(\pi^{(\mathbf{A})})}(\mathbf{X}, E) = \delta_{e^{-t}(\exp(t\mathbf{A}))\mathbf{X}}(E) = \mathbf{1}_E(e^{-t}(\exp(t\mathbf{A}))\mathbf{X})$$

$$= \mathbf{1}_{(e^{-t}(\exp(t\mathbf{A})))^{-1}E}(\mathbf{X})$$

for every $t \in [0, +\infty)$, $\mathbf{X} \in \mathbb{S}_n$, and every Borel subset E of $\mathbb{S}_n$, where, of course, the subset $(e^{-t}(\exp(t\mathbf{A})))^{-1}E$ of $\mathbb{S}_n$ is defined using formula (A.1.4); that is,

$$(e^{-t}\exp(t\mathbf{A}))^{-1}E = \{\mathbf{Y} \in \mathbb{S}_n \mid e^{-t}(\exp(t\mathbf{A}))\mathbf{Y} \in E\}$$

(note that $(e^{-t}\exp(t\mathbf{A}))^{-1}E$ is a Borel subset of $\mathbb{S}_n$ because $\mathbb{S}_n$ is a topological semigroup). We also obtain that $S_t^{(\pi^{(\mathbf{A})})}f(\mathbf{X}) = f(e^{-t}(\exp(t\mathbf{A}))\mathbf{X})$ for every $t \in [0, +\infty)$, $f \in B_b(\mathbb{S}_n)$, and $\mathbf{X} \in \mathbb{S}_n$, and that $T_t^{(\pi^{(\mathbf{A})})}\mu(E) = \mu((e^{-t}(\exp(t\mathbf{A})))^{-1}E)$ for every $t \in [0, +\infty)$, $\mu \in \mathcal{M}(\mathbb{S}_n)$, and $E \in \mathcal{B}(\mathbb{S}_n)$, where, as above, the meaning of the set $(e^{-t}(\exp(t\mathbf{A})))^{-1}E$ is given by the equality (A.1.4).

Note that, given $s \in [0, +\infty)$, if we set $\mathbf{B} = e^{-s}\exp(s\mathbf{A})$, then $P_s^{(\pi^{(\mathbf{A})})} = P_{\mathbf{B}}$, $S_s^{(\pi^{(\mathbf{A})})} = S_{\mathbf{B}}$, and $T_s^{(\pi^{(\mathbf{A})})} = T_{\mathbf{B}}$, where $P_{\mathbf{B}}$ and $(S_{\mathbf{B}}, T_{\mathbf{B}})$ are the transition probability and the Markov pair, respectively, defined by the function $u_{\mathbf{B}}$ discussed in Example 1.1.12.

Since, as pointed out when we defined $\pi^{(\mathbf{A})}$, after Proposition B.4.5, the semiflow $\pi^{(\mathbf{A})}$ is continuous, it follows that $(P_t^{(\pi^{(\mathbf{A})})})_{t \in [0,+\infty)}$ satisfies the s.m.a. and is a $C(\mathbb{S}_n)$-pointwise continuous Feller transition function.

Since, by Proposition B.4.6, the semiflow $\pi^{(\mathbf{A})}$ is equicontinuous with respect to $t \in [0,+\infty)$, using the general discussion about transition functions defined by one-parameter semigroups or groups of elements of $\mathbf{B}(X)$ in this section before Example 2.2.2, we obtain that the transition function $(P_t^{(\pi^{(\mathbf{A})})})_{t \in [0,+\infty)}$ is equicontinuous.

(*b*) Transition Functions of Semiflows on $\mathbb{P}_n$. Let $n \in \mathbb{N}$, $n \geq 2$, and let $\mathbf{A} \in \mathbb{S}_n$. Recall (see the discussion preceding Proposition B.4.7) that $\mathbb{P}_n$ is the compact metric space of all column stochastic vectors in $\mathbb{R}^n$ endowed with the metric d defined by $d(\mathbf{x}, \mathbf{y}) = \sum_{i=1}^{n} |x_i - y_i|$ for every column stochastic vectors $\mathbf{x} = (x_1, x_2, \cdots, x_n)^{\mathrm{T}}$ and $\mathbf{y} = (y_1, y_2, \cdots, y_n)^{\mathrm{T}}$. Let $\varphi^{(\mathbf{A})} = (\varphi_t^{(\mathbf{A})})_{t \in [0,+\infty)}$ be the semiflow discussed in Proposition B.4.7. Thus, $\varphi_t^{(\mathbf{A})}(\mathbf{x}) = e^{-t} \exp(t\mathbf{A})\mathbf{x}$ for every $t \in [0,+\infty)$, and $\mathbf{x} \in \mathbb{P}_n$.

Now, let $(P_t^{(\varphi^{(\mathbf{A})})})_{t \in [0,+\infty)}$ and $((S_t^{(\varphi^{(\mathbf{A})})}, T_t^{(\varphi^{(\mathbf{A})})}))_{t \in [0,+\infty)}$ be the transition function and the family of Markov pairs defined by $\varphi^{(\mathbf{A})}$, respectively.
Then,

$$P_t^{(\varphi^{(\mathbf{A})})}(\mathbf{x}, E) = \delta_{e^{-t} \exp(t\mathbf{A})\mathbf{x}}(E) = \mathbf{1}_E(e^{-t} \exp(t\mathbf{A})\mathbf{x}) = \mathbf{1}_{(e^{-t} \exp(t\mathbf{A}))^{-1}E}(\mathbf{x})$$

for every $t \in [0,+\infty), \mathbf{x} \in \mathbb{P}_n$, and every Borel subset E of $\mathbb{P}_n$, where

$$(e^{-t} \exp(t\mathbf{A}))^{-1}E = \{\mathbf{y} \in \mathbb{P}_n \mid e^{-t} \exp(t\mathbf{A})\mathbf{y} \in E\};$$

$$S_t^{(\varphi^{(\mathbf{A})})} f(\mathbf{x}) = f(e^{-t} \exp(t\mathbf{A})\mathbf{x})$$

for every $t \in [0,+\infty)$, $f \in C(\mathbb{P}_n)$ and $\mathbf{x} \in \mathbb{P}_n$; $T_t^{(\varphi^{(\mathbf{A})})}\mu(E) = \mu((e^{-t} \exp(t\mathbf{A}))^{-1}E)$ for every $t \in [0,+\infty)$; $\mu \in \mathcal{M}(\mathbb{P}_n)$, and $E \in \mathcal{B}(\mathbb{P}_n)$, where $(e^{-t} \exp(t\mathbf{A}))^{-1}E$ has the same meaning as in the above definition of $P_t^{(\varphi^{(\mathbf{A})})}(\mathbf{x}, E)$.

Observe that if we set $\mathbf{B} = e^{-s} \exp(s\mathbf{A})$ for some $s \in [0,+\infty)$, then $P_s^{(\varphi^{(\mathbf{A})})} = P_{(\mathbf{B})}$, $S_s^{(\varphi^{(\mathbf{A})})} = S_{(\mathbf{B})}$, and $T_s^{(\varphi^{(\mathbf{A})})} = T_{(\mathbf{B})}$, where $P_{(\mathbf{B})}$ and $(S_{(\mathbf{B})}, T_{(\mathbf{B})})$ are the transition probability and the Markov pair, respectively, defined by the function $v_{\mathbf{B}}$ discussed in Example 1.1.13.

Taking into consideration that, by Proposition B.4.7, the semiflow $\varphi^{(\mathbf{A})}$ is continuous, we obtain that $(P_t^{(\varphi^{(\mathbf{A})})})_{t \in [0,+\infty)}$ satisfies the s.m.a. and is a $C(\mathbb{P}_n)$-pointwise continuous Feller transition function. Using Proposition B.4.7 again, we obtain that $\varphi^{(\mathbf{A})}$ is equicontinuous with respect to $t \in [0,+\infty)$, so the transition function $(P_t^{(\varphi^{(\mathbf{A})})})_{t \in [0,+\infty)}$ is equicontinuous.   ∎

2.2.3 Transition Functions Defined by One-Parameter Convolution Semigroups of Probability Measures

As mentioned at the beginning of the section, we will now discuss transition functions defined by one-parameter convolution semigroups of probability measures.

To this end, let $(H, \cdot, d)$ be a locally compact separable metric semigroup and assume that H has a neutral element. We will denote by e the neutral element of H.

Let $(\mu_t)_{t \in [0, +\infty)}$ be a one-parameter convolution semigroup of probability measures defined on $(H, \mathcal{B}(H))$.

For every $t \in [0, +\infty)$ let P_t and (S_t, T_t) be the transition probability and the Markov pair defined by μ_t (see Example 1.1.16). Thus, for every $t \in [0, +\infty)$, P_t, S_t and T_t are defined as follows: $P_t(x, A) = (\mu_t * \delta_x)(A)$ for every $x \in H$ and $A \in \mathcal{B}(H)$, $S_t f(x) = \int_H f(zx) \, d\mu_t(z)$ for every $f \in B_b(H)$ and $x \in H$, and $T_t \nu = \mu_t * \nu$ for every $\nu \in \mathcal{M}(H)$.

As expected, it turns out that the family $(P_t)_{t \in [0, +\infty)}$ is a transition function and, therefore, $((S_t, T_t))_{t \in [0, +\infty)}$ is the family of Markov pairs defined by $(P_t)_{t \in [0, +\infty)}$. We discuss the details in the next proposition.

Proposition 2.2.10. *The families $(T_t)_{t \in [0, +\infty)}$ and $(S_t)_{t \in [0, +\infty)}$ are one-parameter semigroups of operators, and $(P_t)_{t \in [0, +\infty)}$ is a transition function.*

Proof. The fact that $(T_t)_{t \in [0, +\infty)}$ is a one-parameter semigroup of operators was discussed at the end of Example 1.1.16 and is easy to see since

$$T_{u+t}\nu = \mu_{u+t} * \nu = \mu_u * \mu_t * \nu$$
$$= T_u(\mu_t * \nu) = T_u T_t \nu$$

for every $u \in [0, +\infty)$, $t \in [0, +\infty)$, and $\nu \in \mathcal{M}(H)$.

Since

$$S_{u+t} f(x) = \langle S_{u+t} f, \delta_x \rangle = \langle f, T_{u+t}\delta_x \rangle$$
$$= \langle f, T_u T_t \delta_x \rangle = \langle f, T_t T_u \delta_x \rangle$$
$$= \langle S_u S_t f, \delta_x \rangle$$
$$= S_u S_t f(x)$$

for every $u \in [0, +\infty)$, $t \in [0, +\infty)$, $f \in B_b(H)$, and $x \in H$, it follows that $(S_t)_{t \in [0, +\infty)}$ is also a one-parameter semigroups of operators.

Since $S_t \mathbf{1}_X = \mathbf{1}_X$ for every $t \in [0, +\infty)$ and since $\langle S_t f, \mu \rangle = \langle f, T_t \mu \rangle$ for every $t \in [0, +\infty)$, $f \in B_b(H)$, and $\mu \in \mathcal{M}(H)$, it follows that we can use Proposition 2.1.2 in order to infer that there exists a unique transition function that defines the family of Markov pairs $((S_t, T_t))_{t \in [0, +\infty)}$.

Taking into consideration that, as pointed out after Lemma 1.1.1, a Markov pair cannot be defined by two distinct transition probabilities, we obtain that $(P_t)_{t \in [0,+\infty)}$ is a transition function and defines the family of Markov pairs $((S_t, T_t))_{t \in [0,+\infty)}$. $\square$

We will often refer to the transition function $(P_t)_{t \in [0,+\infty)}$ and the family $((S_t, T_t))_{t \in [0,+\infty)}$ discussed in Proposition 2.2.10 as the *transition function* and the *family of Markov pairs defined by* $(\mu_t)_{t \in [0,+\infty)}$.

As pointed out in Example 1.1.16, the transition probabilities constructed there define Markov-Feller pairs; consequently, if $(P_t)_{t \in [0,+\infty)}$ is the transition function defined by $(\mu_t)_{t \in [0,+\infty)}$, then $(P_t)_{t \in [0,+\infty)}$ is a Feller transition function.

Our goal now is to discuss a fairly general condition for the transition function $(P_t)_{t \in [0,+\infty)}$ defined by $(\mu_t)_{t \in [0,+\infty)}$ to satisfy the s.m.a. and to be $C_0(H)$-pointwise continuous. To this end, we need the following two lemmas.

Lemma 2.2.11. *Let K be a compact subset of H, let $(x_n)_{n \in \mathbb{N}}$ be a convergent sequence of elements of H, and set $x = \lim_{n \to \infty} x_n$. Then, for every $\varepsilon \in \mathbb{R}$, $\varepsilon > 0$, there exists an $n_\varepsilon \in \mathbb{N}$ such that $d(yx_n, yx) < \varepsilon$ for every $n \in \mathbb{N}$, $n \geq n_\varepsilon$, and every $y \in K$.*

For a proof of the lemma, see Lemma 4.1 of [144].

Lemma 2.2.12. *Let (X, d) be a locally compact separable metric space, let $(\mu_n)_{n \in \mathbb{N}}$ be a sequence of probability measures, $\mu_n \in \mathcal{M}(X)$ for every $n \in \mathbb{N}$, and assume that $(\mu_n)_{n \in \mathbb{N}}$ converges in the weak* topology of $\mathcal{M}(X)$ to a probability measure μ', $\mu' \in \mathcal{M}(X)$. Then the sequence $(\mu_n)_{n \in \mathbb{N}}$ is tight and $\{\mu'\} \cup \{\mu_n \,|\, n \in \mathbb{N}\}$ is a tight set of probability measures.*

For a proof of the above lemma, see, for instance, Observation on p. 96 of [143]. Note that, in the lemma, (X, d) is not necessarily a topological semigroup.

Proposition 2.2.13. *Let $(\mu_t)_{t \in [0,+\infty)}$ be a weak* continuous one-parameter convolution semigroup of probability measures defined on $(H, \mathcal{B}(H))$ (for the definition of the weak* continuity of such one-parameter convolution semigroups, see the beginning of Sect. B.3), and let $(P_t)_{t \in [0,+\infty)}$ and $((S_t, T_t))_{t \in [0,+\infty)}$ be the transition function and the family of Markov pairs defined by $(\mu_t)_{t \in [0,+\infty)}$, respectively. Then $(P_t)_{t \in [0,+\infty)}$ satisfies the s.m.a. and is $C_0(H)$-pointwise continuous.*

Proof. We will first prove that, for every $f \in C_0(H)$, the real valued mapping $(t, x) \mapsto S_t f(x)$, $(t, x) \in [0, +\infty) \times H$, is continuous with respect to the standard topology on $\mathbb{R}$ and the product topology $\mathcal{T}([0, +\infty)) \otimes \mathcal{T}(H)$ on $[0, +\infty) \times H$, where $\mathcal{T}([0, +\infty))$ and $\mathcal{T}(H)$ are the standard topology on $[0, +\infty)$ (the topology induced on $[0, +\infty)$ by the standard topology on $\mathbb{R}$) and the metric topology on H, respectively.

To this end, let $f \in C_0(H)$. Since the assertion is obviously true if $f = 0$, we may and do assume that $f \neq 0$.

Since, by Lemma B.2.5, the topology $\mathcal{T}([0, +\infty)) \otimes \mathcal{T}(H)$ is metrizable, it follows that in order to prove the continuity of the map $(t, x) \mapsto S_t f(x)$, $(t, x) \in [0, +\infty) \times H$, it is enough to prove that if $(t_n)_{n \in \mathbb{N}}$ and $(x_n)_{n \in \mathbb{N}}$ are convergent sequences of elements of $[0, +\infty)$ and H, respectively, then the sequence $(S_{t_n} f(x_n))_{n \in \mathbb{N}}$ converges to $S_{t'} f(x')$, where $t' = \lim_{n \to \infty} t_n$ and $x' = \lim_{n \to \infty} x_n$.

Thus, let $(t_n)_{n \in \mathbb{N}}$ and $(x_n)_{n \in \mathbb{N}}$ be two convergent sequences of elements of $[0, +\infty)$ and H, respectively, and set $t' = \lim_{n \to \infty} t_n$ and $x' = \lim_{n \to \infty} x_n$.

We have to prove that for every $\varepsilon \in \mathbb{R}$, $\varepsilon > 0$, there exists an $n_\varepsilon \in \mathbb{N}$ such that $|S_{t_n} f(x_n) - S_{t'} f(x')| < \varepsilon$ for every $n \in \mathbb{N}$, $n \geq n_\varepsilon$.

Accordingly, let $\varepsilon \in \mathbb{R}$, $\varepsilon > 0$.

Since $(\mu_t)_{t \in [0, +\infty)}$ is a weak* continuous one-parameter convolution semigroup, it follows that the sequence $(\mu_{t_n})_{n \in \mathbb{N}}$ converges in the weak* topology of $\mathcal{M}(H)$ to $\mu_{t'}$. Thus, using Lemma 2.2.12, we obtain that the set $\{\mu_{t'}\} \cup \{\mu_{t_n} \mid n \in \mathbb{N}\}$ is tight. Accordingly, there exists a compact subset K_1 of H such that $\mu_{t'}(H \setminus K_1) < \frac{\varepsilon}{4\|f\|}$ and $\mu_{t_n}(H \setminus K_1) < \frac{\varepsilon}{4\|f\|}$ for every $n \in \mathbb{N}$.

Let $K_2 = \{x'\} \cup \{x_n \mid n \in \mathbb{N}\}$.

Since K_2 is a compact subset of H, it follows that $K_1 K_2 = \{yx \mid y \in K_1, x \in K_2\}$ is also a compact subset of H because the algebraic operation that defines the metric semigroup structure on H is continuous.

Since f is a real-valued continuous function on H, it follows that the restriction of f to $K_1 K_2$ is uniformly continuous, so there exists a $\delta \in \mathbb{R}$, $\delta > 0$, such that $|f(z_1) - f(z_2)| < \frac{\varepsilon}{4}$ whenever $z_1 \in K_1 K_2$, $z_2 \in K_1 K_2$, and $d(z_1, z_2) < \delta$.

By Lemma 2.2.11, there exists an $n'_\varepsilon \in \mathbb{N}$ such that $d(yx_n, yx') < \delta$ for every $n \in \mathbb{N}$, $n \geq n_\varepsilon$ and every $y \in K_1$.

Let $f_{x'} : H \to \mathbb{R}$ be defined by $f_{x'}(y) = f(yx')$ for every $y \in H$. Clearly, $f_{x'}$ is a continuous bounded function; that is, $f_{x'} \in C_b(H)$. Since the sequence $(\mu_{t_n})_{n \in \mathbb{N}}$ weak* converges to $\mu_{t'}$, and since $\mu_{t'}$ is a probability measure, it follows that $(\mu_{t_n})_{n \in \mathbb{N}}$ is also $C_b(H)$-weak* convergent to $\mu_{t'}$ (see the discussion preceding Proposition 1.1.5, or p. 71 of Högnäs and Mukherjea [48]). Accordingly, the sequence $(\langle f_{x'}, \mu_{t_n}\rangle)_{n \in \mathbb{N}}$ converges to $\langle f_{x'}, \mu_{t'}\rangle$, so there exists an $n_\varepsilon \in \mathbb{N}$, $n_\varepsilon \geq n'_\varepsilon$, such that $|\langle f_{x'}, \mu_{t_n}\rangle - \langle f_{x'}, \mu_{t'}\rangle| < \frac{\varepsilon}{4}$ for every $n \in \mathbb{N}$, $n \geq n_\varepsilon$.

Using the definitions of $f_{x'}$, n_ε, K_1, δ and n'_ε, we obtain that

$$\left| S_{t_n} f(x_n) - S_{t'} f(x') \right| \leq \left| S_{t_n} f(x_n) - S_{t_n} f(x') \right| + \left| S_{t_n} f(x') - S_{t'} f(x') \right|$$

$$= \left| \int_H f(yx_n)\, d\mu_{t_n}(y) - \int_H f(yx')\, d\mu_{t_n}(y) \right|$$

$$+ \left| \int_H f(yx')\, d\mu_{t_n}(y) - \int_H f(yx')\, d\mu_{t'}(y) \right|$$

$$\leq \left| \int_{K_1} f(yx_n)\, d\mu_{t_n}(y) - \int_{K_1} f(yx')\, d\mu_{t_n}(y) \right|$$

$$+ \left| \int_{H \setminus K_1} f(yx_n)\, d\mu_{t_n}(y) - \int_{H \setminus K_1} f(yx')\, d\mu_{t_n}(y) \right| + |\langle f_{x'}, \mu_{t_n} \rangle - \langle f_{x'}, \mu_{t'} \rangle|$$

$$< \int_{K_1} \left| f(yx_n) - f(yx') \right| d\mu_{t_n}(y) + \int_{H \setminus K_1} \left| f(yx_n) - f(yx') \right| d\mu_{t_n}(y) + \frac{\varepsilon}{4}$$

$$\leq \frac{\varepsilon}{4} + 2 \|f\| \, \mu_{t_n}(H \setminus K_1) + \frac{\varepsilon}{4}$$

$$< \frac{\varepsilon}{4} + 2 \|f\| \frac{\varepsilon}{4\|f\|} + \frac{\varepsilon}{4} = \frac{\varepsilon}{4} + \frac{\varepsilon}{2} + \frac{\varepsilon}{4} = \varepsilon$$

for every $n \in \mathbb{N}$, $n \geq n_\varepsilon$.

We have therefore proved that, for every $f \in C_0(H)$, the map $(t, x) \mapsto S_t f(x)$, $(t, x) \in [0, +\infty) \times H$ is continuous with respect to the standard topology on $\mathbb{R}$ and the topology $\mathcal{T}([0, +\infty)) \otimes \mathcal{T}(H)$ on $[0, +\infty) \times H$. Thus, it is easy to see now that $(P_t)_{t \in [0,+\infty)}$ is $C_0(H)$-pointwise continuous.

In order to prove that $(P_t)_{t \in [0,+\infty)}$ satisfies the s.m.a., we will use a type of argument similar to that used in the proof of Proposition B.2.1. That is, using D32, p. 348 of Cohn [20] as in Proposition B.2.1, we obtain that H has a countable basis for its topology; since $[0, +\infty)$ has a countable basis for its topology, as well, it follows that we can use Proposition 7.6.2, p. 242 of Cohn [20] in order to infer that the σ-algebra, say $\mathcal{A}$, generated by $\mathcal{B}([0, +\infty)) \times \mathcal{B}(H)$ is equal to the Borel σ-algebra $\mathcal{B}([0, +\infty) \times H)$ generated by the open subsets of $[0, +\infty) \times H$ in the product topology of $[0, +\infty) \times H$. Since we have just proved that, for every $f \in C_0(H)$, the map $(t, x) \mapsto S_t f(x)$, $(t, x) \in [0, +\infty) \times H$, is continuous with respect to the product topology of $[0, +\infty) \times H$ (and the standard topology on $\mathbb{R}$), we obtain (using Proposition 2.1.1, p. 31 of Neveu [81]) that the map is measurable with respect to $\mathcal{B}([0, +\infty) \times H)$ (and $\mathcal{B}(\mathbb{R})$), so it is measurable with respect to $\mathcal{A}$ and $\mathcal{B}(\mathbb{R})$. Since $\mathcal{A} \subseteq \mathcal{L}([0, +\infty)) \otimes \mathcal{B}(H)$ (because $\mathcal{B}([0, +\infty)) \subseteq \mathcal{L}([0, +\infty))$), and using Proposition 2.1.5, we obtain that $(P_t)_{t \in [0,+\infty)}$ satisfies the s.m.a. $\square$

We will now discuss a family of one-parameter convolution semigroups of probability measures and the transition functions that these one-parameter convolution semigroups define. We will often use these transition functions to illustrate various results in the book, especially in Chap. 7.

Example 2.2.14. Let $\mu \in \mathcal{M}(H)$ be a probability measure.

For every $t \in [0, +\infty)$, set $\mu_t = \exp_s(t\mu)$; that is, $\mu_t = e^{-t} \exp(t\mu) = e^{-t} \sum_{k=0}^{\infty} \frac{t^k}{k!} \mu^k$.

By Propositions B.3.2 and B.3.3, $(\mu_t)_{t \in [0,+\infty)}$ is a weak* continuous one-parameter convolution semigroup of probability measures. Therefore, for every $t \in [0, +\infty)$, we can consider the transition probability P_t and the Markov pair (S_t, T_t) defined by μ_t as we did after Example 2.2.9 (see also Example 1.1.16), and the resulting families $(P_t)_{t \in [0,+\infty)}$ and $((S_t, T_t))_{t \in [0,+\infty)}$ are the transition function and the family of Markov pairs defined by $(\mu_t)_{t \in [0,+\infty)}$.

As pointed out before Lemma 2.2.11, $(P_t)_{t\in[0,+\infty)}$ is a Feller transition function and, by Proposition 2.2.13, $(P_t)_{t\in[0,+\infty)}$ satisfies the s.m.a. and is $C_0(H)$-pointwise continuous.   ∎

In view of the notions defined in Sect. 2.1 and of the manner in which we analyzed the examples discussed so far in this section, a natural question comes to mind: can one find a "nice enough" condition on a probability measure $\mu \in \mathcal{M}(H)$ that will guarantee that the transition function defined by $(\mu_t)_{t\in[0,+\infty)}$ is $C_0(H)$-equicontinuous, where $(\mu_t)_{t\in[0,+\infty)}$ is the exponential one-parameter convolution semigroup of probability measures defined by μ and discussed in Example 2.2.14? In turns out that if μ is an equicontinuous probability measure as defined before Example 1.4.29, then $(P_t)_{t\in[0,+\infty)}$ is a $C_0(H)$-equicontinuous transition function. We discuss the details in the next proposition.

Proposition 2.2.15. *Let $\mu \in \mathcal{M}(H)$ be an equicontinuous probability measure, let $(\mu_t)_{t\in[0,+\infty)}$ be the exponential one-parameter convolution semigroup of probability measures defined by μ, and let $(P_t)_{t\in[0,+\infty)}$ be the transition function defined by $(\mu_t)_{t\in[0,+\infty)}$. Then $(P_t)_{t\in[0,+\infty)}$ is a $C_0(H)$-equicontinuous transition function.*

Proof. Let $\mu \in \mathcal{M}(H)$, $(\mu_t)_{t\in[0,+\infty)}$, and $(P_t)_{t\in[0,+\infty)}$ be as in the proposition. Also let $((S_t, T_t))_{t\in[0,+\infty)}$ be the family of Markov pairs defined by $(P_t)_{t\in[0,+\infty)}$.

We have to prove that for every $f \in C_0(H)$, for every convergent sequence $(x_n)_{n\in\mathbb{N}}$ of elements of H, and for every $\varepsilon \in \mathbb{R}$, $\varepsilon > 0$, there exists an $n_\varepsilon \in \mathbb{N}$ such that $|S_t f(x_n) - S_t f(x)| < \varepsilon$ for every $n \in \mathbb{N}$, $n \geq n_\varepsilon$, and every $t \in [0, +\infty)$, where $x = \lim_{n\to\infty} x_n$.

To this end, let $f \in C_0(H)$, let $(x_n)_{n\in\mathbb{N}}$ be a convergent sequence of elements of H, set $x = \lim_{n\to\infty} x_n$, and let $\varepsilon \in \mathbb{R}$, $\varepsilon > 0$. Also, let P_μ and (S_μ, T_μ) be the transition probability and the Markov-Feller pair defined by μ (see Example 1.1.16).

Since we assume that μ is an equicontinuous probability measure, we obtain that there exists an $n_\varepsilon \in \mathbb{N}$ such that

$$\left| S_\mu^k f(x_n) - S_\mu^k f(x) \right| < \frac{\varepsilon}{2} \tag{2.2.4}$$

for every $n \in \mathbb{N}$, $n \geq n_\varepsilon$, and every $k \in \mathbb{N} \cup \{0\}$.

Using the definition of T_μ we obtain that

$$S_\mu^k f(y) = \left\langle S_\mu^k f, \delta_y \right\rangle = \left\langle f, T_\mu^k \delta_y \right\rangle = \left\langle f, \mu^k * \delta_y \right\rangle \tag{2.2.5}$$

for every $k \in \mathbb{N} \cup \{0\}$ and $y \in H$.

Since the sequence $(e^{-t} \sum_{k=0}^{l} \frac{t^k}{k!} \mu^k)_{l\in\mathbb{N}\cup\{0\}}$ converges in the norm topology of $\mathcal{M}(H)$ to μ_t, since the convolution operation defines a structure of topological semigroup on $\mathcal{M}(H)$ with respect to the norm topology of $\mathcal{M}(H)$ (so the operation of convolution is continuous with respect to the norm topology of $\mathcal{M}(H)$), and since the norm convergence implies the weak* convergence of a sequence of elements of $\mathcal{M}(H)$, we obtain that

$$S_t f(y) = \langle S_t f, \delta_y \rangle = \langle f, T_t \delta_y \rangle = \langle f, \mu_t * \delta_y \rangle \tag{2.2.6}$$

$$= \left\langle f, (e^{-t} \sum_{k=0}^{\infty} \frac{t^k}{k!} \mu^k) * \delta_y \right\rangle = \lim_{l \to \infty} \left\langle f, (e^{-t} \sum_{k=0}^{l} \frac{t^k}{k!} \mu^k) * \delta_y \right\rangle$$

$$= \lim_{l \to \infty} e^{-t} \sum_{k=0}^{l} \frac{t^k}{k!} \langle f, \mu^k * \delta_y \rangle = \lim_{l \to \infty} e^{-t} \sum_{k=0}^{l} \frac{t^k}{k!} S_\mu^k f(y)$$

for every $y \in H$ and $t \in [0, +\infty)$.

Using the above expressions (2.2.6), (2.2.5) and (2.2.3) (in this order), we obtain that:

$$|S_t f(x_n) - S_t f(x)| = \left| \lim_{l \to \infty} e^{-t} \sum_{k=0}^{l} \frac{t^k}{k!} S_\mu^k f(x_n) - \lim_{l \to \infty} e^{-t} \sum_{k=0}^{l} \frac{t^k}{k!} S_\mu^k f(x) \right|$$

$$= \lim_{l \to \infty} e^{-t} \left| \sum_{k=0}^{l} \frac{t^k}{k!} (S_\mu^k f(x_n) - S_\mu^k f(x)) \right|$$

$$\leq \limsup_{l \to \infty} e^{-t} \sum_{k=0}^{l} \frac{t^k}{k!} \left| S_\mu^k f(x_n) - S_\mu^k f(x) \right|$$

$$\leq \frac{\varepsilon}{2} e^{-t} \limsup_{l \to \infty} \sum_{k=0}^{l} \frac{t^k}{k!} = \frac{\varepsilon}{2} e^{-t} \lim_{l \to \infty} \sum_{k=0}^{l} \frac{t^k}{k!}$$

$$= \frac{\varepsilon}{2} e^{-t} e^t = \frac{\varepsilon}{2} < \varepsilon$$

for every $n \in \mathbb{N}$, $n \geq n_\varepsilon$, and every $t \in [0, +\infty)$.

Accordingly, $(P_t)_{t \in [0,+\infty)}$ is an equicontinuous transition function. $\qquad \square$

It is important to realize that the transition functions discussed in this subsection are significantly different from the transition functions defined by one-parameter semigroups or groups of measurable functions.

For instance, if $(P_t)_{t \in \mathbb{T}}$ and $((S_t, T_t))_{t \in \mathbb{T}}$ are the transition function and the family of Markov pairs defined by a one-parameter semigroup or group $(w_t)_{t \in \mathbb{T}}$ of elements of $\mathbf{B}(X)$ for some locally compact separable metric space (X, d), respectively, then the probability measures $T_t \delta_x$, $t \in \mathbb{T}$, $x \in X$, are all Dirac measures and their supports are singletons (sets that contain exactly one element each). By contrast, if $(P_t)_{t \in [0,+\infty)}$ and $((S_t, T_t))_{t \in [0,+\infty)}$ are the transition function and the family of Markov pairs defined by an exponential one-parameter convolution semigroup $(\mu_t)_{t \in [0,+\infty)}$ of probability measures defined by a probability measure

$\mu \in \mathcal{M}(H)$ where H is a locally compact separable metric semigroup with unit, and if $A = \overline{\bigcup_{n=0}^{\infty} \mathrm{supp}(\mu^n)}$, then, for every $t \in [0, +\infty)$ and $x \in H$, the support of $T_t \delta_x$ is $\overline{Ax}$.

2.3 Invariant Probability Measures

After having introduced the transition functions in Sect. 2.1 and after having discussed several examples in Sect. 2.2, our goal in this section is to introduce the second main object studied in this book, namely the invariant probability measures of transition functions.

Let (X, d) be a locally compact separable metric space, and, as always in this book, let $\mathbb{T}$ stand for $\mathbb{R}$ or the interval $[0, +\infty)$.

Let $(P_t)_{t \in \mathbb{T}}$ be a transition function on (X, d), and let $((S_t, T_t))_{t \in \mathbb{T}}$ be the family of Markov pairs defined by $(P_t)_{t \in \mathbb{T}}$.

Given $\mu \in \mathcal{M}(X)$, we say that μ is an *invariant element for* $(P_t)_{t \in \mathbb{T}}$ (or *for* $((S_t, T_t))_{t \in \mathbb{T}}$, or *for* $(T_t)_{t \in \mathbb{T}}$) if $T_t \mu = \mu$ for every $t \in \mathbb{T}$.

Since the zero measure in $\mathcal{M}(X)$ is always an invariant element for $(P_t)_{t \in \mathbb{T}}$, the interesting situation is when $(P_t)_{t \in \mathbb{T}}$ also has nonzero invariant elements. Using arguments perfectly similar to those used for a Markov-Feller pair (and the associated Feller transition probability) at the beginning of the subsection *Invariant Probabilities of Markov-Feller Operators* of Section 1.2 on p. 17 of [143], we obtain that $(P_t)_{t \in \mathbb{T}}$ has nonzero invariant elements if and only if $(P_t)_{t \in \mathbb{T}}$ has invariant probability measures; that is, if and only if there exists a $\mu \in \mathcal{M}(X)$, $\mu \geq 0$, $\|\mu\| = 1$, such that $T_t \mu = \mu$ for every $t \in \mathbb{T}$.

As in the case of Markov-Feller pairs, when dealing with a transition function $(P_t)_{t \in \mathbb{T}}$, in order to understand the structure of the set of all invariant elements of $(P_t)_{t \in \mathbb{T}}$, in most cases it is enough to understand the structure of the set of invariant probabilities for $(P_t)_{t \in \mathbb{T}}$. Thus, in this book, we will mostly be interested in studying invariant probability measures for transition functions.

Note that if $\mathbf{w} = (w_t)_{t \in \mathbb{T}}$ is a semiflow or a flow on (X, d), and if $(P_t^{(\mathbf{w})})_{t \in \mathbb{T}}$ is the transition function defined by $\mathbf{w}$ (see the beginning of Sect. 2.2.1 for details), then $\mu \in \mathcal{M}(X)$ is an invariant element for $(P_t^{(\mathbf{w})})_{t \in \mathbb{T}}$ if and only if μ is an invariant element for $\mathbf{w}$ as defined in Appendix B before Example B.1.2.

Let us use the above observation to list a few examples of invariant probabilities.

Example 2.3.1. Let $\mathbf{w} = (w_t)_{t \in \mathbb{R}}$ be the flow of the rotations of the unit circle $\mathbb{R}/\mathbb{Z}$, and let $(P_t^{(\mathbf{w})})_{t \in \mathbb{R}}$ be the transition function of $\mathbf{w}$. Then the Haar-Lebesgue probability measure on $\mathbb{R}/\mathbb{Z}$ is an invariant probability for $(P_t^{(\mathbf{w})})_{t \in \mathbb{R}}$ because, as pointed out in Example B.1.6, the Haar-Lebesgue measure on $\mathbb{R}/\mathbb{Z}$ is an invariant probability measure for $\mathbf{w}$. ∎

Example 2.3.2. Let $n \in \mathbb{N}$, $n \geq 2$, let $\mathbf{v} \in \mathbb{R}^n$, let $\mathbf{w} = (w_t)_{t \in \mathbb{R}}$ be the rectilinear flow on the n-dimensional torus $\mathbb{R}^n/\mathbb{Z}^n$ with velocity $\mathbf{v}$, and let $(P_t^{(\mathbf{w})})_{t \in \mathbb{R}}$ be the transition function defined by $\mathbf{w}$ (see Example 2.2.5). Then the Haar-Lebesgue probability measure on $\nu_{\mathbb{R}^n/\mathbb{Z}^n}$ on $\mathbb{R}^n/\mathbb{Z}^n$ is an invariant probability measure for $(P_t^{(\mathbf{w})})_{t\in\mathbb{R}}$ because, as discussed in Example B.1.7, $\nu_{\mathbb{R}^n/\mathbb{Z}^n}$ is an invariant probability for $\mathbf{w}$.   ∎

Example 2.3.3. Let Γ be a lattice in $\mathrm{PSL}(2,\mathbb{R})$, let $\mathbf{w}^{(\Gamma)} = (w_t^{(\Gamma)})_{t\in\mathbb{R}}$ be the geodesic flow on $(\mathrm{PSL}(2,\mathbb{R})/\Gamma)_{\mathrm{R}}$ defined in Example B.1.8, and let $(P_t^{(\mathbf{w}^{(\Gamma)})})_{t\in\mathbb{R}}$ be the transition function defined by $\mathbf{w}^{(\Gamma)}$ (see (*b*) of Example 2.2.6). Then the standard $\mathrm{PSL}(2,\mathbb{R})$-invariant probability measure $\nu_{(\mathrm{PSL}(2,\mathbb{R})/\Gamma)_{\mathrm{R}}}$ is an invariant measure for $\mathbf{w}^{(\Gamma)}$, so $\nu_{(\mathrm{PSL}(2,\mathbb{R})/\Gamma)_{\mathrm{R}}}$ is also an invariant probability for $(P_t^{(\mathbf{w}^{(\Gamma)})})_{t\in\mathbb{R}}$. If $\overline{\mathbf{v}}^{(j\Gamma)} = (v_t^{(j\Gamma)})_{t\in\mathbb{R}}$, $j = 1,2$, are the horocycle flows defined in (*a*) of Example B.1.9 and $(P_t^{(\overline{\mathbf{v}}^{(j\Gamma)})})_{t\in\mathbb{R}}$, $j = 1,2$, are the two corresponding transition functions (see (*b*) of Example 2.2.7), then $\nu_{(\mathrm{PSL}(2,\mathbb{R})/\Gamma)_{\mathrm{R}}}$ is an invariant probability for both transition functions $(P_t^{(\overline{\mathbf{v}}^{(j\Gamma)})})_{t\in\mathbb{R}}$, $j = 1,2$, because $\nu_{(\mathrm{PSL}(2,\mathbb{R})/\Gamma)_{\mathrm{R}}}$ is an invariant measure for $\mathbf{v}^{(j\Gamma)}$, $j = 1,2$.   ∎

Example 2.3.4. Let $n \in \mathbb{N}$, $n \geq 2$, and let Γ be a lattice in $\mathrm{SL}(n,\mathbb{R})$. Let $\mathbf{w} = (w_t)_{t\in\mathbb{R}}$ be one of the four horocycle flows defined on $(\mathrm{SL}(n,\mathbb{R})/\Gamma)_{\mathrm{S}}$ where S stands for L or R, and $n = 2$ (see (*b*) of Example B.1.9 and (*c*) of Example 2.2.7), or assume that $\mathbf{w} = (w_t)_{t\in\mathbb{R}}$ is one of the exponential flows defined on $(\mathrm{SL}(n,\mathbb{R})/\Gamma)_{\mathrm{S}}$, where, again, S stands for L or R (see Sect. B.4.2 and Example 2.2.8). Now, let $(P_t^{(\mathbf{w})})_{t\in\mathbb{R}}$ be the transition function defined by $\mathbf{w}$ on the corresponding space $(\mathrm{SL}(n,\mathbb{R})/\Gamma)_{\mathrm{S}}$; thus, $(P_t^{(\mathbf{w})})_{t\in\mathbb{R}}$ is defined as in (*c*) of Example 2.2.7 if $\mathbf{w}$ is a horocycle flow, or $(P_t^{(\mathbf{w})})_{t\in\mathbb{R}}$ is defined as in Example 2.2.8 if $\mathbf{w}$ is one of the exponential flows discussed in Sect. B.4.2. In all these cases the corresponding $\mathrm{SL}(n,\mathbb{R})$-invariant probability measure $\nu_{(\mathrm{SL}(n,\mathbb{R})/\Gamma)_{\mathrm{S}}}$ is an invariant probability for $(P_t^{(\mathbf{w})})_{t\in\mathbb{R}}$ because $\nu_{(\mathrm{SL}(n,\mathbb{R})/\Gamma)_{\mathrm{S}}}$ is an invariant measure for $\mathbf{w}$, where, of course, $(\mathrm{SL}(n,\mathbb{R})/\Gamma)_{\mathrm{S}}$ is the space on which $\mathbf{w}$ and $(P_t^{(\mathbf{w})})_{t\in\mathbb{R}}$ are defined.   ∎

Let $\mathbf{w} = (w_t)_{t\in\mathbb{T}}$ be a one-parameter semigroup or a one-parameter group of elements of $\mathbf{B}(X)$. As expected, an element μ of $\mathcal{M}(X)$ is said to be an *invariant element for* (or *of*) $\mathbf{w}$ if $\mu(w_t^{-1}(A)) = \mu(A)$ for every $t \in \mathbb{T}$ and $A \in \mathcal{B}(X)$. Clearly, the notion of an invariant element of a one-parameter semigroup or one-parameter group of elements of $\mathbf{B}(X)$ is a natural extension of the corresponding notion for a semiflow or a flow in the sense that if $\mathbf{w}$ is a semiflow or a flow and $\mu \in \mathcal{M}(X)$, then μ is an invariant element of $\mathbf{w}$ in the sense of the definition given in Appendix B before Example B.1.2, where we think of $\mathbf{w}$ as a semiflow or a flow, if and only if μ is an invariant element of $\mathbf{w}$, thought of as a one-parameter semigroup or a one-parameter group of elements of $\mathbf{B}(X)$, respectively.

Note that if $\mathbf{w} = (w_t)_{t\in\mathbb{T}}$ is a one-parameter semigroup or a one-parameter group of elements of $\mathbf{B}(X)$, if $(P_t^{(\mathbf{w})})_{t\in\mathbb{T}}$ is the transition function defined by $\mathbf{w}$ and if $\mu \in \mathcal{M}(X)$, then μ is an invariant element of $\mathbf{w}$ if and only if μ is an invariant element of $(P_t^{(\mathbf{w})})_{t\in\mathbb{T}}$. The next example deals with such a situation.

Example 2.3.5. Let $\mathbf{w} = (w_t)_{t \in [0,+\infty)}$ be the one-parameter semigroup of elements of $\mathbf{B}([0,1])$ defined in Example 2.2.3 (note that, as pointed out there, $\mathbf{w}$ is not a semiflow), and let $(P_t^{(\mathbf{w})})_{t \in [0,+\infty)}$ be the transition function defined by $\mathbf{w}$. Then the Dirac measure δ_1 concentrated at 1 is an invariant probability measure for $(P_t^{(\mathbf{w})})_{t \in [0,+\infty)}$ because δ_1 is an invariant probability for $\mathbf{w}$. ∎

We conclude this discussion of examples of invariant probabilities for transition functions by discussing certain invariant probabilities of transition functions defined by exponential one-parameter convolution semigroups of probability measures.

Example 2.3.6. Let $(H, \cdot, d)$ be a locally compact separable metric semigroup, and assume that H has a neutral element. Also, let $\mu \in \mathcal{M}(H)$ be a probability measure, let $(\mu_t)_{t \in [0,+\infty)}$ be the exponential one-parameter convolution semigroup of probability measures defined by μ (see the first paragraph after Proposition B.3.2), and let $(P_t)_{t \in [0,+\infty)}$ and $((S_t, T_t))_{t \in [0,+\infty)}$ be the transition function and the family of Markov pairs defined by $(\mu_t)_{t \in [0,+\infty)}$ (see Sect. 2.2.3).

Now let $\nu \in \mathcal{M}(H)$ be a probability measure which satisfies the Choquet-Deny equation $\mu * \nu = \nu$ (see the equality (1.4.1) and Section 1 of [144]). Then using the fact that $\mathcal{M}(H)$ is a Banach algebra when endowed with the operation of convolution (see Proposition B.2.4), we obtain that $e^{-t} \exp(t\mu) * \nu = \nu$ for every $t \in [0, +\infty)$. Accordingly, $T_t \nu = \nu$ for every $t \in [0, +\infty)$; that is, ν is an invariant probability measure for $(P_t)_{t \in [0,+\infty)}$. ∎

If $\mu \in \mathcal{M}(X)$ is an invariant element for a transition function $(P_t)_{t \in \mathbb{T}}$, then, for every $t \in \mathbb{T}$, μ is an invariant element for the transition probability P_t, but if μ is an invariant element of one of the transition probabilities, say P_{t_0}, $t_0 \in \mathbb{T}$, that make up a transition function $(P_t)_{t \in \mathbb{T}}$, then, in general, μ is not an invariant element for the transition function (for instance, if $\mathbf{w} = (w_t)_{t \in \mathbb{R}}$ is the flow of the rotations of the unit circle $\mathbb{R}/\mathbb{Z}$ defined in Example A.3.4, and if $(P_t^{(\mathbf{w})})_{t \in \mathbb{R}}$ and $((S_t, T_t))_{t \in \mathbb{R}}$ are the transition function and the family of Markov pairs defined by $\mathbf{w}$, respectively (see Example 2.2.4), then, for every $s \in \mathbb{Z}$, $T_s^{(\mathbf{w})}$ is the identity operator on $\mathcal{M}(\mathbb{R}/\mathbb{Z})$, so the Dirac measure $\delta_{\hat{0}}$ concentrated at $\hat{0}$ is an invariant probability measure for the transition probability $P_s^{(\mathbf{w})}$ even though $\delta_{\hat{0}}$ is not an invariant element for the transition function $(P_t^{(\mathbf{w})})_{t \in \mathbb{R}}$ because if $s \in \mathbb{R} \setminus \mathbb{Z}$, then $\delta_{\hat{0}}$ is not an invariant measure for $P_s^{(\mathbf{w})}$). However, if $\mathbb{T} = \mathbb{R}$ and μ is an invariant element of the restriction $(P_t)_{t \in [0,+\infty)}$ of a transition function $(P_t)_{t \in \mathbb{R}}$ to $[0, +\infty)$, then μ is an invariant element for $(P_t)_{t \in \mathbb{R}}$. For future reference, we discuss this fact in detail in the next proposition.

Proposition 2.3.7. *Let $(P_t)_{t \in \mathbb{R}}$ be a transition function defined on (X, d) and let $\mu \in \mathcal{M}(X)$. The following assertions are equivalent:*

(a) μ is an invariant element for $(P_t)_{t \in \mathbb{R}}$.
(b) μ is an invariant element of the restriction $(P_t)_{t \in [0,+\infty)}$ of $(P_t)_{t \in \mathbb{R}}$ to $[0, +\infty)$.

Proof.

$(a) \Rightarrow (b)$ is obvious.

$(b) \Rightarrow (a)$.   Let $((S_t, T_t))_{t \in \mathbb{R}}$ be the family of Markov pairs defined by $(P_t)_{t \in \mathbb{R}}$ and let $((S_t, T_t))_{t \in [0, +\infty)}$ be the restriction of $((S_t, T_t))_{t \in \mathbb{R}}$ to $[0, +\infty)$. Taking into consideration that we assume that $T_t \mu = \mu$ for every $t \geq 0$ and using the fact that $(T_t)_{t \in \mathbb{R}}$ is a one-parameter group of operators (see Proposition 2.1.1), we obtain that $T_t \mu = T_t(T_{-t}\mu) = T_0 \mu = \mu$ for every $t \in \mathbb{R}$, $t < 0$.

Thus, μ is an invariant element for $(P_t)_{t \in \mathbb{R}}$. $\square$

Note that a calculation similar to that used in the proof of $(b) \Rightarrow (a)$ in the above proposition can be used to show that if $(P_t)_{t \in \mathbb{R}}$ is a transition function which has the property that T_0 is the identity operator on $\mathcal{M}(X)$, where (S_0, T_0) is the Markov pair defined by the transition probability P_0, and if μ is an element of $\mathcal{M}(X)$, then, μ is an invariant element of the transition probability P_t if and only if μ is an invariant element of P_{-t} for every $t \in \mathbb{R}$. Therefore, for such a transition function $(P_t)_{t \in \mathbb{R}}$ and for $\mu \in \mathcal{M}(X)$, the following assertions are equivalent:

(a) μ is an invariant element for $(P_t)_{t \in \mathbb{R}}$.

(b) μ is an invariant element for $(P_t)_{t \in [0, +\infty)}$.

(c) For every $t \in \mathbb{R}$, $t > 0$, μ is an invariant element for the transition probability P_t.

(d) For every $t \in \mathbb{R}$, $t \leq 0$, μ is an invariant element for the transition probability P_t.

(e) For every $t \in \mathbb{R}$, $t < 0$, μ is an invariant element for the transition probability P_t.

Note that among the transition functions for which the above discussion makes sense are the transition functions defined by flows.

As in the case of transition probabilities, given a transition function $(P_t)_{t \in \mathbb{T}}$ defined on (X, d) and the family of Markov pairs $((S_t, T_t))_{t \in \mathbb{T}}$ generated by $(P_t)_{t \in \mathbb{T}}$, we say that $(P_t)_{t \in \mathbb{T}}$ (or $((S_t, T_t))_{t \in \mathbb{T}}$, or $(T_t)_{t \in \mathbb{T}}$) is *uniquely ergodic* if $(P_t)_{t \in \mathbb{T}}$ has nonzero invariant elements and has only one invariant probability measure. We say that $(P_t)_{t \in \mathbb{T}}$ (or $((S_t, T_t))_{t \in \mathbb{T}}$, or $(T_t)_{t \in \mathbb{T}}$) is *strictly ergodic* if $(P_t)_{t \in \mathbb{T}}$ is uniquely ergodic and the support of the unique invariant probability measure is the entire space X. Note that if $(P_t)_{t \in \mathbb{T}}$ is uniquely ergodic and if $\mu \in \mathcal{M}(X)$ is the unique invariant probability of $(P_t)_{t \in \mathbb{T}}$, then any invariant element $\nu \in \mathcal{M}(X)$ of $(P_t)_{t \in \mathbb{T}}$ is of the form $\nu = a\mu$ for some $a \in \mathbb{R}$.

As in the case of semiflows and flows (see the discussion preceding Example B.1.6), given a one-parameter semigroup or a one-parameter group $\mathbf{w} = (w_t)_{t \in \mathbb{T}}$ of elements of $\mathbf{B}(X)$, we say that $\mathbf{w}$ is *uniquely ergodic* if $\mathbf{w}$ has exactly one invariant probability measure. Clearly, the notion of unique ergodicity of one-parameter semigroups or groups of elements of $\mathbf{B}(X)$ is a natural extension of the corresponding notion for semiflows and flows, respectively. Also obvious is the fact that $\mathbf{w}$ is uniquely ergodic if and only if the transition function $(P_t^{(\mathbf{w})})_{t \in \mathbb{T}}$ defined by $\mathbf{w}$ is uniquely ergodic. The one-parameter semigroup or group $\mathbf{w}$ of elements of $\mathbf{B}(X)$ is said to be *strictly ergodic* if $\mathbf{w}$ is uniquely ergodic and the support of the unique invariant probability of $\mathbf{w}$ is the entire space X. Plainly, $\mathbf{w}$ is strictly ergodic if and only if the transition function $(P_t^{(\mathbf{w})})_{t \in \mathbb{T}}$ defined by $\mathbf{w}$ is strictly ergodic.

Example 2.3.8. Let $\mathbf{w} = (w_t)_{t \in \mathbb{R}}$ be the flow of the rotations of the unit circle $\mathbb{R}/\mathbb{Z}$, and let $(P_t^{(\mathbf{w})})_{t \in \mathbb{R}}$ be the transition function defined by $\mathbf{w}$. Then, as pointed out in Example B.1.6, $\mathbf{w}$ is a uniquely ergodic flow and the Haar-Lebesgue measure on $\mathbb{R}/\mathbb{Z}$ is the unique invariant probability of $\mathbf{w}$. Using the observations made before this example, we obtain that $\mathbf{w}$ and $(P_t^{(\mathbf{w})})_{t \in \mathbb{R}}$ are strictly ergodic and their unique invariant probability is the Haar-Lebesgue measure on $\mathbb{R}/\mathbb{Z}$ (note that the fact that the Haar-Lebesgue measure on $\mathbb{R}/\mathbb{Z}$ is invariant for $(P_t^{(\mathbf{w})})_{t \in \mathbb{R}}$ was already pointed out in Example 2.3.1). ∎

Example 2.3.9. Let $n \in \mathbb{N}$, $n \geq 2$, let $\mathbf{v} \in \mathbb{R}^n$, $\mathbf{v} = (v_1, v_2, \cdots, v_n)$, let $\mathbf{w} = (w_t)_{t \in \mathbb{R}}$ be the rectilinear flow on the n-dimensional torus $\mathbb{R}^n/\mathbb{Z}^n$ with velocity $\mathbf{v}$ defined in Example A.3.5, and let $(P_t^{(\mathbf{w})})_{t \in \mathbb{R}}$ be the transition function defined by $\mathbf{w}$ (see Example 2.2.5).

As pointed out in Example B.1.7, the Haar-Lebesgue measure $\nu_{\mathbb{R}^n/\mathbb{Z}^n}$ on $\mathbb{R}^n/\mathbb{Z}^n$ is an invariant probability measure for $\mathbf{w}$, and $\mathbf{w}$ is uniquely ergodic if and only if the numbers $v_1, v_2, \cdots, v_n$ are rationally independent (see also Section 3.1 of Cornfeld, Fomin and Sinai's monograph [22]). Consequently, $(P_t^{(\mathbf{w})})_{t \in \mathbb{R}}$ is uniquely ergodic if and only if the numbers $v_1, v_2, \cdots, v_n$ are rationally independent, and in this case $(P_t^{(\mathbf{w})})_{t \in \mathbb{R}}$ is strictly ergodic because the unique invariant probability measure of $(P_t^{(\mathbf{w})})_{t \in \mathbb{R}}$ is $\nu_{\mathbb{R}^n/\mathbb{Z}^n}$. ∎

Example 2.3.10. Let Γ be a cocompact lattice in $\mathrm{SL}(2, \mathbb{R})$, let $\mathbf{v}^{(j\Gamma S)} = (v_t^{(j\Gamma S)})_{t \in \mathbb{R}}$ be one of the four horocycle flows considered at (c) of Example 2.2.7, where $j = 1$ or 2, and $S = L$ or R, and let $(P_t^{(\mathbf{v}^{(j\Gamma S)})})_{t \in \mathbb{R}}$ be the transition function defined by $\mathbf{v}^{(j\Gamma S)}$ (see (c) of Example 2.2.7). Then using the result of Furstenberg [37] discussed in (b) of Example B.1.9, we obtain that $(P_t^{(\mathbf{v}^{(j\Gamma S)})})_{t \in \mathbb{R}}$ is uniquely ergodic because, by the above-mentioned result of Furstenberg, the flow $\mathbf{v}^{(j\Gamma S)}$ is uniquely ergodic. Since, as pointed out in (b) of Example B.1.9, the standard $\mathrm{SL}(2, \mathbb{R})$-invariant probability measure $\nu_{(\mathrm{SL}(2,\mathbb{R})/\Gamma)_S}$ is an invariant measure for $\mathbf{v}^{(j\Gamma S)}$, it follows that $\nu_{(\mathrm{SL}(2,\mathbb{R})/\Gamma)_S}$ is an invariant probability for $(P_t^{(\mathbf{v}^{(j\Gamma S)})})_{t \in \mathbb{R}}$, as well, so $(P_t^{(\mathbf{v}^{(j\Gamma S)})})_{t \in \mathbb{R}}$ is strictly ergodic. ∎

Among the nonzero invariant measures of transition functions one can single out various kinds of such measures. For our purposes in this book, the most important type of nonzero invariant measures are the ergodic invariant probability measures.

In a similar manner as in the case of transition probabilities, given a transition function $(P_t)_{t \in \mathbb{T}}$ and a finite nonzero Borel measure μ on (X, d) such that μ is invariant for $(P_t)_{t \in \mathbb{T}}$, we say that μ is an *ergodic measure* if there is *no* Borel measurable subset A of X such that $\mu(A) > 0$ and $\mu(X \setminus A) > 0$, and such that the measures $\mu_1 : \mathcal{B}(X) \to \mathbb{R}$ and $\mu_2 : \mathcal{B}(X) \to \mathbb{R}$ defined by $\mu_1(B) = \mu(A \cap B)$ and $\mu_2(B) = \mu((X \setminus A) \cap B)$ for every $B \in \mathcal{B}(X)$ are both invariant for $(P_t)_{t \in \mathbb{T}}$.

If $\mathbf{w} = (w_t)_{t \in \mathbb{T}}$ is a one-parameter semigroup or a one-parameter group of elements of $\mathbf{B}(X)$, then a nonzero finite measure μ on $(X, \mathcal{B}(X))$ which is invariant for $\mathbf{w}$ is said to be *ergodic* if μ is ergodic for $(P_t^{(\mathbf{w})})_{t \in \mathbb{T}}$, where, as usual, $(P_t^{(\mathbf{w})})_{t \in \mathbb{R}}$ is the transition function defined by $\mathbf{w}$. Note that the notion of ergodic measure defined here for one-parameter semigroups and one-parameter groups of elements of $\mathbf{B}(X)$ is a natural extension of the corresponding notion for semiflows and flows, respectively, defined in Appendix B before Example B.1.6.

It is possible to define a notion of ergodicity that is valid for measures that are not necessarily finite or invariant (see, for instance, p. 2 of Bachir Bekka and Mayer's monograph [10]). The more general notion is a natural extension of the notion of ergodic measure defined above. However, in this book we deal only with nonzero finite invariant ergodic measures, which will be probability measures most of the time.

There are many ways to define the ergodicity of a nonzero finite invariant measure and, generally, these definitions are equivalent if the settings are the same. For instance, we will see in Sect. 6.1 that if $\mathbf{w} = (w_t)_{t \in \mathbb{T}}$ is either a measurable flow, or a measurable semiflow which has the property that w_t is a surjective function for every $t \in [0, +\infty)$, and if $\mu \in \mathcal{M}(X)$ is an invariant probability measure for the transition function $(P_t^{(\mathbf{w})})_{t \in \mathbb{T}}$ defined by $\mathbf{w}$, then μ is an ergodic measure in the sense of the definition given in this section if and only if, in the terminology of Cornfeld, Fomin and Sinai's monograph [22], the dynamical system defined by $\mathbf{w}$ and μ is ergodic (see p. 14 of [22]); also in Sect. 6.1 it will be shown that if $\mathbf{w} = (w_t)_{t \in [0,+\infty)}$ is a measurable semiflow and $\mu \in \mathcal{M}(X)$ is a nonzero invariant measure, then μ is an ergodic measure as defined in this section if and only if, in the words of Stroock's book [119], p. 315, $\mathbf{w}$ is ergodic (with respect to μ).

Note that if $(P_t)_{t \in \mathbb{T}}$ is a transition function defined on (X, d), and if, for some $x \in X$, the Dirac measure δ_x is an invariant measure for $(P_t)_{t \in \mathbb{T}}$, then δ_x is ergodic. For instance, if $\mathbf{w}$ is the one-parameter semigroup defined in Example 2.2.3, and if $(P_t^{(\mathbf{w})})_{t \in [0,+\infty)}$ is the transition function generated by $\mathbf{w}$, then, as pointed out in Example 2.3.5, δ_1 is an invariant probability for $(P_t^{(\mathbf{w})})_{t \in [0,+\infty)}$, so δ_1 is an invariant ergodic measure for $(P_t^{(\mathbf{w})})_{t \in [0,+\infty)}$.

Observe that using the notion of ergodicity of a measure as defined in this section, we obtain that the unique invariant probability measure of a uniquely ergodic transition function is an ergodic measure. Actually, in Chap. 6 we will see that the following converse of the above observation holds true, as well: if a transition function $(P_t)_{t \in \mathbb{T}}$ has exactly one invariant ergodic probability measure, then $(P_t)_{t \in \mathbb{T}}$ is uniquely ergodic. Thus, since the invariant probability measure of a uniquely ergodic transition function is an ergodic measure, it follows that all the invariant probabilities of the uniquely ergodic transition functions discussed in Examples 2.3.8–2.3.10 are ergodic.

Of course, transition functions that are not uniquely ergodic can have invariant ergodic probability measures, as well (actually, as a straightforward consequence of the results obtained in Chaps. 5 and 6, we will see that every transition function that has nonzero invariant elements, also has invariant ergodic probability measures). Below are a few notable examples of invariant ergodic probabilities of transition functions that are not uniquely ergodic.

Example 2.3.11. Let Γ be a lattice in $\mathrm{PSL}(2, \mathbb{R})$, let $\mathbf{w}^{(\Gamma)} = (w_t^{(\Gamma)})_{t \in \mathbb{R}}$ be the geodesic flow on $(\mathrm{PSL}(2, \mathbb{R})/\Gamma)_{\mathrm{R}}$ (see Example B.1.8), and let $(P_t^{(\mathbf{w}^{(\Gamma)})})_{t \in \mathbb{R}}$ be the transition function defined by $\mathbf{w}^{(\Gamma)}$. Then, as pointed out in Example B.1.8, the standard $\mathrm{PSL}(2, \mathbb{R})$-invariant probability measure $\nu_{(\mathrm{PSL}(2,\mathbb{R})/\Gamma)_{\mathrm{R}}}$ is an invariant ergodic measure for $\mathbf{w}^{(\Gamma)}$ by a result of Hedlund [41]; therefore, $\nu_{(\mathrm{PSL}(2,\mathbb{R})/\Gamma)_{\mathrm{R}}}$ is an invariant ergodic probability measure for $(P_t^{(\mathbf{w}^{(\Gamma)})})_{t \in \mathbb{R}}$, as well. Note that, as mentioned in Example B.1.8, the geodesic flow is not uniquely ergodic, so $(P_t^{(\mathbf{w}^{(\Gamma)})})_{t \in \mathbb{R}}$ is not uniquely ergodic either. ∎

Example 2.3.12. Let Γ be a lattice in $\mathrm{SL}(2, \mathbb{R})$, let $\mathbf{v}^{(j\Gamma L)}$ and $\mathbf{v}^{(j\Gamma R)}$, $j = 1, 2$, be the four horocycle flows defined in (b) of Example B.1.9, and let $(P_t^{(\mathbf{v}^{(j\Gamma L)})})_{t \in \mathbb{R}}$ and $(P_t^{(\mathbf{v}^{(j\Gamma R)})})_{t \in \mathbb{R}}$, $j = 1, 2$, be the corresponding four transition functions defined by these flows. Assume also that the four flows have periodic points (this happens if $\Gamma = \mathrm{SL}(2, \mathbb{Z})$, for instance; for details, see (b) of Example B.1.9). It follows that the invariant ergodic probability measures of the four transition functions are the invariant ergodic probabilities of the corresponding flows; that is, using a result of Dani [23] discussed in (b) of Example B.1.9, we obtain that these invariant ergodic probability measures are precisely the standard $\mathrm{SL}(2, \mathbb{R})$-invariant probabilities $\nu_{(\mathrm{SL}(2,\mathbb{R})/\Gamma)_{\mathrm{L}}}$ (for $(P_t^{(\mathbf{v}^{(j\Gamma L)})})_{t \in \mathbb{R}}$, $j = 1, 2$), and $\nu_{(\mathrm{SL}(2,\mathbb{R})/\Gamma)_{\mathrm{R}}}$ (for $(P_t^{(\mathbf{v}^{(j\Gamma R)})})_{t \in \mathbb{R}}$, $j = 1, 2$), and the invariant ergodic probability measures whose supports are the orbits of the periodic points of the corresponding flows. ∎

In the next proposition, we discuss a property of (nonzero) ergodic measures that is similar to the property of the invariant elements of an $\mathbb{R}$-transition function discussed in Proposition 2.3.7.

Proposition 2.3.13. *Let $(P_t)_{t \in \mathbb{R}}$ be a transition function defined on (X, d), and let $\mu \in \mathcal{M}(X)$, $\mu \geq 0$, $\mu \neq 0$. The following assertions are equivalent:*

(a) μ *is an invariant ergodic measure for* $(P_t)_{t \in \mathbb{R}}$.
(b) μ *is an invariant ergodic measure for the restriction* $(P_t)_{t \in [0,+\infty)}$ *of* $(P_t)_{t \in \mathbb{R}}$ *to* $[0, +\infty)$.

Proof.

$(a) \Rightarrow (b)$ Since we assume that μ is an invariant measure for $(P_t)_{t \in \mathbb{R}}$, we obtain that μ is an invariant measure for $(P_t)_{t \in [0,+\infty)}$, as well.

Now, if we assume that μ is not an ergodic measure for $(P_t)_{t \in [0,+\infty)}$, then there exist two mutually singular nonzero measures ν_1 and ν_2 such that $\mu = \nu_1 + \nu_2$ and such that ν_1 and ν_2 are invariant measures for $(P_t)_{t \in [0,+\infty)}$. By Proposition 2.3.7,

ν_1 and ν_2 are also invariant measures for $(P_t)_{t\in\mathbb{R}}$. We have obtained a contradiction because we assume that (a) holds true.

$(b) \Rightarrow (a)$ Since we assume that μ is an invariant measure for $(P_t)_{t\in[0,+\infty)}$, using Proposition 2.3.7 again, we obtain that μ is also invariant for $(P_t)_{t\in\mathbb{R}}$. The proof of the implication is completed by noting that if we assume that μ fails to be an ergodic measure for $(P_t)_{t\in\mathbb{R}}$, then, obviously, μ fails to be ergodic for $(P_t)_{t\in[0,+\infty)}$, as well. $\qquad\square$

We conclude the section (and the chapter) with a brief discussion of a notion that will appear frequently in Chaps. 5 and 6.

Let $(P_t)_{t\in\mathbb{T}}$ be a transition function defined on (X,d), and let $((S_t,T_t))_{t\in\mathbb{T}}$ be the family of Markov pairs defined by $(P_t)_{t\in\mathbb{T}}$.

A Borel measurable subset A of X is said to be a *set of maximal probability for* $(P_t)_{t\in\mathbb{T}}$ (or *for* $((S_t,T_t))_{t\in\mathbb{T}}$, or *for* $(T_t)_{t\in\mathbb{T}}$) if either $(P_t)_{t\in\mathbb{T}}$ does not have invariant probability measures, or else every invariant probability measure of $(P_t)_{t\in\mathbb{T}}$ is concentrated on A.

Thus, when studying the set of all invariant probabilities of a transition function $(P_t)_{t\in\mathbb{T}}$ in connection with the structure of the space (X,d) on which $(P_t)_{t\in\mathbb{T}}$ is defined, the sets of maximal probability are the sets "where the action is" and, naturally, we would like to find sets of maximal probability as "small" as possible. The relation of inclusion $\subseteq$ defined on $\mathcal{B}(X)$ is an order relation and $(\mathcal{B}(X),\subseteq)$ is a lattice; therefore, it makes sense to ask if the collection

$$\mathcal{A} = \{A \in \mathcal{B}(X) \mid A \text{ is a set of maximal probability for } (P_t)_{t\in\mathbb{T}}\}$$

has a minimum or at least an infimum in $\mathcal{B}(X)$. However, after a moment of reflection we realize that usually $\mathcal{A}$ has neither a minimum nor an infimum. In Chaps. 5 and 6 we will obtain various sets of maximal probability for transition functions that have invariant probabilities, sets which, for most purposes, are small enough.

Chapter 3
Preliminaries on Vector Integrals and Almost Everywhere Convergence

Our goal in this chapter is to discuss several tools that we need in order to obtain the results presented in Chaps. 5–7.

In the first section we briefly review the Bochner integral following Appendix E of Cohn [20] and the Dunford-Schwartz integral defined in [30]. Before discussing the Dunford and Schwartz integral, we go over several basic facts about complete measure spaces because in this book we consider Dunford-Schwartz integrals only when dealing with complete measure spaces.

When obtaining the KBBY decomposition for transition functions and various related results, the role played by the Hopf ergodic theorem (Theorem 1.2.1) in dealing with the decomposition for transition probabilities is played by a result of Dunford and Schwartz; namely, by Theorem 8.7.5, p. 690 of the monograph [30]. In Sect. 3.2 we restate the above-mentioned theorem of Dunford and Schwartz in a form suitable for our purposes and obtain various nontrivial and new consequences of the theorem, consequences that will be used in the remaining chapters of the book (especially in Chaps. 5 and 6).

Finally, in Sect. 3.3 we define a vector integral, which we call the pointwise integral and which, in all likelihood, has appeared earlier in the literature, but we are unable to provide a reference. Also in this section, as a warm up application of the pointwise integral, we use the integral to study the existence of invariant probability measures for transition functions.

3.1 The Bochner and the Dunford-Schwartz Integrals

This section is formed of three subsections. In the first subsection we discuss the Bochner integral. In Sect. 3.1.2 we briefly review several facts about complete measure spaces and the completion of a measure space. Finally, in the last subsection (Sect. 3.1.3) we discuss the Dunford-Schwartz integral. The reason for the review of measure theoretical completeness in the second subsection is that, in

R. Zaharopol, *Invariant Probabilities of Transition Functions*, Probability and Its Applications 44, DOI 10.1007/978-3-319-05723-1_3,
© Springer International Publishing Switzerland 2014

the last subsection and throughout the book, we deal only with Dunford-Schwartz integrals of Banach space-valued functions defined on complete measure spaces and many results for not necessarily complete measure spaces are proved using the completions of these measure spaces.

3.1.1 The Bochner Integral

Our goal in this subsection is to briefly go over several basic facts about the Bochner integral. We will follow Appendix E of Cohn's book [20].

Let $(Y, \mathcal{Y}, \mu)$ be a (finite or) σ-finite measure space, let $(E, \|\cdot\|)$ be a real Banach space, and let $\mathcal{B}(E)$ be the σ-algebra of all the Borel subsets of E.

As usual, given a function $f : Y \to E$ we say that f is *measurable* if $f^{-1}(A) \in \mathcal{Y}$ for every $A \in \mathcal{B}(E)$.

We say that a function $f : Y \to E$ is *strongly measurable* or *Bochner measurable* if f is measurable and the range $f(Y)$ of f is a separable subset of E. Naturally, if E is a separable Banach space, then any measurable function $g : Y \to E$ is strongly measurable. In particular, any real-valued measurable function on $\mathbb{R}$ is strongly measurable.

A function $f : Y \to E$ is said to be *simple* if the range $f(Y)$ of f is a finite subset of E. Thus, $f : Y \to E$ is a simple function if and only if there exist a natural number m, m elements $y_1, y_2, \ldots, y_m$ of E and m subsets $B_1, B_2, \ldots, B_m$ of Y such that $f = \sum_{i=1}^{m} y_i \mathbf{1}_{B_i}$. Plainly, a simple function is measurable if and only if the function is Bochner measurable.

Given a simple measurable function $f : Y \to E$, it is easy to see that there exist $n \in \mathbb{N}$, n measurable subsets $A_1, A_2, \ldots, A_n$ of Y such that $A_i \cap A_j = \emptyset$ whenever $i \neq j$, $i = 1, 2, \ldots, n$, $j = 1, 2, \ldots, n$, and n elements $x_1, x_2, \ldots, x_n$ of E such that $f = \sum_{i=1}^{n} x_i \mathbf{1}_{A_i}$. We call the sum $\sum_{i=1}^{n} x_i \mathbf{1}_{A_i}$ a *disjoint representation of* f. We say that f is *Bochner integrable* if $\mu(A_i) < +\infty$ or $x_i = \mathbf{0}$ (the zero element of E) for every $i = 1, 2, \ldots, n$. If f is Bochner integrable, the sum $\sum_{i=1}^{n} x_i \mu(A_i)$ which is a well-defined element of E, is called the *Bochner integral of* f and we will denote it by B-$\int_Y f(y) \, d\mu(y)$. Other notations that we will use are B-$\int f(y) \, d\mu(y)$, or B-$\int f \, d\mu$, or B-$\int_Y f \, d\mu$.

Note that the definition of the Bochner integrability of a simple Bochner measurable function is correct in the following sense: if $f : Y \to E$, $f = \sum_{i=1}^{n} x_i \mathbf{1}_{A_i}$, $n \in \mathbb{N}$, $x_i \in E$ and $A_i \in \mathcal{Y}$ for every $i = 1, 2, \ldots, n$, and $A_i \cap A_j = \emptyset$ for every $i \neq j$, $i = 1, 2, \ldots, n$, $j = 1, 2, \ldots, n$, is a Bochner integrable function and $f = \sum_{j=1}^{m} z_j \mathbf{1}_{B_j}$ for some $m \in \mathbb{N}$, $z_j \in E$, $B_j \in \mathcal{Y}$, $j = 1, 2, \ldots, m$, and $B_{j_1} \cap B_{j_2} = \emptyset$ for every $j_1 \neq j_2$, $j_1 = 1, 2, \ldots, m$, $j_2 = 1, 2, \ldots, m$, then $\mu(B_j) < +\infty$ or $z_j = \mathbf{0}$ for every $j = 1, 2, \ldots, m$, and $\sum_{i=1}^{n} x_i \mu(A_i) = \sum_{j=1}^{m} z_j \mu(B_j)$.

Now let $f : Y \to E$ be a strongly measurable function. We say that f is *Bochner integrable* if there exists a sequence $(f_n)_{n\in\mathbb{N}}$ of E-valued simple Bochner integrable functions defined on Y such that the following three conditions are satisfied:

(B1) The sequence $(f_n(y))_{n\in\mathbb{N}}$ converges to $f(y)$ for every $y \in Y$ (that is, the sequence $(f_n)_{n\in\mathbb{N}}$ converges pointwise to f (on Y)).

(B2) $\|f_n(y)\| \le \|f(y)\|$ for every $y \in Y$ and $n \in \mathbb{N}$.

(B3) The sequence $(f_n)_{n\in\mathbb{N}}$ is Cauchy in the mean in the following sense: for every $\varepsilon \in \mathbb{R}$, $\varepsilon > 0$, there exists an $n_\varepsilon \in \mathbb{N}$ such that

$$\int_Y \|f_n(y) - f_m(y)\|\ \mathrm{d}\mu(y) < \varepsilon \tag{3.1.1}$$

for every $n \ge n_\varepsilon$ and $m \ge n_\varepsilon$ (note that the integral that appears on the left-hand side of (3.1.1) is meaningful in the sense that it can be shown (see Appendix E of Cohn [20]) that $f_n - f_m$ is a simple Bochner integrable function, so the mapping $y \mapsto \|f_n(y) - f_m(y)\|$, $y \in Y$, is a simple real-valued function integrable in the sense of classical integration theory (in this book by *classical integration theory* or *classical measure theory* we mean the integration theory or the measure theory developed for real or complex-valued functions and real or complex-valued measures that can be found, for instance, in the first six chapters of Cohn [20]; if f is a real-valued or extended real-valued function defined on Y, then the notations $\int f\ \mathrm{d}\mu$, or $\int_Y f\ \mathrm{d}\mu$, or $\int f(y)\ \mathrm{d}\mu(y)$, or $\int_Y f(y)\ \mathrm{d}\mu(y)$ stand for the integral of f in the sense of classical integration theory)).

Proposition 3.1.1. (*a*) *If* $(f_n)_{n\in\mathbb{N}}$ *is a sequence of E-valued simple Bochner integrable functions defined on Y that satisfies condition* $(B3)$ *in the definition of Bochner integrability of a strongly measurable function, then the sequences* $(B\text{-}\int_Y f_n\ \mathrm{d}\mu)_{n\in\mathbb{N}}$ *and* $(\int_Y \|f_n(y)\|\ \mathrm{d}\mu(y))_{n\in\mathbb{N}}$ *are both Cauchy sequences (of elements of E and of real numbers, respectively).*

(*b*) *A strongly measurable function* $f : Y \to E$ *is Bochner integrable if and only if the function* $y \mapsto \|f(y)\|$, $y \in Y$, *is integrable in the classical sense.*

Proof. (*a*) It is easy to see that $\left\| B\text{-}\int_Y h\ \mathrm{d}\mu \right\| \le \int_Y \|h(y)\|\ \mathrm{d}\mu(y)$ for every E-valued simple Bochner integrable function h defined on Y. Thus,

$$\left\| B\text{-}\int_Y f_n\ \mathrm{d}\mu - B\text{-}\int_Y f_m\ \mathrm{d}\mu \right\| = \left\| B\text{-}\int_Y (f_n - f_m)\ \mathrm{d}\mu \right\|$$

$$\le \int_Y \|f_n(y) - f_m(y)\|\ \mathrm{d}\mu(y)$$

for every $n \in \mathbb{N}$ and $m \in \mathbb{N}$, so the sequence $(B\text{-}\int_Y f_n\ \mathrm{d}\mu)_{n\in\mathbb{N}}$ is Cauchy because $(f_n)_{n\in\mathbb{N}}$ satisfies condition (B3).

Since

$$\left| \int_Y \| f_n(y) \|\, d\mu(y) - \int_Y \| f_m(y) \|\, d\mu(y) \right| \le \int_Y \left| \| f_n(y) \| - \| f_m(y) \| \right| d\mu(y)$$

$$\le \int_Y \| f_n(y) - f_m(y) \|\, d\mu(y)$$

for every $n \in \mathbb{N}$ and $m \in \mathbb{N}$, and using again the assumption that $(f_n)_{n\in\mathbb{N}}$ satisfies (B3), we obtain that $(\int_Y \| f_n(y) \|\, d\mu(y))_{n\in\mathbb{N}}$ is a Cauchy sequence.

(b) Let $f : Y \to E$ be a strongly measurable function.

Assume that f is Bochner integrable, and let $(f_n)_{n\in\mathbb{N}}$ be a sequence of E-valued simple Bochner integrable functions on Y which satisfies the conditions (B1), (B2) and (B3) of the definition of Bochner integrability.

For every $n \in \mathbb{N}$, let $g_n : Y \to \mathbb{R}$ be defined by $g_n(y) = \| f_n(y) \|$ for every $y \in Y$. Using (a) we obtain that the sequence $(g_n)_{n\in\mathbb{N}}$ is a Cauchy sequence in $L^1(Y, \mathcal{Y}, \mu)$, where $L^1(Y, \mathcal{Y}, \mu)$ is the usual Banach space in classical measure theory. Using standard results of classical measure theory (see, for instance, Propositions 3.1.4 and 3.1.2 of Cohn [20]), and since $(g_n)_{n\in\mathbb{N}}$ converges to the function $g : Y \to \mathbb{R}$ defined by $g(y) = \| f(y) \|$ for every $y \in Y$, we obtain that g is the limit of $(g_n)_{n\in\mathbb{N}}$ in the mean; that is, the sequence $(\int_Y |g_n - g|\, d\mu)_{n\in\mathbb{N}}$ converges to zero, so g is integrable.

The fact that the integrability of g implies that f is Bochner integrable is proved on p. 352 in Cohn [20]. $\square$

Let $f : Y \to E$ be a not necessarily simple Bochner integrable function. It is shown on p. 352 of Cohn [20] that for every sequence $(f_n)_{n\in\mathbb{N}}$ of E-valued simple Bochner integrable functions defined on Y that satisfies conditions (B1), (B2) and (B3) relative to f, the limit of the sequence $(\text{B-}\int_Y f_n\, d\mu)_{n\in\mathbb{N}}$ which exists by (a) of Proposition 3.1.1, is independent of the particular sequence $(f_n)_{n\in\mathbb{N}}$ under consideration. This limit is called the *Bochner integral of f*, and, as in the case of simple Bochner integrable functions, is denoted by $\text{B-}\int_Y f(y)\, d\mu(y)$, or $\text{B-}\int_Y f\, d\mu$, or $\text{B-}\int f(y)\, d\mu(y)$, or $\text{B-}\int f\, d\mu$.

For future reference, we state several useful facts about Bochner integrable functions in the next proposition.

Proposition 3.1.2. (a) $\left\| \text{B-}\int f\, d\mu \right\| \le \int_Y \| f(y) \|\, d\mu(y)$ *for every Bochner integrable function* $f : Y \to E$.

(b) *The set* $\mathcal{L}_B^1(Y, \mathcal{Y}, \mu, E)$ *of all Bochner integrable functions* $f : Y \to E$ *is a (real) vector space when endowed with pointwise addition and scalar multiplication, and the map* $f \mapsto \int_Y \| f(y) \|\, d\mu(y)$, $f \in \mathcal{L}_B^1(Y, \mathcal{Y}, \mu, E)$, *is a seminorm on* $\mathcal{L}_B^1(Y, \mathcal{Y}, \mu, E)$.

For a proof of (a) of the above proposition, see Proposition E5 on p. 353 of Cohn [20]. Note that if f is a simple Bochner integrable function, the inequality stated in (a) is easy to prove, and was used in the proof of (a) of Proposition 3.1.1. A proof that $\mathcal{L}_B^1(Y, \mathcal{Y}, \mu, E)$ is a vector space is given in Proposition E4 on p. 353 of

Cohn [20]. The fact that the map $f \mapsto \int_Y \|f(y)\| \, d\mu(y)$, $f \in \mathcal{L}^1_B(Y, \mathcal{Y}, \mu, E)$, is a seminorm on $\mathcal{L}^1_B(Y, \mathcal{Y}, \mu, E)$ is discussed on p. 354 of Cohn [20]. The seminorm will be denoted by $\||\cdot\||^{(B)}_{1E}$; thus, $\||f\||^{(B)}_{1E} = \int_Y \|f(y)\| \, d\mu(y)$, $f \in \mathcal{L}^1_B(Y, \mathcal{Y}, \mu, E)$.

As in classical measure theory we can define the almost everywhere equivalence relation on $\mathcal{L}^1_B(Y, \mathcal{Y}, \mu, E)$ as follows: $f \sim g$ if, by definition, $f = g$ μ-a.e., $f \in \mathcal{L}^1_B(Y, \mathcal{Y}, \mu, E)$, $g \in \mathcal{L}^1_B(Y, \mathcal{Y}, \mu, E)$; clearly, $\sim$ is an equivalence relation. The collection $L^1_B(Y, \mathcal{Y}, \mu, E)$ of all the equivalence classes defined by $\sim$ can be endowed with a vector space structure, and the seminorm $\||\cdot\||^{(B)}_{1E}$ on $\mathcal{L}^1_B(Y, \mathcal{Y}, \mu, E)$ can be used to define a norm on $L^1_B(Y, \mathcal{Y}, \mu, E)$ in a similar manner as in classical measure theory; we will use the same notation $\||\cdot\||^{(B)}_{1E}$ for both the seminorm on $\mathcal{L}^1_B(Y, \mathcal{Y}, \mu, E)$ and the norm on $L^1_B(Y, \mathcal{Y}, \mu, E)$. Finally, it can be shown that, when endowed with $\||\cdot\||^{(B)}_{1E}$ the normed vector space $L^1_B(Y, \mathcal{Y}, \mu, E)$ is a Banach space.

Again, as in classical measure theory, we can define

$$\mathcal{L}^\infty_B(Y, \mathcal{Y}, \mu, E) = \left\{ f : Y \to E \;\middle|\; \begin{array}{l} f \text{ is Bochner measurable} \\ \text{and the map } y \mapsto \|f(y)\|, \\ y \in Y, \text{ belongs to } \mathcal{L}^\infty(Y, \mathcal{Y}, \mu) \end{array} \right\},$$

where $\mathcal{L}^\infty(Y, \mathcal{Y}, \mu)$ is the usual vector space of all real-valued essentially bounded (measurable) functions that is defined in classical measure theory. Clearly, $\mathcal{L}^\infty_B(Y, \mathcal{Y}, \mu, E)$ is a vector space and the map $\||\cdot\||^{(B)}_{\infty E} : \mathcal{L}^\infty_B(Y, \mathcal{Y}, \mu, E) \to \mathbb{R}$ defined by $\||f\||^{(B)}_{\infty E} = \|g_f\|_\infty$ for every $f \in \mathcal{L}^\infty_B(Y, \mathcal{Y}, \mu, E)$ is a seminorm on $\mathcal{L}^\infty_B(Y, \mathcal{Y}, \mu, E)$, where $g_f : Y \to E$ is defined by $g_f(y) = \|f(y)\|$ for every $y \in Y$ and $f \in \mathcal{L}^\infty_B(Y, \mathcal{Y}, \mu, E)$, and $\|\cdot\|_\infty$ is the usual seminorm on $\mathcal{L}^\infty(Y, \mathcal{Y}, \mu)$. As in the case of $\mathcal{L}^1_B(Y, \mathcal{Y}, \mu, E)$, the almost everywhere equality relation can be defined on $\mathcal{L}^\infty_B(Y, \mathcal{Y}, \mu, E)$, as well, and the resulting set of equivalence classes, denoted $L^\infty_B(Y, \mathcal{Y}, \mu, E)$, can be endowed with a vector space structure. The seminorm $\||\cdot\||^{(B)}_{\infty E}$ can be used in a natural manner to define a norm on $L^\infty_B(Y, \mathcal{Y}, \mu, E)$ denoted again by $\||\cdot\||^{(B)}_{\infty E}$, and the norm $\||\cdot\||^{(B)}_{\infty E}$ defines a Banach space structure on $L^\infty_B(Y, \mathcal{Y}, \mu, E)$.

Note that if $E = \mathbb{R}$, then the spaces $\mathcal{L}^1_B(Y, \mathcal{Y}, \mu, E)$, $L^1_B(Y, \mathcal{Y}, \mu, E)$, $\mathcal{L}^\infty_B(Y, \mathcal{Y}, \mu, E)$ and $L^\infty_B(Y, \mathcal{Y}, \mu, E)$ are precisely the spaces $\mathcal{L}^1(Y, \mathcal{Y}, \mu)$, $L^1(Y, \mathcal{Y}, \mu)$, $\mathcal{L}^\infty(Y, \mathcal{Y}, \mu)$ and $L^\infty(Y, \mathcal{Y}, \mu)$, respectively, that are defined in classical measure theory.

Let V be a vector space of measurable functions $f : Y \to E$, where the algebraic operations that define the vector space structure on V are pointwise addition and scalar multiplication. Note that the functions in V are not necessarily strongly measurable.

As in classical measure theory, given a sequence $(f_n)_{n \in \mathbb{N}}$ of elements of V and $f \in V$, we say that $(f_n)_{n \in \mathbb{N}}$ *converges in measure to f* if, for every $\varepsilon \in \mathbb{R}$, $\varepsilon > 0$, the sequence $(\mu(\{y \in Y \mid \|f_n(y) - f(y)\| > \varepsilon\}))_{n \in \mathbb{N}}$ converges to zero. The convergence to zero of $(\mu(\{y \in Y \mid \|f_n(y) - f(y)\| > \varepsilon\}))_{n \in \mathbb{N}}$ means, of course, that all the terms of the sequence, except possibly a finite number of them, are real numbers, and that, by deleting the terms that are equal to $+\infty$, the resulting sequence of real numbers converges to zero.

Note that the above definition makes sense because, for every $n \in \mathbb{N}$, the function $f_n - f$ is measurable, so the set $\{y \in Y \mid \|f_n(y) - f(y)\| > \varepsilon\}$ belongs to $\mathcal{Y}$. The reason for introducing the vector space V stems from the fact that, in general, pointwise addition and scalar multiplication do not define a vector space structure on the set of all measurable functions $g : Y \to E$ because pointwise addition is not a well defined algebraic operation on this set.

In a similar way as in classical measure theory, convergence in the seminorm $\|\|\cdot\|\|_{1E}^{(B)}$ implies convergence in measure. For future reference, we discuss this fact in the next proposition.

Proposition 3.1.3. *Let $f \in \mathcal{L}_B^1(Y, \mathcal{Y}, \mu, E)$ and let $(f_n)_{n \in \mathbb{N}}$ be a sequence of elements of $\mathcal{L}_B^1(Y, \mathcal{Y}, \mu, E)$. If $(f_n)_{n \in \mathbb{N}}$ converges to f with respect to the seminorm $\|\|\cdot\|\|_{1E}^{(B)}$, then $(f_n)_{n \in \mathbb{N}}$ converges to f in measure. In particular, if $(f_n)_{n \in \mathbb{N}}$ is a sequence of simple Bochner integrable functions such that f and $(f_n)_{n \in \mathbb{N}}$ satisfy the conditions (B1), (B2) and (B3) of the definition of the Bochner integrability of a not necessarily simple function, then $(f_n)_{n \in \mathbb{N}}$ converges to f in measure.*

The proof of the proposition is very similar to the proof of the corresponding result in classical measure theory (see, for instance, Proposition 3.1.4 of Cohn [20]). Indeed, it is easy to see that, for every $\varepsilon \in \mathbb{R}$, $\varepsilon > 0$, and every $n \in \mathbb{N}$, the inequalities

$$\mu(A_\varepsilon^{(n)}) \le \frac{1}{\varepsilon} \int_Y \mathbf{1}_{A_\varepsilon^{(n)}} \|f_n(y) - f(y)\| \, \mathrm{d}\mu(y)$$

$$\le \frac{1}{\varepsilon} \int_Y \|f_n(y) - f(y)\| \, \mathrm{d}\mu(y)$$

hold true, where $A_\varepsilon^{(n)} = \{y \in Y \mid \|f_n(y) - f(y)\| > \varepsilon\}$. Thus, the main assertion of the proposition and, therefore, the entire proposition are true.

3.1.2 Complete Measure Spaces

As mentioned at the beginning of the section, in this subsection we discuss several facts about measure theoretical completeness that will be used in the next subsection and throughout the book.

Let $(Y, \mathcal{Y}, \mu)$ be a measure space.

As usual (see, for instance, Cohn's book [20]) we say that the measure space $(Y, \mathcal{Y}, \mu)$ is *complete* (or that the measurable space $(Y, \mathcal{Y})$, or the σ-algebra $\mathcal{Y}$ is *complete with respect to* μ) if A belongs to $\mathcal{Y}$ for every subset A of Y which has the property that there exists an $N \in \mathcal{Y}$ such that $A \subseteq N$ and $\mu(N) = 0$.

Assume that $(Y, \mathcal{Y}, \mu)$ is a not necessarily complete measure space, and set $\tilde{\mathcal{Y}} = \{A \subseteq Y \mid$ there exist $B \in \mathcal{Y}$ and $C \in \mathcal{Y}$ such that $B \subseteq A \subseteq C$ and $\mu(C \setminus B) = 0\}$. It is well-known (and not difficult to see) that $\tilde{\mathcal{Y}}$ is a σ-algebra.

Let $\tilde{\mu} : \tilde{\mathcal{Y}} \to \mathbb{R} \cup \{\infty\}$ be defined as follows: $\tilde{\mu}(A) = \mu(C)$, where $C \in \mathcal{Y}$ is such that $A \subseteq C$ and there exists an $N \in \mathcal{Y}$ such that $\mu(N) = 0$ and $C \setminus A \subseteq N$ (the existence of C and N is a consequence of the manner in which $\tilde{\mathcal{Y}}$ is defined). It is easy to see that the definition of $\tilde{\mu}(A)$ is correct, in the sense that it does not depend on the particular set C used to define $\tilde{\mu}(A)$; that is, if $D \in \mathcal{Y}$ is such that $A \subseteq D$ and $D \setminus A \subseteq M$ for some $M \in \mathcal{Y}$, $\mu(M) = 0$, then $\mu(C) = \mu(D)$. It is also easy to see that $\tilde{\mu}$ is a measure on $(Y, \tilde{\mathcal{Y}})$, and that the measure space $(Y, \tilde{\mathcal{Y}}, \tilde{\mu})$ is complete. As usual, we say that $(Y, \tilde{\mathcal{Y}}, \tilde{\mu})$ is the *completion of* $(Y, \mathcal{Y}, \mu)$. The σ-algebra $\tilde{\mathcal{Y}}$ is called the *completion of* $\mathcal{Y}$ *with respect to* μ and we say that $\tilde{\mu}$ is the *completion of* μ. We will use the notation $\tilde{\mathcal{Y}}^{(\mu)}$ for $\tilde{\mathcal{Y}}$ in order to emphasize that the completion of $\mathcal{Y}$ under consideration is with respect to μ.

Now let $(Y, \mathcal{Y}, \mu)$ be a σ-finite measure space and let $(Y, \tilde{\mathcal{Y}}, \tilde{\mu})$ be the completion of $(Y, \mathcal{Y}, \mu)$ (even though we do not have to assume that μ is σ-finite in the discussion which follows, we impose the σ-finiteness condition on μ because in this book, the measures under consideration are σ-finite (or even finite) when dealing with complete (or completions of) measure spaces, and we want to avoid unnecessary technicalities that appear when dealing with L^∞-spaces defined by measures that are not σ-finite).

Now let $p = +\infty$, or $p \in \mathbb{R}$, $p \geq 1$, let $L^p(Y, \mathcal{Y}, \mu)$ and $L^p(Y, \tilde{\mathcal{Y}}, \tilde{\mu})$ be the usual Banach spaces defined and studied in classical integration theory, and let $\varphi_p : L^p(Y, \mathcal{Y}, \mu) \to L^p(Y, \tilde{\mathcal{Y}}, \tilde{\mu})$ be defined as follows: given $\bar{f} \in L^p(Y, \mathcal{Y}, \mu)$, we consider a real-valued function g on Y which is measurable with respect to $\mathcal{Y}$ and belongs to the equivalence class $\bar{f}$; then g is obviously measurable with respect to $\tilde{\mathcal{Y}}$; also, $|g|^p$ is integrable with respect to $\tilde{\mu}$ if $p \in \mathbb{R}$, or g is essentially bounded with respect to $\tilde{\mu}$ if $p = +\infty$; therefore, g defines an element $\tilde{\bar{g}}$ of $L^p(Y, \tilde{\mathcal{Y}}, \tilde{\mu})$; set $\varphi_p(\bar{f}) = \tilde{\bar{g}}$. It is easy to see that the definition of $\varphi_p(\bar{f})$ is correct in the sense that it does not depend on our particular choice of the function g in $\bar{f}$; that is, if h is another real-valued $\mathcal{Y}$-measurable function on Y that belongs to $\bar{f}$, then $\tilde{\bar{g}} = \tilde{\bar{h}}$.

Clearly, the mapping φ_p is one-to-one. It is also not difficult to see that φ_p is a surjection, as well. Indeed, if $\tilde{A} \in \tilde{\mathcal{Y}}$ is such that $\tilde{\mu}(\tilde{A}) < +\infty$, then taking into consideration the manner in which $\tilde{\mathcal{Y}}$ and $\tilde{\mu}$ are defined, we obtain that there exists a $B \in \mathcal{Y}$ such that $\mu(B) = \tilde{\mu}(\tilde{A})$, so $\varphi_p(\bar{\mathbf{1}}_B) = \tilde{\bar{\mathbf{1}}}_{\tilde{A}}$; accordingly, if $\tilde{\bar{f}} \in L^p(Y, \tilde{\mathcal{Y}}, \tilde{\mu})$ has the property that there exists a simple $\tilde{\mathcal{Y}}$-measurable real-valued function $\tilde{g}$ that belongs to $\tilde{\bar{f}}$, then there exists a real-valued simple $\mathcal{Y}$-measurable function h such that $\bar{h} \in L^p(Y, \mathcal{Y}, \mu)$ and $\varphi_p(\bar{h}) = \tilde{\bar{g}}$, so $\varphi_p(\bar{h}) = \tilde{\bar{f}}$ because $\tilde{\bar{g}} = \tilde{\bar{f}}$; if $\tilde{\bar{f}} \in L^p(Y, \tilde{\mathcal{Y}}, \tilde{\mu})$ is such that $\tilde{\bar{f}} \geq 0$, then there exists a real-valued $\tilde{\mathcal{Y}}$-measurable function g in the class $\tilde{\bar{f}}$ such that $g(y) \geq 0$ for every $y \in Y$; moreover, there exists a sequence $(g_n)_{n \in \mathbb{N}}$ of real-valued simple $\tilde{\mathcal{Y}}$-measurable functions such that $0 \leq g_n \leq g$ for every $n \in \mathbb{N}$, and such that $(g_n)_{n\in\mathbb{N}}$ converges pointwise to g on Y; in view of our discussion so far, for every $n \in \mathbb{N}$, there exists a real-valued $\mathcal{Y}$-measurable function h_n such that $h_n \geq 0$ and $h_n = g_n$ $\tilde{\mu}$-a.e.; since $(h_n)_{n\in\mathbb{N}}$ converges μ-a.e. to a real-valued $\mathcal{Y}$-measurable function, say h, it follows that $h = g$ $\tilde{\mu}$-a.e., h defines an equivalence class $\bar{h} \in L^p(Y, \mathcal{Y}, \mu)$, and $\varphi_p(\bar{h}) = \bar{\tilde{g}} = \bar{\tilde{f}}$; finally, if $\bar{\tilde{f}}$ is not necessarily a positive element of $L^p(Y, \tilde{\mathcal{Y}}, \tilde{\mu})$ and if g is a real-valued $\tilde{\mathcal{Y}}$-measurable function in the class $\bar{\tilde{f}}$, then $g = g^+ - g^-$, where g^+ and g^- are positive $\tilde{\mathcal{Y}}$-measurable functions such that $\bar{\tilde{g}}^+$ and $\bar{\tilde{g}}^-$ are elements of $L^p(Y, \tilde{\mathcal{Y}}, \tilde{\mu})$; thus, by applying the preceding discussion to $\bar{\tilde{g}}^+$ and $\bar{\tilde{g}}^-$, we obtain that there exists an $\bar{h} \in L^p(Y, \mathcal{Y}, \mu)$ such that $\varphi_p(\bar{h}) = \bar{\tilde{f}}$.

Observation. Using the above arguments, we also obtain that if f is a real-valued $\tilde{\mathcal{Y}}$-measurable function on Y, then there exists a real-valued $\mathcal{Y}$-measurable function g on Y such that $f = g$ $\tilde{\mu}$-a.e. $\blacktriangle$

It is easy to see that φ_p is a linear isometry, and that both φ_p and φ_p^{-1} are positive operators in the sense that $\varphi_p(\bar{f}) \geq 0$ if and only if $\bar{f} \geq 0$, whenever $\bar{f} \in L^p(Y, \mathcal{Y}, \mu)$.

For future reference we state all the information obtained so far about φ_p in the next proposition.

Proposition 3.1.4. *The mapping $\varphi_p : L^p(Y, \mathcal{Y}, \mu) \to L^p(Y, \mathcal{Y}, \mu)$ is a linear isometry onto $L^p(Y, \tilde{\mathcal{Y}}, \tilde{\mu})$. Moreover, both φ_p and φ_p^{-1} are positive operators.*

We call the operator φ_p the *standard isometry from $L^p(Y, \mathcal{Y}, \mu)$ onto $L^p(Y, \tilde{\mathcal{Y}}, \tilde{\mu})$.*

An important feature of φ_p is that, as shown in the next proposition, the mapping preserves the almost everywhere convergence of sequences of elements of the L^p-spaces under consideration.

Proposition 3.1.5. (a) *If $(\bar{f}_n)_{n\in\mathbb{N}}$ is a sequence of elements of $L^p(Y, \mathcal{Y}, \mu)$ which converges μ-a.e., then the sequence $(\varphi_p(\bar{f}_n))_{n\in\mathbb{N}}$ converges $\tilde{\mu}$-a.e.*

(b) *Conversely, if $(\bar{\tilde{f}}_n)_{n\in\mathbb{N}}$ is a sequence of elements of $L^p(Y, \tilde{\mathcal{Y}}, \tilde{\mu})$ which converges $\tilde{\mu}$-a.e., then the sequence $(\varphi_p^{-1}(\bar{\tilde{f}}_n))_{n\in\mathbb{N}}$ converges μ-a.e.*

Proof. (a) Let $(\bar{f}_n)_{n\in\mathbb{N}}$ be a sequence of elements of $L^p(Y, \mathcal{Y}, \mu)$ and assume that $(\bar{f}_n)_{n\in\mathbb{N}}$ converges μ-a.e. Thus, if $(g_n)_{n\in\mathbb{N}}$ is a sequence of real-valued $\mathcal{Y}$-measurable functions such that g_n belongs to the equivalence class $\bar{f}_n$ for every $n \in \mathbb{N}$, then there exists a $\mathcal{Y}$-measurable subset N of Y such that $\mu(N) = 0$ and such that $(g_n(x))_{n\in\mathbb{N}}$ converges for every $x \in Y \setminus N$. Taking into consideration the way in which φ_p is defined, we obtain that g_n belongs to $\varphi_p(\bar{f}_n)$ for every $n \in \mathbb{N}$. Since $N \in \tilde{\mathcal{Y}}$ and $\tilde{\mu}(N) = 0$, it follows that $(\varphi_p(\bar{f}_n))_{n\in\mathbb{N}}$ converges $\tilde{\mu}$-a.e.

(b) Let $(\bar{\tilde{f}}_n)_{n\in\mathbb{N}}$ be a sequence of elements of $L^p(Y, \tilde{\mathcal{Y}}, \tilde{\mu})$ that converges $\tilde{\mu}$-a.e. Since φ_p is a surjection, it follows that, for every $n \in \mathbb{N}$, there exists a real-valued $\mathcal{Y}$-measurable function h_n that belongs to $\bar{\tilde{f}}_n$. It follows that there exists a $\tilde{\mu}$-negligible subset M of Y such that $(h_n(x))_{n \in \mathbb{N}}$ converges for every $x \in Y \setminus M$. Taking into consideration the manner in which $\tilde{\mathcal{Y}}$ and $\tilde{\mu}$ are defined, we obtain that there exists an $N \in \mathcal{Y}$ such that $\mu(N) = 0$ and $M \subseteq N$. Consequently, the sequence $(h_n)_{n \in \mathbb{N}}$ converges μ-a.e. Since $h_n \in \varphi_p^{-1}(\tilde{\tilde{f}}_n)$ for every $n \in \mathbb{N}$, it follows that $(\varphi_p^{-1}(\tilde{\tilde{f}}_n))_{n \in \mathbb{N}}$ converges μ-a.e. $\square$

Propositions 3.1.4 and 3.1.5 will be used in order to significantly simplify the presentation of certain topics by using completions of measure spaces whenever suitable and convenient. In particular, we will be able to use a simplified version of the Dunford-Schwartz integral discussed in the next subsection.

We will often denote arbitrary elements of $L^p(X, \widetilde{\mathcal{B}(X)}, \tilde{\mu})$, $p \in [1, +\infty)$ or $p = +\infty$, by $\tilde{\tilde{f}}$ in order to stress the fact that any such element can be obtained by starting with a real-valued $\mathcal{B}(X)$-measurable function f and by thinking of f as a $\widetilde{\mathcal{B}(X)}$-measurable function $\tilde{f}$. In view of the Observation that appears before Proposition 3.1.4, the notation $\tilde{\tilde{f}}$ will be used for elements of $\mathbb{M}(X, \widetilde{\mathcal{B}(X)}, \tilde{\mu})$, as well.

We will now discuss the behavior of the Bochner integral when dealing with completions of measure spaces.

As before, let $(Y, \mathcal{Y}, \mu)$ be a σ-finite measure space. Also, let $(Z, \mathcal{Z}, \nu)$ be a σ-finite measure space, let $(Z, \tilde{\mathcal{Z}}, \tilde{\nu})$ be the completion of $(Z, \mathcal{Z}, \nu)$, let $p \in \mathbb{R} \cup \{+\infty\}$, $p \geq 1$, and let $\varphi_p : L^p(Z, \mathcal{Z}, \nu) \to L^p(Z, \tilde{\mathcal{Z}}, \tilde{\nu})$ be the standard isometry.

We can use φ_p to establish a one-to-one correspondence from the set of all $L^p(Z, \mathcal{Z}, \nu)$-valued functions defined on Y onto the set of all $L^p(Z, \tilde{\mathcal{Z}}, \tilde{\nu})$-valued functions defined on Y, as follows: for every $\zeta : Y \to L^p(Z, \mathcal{Z}, \nu)$, let $\tilde{\zeta} : Y \to L^p(Z, \tilde{\mathcal{Z}}, \tilde{\nu})$ be defined by $\tilde{\zeta}(y) = \varphi_p(\zeta(y))$ for every $y \in Y$; that is, $\tilde{\zeta} = \varphi_p \circ \zeta$.

Proposition 3.1.6. *A function $\zeta : Y \to L^p(Z, \mathcal{Z}, \nu)$ is Bochner integrable if and only if $\tilde{\zeta}$ is Bochner integrable. If ζ is Bochner integrable, then $B\text{-}\int_Y \zeta(y)\, d\mu(y) = B\text{-}\int_Y \tilde{\zeta}(y)\, d\mu(y)$.*

Proof. Since φ_p is a surjective isometry, it is obvious that ζ is Bochner measurable if and only if $\tilde{\zeta}$ is Bochner measurable.

The proof of the proposition is completed by observing that ζ is a simple Bochner integrable function if and only if $\tilde{\zeta}$ is a simple Bochner integrable function, and that in this case $B\text{-}\int_Y \zeta(y)\, d\mu(y) = B\text{-}\int_Y \tilde{\zeta}(y)\, d\mu(y)$ and $\int_Y \|\zeta(y)\|\, d\mu(y) = \int_Y \left\|\tilde{\zeta}(y)\right\|\, d\mu(y)$. $\square$

3.1.3   The Dunford-Schwartz Integral

Let $(Y, \mathcal{Y}, \mu)$ be a complete σ-finite measure space, let $(E, \|\cdot\|)$ be a real Banach space, and let $\mathcal{B}(E)$ be the Borel σ-algebra on E.

Let $f : Y \to E$.

We say that f is μ-*essentially separably valued* if there exists a μ-negligible subset N of Y such that $f(Y \setminus N) = \{z \in E \mid z = f(y) \text{ for some } y \in Y \setminus N\}$ is a separable subset of E. Note that in our setting here, the function f is μ-essentially separably valued in the sense of the above definition if and only if f is μ-essentially separably valued in the sense of the definition stated on p. 101 of Dunford and Schwartz [30].

We say that f is *DS-strongly measurable* or *DS-measurable* if f is measurable and μ-essentially separably valued. Note that using Theorem 3.6.10 on p. 148 of [30], we obtain that in our setting, f is DS-measurable if and only if f is measurable in the sense of Definition 3.2.10 on p. 106 of [30].

Clearly, a Bochner measurable function is DS-measurable. As expected, the converse is not true; that is, there exist DS-measurable functions that are not Bochner measurable. Here is a "hands-on" example:

Example 3.1.7. Let $(\mathbb{R}, \mathcal{L}(\mathbb{R}), \lambda)$ be the (complete) measure space defined by the σ-algebra $\mathcal{L}(\mathbb{R})$ of all Lebesgue measurable subsets of $\mathbb{R}$ and the Lebesgue measure λ. Also, as usual, let $\mathcal{M}(\mathbb{R})$ be the Banach space of all real-valued signed Borel measures on $\mathbb{R}$.

Now, let Δ be the standard Cantor set, which is a λ-negligible and uncountable subset of $[0,1]$, and let $f : \mathbb{R} \to \mathcal{M}(\mathbb{R})$ be defined by

$$f(x) = \begin{cases} \delta_x & \text{if } x \in \Delta, \\ 0 & \text{if } x \in \mathbb{R} \setminus \Delta. \end{cases}$$

Then f is DS-measurable but is not Bochner measurable.

Note that many $\mathcal{M}(\mathbb{R})$-valued functions defined on $\mathbb{R}$ are neither Bochner measurable nor DS-measurable. For instance, $g : \mathbb{R} \to \mathcal{M}(\mathbb{R})$, $g(x) = \delta_x$ for every $x \in \mathbb{R}$, is not DS-measurable.   ∎

A function $g : Y \to E$ is called a *DS-null function* if there exists a μ-negligible subset N of Y such that $g(y) = 0$ for every $y \in Y \setminus N$. Note that, in our setting, the DS-null functions are precisely the μ-null functions defined on p. 103, Definition 3.2.3 of [30]. Clearly, a DS-null function is DS-measurable.

A function $f : Y \to E$ is called a *simple measurable function in the sense of Dunford and Schwartz*, or a *DS-simple measurable function* if $f = g + h$ for some DS-null function g and some simple (Bochner) measurable function h. Note that, in our setting a function is a simple measurable function in the sense of Dunford and Schwartz if and only if the function is a μ-simple function in the sense of Definition 3.2.9 on p. 105 of [30].

Let $f : Y \to E$ be a simple measurable function in the sense of Dunford and Schwartz, so $f = g + h$ for some DS-null function g and some simple measurable function h. We say that f is *DS-integrable* if h is Bochner integrable. If f is DS-integrable, then B-$\int_Y h(y)\,d\mu(y)$ is called the *DS-integral of f (on Y with respect to μ)* and is denoted DS-$\int_Y f(y)\,d\mu(y)$. Other notations used in this book are DS-$\int_Y f\,d\mu$, or DS-$\int f\,d\mu$, or DS-$\int f(y)\,d\mu(y)$. Note that the definitions of

DS-integrability and of the DS-integral are correct in the sense that if $f = g' + h'$ is another representation of f as a sum of a DS-null function g' and a simple measurable function h', then h is Bochner integrable if and only if h' is Bochner integrable, and if f is DS-integrable, then B-$\int h \, d\mu = $ B-$\int h' \, d\mu$. Note also that in our setting, f is DS-integrable if and only if f is μ-integrable in the sense of Definition 3.2.13, p. 108 of [30], and if f is DS-integrable, then DS-$\int f \, d\mu$ is the same as the integral defined on p. 108 of [30].

It can be shown that the set of all DS-measurable functions $f : Y \to E$ is a vector space of measurable functions when endowed with pointwise addition and scalar multiplication (see Lemma 3.2.11 on p. 106 of Dunford and Schwartz [30]). We will denote this vector space by $M_{DS}(Y, E)$.

A function $f : Y \to E$ is said to be a *DS-integrable function* if f is DS-measurable and there exists a sequence $(f_n)_{n \in \mathbb{N}}$ of simple measurable functions in the sense of Dunford and Schwartz, $f_n : Y \to E$ for every $n \in \mathbb{N}$, such that the following three conditions are satisfied:

(DS1) f_n is DS-integrable for every $n \in \mathbb{N}$.

(DS2) The sequence $(f_n)_{n \in \mathbb{N}}$ converges in measure to f (note that it makes sense to consider the convergence in measure of $(f_n)_{n \in \mathbb{N}}$ to f because $f \in M_{DS}(Y, E)$ and $f_n \in M_{DS}(Y, E)$ for every $n \in \mathbb{N}$, so $f_n - f, n \in \mathbb{N}$, are measurable functions).

(DS3) The sequence $(f_n)_{n \in \mathbb{N}}$ is Cauchy in the mean in the following sense: for every $\varepsilon \in \mathbb{R}$, $\varepsilon > 0$, there exists a $k_\varepsilon \in \mathbb{N}$ such that

$$\int_Y \| f_k(y) - f_l(y) \| \, d\mu(y) < \varepsilon \tag{3.1.2}$$

for every $k \geq k_\varepsilon$ and $l \geq k_\varepsilon$ (it is easy to see that the integral that appears on the left-hand side of (3.1.2) is well-defined; that is, the real-valued function $y \mapsto \| f_k(y) - f_l(y) \|$, $y \in Y$, is integrable in the sense of classical integration theory because $f_k - f_l$ is the sum of a DS-null function and a simple measurable Bochner integrable function for every $k \in \mathbb{N}$ and $l \in \mathbb{N}$).

Let $f : Y \to E$ be a DS-integrable function, and let $(f_n)_{n \in \mathbb{N}}$ be a sequence of simple measurable functions in the sense of Dunford and Schwartz, $f_n : Y \to E$ for every $n \in \mathbb{N}$, such that f and $(f_n)_{n \in \mathbb{N}}$ satisfy the conditions (DS1), (DS2) and (DS3) of the definition of DS-integrability. Using Lemma 3.2.15 on pp. 109–110 of Dunford and Schwartz [30], we obtain that $\| \text{DS-}\int_Y f_k \, d\mu - \text{DS-}\int_Y f_l \, d\mu \| \leq \int_Y \| f_k - f_l \| \, d\mu$ for every $k \in \mathbb{N}$ and $l \in \mathbb{N}$; therefore, $(\text{DS-}\int_Y f_n \, d\mu)_{n \in \mathbb{N}}$ is a Cauchy sequence of elements of E with respect to the norm on E, so the sequence $(\text{DS-}\int_Y f_n \, d\mu)_{n \in \mathbb{N}}$ converges in the norm topology of E. The limit $\lim_{n \to \infty} \text{DS-}\int_Y f_n \, d\mu$ is called the *Dunford and Schwartz integral*, or the *DS-integral of f (with respect to μ)*, and is denoted by DS-$\int_Y f(y) \, d\mu(y)$, or DS-$\int f(y) \, d\mu(y)$, or DS-$\int_Y f \, d\mu$, or DS-$\int f \, d\mu$. Using Lemma 3.2.16 on pp. 111–112 of Dunford and Schwartz [30], we obtain that the definition of the DS-integral of f is correct in the sense that DS-$\int f \, d\mu$ does not depend on the particular sequence $(f_n)_{n \in \mathbb{N}}$ used to calculate DS-$\int f \, d\mu$; that is, if $(f'_n)_{n \in \mathbb{N}}$ is another sequence of simple measurable functions in the sense of Dunford and Schwartz, $f'_n : Y \to E$ for every $n \in \mathbb{N}$, such that the conditions (DS1), (DS2) and (DS3) are also satisfied by $(f'_n)_{n \in \mathbb{N}}$ and f, then the sequence (DS-$\int f'_n \, d\mu)_{n \in \mathbb{N}}$ converges, as well, and $\lim_{n \to \infty}$ DS-$\int f_n \, d\mu = \lim_{n \to \infty}$ DS-$\int f'_n \, d\mu$.

Using Proposition 3.1.3, we obtain that, in the setting of this subsection, if $f : Y \to E$ is a Bochner integrable function, then f is DS-integrable. For future reference, we state this fact in the next proposition:

Proposition 3.1.8. *If $f : Y \to E$ is Bochner integrable, then f is also DS-integrable and DS-$\int_Y f \, d\mu = B\text{-}\int_Y f \, d\mu$.*

Note that the DS-integrability of a function $f : Y \to E$ does not imply that f is Bochner integrable because there exist DS-measurable functions that fail to be Bochner measurable (see Example 3.1.7).

We will need the following lemma:

Lemma 3.1.9. *Under the assumption that $E = \mathbb{R}$, a function $f : Y \to \mathbb{R}$ is DS-integrable if and only if f is integrable in the sense of classical measure theory. If f is DS-integrable, then DS-$\int_Y f \, d\mu = \int_Y f \, d\mu$.*

Proof. Assume that $f : Y \to \mathbb{R}$ is integrable in the classical sense. Then, as pointed out in Sect. 3.1.1 before recalling the definition of convergence in measure, we obtain that f is Bochner integrable, so, by Proposition 3.1.8, f is DS-integrable and DS-$\int f \, d\mu = \int f \, d\mu$.

Now assume that f is DS-integrable, and let $(f_n)_{n \in \mathbb{N}}$ be a sequence of simple measurable functions in the sense of Dunford and Schwartz such that f and $(f_n)_{n \in \mathbb{N}}$ satisfy the conditions (DS1), (DS2) and (DS3) of the definition of DS-integrability of a function. Then $(f_n)_{n \in \mathbb{N}}$ is a sequence of integrable functions in the classical sense. Since $(f_n)_{n \in \mathbb{N}}$ satisfies condition (DS3), it follows that there exists an integrable function $g : Y \to \mathbb{R}$ such that $(f_n)_{n \in \mathbb{N}}$ converges to g in the mean (that is, such that the sequence $(\int_Y |f_n - g| \, d\mu)_{n \in \mathbb{N}}$ converges to zero). Using Propositions 3.1.4 and 3.1.2 of Cohn [20] we obtain that there exists a subsequence $(f_{n_k})_{k \in \mathbb{N}}$ of $(f_n)_{n \in \mathbb{N}}$ such that $(f_{n_k})_{k \in \mathbb{N}}$ converges to g, μ-a.e. Using Proposition 3.1.2 of Cohn [20] again, we obtain that there exists a further subsequence $(f_{n_{k_l}})_{l \in \mathbb{N}}$ of $(f_{n_k})_{k \in \mathbb{N}}$ such that $(f_{n_{k_l}})_{l \in \mathbb{N}}$ converges μ-a.e. to f. Thus, $f = g$, μ-a.e. Since g is integrable in the classical sense, it follows that f is integrable in the classical sense, as well. $\square$

For every $f : Y \to E$, let $g_f : Y \to \mathbb{R}$ be defined by $g_f(y) = \|f(y)\|$ for every $y \in Y$. Also, let $\mathcal{L}^1_{DS}(Y, \mathcal{Y}, \mu, E)$ be the set of all DS-integrable functions $f : Y \to E$. On $\mathcal{L}^1_{DS}(Y, \mathcal{Y}, \mu, E)$ we can define the almost everywhere equality relation $\sim$ in the usual manner: $f_1 \sim f_2$ if, by definition, $f_1 = f_2$ μ-a.e.; obviously, $\sim$ is an equivalence relation, and we denote by $L^1_{DS}(Y, \mathcal{Y}, \mu, E)$ the set of all the equivalence classes of elements of $\mathcal{L}^1_{DS}(Y, \mathcal{Y}, \mu, E)$ defined by the relation $\sim$.

Proposition 3.1.10. (a) *For every $f \in \mathcal{L}^1_{DS}(Y, \mathcal{Y}, \mu, E)$, the function $g_f : Y \to$*
$\mathbb{R}$, $g_f(y) = \|f(y)\|$ for every $y \in Y$ is both DS-integrable and integrable in
the sense of classical measure theory.

(b) *The map $\|\|\cdot\|\|^{(DS)}_{1E} : \mathcal{L}^1_{DS}(Y, \mathcal{Y}, \mu, E) \to \mathbb{R}$, $\|\|f\|\|^{(DS)}_{1E} = \int g_f \, d\mu$ for every*
$f \in \mathcal{L}^1_{DS}(Y, \mathcal{Y}, \mu, E)$, is well-defined and $(\mathcal{L}^1_{DS}(Y, \mathcal{Y}, \mu, E), \|\|\cdot\|\|^{(DS)}_{1E})$ is a
seminormed vector space.

(c) *If we use the seminorm $\|\|\cdot\|\|^{(DS)}_{1E}$ on $\mathcal{L}^1_{DS}(Y, \mathcal{Y}, \mu, E)$ to define a real-*
valued function on $L^1_{DS}(Y, \mathcal{Y}, \mu, E)$, also denoted by $\|\|\cdot\|\|^{(DS)}_{1E}$, as follows:
$\|\|\bar{f}\|\|^{(DS)}_{1E} = \int g_f \, d\mu$ for every $\bar{f} \in L^1_{DS}(Y, \mathcal{Y}, \mu, E)$, where $f \in$
$\mathcal{L}^1_{DS}(Y, \mathcal{Y}, \mu, E)$ is an element in the equivalence class $\bar{f}$, then this new
function is a well-defined norm on $L^1_{DS}(Y, \mathcal{Y}, \mu, E)$. Moreover, when endowed
with this norm, $L^1_{DS}(Y, \mathcal{Y}, \mu, E)$ is a Banach space.

Proof. (a) Let $f \in \mathcal{L}^1_{DS}(Y, \mathcal{Y}, \mu, E)$. Then using (a) of Theorem 3.2.22 on
pp. 117–118 of Dunford and Schwartz [30], we obtain that g_f is DS-integrable.
In view of Lemma 3.1.9, it follows that g_f is also integrable in the sense of
classical measure theory.

Note that using (a) of Theorem 3.2.22 on pp. 117–118 of [30], we obtain
that a DS-measurable function $f : Y \to E$ belongs to $\mathcal{L}^1_{DS}(Y, \mathcal{Y}, \mu, E)$
if and only if g_f is DS-integrable. Thus, $\mathcal{L}^1_{DS}(Y, \mathcal{Y}, \mu, E)$ is precisely the
space $L^0_1(Y, \mathcal{Y}, \mu, E)$ introduced in Definition 3.3.1 on p. 119 of Dunford and
Schwartz [30].

(b) Use (a) and (b) of Lemma 3.3.3 on pp. 120–121 of Dunford and Schwartz [30].

(c) See Theorems 3.3.5 and 3.6.6 on p. 121 and pp. 146–147 of [30], respectively.

$\square$

Note that using Lemma 3.1.8, we obtain that if $E = \mathbb{R}$, then $\mathcal{L}^1_{DS}(Y, \mathcal{Y}, \mu, \mathbb{R}) =$
$\mathcal{L}^1(Y, \mathcal{Y}, \mu)$ and $L^1_{DS}(Y, \mathcal{Y}, \mu, \mathbb{R}) = L^1(Y, \mathcal{Y}, \mu)$.

We will now discuss a useful relationship between linear bounded operators and
DS-integrable functions.

Let $T : E \to E$ be a linear bounded operator. For every $f \in \mathcal{L}^1_{DS}(Y, \mathcal{Y}, \mu, E)$,
let $h^{(T)}_f : Y \to E$ be defined by $h^{(T)}_f(y) = T(f(y))$ for every $y \in Y$.

Theorem 3.1.11. *For every $f \in \mathcal{L}^1_{DS}(Y, \mathcal{Y}, \mu, E)$, the function $h^{(T)}_f$ is DS-*
integrable, and

$$T\left(DS\text{-}\int_Y f \, d\mu\right) = DS\text{-}\int_Y h^{(T)}_f \, d\mu. \tag{3.1.3}$$

For a proof, see (c) of Theorem 3.2.19, pp. 113–114 of Dunford and Schwartz
[30].

Informally, it is the custom to use the notation $Tf(\cdot)$ for $h^{(T)}_f$, and in this case the
equality (3.1.3) becomes $T(DS\text{-}\int_Y f(y) \, d\mu(y)) = DS\text{-}\int_Y Tf(y) \, d\mu(y)$.

3.2   Almost Everywhere Convergence and the Dunford-Schwartz Theorem

Our goal in this section is to discuss an application of a result of Dunford and Schwartz (Theorem 8.7.5, p. 690 of [30]) to a certain one-parameter semigroup of operators (for a systematic study of one-parameter semigroups of operators, see Engel and Nagel's monograph [34]) defined in terms of an invariant probability measure of a transition function (assuming, of course, that the transition function has invariant probabilities). The application has several consequences, also discussed in this section.

The section is organized into two subsections. In the first subsection, we discuss several basic facts about pointwise and almost everywhere convergence. In the second subsection, we define the above-mentioned semigroup of operators, discuss some of its properties, and apply Theorem 8.7.5, p. 690 of Dunford and Schwartz [30], to the semigroup of operators in order to obtain various useful results.

3.2.1   *Pointwise and Almost Everywhere Convergence*

Let $c \in \mathbb{R}$, let $A \subseteq \mathbb{R}$ be such that $(c, +\infty) \subseteq A$, and let $f : A \to \mathbb{R}$.

As usual, we say that the *limit of f as t tends to $+\infty$ exists and is a real number* if there exists an $L \in \mathbb{R}$ such that the following condition is satisfied: for every $\varepsilon \in \mathbb{R}$, $\varepsilon > 0$, there exists a $t_\varepsilon \in A$ such that $|f(t) - L| < \varepsilon$ for every $t \in A, t \geq t_\varepsilon$. The number L is called the *limit of f as t tends to $+\infty$* and is usually denoted $\lim_{t \to +\infty} f(t)$.

Given a dense subset D of A, we say that the *limit of f as t tends to $+\infty$ along D exists and is a real number* if there exists a real number L_D that has the following property: for every $\varepsilon \in \mathbb{R}$, $\varepsilon > 0$, there exists a $t_\varepsilon \in D$ such that $|f(t) - L_D| < \varepsilon$ for every $t \in D, t \geq t_\varepsilon$. The number L_D is denoted $\displaystyle\lim_{\substack{t \to +\infty \\ t \in D}} f(t)$ and is called the *limit of f as t tends to $+\infty$ along D*.

In the next lemma and throughout the book we will use the following notation:

$$\liminf_{\substack{t \to +\infty \\ t \in A}} f(t) = \sup_{s \in A} \inf_{\substack{t \geq s \\ t \in A}} f(t),$$

$$\limsup_{\substack{t \to +\infty \\ t \in A}} f(t) = \inf_{s \in A} \sup_{\substack{t \geq s \\ t \in A}} f(t),$$

$$\liminf_{\substack{t \to +\infty \\ t \in D}} f(t) = \sup_{s \in D} \inf_{\substack{t \geq s \\ t \in D}} f(t),$$

and

$$\limsup_{\substack{t \to +\infty \\ t \in D}} f(t) = \inf_{s \in D} \sup_{\substack{t \geq s \\ t \in D}} f(t),$$

where D is a dense subset of the domain A of f.

Lemma 3.2.1. *Let $c \in \mathbb{R}$, let A be a subset of $\mathbb{R}$ such that $(c, +\infty) \subseteq A$, let D be a dense subset of A, and let $f : A \to \mathbb{R}$ be a continuous function. The following assertions are equivalent:*

(a) $\lim_{t \to +\infty} f(t)$ exists and is a real number.

(b) $\lim_{\substack{d \to +\infty \\ d \in D}} f(d)$ exists and is a real number.

(c) $\liminf_{\substack{d \to +\infty \\ d \in D}} f(d) = \limsup_{\substack{d \to +\infty \\ d \in D}} f(d)$ and $\liminf_{\substack{d \to +\infty \\ d \in D}} f(d)$ is a real number.

If $\lim_{t \to +\infty} f(t)$ exists and is a real number L, then

$$L = \lim_{\substack{d \to +\infty \\ d \in D}} f(d) = \liminf_{\substack{d \to +\infty \\ d \in D}} f(d) = \limsup_{\substack{d \to +\infty \\ d \in D}} f(d).$$

Proof. The implication $(a) \Rightarrow (b)$ and the fact that the equality $\lim_{t \to +\infty} f(t) = \lim_{\substack{d \to +\infty \\ d \in D}} f(d)$ holds true whenever assertion (a) is true are obvious, even if f is not continuous.

$(b) \Rightarrow (a)$ Assume that $\lim_{\substack{d \to +\infty \\ d \in D}} f(d)$ exists and is a real number L. Also, let $\varepsilon \in \mathbb{R}$, $\varepsilon > 0$. Then there exists a $d_\varepsilon \in D$ such that $|f(d) - L| < \frac{\varepsilon}{2}$ for every $d \in D, d \geq d_\varepsilon$.

Let $t \in A, t > d_\varepsilon$. Since f is continuous and D is dense in A, there exists a $d \in D, d > t$, such that $|f(t) - f(d)| < \frac{\varepsilon}{2}$. Since $d > d_\varepsilon$, it follows that $|f(t) - L| \leq |f(t) - f(d)| + |f(d) - L| < \frac{\varepsilon}{2} + \frac{\varepsilon}{2} = \varepsilon$.

We conclude that $\lim_{t \to +\infty} f(t) = L$ because we have shown that for every $\varepsilon \in \mathbb{R}$, $\varepsilon > 0$, there exists a $d_\varepsilon \in A$ such that $|f(t) - L| < \varepsilon$ for every $t \in A$, $t \geq d_\varepsilon$.

$(b) \Rightarrow (c)$ Assume that $\lim_{\substack{d \to +\infty \\ d \in D}} f(d)$ exists and is a real number L. We will prove that

$$\liminf_{\substack{d\to+\infty\\ d\in D}} f(d) = \limsup_{\substack{d\to+\infty\\ d\in D}} f(d) = L.$$

To this end, let $\varepsilon \in \mathbb{R}$, $\varepsilon > 0$. Then there exists a $d_\varepsilon \in D$ such that $|f(d) - L| < \frac{\varepsilon}{2}$ for every $d \in D$, $d \geq d_\varepsilon$. Thus, $-\frac{\varepsilon}{2} + L < f(d) < \frac{\varepsilon}{2} + L$ for every $d \in D$, $d \geq d_\varepsilon$. Consequently,

$$-\frac{\varepsilon}{2} + L \leq \inf_{\substack{d\in D\\ d\geq c}} f(d) < \frac{\varepsilon}{2} + L \quad \text{and} \quad -\frac{\varepsilon}{2} + L < \sup_{\substack{d\in D\\ d\geq c}} f(d) \leq \frac{\varepsilon}{2} + L$$

for every $c \in D$, $c \geq d_\varepsilon$. Since the maps $c \mapsto \inf_{\substack{d\in D\\ d\geq c}} f(d)$, $c \in D$, and $c \mapsto \sup_{\substack{d\in D\\ d\geq c}} f(d)$, $c \in D$, are increasing and decreasing, respectively, it follows that

$$-\frac{\varepsilon}{2} + L \leq \sup_{c\in D}\inf_{\substack{d\in D\\ d\geq c}} f(d) \leq \frac{\varepsilon}{2} + L \quad \text{and} \quad -\frac{\varepsilon}{2} + L \leq \inf_{c\in D}\sup_{\substack{d\in D\\ d\geq c}} f(d) \leq \frac{\varepsilon}{2} + L.$$

We obtain that the implication holds true because we have proved that

$$\left|\liminf_{\substack{d\to+\infty\\ d\in D}} f(d) - L\right| \leq \frac{\varepsilon}{2} < \varepsilon \quad \text{and} \quad \left|\limsup_{\substack{d\to+\infty\\ d\in D}} f(d) - L\right| \leq \frac{\varepsilon}{2} < \varepsilon$$

for every $\varepsilon \in \mathbb{R}$, $\varepsilon > 0$.

$(c) \Rightarrow (b)$ Assume that there exists a real number L such that

$$L = \liminf_{\substack{d\to+\infty\\ d\in D}} f(d) = \limsup_{\substack{d\to+\infty\\ d\in D}} f(d),$$

and let $\varepsilon \in \mathbb{R}$, $\varepsilon > 0$.

Again using the fact that the maps $c \mapsto \inf_{\substack{d \in D \\ d \geq c}} f(d)$, $c \in D$, and $c \mapsto \sup_{\substack{d \in D \\ d \geq c}} f(d)$,

$c \in D$, are increasing and decreasing, respectively, and since, by our assumption, $L - \varepsilon < \sup_{c \in D} \inf_{\substack{d \in D \\ d \geq c}} f(d) = L$ and $L = \inf_{c \in D} \sup_{\substack{d \in D \\ d \geq c}} f(d) < L + \varepsilon$, we obtain that there

exist $d_\varepsilon^{(1)} \in D$ and $d_\varepsilon^{(2)} \in D$ such that $L - \varepsilon < \inf_{\substack{d \in D \\ d \geq c}} f(d) \leq L < L + \varepsilon$ for every

$c \in D, c \geq d_\varepsilon^{(1)}$, and such that $L - \varepsilon < L \leq \sup_{\substack{d \in D \\ d \geq c}} f(d) < L + \varepsilon$ for every $c \in D$,

$c \geq d_\varepsilon^{(2)}$.

If we set $d_\varepsilon = \max\{d_\varepsilon^{(1)}, d_\varepsilon^{(2)}\}$, then $L - \varepsilon < f(d) < L + \varepsilon$ for every $d \in D$, $d \geq d_\varepsilon$.

We have therefore proved that $\lim_{\substack{d \to +\infty \\ d \in D}} f(d)$ exists and is equal to L. $\square$

Let (X, d) be a locally compact separable metric space, let $\mathbb{T}$ stand for $\mathbb{R}$ or for $[0, +\infty)$, let $(P_t)_{t \in \mathbb{T}}$ be a transition function that satisfies the s.m.a., and let $((S_t, T_t))_{t \in \mathbb{T}}$ be the family of Markov pairs defined by $(P_t)_{t \in \mathbb{T}}$.

Now, let $f \in B_b(X)$ and $x \in X$. By Corollary 2.1.6, the function $\psi_f^{(x)} : \mathbb{T} \to \mathbb{R}$ defined by $\psi_f^{(x)}(t) = S_t f(x)$ for every $t \in \mathbb{T}$ is measurable (with respect to the Borel σ-algebra on $\mathbb{R}$ and the σ-algebra $\mathcal{L}(\mathbb{T})$ of all Lebesgue measurable subsets of $\mathbb{T}$). Since $\psi_f^{(x)}$ is also bounded (because $|S_t f(x)| \leq \|f\|$ for every $t \in \mathbb{T}$), it follows that $\psi_f^{(x)}$ is integrable over any interval of the form $[0, s]$, $s \in \mathbb{T}$, $s > 0$. Therefore, the function $\phi_f^{(x)} : (0, +\infty) \to \mathbb{R}$, $\phi_f^{(x)}(s) = \frac{1}{s} \int_0^s S_t f(x)\, dt$ for every $s \in (0, +\infty)$, is well defined.

Lemma 3.2.2. *For every $f \in B_b(X)$ and $x \in X$, the function $\phi_f^{(x)}$ is continuous.*

Proof. Let $f \in B_b(X)$ and $x \in X$.

Clearly, the lemma is true if f is the constant zero function, so we may and do assume that $f \neq 0$.

Let $s \in (0, +\infty)$. We have to prove that for every sequence $(s_n)_{n \in \mathbb{N}}$ of elements of $(0, +\infty)$ that converges to s, the sequence $(\frac{1}{s_n} \int_0^{s_n} S_t f(x)\, dt)_{n \in \mathbb{N}}$ converges to $\frac{1}{s} \int_0^s S_t f(x)\, dt$. Clearly, we may assume that the sequence $(s_n)_{n \in \mathbb{N}}$ has the property that $s_n \geq \frac{s}{2}$ for every $n \in \mathbb{N}$.

Thus, let $(s_n)_{n \in \mathbb{N}}$ be such a sequence, let $\varepsilon \in \mathbb{R}$, $\varepsilon > 0$, and let n_ε be such that $|s - s_n| < \frac{s\varepsilon}{4\|f\|}$ for every $n \geq n_\varepsilon$.

Using the fact that S_t, $t \in \mathbb{T}$, are contractions of $B_b(X)$, we obtain that

$$\left| \frac{1}{s_n} \int_0^{s_n} S_t f(x) \, dt - \frac{1}{s} \int_0^s S_t f(x) \, dt \right|$$

$$\leq \left| \frac{1}{s_n} \int_0^{s_n} S_t f(x) \, dt - \frac{1}{s_n} \int_0^s S_t f(x) \, dt \right| + \left| \frac{s - s_n}{s s_n} \int_0^s S_t f(x) \, dt \right|$$

$$\leq \frac{1}{s_n} \int_{I_n} |S_t f(x)| \, dt + \frac{|s - s_n|}{s s_n} \int_0^s |S_t f(x)| \, dt \leq \frac{|s - s_n|}{\frac{s}{2}} \|f\| + \frac{|s - s_n|}{\frac{s^2}{2}} s \|f\| < \varepsilon$$

for every $n \in \mathbb{N}$, $n \geq n_\varepsilon$, where I_n is the closed interval with endpoints s_n and s.

We have therefore proved that the sequence $(\frac{1}{s_n} \int_0^{s_n} S_t f(x) \, dt)_{n \in \mathbb{N}}$ converges to $\frac{1}{s} \int_0^s S_t f(x) \, dt$. $\qquad\square$

Using Lemmas 3.2.1 and 3.2.2, we obtain the following proposition:

Proposition 3.2.3. *Let $f \in B_b(X)$ and $x \in X$. Also, let D be the set of all rational numbers in $(0, +\infty)$. The following assertions are equivalent:*

(a) The limit $\lim_{s \to +\infty} \frac{1}{s} \int_0^s S_t f(x) \, dt$ exists and is a real number.

(b) The limit $\lim\limits_{\substack{d \to +\infty \\ d \in D}} \frac{1}{d} \int_0^d S_t f(x) \, dt$ exists and is a real number.

(c) $\liminf\limits_{\substack{d \to +\infty \\ d \in D}} \frac{1}{d} \int_0^d S_t f(x) \, dt = \limsup\limits_{\substack{d \to +\infty \\ d \in D}} \frac{1}{d} \int_0^d S_t f(x) \, dt$ and

$\liminf\limits_{\substack{d \to +\infty \\ d \in D}} \frac{1}{d} \int_0^d S_t f(x) \, dt$ *is a real number.*

If (a) holds true and we set $L = \lim_{s \to +\infty} \frac{1}{s} \int_0^s S_t f(x) \, dt$, then

$$L = \lim\limits_{\substack{d \to +\infty \\ d \in D}} \frac{1}{d} \int_0^d S_t f(x) \, dt = \liminf\limits_{\substack{d \to +\infty \\ d \in D}} \frac{1}{d} \int_0^d S_t f(x) \, dt$$

$$= \limsup\limits_{\substack{d \to +\infty \\ d \in D}} \frac{1}{d} \int_0^d S_t f(x) \, dt.$$

The proof of the proposition is obvious.

For every $f \in B_b(X)$ and $s \in \mathbb{T}$, $s > 0$, let $\phi_f^{(s)} : X \to \mathbb{R}$ be defined by $\phi_f^{(s)}(x) = \frac{1}{s} \int_0^s S_t f(x) \, dt$ for every $x \in X$.

Lemma 3.2.4. *The functions* $\phi_f^{(s)}$, $f \in B_b(X)$, $s \in \mathbb{T}$, $s > 0$, *belong to* $B_b(X)$.

Proof. Since S_t, $t \in \mathbb{T}$, are contractions of $B_b(X)$, it follows that $\left|\phi_f^{(s)}(x)\right| \le \frac{1}{s}\int_0^s |S_t f(x)|\, dt \le \frac{1}{s} s \|f\| = \|f\|$ for every $f \in B_b(X)$, $s \in \mathbb{T}$, $s > 0$, and $x \in X$; therefore, $\phi_f^{(s)}$ is a bounded function whenever $f \in B_b(X)$ and $s \in \mathbb{T}$, $s > 0$.

It remains to prove that the functions $\phi_f^{(s)}$, $f \in B_b(X)$, $s \in \mathbb{T}$, $s > 0$, are measurable.

To this end, let $s \in \mathbb{T}$, $s > 0$.

Using (i) of Lemma 1.1.1, we obtain that

$$\phi_{\mathbf{1}_A}^{(s)} = \frac{1}{s}\int_0^s S_t \mathbf{1}_A(x)\, dt = \frac{1}{s}\int_0^s P_t(x, A)\, dt$$

whenever $f = \mathbf{1}_A$ for some $A \in \mathcal{B}(X)$. Since $(P_t)_{t\in\mathbb{T}}$ satisfies the s.m.a., by (a) of Proposition 5.2.1 on p. 159 of Cohn [20], we obtain that $\phi_{\mathbf{1}_A}^{(s)}$ is a measurable function for every $A \in \mathcal{B}(X)$. (Note that even though in Proposition 5.2.1 of [20] one deals with two measure spaces, the assertion that is of interest to us is true even if we consider a measure space and a measurable space rather than two measure spaces.)

In view of the previous paragraph, it is easy to see that if f is a real-valued simple measurable function, $f = \sum_{i=1}^n a_i \mathbf{1}_{A_i}$ for some $n \in \mathbb{N}$, n real numbers $a_1, a_2, \cdots, a_n$, and n measurable subsets $A_1, A_2, \cdots, A_n$ of X, then $\phi_f^{(s)}$ is measurable.

Finally, if $f \in B_b(X)$ is not necessarily a simple function, then there exists a sequence $(f_k)_{k\in\mathbb{N}}$ of real-valued simple measurable functions such that $(f_k)_{k\in\mathbb{N}}$ converges to f in the norm topology of $B_b(X)$. Thus, $(f_k)_{k\in\mathbb{N}}$ is a bounded sequence with respect to the norm of $B_b(X)$. Since S_t, $t \in [0, 1]$, are contractions of $B_b(X)$, there exists an $M \in \mathbb{R}$, $M > 0$, such that for every $k \in \mathbb{N}$, $|S_t f_k(x)| \le M$ and $|S_t f(x)| \le M$ for every $t \in [0, s]$ and $x \in X$.

Now, given $x \in X$, the sequence of functions $(g_k^{(x)})_{k\in\mathbb{N}}$, $g_k^{(x)} : [0, s] \to \mathbb{R}$, $g_k^{(x)}(t) = S_t f_k(x)$ for every $t \in [0, s]$, $k \in \mathbb{N}$, converges pointwise (everywhere) on $[0, s]$ to the function $g^{(x)} : [0, s] \to \mathbb{R}$ defined by $g^{(x)}(t) = S_t f(x)$ for every $t \in [0, s]$, and $\left|g_k^{(x)}\right| \le M\mathbf{1}_{[0,s]}$ for every $k \in \mathbb{N}$, so we can apply the dominated convergence theorem to the sequence $(g_k^{(x)})_{k\in\mathbb{N}}$ in order to obtain that the sequence $(\phi_{f_k}^{(s)}(x))_{k\in\mathbb{N}}$ converges to $\phi_f^{(s)}(x)$.

Since the functions $\phi_{f_k}^{(s)}$, $k \in \mathbb{N}$, are measurable and the sequence $(\phi_{f_k}^{(s)})_{k\in\mathbb{N}}$ converges pointwise on X to $\phi_f^{(s)}$, it follows that $\phi_f^{(s)}$ is measurable. $\square$

We will now conclude our discussion about pointwise convergence with a proposition that will be used often throughout the book.

Proposition 3.2.5. *Let $f \in B_b(X)$ and set*

$$A_{\overrightarrow{f}} = \left\{ x \in X \ \middle| \ \lim_{s \to +\infty} \frac{1}{s} \int_0^s S_t f(x)\, dt \ \text{exists} \right\}.$$

Also, let $f^{\to} : X \to \mathbb{R}$ be the function defined by

$$f^{\to}(x) = \begin{cases} \lim_{s \to +\infty} \frac{1}{s} \int_0^s S_t f(x)\, dt & \text{if } x \in A_{\overrightarrow{f}}, \\ 0 & \text{if } x \in X \setminus A_{\overrightarrow{f}}. \end{cases}$$

Then $A_{\overrightarrow{f}} \in \mathcal{B}(X)$ and $f^{\to} \in B_b(X)$.

Proof. Let $f \in B_b(X)$.

Using Proposition 3.2.3, we obtain that

$$A_{\overrightarrow{f}} = \left\{ x \in X \ \middle| \ \liminf_{\substack{q \to +\infty \\ q \in D}} \frac{1}{q} \int_0^q S_t f(x)\, dt = \limsup_{\substack{q \to +\infty \\ q \in D}} \frac{1}{q} \int_0^q S_t f(x)\, dt \right\}$$

$$= \left\{ x \in X \ \middle| \ \sup_{\substack{q \in \mathbb{Q} \\ q > 0}} \inf_{\substack{r \geq q \\ r \in \mathbb{Q}}} \frac{1}{r} \int_0^r S_t f(x)\, dt = \inf_{\substack{q \in \mathbb{Q} \\ q > 0}} \sup_{\substack{r \geq q \\ r \in \mathbb{Q}}} \frac{1}{r} \int_0^r S_t f(x)\, dt \right\},$$

where D stands for the set of all rational numbers in the interval $(0, +\infty)$.

Since by Lemma 3.2.4, the functions $\phi_f^{(q)}$, $q \in \mathbb{Q}$, $q > 0$, are measurable, it follows that $\inf_{\substack{r \geq q \\ r \in \mathbb{Q}}} \phi_f^{(r)}$, $q \in \mathbb{Q}$, $q > 0$, and, consequently, $\sup_{\substack{q \in \mathbb{Q} \\ q > 0}} \inf_{\substack{r \geq q \\ r \in \mathbb{Q}}} \phi_f^{(r)}$ are measurable functions (because we take the inf and the sup of countably many measurable functions, so we can apply Proposition 2.1.4 of Cohn [20]). Thus, the function $x \mapsto \sup_{\substack{q \in \mathbb{Q} \\ q > 0}} \inf_{\substack{r \geq q \\ r \in \mathbb{Q}}} \frac{1}{r} \int_0^r S_t f(x)\, dt$, $x \in X$, is measurable. In a similar manner, we obtain that the function $x \mapsto \inf_{\substack{q \in \mathbb{Q} \\ q > 0}} \sup_{\substack{r \geq q \\ r \in \mathbb{Q}}} \frac{1}{r} \int_0^r S_t f(x)\, dt$, $x \in X$, is also measurable.

It follows that the set $A_{\overrightarrow{f}}$ is measurable because it is the set of all $x \in X$ at which two measurable functions have real values and are equal.

Since from the above discussion it follows that the restriction of $f^{\rightarrow}$ to $A_f^{\rightarrow}$ is measurable, it follows that $f^{\rightarrow}$ is a measurable function. $\qquad\square$

We will now discuss several facts about almost everywhere convergence that we will need in this section.

Let $(Y, \mathcal{Y}, \mu)$ be a measure space, and, as usual, let $\mathbb{T}$ stand for the additive metric group $\mathbb{R}$, or else the additive metric semigroup $[0, +\infty)$.

Given a family $(f_\alpha)_{\alpha \in \mathbb{T}}$ of real-valued $\mathcal{Y}$-measurable functions defined on Y, we say that $(f_\alpha)_{\alpha \in \mathbb{T}}$ *converges almost everywhere as* $\alpha \rightarrow +\infty$ if there exists a real-valued $\mathcal{Y}$-measurable function f on Y and a μ-negligible subset N of Y such that $\lim_{\alpha \rightarrow +\infty} f_\alpha(x)$ exists and is equal to $f(x)$ for every $x \in Y \setminus N$. We say that f is a μ-*a.e. limit of* $(f_\alpha)_{\alpha \in \mathbb{T}}$ *as* $\alpha \rightarrow +\infty$, or that $(f_\alpha)_{\alpha \in \mathbb{T}}$ *converges* μ-*a.e. to* f *as* $\alpha \rightarrow +\infty$. Note that the μ-a.e. limit of $(f_\alpha)_{\alpha \in \mathbb{T}}$ as $\alpha \rightarrow +\infty$ is unique μ-a.e. in the sense that if f and g are μ-a.e. limits of $(f_\alpha)_{\alpha \in \mathbb{T}}$ as $\alpha \rightarrow +\infty$, then $f = g$ μ-a.e.

Naturally, the above definitions can be extended in an obvious way to families $(f_\alpha)_{\alpha \in I}$, where I is a set of the form $[a, +\infty)$ or $(a, +\infty)$ for some $a \in \mathbb{R}$. For instance, in this section, we will deal with the almost everywhere convergence as $\alpha \rightarrow +\infty$ of a family $(f_\alpha)_{\alpha \in (0,+\infty)}$.

In the next lemma we discuss the relationship between the convergence μ-a.e. of a family $(f_\alpha)_{\alpha \in \mathbb{T}}$ of real-valued $\mathcal{Y}$-measurable functions on Y as $\alpha \rightarrow +\infty$ and the convergence $\tilde{\mu}$-a.e. of $(f_\alpha)_{\alpha \in \mathbb{T}}$ as $\alpha \rightarrow +\infty$ in the completion $(Y, \tilde{\mathcal{Y}}, \tilde{\mu})$ of $(Y, \mathcal{Y}, \mu)$.

Lemma 3.2.6. *Let* $(Y, \tilde{\mathcal{Y}}, \tilde{\mu})$ *be the completion of the measure space* $(Y, \mathcal{Y}, \mu)$, *and let* $(f_\alpha)_{\alpha \in \mathbb{T}}$ *be a family of real-valued* $\mathcal{Y}$-*measurable functions defined on* Y. *The following assertions are equivalent:*

(a) *The family* $(f_\alpha)_{\alpha \in \mathbb{T}}$ *converges* μ-*a.e. as* $\alpha \rightarrow +\infty$.
(b) *The family* $(f_\alpha)_{\alpha \in \mathbb{T}}$, *thought of as a family of* $\tilde{\mathcal{Y}}$-*measurable functions on* Y, *converges* $\tilde{\mu}$-*a.e. as* $\alpha \rightarrow +\infty$.

Proof. $(a) \Rightarrow (b)$ is obvious. Indeed, if f is a μ-a.e. limit of $(f_\alpha)_{\alpha \in \mathbb{T}}$, then f is $\tilde{\mathcal{Y}}$-measurable (because f is $\mathcal{Y}$-measurable), and there exists a μ-negligible subset N of Y such that $(f_\alpha(x))_{\alpha \in \mathbb{T}}$ converges as $\alpha \rightarrow +\infty$ for every $x \in Y \setminus N$. Since N is also a $\tilde{\mu}$-negligible subset of Y, it follows that $(f_\alpha)_{\alpha \in \mathbb{T}}$ converges to f $\tilde{\mu}$-a.e.

$(b) \Rightarrow (a)$ Assume that $(f_\alpha)_{\alpha \in \mathbb{T}}$ converges $\tilde{\mu}$-a.e. as $\alpha \rightarrow +\infty$, and let f be a $\tilde{\mu}$-a.e. limit of $(f_\alpha)_{\alpha \in \mathbb{T}}$ as $\alpha \rightarrow +\infty$. Then, there exists a $\tilde{\mu}$-negligible subset N_1 of Y such that $(f_\alpha(x))_{\alpha \in \mathbb{T}}$ converges for every $x \in Y \setminus N_1$. Using the comment made in the Observation before Proposition 3.1.4, we obtain that there exists a real-valued $\mathcal{Y}$-measurable function g on Y such that $f = g$ $\tilde{\mu}$-a.e., so there exists a $\tilde{\mu}$-negligible subset N_2 of Y such that $f(x) = g(x)$ for every $x \in Y \setminus N_2$. Since $N_1 \cup N_2$ is a $\tilde{\mu}$-negligible subset of Y, and taking into consideration the manner in which the completion $\tilde{\mathcal{Y}}$ of $\mathcal{Y}$ is defined (see Sect. 3.1.2), we obtain that there exists a μ-negligible subset N of Y such that $N_1 \cup N_2 \subseteq N$. It follows that $(f_\alpha(x))_{\alpha \in \mathbb{T}}$ converges to $g(x)$ as $\alpha \rightarrow +\infty$ for every $x \in Y \setminus N$, so $(f_\alpha)_{\alpha \in \mathbb{T}}$ converges μ-a.e. (to g) as $\alpha \rightarrow +\infty$. $\qquad\square$

Let $\mathbb{M}(Y, \mathcal{Y}, \mu)$ be the vector space of all equivalence classes of real-valued $\mathcal{Y}$-measurable functions on Y with respect to the equivalence relation defined by the μ-almost everywhere equality of two measurable functions.

Let $(\bar{f}_\alpha)_{\alpha \in \mathbb{T}}$ be a family of elements of $\mathbb{M}(Y, \mathcal{Y}, \mu)$. We say that $(\bar{f}_\alpha)_{\alpha \in \mathbb{T}}$ *converges μ-a.e. as $\alpha \to +\infty$* if there exists an $\bar{f} \in \mathbb{M}(Y, \mathcal{Y}, \mu)$ such that $(\bar{f}_{\alpha_n})_{n \in \mathbb{N}}$ converges μ-a.e. to $\bar{f}$ for every sequence $(\alpha_n)_{n \in \mathbb{N}}$ of elements of $\mathbb{T}$ that tends to $+\infty$. Even though this definition is correct from the formal point of view, it does not have the flavor of the definition of the almost everywhere convergence of a family $(h_\alpha)_{\alpha \in \mathbb{T}}$ of real-valued measurable functions as $\alpha \to +\infty$ that we stated earlier; that is, even if $\mathbb{T} = [0, +\infty)$, if $(\bar{f}_\alpha)_{\alpha \in [0,+\infty)}$ converges μ-a.e. as $\alpha \to +\infty$, there is no guarantee that there exists a family $(h_\alpha)_{\alpha \in [0,+\infty)}$ of real-valued $\mathcal{Y}$-measurable functions on Y such that h_α belongs to the equivalence class $\bar{f}_\alpha$ for every $\alpha \in [0, +\infty)$ and such that $(h_\alpha)_{\alpha \in [0,+\infty)}$ converges μ-a.e. as $\alpha \to +\infty$. In view of this situation, Dunford and Schwartz in [30] obtained an approach to the almost everywhere convergence for families $(\bar{f}_\alpha)_{\alpha \in [0,+\infty)}$ of elements of $\mathbb{M}(Y, \mathcal{Y}, \mu)$ that we will now describe for later use.

From now on, we assume that $(Y, \mathcal{Y}, \mu)$ is a complete σ-finite measure space. Note that, under this assumption, any real-valued $\mathcal{Y}$-measurable function on Y is DS-measurable.

Given a family $(\bar{f}_\alpha)_{\alpha \in [0,+\infty)}$ of elements of $\mathbb{M}(Y, \mathcal{Y}, \mu)$ and a function $\omega : [0, +\infty) \times Y \to \mathbb{R}$, we say that ω is a *measurable representation of $(\bar{f}_\alpha)_{\alpha \in [0,+\infty)}$* if the following two conditions are satisfied:

(MR1) ω is DS-measurable with respect to the completion of the σ-algebra product $\mathcal{L}([0, +\infty)) \otimes \mathcal{Y}$, where $\mathcal{L}([0, +\infty))$ is the σ-algebra of all Lebesgue measurable subsets of $[0, +\infty)$.

(MR2) There exists a Lebesgue measurable subset N of $[0, +\infty)$ of Lebesgue measure zero such that the function $g_\alpha : Y \to \mathbb{R}$ defined by $g_\alpha(y) = \omega(\alpha, y)$ for every $y \in Y$ is $\mathcal{Y}$-measurable and belongs to (the equivalence class) $\bar{f}_\alpha$ for every $\alpha \in [0, +\infty) \setminus N$.

If $(\bar{f}_\alpha)_{\alpha \in [0,+\infty)}$ is a family of elements of $\mathbb{M}(Y, \mathcal{Y}, \mu)$, and if $\omega : [0, +\infty) \times Y \to \mathbb{R}$ is a measurable representation of $(\bar{f}_\alpha)_{\alpha \in [0,+\infty)}$, we say that ω is a *strongly measurable representation of $(\bar{f}_\alpha)_{\alpha \in [0,+\infty)}$* if, instead of condition (MR2), ω satisfies the following stronger condition:

(SMR) For every $\alpha \in (0, +\infty)$, the function $g_\alpha : Y \to \mathbb{R}$ defined by $g_\alpha(y) = \omega(\alpha, y)$ for every $y \in Y$ is $\mathcal{Y}$-measurable and belongs to $\bar{f}_\alpha$.

Let $(\bar{f}_\alpha)_{\alpha \in [0,+\infty)}$ be a family of elements of $\mathbb{M}(Y, \mathcal{Y}, \mu)$, assume that $(\bar{f}_\alpha)_{\alpha \in [0,+\infty)}$ has strongly measurable representations, and let $\omega : [0, +\infty) \times Y \to \mathbb{R}$ be such a strongly measurable representation of $(\bar{f}_\alpha)_{\alpha \in [0,+\infty)}$. We say that $(\bar{f}_\alpha)_{\alpha \in [0,+\infty)}$ *converges μ-almost everywhere as $\alpha \to +\infty$ along ω* if the family $(g_\alpha)_{\alpha \in [0,+\infty)}$, $g_\alpha : Y \to \mathbb{R}$, $g_\alpha(y) = \omega(\alpha, y)$ for every $y \in Y$ and $\alpha \in [0, +\infty)$, converges μ-almost everywhere as $\alpha \to +\infty$ (as a family of $\mathcal{Y}$-measurable real-valued functions).

Let $(\bar{f}_\alpha)_{\alpha \in \mathbb{R}}$ be a family of elements of $\mathbb{M}(Y, \mathcal{Y}, \mu)$. By a *measurable representation*, or a *strongly measurable representation of* $(\bar{f}_\alpha)_{\alpha \in \mathbb{R}}$, we mean a function $\omega : [0, +\infty) \times Y \to \mathbb{R}$ which is a measurable representation, or a strongly measurable representation, of the restriction $(\bar{f}_\alpha)_{\alpha \in [0,+\infty)}$ of $(\bar{f}_\alpha)_{\alpha \in \mathbb{R}}$ to $[0, +\infty)$, respectively. Thus, if $(\bar{f}_\alpha)_{\alpha \in \mathbb{R}}$ has strongly measurable representations (or, equivalently, if $(\bar{f}_\alpha)_{\alpha \in [0,+\infty)}$ has strongly measurable representations), and if ω is such a strongly measurable representation of $(\bar{f}_\alpha)_{\alpha \in \mathbb{R}}$ (and of $(\bar{f}_\alpha)_{\alpha \in [0,+\infty)}$), then we say that $(\bar{f}_\alpha)_{\alpha \in \mathbb{R}}$ *converges μ-almost everywhere as $\alpha \to +\infty$ along ω* if $(\bar{f}_\alpha)_{\alpha \in [0,+\infty)}$ converges μ-almost everywhere as $\alpha \to +\infty$ along ω.

Let $(\bar{f}_\alpha)_{\alpha \in \mathbb{T}}$ be a family of elements of $\mathbb{M}(Y, \mathcal{Y}, \mu)$, and assume that $(\bar{f}_\alpha)_{\alpha \in \mathbb{T}}$ has strongly measurable representations. On the set of all strongly measurable representations of $(\bar{f}_\alpha)_{\alpha \in \mathbb{T}}$ we define a relation $\sim$ as follows: given two strongly measurable representations $\omega : [0, +\infty) \times Y \to \mathbb{R}$ and $\eta : [0, +\infty) \times Y \to \mathbb{R}$, let $g_\alpha : Y \to \mathbb{R}$ and $h_\alpha : Y \to \mathbb{R}$ be defined by $g_\alpha(y) = \omega(\alpha, y)$ and $h_\alpha(y) = \eta(\alpha, y)$ for every $y \in Y$ and $\alpha \in [0, +\infty)$; by definition, $\omega \sim \eta$ if there exists a μ-negligible subset N of Y such that $g_\alpha(y) = h_\alpha(y)$ for every $y \in Y \setminus N$ and every $\alpha \in [0, +\infty)$.

Clearly, $\sim$ is an equivalence relation.

Also easy to see is the fact that if $(\bar{f}_\alpha)_{\alpha \in \mathbb{T}}$ is a family of elements of $\mathbb{M}(Y, \mathcal{Y}, \mu)$ that has strongly measurable representations, and if ω is a strongly measurable representation of $(\bar{f}_\alpha)_{\alpha \in \mathbb{T}}$ such that $(\bar{f}_\alpha)_{\alpha \in \mathbb{T}}$ converges μ-almost everywhere as $\alpha \to +\infty$ along ω, then for every strongly measurable representation η of $(\bar{f}_\alpha)_{\alpha \in \mathbb{T}}$ such that $\omega \sim \eta$ it follows that $(\bar{f}_\alpha)_{\alpha \in \mathbb{T}}$ converges μ-almost everywhere as $\alpha \to +\infty$ along η, as well.

Observation. Let $(\bar{f}_\alpha)_{\alpha \in \mathbb{T}}$ be a family of elements of $\mathbb{M}(Y, \mathcal{Y}, \mu)$, assume that $(\bar{f}_\alpha)_{\alpha \in \mathbb{T}}$ has strongly measurable representations, let $\omega : [0, +\infty) \times Y \to \mathbb{R}$ be such a strongly measurable representation of $(\bar{f}_\alpha)_{\alpha \in \mathbb{T}}$, and assume that $(\bar{f}_\alpha)_{\alpha \in \mathbb{T}}$ converges μ-almost everywhere as $\alpha \to +\infty$ along ω.

Let $(g_\alpha)_{\alpha \in [0,+\infty)}$ be the representation of ω as a family of real-valued $\mathcal{Y}$-measurable functions defined on Y; that is, the family $(g_\alpha)_{\alpha \in [0,+\infty)}$, $g_\alpha : Y \to \mathbb{R}$ for every $\alpha \in [0, +\infty)$, is defined by $g_\alpha(y) = \omega(\alpha, y)$ for every $y \in Y$ and $\alpha \in [0, +\infty)$. Since $(\bar{f}_\alpha)_{\alpha \in \mathbb{T}}$ converges μ-a.e. as $\alpha \to +\infty$ along ω, it follows that there exists a $\mathcal{Y}$-measurable real-valued function f such that $(g_\alpha)_{\alpha \in [0,+\infty)}$ converges μ-almost everywhere to f as $\alpha \to +\infty$.

Let $\bar{f}$ be the element of $\mathbb{M}(Y, \mathcal{Y}, \mu)$ defined by f. Then $(\bar{f}_\alpha)_{\alpha \in \mathbb{T}}$ converges μ-a.e. as $\alpha \to +\infty$ to $\bar{f}$ in the sense that $(\bar{f}_{\alpha_n})_{n \in \mathbb{N}}$ converges μ-a.e. to $\bar{f}$ for every sequence $(\alpha_n)_{n \in \mathbb{N}}$ of elements of $\mathbb{T}$ that tends to $+\infty$. Since the μ-a.e. limit $\bar{f}$ of $(\bar{f}_\alpha)_{\alpha \in \mathbb{T}}$ is unique, it follows that given another strongly measurable representation $\eta : [0, +\infty) \times Y \to \mathbb{R}$ of $(\bar{f}_\alpha)_{\alpha \in \mathbb{T}}$, not necessarily equivalent to ω, then either $(\bar{f}_\alpha)_{\alpha \in \mathbb{T}}$ converges μ-a.e. to $\bar{f}$ as $\alpha \to +\infty$ along η, or else the family $(h_\alpha)_{\alpha \in [0,+\infty)}$, $h_\alpha : Y \to \mathbb{R}, h_\alpha(y) = \eta(\alpha, y)$ for every $y \in Y$ and $\alpha \in [0, +\infty)$, does not converge μ-a.e. as $\alpha \to +\infty$ (as a family of real-valued $\mathcal{Y}$-measurable functions).

We will often refer to the μ-a.e. convergence of $(\bar{f}_\alpha)_{\alpha \in \mathbb{T}}$ to $\bar{f}$ as $\alpha \to +\infty$ as the *absolute μ-a.e. convergence of $(\bar{f}_\alpha)_{\alpha \in \mathbb{T}}$* in order to emphasize that this almost everywhere convergence is independent of measurable representations. Thus, if $(\bar{f}_\alpha)_{\alpha \in \mathbb{T}}$ converges absolutely μ-a.e. to $\bar{f}$ as $\alpha \to +\infty$, we say that $\bar{f}$ is the *absolute μ-a.e. limit of $(\bar{f}_\alpha)_{\alpha \in \mathbb{T}}$ as $\alpha \to +\infty$.*   ▲

3.2.2   Semigroups of Operators Defined by Invariant Probabilities, and a Theorem of Dunford and Schwartz

Our goal in this subsection is to define certain semigroups of operators related to invariant probabilities of transition functions, and to use Theorem 8.7.5, p. 690 of Dunford and Schwartz [30], in order to obtain "continuous-time" versions of two results discussed in [143]: Theorem 1.2.6 and Corollary 1.2.7 on pp. 21–22 (see also Theorem 1.2.2 in this book).

As usual in this monograph, when dealing with transition functions, let (X, d) be a locally compact separable metric space.

Also, let $(P_t)_{t \in [0,+\infty)}$ be a transition function defined on (X, d), assume that $(P_t)_{t \in [0,+\infty)}$ satisfies the s.m.a., and let $((S_t, T_t))_{t \in [0,+\infty)}$ be the family of Markov pairs defined by $(P_t)_{t \in [0,+\infty)}$.

Assume that $(P_t)_{t \in [0,+\infty)}$ has invariant probability measures, and let μ be such an invariant probability.

Let $t \in [0, +\infty)$. Using the same arguments that appear before Theorem 1.2.2 (see also the proof of Theorem 1.2.6, p. 21 of [143]) we note that, by Lemma 5.1 of Lin [67], or by Proposition 1.1 of Chapter 4 of Revuz [97], the Banach subspace B_μ of $\mathcal{M}(X)$ of all elements of $\mathcal{M}(X)$ that are absolutely continuous with respect to μ is T_t-invariant (in the sense that $T_t \nu \in B_\mu$ for every $\nu \in B_\mu$), so we can define an operator $U_t^{(\mu)} : L^1(X, \mathcal{B}(X), \mu) \to L^1(X, \mathcal{B}(X), \mu)$ by letting $U_t^{(\mu)} \bar{f}$ be the equivalence class of a Radon-Nikodým derivative of $T_t(\bar{f}\mu)$ with respect to μ (note that we used here the fact that B_μ is Banach space isomorphic and isometric to $L^1(X, \mathcal{B}(X), \mu)$ in a standard way using the Radon-Nikodým derivatives). Since $U_t^{(\mu)}$ is a (positive) contraction of $L^1(X, \mathcal{B}(X), \mu)$, we may and do consider the dual $U_t^{(\mu)'} : L^\infty(X, \mathcal{B}(X), \mu) \to L^\infty(X, \mathcal{B}(X), \mu)$, and using arguments similar to the arguments used for the operator $U^{(\mu)'}$ in the discussion preceding Theorem 1.2.2, we obtain that $U_t^{(\mu)'}$ can be uniquely extended to a positive contraction $V_t^{(\mu)} : L^1(X, \mathcal{B}(X), \mu) \to L^1(X, \mathcal{B}(X), \mu)$.

Proposition 3.2.7. *The family* $(V_t^{(\mu)})_{t \in [0,+\infty)}$ *is a one-parameter semigroup of positive contractions of* $L^1(X, \mathcal{B}(X), \mu)$.

Proof. We have to prove that $V_{q+t}^{(\mu)} = V_q^{(\mu)} V_t^{(\mu)}$ for every $q \in [0, +\infty)$ and $t \in [0, +\infty)$.

To this end, set $\mathcal{H} = \{\bar{g} \in L^1(X, \mathcal{B}(X), \mu) \mid g \in C_0(X)\}$; thus, $\mathcal{H}$ consists of all elements $\bar{f}$ of $L^1(X, \mathcal{B}(X), \mu)$ which have the property that there exists a real-valued continuous function h on X that vanishes at infinity and belongs to the equivalence class $\bar{f}$.

Now let $q \in [0, +\infty)$, $t \in [0, +\infty)$, $\bar{f} \in L^1(X, \mathcal{B}(X), \mu)$, and $\varepsilon \in \mathbb{R}$, $\varepsilon > 0$.

By Proposition 7.4.2, p. 227 of Cohn [20], $\mathcal{H}$ is a dense vector subspace of $L^1(X, \mathcal{B}(X), \mu)$, so there exists a $\bar{g} \in \mathcal{H}$ such that $\|\bar{f} - \bar{g}\|_1 < \frac{\varepsilon}{2}$. We may and do assume that the function g that defines the equivalence class $\bar{g}$ belongs to $C_0(X)$.

Using the fact that $V_r^{(\mu)} \bar{h} = \overline{S_r h}$ (because $U_r^{(\mu)'} \bar{h} = \overline{S_r h}$ by Proposition 1.4 of Chapter 4 of Revuz [97]) for every $h \in B_b(X)$ and $r \in [0, +\infty)$, as well as the fact that $V_s^{(\mu)}$ is an extension of $U_s^{(\mu)'}$ and $U_s^{(\mu)'}$ is a positive contraction of $L^\infty(X, \mathcal{B}(X), \mu)$ for every $s \in [0, \infty)$, and since (by Proposition 2.1.1) the family $(S_r)_{r \in [0, +\infty)}$ is a one-parameter semigroup of operators, we obtain that

$$V_{q+t}^{(\mu)} \bar{g} = \overline{S_{q+t} g} = \overline{S_q S_t g} = V_q^{(\mu)} \overline{S_t g} = V_q^{(\mu)} V_t^{(\mu)} \bar{g}.$$

Since $V_r^{(\mu)}$, $r \in [0, +\infty)$, are positive contractions of $L^1(X, \mathcal{B}(X), \mu)$, it follows that

$$\left\| V_{q+t}^{(\mu)} \bar{f} - V_q^{(\mu)} V_t^{(\mu)} \bar{f} \right\|_1 \leq \left\| V_{q+t}^{(\mu)} \bar{f} - V_{q+t}^{(\mu)} \bar{g} \right\|_1$$

$$+ \left\| V_{q+t}^{(\mu)} \bar{g} - V_q^{(\mu)} V_t^{(\mu)} \bar{g} \right\|_1 + \left\| V_q^{(\mu)} V_t^{(\mu)} \bar{g} - V_q^{(\mu)} V_t^{(\mu)} \bar{f} \right\|_1$$

$$= \left\| V_{q+t}^{(\mu)} \bar{f} - V_{q+t}^{(\mu)} \bar{g} \right\|_1 + \left\| V_q^{(\mu)} V_t^{(\mu)} \bar{g} - V_q^{(\mu)} V_t^{(\mu)} \bar{f} \right\|_1 < \frac{\varepsilon}{2} + \frac{\varepsilon}{2} = \varepsilon.$$

We have therefore proved that $\left\| V_{q+t}^{(\mu)} \bar{f} - V_q^{(\mu)} V_t^{(\mu)} \bar{f} \right\|_1 < \varepsilon$ for every $\varepsilon \in \mathbb{R}$, $\varepsilon > 0$, so $V_{q+t}^{(\mu)} \bar{f} = V_q^{(\mu)} V_t^{(\mu)} \bar{f}$ for every $\bar{f} \in L^1(X, \mathcal{B}(X), \mu)$, $q \in [0, +\infty)$, and $t \in [0, +\infty)$.   □

Let $(X, \widetilde{\mathcal{B}(X)}, \tilde{\mu})$ be the completion of the probability space $(X, \mathcal{B}(X), \mu)$.

A family $(Q_t)_{t \in [0, +\infty)}$ of linear bounded operators, $Q_t : L^1(X, \widetilde{\mathcal{B}(X)}, \tilde{\mu}) \to L^1(X, \widetilde{\mathcal{B}(X)}, \tilde{\mu})$ for every $t \in [0, +\infty)$, is said to be *strongly DS-measurable* if the function $t \mapsto Q_t \bar{\bar{f}}$, $t \in [0, +\infty)$, is DS-measurable with respect to the Borel σ-algebra generated by the open sets in the norm topology of $L^1(X, \widetilde{\mathcal{B}(X)}, \tilde{\mu})$ and the σ-algebra $\mathcal{L}([0, +\infty))$ of all Lebesgue measurable subsets of $[0, +\infty)$ for every $\bar{\bar{f}} \in L^1(X, \widetilde{\mathcal{B}(X)}, \tilde{\mu})$. The family $(Q_t)_{t \in [0, +\infty)}$ is said to be *strongly DS-integrable over every finite interval* if for every $t \in (0, +\infty)$ and every $\bar{\bar{f}} \in L^1(X, \widetilde{\mathcal{B}(X)}, \tilde{\mu})$, the function $s \mapsto Q_s \bar{\bar{f}}$, $s \in [0, t]$, is DS-integrable with respect to the Lebesgue measure on $[0, t]$. Note that if $(Q_t)_{t \in [0, +\infty)}$ is strongly DS-integrable over every finite interval, then $(Q_t)_{t \in [0, +\infty)}$ is also strongly DS-measurable.

Now assume that $(Q_t)_{t\in[0,+\infty)}$ is a semigroup of linear bounded operators, $Q_t : L^1(X,\widetilde{\mathcal{B}(X)},\tilde\mu) \to L^1(X,\widetilde{\mathcal{B}(X)},\tilde\mu)$ for every $t \in [0,+\infty)$, and assume that $(Q_t)_{t\in[0,+\infty)}$ is strongly DS-integrable over every finite interval.

Under the above conditions, it makes sense to consider, for every $t \in (0,+\infty)$, a map $A_t : L^1(X,\widetilde{\mathcal{B}(X)},\tilde\mu) \to L^1(X,\widetilde{\mathcal{B}(X)},\tilde\mu)$ defined by $A_t\bar{\bar{f}} = \frac{1}{t}\text{DS-}\int_0^t Q_s\bar{\bar{f}}\,\mathrm{d}s$ for every $\bar{\bar{f}} \in L^1(X,\widetilde{\mathcal{B}(X)},\tilde\mu)$, where, DS-$\int_0^t Q_s\bar{\bar{f}}\,\mathrm{d}s$ stands for the Dunford-Schwartz integral DS-$\int_{[0,t]} Q_s\bar{\bar{f}}\,\mathrm{d}s$ (over the interval $[0,t]$, of course). It is easy to see that A_t is a linear operator and it can be shown (see p. 685 of Dunford and Schwartz [30]) that A_t is also continuous for every $t \in (0,+\infty)$. The linear bounded operator A_t is called the *operator average of* $(Q_s)_{s\in[0,+\infty)}$ (or *defined by* $(Q_s)_{s\in[0,+\infty)}$) *over the interval* $[0,t]$. As expected, we call $(A_t)_{t\in[0,+\infty)}$ the *family of operator averages defined by* $(Q_t)_{t\in[0,+\infty)}$, where $A_0 = Q_0$.

Since $(Q_t)_{t\in[0,+\infty)}$ is strongly DS-integrable over every finite interval, using Theorem 3.11.17 on pp. 198–200 of Dunford and Schwartz [30], we obtain that for every $\bar{\bar{f}} \in L^1(X,\widetilde{\mathcal{B}(X)},\tilde\mu)$, there exists a measurable representation of the family $(Q_t\bar{\bar{f}})_{t\in[0,+\infty)}$.

Let $\bar{\bar{f}} \in L^1(X,\widetilde{\mathcal{B}(X)},\tilde\mu)$ and let $\omega : [0,+\infty) \times X \to \mathbb{R}$ be a measurable representation of $(Q_t\bar{\bar{f}})_{t\in[0,+\infty)}$. It can be shown (see p. 686 of Dunford and Schwartz [30]) that there exists a $\tilde\mu$-negligible subset N of X such that, for every $x \in X \setminus N$ and every $t \in (0,+\infty)$, the map $s \mapsto \omega(s,x)$, $s \in [0,t]$, is integrable, and such that, if we define $\eta : [0,+\infty) \times X \to \mathbb{R}$ by

$$\eta(t,x) = \begin{cases} 0 & \text{if } t = 0 \text{ or } x \in N \\ \frac{1}{t}\int_0^t \omega(s,x)\,\mathrm{d}s & \text{if } t > 0 \text{ and } x \in X \setminus N, \end{cases}$$

then, for every $t \in (0,+\infty)$, the function $x \mapsto \eta(t,x)$, $x \in X$, belongs to the equivalence class $A_t\bar{\bar{f}}$; that is, η is a strongly measurable representation of $(A_t\bar{\bar{f}})_{t\in[0,+\infty)}$. Any measurable representation η of $(A_t\bar{\bar{f}})_{t\in[0,+\infty)}$ for which there exists a measurable representation ω of $(Q_t\bar{\bar{f}})_{t\in[0,+\infty)}$ such that η is obtained from ω by the procedure described above is called a *standard measurable representation of* $(A_t\bar{\bar{f}})_{t\in[0,+\infty)}$. Note that any standard measurable representation of $(A_t)_{t\in[0,+\infty)}$ is necessarily strongly measurable. It can be shown (see p. 686 of Dunford and Schwartz [30]) that any two standard measurable representations of $(A_t\bar{\bar{f}})_{t\in[0,+\infty)}$ are equivalent.

Let $\varphi_1 : L^1(X,\mathcal{B}(X),\mu) \to L^1(X,\widetilde{\mathcal{B}(X)},\tilde\mu)$ be the isometry (for $p = 1$, of course) discussed in Proposition 3.1.4.

Let $t \in [0,+\infty)$. Also, let $\widetilde{V_t^{(\mu)}} : L^1(X,\widetilde{\mathcal{B}(X)},\tilde\mu) \to L^1(X,\widetilde{\mathcal{B}(X)},\tilde\mu)$ be defined by $\widetilde{V_t^{(\mu)}}\bar{\bar{f}} = \varphi_1 V_t^{(\mu)}\varphi_1^{-1}(\bar{\bar{f}})$ for every $\bar{\bar{f}} \in L^1(X,\widetilde{\mathcal{B}(X)},\tilde\mu)$.

Since φ_1 is an isometry onto $L^1(X,\widetilde{\mathcal{B}(X)},\tilde\mu)$ and $V_t^{(\mu)}$ is a positive contraction of $L^1(X,\mathcal{B}(X),\mu)$, it follows that $\widetilde{V_t^{(\mu)}}$ is a positive contraction of $L^1(X,\widetilde{\mathcal{B}(X)},\tilde\mu)$.

Taking into consideration that μ and $\tilde{\mu}$ are probability measures, we obtain that $L^\infty(X,\mathcal{B}(X),\mu) \subseteq L^1(X,\mathcal{B}(X),\mu)$ and $L^\infty(X,\widetilde{\mathcal{B}(X)},\tilde{\mu}) \subseteq L^1(X,\widetilde{\mathcal{B}(X)},\tilde{\mu})$, so $\varphi_\infty(\bar{f}) = \varphi_1(\bar{f})$ for every $\bar{f} \in L^\infty(X,\mathcal{B}(X),\mu)$; thus, using also the fact that φ_∞ is an isometry onto $L^\infty(X,\widetilde{\mathcal{B}(X)},\tilde{\mu})$, and since $V_t^{(\mu)}\bar{f} \in L^\infty(X,\mathcal{B}(X),\mu)$ and $\left\| V_t^{(\mu)}\bar{f} \right\|_\infty \le \left\| \bar{f} \right\|_\infty$ for every $\bar{f} \in L^\infty(X,\mathcal{B}(X),\mu)$, we obtain that $\widetilde{V_t^{(\mu)}}\tilde{\bar{f}} \in L^\infty(X,\widetilde{\mathcal{B}(X)},\tilde{\mu})$ and $\left\| \widetilde{V_t^{(\mu)}}\tilde{\bar{f}} \right\|_\infty \le \left\| \tilde{\bar{f}} \right\|_\infty$ for every $\tilde{\bar{f}} \in L^\infty(X,\widetilde{\mathcal{B}(X)},\tilde{\mu})$.

Finally, we note that $(\widetilde{V_t^{(\mu)}})_{t \in [0,+\infty)}$ is a one-parameter semigroup of operators because

$$\widetilde{V_{r+s}^{(\mu)}}\tilde{\bar{f}} = \varphi_1 V_{r+s}^{(\mu)} \varphi_1^{-1}(\tilde{\bar{f}}) = \varphi_1 V_r^{(\mu)} V_s^{(\mu)} \varphi_1^{-1}(\tilde{\bar{f}})$$

$$= \varphi_1 V_r^{(\mu)} \varphi_1^{-1} \varphi_1 V_s^{(\mu)} \varphi_1^{-1}(\tilde{\bar{f}}) = \widetilde{V_r^{(\mu)}}\, \widetilde{V_s^{(\mu)}}(\tilde{\bar{f}})$$

for every $r \in [0,+\infty)$, $s \in [0,+\infty)$, and $f \in L^1(X,\widetilde{\mathcal{B}(X)},\tilde{\mu})$.

For future reference we summarize the properties of the family of operators $(\widetilde{V_t^{(\mu)}})_{t \in [0,+\infty)}$ that we have discussed so far in the next proposition.

Proposition 3.2.8. *The family $(\widetilde{V_t^{(\mu)}})_{t \in [0,+\infty)}$ is a one-parameter semigroup of positive contractions of $L^1(X,\widetilde{\mathcal{B}(X)},\tilde{\mu})$, and the operators $\widetilde{V_t^{(\mu)}}$, $t \in [0,+\infty)$, have the property that $\widetilde{V_t^{(\mu)}}(\tilde{\bar{f}}) \in L^\infty(X,\widetilde{\mathcal{B}(X)},\tilde{\mu})$ and $\left\| \widetilde{V_t^{(\mu)}}(\tilde{\bar{f}}) \right\|_\infty \le \left\| \tilde{\bar{f}} \right\|_\infty$ for every $\tilde{\bar{f}} \in L^\infty(X,\widetilde{\mathcal{B}(X)},\tilde{\mu})$.*

Proposition 3.2.9. *The family $(\widetilde{V_t^{(\mu)}})_{t \in [0,+\infty)}$ is strongly DS-integrable over every finite interval.*

Proof. We have to prove that for every $t \in (0,+\infty)$ and for every $\tilde{\bar{f}} \in L^1(X,\widetilde{\mathcal{B}(X)},\tilde{\mu})$, the function $s \mapsto \widetilde{V_s^{(\mu)}}(\tilde{\bar{f}})$, $s \in [0,t]$, is DS-integrable with respect to the Lebesgue measure on $[0,t]$.

To this end, let $t \in (0,+\infty)$. We will prove the DS-integrability of the functions $s \mapsto \widetilde{V_s^{(\mu)}}(\tilde{\bar{f}})$, $s \in [0,t]$, where $\tilde{\bar{f}} \in L^1(X,\widetilde{\mathcal{B}(X)},\tilde{\mu})$, by studying first the case when $\tilde{\bar{f}} \in L^\infty(X,\widetilde{\mathcal{B}(X)},\tilde{\mu})$, and then the general case.

Thus, let $\tilde{\bar{f}} \in L^\infty(X,\widetilde{\mathcal{B}(X)},\tilde{\mu})$. Then there exists a function $g \in B_b(X)$ such that $g \in \varphi_\infty^{-1}(\tilde{\bar{f}}) = \varphi_1^{-1}(\tilde{\bar{f}})$. Accordingly, $\tilde{\bar{g}} = \tilde{\bar{f}}$, where $\tilde{\bar{g}}$ is the class defined by g in $L^1(X,\widetilde{\mathcal{B}(X)},\tilde{\mu})$.

Let $r \in [0,+\infty)$. As pointed out in the proof of Proposition 3.2.7, $\overline{S_r g} = V_r^{(\mu)}\bar{g}$, where $\bar{h}$ denotes the element of $L^1(X,\mathcal{B}(X),\mu)$ defined by a function $h \in B_b(X)$; therefore,

$$S_r g \in V_r^{(\mu)}(\bar{g}) \subseteq \varphi_1(V_r^{(\mu)}(\bar{g})) = \varphi_1 V_r^{(\mu)} \varphi_1^{-1}(\bar{g}) = \widetilde{V_r^{(\mu)}}(\bar{g}) = \widetilde{V_r^{(\mu)}}(\bar{\bar{f}}),$$

where we think of $V_r^{(\mu)}(\bar{g})$ and $\varphi_1(V_r^{(\mu)}(\bar{g}))$ as sets of real-valued measurable functions on $(X, \mathcal{B}(X), \tilde{\mu})$ here.

We have therefore proved that $S_r g$ belongs to $\widetilde{V_r^{(\mu)}}(\bar{\bar{f}})$ for every $r \in [0, +\infty)$.

Since we assume that $(P_t)_{t \in [0,+\infty)}$ satisfies the s.m.a., we obtain that the function $\omega : [0, +\infty) \times X \to \mathbb{R}$, defined by $\omega(r, x) = S_r g(x)$ for every $(r, x) \in [0, +\infty) \times X$, is measurable with respect to the product σ-algebra $\mathcal{L}([0, +\infty)) \otimes \mathcal{B}(X)$ of $\mathcal{L}([0, +\infty))$ and $\mathcal{B}(X)$, so, obviously, ω is also measurable with respect to the completion of $\mathcal{L}([0, +\infty)) \otimes \widetilde{\mathcal{B}(X)}$, the completion being taken with respect to the product of the Lebesgue measure on $[0, +\infty)$ and the probability measure $\tilde{\mu}$.

Thus, it follows that ω is a measurable representation of the family $(\widetilde{V_r^{(\mu)}} \bar{\bar{f}})_{r \in [0,+\infty)}$ (actually, it follows that ω is a strongly measurable representation of $(\widetilde{V_r^{(\mu)}} \bar{\bar{f}})_{r \in [0,+\infty)}$). Using (b) of Lemma 3.11.16, pp. 196–198 of Dunford and Schwartz [30], we obtain that the family $(\widetilde{V_r^{(\mu)}} \bar{\bar{f}})_{r \in [0,+\infty)}$ is DS-measurable in the sense that the vector-valued function $r \mapsto \widetilde{V_r^{(\mu)}} \bar{\bar{f}}$, $r \in [0, +\infty)$, is DS-measurable. In particular, the restriction $(\widetilde{V_r^{(\mu)}} \bar{\bar{f}})_{r \in [0,t]}$ of $(\widetilde{V_r^{(\mu)}} \bar{\bar{f}})_{r \in [0,+\infty)}$ to the interval $[0, t]$ is DS-measurable. Therefore, the function $r \mapsto \left\| \widetilde{V_r^{(\mu)}} \bar{\bar{f}} \right\|_{L^1(X, \widetilde{\mathcal{B}(X)}, \tilde{\mu})}$, $r \in [0, t]$, is DS-measurable. Since the constant function $\left\| \bar{\bar{f}} \right\|_{L^1(X, \widetilde{\mathcal{B}(X)}, \tilde{\mu})} \mathbf{1}_{[0,t]}$ is obviously DS-integrable with respect to the Lebesgue measure on $[0, t]$ and since $\left\| \widetilde{V_r^{(\mu)}} \bar{\bar{f}} \right\|_{L^1(X, \widetilde{\mathcal{B}(X)}, \tilde{\mu})} \leq \left\| \bar{\bar{f}} \right\|_{L^1(X, \widetilde{\mathcal{B}(X)}, \tilde{\mu})}$ for every $r \in [0, t]$ (because $\widetilde{V_r^{(\mu)}}$, $r \in [0, t]$, are positive contractions of $L^1(X, \mathcal{B}(X), \tilde{\mu})$, using Theorem 3.2.22 on pp. 117–118 of Dunford and Schwartz [30], we obtain that $(\widetilde{V_r^{(\mu)}} \bar{\bar{f}})_{r \in [0,t]}$ is DS-integrable).

Now let $\bar{\bar{f}} \in L^1(X, \widetilde{\mathcal{B}(X)}, \tilde{\mu})$, where $\bar{\bar{f}}$ does not necessarily belong to $L^\infty(X, \widetilde{\mathcal{B}(X)}, \tilde{\mu})$. Also, let $(\widetilde{f_n})_{n \in \mathbb{N}}$ be a sequence of elements of $L^\infty(X, \widetilde{\mathcal{B}(X)}, \tilde{\mu})$ that converges to $\bar{\bar{f}}$ in the norm topology of $L^1(X, \widetilde{\mathcal{B}(X)}, \tilde{\mu})$ (obviously, there exists such a sequence because $L^\infty(X, \widetilde{\mathcal{B}(X)}, \tilde{\mu})$ is dense in $L^1(X, \widetilde{\mathcal{B}(X)}, \tilde{\mu})$).

In view of our discussion so far, for every $n \in \mathbb{N}$, the $L^1(X, \widetilde{\mathcal{B}(X)}, \tilde{\mu})$-valued function $(\widetilde{V_r^{(\mu)}} \bar{\bar{f}}_n)_{r \in [0,t]}$ defined on $[0, t]$ is DS-integrable.

Since $(\bar{\bar{f}}_n)_{n \in \mathbb{N}}$ converges to $\bar{\bar{f}}$ in the norm topology of $L^1(X, \widetilde{\mathcal{B}(X)}, \tilde{\mu})$ and since $\widetilde{V_r^{(\mu)}}$, $r \in [0, t]$, are contractions of $L^1(X, \widetilde{\mathcal{B}(X)}, \tilde{\mu})$, it follows that the sequence of functions $(\widetilde{V_r^{(\mu)}} \bar{\bar{f}}_n)_{r \in [0,t]}$ converges everywhere to $(\widetilde{V_r^{(\mu)}} \bar{\bar{f}})_{r \in [0,t]}$ on the interval $[0, t]$, in the sense that the sequence $(\widetilde{V_r^{(\mu)}} \widetilde{f_n})_{n \in \mathbb{N}}$ of elements of $L^1(X, \widetilde{\mathcal{B}(X)}, \tilde{\mu})$ converges in the norm topology of $L^1(X, \widetilde{\mathcal{B}(X)}, \tilde{\mu})$ to $\widetilde{V_r^{(\mu)}} \bar{\bar{f}}$ for every $r \in [0, t]$.

Again using the fact that $(\widetilde{\bar{f}}_n)_{n\in\mathbb{N}}$ is a convergent sequence (so, $(\widetilde{\bar{f}}_n)_{n\in\mathbb{N}}$ is bounded) in the norm topology of $L^1(X, \widetilde{\mathcal{B}(X)}, \tilde{\mu})$ and the fact that $\widetilde{V_r^{(\mu)}}$, $r \in [0, t]$, are contractions of $L^1(X, \widetilde{\mathcal{B}(X)}, \tilde{\mu})$, we obtain that there exists an $M \in \mathbb{R}$, $M > 0$, such that $\left\| \widetilde{V_r^{(\mu)}} \bar{f} \right\| \leq M$ and $\left\| \widetilde{V_r^{(\mu)}} \widetilde{\bar{f}}_n \right\| \leq M$ for every $n \in \mathbb{N}$ for every $r \in [0, t]$.

Since the real-valued constant function $M\mathbf{1}_{[0,t]}$ is obviously DS-integrable with respect to the Lebesgue measure on $[0, t]$, it follows that we can use Corollary 3.6.16 (the Lebesgue dominated convergence theorem for the DS-integral), p. 151 of Dunford and Schwartz [30], in order to conclude that $(\widetilde{V_r^{(\mu)}} \bar{f})_{r\in[0,t]}$ is DS-integrable. $\quad\square$

In view of the above proposition, it makes sense to consider the operator average $\widetilde{A}_t$ of $(\widetilde{V_r^{(\mu)}})_{r\in[0,+\infty)}$ over the interval $[0, t]$ for every $t \in (0, +\infty)$; thus, $\widetilde{A}_t \bar{f} = \frac{1}{t}\text{DS-}\int_0^t \widetilde{V_s^{(\mu)}} \bar{f}\, ds$ for every $\bar{f} \in L^1(X, \widetilde{\mathcal{B}(X)}, \tilde{\mu})$ and $t \in (0, +\infty)$. For $t = 0$, set $\widetilde{A}_0 = \widetilde{V_0^{(\mu)}}$. Using the comments made on pp. 689–690 of Dunford and Schwartz [30], we obtain that $\widetilde{A}_t$, $t \in [0, +\infty)$, are linear contractions of $L^1(X, \widetilde{\mathcal{B}(X)}, \tilde{\mu})$. Also, using the fact that (by Proposition 3.2.8) the operators $\widetilde{V_s^{(\mu)}}$, $s \in [0, +\infty)$, are positive and applying (a) of Lemma 3.11.16 of Dunford and Schwartz [30] to the $L^1(X, \widetilde{\mathcal{B}(X)}, \tilde{\mu})$-valued function $(\widetilde{V_s^{(\mu)}} \bar{f})_{s\in[0,t]}$ defined on $[0, t]$, we obtain that $\widetilde{A}_t \bar{f} \geq 0$ for every $\bar{f} \in L^1(X, \widetilde{\mathcal{B}(X)}, \tilde{\mu})$, $\bar{f} \geq 0$, and every $t \in (0, +\infty)$; since $\widetilde{A}_0$ is equal to $\widetilde{V_0^{(\mu)}}$, we conclude that all the operators $\widetilde{A}_t$, $t \in [0, +\infty)$, are positive.

Lemma 3.2.10. *The operators $\widetilde{A}_t$, $t \in [0, +\infty)$, are Markov operators; that is, the operators $\widetilde{A}_t$, $t \in [0, +\infty)$, have the property that* $\text{DS-}\int_X \widetilde{A}_t \bar{f}\, d\tilde{\mu} = \text{DS-}\int_X \bar{f}\, d\tilde{\mu}$ *for every $t \in [0, +\infty)$ and $\bar{f} \in L^1(X, \widetilde{\mathcal{B}(X)}, \tilde{\mu})$.*

Observation. Note that using the comment made at the end of Sect. 3.1.3 (after Theorem 3.1.11), we obtain that the DS-integrals that appear in the above lemma are also integrals in the sense of classical measure theory. We prefer the DS-integrals because the proof of the lemma and its use are in the context of the integration theory developed by Dunford and Schwartz in [30]. $\blacktriangle$

Proof (of Lemma 3.2.10). It is easy to see that $\widetilde{V_t^{(\mu)}}$, $t \in [0, +\infty)$, are Markov operators, so, since $\widetilde{A}_0 = \widetilde{V_0^{(\mu)}}$, it is enough to prove the lemma under the assumption that $t > 0$.

Thus, let $t \in (0, +\infty)$.

Also, let $\bar{f} \in L^1(X, \widetilde{\mathcal{B}(X)}, \tilde{\mu})$, and assume that $\bar{f} \geq 0$.

Since (by Proposition 3.2.9) the function $(\widetilde{V_s^{(\mu)}} \bar{f})_{s\in[0,t]}$ is DS-integrable with respect to the Lebesgue measure $\tilde{\lambda}_t$ on $[0, t]$, by (a) of Lemma 3.11.16, pp. 196–198, of Dunford and Schwartz [30], there exists a function $\omega : [0, t] \times X \to \mathbb{R}$ which is

DS-integrable with respect to the completion $\widetilde{\lambda_t \otimes \tilde{\mu}}$ of the product measure $\lambda_t \otimes \tilde{\mu}$ such that the function $x \mapsto \omega(s, x)$, $x \in X$, belongs to the equivalence class $\widetilde{V_s^{(\mu)}} \bar{\tilde{f}}$ for λ_t-almost all $s \in [0, t]$, and such that the integral DS-$\int_0^t \omega(s, x)\, \mathrm{d}\lambda_t(s)$ exists for $\tilde{\mu}$-almost all $x \in X$ and, as a function of x, the integral belongs to the equivalence class DS-$\int_0^t \widetilde{V_s^{(\mu)}} \bar{\tilde{f}}\, \mathrm{d}s$.

Using Fubini's theorem for the DS-integral (Theorem 3.11.9, pp. 190–191 of Dunford and Schwartz [30]) and the fact that $\widetilde{V_s^{(\mu)}}$, $s \in [0, t]$, are Markov operators, we obtain that

$$
\begin{aligned}
\text{DS-}\int_X \widetilde{A}_t\, \bar{\tilde{f}}(x)\, \mathrm{d}\tilde{\mu}(x) &= \text{DS-}\int_X \frac{1}{t}\Big(\text{DS-}\int_0^t \widetilde{V_s^{(\mu)}}\, \bar{\tilde{f}}(x)\, \mathrm{d}s\Big)\, \mathrm{d}\tilde{\mu}(x) \\
&= \frac{1}{t}\Big(\text{DS-}\int_X \big(\text{DS-}\int_0^t \omega(s, x)\, \mathrm{d}\lambda_t(s)\big)\, \mathrm{d}\tilde{\mu}(x)\Big) \\
&= \frac{1}{t}\Big(\text{DS-}\int_0^t \big(\text{DS-}\int_X \omega(s, x)\, \mathrm{d}\tilde{\mu}(x)\big)\, \mathrm{d}\lambda_t(s)\Big) \\
&= \frac{1}{t}\Big(\text{DS-}\int_0^t \big(\text{DS-}\int_X \widetilde{V_s^{(\mu)}}\, \bar{\tilde{f}}(x)\, \mathrm{d}\tilde{\mu}(x)\big)\, \mathrm{d}\lambda_t(s)\Big) \\
&= \frac{1}{t}\Big(\text{DS-}\int_0^t \big(\text{DS-}\int_X \bar{\tilde{f}}(x)\, \mathrm{d}\tilde{\mu}(x)\big)\, \mathrm{d}\lambda_t(s)\Big) \\
&= \frac{1}{t}t\Big(\text{DS-}\int_X \bar{\tilde{f}}(x)\, \mathrm{d}\tilde{\mu}(x)\Big) = \text{DS-}\int_X \bar{\tilde{f}}(x)\, \mathrm{d}\tilde{\mu}(x).
\end{aligned}
$$

Now, if $\bar{\tilde{f}} \in L^1(X, \widetilde{\mathcal{B}(X)}, \tilde{\mu})$ is not necessarily a positive element of $L^1(X, \widetilde{\mathcal{B}(X)}, \tilde{\mu})$, then $\bar{\tilde{f}} = (\bar{\tilde{f}})^+ - (\bar{\tilde{f}})^-$, where $(\bar{\tilde{f}})^+$ and $(\bar{\tilde{f}})^-$ are the positive and the negative parts of $\bar{\tilde{f}}$, respectively.

Using our discussion so far, we obtain that

$$
\begin{aligned}
\text{DS-}\int_X \widetilde{A}_t\, \bar{\tilde{f}}\, \mathrm{d}\tilde{\mu} &= \text{DS-}\int_X \widetilde{A}_t(\bar{\tilde{f}})^+\, \mathrm{d}\tilde{\mu} - \text{DS-}\int_X \widetilde{A}_t(\bar{\tilde{f}})^-\, \mathrm{d}\tilde{\mu} \\
&= \text{DS-}\int_X (\bar{\tilde{f}})^+\, \mathrm{d}\tilde{\mu} - \text{DS-}\int_X (\bar{\tilde{f}})^-\, \mathrm{d}\tilde{\mu} \\
&= \text{DS-}\int_X \bar{\tilde{f}}\, \mathrm{d}\tilde{\mu}.
\end{aligned}
$$

$\square$

In view of our discussion so far, we are now ready to present the main result of this section. The result is stated in the next theorem.

Theorem 3.2.11 (The Dunford and Schwartz Almost Everywhere Mean Ergodic Theorem for the One-Parameter Semigroup $(\widetilde{V_t^{(\mu)}})_{t\in[0,+\infty)}$).

(a) *For every $\bar{\tilde{f}} \in L^1(X, \widetilde{\mathcal{B}(X)}, \tilde{\mu})$, there exist standard measurable representations of $(\widetilde{A_t}\,\bar{\tilde{f}})_{t\in[0,+\infty)}$, and $(\widetilde{A_t}\,\bar{\tilde{f}})_{t\in[0,+\infty)}$ converges $\tilde{\mu}$-almost everywhere as $t \to +\infty$ along each of its standard measurable representations. Consequently, $(\widetilde{A_t}\,\bar{\tilde{f}})_{t\in[0,+\infty)}$ converges absolutely $\tilde{\mu}$-almost everywhere as $t \to +\infty$.*

(b) *Let $\bar{\tilde{f}} \in L^1(X, \widetilde{\mathcal{B}(X)}, \tilde{\mu})$, and let $\bar{\tilde{g}} \in \mathbb{M}(X, \widetilde{\mathcal{B}(X)}, \tilde{\mu})$ be the absolute $\tilde{\mu}$-a.e. limit of $(\widetilde{A_t}\,\bar{\tilde{f}})_{t\in[0,+\infty)}$ as $t \to +\infty$. Then $\bar{\tilde{g}} \in L^1(X, \widetilde{\mathcal{B}(X)}, \tilde{\mu})$ and $DS\text{-}\int_X \bar{\tilde{f}}\,\mathrm{d}\tilde{\mu} = DS\text{-}\int_X \bar{\tilde{g}}\,\mathrm{d}\tilde{\mu}$.*

Proof. (a) Using Propositions 3.2.7–3.2.9, and the observation that a one-parameter semigroup of linear bounded operators on $L^1(X, \widetilde{\mathcal{B}(X)}, \tilde{\mu})$ that is strongly DS-integrable over every finite interval is also strongly DS-measurable, we obtain that the semigroup $(\widetilde{V_s^{(\mu)}})_{s\in[0,+\infty)}$ satisfies all the conditions of Theorem 8.7.5 on p. 690 of Dunford and Schwartz [30]. By the comments made on pp. 685–686 of Dunford and Schwartz [30], $(\widetilde{A_t}\,\bar{\tilde{f}})_{t\in[0,+\infty)}$ has standard measurable representations whenever $\bar{\tilde{f}} \in L^1(X, \widetilde{\mathcal{B}(X)}, \tilde{\mu})$. By Theorem 8.7.5, p. 690, of Dunford and Schwartz [30], $(\widetilde{A_t}\,\bar{\tilde{f}})_{t\in[0,+\infty)}$ converges $\tilde{\mu}$-a.e. as $t \to +\infty$ along each of its measurable representations for all $\bar{\tilde{f}} \in L^1(X, \widetilde{\mathcal{B}(X)}, \tilde{\mu})$. Using an observation made at the end of Sect. 3.2.1, we obtain that the families $(\widetilde{A_t}\,\bar{\tilde{f}})_{t\in[0,+\infty)}$, $\bar{\tilde{f}} \in L^1(X, \widetilde{\mathcal{B}(X)}, \tilde{\mu})$, converge absolutely $\tilde{\mu}$-a.e. as $t \to +\infty$, as well.

(b) Let $\bar{\tilde{f}} \in L^1(X, \widetilde{\mathcal{B}(X)}, \tilde{\mu})$, let $\bar{\tilde{g}} \in \mathbb{M}(X, \widetilde{\mathcal{B}(X)}, \tilde{\mu})$, and assume that the family of averages $(\tilde{A}_t\,\bar{\tilde{f}})_{t\in[0,+\infty)}$ converges absolutely $\tilde{\mu}$-a.e. to $\bar{\tilde{g}}$ as $t \to +\infty$. It follows that, in particular, the sequence $(\tilde{A}_n\,\bar{\tilde{f}})_{n\in\mathbb{N}}$ converges $\tilde{\mu}$-a.e. to $\bar{\tilde{g}}$.

Note that the equality (VI) that appears in the proof of Theorem 8.7.1 on pp. 687–688 of Dunford and Schwartz [30] holds true for the semigroup $(\widetilde{V_t^{(\mu)}})_{t\in[0,+\infty)}$ and the averages $(\tilde{A}_t)_{t\in[0,+\infty)}$. Using the equality, we obtain that

$$\tilde{A}_n\,\bar{\tilde{f}} = \frac{1}{n}\sum_{k=0}^{n-1}\widetilde{V_k^{(\mu)}}\,\widetilde{A_1}\,\bar{\tilde{f}} \tag{3.2.1}$$

for every $n \in \mathbb{N}$. Since $(\widetilde{V_t^{(\mu)}})_{t\in[0,+\infty)}$ is a one-parameter semigroup of operators, it follows that the equality (3.2.1) can be written as

$$\tilde{A}_n\,\bar{\tilde{f}} = \frac{1}{n}\sum_{k=0}^{n-1}(\widetilde{V_1^{(\mu)}})^k\,\tilde{A}_1\,\bar{\tilde{f}} \tag{3.2.2}$$

for every $n \in \mathbb{N}$.

We now note that the operator $\widetilde{V_1^{(\mu)}}$ satisfies all the conditions of Theorem 1.2.5 (The Hopf Ergodic Theorem) of [143]. Taking into consideration that $(\tilde{A}_n \bar{\tilde{f}})_{n\in\mathbb{N}}$ converges $\tilde{\mu}$-a.e. to $\bar{\tilde{g}}$, and using (b) of the above-mentioned theorem of Hopf, we obtain that $\bar{\tilde{g}} \in L^1(X, \widetilde{\mathcal{B}(X)}, \tilde{\mu})$ and that DS-$\int_X \tilde{A}_1 \bar{\tilde{f}}\, d\tilde{\mu} = $ DS-$\int_X \bar{\tilde{g}}\, d\tilde{\mu}$.

Finally, using Lemma 3.2.10, we obtain that

$$\text{DS-}\int_X \bar{\tilde{f}}\, d\tilde{\mu} = \text{DS-}\int_X \tilde{A}_1 \bar{\tilde{f}}\, d\tilde{\mu} = \text{DS-}\int_X \bar{\tilde{g}}\, d\tilde{\mu}.$$

$\square$

Using Theorem 3.2.11, we can extend Theorem 1.2.6 and Corollary 1.2.7 of [143] to transition functions. We discuss these two extensions in the next corollary.

Corollary 3.2.12. *(a) Let $f \in B_b(X)$. The family $(\frac{1}{\alpha} \int_0^\alpha S_t f(x)\, dt)_{\alpha\in(0,+\infty)}$ of real-valued $\mathcal{B}(X)$-measurable functions defined on X converges μ-a.e. as $\alpha \to +\infty$. If g is a μ-a.e. limit of $(\frac{1}{\alpha} \int_0^\alpha S_t f(x)\, dt)_{\alpha\in(0,+\infty)}$ as $\alpha \to +\infty$, then g is μ-integrable and $\int_X f\, d\mu = \int_X g\, d\mu$.*

(b) Let $f \in B_b(X)$ be such that $f \geq 0$ and $\int_X f(x)\, d\mu(x) > 0$, and set $A_f^{\to+} = \{x \in A_f^{\to} \mid \lim_{s\to+\infty} \frac{1}{s} \int_0^s S_t f(x)\, dt > 0\}$, where $A_f^{\to}$ is the set defined in Proposition 3.2.5. Then $\mu(A_f^{\to+}) > 0$ and $A_f^{\to+} \cap (\text{supp}\mu) \neq \emptyset$.

Proof. (a) Let $f \in B_b(X)$. As shown in the proof of Proposition 3.2.9, the function $(t, x) \mapsto S_t f(x)$, $(t, x) \in [0, +\infty)\times X$, is a measurable representation of the family $(\widetilde{V_t^{(\mu)}} \bar{\tilde{f}})_{t\in[0,+\infty)}$ (actually, a strong measurable representation of $(\widetilde{V_t^{(\mu)}} \bar{\tilde{f}})_{t\in[0,+\infty)}$), where $\bar{\tilde{f}}$ is the element of $L^1(X, \widetilde{\mathcal{B}(X)}, \tilde{\mu})$ defined by f. Accordingly, the function $\eta : [0, +\infty) \times X \to \mathbb{R}$,

$$\eta(t, x) = \begin{cases} \frac{1}{t} \int_0^t S_r f(x)\, dr & \text{if } t > 0 \text{ and } x \in X \\ S_0 f(x) & \text{if } t = 0 \text{ and } x \in X \end{cases}$$

is a standard measurable representation of $(\widetilde{V_t^{(\mu)}} \bar{\tilde{f}})_{t\in[0,+\infty)}$. By (a) of Theorem 3.2.11, the family $(\frac{1}{\alpha} \int_0^\alpha S_t f(x)\, dt)_{\alpha\in(0,+\infty)}$ converges $\tilde{\mu}$-a.e. as $\alpha \to +\infty$. By Lemma 3.2.6, the family $(\frac{1}{\alpha} \int_0^\alpha S_t f(x)\, dt)_{\alpha\in(0,+\infty)}$ converges μ-a.e. as $\alpha \to +\infty$, as well.

Let g be the μ-a.e. limit of $(\frac{1}{\alpha} \int_0^\alpha S_t f(x)\, dt)_{\alpha\in(0,+\infty)}$, and let $\bar{\tilde{g}}$ be the element of $L^1(X, \widetilde{\mathcal{B}(X)}, \tilde{\mu})$ defined by g. Then $\bar{\tilde{g}}$ is the absolute $\tilde{\mu}$-a.e. limit of $(\widetilde{A_\alpha \bar{\tilde{f}}})_{\alpha\in[0,+\infty)}$ as $\alpha \to +\infty$, where $(\widetilde{A_\alpha})_{\alpha\in[0,+\infty)}$ is the family of operator averages defined by $(\widetilde{V_t^{(\mu)}})_{t\in[0,+\infty)}$. Using (b) of Theorem 3.2.11, we obtain that DS-$\int_X \bar{\tilde{f}}\, d\tilde{\mu} = $ DS-$\int_X \bar{\tilde{g}}\, d\tilde{\mu}$. Since, in this case, the DS-integrability is the same as the integrability of classical measure theory, it follows that $\int_X \bar{\tilde{f}} \, d\tilde{\mu} = \int_X \bar{\tilde{g}} \, d\tilde{\mu}$, so $\int_X f \, d\mu = \int_X g \, d\mu$.

(b) Let $f \in B_b(X)$ be such that $f \geq 0$ and $\int_X f \, d\mu > 0$. Also, let g be a μ-a.e. limit of $(\frac{1}{\alpha} \int_0^\alpha S_t f(x) \, dt)_{\alpha \in (0,+\infty)}$ as $\alpha \to +\infty$ whose existence is assured by (a) of this corollary. Then $\int g \, d\mu = \int f \, d\mu > 0$, so $\mu(A_f^{\vec{}+}) > 0$ because $A_f^{\vec{}+} = \{g > 0\}$ μ-a.e. Since $\mu(A_f^{\vec{}+}) > 0$, it is obvious that $A_f^{\vec{}+} \cap \operatorname{supp}\mu \neq \emptyset$. $\qquad\square$

3.3   The Pointwise Integral

In this section, we introduce a vector integral that we call the pointwise integral. The main features of this integral are that it can be defined in very general settings and that the conditions for the integrability of a function in these settings are also quite general. The price that has to be paid for these features is that, in sharp contrast with the Bochner and Dunford-Schwartz integrals, we cannot build a solid measure theoretical theory of integration for the pointwise integral. However, when used together with the Bochner or the Dunford-Schwartz integrals, we can obtain results that, at least at this time, cannot be obtained without the pointwise integral. Even when it appears alone, the pointwise integral can be quite useful as we will see in the second subsection of this section. The most important results that involve the pointwise integral in the study of transition functions are in the last chapter of the book. In view of the above comments and certain results of Chap. 7 in this work, we often like to call the pointwise integral the *Cinderella integral* (see [147]).

We believe that the pointwise integral has appeared in explicit or at least implicit form earlier in the literature. However, we are unable to provide a reference.

The section is organized into two subsections. In the first subsection, we define the pointwise integral in general, and then we single out and study in some detail two types of pointwise integrals that are of interest to us in this book: a pointwise integral on $B_b(X)$ and another one on $\mathcal{M}(X)$.

In the second subsection we use the pointwise integral on $\mathcal{M}(X)$ in order to study the existence and some of the properties of invariant probabilities of a transition function in terms of the invariant probabilities of a single transition probability that belongs to the transition function.

3.3.1   *Definitions and Basic Properties*

Let $(Y, \mathcal{Y}, \nu)$ be a measure space, let Z be a nonempty set, and let $\mathbf{F}$ be a collection of real-valued functions defined on Z (in general, $\mathbf{F}$ is a Banach space of such functions).

We say that a function $\varphi : Y \to \mathbf{F}$ is *pointwise measurable* if the following condition is satisfied:

(PM) For every $z \in Z$, the function $\varphi_z : Y \to \mathbb{R}$ defined by $\varphi_z(y) = \varphi(y)(z)$ for every $y \in Y$ is $\mathcal{Y}$-measurable.

A pointwise measurable function $\varphi : Y \to \mathbf{F}$ is said to be *pointwise integrable* if the two conditions below are satisfied:

(PI1) The function φ_z defined at (PM) is ν-integrable for every $z \in Z$.
(PI2) The function $\mathbf{I} : Z \to \mathbb{R}$ defined by $\mathbf{I}(z) = \int_Y \varphi(y)(z)\, d\nu(y)$ for every $z \in Z$ belongs to $\mathbf{F}$.

If a function $\varphi : Y \to \mathbf{F}$ is pointwise integrable, then the function $\mathbf{I}$ defined at (PI2) in terms of φ is called the *pointwise integral of φ* (*over* or *on Y*) and is denoted by P-$\int_Y \varphi(y)\, d\nu(y)$. For $z \in Z$, we will use the notation P-$\int_Y \varphi(y)(z)\, d\nu(y)$ for $\mathbf{I}(z)$, rather than (P-$\int_Y \varphi(y)\, d\nu(y))(z)$, which is the formally correct notation, but is more complicated.

Depending on the type of set Z and the type of Banach space $\mathbf{F}$ under consideration one can define various types of pointwise integrals. As pointed out at the beginning of this section, in this book we will use two types of pointwise integrals that we introduce in this subsection.

The first pointwise integral that we will discuss is closely related to Fubini's theorem. The simplicity of this type of pointwise integral and its connection to results that are used to prove Fubini's theorem is one of the reasons why we believe that the pointwise integral has been considered earlier in the literature.

As usual in this book, let (X, d) be a locally compact separable metric space.

Also let, as before, $(Y, \mathcal{Y}, \nu)$ be a measure space but assume, in addition, that ν is a finite measure.

Now, let $\mathbf{g} : Y \times X \to \mathbb{R}$ be a bounded function measurable with respect to the product σ-algebra $\mathcal{Y} \otimes \mathcal{B}(X)$.

Since $\mathbf{g}$ is bounded and measurable with respect to $\mathcal{Y} \otimes \mathcal{B}(X)$, it follows that the function $\mathbf{g}_y : X \to \mathbb{R}$ defined by $\mathbf{g}_y(x) = \mathbf{g}(y, x)$ for every $x \in X$ belongs to $B_b(X)$ for every $y \in Y$ (see, for instance, (b) of Lemma 5.1.1, p. 155 of Cohn [20]). Thus, the function $\varphi : Y \to B_b(X)$, $\varphi(y) = \mathbf{g}_y$ for every $y \in Y$ is well-defined in the sense that $\varphi(y)$ belongs to $B_b(X)$ for every $y \in Y$.

It is easy to see that φ satisfies the conditions (PM) and (PI1). Using (a) of Proposition 5.2.1, p. 159, of Cohn [20], we obtain that φ satisfies condition (PI2), as well, so the pointwise integral P-$\int_Y \varphi(y)\, d\nu(y)$ of φ on Y does exist (note that even though, in Proposition 5.2.1, p. 159, of Cohn [20], there are two measure spaces under consideration, if we carefully read the proof of (a) of the proposition, and the proof of Proposition 5.1.2, p. 156, of Cohn [20], we see that our conclusions are true even though no particular measure is under consideration on the measurable space $(X, \mathcal{B}(X)))$. We also remark that in the above arguments, the fact that (X, d) is a locally compact separable metric space is of no importance; we could have used any measurable space $(Z, \mathcal{Z})$ instead of $(X, \mathcal{B}(X))$.

For future reference we summarize the above discussion in the next proposition.

Proposition 3.3.1. *If $(Y, \mathcal{Y}, \nu)$ is a finite measure space, and if $g : Y \times X \to \mathbb{R}$ is a bounded function measurable with respect to the product σ-algebra $\mathcal{Y} \otimes \mathcal{B}(X)$, then the function $\varphi : Y \to B_b(X)$, $\varphi(y) = \mathbf{g}_y$ for every $y \in Y$, where $\mathbf{g}_y : X \to \mathbb{R}$, $\mathbf{g}_y(x) = \mathbf{g}(y, x)$ for every $x \in X$, is well-defined in the sense that $\mathbf{g}_y \in B_b(X)$ for every $y \in Y$ and is pointwise integrable. The pointwise integral $P\text{-}\int_Y \varphi(y)\,d\nu(y)$ is an element of $B_b(X)$ and is defined by*

$$P\text{-}\int_Y \varphi(y)(x)\,d\nu(y) = \int_Y \mathbf{g}_y(x)\,d\nu(y) = \int_Y \mathbf{g}(y, x)\,d\nu(y)$$

for every $x \in X$, where, as mentioned earlier, $P\text{-}\int_Y \varphi(y)(x)\,d\nu(y)$ stands for $(P\text{-}\int_Y \varphi(y)\,d\nu(y))(x)$.

We will refer to the pointwise integral $P\text{-}\int_Y \varphi(y)\,d\nu(y)$ that appears in the above proposition as a $B_b(X)$-*pointwise integral*. The $B_b(X)$-pointwise integral that is of interest to us in this book appears quite naturally when dealing with transition functions, and is described in the next proposition.

Proposition 3.3.2. *Let $(P_t)_{t \in \mathbb{T}}$ be a transition function defined on (X, d), assume that $(P_t)_{t \in \mathbb{T}}$ satisfies the s.m.a., and let $((S_t, T_t))_{t \in \mathbb{T}}$ be the family of Markov pairs defined by $(P_t)_{t \in \mathbb{T}}$. Then, for every $f \in B_b(X)$ and every Lebesgue measurable subset L of $\mathbb{T}$ of finite Lebesgue measure, the function $\varphi_f^{(L)} : L \to B_b(X)$ defined by $\varphi_f^{(L)}(t) = S_t f$ for every $t \in L$ is pointwise integrable.*

Proof. We apply Proposition 3.3.1 in the case in which $Y = L$, ν is the Lebesgue measure on Y, and $\mathbf{g}^{(f)} : L \times X \to \mathbb{R}$ is defined by $\mathbf{g}^{(f)}(t, x) = S_t f(x)$ for every $(t, x) \in L \times X$ and $f \in B_b(X)$. We can apply the proposition because, given $f \in B_b(X)$, the function $\mathbf{g}^{(f)}$ is obviously bounded, and, since $(P_t)_{t \in \mathbb{T}}$ satisfies the s.m.a., using Proposition 2.1.5, we obtain that $\mathbf{g}^{(f)}$ is measurable with respect to the product σ-algebra $\mathcal{L}(L) \otimes \mathcal{B}(X)$, where $\mathcal{L}(L)$ is the σ-algebra of all Lebesgue measurable subsets of L. By Proposition 3.3.1, the functions $\varphi_f^{(L)}$, $f \in B_b(X)$, are pointwise integrable. $\qquad\square$

The $B_b(X)$-pointwise integral of $\varphi_f^{(L)}$ whose existence is discussed in Proposition 3.3.2 will be denoted by $P\text{-}\int_L S_t f\,d\lambda_L(t)$, where λ_L is the Lebesgue measure on L, or simply $P\text{-}\int_L S_t f\,dt$.

Let $\mathcal{J}$ be a Lebesgue measurable subset of $\mathbb{R}$ of finite Lebesgue measure. In the next theorem we discuss sufficient conditions under which a function $\varphi : \mathcal{J} \to B_b(X)$ is pointwise, Bochner, and Dunford-Schwartz integrable, and the three integrals are equal. In order to discuss the theorem we need the following lemma:

Lemma 3.3.3. *Let $\mathcal{J}$ be a Lebesgue measurable subset of $\mathbb{R}$, let $\varphi : \mathcal{J} \to B_b(X)$ be a continuous function (with respect to the norm topology on $B_b(X)$ and the topology induced on $\mathcal{J}$ by the standard topology of $\mathbb{R}$), assume that the range of φ is included in $C_b(X)$, and let $\mathbf{g} : \mathcal{J} \times X \to \mathbb{R}$ be defined by $\mathbf{g}(t, x) = \varphi(t)(x)$ for every $(t, x) \in \mathcal{J} \times X$. Then $\mathbf{g}$ is jointly continuous with respect to t and x (that* *is, $\mathbf{g}$ is continuous with respect to the standard topology $\mathcal{T}(\mathbb{R})$ on $\mathbb{R}$ and the product topology $\mathcal{T}(\mathcal{J}) \otimes \mathcal{T}_d(X)$ on $\mathcal{J} \times X$, where $\mathcal{T}(\mathcal{J})$ is the standard topology on $\mathcal{J}$ inherited from $\mathbb{R}$ and $\mathcal{T}_d(X)$ is the metric topology on X).*

Proof. It is well known (see, for instance, Chapter 8, Section 5, p. 184 of Royden [103]) that since the topologies $\mathcal{T}(\mathcal{J})$ and $\mathcal{T}_d(X)$ are defined by metrics on $\mathcal{J}$ and X, respectively, it follows that the product topology $\mathcal{T}(\mathcal{J}) \otimes \mathcal{T}_d(X)$ is metrizable. Therefore, in order to prove that $\mathbf{g}$ is continuous, it is enough to prove that for every convergent sequence $((t_n, x_n))_{n\in\mathbb{N}}$ of elements of $\mathcal{J} \times X$, the sequence $(\mathbf{g}(t_n, x_n))_{n\in\mathbb{N}}$ converges to $\mathbf{g}(t, x)$, where $(t, x) = \lim_{n\to+\infty}(t_n, x_n)$.

To this end, let $((t_n, x_n))_{n\in\mathbb{N}}$ be a convergent sequence of elements of $\mathcal{J} \times X$, set $(t, x) = \lim_{n\to+\infty}(t_n, x_n)$, and let $\varepsilon \in \mathbb{R}$, $\varepsilon > 0$. Since the sequence $(t_n)_{n\in\mathbb{N}}$ converges to t, and since φ is continuous, it follows that there exists an $m_\varepsilon \in \mathbb{N}$ such that $\|\varphi(t_n) - \varphi(t)\| < \frac{\varepsilon}{2}$ for every $n \in \mathbb{N}$, $n \geq m_\varepsilon$. Since $\varphi(t) \in C_b(X)$ and $(x_n)_{n\in\mathbb{N}}$ converges to x, there exists an $n_\varepsilon \in \mathbb{N}$, $n_\varepsilon \geq m_\varepsilon$, such that $|\mathbf{g}(t, x_n) - \mathbf{g}(t, x)| < \frac{\varepsilon}{2}$ for every $n \in \mathbb{N}$, $n \geq n_\varepsilon$.

We obtain that

$$|\mathbf{g}(t_n, x_n) - \mathbf{g}(t, x)| \leq |\mathbf{g}(t_n, x_n) - \mathbf{g}(t, x_n)| + |\mathbf{g}(t, x_n) - \mathbf{g}(t, x)|$$

$$< \|\varphi(t_n) - \varphi(t)\| + \frac{\varepsilon}{2} < \frac{\varepsilon}{2} + \frac{\varepsilon}{2} = \varepsilon.$$

$\square$

Theorem 3.3.4. *Let $\mathcal{J}$ be a Lebesgue measurable subset of $\mathbb{R}$ of finite Lebesgue measure, let $\varphi : \mathcal{J} \to B_b(X)$ be a continuous bounded function whose range is included in $C_b(X)$. Then:*

(a) φ is pointwise integrable;
(b) φ is Bochner integrable;
(c) φ is DS-integrable;
(d) $P\text{-}\int_{\mathcal{J}} \varphi(t)\,d\lambda_{\mathcal{J}}(t) = B\text{-}\int_{\mathcal{J}} \varphi(t)\,d\lambda_{\mathcal{J}}(t) = DS\text{-}\int_{\mathcal{J}} \varphi(t)\,d\lambda_{\mathcal{J}}(t)$, where $\lambda_{\mathcal{J}}$ is the Lebesgue measure on $\mathcal{J}$.

Proof. (a) Let $\mathbf{g} : \mathcal{J} \times X \to \mathbb{R}$ be the function defined by $\mathbf{g}(t, x) = \varphi(t)(x)$ for every $(t, x) \in \mathcal{J} \times X$. Using Lemma 3.3.3, we obtain that $\mathbf{g}$ is jointly continuous with respect to t and x; therefore, using three well-known results (see, for instance, Proposition 2.6.2, p. 79 of Cohn [20], Section D32 of Appendix D, p. 348 of [20], and Proposition 7.6.2, p. 242 of [20]), we obtain that $\mathbf{g}$ is measurable with respect to the Borel σ-algebra of $\mathbb{R}$ and the product σ-algebra $\mathcal{B}(\mathcal{J}) \otimes \mathcal{B}(X)$, where $\mathcal{B}(\mathcal{J})$ is the σ-algebra generated by $\mathcal{T}(\mathcal{J})$; since $\mathcal{B}(\mathcal{J}) \subseteq \mathcal{L}(\mathcal{J})$, it follows that $\mathbf{g}$ is measurable with respect to $\mathcal{L}(\mathcal{J}) \otimes \mathcal{B}(X)$, as well, where $\mathcal{L}(\mathcal{J})$ is the σ-algebra of all Lebesgue measurable subsets of $\mathcal{J}$. Using Proposition 3.3.1, we obtain that φ is pointwise integrable.

(b) Since φ is continuous, using Proposition 2.6.2, p. 79, of Cohn [20], we obtain that φ is measurable (in the sense that $\varphi^{-1}(A)$ is a Lebesgue measurable subset of $\mathcal{J}$ for every Borel measurable subset A of $B_b(X)$). Using again the continuity of φ, we obtain that the range $\varphi(\mathcal{J})$ of φ is a separable subset of $B_b(X)$. Accordingly, φ is Bochner measurable.

Since $|\,\|\varphi(t)\| - \|\varphi(s)\|\,| \le \|\varphi(t) - \varphi(s)\|$ for every $t \in \mathcal{J}$ and $s \in \mathcal{J}$, and since φ is a continuous bounded function, it follows that the function $h : \mathcal{J} \to \mathbb{R}$ defined by $h(t) = \|\varphi(t)\|$ for every $t \in \mathcal{J}$ is also continuous and bounded. Since $\mathcal{J}$ has finite Lebesgue measure, it follows that h is integrable in the sense of classical measure theory. Using (b) of Proposition 3.1.1, we obtain that φ is Bochner integrable.

(c) Since, as we have just proved, φ is Bochner integrable, and since $\mathcal{L}(\mathcal{J})$ is a complete σ-algebra with respect to the Lebesgue measure on $\mathcal{J}$, using Proposition 3.1.8, we obtain that φ is DS-integrable.

(d) We first note that, by Proposition 3.1.8,

$$\text{B-}\int_{\mathcal{J}} \varphi(t)\,\mathrm{d}\lambda_{\mathcal{J}}(t) = \text{DS-}\int_{\mathcal{J}} \varphi(t)\,\mathrm{d}\lambda_{\mathcal{J}}(t),$$

so we only have to prove that $\text{P-}\int_{\mathcal{J}} \varphi(t)\,\mathrm{d}\lambda_{\mathcal{J}}(t) = \text{B-}\int_{\mathcal{J}} \varphi(t)\,\mathrm{d}\lambda_{\mathcal{J}}(t)$.

To this end, let $\psi : \mathcal{J} \to B_b(X)$ be a simple Bochner measurable function. Then there exist $n \in \mathbb{N}$, n elements $f_1, f_2, \ldots, f_n$ of $B_b(X)$, n mutually disjoint Lebesgue measurable subsets $A_1, A_2, \ldots, A_n$ of $\mathcal{J}$ (the fact that $A_1, A_2, \ldots, A_n$ are mutually disjoint means, of course, that $A_i \cap A_j = \varnothing$ whenever $i \ne j$ for every $i = 1, 2, \ldots, n$, and $j = 1, 2, \ldots, n$) such that $\psi(t) = \sum_{i=1}^{n} \mathbf{1}_{A_i}(t) f_i$ for every $t \in \mathcal{J}$. Since $\mathcal{J}$ has finite Lebesgue measure, it follows that ψ is Bochner integrable and the Bochner integral of ψ is $\text{B-}\int_{\mathcal{J}} \psi(t)\,\mathrm{d}\lambda_{\mathcal{J}}(t) = \sum_{i=1}^{n} \lambda_{\mathcal{J}}(A_i) f_i$ (see the beginning of Sect. 3.1.1 for details). Also, ψ is pointwise integrable because:

- For every $x \in X$, the function $\psi^{(x)} : \mathcal{J} \to \mathbb{R}$ defined by $\psi^{(x)}(t) = \sum_{i=1}^{n} \mathbf{1}_{A_i}(t) f_i(x)$ for every $t \in \mathcal{J}$ is a real-valued simple measurable function, so ψ is pointwise measurable;
- The function $\psi^{(x)}$ defined above is integrable with respect to the Lebesgue measure $\lambda_{\mathcal{J}}$ on $\mathcal{J}$ for every $x \in X$ (this is so since $\psi^{(x)}$, $x \in X$, are bounded measurable functions on $\mathcal{J}$ and $\lambda_{\mathcal{J}}$ is a finite measure);
- The function $x \mapsto \int \psi^{(x)}(t)\,\mathrm{d}\lambda_{\mathcal{J}}(t) = \sum_{i=1}^{n} \lambda_{\mathcal{J}}(A_i) f_i(x)$, $x \in X$, belongs to $B_b(X)$.

It follows that $\text{P-}\int_{\mathcal{J}} \psi(t)\,\mathrm{d}\lambda_{\mathcal{J}}(t) = \sum_{i=1}^{n} \lambda_{\mathcal{J}}(A_i) f_i$, so $\text{P-}\int_{\mathcal{J}} \psi(t)\,\mathrm{d}\lambda_{\mathcal{J}}(t) = \text{B-}\int_{\mathcal{J}} \psi(t)\,\mathrm{d}\lambda_{\mathcal{J}}(t)$.

Since φ is Bochner integrable (as shown in (b)), there exists a sequence $(\psi_n)_{n\in\mathbb{N}}$ of $B_b(X)$-valued simple Bochner integrable functions defined on $\mathcal{J}$ which satisfies conditions (B1), (B2) and (B3) stated before Proposition 3.1.1, and which has the property that the sequence of Bochner integrals $(\text{B-}\int_{\mathcal{J}} \psi_n(t)\,\mathrm{d}\lambda_{\mathcal{J}}(t))_{n\in\mathbb{N}}$ converges in the norm topology of $B_b(X)$ (that is, uniformly on X) to $\text{B-}\int \varphi(t)\,\mathrm{d}\lambda_{\mathcal{J}}(t)$.

Now, for every $x \in X$, the sequence $(\psi_n^{(x)})_{n \in \mathbb{N}}$, $\psi_n^{(x)} : \mathcal{J} \to \mathbb{R}$, $\psi_n^{(x)}(t) = \psi_n(t)(x)$ for every $t \in \mathcal{J}$ and $n \in \mathbb{N}$, converges pointwise on $\mathcal{J}$ to $\varphi^{(x)} : \mathcal{J} \to \mathbb{R}$, $\varphi^{(x)}(t) = \varphi(t)(x)$ for every $t \in \mathcal{J}$; since $\varphi^{(x)}$ is integrable (because, as shown in (a), φ is pointwise integrable) and bounded, since, for every $n \in \mathbb{N}$, the function $\psi_n^{(x)}$ is integrable with respect to $\lambda_{\mathcal{J}}$ (as shown earlier in the proof of (d)), since $\lambda_{\mathcal{J}}$ is a finite measure, and since $\|\psi_n(t)\| \le \|\varphi(t)\|$ for every $t \in \mathcal{J}$ (because $(\psi_m)_{m \in \mathbb{N}}$ and φ satisfy condition (B2) stated before Proposition 3.1.1), we obtain, by applying the dominated convergence theorem to the sequence $(\psi_n^{(x)})_{n \in \mathbb{N}}$, that the sequence $((\text{P-} \int_{\mathcal{J}} \psi_n(t) \, \mathrm{d}\lambda_{\mathcal{J}}(t))(x))_{n \in \mathbb{N}}$ converges to $(\text{P-} \int_{\mathcal{J}} \varphi(t) \, \mathrm{d}\lambda_{\mathcal{J}}(t))(x)$ (note that the fact that the pointwise integrals of ψ_n, $n \in \mathbb{N}$, exist was proved in the first part of the proof of (d)). Thus, the sequence $(\text{P-} \int_{\mathcal{J}} \psi_n(t) \, \mathrm{d}\lambda_{\mathcal{J}}(t))_{n \in \mathbb{N}}$ converges pointwise on X to $\text{P-} \int_{\mathcal{J}} \varphi(t) \, \mathrm{d}\lambda_{\mathcal{J}}(t)$.

Since (as shown in the first part of the proof of (d)) $\text{B-} \int_{\mathcal{J}} \psi_n(t) \, \mathrm{d}\lambda_{\mathcal{J}}(t) = \text{P-} \int_{\mathcal{J}} \psi_n(t) \, \mathrm{d}\lambda_{\mathcal{J}}(t)$ for every $n \in \mathbb{N}$, and since $(\text{B-} \int_{\mathcal{J}} \psi_n(t) \, \mathrm{d}\lambda_{\mathcal{J}}(t))_{n \in \mathbb{N}}$ converges to $\text{B-} \int_{\mathcal{J}} \varphi(t) \, \mathrm{d}\lambda_{\mathcal{J}}(t)$ uniformly on X, it follows that $\text{P-} \int_{\mathcal{J}} \varphi(t) \, \mathrm{d}\lambda_{\mathcal{J}}(t) = \text{B-} \int_{\mathcal{J}} \varphi(t) \, \mathrm{d}\lambda_{\mathcal{J}}(t)$. $\qquad\square$

We will now discuss the second type of pointwise integral that will be of interest to us in this book.

As before, we assume given a locally compact separable metric space (X, d).

Let $(Y, \mathcal{Y}, \nu)$ be a measure space, let $\varphi : Y \to \mathcal{M}(X)$, and let us think of $\mathcal{M}(X)$ as a Banach space of real-valued functions defined on $\mathcal{B}(X)$. Then it makes sense to consider the pointwise integrability of φ. Thus, we say that φ is pointwise measurable, or $\mathcal{M}(X)$-*pointwise measurable*, if the function $\varphi_A : Y \to \mathbb{R}$ defined by $\varphi_A(y) = \varphi(y)(A)$ for every $y \in Y$ is $\mathcal{Y}$-measurable for every $A \in \mathcal{B}(X)$, and we say that φ is pointwise integrable or $\mathcal{M}(X)$-*pointwise integrable* if, for every $A \in \mathcal{B}(X)$, the function φ_A is $\mathcal{Y}$-measurable and ν-integrable, and if the map $\mathbf{I} : \mathcal{B}(X) \to \mathbb{R}$ defined by $\mathbf{I}(A) = \int_Y \varphi_A(y) \, \mathrm{d}\nu(y)$ for every $A \in \mathcal{B}(X)$ belongs to $\mathcal{M}(X)$. If φ is $\mathcal{M}(X)$-pointwise integrable, then we call $\mathbf{I}$ the $\mathcal{M}(X)$-*pointwise integral of φ on Y*, and we use the notation $\text{P-} \int_Y \varphi(y) \, \mathrm{d}\nu(y)$ rather than $\mathbf{I}$. Since Y will usually be a subset of $\mathbb{T}$, the variable y will be denoted by t and we will prefer the notation φ_t rather than $\varphi(t)$. Also, as we generally do when dealing with pointwise integrals, we will prefer the notation $\text{P-} \int_Y \varphi_t(A) \, \mathrm{d}\nu(t)$ rather than $(\text{P-} \int_Y \varphi_t \, \mathrm{d}\nu(t))(A)$, which is the formally correct notation, but is cumbersome and less intuitive.

Since the first concern when dealing with an integral is to find conditions for integrability, we discuss such conditions in the next proposition.

Proposition 3.3.5. *Let $\varphi : Y \to \mathcal{M}(X)$ be an $\mathcal{M}(X)$-pointwise measurable function, and assume that there exists a ν-integrable function $\rho : Y \to \mathbb{R}$ such that $|\varphi_t(A)| \le \rho(t)$ for every $t \in Y$ and $A \in \mathcal{B}(X)$. Then φ is $\mathcal{M}(X)$-pointwise integrable. In particular, φ is $\mathcal{M}(X)$-pointwise integrable whenever the measure ν is finite and there exists an $M \in \mathbb{R}$ such that $|\varphi_t(A)| \le M$ for every $t \in Y$ and $A \in \mathcal{B}(X)$.*

Proof. Let $\varphi : Y \to \mathcal{M}(X)$ be a function that is $\mathcal{M}(X)$-pointwise measurable, and assume that there exists a ν-integrable map $\rho : Y \to \mathbb{R}$ such that $\sup_{A \in \mathcal{B}(X)} |\varphi_t(A)| \le \rho(t)$ for every $t \in Y$.

Since φ is $\mathcal{M}(X)$-pointwise measurable, given $A \in \mathcal{B}(X)$, it follows that the function φ_A defined before the proposition ($\varphi_A : Y \to \mathbb{R}$, $\varphi_A(t) = \varphi_t(A)$ for every $t \in Y$) is $\mathcal{Y}$-measurable, and, since $|\varphi_A(t)| \le \rho(t)$ for all $t \in Y$, it follows that φ_A is also ν-integrable. Therefore, the function $\mathbf{I} : \mathcal{B}(X) \to \mathbb{R}$, $\mathbf{I}(A) = \int_Y \varphi_A(t)\,d\nu(t)$ for every $A \in \mathcal{B}(X)$, is well-defined, in the sense that the integrals that appear in the definition of $\mathbf{I}(A)$, $A \in \mathcal{B}(X)$, exist. Clearly, $\mathbf{I}(\emptyset) = 0$; therefore, in order to prove that $\mathbf{I} \in \mathcal{M}(X)$, it only remains to prove that $\mathbf{I}$ is σ-additive.

To this end, let $(A_n)_{n \in \mathbb{N}}$ be a sequence of mutually disjoint $\mathcal{B}(X)$-measurable subsets of X. We have to prove that the series $\sum_{n=1}^{\infty} \mathbf{I}(A_n)$ is convergent and that

$$\sum_{n=1}^{\infty} \mathbf{I}(A_n) = \mathbf{I}(\cup_{n=1}^{\infty} A_n). \tag{3.3.1}$$

Since φ_t is a signed measure for every $t \in Y$, it follows that

$$\mathbf{I}(\cup_{i=1}^{n} A_i) = \int_Y \varphi_t(\cup_{i=1}^{n} A_i)\,d\nu(t) = \sum_{i=1}^{n} \int_Y \varphi_t(A_i)\,d\nu(t) = \sum_{i=1}^{n} \mathbf{I}(A_i)$$

for every $n \in \mathbb{N}$. Accordingly, if the sequence $(\mathbf{I}(\cup_{i=1}^{n} A_i))_{n \in \mathbb{N}}$ converges, then the series $\sum_{n=1}^{\infty} \mathbf{I}(A_n)$ converges, as well, and $\sum_{n=1}^{\infty} \mathbf{I}(A_n) = \lim_{n \to \infty} \mathbf{I}(\cup_{i=1}^{n} A_i)$. Therefore, in order to prove that the series $\sum_{n=1}^{\infty} \mathbf{I}(A_n)$ converges and that the equality (3.3.1) holds true, it is enough to prove that $\lim_{n \to \infty} \mathbf{I}(\cup_{i=1}^{n} A_i)$ exists and that

$$\lim_{n \to \infty} \mathbf{I}(\cup_{i=1}^{n} A_i) = \mathbf{I}(\cup_{n=1}^{\infty} A_n). \tag{3.3.2}$$

Since the sequence $(\varphi_{(\cup_{i=1}^{n} A_i)})_{n \in \mathbb{N}}$ (which is a sequence of real-valued ν-integrable functions defined on Y) converges to $\varphi_{(\cup_{n=1}^{\infty} A_n)}$ everywhere on Y, since $\left| \varphi_{(\cup_{i=1}^{n} A_i)}(t) \right| \le \rho(t)$ for every $n \in \mathbb{N}$ and $t \in Y$, and since ρ is a ν-integrable function, it follows that we can apply the Lebesgue dominated convergence theorem to the sequence $(\varphi_{(\cup_{i=1}^{n} A_i)})_{n \in \mathbb{N}}$ in order to conclude that the sequence of integrals $(\int_Y \varphi_{(\cup_{i=1}^{n} A_i)}(t)\,d\nu(t))_{n \in \mathbb{N}}$ converges and $\lim_{n \to \infty} \int_Y \varphi_{(\cup_{i=1}^{n} A_i)}(t)\,d\nu(t) = \int_Y \varphi_{(\cup_{i=1}^{\infty} A_i)}(t)\,d\nu(t)$.

We have therefore proved that the equality (3.3.2) holds true, so $\mathbf{I}$ is a real-valued signed Borel measure; that is, $\mathbf{I} \in \mathcal{M}(X)$. Accordingly, φ is $\mathcal{M}(X)$-pointwise integrable.

Assume now that ν is a finite measure and there exists an $M \in \mathbb{R}$ such that $|\varphi_t(A)| \le M$ for every $t \in Y$ and $A \in \mathcal{B}(X)$. Then, if we set $\rho = M \mathbf{1}_Y$ and use the above arguments, we obtain that φ is $\mathcal{M}(X)$-pointwise integrable in this case, as well.   □

As expected, we have defined the $\mathcal{M}(X)$-pointwise integral because we need it in the study of transition functions. In the following consequence of Proposition 3.3.5, we describe the kind of $\mathcal{M}(X)$-pointwise integral that will be used in the book:

Corollary 3.3.6. *Let $(P_t)_{t\in\mathbb{T}}$ be a transition function defined on (X,d) and let $((S_t, T_t))_{t\in\mathbb{T}}$ be the family of Markov pairs defined by $(P_t)_{t\in\mathbb{T}}$. Assume that $(P_t)_{t\in\mathbb{T}}$ satisfies the s.m.a., and let $\mathcal{J}$ be a Lebesgue measurable subset of $\mathbb{T}$ of finite Lebesgue measure. Then, for every $\mu \in \mathcal{M}(X)$, the function $\varphi^{(\mu)} : \mathcal{J} \to \mathcal{M}(X)$ defined by $\varphi_t^{(\mu)} = T_t\mu$ for every $t \in \mathcal{J}$ is $\mathcal{M}(X)$-pointwise integrable.*

Proof. We start by showing that the map $\varphi^{(\mu)}$ defined in the corollary is $\mathcal{M}(X)$-pointwise measurable.

To this end, assume first that $\mu \geq 0$ and let $A \in \mathcal{B}(X)$. Since $(P_t)_{t\in\mathbb{T}}$ satisfies the s.m.a., we can use (a) of Proposition 5.2.1, p. 159 of Cohn [20] applied to the measure spaces $(\mathcal{J}, \mathcal{L}(\mathcal{J}), \lambda_{\mathcal{J}})$ and $(X, \mathcal{B}(X), \mu)$, where $\mathcal{L}(\mathcal{J})$ is the σ-algebra of all Lebesgue measurable subsets of $\mathcal{J}$ and $\lambda_{\mathcal{J}}$ is the Lebesgue measure on $\mathcal{J}$, and to the map $(t, x) \mapsto P_t(x, A)$, $(t, x) \in \mathcal{J} \times X$, in order to obtain that the function $t \mapsto \varphi_t^{(\mu)}(A) = T_t\mu(A) = \int_X P_t(x, A)\, d\mu(x)$, $t \in \mathcal{J}$ is measurable. Since the map $t \mapsto \varphi_t^{(\mu)}(A)$, $t \in \mathcal{J}$, is measurable for every $A \in \mathcal{B}(X)$, it follows that $\varphi^{(\mu)}$ is $\mathcal{M}(X)$-pointwise measurable whenever $\mu \in \mathcal{M}(X)$, $\mu \geq 0$. If $\mu \in \mathcal{M}(X)$ is not necessarily a positive element of $\mathcal{M}(X)$, then $\varphi^{(\mu)} = \varphi^{(\mu^+)} - \varphi^{(\mu^-)}$, where μ^+ and μ^- are the positive measures that appear in the Jordan decomposition of μ ($\mu = \mu^+ - \mu^-$). Using our discussion so far, we obtain that both $\varphi^{(\mu^+)}$ and $\varphi^{(\mu^-)}$ are $\mathcal{M}(X)$-pointwise measurable, so $\varphi^{(\mu)}$ is $\mathcal{M}(X)$-pointwise measurable, as well.

Now, in order to prove that $\varphi^{(\mu)}$, $\mu \in \mathcal{M}(X)$, are $\mathcal{M}(X)$-pointwise integrable functions, we note that since T_t, $t \in \mathbb{T}$, are Markov operators (so, they are also positive contractions of $\mathcal{M}(X)$), we obtain that

$$\left|\varphi_t^{(\mu)}(A)\right| \leq \left|\varphi_t^{(\mu)}\right|(A) \leq \left|\varphi_t^{(\mu)}\right|(X) = |T_t\mu|(X)$$

$$\leq T_t|\mu|(X) = \|T_t|\mu|\| = \||\mu|\| = \|\mu\|$$

for every $t \in \mathcal{J}$, $A \in \mathcal{B}(X)$, and $\mu \in \mathcal{M}(X)$. Since $\mathcal{J}$ has finite Lebesgue measure, using Proposition 3.3.5 for $M = \|\mu\|$, we obtain that $\varphi^{(\mu)}$ is $\mathcal{M}(X)$-pointwise integrable for every $\mu \in \mathcal{M}(X)$.   □

In a similar manner as in the case of the $B_b(X)$-pointwise integral P-$\int_L S_t f\, dt$ whose notation was established after Proposition 3.3.2, we will denote the $\mathcal{M}(X)$-pointwise integral of the function $\varphi^{(\mu)}$ discussed in Corollary 3.3.6 by P-$\int_{\mathcal{J}} T_t\mu\, d\lambda_{\mathcal{J}}(t)$ or by P-$\int_{\mathcal{J}} T_t\mu\, dt$.

There is a useful and interesting relationship between the $B_b(X)$-pointwise integral considered in Proposition 3.3.2 and the $\mathcal{M}(X)$-pointwise integral that is studied in Corollary 3.3.6. We discuss this relationship in the next proposition.

Proposition 3.3.7. *Let $(P_t)_{t \in \mathbb{T}}$ be a transition function defined on (X, d) and assume that $(P_t)_{t \in \mathbb{T}}$ satisfies the s.m.a. Let $((S_t, T_t))_{t \in \mathbb{T}}$ be the family of Markov pairs defined by $(P_t)_{t \in \mathbb{T}}$, and let L be a Lebesgue measurable subset of $\mathbb{T}$ of finite Lebesgue measure. Then*

$$\left\langle P\text{-}\int_L S_t f \, dt, \mu \right\rangle = \left\langle f, P\text{-}\int_L T_t \mu \, dt \right\rangle \tag{3.3.3}$$

for every $f \in B_b(X)$ and $\mu \in \mathcal{M}(X)$.

Note that the equality (3.3.3) makes sense because the pointwise integrals P-$\int_L S_t f \, dt$ and P-$\int_L T_t \mu \, dt$ exist by Proposition 3.3.2 and Corollary 3.3.6, respectively.

Proof. We first note that it is easy to see that it is enough to prove the proposition in the case when $\mu \geq 0$, so we will assume that $\mu \geq 0$.

We will prove the proposition in two steps: first, under the assumption that f is a $\mathcal{B}(X)$-measurable simple function, and then for the general case when $f \in B_b(X)$ is not necessarily a simple function.

Step 1. It is easy to see that in order to prove (3.3.3) in the case when f is a $\mathcal{B}(X)$-measurable simple function (that is, when $f = \sum_{i=1}^m a_i \mathbf{1}_{A_i}$ for some $m \in \mathbb{N}$, m Borel measurable subsets $A_1, A_2, \ldots, A_m$ of X, and m real numbers $a_1, a_2, \ldots, a_m$) it is enough to prove that (3.3.3) holds true for $f = \mathbf{1}_A$, $A \in \mathcal{B}(X)$.
To this end, let $A \in \mathcal{B}(X)$.
Since $(P_t)_{t \in \mathbb{T}}$ satisfies the s.m.a., we can use (b) of Proposition 5.2.1 on p. 159 of Cohn [20], in order to obtain that

$$\left\langle P\text{-}\int_L S_t \mathbf{1}_A \, dt, \mu \right\rangle = \int_X (P\text{-}\int_L S_t \mathbf{1}_A(x) \, dt) \, d\mu(x)$$

$$= \int_X (\int_L P_t(x, A) \, dt) \, d\mu(x) = \int_L (\int_X P_t(x, A) \, d\mu(x)) \, dt$$

$$= P\text{-}\int_L T_t \mu(A) \, dt = \left\langle \mathbf{1}_A, P\text{-}\int_L T_t \mu \, dt \right\rangle.$$

Step 2. We now prove that (3.3.3) is true for every $f \in B_b(X)$. To this end, note that it is enough to prove (3.3.3) under the assumption that $f \geq 0$.
Thus, let $f \in B_b(X)$, $f \geq 0$.
Then there exists a monotone nondecreasing sequence $(f_n)_{n \in \mathbb{N}}$ of real-valued $\mathcal{B}(X)$-measurable simple positive functions on X such that $(f_n)_{n \in \mathbb{N}}$ converges uniformly to f. Accordingly, the sequence $(\langle f_n, P\text{-}\int_L T_t \mu \, dt \rangle)_{n \in \mathbb{N}}$ converges to $\langle f, P\text{-}\int_L T_t \mu \, dt \rangle$.
Since S_t is a positive contraction of $B_b(X)$, $(S_t f_n)_{n \in \mathbb{N}}$ is a monotone nondecreasing sequence of positive elements of $B_b(X)$ and converges

uniformly on X to $S_t f$ for every t. Therefore, for every $x \in X$, the sequence $(\xi_n^{(x)})_{n \in \mathbb{N}}$, $\xi_n^{(x)} : L \to \mathbb{R}$, $\xi_n^{(x)}(t) = S_t f_n(x)$ for every $t \in L$ and $n \in \mathbb{N}$ converges pointwise on L to $\xi^{(x)} : L \to \mathbb{R}$, $\xi^{(x)}(t) = S_t f(x)$ for every $t \in L$; since $(\xi_n^{(x)})_{n \in \mathbb{N}}$ is a monotone nondecreasing sequence of positive functions, by the monotone convergence theorem, it follows that the sequence $(\int_L \xi_n^{(x)}(t)\,dt)_{n \in \mathbb{N}}$ converges to $\int_L \xi^{(x)}(t)\,dt$. The convergence of $(\int_L \xi_n^{(x)}(t)\,dt)_{n \in \mathbb{N}}$ to $\int_L \xi^{(x)}(t)\,dt$ for every $x \in X$ means that the sequence of functions $(\text{P-}\int_L S_t f_n\,dt)_{n \in \mathbb{N}}$ converges pointwise on X to $\text{P-}\int_L S_t f\,dt$. Since $(\text{P-}\int_L S_t f_n\,dt)_{n \in \mathbb{N}}$ is a monotone nondecreasing sequence of positive $\mathcal{B}(X)$-measurable functions that converges pointwise on X to $\text{P-}\int_L S_t f\,dt$, we can use the monotone convergence theorem in the space $(X, \mathcal{B}(X), \mu)$ in order to infer that the sequence $(\int_X (\text{P-}\int_L S_t f_n(x)\,dt)\,d\mu(x))_{n \in \mathbb{N}}$ converges to $\int_X (\text{P-}\int_L S_t f(x)\,dt)\,d\mu(x)$; that is, the sequence $(\langle \text{P-}\int_L S_t f_n\,dt, \mu \rangle)_{n \in \mathbb{N}}$ converges to $\langle \text{P-}\int_L S_t f\,dt, \mu \rangle$.

Since by Step 1 $\langle \text{P-}\int_L S_t f_n\,dt, \mu \rangle = \langle f_n, \text{P-}\int_L T_t \mu\,dt \rangle$ for every $n \in \mathbb{N}$, it follows that $\langle \text{P-}\int_L S_t f\,dt, \mu \rangle = \langle f, \text{P-}\int_L T_t \mu\,dt \rangle$. $\qquad\square$

If L is an interval in $\mathbb{R}$ with endpoints a and b, $a \leq b$, and if $\varphi : L \to B_b(X)$, then we will often denote the $B_b(X)$-pointwise integral $\text{P-}\int_L \varphi(t)\,d\lambda_L(t)$, the Bochner integral $\text{B-}\int_L \varphi(t)\,d\lambda_L(t)$, and the DS-integral $\text{DS-}\int_L \varphi(t)\,d\lambda_L(t)$ (where λ_L is the Lebesgue measure on L) simply by $\text{P-}\int_a^b \varphi_t\,dt$, $\text{B-}\int_a^b \varphi_t\,dt$, and $\text{DS-}\int_a^b \varphi_t\,dt$, respectively; also, if $\psi : L \to \mathcal{M}(X)$, we will use the notation $\text{P-}\int_a^b \psi_t\,dt$ for the $\mathcal{M}(X)$-pointwise integral of ψ with respect to the Lebesgue measure on L.

3.3.2 *An Application: The Existence of Invariant Probabilities for Transition Functions*

An interesting approach in the study of a transition function $(P_t)_{t \in \mathbb{T}}$ is to study the relationship between the properties of an individual transition probability P_{t_0}, $t_0 \in \mathbb{T}$, $t_0 \neq 0$ that belongs to the collection $\{P_t \mid t \in \mathbb{T}\}$ of transition probabilities that define $(P_t)_{t \in \mathbb{T}}$, and the properties of the transition function $(P_t)_{t \in \mathbb{T}}$ as a whole. Although we will not pursue this approach systematically in the book, in this subsection, in order to illustrate the use of the pointwise integrals introduced in the previous subsection, we will briefly discuss the relationship between the existence and the support of an invariant probability for the transition function $(P_t)_{t \in \mathbb{T}}$, on the one hand, and the existence and the support of an invariant probability of a transition probability P_{t_0}, $t_0 \in \mathbb{T}$, $t_0 \neq 0$, on the other.

Let us mention that a rather surprising and interesting approach to the study of a transition function using a transition probability was taken by Worm and Hille in [132], where the authors obtain quite significant results about a transition function

$(P_t)_{t\in[0,+\infty)}$ by associating to $(P_t)_{t\in[0,+\infty)}$ a transition probability P which is related to $(P_t)_{t\in[0,+\infty)}$, but is not one of the transition probabilities P_t, $t \in [0, +\infty)$.

Let, as usual, in this book, $(P_t)_{t\in\mathbb{T}}$ be a transition function defined on a locally compact separable metric space (X, d), assume that $(P_t)_{t\in\mathbb{T}}$ satisfies the s.m.a., and let $((S_t, T_t))_{t\in\mathbb{T}}$ be the family of Markov pairs defined by $(P_t)_{t\in\mathbb{T}}$.

Let $\alpha \in \mathbb{T}$, $\alpha > 0$, assume that T_α has invariant probabilities (that is, assume that there exists a probability measure $\mu \in \mathcal{M}(X)$ such that $T_\alpha\mu = \mu$), and let μ_α be an invariant probability measure for T_α. (Note that, in general, μ_α is not an invariant probability for $(T_t)_{t\in\mathbb{T}}$.)

Now set $\nu_\alpha = \frac{1}{\alpha}(\text{P-}\int_0^\alpha T_t\mu_\alpha\, dt)$. Observe that, by Corollary 3.3.6, the pointwise integral that appears in the definition of ν_α does exist, so ν_α is a well defined element of $\mathcal{M}(X)$. Moreover, using the definition of an $\mathcal{M}(X)$-pointwise integral and the fact that μ_α is a probability measure, we obtain that ν_α is a probability measure, as well.

Our goal now is to prove that ν_α is an invariant probability measure for the transition function $(P_t)_{t\in\mathbb{T}}$. To this end, we need the following technical lemma in which we use the notation introduced so far.

Lemma 3.3.8. *Let $\mathcal{J}$ be a Lebesgue measurable subset of $\mathbb{T}$, assume that $\mathcal{J}$ has finite Lebesgue measure, let $c \in \mathbb{T}$, and let $A \in \mathcal{B}(X)$. Then:*

(a) $\int_{\mathcal{J}} S_{u+c}\mathbf{1}_A(x)\, du = \int_{\mathcal{J}+c} S_u\mathbf{1}_A(x)\, du$ *for every $x \in X$.*
(b) $\int_{\mathcal{J}} T_{u+c}\mu_\alpha(A)\, du = \int_{\mathcal{J}+c} T_u\mu_\alpha(A)\, du$.

Proof. For both (a) and (b) we will use the fact that if $f : \mathbb{R} \to \mathbb{R}$ is a Lebesgue integrable function, then, for every $c \in \mathbb{R}$, the function $u \mapsto f(u + c)$, $u \in \mathbb{R}$, is also Lebesgue integrable and $\int_{\mathbb{R}} f(u + c)\, du = \int_{\mathbb{R}} f(u)\, du$; the assertions about f are true because the Lebesgue measure is translation invariant.

(a) Let $x \in X$ and let $g : \mathbb{R} \to \mathbb{R}$ be defined by

$$g(u) = \begin{cases} S_u\mathbf{1}_A(x) & \text{if } u \in \mathbb{T} \\ 0 & \text{if } u \in \mathbb{R} \setminus \mathbb{T}, \end{cases}$$

and let $f : \mathbb{R} \to \mathbb{R}$ be defined by $f(u) = \mathbf{1}_{\mathcal{J}+c}(u)g(u)$ for every $u \in \mathbb{R}$. Since $(P_t)_{t\in\mathbb{T}}$ satisfies the s.m.a., it follows that g is measurable, so using the fact that $\mathcal{J} + c$ is a measurable subset of $\mathbb{R}$ of finite Lebesgue measure, we obtain that f is Lebesgue integrable.

It follows that

$$\int_{\mathcal{J}} S_{u+c}\mathbf{1}_A(x)\, du = \int_{\mathbb{R}} \mathbf{1}_{\mathcal{J}}(u)S_{u+c}\mathbf{1}_A(x)\, du = \int_{\mathbb{R}} \mathbf{1}_{\mathcal{J}+c}(u + c)g(u + c)\, du$$

$$= \int_{\mathbb{R}} f(u + c)\, du = \int_{\mathbb{R}} f(u)\, du = \int_{\mathbb{R}} \mathbf{1}_{\mathcal{J}+c}(u)S_u\mathbf{1}_A(x)\, du = \int_{\mathcal{J}+c} S_u\mathbf{1}_A(x)\, du.$$

Note that in the above equalities we used the fact that both $\mathcal{J}$ and $\mathcal{J}+c$ are subsets of $\mathbb{T}$.

(b) In this case, let $g : \mathbb{R} \to \mathbb{R}$ be defined by

$$g(u) = \begin{cases} T_u \mu_\alpha(A) & \text{if } u \in \mathbb{T} \\ 0 & \text{if } u \in \mathbb{R} \setminus \mathbb{T}. \end{cases}$$

Since $(P_t)_{t\in\mathbb{T}}$ satisfies the s.m.a., given $u \in \mathbb{T}$, the map $x \mapsto P_u(x, A)$, $x \in X$, is $\mathcal{B}(X)$-measurable; since $x \mapsto P_u(x, A)$, $x \in X$, is also a positive bounded function, and since μ_α is a probability measure, it follows that $\int_X P_u(x, A) \, d\mu_\alpha(x)$ exists and is a real number. Using (a) of Proposition 5.2.1, p. 159, of Cohn [20], we obtain that the map $u \mapsto \int_X P_u(x, A) \, d\mu_\alpha(x) = T_u \mu_\alpha(A)$, $u \in \mathbb{T}$, is measurable, so g is measurable.

Let $f : \mathbb{R} \to \mathbb{R}$ be defined by $f(u) = \mathbf{1}_{\mathcal{J}+c}(u) g(u)$ for all $u \in \mathbb{R}$. Clearly, f is measurable, and, since the set $\mathcal{J} + c$ has finite Lebesgue measure and f is a bounded function, it follows that f is also integrable.

Using the fact that, as pointed out at (a), both $\mathcal{J}$ and $\mathcal{J} + c$ are subsets of $\mathbb{T}$, we obtain that

$$\int_{\mathcal{J}} T_{u+c} \mu_\alpha(A) \, du = \int_{\mathbb{R}} \mathbf{1}_{\mathcal{J}}(u) T_{u+c} \mu_\alpha(A) \, du = \int_{\mathbb{R}} \mathbf{1}_{\mathcal{J}+c}(u+c) T_{u+c} \mu_\alpha(A) \, du$$

$$= \int_{\mathbb{R}} f(u+c) \, du = \int_{\mathbb{R}} f(u) \, du = \int_{\mathbb{R}} \mathbf{1}_{\mathcal{J}+c}(u) T_u \mu_\alpha(A) \, du = \int_{\mathcal{J}+c} T_u \mu_\alpha(A) \, du.$$

$\square$

Theorem 3.3.9. *The measure ν_α is an invariant probability measure for the transition function $(P_t)_{t\in\mathbb{T}}$; that is, ν_α has the property that $T_\beta \nu_\alpha = \nu_\alpha$ for every $\beta \in \mathbb{T}$.*

Proof. Let $\beta \in \mathbb{T}$. We have to prove that

$$\left(T_\beta\left(\frac{1}{\alpha}(\text{P-}\int_0^\alpha T_t \mu_\alpha \, dt)\right)\right)(A) = \frac{1}{\alpha}(\text{P-}\int_0^\alpha T_t \mu_\alpha \, dt)(A),$$

or, equivalently, that $\left\langle \mathbf{1}_A, T_\beta(\frac{1}{\alpha}(\text{P-}\int_0^\alpha T_t \mu_\alpha \, dt)) \right\rangle = \left\langle \mathbf{1}_A, \frac{1}{\alpha}(\text{P-}\int_0^\alpha T_t \mu_\alpha \, dt) \right\rangle$ for every $A \in \mathcal{B}(X)$.

We first note that, using the implication $(b) \Rightarrow (a)$ of Proposition 2.3.7, we obtain that, if $\mathbb{T} = \mathbb{R}$, it is enough to prove that $T_\beta \nu_\alpha = \nu_\alpha$ whenever $\beta \in \mathbb{T}$, $\beta \geq 0$. Thus, we may and do assume that $\beta \geq 0$. Consequently, there exist a unique integer $l \in \mathbb{N} \cup \{0\}$ and a unique $r \in \mathbb{R}$, $0 \leq r < \alpha$, such that $\beta = l\alpha + r$.

Now, let $A \in \mathcal{B}(X)$.

Using the equality (3.3.3) of Proposition 3.3.7 and Lemma 3.3.8, we obtain that

$$\left\langle \mathbf{1}_A, T_\beta\big(\frac{1}{\alpha}(\text{P-}\int_0^\alpha T_t\mu_\alpha\,\mathrm{d}t)\big)\right\rangle = \frac{1}{\alpha}\left\langle S_\beta\mathbf{1}_A, \text{P-}\int_0^\alpha T_t\mu_\alpha\,\mathrm{d}t\right\rangle$$

$$= \frac{1}{\alpha}\left\langle \text{P-}\int_0^\alpha S_t S_\beta\mathbf{1}_A\,\mathrm{d}t, \mu_\alpha\right\rangle = \frac{1}{\alpha}\Big(\int_X\Big(\int_0^\alpha S_{t+\beta}\mathbf{1}_A(x)\,\mathrm{d}t\Big)\,\mathrm{d}\mu_\alpha(x)\Big)$$

$$= \frac{1}{\alpha}\Big(\int_X\Big(\int_\beta^{\alpha+\beta} S_t\mathbf{1}_A(x)\,\mathrm{d}t\Big)\,\mathrm{d}\mu_\alpha(x)\Big) = \frac{1}{\alpha}\left\langle \text{P-}\int_\beta^{\alpha+\beta} S_t\mathbf{1}_A\,\mathrm{d}t, \mu_\alpha\right\rangle$$

$$= \frac{1}{\alpha}\left\langle \mathbf{1}_A, \Big(\text{P-}\int_\beta^{\alpha+\beta} T_t\mu_\alpha\,\mathrm{d}t\Big)\right\rangle = \frac{1}{\alpha}\left\langle \mathbf{1}_A, \Big(\text{P-}\int_0^\alpha T_{t+\beta}\mu_\alpha\,\mathrm{d}t\Big)\right\rangle$$

$$= \frac{1}{\alpha}\left\langle \mathbf{1}_A, \Big(\text{P-}\int_0^\alpha T_t T_\beta\mu_\alpha\,\mathrm{d}t\Big)\right\rangle = \frac{1}{\alpha}\left\langle \mathbf{1}_A, \Big(\text{P-}\int_0^\alpha T_t T_r T_\alpha^l\mu_\alpha\,\mathrm{d}t\Big)\right\rangle$$

$$= \frac{1}{\alpha}\left\langle \mathbf{1}_A, \Big(\text{P-}\int_0^\alpha T_{t+r}\mu_\alpha\,\mathrm{d}t\Big)\right\rangle = \frac{1}{\alpha}\left\langle \mathbf{1}_A, \Big(\text{P-}\int_r^{\alpha+r} T_t\mu_\alpha\,\mathrm{d}t\Big)\right\rangle$$

$$= \frac{1}{\alpha}\left\langle \mathbf{1}_A, \Big(\big(\text{P-}\int_r^\alpha T_t\mu_\alpha\,\mathrm{d}t\big) + \big(\text{P-}\int_\alpha^{\alpha+r} T_t\mu_\alpha\,\mathrm{d}t\big)\Big)\right\rangle$$

$$= \frac{1}{\alpha}\left\langle \mathbf{1}_A, \Big(\big(\text{P-}\int_r^\alpha T_t\mu_\alpha\,\mathrm{d}t\big) + \big(\text{P-}\int_0^r T_{t+\alpha}\mu_\alpha\,\mathrm{d}t\big)\Big)\right\rangle$$

$$= \frac{1}{\alpha}\left\langle \mathbf{1}_A, \Big(\big(\text{P-}\int_r^\alpha T_t\mu_\alpha\,\mathrm{d}t\big) + \big(\text{P-}\int_0^r T_t T_\alpha\mu_\alpha\,\mathrm{d}t\big)\Big)\right\rangle$$

$$= \frac{1}{\alpha}\left\langle \mathbf{1}_A, \Big(\big(\text{P-}\int_r^\alpha T_t\mu_\alpha\,\mathrm{d}t\big) + \big(\text{P-}\int_0^r T_t\mu_\alpha\,\mathrm{d}t\big)\Big)\right\rangle$$

$$= \frac{1}{\alpha}\left\langle \mathbf{1}_A, \Big(\text{P-}\int_0^\alpha T_t\mu_\alpha\,\mathrm{d}t\Big)\right\rangle.$$

Since the above equalities hold true for every $A \in \mathcal{B}(X)$, it follows that $T_\beta\nu_\alpha = \nu_\alpha$ for every $\beta \in \mathbb{T}$. $\qquad\square$

Using the fact that an invariant probability measure for the transition function $(P_t)_{t\in\mathbb{T}}$ is an invariant probability for every transition probability P_t, $t \in \mathbb{T}$, and the comments that follow Proposition 2.3.7, we obtain the following obvious consequence of Theorem 3.3.9:

Corollary 3.3.10. *For the transition function $(P_t)_{t\in\mathbb{T}}$ that is defined on (X, d) and satisfies the s.m.a., the following statements hold true:*

(a) If $\mathbb{T} = [0, +\infty)$, then the following assertions are equivalent:

(i) There exists a probability measure $\nu \in \mathcal{M}(X)$ such that ν is an invariant measure for $(P_t)_{t\in\mathbb{T}}$ (that is, $(P_t)_{t\in\mathbb{T}}$ has invariant probability measures).

(ii) *There exists a $t_0 \in \mathbb{T}$, $t_0 > 0$, such that the transition probability P_{t_0} has invariant probability measures (that is, there exists a probability measure $\mu \in \mathcal{M}(X)$ such that μ is invariant for P_{t_0}).*

(b) *If $\mathbb{T} = \mathbb{R}$ and if the transition probability P_0 is defined by $P_0(x, A) = \mathbf{1}_A(x)$ for every $x \in X$ and $A \in \mathcal{B}(X)$ (that is, if P_0 has the property that the operators S_0 and T_0 that appear in the Markov pair (S_0, T_0) defined by P_0 are the identity operators on $B_b(X)$ and $\mathcal{M}(X)$, respectively), then the following assertions are equivalent:*

(1) *The transition function $(P_t)_{t \in \mathbb{R}}$ has at least one invariant probability measure.*

(2) *There exists a $t_0 \in \mathbb{R}$, $t_0 \neq 0$, such that the transition probability P_{t_0} has at least one invariant probability measure.*

Observation. The condition that $P_0(x, A) = \mathbf{1}_A(x)$ for every $x \in X$ and $A \in \mathcal{B}(X)$ imposed on the transition probability P_0 in (b) of the above corollary is needed in order to make sure that a probability measure $\mu \in \mathcal{M}(X)$ is invariant for the transition probability P_t if and only if μ is invariant for P_{-t} whenever $t \in \mathbb{R}$ (see the comments that appear after Proposition 2.3.7); the fact that μ is invariant for P_t if and only if μ is invariant for $P_{|t|}$ whenever $t \in \mathbb{R}$, $t \neq 0$, is then used in order to be able to apply Theorem 3.3.9 when proving the implication $(2) \Rightarrow (1)$ in the corollary.

Note that the above-mentioned condition on P_0 is rather mild and is satisfied quite often; for instance, it is satisfied whenever $(P_t)_{t \in \mathbb{R}}$ is defined by a flow.   ▲

It is important to be aware that the probability measures μ_α and ν_α defined just before Lemma 3.3.8 are distinct, in general. For instance, let $\mathbf{w} = (w_t)_{t \in \mathbb{R}}$ be the flow of the rotations of the unit circle (the flow is defined in Example A.3.4), let $(P_t^{(\mathbf{w})})_{t \in \mathbb{R}}$ be the transition function defined by $\mathbf{w}$, and let $((S_t^{(\mathbf{w})}, T_t^{(\mathbf{w})}))_{t \in \mathbb{R}}$ be the family of Markov pairs defined by $\mathbf{w}$ (see Example 2.2.4); it is easy to see that $T_1^{(\mathbf{w})}$ is the identity operator on $\mathcal{M}(\mathbb{R}/\mathbb{Z})$, so any probability measure in $\mathcal{M}(\mathbb{R}/\mathbb{Z})$ is invariant for $T_1^{(\mathbf{w})}$; in particular the Dirac measure $\delta_{\hat{0}}$ is such an invariant probability measure; now, let $\alpha = 1$ and set $\mu_\alpha = \delta_{\hat{0}}$; then ν_α is the Haar-Lebesgue measure on $\mathbb{R}/\mathbb{Z}$ because, by Theorem 3.3.9, ν_α is an invariant probability measure for the transition function $(P_t^{(\mathbf{w})})_{t \in \mathbb{R}}$, and, as pointed out in Example 2.3.8, $(P_t^{(\mathbf{w})})_{t \in \mathbb{R}}$ is strictly ergodic and the unique invariant probability for $(P_t^{(\mathbf{w})})_{t \in \mathbb{R}}$ is the Haar-Lebesgue measure on $\mathbb{R}/\mathbb{Z}$. Thus, $\mu_\alpha \neq \nu_\alpha$ in this case.

Looking at the transition function $(P_t^{(\mathbf{w})})_{t \in \mathbb{R}}$ discussed above (where $\mathbf{w}$ is the flow of the rotations of the unit circle), we note that, even though $(P_t^{(\mathbf{w})})_{t \in \mathbb{R}}$ is uniquely ergodic, some of the transition probabilities $P_t^{(\mathbf{w})}$, $t \in \mathbb{R}$, fail to be uniquely ergodic (for instance, the transition probability $P_1^{(\mathbf{w})}$ is not uniquely ergodic), so, the unique ergodicity of a transition function does not imply the unique ergodicity of the transition probabilities that define the transition function; on the other hand, it follows from Example 1.4.27 that if $t \notin \mathbb{Q}$, then the transition probability $P_t^{(\mathbf{w})}$ is uniquely ergodic. Thus, a natural question is whether or not a transition function $(P_t)_{t \in \mathbb{T}}$ (that satisfies the s.m.a.) is uniquely ergodic whenever there exists a $t_0 \in \mathbb{T}$, $t_0 \neq 0$, such that the transition probability P_{t_0} is uniquely ergodic. Under the rather general conditions of Corollary 3.3.10, the answer is yes and is discussed in the next corollary.

Corollary 3.3.11. *Let, as in Corollary 3.3.10, $(P_t)_{t \in \mathbb{R}}$ be a transition function defined on (X, d) and assume that $(P_t)_{t \in \mathbb{R}}$ satisfies the s.m.a.*

(a) *Assume that $\mathbb{T} = [0, +\infty)$ and that there exists a $t_0 \in [0, +\infty)$, $t_0 > 0$, such that the transition probability P_{t_0} is uniquely ergodic. Then the transition function $(P_t)_{t \in [0,+\infty)}$ is also uniquely ergodic.*
(b) *Assume that $\mathbb{T} = \mathbb{R}$, and that the transition probability P_0 is defined by $P_0(x, A) = \mathbf{1}_A(x)$ for every $x \in X$ and $A \in \mathcal{B}(X)$. If there exists a $t_0 \in \mathbb{R}$, $t_0 \neq 0$, such that the transition probability P_{t_0} is uniquely ergodic, then the transition function $(P_t)_{t \in \mathbb{R}}$ is uniquely ergodic, as well.*

Proof. (a) Let $\mathbb{T} = [0, +\infty)$ and assume that for some $t_0 \in [0, +\infty)$, $t_0 > 0$, the transition probability P_{t_0} is uniquely ergodic. Then, using (a) of Corollary 3.3.10, we obtain that the transition function $(P_t)_{t \in [0,+\infty)}$ has at least one invariant probability measure. However, $(P_t)_{t \in [0,+\infty)}$ cannot have more than one invariant probability measure because, if we assume that $(P_t)_{t \in [0,+\infty)}$ has more than one invariant probability measure, then the transition probability P_{t_0} fails to be uniquely ergodic.
(b) The proof is similar to the proof of (a), but using (b) of Corollary 3.3.10, of course.   □

A Feller transition function that satisfies the s.m.a. and is defined on a compact metric space always has invariant probability measures. This fact, which is a consequence of Theorem 3.3.9, is discussed in the next corollary.

Corollary 3.3.12. *Assume that (X, d) is a compact metric space, and let $(P_t)_{t \in \mathbb{T}}$ be a Feller transition function that satisfies the s.m.a. Then $(P_t)_{t \in \mathbb{T}}$ has at least one invariant probability measure.*

Proof. Using Theorem 1.4.2, we obtain that a Feller transition probability defined on a compact metric space has invariant probability measures. Therefore, the transition probability P_1 has at least one invariant probability measure. Using Theorem 3.3.9, we obtain that the transition function $(P_t)_{t \in \mathbb{T}}$ has invariant probability measures, as well.   □

Chapter 4
Special Topics

In this chapter, we present several topics in measure theory, general topology, and functional analysis that will be used to obtain the results discussed in the rest of the book. Although necessary, the topics discussed in the chapter do not have such a dominant influence on the theory developed in the remaining part of the work as the topics discussed in Chap. 3.

In Sect. 4.1 we discuss certain measure theoretical facts about functions constant almost everywhere, in Sect. 4.2 we review briefly conditional expectations, in Sect. 4.3 we discuss in some detail continuous-time limit supports in the mean (the continuous-time versions of the corresponding notions for sequences of probability measures discussed in Sect. 1.3), in Sects. 4.4 and 4.5 we discuss continuous-time Banach limits and the Ascoli-Arzelà theorem, respectively, and finally, in Sect. 4.6, we go over several topics in ordered vector lattices, Banach lattices, and positive operators.

4.1 Functions Constant Almost Everywhere

Let $(Y, \mathcal{Y}, v)$ be a probability space.

We will denote by $\tilde{B}(Y, \mathcal{Y}, v)$ the set of all $\mathcal{Y}$-measurable functions $f : Y \to \mathbb{R} \cup \{+\infty, -\infty\}$. As usual, we say that two elements f and g of $\tilde{B}(Y, \mathcal{Y}, v)$ are *equal v-almost everywhere* (*v-a.e.*) if there exists a $\mathcal{Y}$-measurable subset N of Y such that $v(N) = 0$ and $f(x) = g(x)$ for every $x \in Y \setminus N$. Clearly, the equality v-a.e. is an equivalence relation on $\tilde{B}(Y, \mathcal{Y}, v)$ and we will denote by $\tilde{\mathbb{M}}(Y, \mathcal{Y}, v)$ the set of all equivalence classes defined by this equivalence relation.

As usual, given a $\mathcal{Y}$-measurable function $f : Y \to \mathbb{R} \cup \{+\infty, -\infty\}$, we say that f is *constant v-a.e.* if there exist a $\mathcal{Y}$-measurable subset N of Y such that $v(N) = 0$, and $\alpha \in \mathbb{R} \cup \{+\infty, -\infty\}$ such that $f(x) = \alpha$ for every $x \in Y \setminus N$. If $\tilde{f} \in \tilde{\mathbb{M}}(Y, \mathcal{Y}, v)$, we say that $\tilde{f}$ is *constant v-a.e.* if there exists an element g of $\tilde{B}(Y, \mathcal{Y}, v)$ in the equivalence class $\tilde{f}$ such that g is constant v-a.e. Plainly, $\tilde{f} \in$

R. Zaharopol, *Invariant Probabilities of Transition Functions*, Probability and Its Applications 44, DOI 10.1007/978-3-319-05723-1_4,
© Springer International Publishing Switzerland 2014

$\tilde{\mathbb{M}}(Y, \mathcal{Y}, \nu)$ is constant ν-a.e. if and only if every $\mathcal{Y}$-measurable function g in the equivalence class $\bar{f}$ is constant ν-a.e.

Our goal in this section is to discuss necessary and sufficient conditions for an element of $\tilde{B}(Y, \mathcal{Y}, \nu)$ or of $\tilde{\mathbb{M}}(Y, \mathcal{Y}, \nu)$ to be constant ν-a.e.

Proposition 4.1.1. *Let $f \in \tilde{B}(Y, \mathcal{Y}, \nu)$. The following assertions are equivalent:*

(1) *f is constant ν-a.e.*
(2) *$\nu(\{x \in Y \mid f(x) > a\}) = 0$ or 1 for every $a \in \mathbb{R}$.*
(3) *$\nu(\{x \in Y \mid f(x) < a\}) = 0$ or 1 for every $a \in \mathbb{R}$.*
(4) *$\nu(\{x \in Y \mid f(x) \geq a\}) = 0$ or 1 for every $a \in \mathbb{R}$.*
(5) *$\nu(\{x \in Y \mid f(x) \leq a\}) = 0$ or 1 for every $a \in \mathbb{R}$.*

Proof. $(1) \Rightarrow (2)$ is obvious.
$(2) \Rightarrow (1)$ Under the assumption that (2) holds true, we have to prove that f is constant ν-a.e. in the following three cases:

$\quad$ (a) $\nu(\{x \in Y \mid f(x) > \alpha\}) = 0$ for every $\alpha \in \mathbb{R}$.
$\quad$ (b) $\nu(\{x \in Y \mid f(x) > \alpha\}) = 1$ for every $\alpha \in \mathbb{R}$.
$\quad$ (c) There exist $a \in \mathbb{R}$ and $b \in \mathbb{R}$ such that $\nu(\{x \in Y \mid f(x) > a\}) = 1$ and $\nu(\{x \in Y \mid f(x) > b\}) = 0$.

(a) Since we assume that $\nu(\{x \in Y \mid f(x) > \alpha\}) = 0$ for every $\alpha \in \mathbb{R}$, it follows that if we set $A = \bigcap_{n \in \mathbb{Z}} \nu(\{x \in Y \mid f(x) \leq n\})$, then $f(x) = -\infty$ for every $x \in A$, and $\nu(A) = 1 - \nu(Y \setminus A) = 1 - \nu(\cup_{n \in \mathbb{Z}}\{x \in Y \mid f(x) > n\}) = 1 - \sum_{n \in \mathbb{Z}} \nu(\{x \in Y \mid f(x) > n\}) = 1$.
$\quad$ Thus, $f = -\infty$ ν-a.e.
(b) In a similar manner as in (a), if $\nu(\{x \in Y \mid f(x) > \alpha\}) = 1$ for every $\alpha \in \mathbb{R}$, then $f = +\infty$ ν-a.e. because if we set $A = \cap_{n \in \mathbb{Z}}\{x \in Y \mid f(x) > n\}$, then $f(x) = +\infty$ for every $x \in A$ and

$$\nu(A) = \nu(Y \setminus (\cup_{n \in \mathbb{Z}}\{x \in Y \mid f(x) \leq n\}))$$

$$= 1 - \nu(\cup_{n \in \mathbb{Z}}\{x \in Y \mid f(x) \leq n\})$$

$$\geq 1 - \sum_{n \in \mathbb{Z}} \nu(\{x \in Y \mid f(x) \leq n\}) = 1$$

$\quad$ since $\nu(\{x \in Y \mid f(x) \leq n\}) = 0$ for every $n \in \mathbb{Z}$.
(c) Let $a \in \mathbb{R}$ and $b \in \mathbb{R}$ be such that $\nu(\{x \in Y \mid f(x) > a\}) = 1$ and $\nu(\{x \in Y \mid f(x) > b\}) = 0$.

Taking into consideration that for every $c \in \mathbb{R}$, $d \in \mathbb{R}$, $c \leq d$, we have $\{x \in Y \mid f(x) > d\} \subseteq \{x \in Y \mid f(x) > c\}$, and since ν is a (positive) measure, we obtain that $a < b$.

Set $\alpha = \inf\{c \in \mathbb{R} \mid \nu(\{x \in Y \mid f(x) > c\}) = 0\}$.
Then $\alpha \in \mathbb{R}$ and $a \leq \alpha \leq b$.

Let $B_n = \{x \in Y \mid f(x) > \alpha - \frac{1}{n}\}$ and $C_n = \{x \in Y \mid f(x) \leq \alpha - \frac{1}{n}\}$, and set $A_n = B_n \cap C_n$ for every $n \in \mathbb{N}$.

Then, since we assume that (2) holds true and using the definition of α, we obtain that $v(B_n) = 1$; again using the definition of α we obtain that $v(C_n) = 1$ because $v(Y \setminus C_n) = v(\{x \in Y \mid f(x) > \alpha - \frac{1}{n}\}) = 0$; thus, $v(A_n) = v(B_n \cap C_n) = 1$ for every $n \in \mathbb{N}$.

Now set $A = \bigcap_{n \in \mathbb{N}} A_n$.

Then $v(A) = 1 - v(Y \setminus (\cap_{n \in \mathbb{N}} A_n)) = 1 - v(\cup_{n \in \mathbb{N}}(Y \setminus A_n)) \geq 1 - \sum_{n=1}^{\infty} v(Y \setminus A_n) = 1$; therefore $v(A) = 1$.

Since $A = \{x \in Y \mid \alpha - \frac{1}{n} < f(x) \leq \alpha - \frac{1}{n}$ for every $n \in \mathbb{N}\} = \{x \in Y \mid f(x) = \alpha\}$, it follows that f is constant v-a.e.

The implications (2) $\Leftrightarrow$ (5) are obvious because $\{x \in Y \mid f(x) > a\} = Y \setminus \{x \in Y \mid f(x) \leq a\}$ for every $a \in \mathbb{R}$.

(3) $\Rightarrow$ (5) is true because $\{x \in Y \mid f(x) \leq a\} = \cap_{n \in \mathbb{N}}\{x \in Y \mid f(x) < a + \frac{1}{n}\}$ for every $a \in \mathbb{R}$.

(5) $\Rightarrow$ (3) because $\{x \in Y \mid f(x) < a\} = \cup_{n \in \mathbb{N}}\{x \in Y \mid f(x) \leq a - \frac{1}{n}\}$ for every $a \in \mathbb{R}$.

The implications (3) $\Leftrightarrow$ (4) are true because $\{x \in Y \mid f(x) < a\} = Y \setminus \{x \in Y \mid f(x) \geq a\}$ for every $a \in \mathbb{R}$. $\qquad\square$

The above proposition has the following obvious consequence:

Corollary 4.1.2. *Let $\bar{f} \in \tilde{\mathbb{M}}(Y, \mathcal{Y}, v)$. The following assertions are equivalent:*

(a) *$\bar{f}$ is constant v-a.e.*
(b) *There exists a $g \in \tilde{B}(Y, \mathcal{Y}, v)$ such that g belongs to the equivalence class $\bar{f}$ and g satisfies any of the five equivalent assertions of Proposition 4.1.1.*
(c) *Every $g \in \tilde{B}(Y, \mathcal{Y}, v)$ that belongs to $\bar{f}$ satisfies any of the five equivalent assertions of Proposition 4.1.1.*

4.2 Conditional Expectation

Our goal in this section is to briefly review several facts about conditional expectations that will be used in Chap. 6. In our discussion here, we follow Section 4.3 of Neveu's monograph [81] (see also Chapter 6, Section 34 of Billingsley's book [12]).

Let $(Y, \mathcal{Y}, v)$ be a probability space and let $L^1(Y, \mathcal{Y}, v)$ be the usual Banach space of all equivalence classes generated by the equivalence relation defined by the equality v-a.e. of two real-valued v-integrable functions (these equivalence classes are often called *equivalence classes modulo v* and we will use this terminology, as well). Also, let $\mathcal{S}$ be a sub-σ-algebra of $\mathcal{Y}$, and let $v_\mathcal{S}$ be the restriction of v to $\mathcal{S}$. Then $(Y, \mathcal{S}, v_\mathcal{S})$ is a probability space in its own right.

Now let $\bar{f} \in L^1(Y, \mathcal{Y}, v)$, $\bar{f} \geq 0$. Then $\bar{f}v$ is a finite (positive) measure on $(Y, \mathcal{Y})$, and, clearly, $\bar{f}v$ is absolutely continuous with respect to v.

Let $v_{\mathcal{S}}^{(\bar{f})}$ be the restriction of $\bar{f}v$ to $\mathcal{S}$. It is easy to see that $v_{\mathcal{S}}^{(\bar{f})}$ is absolutely continuous with respect to $v_{\mathcal{S}}$. Thus, by the Radon-Nikodým theorem, there exists an $\mathcal{S}$-measurable real-valued function g on Y such that $v_{\mathcal{S}}^{(\bar{f})} = gv_{\mathcal{S}}$. It can be shown that g is unique $v_{\mathcal{S}}$-a.e. in the sense that if g' is another $\mathcal{S}$-measurable real-valued function on Y such that $v_{\mathcal{S}}^{(\bar{f})} = g'v_{\mathcal{S}}$, then $g = g'$ $v_{\mathcal{S}}$-a.e.; in other words, given $v_{\mathcal{S}}^{(\bar{f})}$ and $v_{\mathcal{S}}$ as above, there exists a unique element $\bar{g} \in L^1(Y, \mathcal{S}, v_{\mathcal{S}})$ such that $v_{\mathcal{S}}^{(\bar{f})} = \bar{g}v_{\mathcal{S}}$. The element $\bar{g}$ is usually denoted $E(\bar{f}|\mathcal{S})$. Given a measurable function f in the class $\bar{f}$, we might also use the notation $v_{\mathcal{S}}^{(f)}$ for $v_{\mathcal{S}}^{(\bar{f})}$, and, in this case, a real-valued $\mathcal{S}$-measurable function g such that $v_{\mathcal{S}}^{(f)} = gv_{\mathcal{S}}$ is denoted by $E(f|\mathcal{S})$. The function $E(f|\mathcal{S})$ or the equivalence class $E(\bar{f}|\mathcal{S})$ are both called *conditional expectations of f or of $\bar{f}$ with respect to $\mathcal{S}$*, respectively.

If $\bar{f} \in L^1(Y, \mathcal{Y}, v)$ is not necessarily a positive element of $L^1(Y, \mathcal{Y}, v)$, $\bar{f} = \bar{f}^+ - \bar{f}^-$, where $\bar{f}^+ = \bar{f} \vee \bar{0}$ and $\bar{f}^- = (-\bar{f}) \vee \bar{0}$, and $\bar{0}$ is the zero element of $L^1(Y, \mathcal{Y}, v)$, then the conditional expectation of $\bar{f}$ with respect to $\mathcal{S}$ is denoted by $E(\bar{f}|\mathcal{S})$ and is defined as the difference $E(\bar{f}^+|\mathcal{S}) - E(\bar{f}^-|\mathcal{S})$. Similarly, if f is a not necessarily positive real-valued $\mathcal{Y}$-measurable v-integrable function on Y, $f = f^+ - f^-$ where $f^+ = f \vee 0$ and $f^- = (-f) \vee 0$, then the conditional expectation of f with respect to $\mathcal{S}$ is denoted by $E(f|\mathcal{S})$ and is defined by $E(f^+|\mathcal{S}) - E(f^-|\mathcal{S})$.

The following theorem is often useful:

Theorem 4.2.1. *Let f be a real-valued $\mathcal{Y}$-measurable v-integrable function on Y, and let g be a real-valued $\mathcal{S}$-measurable function on Y. The following two assertions are equivalent:*

(a) $g = E(f|\mathcal{S})$.
(b) $\int_A g \, dv = \int_A f \, dv$ *for every $A \in \mathcal{S}$.*

For a proof of Theorem 4.2.1, see p. 466 of Billingsley's book [12].

Finally, we note that if $f : Y \to \mathbb{R}$ is v-integrable and happens to be not just $\mathcal{Y}$-measurable, but also $\mathcal{S}$-measurable, as well, then $\bar{f} = E(\bar{f}|\mathcal{S})$; that is, $\bar{f}$ is equal to its own conditional expectation with respect to $\mathcal{S}$.

4.3 Weak* Convergence and Continuous-Time Limit Supports

In this section, we discuss certain aspects of the behavior of families $(\mu_t)_{t \in [0, +\infty)}$ of real-valued signed measures as t tends to $+\infty$. Thus, we start by discussing the weak* convergence of such families. Next, we define the upper and lower limit support of a family $(\mu_t)_{t \in [0, +\infty)}$ of Borel probability measures, and define the limit support of such a family. As the terminology suggests, the above notions are "continuous-time" versions of the corresponding concepts for sequences of probability measures discussed in Sect. 1.3 (see also Section 4 of [146]). As in the case of sequences of probability measures, we prove that if the family $(\mu_t)_{t\in[0,+\infty)}$ of probability measures converges in the weak* topology of $\mathcal{M}(X)$, then $(\mu_t)_{t\in[0,+\infty)}$ has a limit support, and the limit support of $(\mu_t)_{t\in[0,+\infty)}$ is supp μ, where μ is the weak* limit of $(\mu_t)_{t\in[0,+\infty)}$; we also discuss two examples in which we show that if $(\mu_t)_{t\in[0,+\infty)}$ does not converge in the weak* topology of $\mathcal{M}(X)$, then $(\mu_t)_{t\in[0,+\infty)}$ may or may not have a limit support. We conclude the section with an application of the (upper) limit supports to the study of transition functions and the orbit-closures under the action of these transition functions.

Let (X, d) be a locally compact separable metric space.

Given a family $(\mu_t)_{t\in[0,+\infty)}$ of elements of $\mathcal{M}(X)$, in agreement with the usual definition of the weak* topology (see the discussion preceding Proposition 1.1.5), we say that $(\mu_t)_{t\in[0,+\infty)}$ *converges in the weak* topology of* $\mathcal{M}(X)$ *as* $t \to +\infty$ if there exists a $v \in \mathcal{M}(X)$ such that, for every $f \in C_0(X)$, the real-valued function $(\langle f, \mu_t\rangle)_{t\in[0,+\infty)}$ defined on $[0, +\infty)$ converges to $\langle f, v\rangle$ as $t \to +\infty$. In this case, we call v the *weak* limit of* $(\mu_t)_{t\in[0,+\infty)}$ *as* $t \to +\infty$, or, we can also say that $(\mu_t)_{t\in[0,+\infty)}$ *converges to* v *in the weak* topology of* $\mathcal{M}(X)$ *as* $t \to +\infty$, and we use the notation $\lim_{t\to+\infty}{}_{w^*}\mu_t$ or w^*-$\lim_{t\to+\infty}\mu_t$ for v.

If $(\mu_t)_{t\in[0,+\infty)}$ converges in the weak* topology of $\mathcal{M}(X)$ as $t \to +\infty$, then the weak* limit is, of course, unique because the weak* topology of $\mathcal{M}(X)$ is Hausdorff.

The next proposition is an extension of Proposition 4.2.1 of [143] to the "continuous-time" case.

Proposition 4.3.1. *Let* $(\mu_t)_{t\in[0,+\infty)}$ *be a family of elements of* $\mathcal{M}(X)$. *The following assertions are equivalent:*

(a) $(\mu_t)_{t\in[0,+\infty)}$ *converges in the weak* topology of* $\mathcal{M}(X)$ *as* $t \to +\infty$.
(b) *There exists a* $v_0 \in \mathcal{M}(X)$ *such that the following condition is satisfied: the sequence* $(\mu_{t_n})_{n\in\mathbb{N}}$ *converges to* v_0 *in the weak* topology of* $\mathcal{M}(X)$ *for every sequence* $(t_n)_{n\in\mathbb{N}}$ *of elements of* $[0, +\infty)$ *that diverges to* $+\infty$.
(c) *There exists a* $v_1 \in \mathcal{M}(X)$ *such that the following condition is satisfied: for every sequence* $(t_n)_{n\in\mathbb{N}}$ *of elements of* $[0, +\infty)$, *there exists a subsequence* $(t_{n_k})_{k\in\mathbb{N}}$ *of* $(t_n)_{n\in\mathbb{N}}$ *such that* $(\mu_{t_{n_k}})_{k\in\mathbb{N}}$ *converges to* v_1 *in the weak* topology of* $\mathcal{M}(X)$.

If the above assertions hold true for $(\mu_t)_{t\in[0,+\infty)}$, *then* $\lim_{t\to+\infty}{}_{w^*}\mu_t = v_0 = v_1$.

Proof. $(a) \Rightarrow (b)$ Since we assume that (a) holds true, $\lim_{t\to+\infty}{}_{w^*}\mu_t$ exists.

Set $v_0 = \lim_{t\to+\infty}{}_{w^*}\mu_t$. Our goal is to prove that v_0 satisfies the condition stated at (b).

To this end, let $(t_n)_{n\in\mathbb{N}}$ be a sequence of elements of $[0, +\infty)$ that diverges to $+\infty$.

Now let $f \in C_0(X)$ and let $\varepsilon \in \mathbb{R}, \varepsilon > 0$. Since we assume that (a) is true, and using the definition of v_0, we obtain that $\lim_{t\to+\infty}\langle f, \mu_t\rangle$ exists and is equal to $\langle f, v_0\rangle$.

Thus, there exists a $t_\varepsilon \in [0, +\infty)$ such that $|\langle f, \mu_t \rangle - \langle f, \nu_0 \rangle| < \varepsilon$ for every $t \in [0, +\infty), t \geq t_\varepsilon$.

Since $(t_n)_{n \in \mathbb{N}}$ diverges to $+\infty$, there exists a $n_\varepsilon \in \mathbb{N}$ such that $t_n \geq t_\varepsilon$ for every $n \geq n_\varepsilon$. Then $|\langle f, \mu_{t_n} \rangle - \langle f, \nu_0 \rangle| < \varepsilon$ for every $n \geq n_\varepsilon$.

We have therefore proved that the sequence $((\langle f, \mu_{t_n} \rangle))_{n \in \mathbb{N}}$ converges to $\langle f, \nu_0 \rangle$ whenever $f \in C_0(X)$.

$(b) \Rightarrow (c)$ Assume that (b) holds true, and set $\nu_1 = \nu_0$. Then, clearly, ν_1 satisfies the condition stated at (c).

$(c) \Rightarrow (a)$ Assume that (c) holds true, and let $\nu_1 \in \mathcal{M}(X)$ be such that ν_1 satisfies the condition stated at (c).

Now, assume that (a) is false; that is, assume that $(\mu_t)_{t \in [0, +\infty)}$ does not converge in the weak* topology of $\mathcal{M}(X)$ as $t \to +\infty$. Then, in particular, $(\mu_t)_{t \in [0, +\infty)}$ does not converge to ν_1 in the weak* topology of $\mathcal{M}(X)$ as $t \to +\infty$. Therefore, there exists an $f \in C_0(X)$ such that $\lim_{t \to +\infty} \langle f, \mu_t \rangle$ either does not exist, or else, (if it exists) it is not equal to $\langle f, \nu_1 \rangle$. Thus, there exists an $\varepsilon_0 \in \mathbb{R}, \varepsilon_0 > 0$, such that, for every $t \in [0, +\infty)$, there exists a $t' \in [0, +\infty), t' \geq t$, such that $|\langle f, \mu_{t'} \rangle - \langle f, \nu_1 \rangle| \geq \varepsilon_0$. In particular, for every $n \in \mathbb{N}$, there exists a $t_n \in [0, +\infty)$, $t_n \geq n$, such that $|\langle f, \mu_{t_n} \rangle - \langle f, \nu_1 \rangle| \geq \varepsilon_0$. Clearly, the sequence $(t_n)_{n \in \mathbb{N}}$ diverges to $+\infty$, and there is no subsequence $(t_{n_k})_{k \in \mathbb{N}}$ of $(t_n)_{n \in \mathbb{N}}$ such that $\left(\left\langle f, \mu_{t_{n_k}} \right\rangle \right)_{k \in \mathbb{N}}$ converges to $\langle f, \nu_1 \rangle$. $\qquad\square$

Let $(\mu_t)_{t \in [0, +\infty)}$ be a family of probability measures defined on $(X, \mathcal{B}(X))$. Set

$$G^\sim((\mu_t)_{t \in [0, +\infty)}) = \left\{ x \in X \ \middle| \ \begin{array}{l} \text{there exists an open neighborhood } U \\ \text{of } x \text{ in } X \text{ such that } \lim\sup_{t \to +\infty} \mu_t(U) = 0 \end{array} \right\}$$

and

$$G_\sim((\mu_t)_{t \in [0, +\infty)}) = \left\{ x \in X \ \middle| \ \begin{array}{l} \text{there exists an open neighborhood } U \\ \text{of } x \text{ in } X \text{ such that } \lim\inf_{t \to +\infty} \mu_t(U) = 0 \end{array} \right\}.$$

Also set $\overline{L\mathrm{supp}}_{t \to +\infty}\, \mu_t = X \setminus G^\sim((\mu_t)_{t \in [0, +\infty)})$ and $\underline{L\mathrm{supp}}_{t \to +\infty}\, \mu_t = X \setminus G_\sim((\mu_t)_{t \in [0, +\infty)})$.

Note that both $G^\sim((\mu_t)_{t \in [0, +\infty)})$ and $G_\sim((\mu_t)_{t \in [0, +\infty)})$ are open subsets of X. Indeed, if $x \in G^\sim((\mu_t)_{t \in [0, +\infty)})$ and U is an open neighborhood of x such that $\lim\sup_{t \to +\infty} \mu_t(U) = 0$, then every $y \in U$ belongs to $G^\sim((\mu_t)_{t \in [0, +\infty)})$; in a similar manner, we obtain that $G_\sim((\mu_t)_{t \in [0, +\infty)})$ is an open subset of X, as well.

Since $G^\sim((\mu_t)_{t \in [0, +\infty)})$ and $G_\sim((\mu_t)_{t \in [0, +\infty)})$ are open sets, it follows that $\overline{L\mathrm{supp}}_{t \to +\infty}\mu_t$ and $\underline{L\mathrm{supp}}_{t \to +\infty}\mu_t$ are closed subsets of X.

We call $\overline{L\mathrm{supp}}_{t \to +\infty}\, \mu_t$ and $\underline{L\mathrm{supp}}_{t \to +\infty}\, \mu_t$ the *upper limit support (u.l.s.)* and the *lower limit support (l.l.s.) of* $(\mu_t)_{t \in [0, +\infty)}$ *as* $t \to +\infty$ because we may think of $\overline{L\mathrm{supp}}_{t \to +\infty}\, \mu_t$ and $\underline{L\mathrm{supp}}_{t \to +\infty}\, \mu_t$ as a kind of upper and lower limit of the supports of $\mu_t, t \in [0, +\infty)$, as $t \to +\infty$, respectively.

We say that $(\mu_t)_{t \in [0,+\infty)}$ has a *limit support as* $t \to +\infty$ if $\overline{L\text{supp}}_{t \to +\infty} \mu_t = \underline{L\text{supp}}_{t \to +\infty} \mu_t$. In this case we denote by $L\text{supp}_{t \to +\infty} \mu_t$ any of the sets $\overline{L\text{supp}}_{t \to +\infty} \mu_t$ or $\underline{L\text{supp}}_{t \to +\infty} \mu_t$, and we call $L\text{supp}_{t \to +\infty} \mu_t$ the *limit support of* $(\mu_t)_{t \in [0,+\infty)}$ *as* $t \to +\infty$.

Observation. (1) Note that the u.l.s., the l.l.s., or the l.s., if it exists, may well be equal to the empty set.

(2) Note also that always $\underline{L\text{supp}}_{t \to +\infty} \mu_t \subseteq \overline{L\text{supp}}_{t \to +\infty} \mu_t$.   ▲

Theorem 4.3.2. *Assume that* $(\mu_t)_{t \in [0,+\infty)}$ *converges in the weak* topology of* $\mathcal{M}(X)$ *as* $t \to +\infty$ *and let* $\mu = \lim_{t \to +\infty} w^* \mu_t$. *Then* $(\mu_t)_{t \in [0,+\infty)}$ *has a limit support as* $t \to +\infty$ *and* $L\text{supp}_{t \to +\infty} \mu_t = \text{supp}\, \mu$.

Proof. Assume, as in the theorem, that the limit of $(\mu_t)_{t \in [0,+\infty)}$ in the weak* topology of $\mathcal{M}(X)$ does exist as $t \to +\infty$ and let μ be this limit.

We have to prove that $\text{supp}\, \mu = \overline{L\text{supp}}_{t \to +\infty} \mu_t$ and that $\text{supp}\, \mu = \underline{L\text{supp}}_{t \to +\infty} \mu_t$. However, in view of Observation (2) stated before the theorem, it is enough to prove that:

(a) $\overline{L\text{supp}}_{t \to +\infty} \mu_t \subseteq \text{supp}\, \mu$ and (b) $\text{supp}\, \mu \subseteq \underline{L\text{supp}}_{t \to +\infty} \mu_t$.

(a) Clearly, the inclusion is true if $\overline{L\text{supp}}_{t \to +\infty} \mu_t$ is the empty set. Thus, assume that $\overline{L\text{supp}}_{t \to +\infty} \mu_t$ is nonempty. We have to prove that for every $x \in \overline{L\text{supp}}_{t \to +\infty} \mu_t$ and every open neighborhood U of x in X, we have that $\mu(U) > 0$.

To this end, let $x \in \overline{L\text{supp}}_{t \to +\infty} \mu_t$ and let U be an open neighborhood of x.

Since X is a locally compact metric space, there exists a $\rho \in \mathbb{R}$, $\rho > 0$, such that the closed ball $\overline{B(x, \rho)}$ is a compact subset of X and such that $\overline{B(x, \rho)} \subseteq U$. By Proposition 7.1.8, p. 199, of Cohn's book [20], there exists an $f \in C_0(X)$ such that $1_{\overline{B(x,\rho)}} \leq f \leq 1_U$ (actually, we can choose a function f whose support is compact).

Using the fact that $(\mu_t)_{t \in [0,+\infty)}$ converges in the weak* topology of $\mathcal{M}(X)$ to μ as $t \to +\infty$, and the fact that

$$\limsup_{t \to +\infty} \left\langle 1_{\overline{B(x,\rho)}}, \mu_t \right\rangle = \limsup_{t \to +\infty} \mu_t(\overline{B(x, \rho)}) > 0$$

(because $x \in \overline{L\text{supp}}_{t \to +\infty} \mu_t$ and $\overline{B(x, \rho)}$ is a neighborhood of x), we obtain that

$$\mu(U) = \langle 1_U, \mu \rangle \geq \langle f, \mu \rangle = \lim_{t \to +\infty} \langle f, \mu_t \rangle \geq \limsup_{t \to +\infty} \left\langle 1_{\overline{B(x,\rho)}}, \mu_t \right\rangle > 0.$$

We have therefore proved that $\overline{L\text{supp}}_{t \to +\infty} \mu_t \subseteq \text{supp}\, \mu$.

(b) Obviously, the inclusion is true if $\mu = 0$ because in this case supp μ is the empty set.

Thus, assume that $\mu \neq 0$, so supp $\mu \neq \emptyset$.

In order to prove that $\operatorname{supp} \mu \subseteq \underline{L\operatorname{supp}}_{t \to +\infty} \mu_t$, we have to prove that for every $x \in \operatorname{supp} \mu$ and every open neighborhood U of x in X, we have that $\liminf_{t \to +\infty} \mu_t(U) > 0$.

To this end, let $x \in \operatorname{supp} \mu$ and let U be an open neighborhood of x.

As in the proof of the inclusion at (a), we can find $\rho \in \mathbb{R}$, $\rho > 0$, and $f \in C_0(X)$ such that $\overline{B(x, \rho)}$ is a compact subset of X, $\overline{B(x, \rho)} \subseteq U$, and $1_{\overline{B(x,\rho)}} \leq f \leq 1_U$.

Taking into consideration that $\mu(\overline{B(x, \rho)}) > 0$ (because $x \in \operatorname{supp} \mu$), and using the fact that $(\mu_t)_{t \in [0,+\infty)}$ converges in the weak* topology of $\mathcal{M}(X)$ to μ as $t \to +\infty$, we obtain that

$$0 < \left\langle 1_{\overline{B(x,\rho)}}, \mu \right\rangle \leq \langle f, \mu \rangle = \lim_{t \to +\infty} \langle f, \mu_t \rangle \leq \liminf_{t \to +\infty} \langle 1_U, \mu_t \rangle = \liminf_{t \to +\infty} \mu_t(U).$$

$\square$

In general, if a family $(\nu_t)_{t \in [0,+\infty)}$ of probability measures defined on $(X, \mathcal{B}(X))$ does not converge in the weak* topology of $\mathcal{M}(X)$ as $t \to +\infty$, then $(\nu_t)_{t \in [0,+\infty)}$ may or may not have a limit support. The next two examples prove this fact.

Example 4.3.3. This example is a "continuous-time" version of Example B.3 of [146]. We construct here a family $(\nu_t)_{t \in [0,+\infty)}$ of Borel probability measures on $\mathbb{R}$ such that $(\nu_t)_{t \in [0,+\infty)}$ does not converge in the weak* topology of $\mathcal{M}(X)$ as $t \to +\infty$, but has a limit support.

In order to define the measures ν_t, $t \in [0, +\infty)$, we will use the Dirac measures μ_n, $n \in \mathbb{N}$, defined in Example B.3 of [146]. Let us briefly recall the manner in which we defined the measures μ_n, $n \in \mathbb{N}$, in [146]. Thus, let $(l_k)_{k \in \mathbb{N}}$ be the sequence of natural numbers defined as follows: set $l_1 = l_2 = l_3 = l_4 = 1$; next set $l_5 = l_1 + l_3 = 2$, $l_6 = l_5$, $l_7 = 2l_5$, and so on. In general, for $m \in \mathbb{N}$, $m \geq 2$, if we have already defined $l_1, l_2, l_3, \cdots, l_{2m}$, set $l_{2m+1} = 2l_{2m-1}$ and $l_{2m+2} = l_{2m+1}$. Note that $l_{2m+1} = \sum_{i=1}^{m} l_{2i-1}$ for every $m \in \mathbb{N}$.

The measures μ_n, $n \in \mathbb{N}$, are defined as follows: set $\mu_1 = \delta_0$, $\mu_2 = \delta_2$, $\mu_3 = \delta_0$, and $\mu_4 = \delta_4$. Now, set $\mu_5 = \mu_6 = \delta_0$, $\mu_7 = \delta_7$ and $\mu_8 = \delta_8$; that is, set $\mu_{\sum_{i=1}^{4} l_i + 1} = \mu_{\sum_{i=1}^{4} l_i + l_5} = \delta_0$, $\mu_{\sum_{i=1}^{5} l_i + 1} = \delta_{\sum_{i=1}^{5} l_i + 1}$, and $\mu_{\sum_{i=1}^{5} l_i + l_6} = \delta_{\sum_{i=1}^{5} l_i + l_6}$. In general, if we assume that for $m \in \mathbb{N}$, $m \geq 2$, the measures $\mu_1, \mu_2, \mu_3, \ldots, \mu_{\sum_{i=1}^{2m} l_i}$ have already been defined, then set $\mu_{\sum_{i=1}^{2m} l_i + 1} = \mu_{\sum_{i=1}^{2m} l_i + 2} = \cdots = \mu_{\sum_{i=1}^{2m} l_i + l_{2m+1}} = \delta_0$ and $\mu_{\sum_{i=1}^{2m+1} l_i + j} = \delta_{\sum_{i=1}^{2m+1} l_i + j}$ for every $j = 1, 2, 3, \ldots, l_{2m+2}$.

We now define the probability measures ν_t, $t \in [0, +\infty)$, as follows: $\nu_t = \delta_0$ for every $t \in [0, 2)$, and $\nu_t = \frac{1}{[t]} \sum_{i=1}^{[t]} \mu_i$ for every $t \in [2, +\infty)$, where $[t]$ stands for the integer part of t (that is, $[t]$ is the largest integer less than or equal to t).

We first prove that $(\nu_t)_{t\in[0,+\infty)}$ has a limit support. To this end, we will prove that $\{0\} = \underline{L\text{supp}}_{t\to+\infty}\,\nu_t = \overline{L\text{supp}}_{t\to+\infty}\,\nu_t$.

Note that if U is an open interval of length less than 1 and such that $0 \notin U$, then $\sum_{i=1}^{[t]} \mu_i(U) \leq 1$ for every $t \in [2,+\infty)$, so $\lim_{t\to+\infty}\nu_t(U)$ exists and is equal to 0 because $\nu_t(U) = \frac{1}{[t]}\sum_{i=1}^{[t]}\mu_i(U)$ for every $t \in [2,+\infty)$. We have therefore proved that $\overline{L\text{supp}}_{t\to+\infty}\,\nu_t \subseteq \{0\}$. Since $\underline{L\text{supp}}_{t\to+\infty}\,\nu_t \subseteq \overline{L\text{supp}}_{t\to+\infty}\,\nu_t$, it follows that in order to prove that $\{0\}$ is the limit support of $(\nu_t)_{t\in[0,+\infty)}$ as $t \to +\infty$, it is enough to prove that $0 \in \underline{L\text{supp}}_{t\to+\infty}\,\nu_t$. Consequently, it is enough to prove that there exists an $\alpha \in \mathbb{R}, \alpha > 0$, such that $\nu_t(\{0\}) \geq \alpha$ for every $t \in [0,+\infty)$, because in this case $\liminf_{t\to+\infty}\nu_t(U) \geq \liminf_{t\to+\infty}\nu_t(\{0\}) = \sup_{s\in[0,+\infty)}\inf_{t\geq s}\nu_t(\{0\}) \geq \alpha$ for every open neighborhood U of zero.

We will prove that $\alpha = \frac{1}{3}$ satisfies the required condition. To this end, we note that it is enough to prove that $\nu_t(\{0\}) \geq \frac{1}{3}$ for every $t \in [2,+\infty)$ because $\nu_t(\{0\}) = 1 > \frac{1}{3}$ for every $t \in [0,2)$.

Thus, let $t \in [2,+\infty)$. Then t is in one and only one of the following two situations:

(1) There exists an $m \in \mathbb{N}, m \geq 2$, such that $\sum_{i=1}^{2m-1} l_i \leq t < \sum_{i=1}^{2m} l_i$.

(2) There exists an $m \in \mathbb{N}$ such that $\sum_{i=1}^{2m} l_i \leq t < \sum_{i=1}^{2m+1} l_i$.

If t is in case (1), set $r = [t] - \sum_{i=1}^{2m-1} l_i$.

Using the fact that $l_{2i} = l_{2i-1}$ for every $i \in \mathbb{N}$, taking into consideration that $l_{2k+1} = \sum_{i=1}^{k} l_{2i-1}$ for every $k \in \mathbb{N}$, and since $0 \leq r < l_{2m}$, we obtain that

$$\nu_t(\{0\}) = \frac{1}{[t]}\sum_{i=1}^{[t]}\mu_i(\{0\}) = \frac{1}{\sum_{i=1}^{2m-1} l_i + r}\sum_{i=1}^{m} l_{2i-1}$$

$$= \frac{\sum_{i=1}^{m} l_{2i-1}}{\sum_{i=1}^{m} l_{2i-1} + \sum_{i=1}^{m-1} l_{2i} + r} = \frac{\sum_{i=1}^{m-1} l_{2i-1} + l_{2m-1}}{2\sum_{i=1}^{m-1} l_{2i-1} + l_{2m-1} + r}$$

$$> \frac{\sum_{i=1}^{m-1} l_{2i-1} + l_{2m-1}}{2\sum_{i=1}^{m-1} l_{2i-1} + \sum_{i=1}^{m-1} l_{2i-1} + l_{2m}} = \frac{2\sum_{i=1}^{m-1} l_{2i-1}}{3\sum_{i=1}^{m-1} l_{2i-1} + l_{2m-1}}$$

$$= \frac{2\sum_{i=1}^{m-1} l_{2i-1}}{3\sum_{i=1}^{m-1} l_{2i-1} + \sum_{i=1}^{m-1} l_{2i-1}} = \frac{2}{4} = \frac{1}{2} > \frac{1}{3}.$$

If t is in case (2), set $q = [t] - \sum_{i=1}^{2m} l_i$.
Then

$$\nu_t(\{0\}) = \frac{1}{[t]}\sum_{i=1}^{[t]}\mu_i(\{0\}) = \frac{\sum_{i=1}^{m} l_{2i-1} + q}{\sum_{i=1}^{2m} l_i + q}$$

$$> \frac{\sum_{i=1}^m l_{2i-1}}{\sum_{i=1}^m l_{2i-1} + \sum_{i=1}^m l_{2i} + l_{2m+1}}$$

$$= \frac{\sum_{i=1}^m l_{2i-1}}{2\sum_{i=1}^m l_{2i-1} + \sum_{i=1}^m l_{2i-1}} = \frac{1}{3}.$$

We have therefore proved that if t is in either of the cases (1) or (2), then $v_t(\{0\}) \geq \frac{1}{3}$; therefore, $(v_t)_{t \in [0,+\infty)}$ has a limit support, namely, $L\mathrm{supp}_{t \to +\infty} v_t = \{0\}$.

We now prove that $(v_t)_{t \in [0,+\infty)}$ does not converge in the weak* topology of $\mathcal{M}(\mathbb{R})$ as $t \to +\infty$. Clearly, in order to prove the above assertion, it is enough to show that there exists an $f \in C_0(\mathbb{R})$ such that $\lim_{t \to +\infty} \langle f, v_t \rangle$ does not exist.

To this end, let $f \in C_0(\mathbb{R})$, $f \geq 0$, be a function supported on $[-\frac{1}{2}, \frac{1}{2}]$ such that $f(0) = 1$. Since $\langle f, v_t \rangle = v_t(\{0\})$ for every $t \in [0, +\infty)$, in order to prove that $\lim_{t \to +\infty} \langle f, v_t \rangle$ does not exist, it is enough to find two strictly increasing sequences $(t_k)_{k \in \mathbb{N}}$ and $(s_k)_{k \in \mathbb{N}}$ of elements of $[0, +\infty)$ such that both sequences diverge to $+\infty$ and such that the sequences $(v_{t_k}(\{0\}))_{k \in \mathbb{N}}$ and $(v_{s_k}(\{0\}))_{k \in \mathbb{N}}$ both converge, but have different limits.

Let $t_k = \sum_{j=1}^{2k-1} l_j$ and $s_k = \sum_{i=1}^{2k} l_i$ for every $k \in \mathbb{N}$. Clearly, both sequences $(t_k)_{k \in \mathbb{N}}$ and $(s_k)_{k \in \mathbb{N}}$ are strictly increasing and diverge to $+\infty$.

Since

$$v_{t_k}(\{0\}) = \frac{1}{[t_k]} \sum_{i=1}^{[t_k]} \mu_i(\{0\}) = \frac{1}{\sum_{j=1}^{2k-1} l_j} \sum_{i=1}^{t_k} \mu_i(\{0\})$$

$$= \frac{1}{\sum_{i=1}^k l_{2i-1} + \sum_{j=1}^{k-1} l_{2j}} \sum_{j=1}^k l_{2j-1}$$

$$= \frac{2\sum_{j=1}^k l_{2j-1}}{2\sum_{j=1}^k l_{2j-1} + l_{2k-1}} = \frac{2}{3}$$

for every $k \in \mathbb{N}$, $k \geq 2$, it follows that the sequence $(v_{t_k}(\{0\}))_{k \in \mathbb{N}}$ converges to $\frac{2}{3}$.

On the other hand,

$$v_{s_k}(\{0\}) = \frac{1}{[s_k]} \sum_{i=0}^{[s_k]} \mu_i(\{0\}) = \frac{1}{\sum_{j=1}^{2k} l_j} \sum_{i=0}^{s_k} \mu_i(\{0\}) = \frac{\sum_{j=1}^k l_{2j-1}}{\sum_{j=1}^k l_{2j-1} + \sum_{j=1}^k l_{2j}} = \frac{1}{2}$$

for every $k \in \mathbb{N}$, $k \geq 2$; hence, the sequence $(v_{s_k}(\{0\}))_{k \in \mathbb{N}}$ converges to $\frac{1}{2}$. ∎

Example 4.3.4. We now define a family $(\eta_t)_{t \in [0,+\infty)}$ of Borel probability measures on $\mathbb{R}$ which does not have a limit support and which obviously (by Theorem 4.3.2) does not converge in the weak* topology of $\mathcal{M}(X)$ as $t \to +\infty$. The family $(\eta_t)_{t \in [0,+\infty)}$ is a "continuous-time" version of the sequence $(\mu_n)_{n \in \mathbb{N}}$ constructed in Example B.2 of [146].

In a similar manner as in Example 4.3.3, we will use the sequence $(\mu_n)_{n\in\mathbb{N}}$ obtained in Example B.2 of [146] in order to construct the family $(\eta_t)_{t\in[0,+\infty)}$. Thus, let us briefly recall the construction of the sequence $(\mu_n)_{n\in\mathbb{N}}$ in Example B.2 of [146].

Let $(l_k)_{k\in\mathbb{N}}$ be a sequence of natural numbers constructed as follows: set $l_1 = 1$, let $l_2 \in \mathbb{N}$ be large enough such that $\frac{l_2}{l_1+l_2} \geq \frac{1}{2}$, and let $l_3 \in \mathbb{N}$ be large enough such that $\frac{l_3}{l_1+l_2+l_3} \geq \frac{2}{3}$. In general, for $k \in \mathbb{N}$, $k \geq 2$, if we assume that we have already chosen $l_1, l_2, \ldots, l_{k-1}$, pick $l_k \in \mathbb{N}$ large enough such that $\frac{l_k}{l_1+l_2+\cdots+l_k} \geq \frac{k-1}{k}$.

The sequence of probability measures $(\mu_n)_{n\in\mathbb{N}}$ defined in Example B.2 of [146] is obtained as follows:

Set $\mu_1(=\mu_{l_1}) = \delta_0$, $\mu_{l_1+1} = \mu_{l_1+2} = \cdots = \mu_{l_1+l_2} = \delta_1$, $\mu_{l_1+l_2+1} = \mu_{l_1+l_2+2} = \cdots = \mu_{l_1+l_2+l_3} = \delta_0$, and so on. In general, if $k \in \mathbb{N}$ is even, then set $\mu_{\sum_{i=1}^{k} l_i+1} = \mu_{\sum_{i=1}^{k} l_i+2} = \cdots = \mu_{\sum_{i=1}^{k} l_i+l_{k+1}} = \delta_0$, and for k odd, $k \in \mathbb{N}$, set $\mu_{\sum_{i=1}^{k} l_i+1} = \mu_{\sum_{i=1}^{k} l_i+2} = \cdots = \mu_{\sum_{i=1}^{k} l_i+l_{k+1}} = \delta_1$.

We define the measures η_t, $t \in [0,+\infty)$, in a similar manner as the measures ν_t, $t \in [0,+\infty)$, of Example 4.3.3; that is, let $\eta_t = \delta_0$ for every $t \in [0,2)$, and for $t \in [2,+\infty)$, set $\eta_t = \frac{1}{[t]}\sum_{i=1}^{[t]} \mu_i$, where $[t]$ is the integer part of t.

We now show that $(\eta_t)_{t\in[0,+\infty)}$ does not have a limit support as $t \to \infty$. To this end, we will prove that:

(i) $\overline{L\mathrm{supp}_{t\to+\infty}\, \eta_t} = \emptyset$; and

(ii) $0 \in \overline{L\mathrm{supp}_{t\to+\infty}\, \eta_t}$.

(i) In order to prove that $\overline{L\mathrm{supp}_{t\to+\infty}\, \eta_t} = \emptyset$, we will show that:

(i-a) $x \notin \overline{L\mathrm{supp}_{t\to+\infty}\, \eta_t}$ whenever $x \in \mathbb{R} \setminus \{0,1\}$.

(i-b) $0 \notin \overline{L\mathrm{supp}_{t\to+\infty}\, \eta_t}$.

(i-c) $1 \notin \overline{L\mathrm{supp}_{t\to+\infty}\, \eta_t}$.

(i-a) Let $x \in \mathbb{R} \setminus \{0,1\}$. Clearly, there exists an open neighborhood U of x in $\mathbb{R}$ such that $U \cap \{0,1\} = \emptyset$. Then $\eta_t(U) = \delta_0(U) = 0$ if $t \in [0,2)$, and $\eta_t(U) = \frac{1}{[t]}\sum_{i=1}^{[t]} \mu_i(U) = 0$ for every $t \in [2,+\infty)$ because $\mu_n(U) = 0$ for every $n \in \mathbb{N}$.

Accordingly, $x \in G_\sim((\eta_t)_{t\in[0,+\infty)})$.

(i-b) In order to prove that $0 \notin \overline{L\mathrm{supp}_{t\to+\infty}\, \eta_t}$, we will show that there exist an open neighborhood U of 1 and a sequence $(t_k)_{k\in\mathbb{N}}$ of elements of $[0,+\infty)$ such that $(t_k)_{k\in\mathbb{N}}$ diverges to $+\infty$ and such that the sequence $(\eta_{t_k}(U))_{k\in\mathbb{N}}$ converges to zero.

To this end, let $U = (-\frac{1}{2}, \frac{1}{2})$, and set $t_k = \sum_{i=1}^{2k} l_i$. Clearly, U is an open neighborhood of zero and the sequence $(t_k)_{k\in\mathbb{N}}$ diverges to $+\infty$.

Since

$$\eta_{t_k}(U) = \frac{1}{[t_k]} \sum_{i=1}^{[t_k]} \mu_i(U) = \frac{1}{\sum_{i=1}^{2k} l_i} \sum_{i=1}^{\sum_{j=1}^{2k} l_j} \mu_i(U)$$

$$= \frac{l_1 + l_3 + \cdots + l_{2k-1}}{l_1 + l_2 + l_3 + \cdots + l_{2k}}$$

$$= \frac{l_1 + l_3 + \cdots + l_{2k-1} + l_{2k}}{l_1 + l_2 + l_3 + \cdots + l_{2k}} - \frac{l_{2k}}{l_1 + l_2 + l_3 + \cdots + l_{2k}}$$

$$\leq 1 - \frac{2k-1}{2k} = \frac{1}{2k}$$

for every $k \in \mathbb{N}$, it follows that $(\eta_{t_k}(U))_{k \in \mathbb{N}}$ converges to zero.

(i-c) The basic idea in proving that $1 \notin \underline{L}\mathrm{supp}_{t \to +\infty} \eta_t$ is the same as in (i-b), namely, we will find an open neighborhood V of 1 and a sequence $(t_s)_{s \in \mathbb{N}}$ of elements of $[0, +\infty)$ such that $(t_s)_{s \in \mathbb{N}}$ diverges to $+\infty$ and such that $(\eta_{t_s}(V))_{s \in \mathbb{N}}$ converges to zero.

Thus, let $V = (\frac{1}{2}, \frac{3}{2})$, and set $t_s = \sum_{i=1}^{2s-1} l_i$ for every $s \in \mathbb{N}$. Obviously, V is an open neighborhood of zero and $(t_s)_{s \in \mathbb{N}}$ is a sequence of elements of $[0, +\infty)$ that diverges to $+\infty$.

Taking into consideration that

$$\eta_{t_s}(V) = \frac{1}{t_s} \sum_{i=1}^{t_s} \mu_i(V) = \frac{1}{\sum_{i=1}^{2s-1} l_i} \sum_{i=1}^{\sum_{j=1}^{2s-1} l_j} \mu_i(V)$$

$$= \frac{l_2 + l_4 + \cdots + l_{2s-2}}{l_1 + l_2 + l_3 + \cdots + l_{2s-1}}$$

$$= \frac{l_2 + l_4 + \cdots + l_{2s-2} + l_{2s-1}}{l_1 + l_2 + l_3 + \cdots + l_{2s-2} + l_{2s-1}} - \frac{l_{2s-1}}{l_1 + l_2 + l_3 + \cdots + l_{2s-1}}$$

$$\leq 1 - \frac{2s-2}{2s-1} = \frac{1}{2s-1}$$

for every $s \in \mathbb{N}$, we obtain that $(\eta_{t_s}(V))_{s \in \mathbb{N}}$ converges to zero.

(ii) In order to prove that $0 \in \underline{L}\mathrm{supp}_{t \to +\infty} \eta_t$ we have to prove that $\limsup_{t \to +\infty} \eta_t(W) > 0$ for every neighborhood W of 0. Thus, we have to prove that for every neighborhood W of 0, there exists a sequence $(t_k)_{k \in \mathbb{N}}$ of elements of $[0, +\infty)$ such that $(t_k)_{k \in \mathbb{N}}$ diverges to $+\infty$ and such that $\limsup_{k \to +\infty} \eta_{t_k}(W) > 0$.

To this end, let W be a neighborhood of zero and set $t_k = \sum_{i=1}^{2k-1} l_i$ for every $k \in \mathbb{N}$.

Clearly, $(t_k)_{k \in \mathbb{N}}$ is a sequence of elements of $[0, +\infty)$ that diverges to $+\infty$.

Since

$$\eta_{t_k}(W) = \frac{1}{t_k} \sum_{i=1}^{t_k} \mu_i(W) = \frac{1}{\sum_{s=1}^{2k-1} l_s} \sum_{i=1}^{\sum_{j=1}^{2k-1} l_j} \mu_i(W)$$

$$\geq \frac{l_1 + l_3 + l_5 + \cdots + l_{2k-1}}{l_1 + l_2 + l_3 + l_4 + \cdots + l_{2k-1}} \geq \frac{l_{2k-1}}{l_1 + l_2 + l_3 + l_4 + \cdots + l_{2k-1}} \geq \frac{2k-2}{2k-1}$$

for every $k \in \mathbb{N}$, it follows that $\limsup_{k \to +\infty} \eta_{t_k}(W) = 1$ (actually, the sequence $(\eta_{t_k}(W))_{k \in \mathbb{N}}$ converges to 1 because η_{t_k}, $k \in \mathbb{N}$, are probability measures).

Accordingly, $0 \in \overline{L\mathrm{supp}}_{t \to +\infty} \eta_t$.

Remark 4.3.5. Using a similar approach, we can prove that $1 \in \overline{L\mathrm{supp}}_{t \to +\infty} \eta_t$, as well. Also, we can show that $\overline{L\mathrm{supp}}_{t \to +\infty} \eta_t$ is equal to $\{0, 1\}$.   ∎

We will now conclude the section with an application of the upper limit supports to the study of transition functions. The application is discussed in the next proposition and involves orbit-closures of elements under the action of these transition functions (for the definition of the orbit-closures, see the discussion preceding Proposition 2.1.7).

Proposition 4.3.6. *Let $(P_t)_{t \in [0, +\infty)}$ be a transition function defined on (X, d), and let $((S_t, T_t))_{t \in [0, +\infty)}$ be the family of Markov pairs defined by $(P_t)_{t \in [0, +\infty)}$. If $x \in X$, then $\overline{L\mathrm{supp}}_{t \to +\infty}(T_t \delta_x) \subseteq \overline{\mathcal{O}(x)}$.*

Proof. Let $x \in X$. Since the inclusion $\overline{L\mathrm{supp}}_{t \to +\infty}(T_t \delta_x) \subseteq \overline{\mathcal{O}(x)}$ is true if $\overline{L\mathrm{supp}}_{t \to +\infty}(T_t \delta_x)$ is equal to the empty set, assume that $\overline{L\mathrm{supp}}_{t \to +\infty}(T_t \delta_x) \neq \emptyset$.

Now, let $y \in \overline{L\mathrm{supp}}_{t \to +\infty}(T_t \delta_x)$. Then $\limsup_{t \to +\infty}(T_t \delta_x)(U) > 0$ for every neighborhood U of y. In particular, $\limsup_{t \to +\infty}(T_t \delta_x)(B(y, \frac{1}{n})) > 0$ for every $n \in \mathbb{N}$. Thus, for every $n \in \mathbb{N}$, there exists a $t_n \in [0, +\infty)$ such that $(T_{t_n} \delta_x)(B(y, \frac{1}{n})) > 0$; hence $(\mathrm{supp}(T_{t_n} \delta_x)) \cap (B(y, \frac{1}{n})) \neq \emptyset$, so there exists a $y_n \in (\mathrm{supp}(T_{t_n} \delta_x)) \cap (B(y, \frac{1}{n}))$.

Clearly, $y_n \in \mathcal{O}(x)$ for every $n \in \mathbb{N}$, and the sequence $(y_n)_{n \in \mathbb{N}}$ converges in the metric topology of X to y. Therefore, $y \in \overline{\mathcal{O}(x)}$.

We have therefore proved that $y \in \overline{\mathcal{O}(x)}$ whenever $y \in \overline{L\mathrm{supp}}_{t \to +\infty}(T_t \delta_x)$.   □

Note that, in the above proposition, we do not impose any conditions on the transition function $(P_t)_{t \in [0, +\infty)}$, not even that $(P_t)_{t \in [0, +\infty)}$ satisfy the s.m.a.

4.4   Continuous-Time Banach Limits

Our goal in this section is to discuss certain facts about continuous-time Banach limits. We start with the definition of these Banach limits (here we follow Exercise 23 on p. 73 of Dunford and Schwartz's monograph [30]). Next, we discuss a method of obtaining a new continuous-time Banach limit from a given one (the method is a continuous-time version of the construction for discrete-time Banach limits described in Section 2.1 on p. 39 before Theorem 2.1.2 in [143]). Finally, we conclude the section with a result involving the use of continuous-time Banach limits in the study of certain Feller transition functions (more important uses of continuous-time Banach limits when studying Feller transition functions appear in Chap. 7).

Let $B_b^{(L)}([0, +\infty))$ be the Banach space of all real-valued bounded Lebesgue measurable functions defined on $[0, +\infty)$, where the norm on $B_b^{(L)}([0, +\infty))$ is the usual uniform (sup) norm.

Given $f \in B_b^{(L)}([0, +\infty))$ and $t \in [0, +\infty)$, we will denote by f_t the real-valued function on $[0, +\infty)$ defined by $f_t(u) = f(u + t)$ for every $u \in [0, +\infty)$. Clearly, $f_t \in B_b^{(L)}([0, +\infty))$ for every $f \in B_b^{(L)}([0, +\infty))$ and $t \in [0, +\infty)$.

A *continuous-time Banach limit* (or a *nondiscrete Banach limit*) is a map $x^* : B_b^{(L)}([0, +\infty)) \to \mathbb{R}$ such that:

(a) x^* is a positive linear (bounded) functional (the positivity of x^* means, of course, that $x^*(f) \geq 0$ whenever $f \geq 0$, $f \in B_b^{(L)}([0, +\infty))$); note that a positive linear functional on $B_b^{(L)}([0, +\infty))$ is necessarily bounded).

(b) $x^*(f) = x^*(f_t)$ for every $f \in B_b^{(L)}([0, +\infty))$ and every $t \in [0, +\infty)$.

(c) $\liminf_{t \to +\infty} \frac{1}{t} \int_0^t f(u)\, du \leq x^*(f) \leq \limsup_{t \to +\infty} \frac{1}{t} \int_0^t f(u)\, du$ for every $f \in B_b^{(L)}([0, +\infty))$.

(d) $x^*(f) = \lim_{u \to +\infty} f(u)$ for every $f \in B_b^{(L)}([0, +\infty))$ for which $\lim_{u \to +\infty} f(u)$ exists.

Naturally, the most pressing task concerning the continuous-time Banach limits just defined is to prove their existence. To this end, we need several lemmas that we discuss next.

Lemma 4.4.1. *Let $f \in B_b^{(L)}([0, +\infty))$ be such that $\lim_{x \to +\infty} f(x)$ exists and is a real number, say L. Then $\lim_{t \to +\infty} \frac{1}{t} \int_0^t f(x)\, dx$ exists and is equal to L.*

Proof. Let $f \in B_b^{(L)}([0, +\infty))$, assume that $\lim_{x \to +\infty} f(x)$ exists and is a real number, and set $L = \lim_{x \to +\infty} f(x)$.

Now, let $\varepsilon \in \mathbb{R}$, $\varepsilon > 0$.

Since $\lim_{x \to +\infty} f(x) = L$, it follows that there exists an $M_\varepsilon \in \mathbb{R}$, $M_\varepsilon > 0$, such that $|f(x) - L| < \frac{\varepsilon}{2}$ for every $x \in [0, +\infty)$, $x \geq M_\varepsilon$.

Next, let $t_\varepsilon \in (0, +\infty)$ be large enough such that $M_\varepsilon \leq t_\varepsilon$ and $\frac{M_\varepsilon}{t_\varepsilon}(\|f\| + |L|) < \frac{\varepsilon}{2}$, where $\|f\|$ is the norm of f in $B_b^{(L)}([0, +\infty))$.

Then, for $t \geq t_\varepsilon$, we have

$$\left| \frac{1}{t} \int_0^t f(x)\, dx - L \right| = \left| \frac{1}{t} \int_0^t f(x)\, dx - \frac{1}{t} \int_0^t L\, dx \right|$$

$$\leq \frac{1}{t}\left(\left|\int_0^{M_\varepsilon} f(x)\,dx - \int_0^{M_\varepsilon} L\,dx\right| + \left|\int_{M_\varepsilon}^t f(x)\,dx - \int_{M_\varepsilon}^t L\,dx\right|\right)$$

$$\leq \frac{1}{t}\left(\int_0^{M_\varepsilon} |f(x)|\,dx + \int_0^{M_\varepsilon} |L|\,dx + \int_{M_\varepsilon}^t |f(x) - L|\,dx\right)$$

$$\leq \frac{1}{t} M_\varepsilon \left(\|f\| + |L|\right) + \frac{t - M_\varepsilon}{t}\frac{\varepsilon}{2} < \frac{\varepsilon}{2} + \frac{\varepsilon}{2} < \varepsilon.$$

We have therefore proved that for every $\varepsilon \in \mathbb{R}$, $\varepsilon > 0$, there exists a $t_\varepsilon \in [0, +\infty)$ such that $\left|\frac{1}{t}\int_0^t f(x)\,dx - L\right| < \varepsilon$ for every $t \in [0, +\infty)$, $t \geq t_\varepsilon$. $\qquad\qquad\square$

Lemma 4.4.2. *Let* $f \in B_b^{(\mathrm{L})}([0, +\infty))$, $s \in [0, +\infty)$. *Then* $\int_0^s f(x + t)\,dx = \int_t^{s+t} f(x)\,dx$ *for every* $t \in [0, +\infty)$.

Proof. Let $f \in B_b^{(\mathrm{L})}([0, +\infty))$, $s \in [0, +\infty)$, and $t \in [0, +\infty)$. Now, let $h : \mathbb{R} \to \mathbb{R}$ be defined by

$$h(x) = \begin{cases} f(x) & \text{if } x \in [t, s + t] \\ 0 & \text{if } x \notin [t, s + t]. \end{cases}$$

Then, the function $g : \mathbb{R} \to \mathbb{R}$ defined by

$$g(x) = \begin{cases} f(x + t) & \text{if } x \in [0, s] \\ 0 & \text{if } x \notin [0, s] \end{cases}$$

has the property that $g(x) = h(x + t)$ for every $x \in \mathbb{R}$. Since f is Lebesgue integrable over every bounded subinterval of $[0, +\infty)$, it follows that g and h are Lebesgue integrable. Using the fact that the Lebesgue measure is translation invariant, we obtain that $\int_{\mathbb{R}} g(x)\,dx = \int_{\mathbb{R}} h(x)\,dx$, so $\int_0^s f(x + t)\,dx = \int_t^{s+t} f(x)\,dx$. $\qquad\square$

In discussing the existence of the continuous-time Banach limits, we will use the map $p : B_b^{(\mathrm{L})}([0, +\infty)) \to \mathbb{R}$ defined by $p(f) = \limsup_{t \to +\infty}(\frac{1}{t}\int_0^t f(x)\,dx)$ for every $f \in B_b^{(\mathrm{L})}([0, +\infty))$. Note that the definition of p is correct in the sense that, given $f \in B_b^{(\mathrm{L})}([0, +\infty))$, the function f is Lebesgue integrable on every interval $[0, t]$, $t \in [0, +\infty)$, and since f is bounded, the set $\{\frac{1}{t}\int_0^t f(x)\,dx \mid t \in (0, +\infty)\}$ is a bounded subset of $\mathbb{R}$, so $\limsup_{t \to +\infty} \frac{1}{t}\int_0^t f(x)\,dx$ is a real number.

The function p has two properties that we will use soon. We discuss these two properties in the next lemma.

Lemma 4.4.3. *(a)* $p(f + g) \leq p(f) + p(g)$ *for every* $f \in B_b^{(\mathrm{L})}([0, +\infty))$ *and*
   $g \in B_b^{(\mathrm{L})}([0, +\infty))$.
(b) $p(af) = ap(f)$ *for every* $f \in B_b^{(\mathrm{L})}([0, +\infty))$ *and* $a \in \mathbb{R}$, $a \geq 0$.

Proof. (*a*) If $f \in B_b^{(L)}([0, +\infty))$ and $g \in B_b^{(L)}([0, +\infty))$, then

$$p(f + g) = \inf_{s \in (0,+\infty)} \sup_{t \geq s} \frac{1}{t} \int_0^t (f(x) + g(x)) \, dx$$

$$\leq \inf_{s \in (0,+\infty)} \left(\left(\sup_{t \geq s} \frac{1}{t} \int_0^t f(x) \, dx \right) + \left(\sup_{t \geq s} \frac{1}{t} \int_0^t g(x) \, dx \right) \right)$$

$$= \inf_{s \in (0,+\infty)} \left(\sup_{t \geq s} \left(\frac{1}{t} \int_0^t f(x) \, dx \right) \right) + \inf_{s \in (0,+\infty)} \left(\sup_{t \geq s} \left(\frac{1}{t} \int_0^t g(x) \, dx \right) \right)$$

$$= p(f) + p(g).$$

(*b*) Let $f \in B_b^{(L)}([0, +\infty))$ and $a \in \mathbb{R}$, $a \geq 0$. Since $\sup_{t \geq s} \frac{1}{t} \int_0^t a f(x) \, dx = a \sup_{t \geq s} \frac{1}{t} \int_0^t f(x) \, dx$ for every $s \in (0, +\infty)$, and since

$$\inf_{s \in (0,+\infty)} \left(a \sup_{t \geq s} \frac{1}{t} \int_0^t f(x) \, dx \right) = a \left(\inf_{s \in (0,+\infty)} \sup_{t \geq s} \frac{1}{t} \int_0^t f(x) \, dx \right)$$

it follows that $p(af) = ap(f)$.   □

Observation. For the next lemma, we need the following simple remark: If A is a nonempty set of real numbers, then $|\sup_{a \in A} a| \leq \sup_{a \in A} |a|$. Indeed, since $a \leq |a|$ for all $a \in A$, it follows that $\sup_{a \in A} a \leq \sup_{a \in A} |a|$, and since $-a \leq |a|$ for all $a \in A$, we obtain that

$$- \sup_{a \in A} a = \inf_{a \in A} (-a) \leq \inf_{a \in A} |a| \leq \sup_{a \in A} |a|.$$

▲

Lemma 4.4.4. *Let* $f \in B_b^{(L)}([0, +\infty))$ *and let* $t \in [0, +\infty)$. *Then:*

(*a*) $p(f - f_t) = 0$.
(*b*) $p(f_t - f) = 0$.

Proof. Set $M = \sup_{x \in [0,+\infty)} |f(x)|$. Clearly, $M < +\infty$ because f is a bounded function.

We will show that (*a*) and (*b*) hold true by proving that $|p(f - f_t)| = 0$ and $|p(f_t - f)| = 0$, respectively.

(*a*) Using the fact that $p(f - f_t) = \lim_{\substack{u \to +\infty \\ u > 0}} \sup_{s \geq u} \frac{1}{s} \int_0^s (f(x) - f(x + t)) \, dx$

(because the map $u \mapsto \sup_{s \geq u} \frac{1}{s} \int_0^s (f(x) - f(x + t)) \, dx$, $u \in (0, +\infty)$, is monotone nonincreasing), and using Lemma 4.4.2 and the Observation after Lemma 4.4.3, we obtain that

$$
|p(f - f_t)| = \left| \lim_{\substack{u \to +\infty \\ u>0}} \sup_{s \geq u} \frac{1}{s} \int_0^s (f(x) - f(x+t))\, dx \right|
$$

$$
= \lim_{\substack{u \to +\infty \\ u>0}} \left| \sup_{s \geq u} \left(\frac{1}{s} \int_0^s f(x)\, dx - \frac{1}{s} \int_0^s f(x+t)\, dx \right) \right|
$$

$$
= \lim_{\substack{u \to +\infty \\ u>0}} \left| \sup_{s \geq u} \left(\frac{1}{s} \int_0^s f(x)\, dx - \frac{1}{s} \int_t^{s+t} f(x)\, dx \right) \right|
$$

$$
= \lim_{\substack{u \to +\infty \\ u>\max\{0,t\}}} \left| \sup_{s \geq u} \left(\frac{1}{s} \int_0^t f(x)\, dx + \frac{1}{s} \int_t^s f(x)\, dx - \frac{1}{s} \int_t^s f(x)\, dx \right. \right.
$$

$$
\left. \left. - \frac{1}{s} \int_s^{s+t} f(x)\, dx \right) \right| = \lim_{\substack{u \to +\infty \\ u>\max\{0,t\}}} \left| \sup_{s \geq u} \left(\frac{1}{s} \int_0^t f(x)\, dx - \frac{1}{s} \int_s^{s+t} f(x)\, dx \right) \right|
$$

$$
\leq \inf_{u>\max\{0,t\}} \sup_{s \geq u} \left| \frac{1}{s} \int_0^t f(x)\, dx - \frac{1}{s} \int_s^{s+t} f(x)\, dx \right|
$$

$$
\leq \inf_{u>\max\{0,t\}} \sup_{s \geq u} \left(\frac{1}{s} \int_0^t |f(x)|\, dx + \frac{1}{s} \int_s^{s+t} |f(x)|\, dx \right)
$$

$$
\leq \inf_{u>\max\{0,t\}} \sup_{s \geq u} \left(\frac{Mt}{s} + \frac{Mt}{s} \right) = \inf_{u>\max\{0,t\}} \frac{2Mt}{u} = 0.
$$

(b) The proof is similar to the proof of (a). Here we use the fact that $p(f_t - f) = \lim_{u \to +\infty} \sup_{s \geq u} \frac{1}{s} \int_0^s (f(x+t) - f(x))\, dx$ (because the function $u \mapsto \sup_{s \geq u} \frac{1}{s} \int_0^s (f(x+t) - f(x))\, dx$, $u \in (0, +\infty)$, is monotone nonincreasing), Lemma 4.4.2, and the Observation after Lemma 4.4.3 in order to obtain that

$$
|p(f_t - f)| = \lim_{\substack{u \to +\infty \\ u>0}} \left| \sup_{s \geq u} \left(\frac{1}{s} \int_0^s f(x+t)\, dx - \frac{1}{s} \int_0^s f(x)\, dx \right) \right|
$$

$$
= \lim_{\substack{u \to +\infty \\ u>\max\{0,t\}}} \left| \sup_{s \geq u} \left(\frac{1}{s} \int_s^{s+t} f(x)\, dx - \frac{1}{s} \int_0^t f(x)\, dx - \frac{1}{s} \int_t^s f(x)\, dx \right) \right|
$$

$$
= \lim_{\substack{u \to +\infty \\ u > \max\{0,t\}}} \left| \sup_{s \geq u} \left(\frac{1}{s} \int_s^{s+t} f(x)\, dx - \frac{1}{s} \int_0^t f(x)\, dx \right) \right|
$$

$$
\leq \inf_{u > \max\{0,t\}} \sup_{s \geq u} \left(\frac{1}{s} \int_s^{s+t} |f(x)|\, dx + \frac{1}{s} \int_0^t |f(x)|\, dx \right)
$$

$$
\leq \inf_{u > \max\{0,t\}} \sup_{s \geq u} \left(\frac{Mt}{s} + \frac{Mt}{s} \right) = 0.
$$

$\square$

Our discussion so far allows us to prove the existence of continuous-time Banach limits. We do this in the next theorem.

Theorem 4.4.5. *Continuous-time Banach limits exist.*

Proof. Set $\mathcal{J} = \{ f \in B_b^{(\mathrm{L})}([0, +\infty)) \mid \lim_{x \to +\infty} f(x) \text{ exists}\}$. Clearly, $\mathcal{J}$ is a linear subspace of $B_b^{(\mathrm{L})}([0, +\infty))$.

Let $\varphi : \mathcal{J} \to \mathbb{R}$ be defined by $\varphi(f) = \lim_{x \to +\infty} f(x)$ for every $f \in \mathcal{J}$. Clearly, φ is well-defined because for every $f \in \mathcal{J}$ the limit $\lim_{x \to +\infty} f(x)$ exists and is a real number. Also obvious is the fact that φ is a linear functional.

Using Lemma 4.4.1, we obtain that $\varphi(f) = p(f)$ for every $f \in \mathcal{J}$.

Since $\varphi(f) \leq p(f)$ for every $f \in \mathcal{J}$ and using Lemma 4.4.3, we obtain that φ and p satisfy the conditions of Theorem 2.3.10, pp. 62–63, of Dunford and Schwartz's monograph [30], a very general version of the Hahn-Banach theorem. Thus, using the theorem, we can extend φ to a linear functional $x^* : B_b^{(\mathrm{L})}([0, +\infty)) \to \mathbb{R}$ such that $x^*(f) \leq p(f)$ for every $f \in B_b^{(\mathrm{L})}([0, +\infty))$.

We now prove that x^* is a continuous-time Banach limit. Thus, we have to show that x^* satisfies the conditions (a), (b), (c) and (d) in the definition of the continuous-time Banach limits.

We first note that, since x^* is an extension of φ and using the definition of φ, we obtain that x^* satisfies condition (d).

Now,

$$
x^*(-f) \leq p(-f) = \inf_{s \in (0,+\infty)} \sup_{t \geq s} \frac{1}{t} \int_0^t (-f(x))\, dx
$$

$$
= - \sup_{s \in (0,+\infty)} \inf_{t \geq s} \frac{1}{t} \int_0^t f(x)\, dx,
$$

so $x^*(f) \geq \liminf_{t \to +\infty} \frac{1}{t} \int_0^t f(x)\, dx$ for every $f \in B_b^{(\mathrm{L})}([0, +\infty))$. Therefore, the first inequality in (c) is satisfied whenever $f \in B_b^{(\mathrm{L})}([0, +\infty))$. The second inequality in (c) is obviously satisfied for all $f \in B_b^{(\mathrm{L})}([0, +\infty))$.

Using Lemma 4.4.4, we obtain that $x^*(f - f_t) \leq p(f - f_t) = 0$ and $x^*(f_t - f) \leq p(f_t - f) = 0$, so $x^*(f) = x^*(f_t)$ for every $f \in B_b^{(L)}([0, +\infty))$ and $t \in [0, +\infty)$. Thus, x^* satisfies condition (b), as well.

In order to show that x^* satisfies condition (a), we have to prove only that x^* is positive. However, this fact is obvious because x^* satisfies (c), so if $f \in B_b^{(L)}([0, +\infty))$ is such that $f \geq 0$, then $0 \leq \liminf_{t \to +\infty} \frac{1}{t} \int_0^t f(x)\,dx \leq x^*(f)$.
$\square$

In the next proposition, we discuss a procedure for constructing continuous-time Banach limits from given ones.

In order to state the proposition, we will make the following notational convention that will be used from now on throughout the book: given a real-valued function $(a_t)_{t \in (0,+\infty)}$ defined on $(0, +\infty)$ we will denote by $(a_t)_{\overline{t \in (0,+\infty)}}$ the extension of $(a_t)_{t \in (0,+\infty)}$ to $[0, +\infty)$ obtained by setting $a_0 = 0$.

Proposition 4.4.6. *Let L be a continuous-time Banach limit on $B_b^{(L)}([0, +\infty))$, and let $L_0 : B_b^{(L)}([0, +\infty)) \to \mathbb{R}$ be defined by $L_0(f) = L\left((\frac{1}{t} \int_0^t f(s)\,ds)_{\overline{t \in (0,+\infty)}}\right)$ for every $f \in B_b^{(L)}([0, +\infty))$. Then L_0 is well-defined and is a continuous-time Banach limit, as well.*

Proof. We first note that L_0 is well-defined, in the sense that the function $(\frac{1}{t} \int_0^t f(s)\,ds)_{\overline{t \in (0,+\infty)}}$ belongs to $B_b^{(L)}([0, +\infty))$. Indeed, the function $(\frac{1}{t} \int_0^t f(s)\,ds)_{\overline{t \in (0,+\infty)}}$ is bounded because f is bounded; since the restriction of $(\frac{1}{t} \int_0^t f(s)\,ds)_{\overline{t \in (0,+\infty)}}$ to the interval $(0, +\infty)$ is continuous, it follows that $(\frac{1}{t} \int_0^t f(s)\,ds)_{\overline{t \in (0,+\infty)}}$ is Lebesgue measurable on $[0, +\infty)$.

We now prove that L_0 is a continuous-time Banach limit by proving that L_0 satisfies the conditions (d), (c), (b) and (a) in this order of the definition of a continuous-time Banach limit.

Proof that L_0 satisfies (d). Let $f \in B_b^{(L)}([0, +\infty))$ be such that $\lim_{x \to +\infty} f(x)$ does exist and set $a = \lim_{x \to +\infty} f(x)$. By Lemma 4.4.1, $a = \lim_{t \to +\infty} \frac{1}{t} \int_0^t f(s)\,ds$. Since L is a continuous-time Banach limit, it follows that

$$L_0(f) = L\left((\frac{1}{t} \int_0^t f(s)\,ds)_{\overline{t \in (0,+\infty)}}\right) = a.$$

Proof that L_0 satisfies (c). We first prove that

$$L_0(f) \leq \limsup_{t \to +\infty} \frac{1}{t} \int_0^t f(s)\,ds \tag{4.4.1}$$

for every $f \in B_b^{(L)}([0, +\infty))$.

To this end, let $f \in B_b^{(L)}([0, +\infty))$.

Set $g = (\frac{1}{t} \int_0^t f(s)\,\mathrm{d}s)_{\overline{t \in (0,+\infty)}}$.

Since L is a continuous-time Banach limit, it follows that $L_0(f) = L(g) \leq \limsup_{t \to +\infty} \frac{1}{t} \int_0^t g(s)\,\mathrm{d}s$. Thus, in order to prove the inequality (4.4.1), it is enough to prove that $\limsup \frac{1}{t} \int_0^t g(s)\,\mathrm{d}s \leq \limsup_{t \to +\infty} g(t)$. Accordingly, we will prove
$$\substack{t \to +\infty \\ t > 0}$$

that $\limsup \frac{1}{t} \int_0^t g(s)\,\mathrm{d}s \leq \limsup_{t \to +\infty} g(t) + \varepsilon$ for every $\varepsilon \in \mathbb{R}$, $\varepsilon > 0$.
$$\substack{t \to +\infty \\ t > 0}$$

So, let $\varepsilon \in \mathbb{R}$, $\varepsilon > 0$.

Also, let $t_1 \in (0, +\infty)$.

We may and do pick $t_2 \in \mathbb{R}$, $t_2 > t_1$, large enough such that $\frac{t_1}{t_2} \|g\| < \frac{\varepsilon}{2}$, where $\|g\|$ is the norm of g in $B_b^{(L)}([0, +\infty))$; that is, $\|g\| = \sup_{t \in [0,+\infty)} |g(t)|$.

For every $t \in (0, +\infty)$, $t \geq t_2$, using Lemma 4.4.2, we obtain that

$$\frac{1}{t} \int_0^t g(s)\,\mathrm{d}s = \frac{1}{t} \int_0^{t_1} g(s)\,\mathrm{d}s + \frac{1}{t} \int_{t_1}^t g(s)\,\mathrm{d}s$$

$$= \frac{1}{t} \int_0^{t_1} g(s)\,\mathrm{d}s + \frac{1}{t} \int_0^{t-t_1} g(s + t_1)\,\mathrm{d}s$$

$$= \frac{1}{t} \int_0^{t_1} g(s)\,\mathrm{d}s + \left(\frac{1}{t} - \frac{1}{t - t_1}\right) \int_0^{t-t_1} g(s + t_1)\,\mathrm{d}s + \frac{1}{t - t_1} \int_0^{t-t_1} g(s + t_1)\,\mathrm{d}s$$

$$\leq \frac{t_1}{t} \|g\| + \left|\left(\frac{1}{t} - \frac{1}{t - t_1}\right) \int_0^{t-t_1} g(s + t_1)\,\mathrm{d}s\right| + \frac{1}{t - t_1}(t - t_1) \sup_{s \geq 0} g(s + t_1)$$

$$\leq \frac{t_1}{t_2} \|g\| + \frac{t_1}{t(t - t_1)} \int_0^{t-t_1} |g(s + t_1)|\,\mathrm{d}s + \sup_{s \geq t_1} g(s)$$

$$< \frac{\varepsilon}{2} + \frac{t_1}{t(t - t_1)}(t - t_1) \|g\| + \sup_{s \geq t_1} g(s) \leq \frac{\varepsilon}{2} + \frac{t_1}{t_2} \|g\| + \sup_{s \geq t_1} g(s) < \sup_{s \geq t_1} g(s) + \varepsilon.$$

It follows that $\sup_{t \geq t_2}(\frac{1}{t} \int_0^t g(s)\,\mathrm{d}s) \leq \sup_{s \geq t_1} g(s) + \varepsilon$.

Therefore, $\inf_{t > 0} \sup_{s \geq t}(\frac{1}{s} \int_0^s g(x)\,\mathrm{d}x) \leq \sup_{s \geq t_1} g(s) + \varepsilon$ for every $t_1 > 0$.

Accordingly,

$$\limsup_{t \to +\infty} \frac{1}{t} \int_0^t g(s)\,\mathrm{d}s \leq \inf_{t > 0} \sup_{s \geq t} g(s) + \varepsilon = \limsup_{t \to +\infty} g(s) + \varepsilon.$$

We have therefore proved that the inequality (4.4.1) holds true.

It is easy to see that L_0 is a linear map. Using the fact that L_0 is a linear functional and (4.4.1), we obtain that

$$-L_0(f) = L_0(-f) \leq \limsup_{\substack{t \to +\infty \\ t > 0}} \left(\frac{1}{t} \int_0^t (-f(x))\,dx \right)$$

$$= \limsup_{\substack{t \to +\infty \\ t > 0}} \left(-\frac{1}{t} \int_0^t f(x)\,dx \right) = -\liminf_{\substack{t \to +\infty \\ t > 0}} \left(\frac{1}{t} \int_0^t f(x)\,dx \right),$$

so $\liminf_{t \to +\infty} \frac{1}{t} \int_0^t f(x)\,dx \leq L_0(f)$ for every $f \in B_b^{(L)}([0, +\infty))$.

Thus, L_0 satisfies condition (c).

Proof that L_0 satisfies (b). Let $f \in B_b^{(L)}([0, +\infty))$ and let $t \in [0, +\infty)$. Our goal is to prove that $L_0(f) = L_0(f_t)$.

By applying Lemma 4.4.2, and taking into consideration that L, as a continuous-time Banach limit, is a linear functional, we obtain that

$$L_0(f_t) = L\left(\left(\frac{1}{s} \int_0^s f(x+t)\,dx\right)_{s \in (0,+\infty)} \right) = L\left(\left(\frac{1}{s} \int_t^{s+t} f(x)\,dx\right)_{s \in (0,+\infty)} \right)$$

$$= L\left(\left(\frac{1}{s} \int_0^{s+t} f(x)\,dx - \frac{1}{s} \int_0^t f(x)\,dx\right)_{s \in (0,+\infty)} \right)$$

$$= L\left(\left(\frac{1}{s} \int_0^{s+t} f(x)\,dx\right)_{s \in (0,+\infty)} \right) - L\left(\left(\frac{1}{s} \int_0^t f(x)\,dx\right)_{s \in (0,+\infty)} \right).$$

Taking into consideration that $\lim_{s \to +\infty} \frac{1}{s} \int_0^t f(x)\,dx$ exists and is equal to zero, we obtain that $L\left((\frac{1}{s} \int_0^t f(x)\,dx)_{s \in (0,+\infty)} \right) = 0$ because L is a continuous-time Banach limit.

Therefore, we further obtain that

$$L_0(f_t) = L\left(\left(\frac{1}{s} \int_0^{s+t} f(x)\,dx\right)_{s \in (0,+\infty)} \right)$$

$$= L\left(\left(\frac{1}{s} \int_0^{s+t} f(x)\,dx - \frac{1}{s+t} \int_0^{s+t} f(x)\,dx + \frac{1}{s+t} \int_0^{s+t} f(x)\,dx\right)_{s \in (0,+\infty)} \right)$$

$$= L\left(\left(\frac{t}{s(s+t)} \int_0^{s+t} f(x)\,dx\right)_{s \in (0,+\infty)} \right) + L\left(\left(\frac{1}{s+t} \int_0^{s+t} f(x)\,dx\right)_{s \in (0,+\infty)} \right).$$

Note that $\left| \frac{t}{s(s+t)} \int_0^{s+t} f(x)\,dx \right| \leq \frac{t}{s(s+t)} (s+t) \|f\| = \frac{t}{s} \|f\|$ for every $s \in (0, +\infty)$, where $\|f\|$ is the norm of f in $B_b^{(L)}([0, +\infty))$ (that is, $\|f\| = \sup_{x \in [0,+\infty)} |f(x)|$); therefore, $\lim_{s \to +\infty} \frac{t}{s(s+t)} \int_0^{s+t} f(x)\,dx$ exists and

is equal to zero. Since L is a continuous-time Banach limit, it follows that
$L\left(\left(\frac{t}{s(s+t)} \int_0^{s+t} f(x)\,\mathrm{d}x\right)_{\overline{s\in(0,+\infty)}}\right) = 0$.

Using the function $g = \left(\frac{1}{s}\int_0^s f(x)\,\mathrm{d}x\right)_{\overline{s\in(0,+\infty)}}$ that we introduced earlier, we obtain that

$$L_0(f_t) = L\left(\left(\frac{1}{s+t}\int_0^{s+t} f(x)\,\mathrm{d}x\right)_{\overline{s\in(0,+\infty)}}\right) = L(g_t) = L(g) = L_0(f).$$

Proof that L_0 satisfies (a). As we mentioned earlier, it is easy to see that L_0 is a linear functional. Thus, the proof that L_0 satisfies (a) is completed if we show that L_0 is positive. But the positivity of L_0 is a straightforward consequence of the fact that L_0 satisfies condition (c) of the definition of a continuous-time Banach limit.

□

We will now conclude the section with a result in which we use continuous-time Banach limits in order to study a certain topic in the theory of transition functions.

As usual in this book, let (X, d) be a locally compact separable metric space. Also, let $(P_t)_{t\in\mathbb{T}}$ be a transition function, and let $((S_t, T_t))_{t\in\mathbb{T}}$ be the family of Markov pairs defined by $(P_t)_{t\in\mathbb{T}}$.

A probability measure $\mu_0 \in \mathcal{M}(X)$ is said to be *attractive for* $(P_t)_{t\in\mathbb{T}}$ (or *for* $((S_t, T_t))_{t\in\mathbb{T}}$, or *for* $(T_t)_{t\in\mathbb{T}}$) if the limit $\lim_{t\to+\infty} w*\,T_t\mu$ (in the weak* topology of $\mathcal{M}(X)$) exists and is equal to μ_0 for every probability measure $\mu \in \mathcal{M}(X)$.

Obviously, if $(P_t)_{t\in\mathbb{T}}$ has an attractive probability measure, then this attractive probability is unique (that is, $(P_t)_{t\in\mathbb{T}}$ cannot have two distinct attractive probability measures).

Also easy to see is that if $\mathbb{T} = \mathbb{R}$, then a probability measure $\mu_0 \in \mathcal{M}(X)$ is attractive for $(P_t)_{t\in\mathbb{R}}$ if and only if μ_0 is attractive for the restriction $(P_t)_{t\in[0,+\infty)}$ of $(P_t)_{t\in\mathbb{R}}$ to $[0, +\infty)$.

A natural question concerning the attractive probability measures of transition functions is: if a transition function $(P_t)_{t\in\mathbb{T}}$ has an attractive probability measure, say μ_0, when is μ_0 also an invariant measure for $(P_t)_{t\in\mathbb{T}}$? It turns out that if $(P_t)_{t\in[0,+\infty)}$ is a Feller transition probability, then an attractive probability measure is invariant, as well, for $(P_t)_{t\in\mathbb{T}}$. We discuss this fact in the next proposition.

Proposition 4.4.7. *Let $(P_t)_{t\in\mathbb{T}}$ be a Feller transition function, assume that $(P_t)_{t\in\mathbb{T}}$ has an attractive probability measure, and let μ_0 be this measure. Then $(P_t)_{t\in\mathbb{T}}$ is uniquely ergodic and μ_0 is the unique invariant probability measure for $(P_t)_{t\in\mathbb{T}}$.*

Proof. We first note that in order to prove the proposition, it is enough to prove that μ_0 is invariant for $(P_t)_{t\in\mathbb{T}}$ because if μ_0 is an attractive probability measure, then $(P_t)_{t\in\mathbb{T}}$ cannot have more than one invariant probability, and if it has one, then the invariant probability has to be μ_0.

Next, we note that using the comment made before the proposition and using also Proposition 2.3.7, we obtain that we may assume that $\mathbb{T} = [0, +\infty)$. Therefore, we will prove the proposition under the assumption that $\mathbb{T} = [0, +\infty)$.

Let $((S_t, T_t))_{t \in [0,+\infty)}$ be the family of Markov-Feller pairs defined by $(P_t)_{t \in [0,+\infty)}$.

We have to prove that $T_t \mu_0 = \mu_0$ for every $t \in [0, +\infty)$.

To this end, let $t \in [0, +\infty)$. Also, let $L : B_b^{(L)}([0, +\infty)) \to \mathbb{R}$ be a continuous-time Banach limit.

Let $\phi : C_b(X) \to \mathbb{R}$ be defined by $\phi(f) = L(((\langle S_r f, \mu_0 \rangle)_{r \in [0,+\infty)})$ for every $f \in C_b(X)$.

It is easy to see that ϕ is a positive linear functional on $C_b(X)$.

Taking into consideration that $T_t \mu_0$ is a probability measure and using the fact that μ_0 is an attractive probability, we obtain that $\lim_{r \to +\infty} {}_{w*} T_r T_t \mu_0$ exists and is equal to μ_0. Consequently, since L is a continuous-time Banach limit, we further obtain that $\phi(S_t f) = \phi(f)$ because

$$\phi(S_t f) = L(((\langle S_r S_t f, \mu_0 \rangle)_{r \in [0,+\infty)}) = L(((\langle f, T_r T_t \mu_0 \rangle)_{r \in [0,+\infty)}) = \langle f, \mu_0 \rangle = \phi(f)$$

for every $f \in C_0(X)$.

From the last equality above, it follows that the restriction of ϕ to $C_0(X)$ is equal to μ_0.

Thus, we can apply Theorem 1.2.3 (Lasota-Yorke Lemma) and we conclude that $T_t \mu_0 = \mu_0$. $\qquad \square$

4.5 The Ascoli-Arzelà Theorem

The Ascoli-Arzelà theorem is a well-known result that has been discussed in many textbooks and monographs (see, for instance, Theorem 7.2, pp. 81–82 of Billingsley [13], Theorem 6.3.8, pp. 179–180 of Conway [21], Theorem 4.6.7, pp. 266–267 of Dunford and Schwartz [30], Section 7.10 of Royden [103], and Theorem 19.3, pp. 78–79 of Yosida's book [137]). However, none of the approaches that I am aware of is convenient for our purposes. Thus, we will discuss here a version of the theorem that is better suited to our needs.

Let (E, ρ) be a metric space, and $(f_n)_{n \in \mathbb{N}}$ be a sequence of real-valued functions defined on E.

As usual, we say that $(f_n)_{n \in \mathbb{N}}$ is *uniformly bounded* or *equibounded* if there exists an $M \in \mathbb{R}$, $M \geq 0$, such that $|f_n(x)| \leq M$ for every $n \in \mathbb{N}$ and $x \in E$.

Recall (see the definitions following Proposition 1.1.5) that the sequence $(f_n)_{n \in \mathbb{N}}$ is said to be *equicontinuous* if for every convergent sequence $(x_k)_{k \in \mathbb{N}}$ of elements of E and every $\varepsilon \in \mathbb{R}$, $\varepsilon > 0$, there exists a $k_\varepsilon \in \mathbb{N}$ such that $|f_n(x_k) - f_n(x)| < \varepsilon$ for every $k \in \mathbb{N}$, $k \geq k_\varepsilon$, and for every $n \in \mathbb{N}$, where $x = \lim_{k \to +\infty} x_k$.

In the proof of the next theorem, we will use a notation that we describe in the following paragraph.

Let $(f_{m_{l_1}})_{l_1 \in \mathbb{N}}$ be a subsequence of the sequence $(f_m)_{m \in \mathbb{N}}$ (of real-valued functions defined on E), let $(f_{m_{l_{1_{l_2}}}})_{l_2 \in \mathbb{N}}$ be a subsequence of $(f_{m_{l_1}})_{l_1 \in \mathbb{N}}$, and so on. In general, for $k \in \mathbb{N}$, let $\left(f_{m_{l_1 l_2 l_3 \cdots l_k l_{k+1}}} \right)_{l_{k+1} \in \mathbb{N}}$ be a subsequence of $\left(f_{m_{l_1 l_2 l_3 \cdots l_{k-1} l_k}} \right)_{l_k \in \mathbb{N}}$. We will use the notation $m_{\underline{l_1, l_2, \ldots, l_{k-1}, l_k}}$ for $m_{l_1 l_2 l_3 \cdots l_{k-1} l_k}$, and, of course, the notation $m_{\underline{l_1, l_2, \ldots, l_k, l_{k+1}}}$ for $m_{l_1 l_2 l_3 \cdots l_k l_{k+1}}$, $k \in \mathbb{N}$. Thus, for instance, we prefer the notation $f_{m_{\underline{l_1, l_2, \ldots, l_{k-1}, l_k}}}$ instead of $f_{m_{l_1 l_2 l_3 \cdots l_k l_{k+1}}}$, $k \in \mathbb{N}$.

Clearly, if $k = 1$, we can interchange m_{l_1} and $m_{\underline{l_1}}$. Also, if $k = 1$, we can switch $f_{m_{l_1}}$ and $f_{m_{\underline{l_1}}}$.

Theorem 4.5.1 (Ascoli-Arzelà). *Let (E, ρ) be a separable metric space, and let $(f_n)_{n \in \mathbb{N}}$ be an equicontinuous uniformly bounded sequence of real-valued functions defined on E. Then there exists a subsequence $(f_{n_l})_{l \in \mathbb{N}}$ of $(f_n)_{n \in \mathbb{N}}$ such that $(f_{n_l})_{l \in \mathbb{N}}$ converges pointwise on E (that is, such that $(f_{n_l}(x))_{l \in \mathbb{N}}$ converges for every $x \in E$).*

Proof. Since (E, ρ) is separable, it follows that there exists a sequence $(x_j)_{j \in \mathbb{N}}$ of elements of E such that the range $\{x_j \mid j \in \mathbb{N}\}$ of $(x_j)_{j \in \mathbb{N}}$ is dense in E.

Since $(f_m)_{m \in \mathbb{N}}$ is uniformly bounded, it follows that $(f_m(x_1))_{m \in \mathbb{N}}$ is a bounded sequence of real numbers, so there exists a convergent subsequence $(f_{m_{l_1}}(x_1))_{l_1 \in \mathbb{N}}$ of $(f_m(x_1))_{m \in \mathbb{N}}$; hence, $(f_{m_{l_1}}(x_1))_{l_1 \in \mathbb{N}}$ is a Cauchy sequence; thus, there exists an $\bar{l}_1 \in \mathbb{N}$ such that $\left| f_{m_{l_1'}}(x_1) - f_{m_{l_1''}}(x_1) \right| < 1$ for every $l_1' \in \mathbb{N}$, $l_1' \geq \bar{l}_1$, and $l_1'' \in \mathbb{N}$, $l_1'' \geq \bar{l}_1$. Set $n_1 = m_{\bar{l}_1}$.

Now, the sequence $(f_{m_{l_1}}(x_2))_{\substack{l_1 \in \mathbb{N} \\ l_1 \geq \bar{l}_1}}$ is a bounded sequence of real numbers, so there exists a convergent subsequence $(f_{m_{l_1, l_2}}(x_2))_{l_2 \in \mathbb{N}}$ of $(f_{m_{l_1}}(x_2))_{\substack{l_1 \in \mathbb{N} \\ l_1 \geq \bar{l}_1}}$; therefore, $(f_{m_{l_1, l_2}}(x_2))_{l_2 \in \mathbb{N}}$ is a Cauchy sequence (note that $(f_{m_{l_1, l_2}}(x_1))_{l_2 \in \mathbb{N}}$ is also a Cauchy (convergent) sequence because $(f_{m_{l_1, l_2}}(x_1))_{l_2 \in \mathbb{N}}$ is a subsequence of the convergent sequence $(f_{m_{l_1}}(x_1))_{l_1 \in \mathbb{N}}$); thus, there exists an $\bar{l}_2 \in \mathbb{N}$, $\bar{l}_2 \geq 2$, large enough such that $\left| f_{m_{l_1, l_2'}}(x_i) - f_{m_{l_1, l_2''}}(x_i) \right| < \frac{1}{2}$ for every $l_2' \in \mathbb{N}$, $l_2' \geq \bar{l}_2$, for every $l_2'' \in \mathbb{N}$, $l_2'' \geq \bar{l}_2$, and for every $i = 1, 2$. Set $n_2 = m_{l_1, \bar{l}_2}$. Note that $n_1 < n_2$ because $\bar{l}_2 \geq 2$.

In general, assume that we have constructed the natural numbers $n_1, n_2, \ldots, n_k$, and the corresponding convergent subsequences $\left(f_{m_{l_1, l_2, l_3, \ldots, l_{j-1}, l_j}}(x_i) \right)_{\substack{l_j \in \mathbb{N} \\ l_j \geq \bar{l}_j}}$, $j = 1, 2, \ldots, k$, $1 \leq i \leq j$. Then the sequence $\left(f_{m_{l_1, l_2, l_3, \ldots, l_{k-1}, l_k}}(x_{k+1}) \right)_{\substack{l_k \in \mathbb{N} \\ l_k \geq \bar{l}_k}}$ is a bounded sequence; therefore, there exists a convergent subsequence $\left(f_{m_{l_1, l_2, l_3, \ldots, l_k, l_{k+1}}}(x_{k+1}) \right)_{l_{k+1} \in \mathbb{N}}$ of $\left(f_{m_{l_1, l_2, l_3, \ldots, l_{k-1}, l_k}}(x_{k+1}) \right)_{\substack{l_k \in \mathbb{N} \\ l_k \geq \bar{l}_k}}$; hence, the sequence $\left(f_{m_{l_1, l_2, l_3, \ldots, l_k, l_{k+1}}}(x_{k+1}) \right)_{l_{k+1} \in \mathbb{N}}$ is also a Cauchy sequence. Note that the sequences $\left(f_{m_{l_1, l_2, l_3, \ldots, l_k, l_{k+1}}}(x_j) \right)_{l_{k+1} \in \mathbb{N}}$, $j = 1, 2, \ldots, k$, are also Cauchy (convergent) sequences because for every $j = 1, 2, \ldots, k$, the sequence $\left(f_{m_{l_1, l_2, l_3, \ldots, l_k, l_{k+1}}}(x_j) \right)_{l_{k+1} \in \mathbb{N}}$ is a subsequence of $\left(f_{m_{l_1, l_2, l_3, \ldots, l_{k-1}, l_k}}(x_j) \right)_{\substack{l_k \in \mathbb{N} \\ l_k \geq \bar{l}_k}}$, and the sequence $\left(f_{m_{l_1, l_2, l_3, \ldots, l_{k-1}, l_k}}(x_j) \right)_{\substack{l_k \in \mathbb{N} \\ l_k \geq \bar{l}_k}}$ is convergent by construction. Therefore, there exists an $\bar{l}_{k+1} \in \mathbb{N}$, $\bar{l}_{k+1} \geq 2$, large enough such that
$$\left| f_{m_{l_1, l_2, l_3, \ldots, l_k, l'_{k+1}}}(x_j) - f_{m_{l_1, l_2, l_3, \ldots, l_k, l''_{k+1}}}(x_j) \right| < \frac{1}{k+1} \text{ for every } j = 1, 2, 3, \ldots, k,$$
$k + 1$, and for every $l'_{k+1} \in \mathbb{N}$, $l'_{k+1} \geq \bar{l}_{k+1}$, $l''_{k+1} \in \mathbb{N}$, $l''_{k+1} \geq \bar{l}_{k+1}$. Set $n_{k+1} = m_{l_1, l_2, l_3, \ldots, l_k, \bar{l}_{k+1}}$, and observe that $n_k < n_{k+1}$ because $\bar{l}_{k+1} \geq 2$.

Note that the subsequence $(f_{n_k})_{k \in \mathbb{N}}$ obtained by the above construction has the property that the sequence $(f_{n_k}(x_j))_{k \in \mathbb{N}}$ is convergent for every $j \in \mathbb{N}$. Indeed, let $j \in \mathbb{N}$, and let $\varepsilon \in \mathbb{R}$, $\varepsilon > 0$; also, let $k_\varepsilon \in \mathbb{N}$ be such that $k_\varepsilon \geq j$ and $\frac{1}{k_\varepsilon} < \varepsilon$. Taking into consideration the manner in which the sequence $(n_k)_{k \in \mathbb{N}}$ was constructed, we obtain that $\left| f_{n_{k'}}(x_j) - f_{n_{k''}}(x_j) \right| < \frac{1}{\min\{k', k''\}} \leq \frac{1}{k_\varepsilon} < \varepsilon$ for every $k' \in \mathbb{N}$, $k' \geq k_\varepsilon$, and $k'' \in \mathbb{N}$, $k'' \geq k_\varepsilon$. Thus, the sequences $(f_{n_k}(x_j))_{k \in \mathbb{N}}$, $j \in \mathbb{N}$, are convergent.

We now prove that the subsequence $(f_{n_k})_{k \in \mathbb{N}}$ converges pointwise on E, or, equivalently, that $(f_{n_k}(x))_{k \in \mathbb{N}}$, $x \in E$, are Cauchy sequences.

To this end, let $x \in E$, and let $\varepsilon \in \mathbb{R}$, $\varepsilon > 0$.

Since $\{ x_j \mid j \in \mathbb{N} \}$ is a dense subset of E, it follows that there exists a sequence $(y_l)_{l \in \mathbb{N}}$ of elements of $\{ x_j \mid j \in \mathbb{N} \}$ such that $(y_l)_{l \in \mathbb{N}}$ converges to x.

Clearly, the subsequence $(f_{n_k})_{k \in \mathbb{N}}$ of $(f_n)_{n \in \mathbb{N}}$, as a sequence in its own right, is equicontinuous because $(f_n)_{n \in \mathbb{N}}$ is equicontinuous. Thus, there exists an $l_\varepsilon \in \mathbb{N}$ such that $\left| f_{n_k}(y_{l'}) - f_{n_k}(x) \right| < \frac{\varepsilon}{3}$ for every $l' \in \mathbb{N}$, $l' \geq l_\varepsilon$ and every $k \in \mathbb{N}$.

Since y_{l_ε} belongs to $\{ x_j \mid j \in \mathbb{N} \}$, it follows that $(f_{n_k}(y_{l_\varepsilon}))_{k \in \mathbb{N}}$ is a Cauchy sequence, so there exists a $k_\varepsilon \in \mathbb{N}$ such that $\left| f_{n_{k'}}(y_{l_\varepsilon}) - f_{n_{k''}}(y_{l_\varepsilon}) \right| < \frac{\varepsilon}{3}$ for every $k' \in \mathbb{N}$, $k' \geq k_\varepsilon$, and $k'' \in \mathbb{N}$, $k'' \geq k_\varepsilon$.

We obtain that

$$\left| f_{n_{k'}}(x) - f_{n_{k''}}(x) \right| \leq \left| f_{n_{k'}}(x) - f_{n_{k'}}(y_{l_\varepsilon}) \right| + \left| f_{n_{k'}}(y_{l_\varepsilon}) - f_{n_{k''}}(y_{l_\varepsilon}) \right|$$
$$+ \left| f_{n_{k''}}(y_{l_\varepsilon}) - f_{n_{k''}}(x) \right| < \frac{\varepsilon}{3} + \frac{\varepsilon}{3} + \frac{\varepsilon}{3} = \varepsilon$$

for every $k' \in \mathbb{N}$, $k' \geq k_\varepsilon$, and $k'' \in \mathbb{N}$, $k'' \geq k_\varepsilon$.

We have therefore proved that, for every $x \in X$, the sequence $(f_{n_k}(x))_{k \in \mathbb{N}}$ is Cauchy. $\qquad\square$

4.6 Ordered Vector Spaces and Positive Operators

Our goal in this section is to review very briefly several topics that appear in the theory of Banach lattices and the study of positive operators on these Banach lattices. We discuss these topics because we need some of the facts presented here later on in the book, and because the topics discussed in the present section can be used to obtain a better understanding of various topics discussed in the volume. The material presented here can be found in a significantly more detailed form in any textbook or monograph dealing with vector lattices and positive operators (for instance, in Abramovich and Aliprantis [1], [2], Aliprantis and Burkinshaw [3], Luxemburg and Zaanen [68], Schaefer [105], and Zaanen [139]).

Inevitably, in order to keep this book self-contained, there is a certain overlap of the material discussed here and in the subsection *Vector Lattices, Banach Lattices, and Positive Operators* in Section 1.3 of [143].

Let E be a real vector space and let $\leq$ be an order relation defined on E.

We say that the ordered pair $(E, \leq)$ is an *ordered vector space* if the following two conditions are satisfied:

- If $x \in E$ and $y \in E$ are such that $x \leq y$, then $x + z \leq y + z$ for every $z \in E$;
- If $x \in E$ and $y \in E$ are such that $x \leq y$, then $\lambda x \leq \lambda y$ for every $\lambda \in \mathbb{R}, \lambda \geq 0$.

It will often be the case that the order relation $\leq$ will be clearly understood from the context, so, in such a situation, we will simply refer to E as the ordered vector space $(E, \leq)$.

Let $(E, \leq)$ be an ordered vector space.

Given $u \in E$, we say that u is a *positive element of E* if $0 \leq u$.

As usual, given a nonempty subset A of X, we will use the notation $\sup A$ and $\inf A$ for the supremum (the least upper bound) and the infimum (the greatest lower bound) of A, respectively, whenever $\sup A$ and $\inf A$ exist in E, of course. If $x \in E$ and $y \in E$ (x and y may or may not be distinct) we will use the notation $x \vee y$ and $x \wedge y$ for $\sup\{x, y\}$ and $\inf\{x, y\}$ (again, provided that $\sup\{x, y\}$ and $\inf\{x, y\}$ exist in E).

The following proposition is often useful.

Proposition 4.6.1. *Let $x \in E$, and let A be a nonempty subset of E.*

(a) *If $\sup A$ exists in E, then $\sup(x+A)$ exists as well, and $x+\sup A = \sup(x+A)$.*
(b) *Similarly, if $\inf A$ exists in E, then $\inf(x + A)$ exists as well, and $x + \inf A = \inf(x + A)$.*
(c) *$\sup A$ exists if and only if $\inf(-A)$ exists, and in this case $-\sup(A) = \inf(-A)$.*

The proof of the proposition is easy.

Let $u \in E$. We will use the notation u^+, u^-, and $|u|$ for $u \vee 0$, $(-u) \vee 0$ and $(-u) \vee u$ whenever the suprema exist in E, and we call u^+, u^-, and $|u|$ the *positive part*, the *negative part*, and the *modulus* (or *absolute value*) *of u*, respectively.

The ordered vector space $(E, \leq)$ is called a *vector lattice* (or a *Riesz space*, or a *linear lattice*) if the order relation $\leq$ defines a lattice structure on E; that is, if $x \vee y$ and $x \wedge y$ exist in E whenever $x \in E$ and $y \in E$.

In the next proposition we discuss a useful criterion for deciding whether an ordered vector space is a vector lattice or not.

Proposition 4.6.2. *The ordered vector space* $(E, \leq)$ *is a vector lattice if and only if* x^+ *and* x^- *exist in E for every* $x \in E$.

Proof. Clearly, if $(E, \leq)$ is a vector lattice, then x^+ and x^- exist in E for every $x \in E$.

Conversely, assume that x^+ and x^- exist in E for every $x \in E$, and let $u \in E$ and $v \in E$. Using our assumption, we obtain that $(u - v)^+$ and $(u - v)^-$ exist in E. Using (a) of Proposition 4.6.1, we obtain that $u \vee v$ exists in E (and is equal to $v + (u - v)^+$). Now, using (c) of Proposition 4.6.1, we obtain that $(u - v) \wedge 0$ exists in E and is equal to $-(u - v)^-$; therefore, using (b) of Proposition 4.6.1, we obtain that $u \wedge v$ exists in E, as well (and is equal to $v - (u - v)^-$). $\qquad\square$

Let E be a vector space, and assume that E is endowed with a norm $\|\cdot\|$ and an order relation $\leq$ that define a Banach space structure and a Riesz space structure on E, respectively. We say that E is a *Banach lattice* if the following condition is satisfied: if $u \in E$ and $v \in E$ are such that $|u| \leq |v|$, then $\|u\| \leq \|v\|$.

Let E be a Banach lattice and let F be a Banach subspace of E. Clearly, the restriction $\leq_F$ of the order relation $\leq_E$ on E to F is an order relation on F that defines a structure of ordered vector space on F. We say that F is a *Banach sublattice of E* if the following condition is satisfied: $x \vee_E y \in F$ and $x \wedge_E y \in F$ for every $x \in F$ and $y \in F$, where $x \vee_E y$ and $x \wedge_E y$ are the supremum and the infimum of the set $\{x, y\}$ with respect to the order relation $\leq_E$ in E.

Thus, if F is a Banach sublattice of E, then $x \vee_F y$ and $x \wedge_F y$ exist and are equal to $x \vee_E y$ and $x \wedge_E y$, respectively, where $\vee_F$ and $\wedge_F$ are the supremum and infimum with respect to the order relation $\leq_F$ in F, respectively, and $\vee_E$ and $\wedge_E$ are the supremum and infimum with respect to $\leq_E$ in E, respectively. Therefore, the Banach sublattice F can and often will be thought of as a Banach lattice in its own right.

The next proposition is a useful tool for deciding whether a Banach subspace of a Banach lattice is a Banach sublattice or not.

Proposition 4.6.3. *Let E be a Banach lattice and let F be a Banach subspace of E. Then F is a Banach sublattice of E if and only if* $x \vee_E 0$ *and* $x \wedge_E 0$ *belong to F for every* $x \in F$, *where* $\vee_E$ *and* $\wedge_E$ *are the supremum and the infimum with respect to the order relation* $\leq_E$ *that defines the Banach lattice structure of E.*

Proof. Obviously, if F is a Banach sublattice of E, then $x \vee_E 0$ and $x \wedge_E 0$ belong to F for every $x \in F$.

Conversely, if F is a Banach subspace of E, then the restriction $\leq_F$ of the order relation $\leq_E$ defines a structure of an ordered vector space on F. Since $x \vee_E 0$ and $x \wedge_E 0$ belong to F, it follows that $x \vee_F 0$ and $x \wedge_F 0$ exist in F (and the equalities $x \vee_F 0 = x \vee_E 0$ and $x \wedge_F 0 = x \wedge_E 0$ hold true), where $\vee_F$ and $\wedge_F$ are the supremum and infimum with respect to the order relation $\leq_F$ in F. Thus, using Proposition 4.6.2, we obtain that $(F, \leq_F)$ is a Riesz space.

It is easy to see that $x \vee_F y = x \vee_E y$ and $x \wedge_F y = x \wedge_E y$ for every $x \in F$ and $y \in F$, so F is a Banach sublattice of E. $\qquad\square$

Most Banach spaces that we encounter in mathematics are Banach lattices when endowed with their standard norms and order relations. For instance, given a locally compact separable metric space (X, d), the spaces $C_0(X)$, $C_b(X)$ and $B_b(X)$ are all Banach lattices when endowed with the pointwise order (recall that the pointwise order is defined as follows: $f \leq g$ if, by definition $f(x) \leq g(x)$ for every $x \in X$, where f and g belong to the space under consideration); the space $\mathcal{M}(X)$ is a Banach lattice when endowed with its standard order relation defined as follows: $\mu \leq \nu$ if $\mu(A) \leq \nu(A)$ for every $A \in \mathcal{B}(X)$. Let $(Y, \mathcal{Y}, \mu)$ be a measure space, let $p \in \mathbb{R} \cup \{+\infty\}$, $1 \leq p \leq +\infty$, and let $L^p(Y, \mathcal{Y}, \mu)$ be the usual L^p-space that appears in classical measure theory; then $L^p(Y, \mathcal{Y}, \mu)$ is a Banach lattice with respect to its standard order relation defined as follows: $\bar{f}_1 \leq \bar{f}_2$ if there exist two measurable functions g_1 and g_2 such that g_i is in the equivalence class $\bar{f}_i$, $i = 1, 2$, and such that $g_1 \leq g_2$ μ-a.e., where $\bar{f}_i \in L^p(Y, \mathcal{Y}, \mu)$ for every $i = 1, 2$.

Naturally, one can find a vector space E endowed with an order relation $\leq$ and a norm $\|\cdot\|$ such that E is a Riesz space with respect to $\leq$, and a Banach space for the norm $\|\cdot\|$, but E fails to be a Banach lattice for $\|\cdot\|$ and $\leq$. Moreover, one can find a Riesz space E for which there is no norm that will define a Banach lattice structure on E (for details, see p. 29 of [143]).

Note that in the above examples $C_0(X)$ can be thought of as a Banach sublattice of $C_b(X)$ and of $B_b(X)$, and $C_b(X)$ is a Banach sublattice of $B_b(X)$.

Let us discuss now a slightly more sophisticated example of a Banach sublattice that will be studied in Chap. 6.

Example 4.6.4. Let $(P_t)_{t \in \mathbb{T}}$ be a transition function defined on a locally compact separable metric space (X, d), and let $\mathcal{M}_{\mathrm{inv}}(X)$ be the vector space of all invariant elements for $(P_t)_{t \in \mathbb{T}}$ in $\mathcal{M}(X)$. It is easy to see that the restriction $\|\cdot\|_{\mathrm{inv}}$ to $\mathcal{M}_{\mathrm{inv}}(X)$ of the norm $\|\cdot\|$ on $\mathcal{M}(X)$ defines a Banach space structure on $\mathcal{M}_{\mathrm{inv}}(X)$. The restriction $\leq_{\mathrm{inv}}$ to $\mathcal{M}_{\mathrm{inv}}(X)$ of the order relation $\leq$ on $\mathcal{M}(X)$ defines a structure of ordered vector space on $\mathcal{M}_{\mathrm{inv}}(X)$. Moreover, since an element μ of $\mathcal{M}(X)$ belongs to $\mathcal{M}_{\mathrm{inv}}(X)$ if and only if $\mu^+ \in \mathcal{M}_{\mathrm{inv}}(X)$ and $\mu^- \in \mathcal{M}_{\mathrm{inv}}(X)$, using Proposition 4.6.3, we obtain that $\mathcal{M}_{\mathrm{inv}}(X)$ is a Banach sublattice of $\mathcal{M}(X)$, so we may and whenever convenient do think of $\mathcal{M}_{\mathrm{inv}}(X)$ as a Banach lattice in its own right. $\qquad\blacksquare$

There are two types of Banach lattices that are of special interest to us. Let us discuss them briefly now.

A Banach lattice E is called an *AL-space* (*abstract L^1-space*) if its norm is additive in the sense that $\|x + y\| = \|x\| + \|y\|$ for every $x \in E$, $x \geq 0$, and $y \in E$, $y \geq 0$. If $(Y, \mathcal{Y}, \mu)$ is a measure space, then the space $L^1(Y, \mathcal{Y}, \mu)$ that appears in classical measure theory is an AL-space. If (X, d) is a locally compact separable metric space, $\mathcal{M}(X)$ is also an AL-space. The Banach lattices $C_0(X)$, $C_b(X)$ and $B_b(X)$ with X as above are not AL-spaces.

We say that a Banach lattice E is an *AM-space* if $\|x \vee y\| = \max\{\|x\|, \|y\|\}$ whenever $x \in E$ and $y \in E$ are such that $x \wedge y = 0$. The Banach lattices $C_0(X)$, $C_b(X)$ and $B_b(X)$ that we mentioned above are AM-spaces. Let $(Y, \mathcal{Y}, \mu)$ be a measure space, assume that there exist two disjoint $\mathcal{Y}$-measurable subsets A and B of Y such that $0 < \mu(A) < +\infty$ and $0 < \mu(B) < +\infty$, and let $L^p(Y, \mathcal{Y}, \mu)$, $p \in \mathbb{R} \cup \{+\infty\}$, $1 \leq p \leq +\infty$, be the L^p-spaces that appear in classical measure theory; then $L^\infty(Y, \mathcal{Y}, \mu)$ is an AM-space, and the Banach lattices $L^p(Y, \mathcal{Y}, \mu)$, $1 < p < +\infty$, are neither AM-spaces, nor AL-spaces.

Let E be an AM-space. If the closed unit ball of E has a largest element, (a maximum), then this element is unique and is called a (*strong order*) *unit* of E. If E has a unit, then E is called an *AM-space with unit*. The AM-spaces $C_b(X)$ and $B_b(X)$ that we mentioned earlier are AM-spaces with unit (the function $\mathbf{1}_X$ is the unit for both $C_b(X)$ and $B_b(X)$). By contrast, if X is not compact, then $C_0(X)$ is an AM-space without unit.

Let E and F be two Banach lattices.

A linear operator $T : E \to F$ is said to be *positive* if $Tx \geq 0$ whenever $x \in E$ is such that $x \geq 0$. It can be shown (see, for example, Theorem 5.3, p. 84 of Schaefer's monograph [105]) that a positive operator from a Banach lattice to another one is necessarily bounded (continuous).

As usual, a linear bounded operator $T : E \to F$ is said to be a *contraction* if $\|T\| \leq 1$.

If E is an AL-space, a linear operator $T : E \to E$ is called a *Markov operator* if T is positive and if $\|Tu\| = \|u\|$ for every $u \in E$, $u \geq 0$. It is easy to see that if T is a Markov operator then $\|T\| = 1$ so T is also a contraction.

Example 4.6.5. Let $(P_t)_{t \in \mathbb{T}}$ be a transition function defined on a locally compact separable metric space (X, d) and let $((S_t, T_t))_{t \in \mathbb{T}}$ be the family of Markov pairs generated by $(P_t)_{t \in \mathbb{T}}$. Then S_t is a positive contraction of $B_b(X)$ and T_t is a Markov operator on $\mathcal{M}(X)$ for every $t \in \mathbb{T}$. ∎

Let E and F be two vector lattices.

A linear operator $T : E \to F$ is said to be a *lattice* (or *Riesz*) *homomorphism* if $T(x \vee y) = (T(x)) \vee (T(y))$ for every $x \in E$ and $y \in E$.

Note that, even though we defined the notion of positive operators only in the case when the domains and the codomains of the operators are Banach lattices, the definition makes perfect sense when the domains and the codomains are just vector lattices rather than Banach lattices, and it is easy to see that Riesz homomorphisms are positive operators.

The following proposition is useful (for a proof, see Theorem 7.2, p. 88 of Aliprantis and Burkinshaw's monograph [3]):

Proposition 4.6.6. *Let E and F be two Riesz spaces and let $T : E \to F$ be a linear operator. The following assertions are equivalent:*

(a) T is a lattice homomorphism.
(b) $T(u \wedge v) = (T(u)) \wedge (T(v))$ for every $u \in E$ and $v \in E$.
(c) $(T(u)) \wedge (T(v)) = 0$ whenever $u \in E$ and $v \in E$ are such that $u \wedge v = 0$.

A lattice homomorphism which is also one-to-one is called a *lattice isomorphism.* Two vector lattices E and F are called *Riesz isomorphic* or *vector lattice isomorphic* if there exists a lattice isomorphism from E onto F.

We conclude the section and the chapter with a result on lattice isomorphisms.

Proposition 4.6.7. *Let E and F be two Riesz spaces, let $T : E \to F$ be a linear operator, and assume that T is one-to-one and onto. Then T is a lattice isomorphism if and only if both T and T^{-1} are positive operators.*

For a proof of the proposition, see Theorem 7.3, p. 89 of Aliprantis and Burkinshaw [3].

Chapter 5
The Ergodic Decomposition of Kryloff, Bogoliouboff, Beboutoff and Yosida, Part I

Our goal in this and the next chapter is to discuss an ergodic decomposition for transition functions. The decomposition (which is also valid for transition probabilities (see Sect. 1.3 and the paper [146])) stems from the pioneering works of Kryloff and Bogoliouboff [54], Beboutoff [9], and Yosida [135] and [136], so we call it the KBBY decomposition.

Let (X, d) be a locally compact separable metric space, let $(P_t)_{t \in \mathbb{T}}$ be a transition function, let $((S_t, T_t))_{t \in \mathbb{T}}$ be the family of Markov pairs defined by $(P_t)_{t \in \mathbb{T}}$, and assume that $(P_t)_{t \in \mathbb{T}}$ satisfies the s.m.a.

The KBBY decomposition for $(P_t)_{t \in \mathbb{T}}$ is a splitting of the space X in terms of the behavior, as $s \to +\infty$, of the real-valued functions $(\frac{1}{s} \int_0^s S_t f(x) \, dt)_{s \in (0, +\infty)}$, $f \in C_0(X)$, $x \in X$, defined on $(0, +\infty)$.

The decomposition and its features are surprisingly similar to those for a transition probability (obtained in [146] and described briefly in Sect. 1.3) even though the details involved in the proofs of the results that appear in stating the decomposition are significantly more sophisticated than for a transition probability (most of the topics discussed in Chaps. 3 and 4 are preliminaries for the decomposition for transition functions).

We note that recently Worm and Hille [132] (see also Worm [130]) have obtained an approach to the KBBY decomposition defined by a transition function that is valid in Polish spaces and requires only that the transition function satisfy the s.m.a. The price that has to be paid for this more general approach consists of the fact that one cannot define dissipativeness or the sets Γ_0 and Γ_c. Also, if we consider the case of a locally compact separable metric space (X, d), it is harder to use their approach than ours. Worm and Hille obtained the above-mentioned results after the announcement in Section 7 of [146] of our results that we discuss here.

Essentially, in this chapter we study the aspects of the decomposition that do not directly involve the ergodic invariant probability measures (in the case of transition probabilities, these topics appear in Sections 3–5 of [146]), while in the next chapter we discuss the role of the ergodic invariant probability measures in the decomposition.

R. Zaharopol, *Invariant Probabilities of Transition Functions*, Probability and Its Applications 44, DOI 10.1007/978-3-319-05723-1_5,
© Springer International Publishing Switzerland 2014

The present chapter is organized as follows: in the first section we define various sets that appear in the decomposition, we define the elementary measures, and we discuss the supports of the standard elementary measures; in the second section we prove that certain sets defined in the first section are measurable, and in the last section (Sect. 5.3) we prove that some of the sets defined in Sect. 5.1 are sets of maximal probability.

5.1   Elementary Measures and Their Role
in the Decomposition

In Sect. 1.3 we recalled from [146] the definitions of the sets $\Omega^{(\mathrm{TP})}$, $\Gamma^{(\mathrm{TP})}$, $\mathcal{D}^{(\mathrm{TP})}$, $\Gamma_0^{(\mathrm{TP})}$, $\Gamma_c^{(\mathrm{TP})}$, $\Gamma_{\mathrm{cp}}^{(\mathrm{TP})}$ and $\Gamma_{\mathrm{cpi}}^{(\mathrm{TP})}$ that appear in the ergodic decomposition defined by a transition probability. In this section, we will define and study the corresponding sets that appear in the ergodic decomposition defined by a transition function that satisfies the s.m.a. We will also define and study the elementary measures and the standard elementary measures defined using transition functions.

Let (X, d) be a locally compact separable metric space.

Let $(P_t)_{t \in \mathbb{T}}$ be a transition function defined on (X, d), and let $((S_t, T_t))_{t \in \mathbb{T}}$ be the family of Markov pairs defined by $(P_t)_{t \in \mathbb{T}}$.

Set

$$\Omega = \left\{ x \in X \;\middle|\; \begin{array}{l} L^{(\mathrm{ct})}(((\langle f, T_t \delta_x \rangle)_{t \in [0,+\infty)})) = 0 \text{ for} \\ \text{every continuous-time Banach} \\ \text{limit } L^{(\mathrm{ct})} \text{ and every } f \in C_0(X) \end{array} \right\}$$

and $\Gamma = X \setminus \Omega$. (For the definition and details on continuous-time Banach limits, see Sect. 4.4.)

Note that more complete notations for the sets Ω and Γ are $\Omega((P_t)_{t \in \mathbb{T}})$ and $\Gamma((P_t)_{t \in \mathbb{T}})$, respectively. However, since it will always be clear from the context which transition probability is under consideration, we will prefer the notation Ω and Γ rather than $\Omega((P_t)_{t \in \mathbb{T}})$ and $\Gamma((P_t)_{t \in \mathbb{T}})$. We will use the same convention for all other subsets of X that will appear in the KBBY decomposition defined by $(P_t)_{t \in \mathbb{T}}$.

For every $x \in X$ and every continuous-time Banach limit $L^{(\mathrm{ct})}$, we can define a function $\varepsilon_x^{(L^{(\mathrm{ct})})} : C_0(X) \to \mathbb{R}$ as follows: $\varepsilon_x^{(L^{(\mathrm{ct})})}(f) = L^{(\mathrm{ct})}(((\langle f, T_t \delta_x \rangle)_{t \in [0,+\infty)}))$ for every $f \in C_0(X)$.

Let $x \in X$ and let $L^{(\mathrm{ct})}$ be a continuous-time Banach limit. It is easy to see that $\varepsilon_x^{(L^{(\mathrm{ct})})}$ is a linear functional on $C_0(X)$, and that $\varepsilon_x^{(L^{(\mathrm{ct})})} \geq 0$ for every $f \in C_0(X)$, $f \geq 0$. Thus, $\varepsilon_x^{(L^{(\mathrm{ct})})}$ is continuous (because it is a positive linear functional on $C_0(X)$ and $C_0(X)$ is a Banach lattice); therefore $\varepsilon_x^{(L^{(\mathrm{ct})})}$ is an element of the topological dual $(C_0(X))'$ of $C_0(X)$; since $(C_0(X))'$ can be identified with $\mathcal{M}(X)$ in a standard manner, we may and do think of $\varepsilon_x^{(L^{(\mathrm{ct})})}$ as an element of $\mathcal{M}(X)$.

In a similar manner as in the case of transition probabilities, given $x \in X$ and a continuous-time Banach limit $L^{(\mathrm{ct})}$, we say that $\varepsilon_x^{(L^{(\mathrm{ct})})}$ is an *elementary measure* if $\varepsilon_x^{(L^{(\mathrm{ct})})} \neq 0$; of course, if $\varepsilon_x^{(L^{(\mathrm{ct})})}$ is a probability measure, we call $\varepsilon_x^{(L^{(\mathrm{ct})})}$ an elementary probability measure.

Note that if $\varepsilon_x^{(L^{(\mathrm{ct})})}$ is an elementary measure for some $x \in X$ and some continuous-time Banach limit $L^{(\mathrm{ct})}$, then $x \in \Gamma$. Moreover, if $x \in X$, then $x \in \Gamma$ if and only if there exists a continuous-time Banach limit $L^{(\mathrm{ct})}$ such that $\varepsilon_x^{(L^{(\mathrm{ct})})}$ is an elementary measure.

From now on we will assume that the transition function $(P_t)_{t \in \mathbb{T}}$ satisfies the s.m.a.

Set

$$
\mathcal{D} = \left\{ x \in X \;\middle|\; \begin{array}{l} \text{the limit} \quad \lim_{\substack{s \to +\infty \\ s > 0}} \frac{1}{s} \int_0^s S_t f(x)\, dt \ \text{exists} \\[2ex] \text{and is equal to zero for every } f \in C_0(X) \end{array} \right\}
$$

and $\Gamma_0 = X \setminus \mathcal{D}$.

Note that the set $\mathcal{D}$ is well-defined because $(P_t)_{t \in \mathbb{T}}$ satisfies the s.m.a., so, by Corollary 2.1.6, the integrals $\int_0^s S_t f(x)\, dt$, $s \in \mathbb{R}$, $s > 0$, $f \in C_0(X)$, $x \in X$, exist.

The set $\mathcal{D}$ is called the *dissipative part of X defined* (or *generated*) by $(P_t)_{t \in \mathbb{T}}$, or *by* $((S_t, T_t))_{t \in \mathbb{T}}$. We say that the transition function $(P_t)_{t \in \mathbb{T}}$ (or the family $((S_t, T_t))_{t \in \mathbb{T}}$ of Markov pairs defined by $(P_t)_{t \in \mathbb{T}}$) is *dissipative* if $X = \mathcal{D}$. For instance, it is not difficult to see that the transition function defined in Example 2.2.2 is dissipative. By contrast, if $(P_t)_{t \in \mathbb{T}}$ is defined on a compact metric space (X, d), then the dissipative part of X defined by $(P_t)_{t \in \mathbb{T}}$ is the empty set. Thus, for instance, the dissipative parts defined by the transition functions in Examples 2.2.3–2.2.5 are all equal to the empty set. There exist transition functions defined on noncompact spaces such that the dissipative parts defined by these transition functions are equal to the empty set, as well. The next two examples illustrate such a situation.

Example 5.1.1. Let Γ be a lattice in $\mathrm{SL}(2, \mathbb{R})$ such that the coset space $(\mathrm{SL}(2, \mathbb{R})/\Gamma)_{\mathrm{L}}$ is not a compact space (an example of such a lattice is $\mathrm{SL}(2, \mathbb{Z})$), and let $(P_t^{(\mathbf{v}^{(1\Gamma \mathrm{L})})})_{t \in \mathbb{R}}$ be the transition function defined by the horocycle flow $\mathbf{v}^{(1\Gamma \mathrm{L})}$ (see (c) of Example 2.2.7). Using a result of Dani and Smillie [28] stated in (b) of Example B.1.9, we obtain that the dissipative part defined by $(P_t^{(\mathbf{v}^{(1\Gamma \mathrm{L})})})_{t \in \mathbb{R}}$ is the empty set. ∎

Example 5.1.2. Let $n \in \mathbb{N}$, $n \geq 2$, let Γ be a lattice in $\mathrm{SL}(n, \mathbb{R})$ such that $(\mathrm{SL}(n, \mathbb{R})/\Gamma)_{\mathrm{R}}$ is not compact, let $\mathbf{v}$ be a unipotent flow on $(\mathrm{SL}(n, \mathbb{R})/\Gamma)_{\mathrm{R}}$, and let $(P_t^{(\mathbf{v})})_{t \in \mathbb{R}}$ be the transition function defined by $\mathbf{v}$ (for the terminology used here, see Sect. B.4.2 and Example 2.2.8). Then, using one of Ratner's theorems, Theorem B.4.9, we obtain that the dissipative part defined by $(P_t^{(\mathbf{v})})_{t \in \mathbb{R}}$ is the empty set. ∎

Set

$$\Gamma_{\mathrm{c}} = \left\{ x \in \Gamma_0 \;\middle|\; \begin{array}{l} \text{the limit } \lim_{\substack{s \to +\infty \\ s > 0}} \frac{1}{s} \int_0^s S_t f(x)\, dt \text{ exists} \\[2mm] \text{for every } f \in C_0(X) \end{array} \right\}.$$

As in the case of transition probabilities, the definition of Γ_{c} above suggests that we consider for every $x \in \Gamma_{\mathrm{c}}$ the map $\varepsilon_x : C_0(X) \to \mathbb{R}$ defined by $\varepsilon_x(f) = \lim_{\substack{s \to +\infty \\ s > 0}} \frac{1}{s} \int_0^s S_t f(x)\, dt$ for every $f \in C_0(X)$.

Clearly, the functions ε_x, $x \in \Gamma_{\mathrm{c}}$ are linear and $\varepsilon_x(f) \geq 0$ for every $x \in \Gamma_{\mathrm{c}}$ and every $f \in C_0(X)$, $f \geq 0$. Therefore, ε_x, $x \in \Gamma_{\mathrm{c}}$, are continuous, so these functions belong to the topological dual of $C_0(X)$. Accordingly, we may and do think of ε_x, $x \in \Gamma_{\mathrm{c}}$, as nonzero, positive elements of $\mathcal{M}(X)$.

As in this case of transition probabilities (see p. 56 of [146]), a natural question at this point is: are the measures ε_x, $x \in \Gamma_{\mathrm{c}}$, elementary measures? That is, given $x \in \Gamma_{\mathrm{c}}$, can we find a continuous-time Banach limit $L^{(\mathrm{ct})}$ such that $\varepsilon_x = \varepsilon_x^{(L^{(\mathrm{ct})})}$? In view of our comments about the similarity of the ergodic decomposition for transition probabilities on one hand, and the decomposition for transition functions, on the other hand, it is not surprising that the answer is yes. We discuss the details in the next proposition.

Proposition 5.1.3. *For every $x \in \Gamma_{\mathrm{c}}$, ε_x is an elementary measure.*

Proof. Let $x \in \Gamma_{\mathrm{c}}$. We have to prove that there exists a continuous-time Banach limit $L^{(\mathrm{ct})}$ such that $\varepsilon_x = \varepsilon_x^{L^{(\mathrm{ct})}}$.

To this end, let $L_1^{(\mathrm{ct})}$ be a continuous-time Banach limit (the existence of $L_1^{(\mathrm{ct})}$ is assured by Theorem 4.4.5), and let $L^{(\mathrm{ct})}$ be the continuous-time Banach limit constructed in Proposition 4.4.6. Thus, $L^{(\mathrm{ct})}(h) = L_1^{(\mathrm{ct})}\left(\left(\frac{1}{s} \int_0^s h(t)\, dt\right)_{s \in (0,+\infty)}\right)$ for every $h \in B_b^{(L)}([0,+\infty))$.

Since $\lim_{s \to +\infty} \frac{1}{s} \int_0^s S_t f(x)\, dt$ exists and is equal to $\varepsilon_x(f)$ for every $f \in C_0(X)$ because $x \in \Gamma_{\mathrm{c}}$, and since $L_1^{(\mathrm{ct})}$ is a continuous-time Banach limit, we obtain that

$$\varepsilon_x^{L^{(\mathrm{ct})}}(f) = L^{(\mathrm{ct})}(((\langle f, T_t \delta_x \rangle))_{t \in [0,+\infty)})$$

$$= L^{(\mathrm{ct})}((S_t f(x))_{t \in [0,+\infty)}) = L_1^{(\mathrm{ct})}((\frac{1}{s} \int_0^s S_t f(x)\, dt)_{s \in (0,+\infty)})$$

$$= \lim_{s \to +\infty} \frac{1}{s} \int_0^s S_t f(x)\, dt = \varepsilon_x(f)$$

for every $f \in C_0(X)$.

Thus, ε_x is an elementary measure. $\qquad\qquad\square$

In view of the above proposition, and in line with the terminology used when dealing with transition probabilities, we call ε_x, $x \in \Gamma_c$, *standard elementary measures*.

Note that another proof of Proposition 5.1.3 can be obtained by observing that, if $x \in \Gamma$, then $\varepsilon_x^{(L^{(ct)})} = \varepsilon_x$ for every continuous-time Banach limit $L^{(ct)}$ (use condition (c) in the definition of a continuous-time Banach limit) and that continuous-time Banach limits exist (by Theorem 4.4.5).

Note also that $\Gamma_c \subseteq \Gamma \subseteq \Gamma_0$ (use Proposition 5.1.3 and the definition of a continuous-time Banach limit).

As in the case of transition probabilities, we can find "formulas" for the supports of the standard elementary measures, but here we have to use continuous-time limit supports. More precisely, we have the theorem below. In the theorem, and throughout the book, the notation $\left(\frac{1}{s}\left(\text{P-}\int_0^s T_t\delta_x\,dt\right)\right)_{s\in(0,+\infty)}$ stands for the $\mathcal{M}(X)$-valued function ψ on $[0, +\infty)$ defined as follows:

$$\psi(s) = \begin{cases} T_0\delta_x & \text{if } s = 0 \\ \frac{1}{s}(\text{P-}\int_0^s T_t\delta_x\,dt) & \text{if } s \in (0, +\infty). \end{cases}$$

Theorem 5.1.4. *Let $x \in \Gamma_c$. Then the following assertions hold true:*

(a) *The $\mathcal{M}(X)$-pointwise integrals $\text{P-}\int_0^s T_t\delta_x\,dt$, $s \in (0, +\infty)$, exist.*
(b) *The family $(\frac{1}{s}(\text{P-}\int_0^s T_t\delta_x\,dt))_{s\in(0,+\infty)}$ has a limit support as $s \to +\infty$ and*
$$\operatorname{supp}\varepsilon_x = L\operatorname{supp}_{s\to+\infty}(\tfrac{1}{s}(\text{P-}\int_0^s T_t\delta_x\,dt)).$$

Proof. (a) Using Corollary 3.3.6, we obtain that the $\mathcal{M}(X)$-pointwise integrals $\text{P-}\int_0^s T_t\delta_x\,dt$, $s \in (0, +\infty)$, exist.

(b) By Proposition 3.3.2, the pointwise integrals $\text{P-}\int_0^s S_t f\,dt$, $s \in (0, +\infty)$, $f \in C_0(X)$, exist. Thus, using Proposition 3.3.7 for $\mu = \delta_x$ and the fact that $(\frac{1}{s}\int_0^s S_t f(x)\,dt)_{s\in(0,+\infty)}$ converges to $\varepsilon_x(f)$ for every $f \in C_0(X)$ (because $x \in \Gamma_c$), we obtain that $(\frac{1}{s}\text{P-}\int_0^s T_t\delta_x\,dt)_{s\in(0,+\infty)}$ converges in the weak* topology of $\mathcal{M}(X)$ to ε_x. Using Theorem 4.3.2, we obtain that $L\operatorname{supp}_{s\to+\infty}(\frac{1}{s}(\text{P-}\int_0^s T_t\delta_x\,dt))$ exists and is equal to $\operatorname{supp}\varepsilon_x$. $\qquad\square$

The support of a (not necessarily standard) elementary measure $\varepsilon_x^{(L^{(ct)})}$, where $x \in \Gamma$ and $L^{(ct)}$ is a continuous-time Banach limit, is included in the upper limit support of the family of probability measures $(\frac{1}{t}(\text{P-}\int_0^t T_s\delta_x\,ds))_{t\in(0,+\infty)}$ as $t \to +\infty$. We discuss this fact in the next proposition.

Proposition 5.1.5. *Assume that Γ is nonempty, let $x \in \Gamma$, and let $L^{(ct)}$ be a continuous-time Banach limit such that $\varepsilon_x^{(L^{(ct)})}$ is an elementary measure. Then* $\operatorname{supp}\varepsilon_x^{(L^{(ct)})} \subseteq \overline{L\operatorname{supp}_{t\to+\infty}}(\frac{1}{t}(\text{P-}\int_0^t T_s\delta_x\,ds)).$

Proof. Let $\varepsilon_x^{(L^{(ct)})}$ be an elementary measure, where $x \in \Gamma$ and $L^{(ct)}$ is a continuous-time Banach limit. We have to prove that, for every $y \in \mathrm{supp}\,\varepsilon_x^{(L^{(ct)})}$, the inequality $\limsup_{t \to +\infty} \frac{1}{t} \int_0^t T_s \delta_x(U)\,ds > 0$ holds true for every open neighborhood U of y.

To this end, let $y \in \mathrm{supp}\,\varepsilon_x^{(L^{(ct)})}$, and let U be an open neighborhood of y.

Since X is a locally compact metric space, there exists $\rho \in \mathbb{R}$, $\rho > 0$, such that $\overline{B(y,\rho)}$ is a compact subset of X and such that $\overline{B(y,\rho)} \subseteq U$. Using Proposition 7.1.8, p. 199, of Cohn [20], we obtain that there exists a continuous function $f : X \to \mathbb{R}$ that has compact support (so, $f \in C_0(X)$) such that $\mathbf{1}_{\overline{B(y,\rho)}} \le f \le \mathbf{1}_U$.

Since $y \in \mathrm{supp}\,\varepsilon_x^{(L^{(ct)})}$, since $B(y,\rho)$ is a neighborhood of y, since $\mathbf{1}_{B(y,\rho)} \le \mathbf{1}_{\overline{B(y,\rho)}} \le f$, and using the definition of $\varepsilon_x^{(L^{(ct)})}$ and of a continuous-time Banach limit, we obtain that

$$
\begin{aligned}
0 < \varepsilon_x^{(L^{(ct)})}(B(y,\rho)) &= \left\langle \mathbf{1}_{B(y,\rho)}, \varepsilon_x^{(L^{(ct)})} \right\rangle \le \left\langle f, \varepsilon_x^{(L^{(ct)})} \right\rangle \\
&= L^{(ct)}((\langle f, T_t \delta_x \rangle)_{t \in [0,+\infty)}) \le L^{(ct)}((\langle \mathbf{1}_U, T_t \delta_x \rangle)_{t \in [0,+\infty)}) \\
&= L^{(ct)}((T_t \delta_x(U))_{t \in [0,+\infty)}) \le \limsup_{t \to +\infty} \frac{1}{t} \int_0^t (T_s \delta_x)(U)\,ds.
\end{aligned}
$$

$\square$

Our goal now is to prove that the support of an elementary measure $\varepsilon_x^{(L^{(ct)})}$, where $x \in \Gamma$ and $L^{(ct)}$ is a continuous-time Banach limit, is a subset of the orbit-closure $\overline{\mathcal{O}(x)}$ of x under the action of $(T_t)_{t \in [0,+\infty)}$.

We need the following simple lemma:

Lemma 5.1.6. $\overline{L\mathrm{supp}_{t \to +\infty}}\left(\frac{1}{t}\left(P\text{-} \int_0^t T_s \delta_x\,ds \right) \right) \subseteq \overline{L\mathrm{supp}_{t \to +\infty}}(T_t \delta_x)$ *for every* $x \in X$.

Proof. We will prove that

$$
G^{\sim}((T_t \delta_x)_{t \in [0,+\infty)}) \subseteq G^{\sim}\left(\left(\frac{1}{t}(P\text{-} \int_0^t T_s \delta_x\,ds) \right)_{\overline{t \in (0,+\infty)}} \right).
$$

(For the notation used here, see the discussion preceding Theorem 4.3.2.)

Let $y \in G^{\sim}((T_t \delta_x)_{t \in [0,+\infty)})$.

Then there exists an open neighborhood U of y such that $\limsup_{t \to +\infty}(T_t \delta_x)(U) = 0$. Since $T_t \delta_x$, $t \in [0,+\infty)$, are probability measures, it follows that $\lim_{t \to +\infty}(T_t \delta_x)(U)$ exists and is equal to zero. Since $(T_t \delta_x)(U) = S_t \mathbf{1}_U(x)$ for every $t \in [0,+\infty)$, and since the map $t \mapsto S_t \mathbf{1}_U(x)$, $t \in [0,+\infty)$, belongs to $B_b^{(L)}([0,+\infty))$ (because $(P_t)_{t \in [0,+\infty)}$ satisfies the s.m.a.), it follows that we can apply Lemma 4.4.1, and we obtain that $\lim_{t \to +\infty} \frac{1}{t} \int_0^t T_s \delta_x(U)\,ds$ exists and is equal to zero. Accordingly, $y \in G^{\sim}\left(\left(\frac{1}{t}(P\text{-} \int_0^t T_s \delta_x\,ds) \right)_{\overline{t \in (0,+\infty)}} \right)$. $\square$

Proposition 5.1.7. *Assume that* $\mathbb{T} = [0, +\infty)$*, and let* $\varepsilon_x^{(L^{(\mathrm{ct})})}$ *be an elementary measure, where* $x \in \Gamma$ *and* $L^{(\mathrm{ct})}$ *is a continuous-time Banach limit. Then* $\operatorname{supp} \varepsilon_x^{(L^{(\mathrm{ct})})} \subseteq \overline{\mathcal{O}(x)}$*, where, as usual,* $\overline{\mathcal{O}(x)}$ *stands for the orbit-closure of* x *under the action of* $(P_t)_{t \in [0,+\infty)}$*. In particular, if* ε_x *is a standard elementary measure for some* $x \in \Gamma_\mathrm{c}$*, then* $\operatorname{supp} \varepsilon_x \subseteq \overline{\mathcal{O}(X)}$*.*

Proof. Let $\varepsilon_x^{(L^{(\mathrm{ct})})}$ be an elementary measure, where $x \in \Gamma$ and $L^{(\mathrm{ct})}$ is a continuous-time Banach limit.

Using Proposition 5.1.5, Lemma 5.1.6 and Proposition 4.3.6 in this order, we obtain that

$$\operatorname{supp} \varepsilon_x^{(L^{(\mathrm{ct})})} \subseteq \overline{L\operatorname{supp}_{t \to +\infty}}\left(\frac{1}{t}\left(\text{P-}\int_0^t T_s \delta_x \, \mathrm{d}s\right)\right) \subseteq \overline{L\operatorname{supp}_{t \to +\infty}}(T_t \delta_x) \subseteq \overline{\mathcal{O}(X)}.$$

The assertion about standard elementary measures is now obviously true because, by Proposition 5.1.3, standard elementary measures are particular cases of elementary measures. $\qquad\square$

Note that the elementary measures, not necessarily standard, are nonzero measures of norm less than or equal to 1. Therefore, it makes sense to define the set $\Gamma_{\mathrm{cp}} = \{x \in \Gamma_\mathrm{c} \mid \|\varepsilon_x\| = 1\}$. Thus, Γ_{cp} is the set of all $x \in \Gamma_\mathrm{c}$ such that ε_x is a standard elementary probability measure.

As in the case of transition probabilities, it is a natural question whether the elementary measures are invariant or not, and, again as in the study of the KBBY decomposition for transition probabilities, there exist transition functions that have even standard elementary probability measures that fail to be invariant. Let us discuss such an example.

Example 5.1.8. Let $X = [0, 1]$ and let d be the usual metric on $[0, 1]$ defined by the absolute value.

Now let $\mathbf{w} = (w_t)_{t \in [0,+\infty)}$ be the one-parameter semigroup defined in Example 2.2.3, let $(P_t^{(\mathbf{w})})_{t \in [0,+\infty)}$ and $(S_t^{(\mathbf{w})}, T_t^{(\mathbf{w})})_{t \in [0,+\infty)}$ be the transition function and the family of Markov pairs defined by $\mathbf{w}$.

We have already proved in Example 2.2.3 that $(P_t^{(\mathbf{w})})_{t \in [0,+\infty)}$ satisfies the s.m.a. Thus, in view of the theory developed in this section, it makes sense to consider the set Γ_{cp} defined by $(P_t^{(\mathbf{w})})_{t \in [0,+\infty)}$. Our goal now is to determine the set Γ_{cp} and to show that for some $x \in \Gamma_{\mathrm{cp}}$, the probability measure ε_x is not an invariant measure.

To this end, we first prove the following assertion:

Assertion A. $(0, 1) \subseteq \Gamma_{\mathrm{cp}}$ and $\varepsilon_x = \delta_0$ for every $x \in (0, 1)$.

In order to prove the assertion, we have to show that $\lim_{t \to +\infty} \frac{1}{t} \int_0^t S_r f(x) \, \mathrm{d}r$ exists and is equal to $f(0)$ for every continuous function $f : [0, 1] \to \mathbb{R}$ and every $x \in (0, 1)$.

To this end, let $f : [0, 1] \to \mathbb{R}$ be a continuous function and let $x \in (0, 1)$.

Clearly, the assertion is true if $f = 0$, so we may and do assume that $f \neq 0$. Thus, $\|f\| > 0$.

Now, let $\varepsilon \in \mathbb{R}, \varepsilon > 0$.

Since f is a continuous function and since $\lim_{t \to +\infty} \frac{x}{2^t} = 0$, it follows that there exists a $t_\varepsilon^{(1)} \in \mathbb{R}, t_\varepsilon^{(1)} > 0$, large enough such that $\left| f(\frac{x}{2^t}) - f(0) \right| < \frac{\varepsilon}{2}$ for every $t \in \mathbb{R}, t \geq t_\varepsilon^{(1)}$.

Now, there exists a $t_\varepsilon \in \mathbb{R}, t_\varepsilon \geq t_\varepsilon^{(1)}$, large enough such that $\frac{t_\varepsilon^{(1)}}{t_\varepsilon} < \frac{\varepsilon}{4\|f\|}$.

We obtain that

$$\left| \frac{1}{t} \int_0^t S_r f(x) \, dr - f(0) \right| = \left| \frac{1}{t} \int_0^t (S_r f(x) - f(0)) \, dr \right|$$

$$\leq \frac{1}{t} \int_0^t |S_r f(x) - f(0)| \, dr = \frac{1}{t} \int_0^t \left| f(\frac{x}{2^r}) - f(0) \right| \, dr$$

$$= \frac{1}{t} \int_0^{t_\varepsilon^{(1)}} \left| f(\frac{x}{2^r}) - f(0) \right| \, dr + \frac{1}{t} \int_{t_\varepsilon^{(1)}}^t \left| f(\frac{x}{2^r}) - f(0) \right| \, dr$$

$$< \frac{t_\varepsilon^{(1)}}{t} 2\|f\| + \frac{t - t_\varepsilon^{(1)}}{t} \frac{\varepsilon}{2} \leq \frac{t_\varepsilon^{(1)}}{t_\varepsilon} 2\|f\| + \frac{\varepsilon}{2} < \frac{\varepsilon}{2} + \frac{\varepsilon}{2} = \varepsilon$$

for every $t \in \mathbb{R}, t \geq t_\varepsilon$.

We have therefore proved that for every continuous nonzero function $f :$ $[0, 1] \to \mathbb{R}$, for every $x \in (0, 1)$, and for every $\varepsilon \in \mathbb{R}, \varepsilon > 0$, there exists a $t_\varepsilon \in \mathbb{R}$, $t_\varepsilon > 0$, such that $\left| \frac{1}{t} \int_0^t S_r f(x) \, dr - f(0) \right| < \varepsilon$ for every $t \in \mathbb{R}, t \geq t_\varepsilon$.

Thus, Assertion A holds true.

Using Assertion A, we obtain that for every $x \in (0, 1)$, we can construct a standard elementary probability measure ε_x that is not invariant because $\varepsilon_x = \delta_0$ and $T_t \delta_0 = \delta_{w_t(0)} = \delta_1$ for every $t \in [0, +\infty)$.

Note that the following assertion is also true:

Assertion B. If $x = 0$ or $x = 1$, then $\lim_{t \to +\infty} \frac{1}{t} \int_0^t S_r f(x) \, dr$ exists and is equal to $f(1)$.

Indeed, the assertion is true because $w_r(0) = w_r(1) = 1$ and $S_r f(x) = f(w_r(x)) = f(1)$ for every continuous function $f : [0, 1] \to \mathbb{R}$, every $r \in [0, +\infty)$, and every $x = 0$ or 1.

Combining Assertions A and B, we obtain that for the transition function under consideration, $\Gamma_{cp} = [0, 1]$ (the entire space), $\varepsilon_x, x \in (0, 1)$, are standard elementary probability measures that are not invariant measures (because $\varepsilon_x = \delta_0, x \in (0, 1)$, and δ_0 is not an invariant measure), while ε_0 and ε_1 are invariant standard elementary probability measures because $\varepsilon_0 = \varepsilon_1 = \delta_1$ and δ_1 is an invariant measure for the transition function.   ∎

In view of the above example, it makes sense to consider the set

$$
\Gamma_{cpi} = \left\{ x \in \Gamma_{cp} \;\middle|\; \begin{array}{l} \varepsilon_x \text{ is an invariant} \\ \text{probability measure} \\ \text{for } (P_t)_{t \in \mathbb{T}} \end{array} \right\}.
$$

Our goal now, in the next section, is to prove that the sets $\mathcal{D}$, Γ_0, Γ_c, Γ_{cp} and Γ_{cpi} are measurable.

5.2   The Measurability of $\mathcal{D}$, Γ_0, Γ_c, Γ_{cp} and Γ_{cpi}

Our goal in this section is to state and prove the versions for transition functions of Propositions 5.2–5.5 and Theorem 5.6 of [146] (see also Propositions 1.3.1 and 1.3.5–1.3.8).

We will use the setting, terminology, and notation of Sect. 5.1. Thus, we assume given a transition function $(P_t)_{t \in \mathbb{T}}$ defined on a locally compact separable metric space (X, d), we assume that $(P_t)_{t \in \mathbb{T}}$ satisfies the s.m.a., and we assume that $((S_t, T_t))_{t \in \mathbb{T}}$ is the family of Markov pairs defined by $(P_t)_{t \in \mathbb{T}}$.

We will need the following two lemmas:

Lemma 5.2.1. *Let $a \in \mathbb{R}$, and let $f : (a, +\infty) \to \mathbb{R}$. The following assertions are equivalent:*

(a) $\lim_{t \to +\infty} f(t)$ exists and is a real number.
(b) For every $\varepsilon \in \mathbb{R}$, $\varepsilon > 0$, there exists a $t_\varepsilon \in (a, +\infty)$ such that $|f(t') - f(t'')| < \varepsilon$ for every $t' \geq t_\varepsilon$ and $t'' \geq t_\varepsilon$.

The proof of the lemma is easy, so we omit it.

In the next lemma, we will use the fact that $C_0(X)$ is a separable Banach space (for a proof of the separability of $C_0(X)$, see, for example, Theorem 1.3.3 of [143]).

Lemma 5.2.2. *Let $(g_n)_{n \in \mathbb{N}}$ be a sequence of elements of $C_0(X)$ such that the range $\{g_n \mid n \in \mathbb{N}\}$ of $(g_n)_{n \in \mathbb{N}}$ is dense in $C_0(X)$.*

(a) Let $x \in X$, and assume that $\lim_{t \to +\infty} \frac{1}{t} \int_0^t S_r g_n(x)\, dr$ exists and is a real number for every $n \in \mathbb{N}$. Then $\lim_{t \to +\infty} \frac{1}{t} \int_0^t S_r f(x)\, dr$ exists and is a real number for every $f \in C_0(X)$.
(b) Let $x \in X$, and assume that there exists an $L \in \mathbb{R}$ such that $\lim_{t \to +\infty} \frac{1}{t} \int_0^t S_r g_n(x)\, dr$ exists and is equal to L for every $n \in \mathbb{N}$. Then $\lim_{t \to +\infty} \frac{1}{t} \int_0^t S_r f(x)\, dr$ exists and is equal to L for every $f \in C_0(X)$.

Proof. (a) Let $x \in X$ and assume that $\lim_{t \to +\infty} \frac{1}{t} \int_0^t S_r g_n(x)\, dr$ exists and is a real number for every $n \in \mathbb{N}$.

Using Lemma 5.2.1, we obtain that it is enough to prove that for every $f \in C_0(X)$ and every $\varepsilon \in \mathbb{R}$, $\varepsilon > 0$, there exists a $t_\varepsilon \in \mathbb{R}$, $t_\varepsilon > 0$, such that

$\left| \frac{1}{t'} \int_0^{t'} S_r f(x)\, dr - \frac{1}{t''} \int_0^{t''} S_r f(x)\, dr \right| < \varepsilon$ for every $t' \in \mathbb{R}$, $t' \geq t_\varepsilon$, and $t'' \in \mathbb{R}$, $t'' \geq t_\varepsilon$.

To this end, let $f \in C_0(X)$ and let $\varepsilon \in \mathbb{R}$, $\varepsilon > 0$.

Then there exists an $n_\varepsilon \in \mathbb{N}$ such that $\| g_{n_\varepsilon} - f \| < \frac{\varepsilon}{3}$.

Since we assume that $\lim_{t \to +\infty} \frac{1}{t} \int_0^t S_r g_{n_\varepsilon}(x)\, dr$ exists and is a real number, using Lemma 5.2.1 again, we obtain that there exists a $t_\varepsilon \in \mathbb{R}$, $t_\varepsilon > 0$, such that $\left| \frac{1}{t'} \int_0^{t'} S_r g_{n_\varepsilon}(x)\, dr - \frac{1}{t''} \int_0^{t''} S_r g_{n_\varepsilon}(x)\, dr \right| < \frac{\varepsilon}{3}$ for every $t' \in \mathbb{R}$, $t' \geq t_\varepsilon$, and $t'' \in \mathbb{R}$, $t'' \geq t_\varepsilon$.

Using the fact that S_r, $r \in \mathbb{T}$, are contractions of $B_b(X)$, we obtain that

$$\left| \frac{1}{t'} \int_0^{t'} S_r f(x)\, dr - \frac{1}{t''} \int_0^{t''} S_r f(x)\, dr \right| \leq \left| \frac{1}{t'} \int_0^{t'} S_r f(x)\, dr - \frac{1}{t'} \int_0^{t'} S_r g_{n_\varepsilon}(x)\, dr \right|$$

$$+ \left| \frac{1}{t'} \int_0^{t'} S_r g_{n_\varepsilon}(x)\, dr - \frac{1}{t''} \int_0^{t''} S_r g_{n_\varepsilon}(x)\, dr \right|$$

$$+ \left| \frac{1}{t''} \int_0^{t''} S_r g_{n_\varepsilon}(x)\, dr - \frac{1}{t''} \int_0^{t''} S_r f(x)\, dr \right|$$

$$\leq \frac{1}{t'} \int_0^{t'} |S_r f(x) - S_r g_{n_\varepsilon}(x)|\, dr + \frac{\varepsilon}{3} + \frac{1}{t''} \int_0^{t''} |S_r g_{n_\varepsilon}(x) - S_r f(x)|\, dr$$

$$\leq \frac{1}{t'} \int_0^{t'} \| S_r f - S_r g_{n_\varepsilon} \|\, dr + \frac{\varepsilon}{3} + \frac{1}{t''} \int_0^{t''} \| S_r g_{n_\varepsilon} - S_r f \|\, dr$$

$$\leq \frac{1}{t'} \int_0^{t'} \| f - g_{n_\varepsilon} \|\, dr + \frac{\varepsilon}{3} + \frac{1}{t''} \int_0^{t''} \| g_{n_\varepsilon} - f \|\, dr < \frac{\varepsilon}{3} + \frac{\varepsilon}{3} + \frac{\varepsilon}{3} = \varepsilon$$

for every $t' \in \mathbb{R}$, $t' \geq t_\varepsilon$, and $t'' \in \mathbb{R}$, $t'' \geq t_\varepsilon$.

(b) Let $L \in \mathbb{R}$ and $x \in X$, and assume that $\lim_{t \to +\infty} \frac{1}{t} \int_0^t S_r g_n(x)\, dr$ exists and is equal to L for every $n \in \mathbb{N}$.

We have to prove that for every $f \in C_0(X)$ and $\varepsilon \in \mathbb{R}$, $\varepsilon > 0$, there exists a $t_\varepsilon \in \mathbb{R}$, $t_\varepsilon > 0$, such that $\left| \frac{1}{t} \int_0^t S_r f(x)\, dr - L \right| < \varepsilon$ for every $t \in \mathbb{R}$, $t \geq t_\varepsilon$.

Thus, let $f \in C_0(X)$ and $\varepsilon \in \mathbb{R}$, $\varepsilon > 0$.

Since $\{ g_n \mid n \in \mathbb{N} \}$ is dense in $C_0(X)$, there exists an $n_\varepsilon \in \mathbb{N}$ such that $\| g_{n_\varepsilon} - f \| < \frac{\varepsilon}{2}$.

Since we assume that $\lim_{t \to +\infty} \frac{1}{t} \int_0^t S_r g_{n_\varepsilon}(x)\, dr$ exists and is equal to L, it follows that there exists a $t_\varepsilon \in \mathbb{R}$, $t_\varepsilon > 0$, such that $\left| \frac{1}{t} \int_0^t S_r g_{n_\varepsilon}(x)\, dr - L \right| < \frac{\varepsilon}{2}$ for every $t \in \mathbb{R}$, $t \geq t_\varepsilon$.

Using the fact that S_r, $r \in \mathbb{T}$, are positive contractions of $B_b(X)$, we obtain that

$$\left| \frac{1}{t} \int_0^t S_r f(x)\, dr - L \right| \leq \left| \frac{1}{t} \int_0^t S_r f(x)\, dr - \frac{1}{t} \int_0^t S_r g_{n_\varepsilon}(x)\, dr \right|$$

$$+ \left| \frac{1}{t} \int_0^t S_r g_{n_\varepsilon}(x)\,dr - L \right| < \frac{1}{t} \int_0^t |S_r f(x) - S_r g_{n_\varepsilon}(x)|\,dr + \frac{\varepsilon}{2}$$

$$\leq \frac{1}{t} \int_0^t \|S_r(f - g_{n_\varepsilon})\|\,dr + \frac{\varepsilon}{2} \leq \frac{1}{t} \int_0^t \|f - g_{n_\varepsilon}\|\,dr + \frac{\varepsilon}{2} < \frac{\varepsilon}{2} + \frac{\varepsilon}{2} = \varepsilon$$

for every $t \in \mathbb{R}$, $t \geq t_\varepsilon$. $\qquad\qquad\qquad\qquad\qquad\qquad\qquad\qquad\qquad\square$

We are now in a position to discuss the measurability of $\mathcal{D}$ and Γ_0.

Theorem 5.2.3. *The sets $\mathcal{D}$ and Γ_0 are $\mathcal{B}(X)$-measurable.*

Proof. Clearly, since $\Gamma_0 = X \setminus \mathcal{D}$, it is enough to prove only that $\mathcal{D}$ is measurable.

To this end, let $(g_n)_{n \in \mathbb{N}}$ be a sequence of elements of $C_0(X)$ such that the range $\{g_n \mid n \in \mathbb{N}\}$ of $(g_n)_{n \in \mathbb{N}}$ is dense in $C_0(X)$ (the existence of $(g_n)_{n \in \mathbb{N}}$ is assured by the separability of $C_0(X)$).

Using (b) of Lemma 5.2.2, we obtain that

$$\mathcal{D} = \bigcap_{n=1}^{\infty} \left\{ x \in X \ \middle| \ \begin{array}{l} \lim_{t \to +\infty} \frac{1}{t} \int_0^t S_r g_n(x)\,dr \ \text{exists and is} \\ \text{equal to zero} \end{array} \right\}$$

Set

$$A_n = \left\{ x \in X \ \middle| \ \lim_{t \to +\infty} \frac{1}{t} \int_0^t S_r g_n(x)\,dr \ \text{exists and is equal to zero} \right\}$$

for every $n \in \mathbb{N}$.

Using Proposition 3.2.5 and the notations introduced in the proposition, we obtain that $\overrightarrow{g_n} \in B_b(X)$ and $A_{\overrightarrow{g_n}} \in \mathcal{B}(X)$ for every $n \in \mathbb{N}$. Therefore, taking into consideration that $A_n = A_{\overrightarrow{g_n}} \cap \{x \in X \mid \overrightarrow{g_n}(x) = 0\}$, we obtain that A_n is a $\mathcal{B}(X)$-measurable subset of X for every $n \in \mathbb{N}$.

Since $\mathcal{D} = \cap_{n=1}^{\infty} A_n$, it follows that $\mathcal{D}$ is measurable. $\qquad\qquad\qquad\square$

We now prove that Γ_c is measurable.

Theorem 5.2.4. *The set Γ_c belongs to $\mathcal{B}(X)$.*

Proof. Again using the separability of $C_0(X)$, we let $(g_n)_{n \in \mathbb{N}}$ be a sequence of elements of $C_0(X)$ such that the range $\{g_n \mid n \in \mathbb{N}\}$ of $(g_n)_{n \in \mathbb{N}}$ is dense in $C_0(X)$.

By applying (a) of Lemma 5.2.2, we obtain that $\Gamma_c = \cap_{n=1}^{\infty} A_n$, where

$$A_n = \left\{ x \in \Gamma_0 \ \middle| \ \lim_{\substack{t \to +\infty \\ t > 0}} \frac{1}{t} \int_0^t S_r g_n(x)\,dr \ \text{exists} \right\}$$

for every $n \in \mathbb{N}$.

Clearly, $A_n = A_{\overrightarrow{g_n}} \cap \Gamma_0$, where $A_{\overrightarrow{g_n}}$ is the subset of X corresponding to g_n and defined in Proposition 3.2.5, $n \in \mathbb{N}$.

Since $A_{g_n}^{\rightarrow}$ and Γ_0 are measurable by Proposition 3.2.5 and Theorem 5.2.3, respectively, it follows that A_n is measurable, as well, for every $n \in \mathbb{N}$.

Accordingly, $\Gamma_{\mathrm{c}} \in \mathcal{B}(X)$. $\qquad\qquad\qquad\qquad\qquad\qquad\qquad\qquad\qquad\qquad\square$

Our goal now is to prove that Γ_{cp} is measurable. The approach that we will use is similar to that used for transition probabilities in [146]. However, for transition functions we also have to use some of the tools developed in Chap. 3.

For every $f \in B_b(X)$, let $f^{\square} : X \to \mathbb{R}$ be the function defined as follows:

$$f^{\square}(x) = \begin{cases} \langle f, \varepsilon_x \rangle & \text{if } x \in \Gamma_{\mathrm{c}}, \\ 0 & \text{if } x \notin \Gamma_{\mathrm{c}}. \end{cases}$$

Note that if $f \in C_0(X)$, then

$$f^{\square}(x) = \begin{cases} \lim_{t \to +\infty} \frac{1}{t} \int_0^t S_r f(x)\, dr & \text{if } x \in \Gamma_{\mathrm{c}}, \\ 0 & \text{if } x \notin \Gamma_{\mathrm{c}}. \end{cases} \tag{5.2.1}$$

In order to prove the measurability of Γ_{cp} we need the next lemma. The result of the lemma will soon be significantly extended.

Lemma 5.2.5. $f^{\square} \in B_b(X)$ (that is, $f^{\square}$ is a bounded measurable function) whenever $f \in C_0(X)$.

Proof. Let $f \in C_0(X)$. Using the equality (5.2.1), we obtain that $f^{\square} = f^{\rightarrow} 1_{\Gamma_{\mathrm{c}}}$, where $f^{\rightarrow}$ is the function defined in Proposition 3.2.5.

Clearly, $f^{\square}$ is bounded. Since $\Gamma_{\mathrm{c}} \in \mathcal{B}(X)$ by Theorem 5.2.4, and since $f^{\rightarrow} \in B_b(X)$ by Proposition 3.2.5, it follows that $f^{\square}$ is also measurable. $\qquad\square$

Theorem 5.2.6. $\Gamma_{\mathrm{cp}} \in \mathcal{B}(X)$.

Proof. The proof of the theorem is similar to the proof of Proposition 5.4 of [146], but here we have to use Lemma 5.2.5.

Since $C_0(X)$ is a separable Banach space, it follows that the subset A of $C_0(X)$ defined by $A = \{f \in C_0(X) \mid 0 \le f \le 1\}$ can be thought of as a metric space in its own right when endowed with the restriction to $A \times A$ of the distance function defined on $C_0(X) \times C_0(X)$ by the sup norm on $C_0(X)$. It is easy to see that A, as a metric space, is separable. Thus, we may and do pick a sequence $(g_n)_{n \in \mathbb{N}}$ of elements of A such that the range $\{g_n \mid n \in \mathbb{N}\}$ of $(g_n)_{n \in \mathbb{N}}$ is dense in A.

Now, let $\varphi : X \to \mathbb{R}$ be defined as follows:

$$\varphi(x) = \begin{cases} \|\varepsilon_x\| & \text{if } x \in \Gamma_{\mathrm{c}}, \\ 0 & \text{if } x \notin \Gamma_{\mathrm{c}}. \end{cases}$$

The proof of the theorem will be completed if we show that φ is measurable because $\Gamma_{\mathrm{cp}} = \{x \in X \mid \varphi(x) = 1\}$.

Since ε_x is a positive element of $\mathcal{M}(X)$ for every $x \in \Gamma_c$, we obtain that

$$\varphi(x) = \|\varepsilon_x\| = \sup_{\substack{f \in C_0(X) \\ 0 \le f \le 1}} \langle f, \varepsilon_x \rangle = \sup_{l \in \mathbb{N}} \langle g_l, \varepsilon_x \rangle = \sup_{l \in \mathbb{N}} g_l^{\square}(x)$$

for every $x \in \Gamma_c$.

Accordingly, in view of the equality (5.2.1), we obtain that $\varphi(x) = \sup_{l \in \mathbb{N}} g_l^{\square}(x)$ for every $x \in X$; that is, $\varphi = \sup_{l \in \mathbb{N}} g_l^{\square}$.

Since, by Lemma 5.2.5, $g_l^{\square}$ is a measurable function for every $l \in \mathbb{N}$, it follows that φ is measurable, as well. $\square$

We will conclude this section by discussing the $\mathcal{B}(X)$-measurability of Γ_{cpi}. To this end, we have to extend Lemma 5.2.5, and in order to obtain the extension, we need some preparation.

Let $\mathcal{F}_{\mathbb{R}}(X)$ be the vector space of all real-valued functions defined on X and let $Q : B_b(X) \to \mathcal{F}_{\mathbb{R}}(X)$ be a linear operator.

We say that Q is a *bounded function-valued operator* if Qf is a bounded function for every $f \in B_b(X)$.

The operator Q is said to be *sequentially continuous with respect to pointwise convergence* if the following condition is satisfied: if $(f_n)_{n \in \mathbb{N}}$ is a bounded sequence of elements of $B_b(X)$ that converges pointwise to some $f \in B_b(X)$ on X (that is, if there exists an $M \in \mathbb{R}$, $M \ge 0$, such that $|f_n(x)| \le M$ for every $n \in \mathbb{N}$ and $x \in X$, and if $(f_n(x))_{n \in \mathbb{N}}$ converges to $f(x)$ for every $x \in X$), then $(Qf_n)_{n \in \mathbb{N}}$ converges pointwise to Qf on X.

Lemma 5.2.7. *Let* $Q : B_b(X) \to \mathcal{F}_{\mathbb{R}}(X)$ *be a bounded function-valued linear operator, assume that* Q *is sequentially continuous with respect to pointwise convergence, and assume that* $Qf \in B_b(X)$ *for every* $f \in C_0(X)$. *Then* $Q\mathbf{1}_K \in B_b(X)$ *for every compact subset* K *of* X.

Proof. We will use Step 1 in the proof of Proposition 2.2 of [146] (see p. 75 of [146]).

Let K be a compact subset of X. Clearly, the lemma is true if K is the empty set, so we may and do assume that K is nonempty.

By Step 1 of the proof of Proposition 2.2 of [146], there exists a sequence $(f_n)_{n \in \mathbb{N}}$ of real-valued continuous functions with compact support defined on X such that $\mathbf{1}_K \le f_n$ for every $n \in \mathbb{N}$, such that $(f_n)_{n \in \mathbb{N}}$ converges pointwise to $\mathbf{1}_K$ on X, and such that $(f_n)_{n \in \mathbb{N}}$ is a bounded sequence in $B_b(X)$.

Since Q is sequentially continuous with respect to pointwise convergence, it follows that $(Qf_n)_{n \in \mathbb{N}}$ converges to $Q\mathbf{1}_K$. Since we assume that $Qf_n \in B_b(X)$ for every $n \in \mathbb{N}$, we obtain that $Q\mathbf{1}_K \in B_b(X)$, as well. $\square$

Proposition 5.2.8. *Let* $Q : B_b(X) \to \mathcal{F}_{\mathbb{R}}(X)$ *be a bounded function-valued linear operator and assume that* Q *is sequentially continuous with respect to pointwise* convergence. If $Qf \in B_b(X)$ for every $f \in C_0(X)$, then $Qf \in B_b(X)$ for every $f \in B_b(X)$ (that is, the range of Q is actually included in $B_b(X)$).

Proof. We will prove the proposition in two steps.

Step 1.   At this step, we prove that $Q\mathbf{1}_A \in B_b(X)$ for every $A \in \mathcal{B}(X)$.

To this end, set $\mathcal{A} = \{A \in \mathcal{B}(X) \mid Q\mathbf{1}_A \in B_b(X)\}$.

Since Q is linear it follows that $\mathcal{A}$ satisfies conditions (a) and (b) of Lemma 2.1.3.

Using the fact that Q is sequentially continuous with respect to pointwise convergence, we obtain that $\mathcal{A}$ satisfies condition (c) of Lemma 2.1.3, as well.

Finally, using Lemma 5.2.7, we obtain that all the compact subsets of X belong to $\mathcal{A}$, so, by Lemma 2.1.3, $\mathcal{A} = \mathcal{B}(X)$.

Step 2.   At this step, we conclude the proof of the proposition.

Using Step 1, and the fact that Q is a linear operator, we obtain that $Qf \in B_b(X)$ whenever f is a simple measurable function (that is, whenever $f = \sum_{i=1}^{n} a_i \mathbf{1}_{A_i}$ for some $n \in \mathbb{N}$, and $a_i \in \mathbb{R}$ and $A_i \in \mathcal{B}(X)$, $i = 1, 2, \cdots, n$).

If $f \in B_b(X)$, $f \geq 0$, then there exists a sequence $(f_n)_{n \in \mathbb{N}}$ of simple measurable functions such that $0 \leq f_n \leq f$ for every $n \in \mathbb{N}$, and such that $(f_n)_{n \in \mathbb{N}}$ converges pointwise (even uniformly) to f on X. Since Q is sequentially continuous with respect to pointwise convergence, we obtain that $(Qf_n)_{n \in \mathbb{N}}$ converges to Qf pointwise on X. Since $Qf_n \in B_b(X)$ for every $n \in \mathbb{N}$, it follows that Qf belongs to $B_b(X)$, as well.

Finally, if $f \in B_b(X)$ is not necessarily a positive element of $B_b(X)$, then $f = f^+ - f^-$, where $f^+ = f \vee 0$ and $f^- = (-f) \vee 0$. Since $f^+ \in B_b(X)$ and $f^- \in B_b(X)$, it follows that $Qf^+ \in B_b(X)$ and $Qf^- \in B_b(X)$, so, using the fact that Q is a linear operator, we obtain that $Qf \in B_b(X)$. $\square$

Our discussion so far allows us to obtain the following extension of Lemma 5.2.5:

Proposition 5.2.9. *If $f \in B_b(X)$, then $f^{\square}$ belongs to $B_b(X)$, as well.*

Proof. Note that for every $f \in B_b(X)$, the function $f^{\square}$ belongs to $\mathcal{F}_{\mathbb{R}}(X)$, so it makes sense to define the operator $Q : B_b(X) \to \mathcal{F}_{\mathbb{R}}(X)$, $Qf = f^{\square}$ for every $f \in B_b(X)$.

Clearly, the proof of the proposition will be completed if we show that $Qf \in B_b(X)$ for every $f \in B_b(X)$.

To this end, we first note that Q is clearly a linear operator.

Since $|Qf(x)| = 0$ if $x \notin \Gamma_c$, and $|Qf(x)| = |\langle f, \varepsilon_x \rangle| \leq \int |f(y)| \, d\varepsilon_x(y) \leq \|f\|$ if $x \in \Gamma_c$, it follows that $|Qf| \leq \|f\|$ for every $f \in B_b(X)$, so Q is a bounded function-valued linear operator.

We now note that Q is sequentially continuous with respect to pointwise convergence. Indeed, let $(f_n)_{n \in \mathbb{N}}$ be a bounded sequence of elements of $B_b(X)$, and assume that $(f_n)_{n \in \mathbb{N}}$ converges pointwise to some $f \in B_b(X)$ on X. Let $x \in X$, and assume first that $x \notin \Gamma_c$; then $f_n^{\square}(x) = 0$ for every $n \in \mathbb{N}$ and $f^{\square}(x) = 0$, as well; therefore, $(Q f_n(x))_{n \in \mathbb{N}}$ converges to $Q f(x)$ whenever $x \notin \Gamma_c$; on the other hand, if $x \in \Gamma_c$, then, since $(f_n)_{n \in \mathbb{N}}$ is a bounded sequence of measurable functions, and since ε_x is a finite measure, we obtain that we can use the Lebesgue dominated convergence theorem in order to conclude that the sequence $(\int_X f_n(y) \, d\varepsilon_x(y))_{n \in \mathbb{N}}$ converges to $\int_X f(y) \, d\varepsilon_x(y)$; that is, we can conclude that $(Q f_n(x))_{n \in \mathbb{N}}$ converges to $Q f(x)$.

Since $Q f \in B_b(X)$ whenever $f \in C_0(X)$ by Lemma 5.2.5, we obtain that we can use Proposition 5.2.8. By Proposition 5.2.8, $Q f \in B_b(X)$ for every $f \in B_b(X)$. $\qquad\square$

Observation. The proof of Proposition 5.2.9 is based on Proposition 5.2.8. However, Proposition 5.2.8 can also be used to prove Proposition 1.3.7 (Proposition 5.5 of [146]), which is the assertion for transition probabilities that corresponds to Proposition 5.2.9 (and thus, we obtain a unified approach for proving both Propositions 1.3.7 and 5.2.9). Indeed, let P be a transition probability, and let $f^{\square}$, $f \in B_b(X)$, be the functions defined by P and under consideration in Proposition 1.3.7. If we consider the operator $Q_P : B_b(X) \to \mathcal{F}_{\mathbb{R}}(X)$ defined by $Q_P f = f^{\square}$ for every $f \in B_b(X)$, then it can be shown that Q_P is a bounded function-valued linear operator, that Q_P is sequentially continuous with respect to pointwise convergence, and that $Q_P f \in B_b(X)$ for every $f \in C_0(X)$. Thus, by Proposition 5.2.8, $Q_P f \in B_b(X)$ for every $f \in B_b(X)$; that is, $f^{\square} \in B_b(X)$ for every $f \in B_b(X)$. $\qquad\blacktriangle$

In order to discuss the measurability of Γ_{cpi}, we still need two more lemmas that we discuss next.

Lemma 5.2.10. (a) *Let* $t \in \mathbb{T}$ *and* $\mu \in \mathcal{M}(X)$. *If* $T_t \mu = \mu$, *then* $T_0 \mu = \mu$, *as well (that is, if* μ *is an invariant element for* T_t, *then* μ *is also an invariant element for* T_0).

(b) *Assume that* $\mathbb{T} = \mathbb{R}$, *let* $t \in \mathbb{R}$, *and let* $\mu \in \mathcal{M}(X)$. *If* $T_t \mu = \mu$, *then* $T_{-t} \mu = \mu$ *(that is, if* μ *is an invariant element for* T_t, *then* μ *is an invariant element for* T_{-t}, *as well).*

(c) *An element* μ *of* $\mathcal{M}(X)$ *fails to be an invariant element for* $(P_t)_{t \in \mathbb{T}}$ *if and only if there exists an* $s \in \mathbb{T}$, $s > 0$ *such that* $T_s \mu \neq \mu$.

Proof. (a) Assume that $T_t \mu = \mu$. Then $T_0 \mu = T_0(T_t \mu) = T_{0+t} \mu = T_t \mu = \mu$.

(b) Assume that $T_t \mu = \mu$. Using (a) we obtain that $T_{-t} \mu = T_{-t}(T_t \mu) = T_{-t+t} \mu = T_0 \mu = \mu$.

(c) Clearly, if there exists an $s \in \mathbb{T}$, $s > 0$, such that $T_s \mu \neq \mu$, then μ is not an invariant element for $(P_t)_{t \in \mathbb{T}}$.

Conversely, assume that μ is not an invariant element for $(P_t)_{t \in \mathbb{T}}$. Then there exists an $s \in \mathbb{T}$ such that $T_s \mu \neq \mu$. Clearly, in this case, we may choose $s \neq 0$, because, by (a), it cannot happen that $T_0 \mu \neq \mu$ and $T_r \mu = \mu$ for every $r \in \mathbb{T}$, $r \neq 0$. Using (b) we obtain that we can choose $s > 0$. $\qquad\square$

Observation. Note that the above lemma can be used to show that the comments made immediately after Proposition 2.3.7 are valid even if T_0 is not the identity operator. In particular, the equivalence of the assertions (a), (b), (c), (d) and (e) stated there is also valid when T_0 is not the identity operator. ▲

In the next lemma and the theorem that concludes this section, we assume that, in addition to satisfying the s.m.a., the transition function $(P_t)_{t \in \mathbb{T}}$ is also pointwise continuous.

After we obtained the results on the KBBY decomposition for transition functions that we discuss here and in Chap. 6 (these results were announced in [146] which was submitted in 2006 and appeared in 2008), Worm and Hille [132] (see also [130]) obtained an approach to the KBBY decomposition that is valid in Polish spaces and the transition functions do not have to be pointwise continuous. The drawbacks of their approach consist of the fact that one cannot define the dissipative part of a transition function, and of the fact that it is more difficult to decide if an element x of X belongs to a certain set of the decomposition or not.

Lemma 5.2.11. *Assume that the transition function $(P_t)_{t \in \mathbb{T}}$ is pointwise continuous (and, of course, satisfies the s.m.a.) and let $(g_l)_{l \in \mathbb{N}}$ be a sequence of elements of $C_0(X)$ such that the range $\{g_l \mid l \in \mathbb{N}\}$ of $(g_l)_{l \in \mathbb{N}}$ is dense in $C_0(X)$ (such a sequence $(g_l)_{l \in \mathbb{N}}$ exists because $C_0(X)$ is a separable Banach space). Then a probability measure $\mu \in \mathcal{M}(X)$ fails to be an invariant probability for $(P_t)_{t \in \mathbb{T}}$ if and only if there exist $l \in \mathbb{N}$ and $q \in \mathbb{Q}$, $q > 0$, such that $\langle g_l, T_q \mu \rangle \neq \langle g_l, \mu \rangle$ (or, equivalently, such that $\langle S_q g_l, \mu \rangle \neq \langle g_l, \mu \rangle$).*

Proof. First note that if $\mu \in \mathcal{M}(X)$ is a probability measure such that $\langle g_l, T_q \mu \rangle \neq \langle g_l, \mu \rangle$ for some $l \in \mathbb{N}$ and $q \in \mathbb{Q}$, $q > 0$, then obviously, μ is not an invariant measure for $(P_t)_{t \in \mathbb{T}}$.

Thus, in order to prove the lemma, it is enough to prove that if $\mu \in \mathcal{M}(X)$ is not an invariant probability measure for $(P_t)_{t \in \mathbb{T}}$, then there exist $l \in \mathbb{N}$ and $q \in \mathbb{Q}$, $q > 0$, such that $\langle g_l, T_q \mu \rangle \neq \langle g_l, \mu \rangle$.

To this end, let $\mu \in \mathcal{M}(X)$ be a probability measure that fails to be invariant for $(P_t)_{t \in \mathbb{T}}$.

Using (c) of Lemma 5.2.10, we obtain that there exists a $t_0 \in \mathbb{T}$, $t_0 > 0$, such that $T_{t_0} \mu \neq \mu$. Consequently, there exists an $f \in C_0(X)$ such that $\langle f, T_{t_0} \mu \rangle \neq \langle f, \mu \rangle$.

Set $\varepsilon = |\langle f, T_{t_0} \mu \rangle - \langle f, \mu \rangle|$.

Since $\{g_l \mid l \in \mathbb{N}\}$ is dense in $C_0(X)$, it follows that there exists a subsequence $(g_{l_k})_{k \in \mathbb{N}}$ of $(g_l)_{l \in \mathbb{N}}$ such that $(g_{l_k})_{k \in \mathbb{N}}$ converges uniformly to f; consequently, the sequence $(|\langle g_{l_k}, T_{t_0} \mu \rangle - \langle g_{l_k}, \mu \rangle|)_{k \in \mathbb{N}}$ converges to $|\langle f, T_{t_0} \mu \rangle - \langle f, \mu \rangle|$ because the sequences $(\langle g_{l_k}, T_{t_0} \mu \rangle)_{k \in \mathbb{N}}$ and $(\langle g_{l_k}, \mu \rangle)_{k \in \mathbb{N}}$ converges to $\langle f, T_{t_0} \mu \rangle$ and $\langle f, \mu \rangle$, respectively. Therefore, there exists a $k_0 \in \mathbb{N}$ such that $|\langle g_{l_{k_0}}, T_{t_0} \mu \rangle - \langle g_{l_{k_0}}, \mu \rangle| > \frac{\varepsilon}{2}$, or, equivalently, $|\langle S_{t_0} g_{l_{k_0}}, \mu \rangle - \langle g_{l_{k_0}}, \mu \rangle| > \frac{\varepsilon}{2}$. Set $l = l_{k_0}$.

Now, let $(q_n)_{n \in \mathbb{N}}$ be a sequence of rational numbers such that $q_n > 0$ for every $n \in \mathbb{N}$, and such that $(q_n)_{n \in \mathbb{N}}$ converges to t_0.

Since we assume that $(P_t)_{t \in \mathbb{T}}$ is pointwise continuous, we obtain that $(S_{q_n} g_l(x))_{n \in \mathbb{N}}$ converges to $S_{t_0} g_l(x)$ for every $x \in X$; that is, the sequence of measurable functions $(S_{q_n} g_l)_{n \in \mathbb{N}}$ converges pointwise to $S_{t_0} g_l$ on X. Since S_{q_n}, $n \in \mathbb{N}$, are positive contractions of $B_b(X)$, it follows that $|S_{q_n} g_l(x)| \leq$

$S_{q_n}|g_l|(x) \leq \|g_l\|$ for every $n \in \mathbb{N}$ and $x \in X$; hence, using the fact that μ is a probability measure, we obtain that we can apply the Lebesgue dominated convergence theorem to the sequence of functions $(S_{q_n}g_l)_{n\in\mathbb{N}}$ which converges pointwise to $S_{t_0}g_l$ and is bounded in absolute value by the μ-integrable function $\|g_l\| \cdot \mathbf{1}_X$. Using Lebesgue's dominated convergence theorem, we conclude that the sequence $(\langle S_{q_n}g_l, \mu \rangle)_{n\in\mathbb{N}}$ converges to $\langle S_{t_0}g_l, \mu \rangle$. Since $|\langle S_{t_0}g_l, \mu \rangle - \langle g_l, \mu \rangle| > \frac{\varepsilon}{2}$, it follows that there exists an $n_0 \in \mathbb{N}$ such that $|\langle S_{q_{n_0}}g_l, \mu \rangle - \langle g_l, \mu \rangle| > \frac{\varepsilon}{2}$.

Set $q = q_{n_0}$. Then $q \in \mathbb{Q}$, $q > 0$, and $\langle S_q g_l, \mu \rangle \neq \langle g_l, \mu \rangle$ because $|\langle S_q g_l, \mu \rangle - \langle g_l, \mu \rangle| > \frac{\varepsilon}{2} > 0$. $\qquad\square$

We now have all the tools that we need to discuss the measurability of Γ_{cpi}. In the next theorem we prove that if $(P_t)_{t\in\mathbb{T}}$ is pointwise continuous, then Γ_{cpi} is measurable.

Theorem 5.2.12. *Assume that $(P_t)_{t\in\mathbb{T}}$ satisfies the s.m.a. and is pointwise continuous. Then the set Γ_{cpi} belongs to $\mathcal{B}(X)$.*

Proof. Since, by Theorem 5.2.6, Γ_{cp} is measurable with respect to $\mathcal{B}(X)$, and since $\Gamma_{cpi} \subseteq \Gamma_{cp}$, in order to prove that Γ_{cpi} is $\mathcal{B}(X)$-measurable, it is enough to prove that $\Gamma_{cp} \setminus \Gamma_{cpi}$ belongs to $\mathcal{B}(X)$.

To this end, let $(g_n)_{n\in\mathbb{N}}$ be the sequence of elements of $C_0(X)$ that we considered in Lemma 5.2.11 (that is, $(g_n)_{n\in\mathbb{N}}$ has the property that $\{g_n \mid n \in \mathbb{N}\}$ is dense in $C_0(X)$).

Now let $x \in \Gamma_{cp}$.

Using Lemma 5.2.11, we obtain that $x \in \Gamma_{cp} \setminus \Gamma_{cpi}$ if and only if there exists $l \in \mathbb{N}$ and $q \in \mathbb{Q}$, $q > 0$, such that $\langle S_q g_l, \varepsilon_x \rangle \neq \langle g_l, \varepsilon_x \rangle$; that is, if and only if $(S_q g_l)^{\square}(x) \neq g_l^{\square}(x)$ for some $l \in \mathbb{N}$ and $q \in \mathbb{Q}$, $q > 0$.

Thus,

$$\Gamma_{cp} \setminus \Gamma_{cpi} = \left\{ x \in \Gamma_{cp} \,\middle|\, \begin{array}{l} \text{there exist } q \in \mathbb{Q}, q > 0, \text{ and } l \in \mathbb{N} \text{ such} \\ \text{that } (S_q g_l)^{\square}(x) \neq g_l^{\square}(x) \end{array} \right\}$$

$$= \bigcup_{\substack{q\in\mathbb{Q} \\ q>0}} \bigcup_{l=1}^{\infty} \left\{ x \in \Gamma_{cp} \mid (S_q g_l)^{\square}(x) \neq g_l^{\square}(x) \right\}.$$

Since, by Proposition 5.2.9, the functions $(S_q g_l)^{\square}$, $q \in \mathbb{Q}$, $q > 0$, $l \in \mathbb{N}$, and $g_l^{\square}$, $l \in \mathbb{N}$, belong to $B_b(X)$ and since Γ_{cp} is $\mathcal{B}(X)$-measurable by Theorem 5.2.6, it follows that all the sets $\{x \in \Gamma_{cp} \mid (S_q g_l)^{\square}(x) = g_l^{\square}(x)\}$, $q \in \mathbb{Q}$, $q > 0$, $l \in \mathbb{N}$, are measurable. Accordingly, $\Gamma_{cp} \setminus \Gamma_{cpi}$ is measurable because it is a countable union of measurable subsets of X. $\qquad\square$

5.3   Sets of Maximal Probability

Our goal in this section is to prove that the sets Γ_0, Γ_c, Γ_{cp} and Γ_{cpi} are sets of maximal probability. The situation is similar to the corresponding situation for transition probabilities. From Sect. 5.2 we know that the four sets are measurable, and from their definitions we know that $\Gamma_0 \supseteq \Gamma_c \supseteq \Gamma_{cp} \supseteq \Gamma_{cpi}$, so in an ideal situation it is enough to prove only that Γ_{cpi} is of maximal probability. However, at this time, we do not know how to prove that Γ_{cpi} is of maximal probability without knowing that Γ_{cp} is of maximal probability, and, similarly, the proof of the fact that Γ_{cp} is of maximal probability cannot be obtained without first making sure that Γ_c is of maximal probability (the proof that Γ_c is of maximal probability does not depend on knowing in advance that Γ_0 is a set of maximal probability). Thus, in this section, we obtain versions for transition functions of Propositions 5.8, 5.9 and Theorem 5.12 of [146] (see also Propositions 1.3.9, 1.3.10 and Theorem 1.3.12).

We will use the setting, terminology and notations used in Sects. 5.1 and 5.2. Accordingly, we assume given a transition function $(P_t)_{t \in \mathbb{T}}$ defined on a locally compact separable metric space (X, d), we let $((S_t, T_t))_{t \in \mathbb{T}}$ be the family of Markov pairs defined by $(P_t)_{t \in \mathbb{T}}$, and we assume that $(P_t)_{t \in \mathbb{T}}$ satisfies the s.m.a. The assumption that $(P_t)_{t \in \mathbb{T}}$ be pointwise continuous will be made only whenever necessary.

Recall (see the discussion after Proposition 2.3.13 at the end of Sect. 2.3) that a Borel measurable subset A of X is said to be a *set of maximal probability for* $(P_t)_{t \in \mathbb{T}}$ if either $(P_t)_{t \in \mathbb{T}}$ does not have invariant probability measures, or else every invariant probability measure for $(P_t)_{t \in \mathbb{T}}$ is concentrated on A.

Our goal now is to prove that Γ_c is a set of maximal probability.

To this end, we need the following result:

Proposition 5.3.1. *Assume that $(P_t)_{t \in \mathbb{T}}$ has invariant probability measures, and let μ be such an invariant probability. Then there exists a subset A of X such that $A \in \mathcal{B}(X)$, $\mu(A) = 1$, and $A \subseteq \Gamma_c$.*

Proof. Since $C_0(X)$ is a separable Banach space, there exists a sequence $(g_l)_{l \in \mathbb{N}}$ of elements of $C_0(X)$ such that the range $\{g_l \mid l \in \mathbb{N}\}$ of $(g_l)_{l \in \mathbb{N}}$ is dense in $C_0(X)$.

Using (a) of Corollary 3.2.12, we obtain that, for every $l \in \mathbb{N}$, there exists a subset B_l of X such that $B_l \in \mathcal{B}(X)$, $\mu(B_l) = 1$, and such that $\lim_{\alpha \to +\infty} \frac{1}{\alpha} \int_0^{\alpha} S_t g_l(x)\, dt$ exists (and is a real number) for every $x \in B_l$.

Set $B = \cap_{l \in \mathbb{N}} B_l$.

It is easy to see that $B \in \mathcal{B}(X)$ and $\mu(B) = 1$.

Using (a) of Lemma 5.2.2 (applied to every $x \in B$), we obtain that $\lim_{\alpha \to +\infty} \frac{1}{\alpha} \int_0^{\alpha} S_t f(x)\, dt$ exists (and is a real number) for every $x \in B$ and $f \in C_0(X)$.

Set $B' = \{x \in B \mid \lim_{\alpha \to +\infty} \frac{1}{\alpha} \int_0^{\alpha} S_t f(x)\, dt = 0$ for every $f \in C_0(X)\}$.

Clearly, $B' \in \mathcal{B}(X)$ because $B' = B \cap \mathcal{D}$, and, by Theorem 5.2.3, $\mathcal{D}$ is $\mathcal{B}(X)$-measurable.

For every $f \in C_0(X)$, let $f^{(*)}$ be a μ-integrable function such that $f^{(*)}(x) = \lim_{\alpha \to +\infty} \frac{1}{\alpha} \int_0^\alpha S_t f(x)\, dt$ μ-a.e. (the existence of $f^{(*)}$ is assured by (a) of Corollary 3.2.12).

Taking into consideration that S_t, $t \in \mathbb{T}$, are contractions of $B_b(X)$, we obtain that

$$\left| \lim_{\substack{\alpha \to +\infty \\ \alpha > 0}} \frac{1}{\alpha} \int_0^\alpha S_t f(x)\, dt \right| = \lim_{\substack{\alpha \to +\infty \\ \alpha > 0}} \frac{1}{\alpha} \left| \int_0^\alpha S_t f(x)\, dt \right|$$

$$\leq \lim_{\substack{\alpha \to +\infty \\ \alpha > 0}} \frac{1}{\alpha} \int_0^\alpha |S_t f(x)|\, dt \leq \lim_{\substack{\alpha \to +\infty \\ \alpha > 0}} \frac{1}{\alpha} \int_0^\alpha \|f\|\, dt = \|f\|$$

for every $x \in X$ for which $\lim_{\substack{\alpha \to +\infty \\ \alpha > 0}} \frac{1}{\alpha} \int_0^\alpha S_t f(x)\, dt$ exists. Therefore, $\left| f^{(*)} \right| \leq \|f\|$ μ-a.e. on B.

Using (a) of Corollary 3.2.12, the fact that $\mu(B) = 1$, the fact that $f^{(*)} = 0$ μ-a.e. on B', and since $\left| f^{(*)} \right| \leq \|f\|$ μ-a.e. on B, we obtain that

$$|\langle f, \mu \rangle| = \left| \int f^{(*)}(x)\, d\mu(x) \right| = \left| \int_B f^{(*)}(x)\, d\mu(x) \right|$$

$$= \left| \int_{B'} f^{(*)}(x)\, d\mu(x) + \int_{B \setminus B'} f^{(*)}(x)\, d\mu(x) \right| = \left| \int_{B \setminus B'} f^{(*)}(x)\, d\mu(x) \right|$$

$$\leq \int_{B \setminus B'} \left| f^{(*)}(x) \right|\, d\mu(x) \leq \|f\|\, \mu(B \setminus B') = \|f\|\,(1 - \alpha)$$

for every $f \in C_0(X)$.

Accordingly, $\|\mu\| \leq 1 - \alpha$, so we have obtained a contradiction with the fact that μ is a probability measure. Since the contradiction stems from our assumption that $\mu(B') > 0$, we obtain that $\mu(B') = 0$.

If we set $A = B \setminus B'$, then $A \in \mathcal{B}(X)$, $\mu(A) = 1$, and $A \subseteq \Gamma_c$. $\square$

From the above proposition it is now obvious that Γ_c is a set of maximal probability. For future reference we state this fact as a corollary.

Corollary 5.3.2.   *The set Γ_c is of maximal probability.*

For every $f \in B_b(X)$, let $f^* : X \to \mathbb{R}$ be a function defined as follows:

$$f^*(x) = \begin{cases} \lim_{\alpha \to +\infty} \frac{1}{\alpha} \int_0^\alpha S_t f(x)\, dt & \text{if } x \in A_f^{\to} \cap \Gamma_c, \\ 0 & \text{if } x \notin A_f^{\to} \cap \Gamma_c, \end{cases}$$

where $A_f^{\to}$ is the set defined in Proposition 3.2.5.

Since Γ_{c} is $\mathcal{B}(X)$-measurable (by Theorem 5.2.4), and since $A_f^{\rightarrow} \in \mathcal{B}(X)$ and $f^{\rightarrow} \in B_b(X)$ by Proposition 3.2.5 (where $f^{\rightarrow}$ is defined in Proposition 3.2.5), it follows that f^* is a measurable function for every $f \in B_b(X)$. Since f^*, $f \in B_b(X)$, are obviously bounded functions, we obtain that $f^* \in B_b(X)$ for every $f \in B_b(X)$.

Note that if $f \in C_0(X)$, then $A_f^{\rightarrow} \supseteq \Gamma_{\mathrm{c}}$, so, in this case

$$f^*(x) = \begin{cases} \lim_{\alpha \to +\infty} \frac{1}{\alpha} \int_0^\alpha S_t f(x)\, dt & \text{if } x \in \Gamma_{\mathrm{c}}, \\ 0 & \text{if } x \notin \Gamma_{\mathrm{c}}. \end{cases}$$

Theorem 5.3.3. *The set Γ_{cp} is of maximal probability.*

Proof. The proof is similar to the proof of Theorem 5.9 of [146].

Assume that $(P_t)_{t \in \mathbb{T}}$ has invariant probabilities, and that Γ_{cp} is not a set of maximal probability. Then there exists an invariant probability $\mu \in \mathcal{M}(X)$ for $(P_t)_{t \in \mathbb{T}}$ such that $\mu(\Gamma_{\mathrm{cp}}) < 1$.

Since (by Corollary 5.3.2) Γ_{c} is a set of maximal probability, and since $\Gamma_{\mathrm{cp}} \subseteq \Gamma_{\mathrm{c}}$, it follows that $\mu(\Gamma_{\mathrm{c}} \setminus \Gamma_{\mathrm{cp}}) > 0$.

Now, let $\varphi : X \to \mathbb{R}$ be the function defined in the proof of Theorem 5.2.6. Since

$$\varphi(x) = \begin{cases} \|\varepsilon_x\| & \text{if } x \in \Gamma_{\mathrm{c}}, \\ 0 & \text{if } x \notin \Gamma_{\mathrm{c}}, \end{cases}$$

it follows that $\Gamma_{\mathrm{c}} \setminus \Gamma_{\mathrm{cp}} = \{x \in \Gamma_{\mathrm{c}} \mid \varphi(x) < 1\}$, so $\mu(\{x \in \Gamma_{\mathrm{c}} \mid \varphi(x) < 1\}) > 0$.

Since (as we proved in Theorem 5.2.6) φ is a measurable function, it follows that there exists a $\rho \in \mathbb{R}$, $0 < \rho < 1$, such that $\mu(\{x \in \Gamma_{\mathrm{c}} \mid \varphi(x) < \rho\}) > 0$. Set $A = \{x \in \Gamma_{\mathrm{c}} \mid \varphi(x) < \rho\}$ and $\alpha = \mu(A)$.

Using (a) of Corollary 3.2.12, the fact that Γ_{c} is a set of maximal probability, the fact that $\mu(\Gamma_{\mathrm{c}} \setminus A) = 1 - \alpha$, and the fact that $f^*(x) \le 1$ for every $x \in \Gamma_{\mathrm{c}} \setminus A$ and $f^*(x) \le \rho$ for every $x \in A$, we obtain that

$$\langle f, \mu \rangle = \int_{\Gamma_{\mathrm{c}}} f^*(x)\, d\mu(x) = \int_{\Gamma_{\mathrm{c}} \setminus A} f^*(x)\, d\mu(x) + \int_A f^*(x)\, d\mu(x) \le 1 - \alpha + \alpha\rho$$

for every $f \in C_0(X)$, $0 \le f \le 1$.

It follows that $\|\mu\| = \sup_{\substack{f \in C_0(X) \\ 0 \le f \le 1}} \langle f, \mu \rangle \le 1 - \alpha + \alpha\rho < 1$, so we obtain a contradiction with our assumption that μ is a probability measure. Clearly, the contradiction stems from our assumption that Γ_{cp} is not a set of maximal probability. $\square$

Our goal now is to prove that Γ_{cpi} is a set of maximal probability. To this end, we need some preparation.

Let Π be a property that a function $f \in B_b(X)$ may or may not possess. In the next proposition, we obtain a set of conditions which guarantee that every $f \in B_b(X)$ has property Π.

Proposition 5.3.4. *Assume that property Π satisfies the following three conditions:*

(a) *If $f \in C_0(X)$, then f has property Π.*
(b) *(Linearity) If $f \in B_b(X)$, $g \in B_b(X)$, $a \in \mathbb{R}$, and $b \in \mathbb{R}$, and if f and g have property Π, then $af + bg$ has property Π.*
(c) *(Invariance to Pointwise (Everywhere) Convergence) Let $(f_n)_{n\in\mathbb{N}}$ be a sequence of elements of $B_b(X)$, assume that $(f_n)_{n\in\mathbb{N}}$ converges pointwise (everywhere) to a function $f \in B_b(X)$ and assume that $(f_n)_{n\in\mathbb{N}}$ is uniformly bounded (that is, there exists an $M \in \mathbb{R}$, $M \geq 0$, such that $|f_n(x)| \leq M$ for every $n \in \mathbb{N}$ and $x \in X$). If all the functions f_n, $n \in \mathbb{N}$, have property Π, then f has property Π, as well.*

Then every $f \in B_b(X)$ has property Π.

Proof. Set $\mathcal{F} = \{f \in B_b(X) \mid f \text{ has property } \Pi\}$. Then $\mathcal{F}$ satisfies all the conditions of Proposition 1.1.14, so using the proposition, we obtain that every $f \in B_b(X)$ has property Π. $\square$

Now, assume that the transition function $(P_t)_{t\in\mathbb{T}}$ has invariant probability measures, and let $\mu \in \mathcal{M}(X)$ be such an invariant probability.

The property Π that is of interest to us is defined as follows: we say that $f \in B_b(X)$ has property Π if $f^* = f^\square$ μ-a.e.

Our immediate goal is to show that property Π satisfies the invariance to pointwise convergence (condition (c) of Proposition 5.3.4).

Let $(X, \widetilde{\mathcal{B}(X)}, \tilde{\mu})$ be the completion of the measure space $(X, \mathcal{B}(X), \mu)$. Then using Theorem 3.2.11, we can define an operator $\widetilde{W_\mu} : L^1(X, \widetilde{\mathcal{B}(X)}, \tilde{\mu}) \to L^1(X, \widetilde{\mathcal{B}(X)}, \tilde{\mu})$ as follows: given $\tilde{\tilde{f}} \in L^1(X, \widetilde{\mathcal{B}(X)}, \tilde{\mu})$, we let $\widetilde{W_\mu}\tilde{\tilde{f}}$ be the absolute $\tilde{\mu}$-a.e. limit of $(\widetilde{A_t}\tilde{\tilde{f}})_{t\in[0,+\infty)}$ as $t \to +\infty$, whose existence is guaranteed by (a) of Theorem 3.2.11. The fact that the limit is an element of $L^1(X, \widetilde{\mathcal{B}(X)}, \tilde{\mu})$ is a consequence of (b) of Theorem 3.2.11. Another consequence of (b) of Theorem 3.2.11 is the fact that $\widetilde{W_\mu}$ is a Markov operator on $L^1(X, \widetilde{\mathcal{B}(X)}, \tilde{\mu})$.

Lemma 5.3.5. *Let $(f_n)_{n\in\mathbb{N}}$ be a sequence of elements of $B_b(X)$, let $f \in B_b(X)$, and assume that:*

(i) *The sequence $(f_n)_{n\in\mathbb{N}}$ is bounded in $B_b(X)$ (that is, there exists an $M \in \mathbb{R}$, $M \geq 0$, such that $|f_n(x)| \leq M$ for every $n \in \mathbb{N}$ and $x \in X$).*
(ii) *The sequence $(f_n)_{n\in\mathbb{N}}$ converges pointwise (everywhere on X) to f.*
(iii) *f_n has property Π (that is, $f_n^* = f_n^\square$ μ-a.e.) for every $n \in \mathbb{N}$.*

Then f has property Π (that is, $f^* = f^\square$ μ-a.e.).

Proof. Let $M \in \mathbb{R}$, $M \geq 0$, be a constant whose existence is assured by condition (i).

Taking into consideration that $|f_n(x)| \leq M$ for every $n \in \mathbb{N}$ and $x \in X$, and since $(f_n)_{n \in \mathbb{N}}$ converges to f pointwise, we obtain that $|f(x)| \leq M$ for every $x \in X$.

Since $|f_n - f| \leq 2M\mathbf{1}_X$ for every $n \in \mathbb{N}$, since $2M\mathbf{1}_X$ is a μ-integrable function (because μ is a probability measure), and since $(|f_n - f|)_{n \in \mathbb{N}}$ converges pointwise to zero, it follows that we can apply the Lebesgue dominated convergence theorem in order to infer that $(\int |f_n - f| \, d\mu)_{n \in \mathbb{N}}$ converges to zero. Thus, the sequence $(\widetilde{\tilde{f}_n})_{n \in \mathbb{N}}$ converges to $\widetilde{\tilde{f}}$ in the norm topology of $L^1(X, \widetilde{\mathcal{B}(X)}, \tilde{\mu})$, where $\widetilde{\tilde{g}}$ is the class of g in $L^1(X, \widetilde{\mathcal{B}(X)}, \tilde{\mu})$ (the element of $L^1(X, \widetilde{\mathcal{B}(X)}, \tilde{\mu})$ to which g belongs) for every μ-integrable function g.

Since the operator $\widetilde{W_\mu}$ is a (positive) contraction of $L^1(X, \widetilde{\mathcal{B}(X)}, \tilde{\mu})$, it follows that $(\widetilde{W_\mu} \widetilde{\tilde{f}_n})_{n \in \mathbb{N}}$ converges to $\widetilde{W_\mu} \widetilde{\tilde{f}}$ in the norm topology of $L^1(X, \widetilde{\mathcal{B}(X)}, \tilde{\mu})$.

Using (a) of Corollary 3.2.12, the fact that $\mu(\Gamma_c) = 1$ (by Corollary 5.3.2), and Theorem 3.2.11, we obtain that f^* is an element of the equivalence class $\widetilde{W_\mu} \widetilde{\tilde{f}}$, so $\widetilde{f^*} = \widetilde{W_\mu} \widetilde{\tilde{f}}$; similarly, f_n^* is an element of the equivalence class $\widetilde{W_\mu} \widetilde{\tilde{f}_n}$, so $\widetilde{f^*} = \widetilde{W_\mu} \widetilde{\tilde{f}_n}$ for every $n \in \mathbb{N}$. Thus, the sequence $\left(\widetilde{f_n^*}\right)_{n \in \mathbb{N}}$ converges to $\widetilde{f^*}$ in the norm topology of $L^1(X, \widetilde{\mathcal{B}(X)}, \tilde{\mu})$.

Since $|f_n| \leq M\mathbf{1}_X$ for every $n \in \mathbb{N}$, and since $(f_n)_{n \in \mathbb{N}}$ converges pointwise to f, it follows that for every $x \in \Gamma_c$ we can apply the Lebesgue dominated convergence theorem to the sequence $(f_n)_{n \in \mathbb{N}}$ with respect to the measure ε_x in order to obtain that the sequence $(\langle f_n, \varepsilon_x \rangle)_{n \in \mathbb{N}}$ converges to $\langle f, \varepsilon_x \rangle$. Accordingly, the sequence $(f_n^\square)_{n \in \mathbb{N}}$ converges pointwise to $f^\square$.

Since $\varepsilon_x \geq 0$ and $\|\varepsilon_x\| \leq 1$ for every $x \in \Gamma_c$, it follows that $|f^\square| \leq M\mathbf{1}_X$ and $|f_n^\square| \leq M\mathbf{1}_X$ for every $n \in \mathbb{N}$. Since the sequence $(|f_n^\square - f^\square|)_{n \in \mathbb{N}}$ converges pointwise to zero, we can apply the Lebesgue dominated convergence theorem to $(|f_n^\square - f^\square|)_{n \in \mathbb{N}}$ with respect to the probability measure μ in order to conclude that the sequence $(\int |f_n^\square - f^\square| \, d\mu)_{n \in \mathbb{N}}$ converges to zero. Therefore, the sequence $(\widetilde{f_n^\square})_{n \in \mathbb{N}}$ (which is a sequence of elements of $L^1(X, \widetilde{\mathcal{B}(X)}, \tilde{\mu})$) converges to $\widetilde{f^\square}$ in the norm topology of $L^1(X, \widetilde{\mathcal{B}(X)}, \tilde{\mu})$.

Since we assume that $f_n^* = f_n^\square$ μ-a.e. for every $n \in \mathbb{N}$, it follows that $\widetilde{f_n^*} = \widetilde{f_n^\square}$ μ-a.e. for every $n \in \mathbb{N}$. Taking into consideration that $(\widetilde{f_n^*})_{n \in \mathbb{N}}$ and $(\widetilde{f_n^\square})_{n \in \mathbb{N}}$ converge to $\widetilde{f^*}$ and $\widetilde{f^\square}$, respectively, we obtain that $\widetilde{f^*} = \widetilde{f^\square}$; therefore $f^* = f^\square$ μ-a.e. $\square$

Proposition 5.3.6. $f^* = f^\square$ μ-a.e. for every $f \in B_b(X)$.

Proof. Recall (see the discussion preceding Lemma 5.3.5) that we say that $f \in B_b(X)$ has property Π if $f^* = f^\square$ μ-a.e. By Lemma 5.3.5, property Π satisfies condition (c) (the invariance to pointwise convergence) of Proposition 5.3.4. It is easy to see that Π satisfies conditions (a) and (b) of the proposition, as well. Accordingly, each $f \in B_b(X)$ has property Π. $\square$

We are now in a position to prove that Γ_{cpi} is a set of maximal probability and we will carry out this task in the next theorem.

Theorem 5.3.7. *Assume that $(P_t)_{t \in \mathbb{T}}$ satisfies the s.m.a. and is pointwise continuous. Then Γ_{cpi} is a set of maximal probability.*

Proof. We have to prove that if $(P_t)_{t \in \mathbb{T}}$ has invariant probabilities, and if $\mu \in \mathcal{M}(X)$ is such an invariant probability, then $\mu(\Gamma_{\mathrm{cpi}}) = 1$.

Thus, assume that $(P_t)_{t \in \mathbb{T}}$ has invariant probability measures, and let $\mu \in \mathcal{M}(X)$ be an invariant probability for $(P_t)_{t \in \mathbb{T}}$.

Since, by Theorem 5.3.3, Γ_{cp} is a set of maximal probability, and since $\Gamma_{\mathrm{cpi}} \subseteq \Gamma_{\mathrm{cp}}$, it follows that in order to prove that $\mu(\Gamma_{\mathrm{cpi}}) = 1$, it is enough to prove that $\mu(\Gamma_{\mathrm{cp}} \setminus \Gamma_{\mathrm{cpi}}) = 0$.

Let $(g_l)_{l \in \mathbb{N}}$ be a sequence of elements of $C_0(X)$ such that the range $\{g_l \mid l \in \mathbb{N}\}$ of $(g_l)_{l \in \mathbb{N}}$ is dense in $C_0(X)$ (we can find such a sequence because $C_0(X)$ is a separable Banach space).

Let $x \in \Gamma_{\mathrm{cp}}$.

Using Lemma 5.2.11, we obtain that $x \in \Gamma_{\mathrm{cp}} \setminus \Gamma_{\mathrm{cpi}}$ if and only if there exists $q \in \mathbb{Q}, q > 0$, and $l \in \mathbb{N}$ such that $g_l^{\square}(x) \neq (S_q g_l)^{\square}(x)$.

Therefore, if we set $A_{lq} = \{x \in \Gamma_{\mathrm{cp}} \mid g_l^{\square}(x) \neq (S_q g_l)^{\square}(x)\}$ for every $q \in \mathbb{Q}$, $q > 0$, and $l \in \mathbb{N}$, we obtain that $\Gamma_{\mathrm{cp}} \setminus \Gamma_{\mathrm{cpi}} = \cup_{l \in \mathbb{N}} \cup_{\substack{q \in \mathbb{Q} \\ q > 0}} A_{lq}$.

Note that for every $l \in \mathbb{N}$ and $q \in \mathbb{Q}, q > 0$, the set A_{lq} is measurable because Γ_{cp} is measurable (by Theorem 5.2.6), and, by Proposition 5.2.9, both functions $g_l^{\square}$ and $(S_q g_l)^{\square}$ belong to $B_b(X)$.

We now need the following assertion:

Assertion A. $g_l^* = (S_q g_l)^*$ for every $l \in \mathbb{N}$ and $q \in \mathbb{Q}, q > 0$.

Proof of Assertion A. Let $l \in \mathbb{N}$ and $q \in \mathbb{Q}, q > 0$.

We will first prove that, for every $x \in X$, we have that $\lim_{\alpha \to +\infty} \frac{1}{\alpha} \int_0^\alpha S_t g_l(x)\, dt$ exists and is a real number, say L, if and only if $\lim_{\alpha \to +\infty} \frac{1}{\alpha} \int_0^\alpha S_t S_q g_l(x)\, dt$ exists and is equal to L.

Note that in order to prove the above statement, it is enough to prove that, for every $\varepsilon \in \mathbb{R}, \varepsilon > 0$, there exists an $\alpha_\varepsilon \in \mathbb{R}, \alpha_\varepsilon > 0$, such that

$$\left| \frac{1}{\alpha} \int_0^\alpha S_t g_l(x)\, dt - \frac{1}{\alpha} \int_0^\alpha S_t S_q g_l(x)\, dt \right| < \varepsilon$$

for every $\alpha \in \mathbb{R}, \alpha \geq \alpha_\varepsilon$.

Thus, let $x \in X$ and $\varepsilon \in \mathbb{R}, \varepsilon > 0$.

Now let $\alpha_\varepsilon \in \mathbb{R}, \alpha_\varepsilon > q$, be large enough such that $\frac{2q \|g_l\|}{\alpha_\varepsilon} < \varepsilon$.

Using the fact that S_t, $t \in \mathbb{T}$, are (positive) contractions, we obtain that

$$\left| \frac{1}{\alpha} \int_0^\alpha S_t g_l(x)\, dt - \frac{1}{\alpha} \int_0^\alpha S_t S_q g_l(x)\, dt \right|$$

$$= \frac{1}{\alpha} \left| \int_0^q S_t g_l(x)\, dt + \int_q^\alpha S_t g_l(x)\, dt - \int_q^{\alpha+q} S_t g_l(x)\, dt \right|$$

$$= \frac{1}{\alpha} \left| \int_0^q S_t g_l(x)\, dt + \int_q^\alpha S_t g_l(x)\, dt - \int_q^\alpha S_t g_l(x)\, dt - \int_q^{\alpha+q} S_t g_l(x)\, dt \right|$$

$$= \frac{1}{\alpha} \left| \int_0^q S_t g_l(x)\, dt - \int_q^{\alpha+q} S_t g_l(x)\, dt \right|$$

$$\leq \frac{1}{\alpha} \left(\int_0^q |S_t g_l(x)|\, dt + \int_q^{\alpha+q} |S_t g_l(x)|\, dt \right)$$

$$\leq \frac{1}{\alpha} (q \, \|g_l\| + q \, \|g_l\|) = \frac{2q \, \|g_l\|}{\alpha} < \varepsilon$$

for every $\alpha \in \mathbb{R}$, $\alpha \geq \alpha_\varepsilon$.

We have therefore proved that $\lim_{\alpha \to +\infty} \frac{1}{\alpha} \int_0^\alpha S_t g_l(x)\, dt$ exists if and only if $\lim_{\alpha \to +\infty} \frac{1}{\alpha} \int_0^\alpha S_t S_q g_l(x)\, dt$ exists, and if the limits exist, then they are equal.

Now, if $x \in \Gamma_c$, then $\lim_{\alpha \to +\infty} \frac{1}{\alpha} \int_0^\alpha S_t g_l(x)\, dt$ exists because $g_l \in C_0(X)$; consequently, $\lim_{\alpha \to +\infty} \frac{1}{\alpha} \int_0^\alpha S_t S_q g_l(x)\, dt$ exists, as well, and $\lim_{\alpha \to +\infty} \frac{1}{\alpha} \int_0^\alpha S_t g_l(x)\, dt = \lim_{\alpha \to +\infty} \frac{1}{\alpha} \int_0^\alpha S_t S_q g_l(x)\, dt$; therefore, $g_l^*(x) = (S_q g_l)^*(x)$ whenever $x \in \Gamma_c$.

If $x \notin \Gamma_c$, then using the manner in which the functions h^*, $h \in B_b(X)$, are defined we obtain that $g_l^*(x) = 0$ and $(S_q g_l)^*(x) = 0$.

Thus, Assertion A has been completely proved.

Now, let $l \in \mathbb{N}$ and $q \in \mathbb{Q}$, $q > 0$.

Using Assertion A and the fact that $g_l^* = g_l^\square$ (because $g_l \in C_0(X)$), we obtain that

$$A_{lq} = \{x \in \Gamma_{cp} \mid g_l^*(x) \neq (S_q g_l)^\square(x)\} = \{x \in \Gamma_{cp} \mid (S_q g_l)^*(x) \neq (S_q g_l)^\square(x)\}.$$

Using Proposition 5.3.6, we obtain that $(S_q g_l)^* = (S_q g_l)^\square$ μ-a.e. Accordingly, $\mu(A_{lq}) = 0$.

Since the sets A_{lq}, $l \in \mathbb{N}$, $q \in \mathbb{Q}$, $q > 0$, are μ-negligible, and since the union of these sets is $\Gamma_{cp} \setminus \Gamma_{cpi}$, we obtain that $\mu(\Gamma_{cp} \setminus \Gamma_{cpi}) = 0$. $\square$

Chapter 6
The Ergodic Decomposition of Kryloff, Bogoliouboff, Beboutoff and Yosida, Part II: The Role of the Invariant Ergodic Probability Measures in the Decomposition

As pointed out in the abstract for Chap. 5 at the beginning of the chapter, Chap. 6 is the second and last chapter in which we deal exclusively with the ergodic decomposition.

Let $(P_t)_{t\in\mathbb{T}}$ be a transition function defined on a locally compact separable metric space (X, d), and assume that $(P_t)_{t\in\mathbb{T}}$ satisfies the s.m.a. and is pointwise continuous.

In Sect. 5.1, we defined the set Γ_{cpi} of all $x \in X$ for which we can define standard elementary measures ε_x and these measures are all invariant probabilities for $(P_t)_{t\in\mathbb{T}}$. A natural question at this point is whether or not the measures ε_x, $x \in \Gamma_{\mathrm{cpi}}$, are ergodic. In general, these measures are not necessarily ergodic. So, after the first section (Sect. 6.1) in which we review various equivalent definitions of an invariant ergodic probability measure of a transition function, in Sect. 6.2 we define the set Γ_{cpie} of all $x \in \Gamma_{\mathrm{cpi}}$ which have the property that ε_x is also an ergodic measure, we prove that Γ_{cpie} is measurable and of maximal probability, that every invariant ergodic probability measure is of the form ε_x for some $x \in \Gamma_{\mathrm{cpie}}$, etc.; in general, in Sect. 6.2 we prove versions for transition functions of all the results for transition probabilities obtained in Section 6 of [146]. In the last section of the chapter (Sect. 6.3) we extend (and connect to the KBBY decomposition) various results on the role played by the ergodic invariant probability measures in the set of all invariant real-valued signed Borel measures for $(P_t)_{t\in\mathbb{T}}$. These results appear in Phelps' monograph [86], in Rohlin [101], and in Sections 8 and 9 of Oxtoby [85].

Throughout the chapter, we assume that all the transition functions under consideration satisfy the s.m.a., and, in Sects. 6.2 and 6.3, we assume that these transition functions are pointwise continuous, as well.

6.1 Preliminaries on Ergodic Measures

Our goal in this section is to discuss several equivalent definitions of an invariant ergodic probability measure for a transition function.

R. Zaharopol, *Invariant Probabilities of Transition Functions*, Probability and Its Applications 44, DOI 10.1007/978-3-319-05723-1_6,
© Springer International Publishing Switzerland 2014

Let $(P_t)_{t\in\mathbb{T}}$ be a transition function defined on a locally compact separable metric space (X, d), assume that $(P_t)_{t\in\mathbb{T}}$ satisfies the s.m.a., and let $((S_t, T_t))_{t\in\mathbb{T}}$ be the family of Markov pairs defined by $(P_t)_{t\in\mathbb{T}}$.

Assume that $(P_t)_{t\in\mathbb{T}}$ has invariant probability measures and let μ be such an invariant probability.

Also, let $(V_t^{(\mu)})_{t\in[0,+\infty)}$ be the one-parameter semigroup of positive contractions of $L^1(X, \mathcal{B}(X), \mu)$ defined before Proposition 3.2.7.

A subset A of X, $A \in \mathcal{B}(X)$, is said to be μ-*invariant* if $V_t^{(\mu)}\bar{\mathbf{1}}_A = \bar{\mathbf{1}}_A$ for every $t \in [0, +\infty)$.

In a similar manner, let $(X, \widetilde{\mathcal{B}(X)}, \tilde{\mu})$ be the completion of the probability space $(X, \mathcal{B}(X), \mu)$, and let $(\widetilde{V_t^{(\mu)}})_{t\in[0,+\infty)}$ be the one-parameter semigroup of positive contractions of $L^1(X, \widetilde{\mathcal{B}(X)}, \tilde{\mu})$ defined before Proposition 3.2.8; we say that a subset A of X, $A \in \widetilde{\mathcal{B}(X)}$, is $\tilde{\mu}$-*invariant* if $\widetilde{V_t^{(\mu)}}\bar{\bar{\mathbf{1}}}_A = \bar{\bar{\mathbf{1}}}_A$ for every $t \in [0, +\infty)$.

Set

$$\mathcal{B}_\mu(X) = \{A \in \mathcal{B}(X) \mid A \text{ is } \mu\text{-invariant}\} \text{ and}$$

$$\mathcal{B}_{\tilde{\mu}}(X) = \{A \in \widetilde{\mathcal{B}(X)} \mid A \text{ is } \tilde{\mu}\text{-invariant}\}.$$

Lemma 6.1.1. $\mathcal{B}_\mu(X)$ *and* $\mathcal{B}_{\tilde{\mu}}(X)$ *are sub-σ-algebras of* $\mathcal{B}(X)$ *and* $\widetilde{\mathcal{B}(X)}$, *respectively.*

Proof. The proofs that $\mathcal{B}_\mu(X)$ and $\mathcal{B}_{\tilde{\mu}}(X)$ are sub-σ-algebras of $\mathcal{B}(X)$ and $\widetilde{\mathcal{B}(X)}$ are similar and based upon the fact that $V_t^{(\mu)}$, $t \in [0, +\infty)$, and $\widetilde{V_t^{(\mu)}}$, $t \in [0, +\infty)$, are Markov operators on $L^1(X, \mathcal{B}(X), \mu)$ and $L^1(X, \widetilde{\mathcal{B}(X)}, \tilde{\mu})$, respectively. Therefore, we prove only that $\mathcal{B}_{\tilde{\mu}}(X)$ is a sub-σ-algebra of $\widetilde{\mathcal{B}(X)}$.

If $A \in \mathcal{B}_{\tilde{\mu}}(X)$, then $\widetilde{V_t^{(\mu)}}(\bar{\bar{\mathbf{1}}}_{X\setminus A}) = \widetilde{V_t^{(\mu)}}(\bar{\bar{\mathbf{1}}}_X) - \widetilde{V_t^{(\mu)}}(\bar{\bar{\mathbf{1}}}_A) = \bar{\bar{\mathbf{1}}}_X - \bar{\bar{\mathbf{1}}}_A = \bar{\bar{\mathbf{1}}}_{X\setminus A}$ for every $t \in [0, +\infty)$, so $X \setminus A \in \mathcal{B}_{\tilde{\mu}}(X)$.

Let $(A_n)_{n\in\mathbb{N}}$ be a sequence of elements of $\mathcal{B}_{\tilde{\mu}}(X)$. Since

$$\widetilde{V_t^{(\mu)}}(\bar{\bar{\mathbf{1}}}_{\cup_{n=1}^{\infty}A_n}) = \widetilde{V_t^{(\mu)}}(\sup_{n\in\mathbb{N}} \bar{\bar{\mathbf{1}}}_{A_n}) \leq \sup_{n\in\mathbb{N}} \widetilde{V_t^{(\mu)}}(\bar{\bar{\mathbf{1}}}_{A_n}) = \sup_{n\in\mathbb{N}} \bar{\bar{\mathbf{1}}}_{A_n} = \bar{\bar{\mathbf{1}}}_{\cup_{n=1}^{\infty}A_n},$$

since $\left\|\widetilde{V_t^{(\mu)}}(\bar{\bar{\mathbf{1}}}_{\cup_{n=1}^{\infty}A_n})\right\| = \left\|\bar{\bar{\mathbf{1}}}_{\cup_{n=1}^{\infty}A_n}\right\|$ (because $\widetilde{V_t^{(\mu)}}$ is a Markov operator), and using a property of the norm of $L^1(X, \widetilde{\mathcal{B}(X)}, \tilde{\mu})$, we obtain that $\widetilde{V_t^{(\mu)}}(\bar{\bar{\mathbf{1}}}_{\cup_{n=1}^{\infty}A_n}) = \bar{\bar{\mathbf{1}}}_{\cup_{n=1}^{\infty}A_n}$ for every $t \in [0, +\infty)$; accordingly, $\cup_{n=1}^{\infty}A_n \in \mathcal{B}_{\tilde{\mu}}(X)$.

Since $\mathcal{B}_{\tilde{\mu}}(X)$ is a subset of $\widetilde{\mathcal{B}(X)}$, it follows that $\mathcal{B}_{\tilde{\mu}}(X)$ is a sub-σ-algebra of $\widetilde{\mathcal{B}(X)}$. $\qquad\square$

We will often think of $\mathcal{B}_\mu(X)$ and $\mathcal{B}_{\tilde{\mu}}(X)$ as σ-algebras in their own right.

Recall (see the discussion following Example 2.3.10) that the invariant probability measure μ is said to be *ergodic* if μ cannot be written as a sum of two nonzero invariant measures, say ν_1 and ν_2, such that ν_1 and ν_2 are concentrated on disjoint measurable subsets of X.

As usual, given a probability space $(Y, \mathcal{Y}, \nu)$ and a sub-σ-algebra $\mathcal{A}$ of $\mathcal{Y}$, we say that $\mathcal{A}$ is *trivial* if $\nu(A)$ is either equal to zero, or else $\nu(A) = 1$ for every $A \in \mathcal{A}$.

Note that $\mathcal{B}_\mu(X)$ is trivial with respect to $(X, \mathcal{B}(X), \mu)$ if and only if $\mathcal{B}_{\tilde{\mu}}(X)$ is trivial with respect to $(X, \widetilde{\mathcal{B}(X)}, \tilde{\mu})$.

Proposition 6.1.2. *The following assertions are equivalent:*

(a) The probability measure μ is ergodic.
(b) $\mathcal{B}_\mu(X)$ is trivial.
(c) $\mathcal{B}_{\tilde{\mu}}(X)$ is trivial.

Proof. $(a) \Rightarrow (b)$ Assume that $\mathcal{B}_\mu(X)$ is not trivial. Thus, there exists a $\Theta \in \mathcal{B}(X)$ such that $V_t^{(\mu)} \mathbf{1}_\Theta = \mathbf{1}_\Theta$ for every $t \in [0, +\infty)$, and such that $0 < \mu(\Theta) < 1$.

Let $\nu_i : \mathcal{B}(X) \to \mathbb{R}$, $i = 1, 2$ be two maps defined as follows: $\nu_1(A) = \mu(A \cap \Theta)$ and $\nu_2(A) = \mu(A) - \nu_1(A)$ for every $A \in \mathcal{B}(X)$. Clearly, ν_1 is a positive finite measure on $\mathcal{B}(X)$, so $\nu_1 \in \mathcal{M}(X)$; since $\nu_1 \leq \mu$, it follows that $\nu_2 = \mu - \nu_1$ is a positive measure and belongs to $\mathcal{M}(X)$, as well.

We now prove that ν_1 is an invariant measure for $(P_t)_{t \in \mathbb{T}}$. By Proposition 2.3.7, it is enough to prove that ν_1 is an invariant measure for the restriction $(P_t)_{t \in [0, +\infty)}$ of $(P_t)_{t \in \mathbb{T}}$ to the interval $[0, +\infty)$.

Thus, let $t \in [0, +\infty)$. Since $S_t \mathbf{1}_\Theta = \mathbf{1}_\Theta$ μ-a.e., it follows that there exists a $\mathcal{B}(X)$-measurable subset Θ_t of Θ such that $\mu(\Theta_t) = \mu(\Theta)$ and such that $S_t \mathbf{1}_\Theta(x) = 1$ for every $x \in \Theta_t$; since $S_t \mathbf{1}_\Theta(y) = P_t(y, \Theta)$ for every $y \in X$, it follows that $P_t(x, \Theta) = 1$ for every $x \in \Theta_t$. Since $P_t(y, \cdot)$ is a probability measure for every $y \in X$, it follows that $P_t(x, X \setminus \Theta) = 0$ for every $x \in \Theta_t$, so $P_t(x, A \cap (X \setminus \Theta)) = 0$ for every $x \in \Theta_t$ and $A \in \mathcal{B}(X)$. Accordingly, it follows that

$$P_t(x, A \cap (X \setminus \Theta)) \cdot \mathbf{1}_{\Theta_t}(x) = 0 \tag{6.1.1}$$

for every $x \in X$ and $A \in \mathcal{B}(X)$.

Using the fact that $\nu_1 = \mathbf{1}_\Theta \mu$ (because $\nu_1(A) = \mu(A \cap \Theta) = \int_A \mathbf{1}_\Theta \, d\mu$ for every $A \in \mathcal{B}(X)$), the fact that $\mathbf{1}_{\Theta_t} = \mathbf{1}_\Theta$ μ-a.e., the fact that $P(y, \cdot)$ is a probability measure for every $y \in X$, the equality (6.1.1), the definition of T_t, and the fact that μ is an invariant probability for $(T_s)_{s \in [0, +\infty)}$, we obtain that

$$T_t \nu_1(A) = \int_X P_t(x, A) \, d\nu_1(x) = \int_X P_t(x, A) \mathbf{1}_\Theta(x) \, d\mu(x)$$

$$= \int_X P_t(x, A) \mathbf{1}_{\Theta_t}(x) \, d\mu(x) = \int_X P_t(x, ((A \cap \Theta) \cup (A \cap (X \setminus \Theta)))) \mathbf{1}_{\Theta_t}(x) \, d\mu(x)$$

$$= \int_X P_t(x, (A \cap \Theta)) \mathbf{1}_{\Theta_t}(x) \, d\mu(x) + \int_X P_t(x, (A \cap (X \setminus \Theta))) \mathbf{1}_{\Theta_t}(x) \, d\mu(x)$$

$$= \int_X P_t(x, (A \cap \Theta)) \mathbf{1}_{\Theta_t}(x) \, d\mu(x) \le \int_X P_t(x, (A \cap \Theta)) \, d\mu(x)$$

$$= T_t \mu(A \cap \Theta) = \mu(A \cap \Theta) = \nu_1(A)$$

for every $A \in \mathcal{B}(X)$.

Accordingly, $T\nu_1 \le \nu_1$. Since ν_1 is a positive element of $\mathcal{M}(X)$, and since T_t is a Markov operator, we conclude that $T\nu_1 = \nu_1$.

Since $T\nu_2 = \nu_2$ (because $T\nu_2 = T\mu - T\nu_1 = \mu - \nu_1 = \nu_2$), since ν_1 and ν_2 are concentrated on Θ and $X \setminus \Theta$, respectively, and since $\mu(\Theta) > 0$ and $\mu(X \setminus \Theta) > 0$, we have obtained a contradiction because we assume that μ is an ergodic invariant probability measure.

$(b) \Leftrightarrow (c)$ The equivalence of (b) and (c) is obvious and was pointed out before the proposition.

$(b) \Rightarrow (a)$ Assume that $\mathcal{B}_\mu(X)$ is trivial, but that the probability measure μ is not ergodic.

Since we assume that μ is not an ergodic measure for $(P_t)_{t \in \mathbb{T}}$, by Proposition 2.3.13, we may assume that $\mathbb{T} = [0, +\infty)$, and that μ fails to be an ergodic measure for $(P_t)_{t \in [0,+\infty)}$.

In view of our assumption that μ is not ergodic, we obtain that there exist two nonzero invariant measures ν_1 and ν_2 such that $\mu = \nu_1 + \nu_2$ and such that ν_1 and ν_2 are mutually singular. Thus, there exist two $\mathcal{B}(X)$-measurable subsets C_1 and C_2 of X such that $C_1 \cup C_2 = X$, $C_1 \cap C_2 = \emptyset$, and such that $\nu_1(C_2) = 0$ and $\nu_2(C_1) = 0$.

In order to complete the proof of the implication, it is enough to show that C_1 is μ-invariant because $0 < \mu(C_1) < 1$ (since $\mu(C_1) = \nu_1(X)$), so we obtain a contradiction which stems from our assumption that $\mathcal{B}_\mu(X)$ is trivial.

Thus, let $t \in [0, +\infty)$.

Taking into consideration that ν_i is concentrated on C_i, $i = 1, 2$, that $C_1 \cap C_2 = \emptyset$, and that $T_t \nu_1 = \nu_1$, we obtain that

$$\mu(C_1) = \nu_1(C_1) = T_t \nu_1(C_1) = \int P_t(x, C_1) \, d\nu_1(x)$$

$$= \int \mathbf{1}_{C_1}(x) P_t(x, C_1) \, d\nu_1(x) = \int \mathbf{1}_{C_1}(x) P_t(x, C_1) \, d\mu(x).$$

Since $0 \le P_t(x, C_1) \le 1$ for all $x \in X$, using the above equalities, we obtain that there exists a μ-negligible subset N_1 of C_1 such that $P_t(x, C_1) = 1$ for every $x \in C_1 \setminus N_1$.

Since

$$0 = v_2(C_1) = T_t v_2(C_1) = \int P_t(x, C_1) \, dv_2(x)$$

$$= \int \mathbf{1}_{C_2}(x) P_t(x, C_1) \, dv_2(x) = \int \mathbf{1}_{C_2}(x) P_t(x, C_1) \, d\mu(x),$$

it follows that $\mathbf{1}_{C_2}(x) P_t(x, C_1) = 0$ μ-a.e., so there exists a μ-negligible subset N_2 of C_2 such that $P(x, C_1) = 0$ for every $x \in C_2 \setminus N_2$.

Thus,

$$P_t(x, C_1) = \begin{cases} 1 \text{ if } x \in C_1 \setminus N_1 \\ 0 \text{ if } x \in C_2 \setminus N_2; \end{cases}$$

therefore, $P_t(x, C_1) = \mathbf{1}_{C_1}(x)$ for every $x \in X \setminus (N_1 \cup N_2)$. Since N_1 and N_2 are μ-negligible, it follows that $P_t(x, C_1) = \mathbf{1}_{C_1}(x)$ for μ-a.e. $x \in X$.

Using the fact that $S_t \mathbf{1}_{C_1}(x) = P_t(x, C_1)$ for all $x \in X$, and since $\overline{S_t \mathbf{1}_{C_1}} = V_t^{(\mu)} \bar{\mathbf{1}}_{C_1}$ (as pointed out in the proof of Proposition 3.2.7), we obtain that $V_t^{(\mu)} \bar{\mathbf{1}}_{C_1} = \bar{\mathbf{1}}_{C_1}$.

We have therefore proved that $V_t^{(\mu)} \bar{\mathbf{1}}_{C_1} = \bar{\mathbf{1}}_{C_1}$ for every $t \in [0, +\infty)$, so C_1 is a μ-invariant subset of X. $\square$

An element f of $B_b(X)$ is called a μ-*invariant function* if $S_t f = f$ for every $t \in [0, +\infty)$.

Given an element $\bar{f}$ in $L^1(X, \mathcal{B}(X), \mu)$, we say that $\bar{f}$ is a μ-*invariant element of* $L^1(X, \mathcal{B}(X), \mu)$ if $V_t^{(\mu)} \bar{f} = \bar{f}$ for every $t \in [0, +\infty)$.

Observation. If $\mathbb{T} = \mathbb{R}$, then $f \in B_b(X)$ is μ-invariant if and only if $S_t f = f$ for every $t \in \mathbb{R}$. Indeed, if f is μ-invariant and $t \in [0, +\infty)$, then $S_{-t} f = S_{-t}(S_t f) = S_{-t+t} f = S_0 f = f$, so $S_r f = f$ for every $r \in \mathbb{R}$. Clearly, if $S_r f = f$ for every $r \in \mathbb{R}$, then f is μ-invariant.   ▲

Let $(Y, \mathcal{Y}, v)$ be a measure space and let $\mathcal{F} \subseteq \mathcal{Y}$ be a sub-σ-algebra of $\mathcal{Y}$ that contains all the v-negligible subsets of Y (that belong to $\mathcal{Y}$). Given $\bar{f} \in L^1(Y, \mathcal{Y}, v)$, we say that $\bar{f}$ is $\mathcal{F}$-*measurable* if every $\mathcal{Y}$-measurable function g in the class $\bar{f}$ is $\mathcal{F}$-measurable. Note that since $\mathcal{F}$ contains all the v-negligible subsets of Y, it follows that $\bar{f} \in L^1(Y, \mathcal{Y}, v)$ is $\mathcal{F}$-measurable if and only if there exists a $\mathcal{Y}$-measurable function g in class $\bar{f}$ such that, g is also $\mathcal{F}$-measurable.

Example 6.1.3. The sub-σ-algebras $\mathcal{B}_\mu(X)$ and $\mathcal{B}_{\tilde{\mu}}(X)$ of $\mathcal{B}(X)$ and $\widetilde{\mathcal{B}(X)}$ contain all the μ-negligible and $\tilde{\mu}$-negligible subsets of X, respectively. Indeed, if $A \in \mathcal{B}(X)$ is μ-negligible, then $\overline{\mathbf{1}_A} = \bar{0}$, so $V_t^{(\mu)} \overline{\mathbf{1}_A} = V_t^{(\mu)} \bar{0} = \bar{0} = \overline{\mathbf{1}_A}$ for every $t \in [0, +\infty)$; hence, $A \in \mathcal{B}_\mu(X)$. Similarly, if $A \in \widetilde{\mathcal{B}(X)}$ is $\tilde{\mu}$-negligible, then $\widetilde{\mathbf{1}_A} = \bar{\bar{0}}$, so $\widetilde{V_t^{(\mu)}} \widetilde{\mathbf{1}_A} = \widetilde{V_t^{(\mu)}} \bar{\bar{0}} = \bar{\bar{0}} = \widetilde{\mathbf{1}_A}$ for every $t \in [0, +\infty)$; consequently, $A \in \mathcal{B}_{\tilde{\mu}}(X)$.   ■

Proposition 6.1.4. *(a) A function $g \in B_b(X)$ is μ-invariant if and only if g is $\mathcal{B}_\mu(X)$-measurable.*

(b) An element $\bar{f}$ of $L^1(X, \mathcal{B}(X), \mu)$ is μ-invariant if and only if $\bar{f}$ is $\mathcal{B}_\mu(X)$-measurable.

Proof. Let $g \in B_b(X)$. Since, as mentioned in the proof of Proposition 3.2.7, $\overline{S_t g} = V_t^{(\mu)} \bar{g}$ for every $t \in [0, +\infty)$, it follows that g is μ-invariant if and only if the element $\bar{g}$ of $L^1(X, \mathcal{B}(X), \mu)$ defined by g is μ-invariant. Using the fact that g is $\mathcal{B}_\mu(X)$-measurable if and only if $\bar{g}$ is $\mathcal{B}_\mu(X)$-measurable, we obtain that in order to prove the proposition, we have to prove only (b).

In order to prove (b) we have to prove that the following two assertions hold true for $\bar{f} \in L^1(X, \mathcal{B}(X), \mu)$:

(1) If $\bar{f}$ is $\mathcal{B}_\mu(X)$-measurable, then $\bar{f}$ is μ-invariant.
(2) If $\bar{f}$ is μ-invariant, then $\bar{f}$ is $\mathcal{B}_\mu(X)$-measurable.

(1) First note that if $\bar{f} = \overline{\mathbf{1}_A}$ for some $A \in \mathcal{B}(X)$, and if we assume that $\bar{f}$ is $\mathcal{B}_\mu(X)$-measurable, then $A \in \mathcal{B}_\mu(X)$; therefore, $V_t^{(\mu)} \overline{\mathbf{1}_A} = \overline{\mathbf{1}_A}$ for every $t \in [0, +\infty)$; hence, $\bar{f}$ is μ-invariant.

Next, assume that $\bar{f} \in L^1(X, \mathcal{B}(X), \mu)$ is of the form $\bar{f} = \sum_{i=1}^n a_i \overline{\mathbf{1}_{A_i}}$ for some $n \in \mathbb{N}$, n real numbers $a_1, a_2, \ldots, a_n$, and n $\mathcal{B}(X)$-measurable subsets $A_1, A_2, \ldots, A_n$, and assume that $\bar{f}$ is $\mathcal{B}_\mu(X)$-measurable. Clearly, we can choose the numbers $a_1, a_2, \ldots, a_n$ and the sets $A_1, A_2, \ldots, A_n$ such that $a_i \neq a_j$ and $A_i \cap A_j = \emptyset$ for every $i \in \{1, 2, \ldots, n\}, j \in \{1, 2, \ldots, n\}, i \neq j$. Since we assume that $\bar{f}$ is $\mathcal{B}_\mu(X)$-measurable, it follows that $A_i \in \mathcal{B}_\mu(X)$ for every $i = 1, 2, \ldots, n$, so $\overline{\mathbf{1}_{A_i}}, i = 1, 2, \ldots, n$, are μ-invariant elements of $L^1(X, \mathcal{B}(X), \mu)$; consequently, $\bar{f}$ is μ-invariant, as well.

Now, let $\bar{f} \in L^1(X, \mathcal{B}(X), \mu)$, $\bar{f} \geq 0$, and assume that $\bar{f}$ is $\mathcal{B}_\mu(X)$-measurable. In view of the definition of the $\mathcal{B}_\mu(X)$-measurability, it follows that there exists a real-valued $\mathcal{B}_\mu(X)$-measurable function g, $g \geq 0$, such that g belongs to the class $\bar{f}$. Thus, there also exists an increasing sequence $(g_n)_{n \in \mathbb{N}}$ of positive $\mathcal{B}_\mu(X)$-measurable simple real-valued functions such that $(g_n)_{n \in \mathbb{N}}$ converges everywhere to g. Clearly, the functions g_n, $n \in \mathbb{N}$, are μ-integrable. By the Lebesgue dominated convergence theorem applied to the sequence $(g - g_n)_{n \in \mathbb{N}}$, we obtain that the sequence $(\int_X (g - g_n)\, d\mu)_{n \in \mathbb{N}}$ converges to zero. Therefore, the sequence $(\bar{g}_n)_{n \in \mathbb{N}}$ converges in the norm topology of $L^1(X, \mathcal{B}(X), \mu)$ to $\bar{f}$. Using the fact that g_n, $n \in \mathbb{N}$, are simple $\mathcal{B}_\mu(X)$-measurable μ-integrable functions, and in view of the discussion in the preceding paragraph, we obtain that $\bar{g}_n$, $n \in \mathbb{N}$, are μ-invariant elements of $L^1(X, \mathcal{B}(X), \mu)$. Accordingly, $V_t^{(\mu)} \bar{g}_n = \bar{g}_n$ for every $t \in [0, +\infty)$ and $n \in \mathbb{N}$. Since $V_t^{(\mu)}$, $t \in [0, +\infty)$, are bounded (continuous) operators and since $(\bar{g}_n)_{n \in \mathbb{N}}$ converges in the norm topology of $L^1(X, \mathcal{B}(X), \mu)$ to $\bar{f}$, it follows that $V_t^{(\mu)} \bar{f} = \bar{f}$ for every $t \in [0, +\infty)$, so $\bar{f}$ is μ-invariant in this case, as well.

Finally, let $\bar{f} \in L^1(X, \mathcal{B}(X), \mu)$ be a not necessarily positive element of $L^1(X, \mathcal{B}(X), \mu)$, and assume that $\bar{f}$ is $\mathcal{B}_\mu(X)$-measurable. It is easy to see that in this case $\bar{f}^+ = \bar{f} \vee \bar{0}$ and $\bar{f}^- = (-\bar{f}) \vee \bar{0}$ are $\mathcal{B}_\mu(X)$-measurable functions, as well. Using the discussion of the preceding paragraph, we obtain that $\bar{f}^+$ and $\bar{f}^-$ are μ-invariant elements of $L^1(X, \mathcal{B}(X), \mu)$. Since $\bar{f} = \bar{f}^+ - \bar{f}^-$ and the operators $V_t^{(\mu)}$, $t \in [0, +\infty)$, are linear, it follows that $\bar{f}$ is μ-invariant.

(2) The proof of assertion (2) will be carried out in several steps.

Step 1.   Our goal at this step is to discuss two properties of μ-invariant elements that we will need soon.

Let $\bar{f}$ and $\bar{g}$ be two positive μ-invariant elements of $L^1(X, \mathcal{B}(X), \mu)$. Then $\bar{f} \wedge \bar{g}$ is μ-invariant, as well. Indeed, let $t \in [0, +\infty)$, and note that $V_t^{(\mu)}(\bar{f} \wedge \bar{g}) \leq (V_t^{(\mu)} \bar{f}) \wedge (V_t^{(\mu)} \bar{g}) = \bar{f} \wedge \bar{g}$; since $\bar{f} \wedge \bar{g}$ is a positive element of $L^1(X, \mathcal{B}(X), \mu)$ and since $V_t^{(\mu)}$ is a Markov operator, it follows that $V_t^{(\mu)}(\bar{f} \wedge \bar{g}) = \bar{f} \wedge \bar{g}$. Thus, $V_t^{(\mu)}(\bar{f} \wedge \bar{g}) = \bar{f} \wedge \bar{g}$ for every $t \in [0, +\infty)$; that is, $\bar{f} \wedge \bar{g}$ is μ-invariant.

Let $(\bar{f}_n)_{n \in \mathbb{N}}$ be a sequence of μ-invariant positive elements of $L^1(X, \mathcal{B}(X), \mu)$, and assume that $\sup_{n \in \mathbb{N}} \bar{f}_n$ exists in $L^1(X, \mathcal{B}(X), \mu)$. Then $\sup_{n \in \mathbb{N}} \bar{f}_n$ is a μ-invariant element of $L^1(X, \mathcal{B}(X), \mu)$. Indeed, taking into consideration that $V_t^{(\mu)}$, $t \in [0, +\infty)$, are positive operators, we obtain that $V_t^{(\mu)}(\sup_{n \in \mathbb{N}} \bar{f}_n) \geq \sup_{n \in \mathbb{N}} \bar{f}_n$ for every $t \in [0, +\infty)$. Since $V_t^{(\mu)}$, $t \in [0, +\infty)$, are Markov operators and since $\sup_{n \in \mathbb{N}} \bar{f}_n$ is a positive element of $L^1(X, \mathcal{B}(X), \mu)$, it follows that $V_t^{(\mu)}(\sup_{n \in \mathbb{N}} \bar{f}_n) = \sup_{n \in \mathbb{N}} \bar{f}_n$ for every $t \in [0, +\infty)$; therefore, $\sup_{n \in \mathbb{N}} \bar{f}_n$ is μ-invariant.

Step 2.   At this step, we prove that if $\bar{f}$ is a μ-invariant element of $L^1(X, \mathcal{B}(X), \mu)$ such that $\bar{f} \geq 0$ and such that there exists a bounded $\mathcal{B}(X)$-measurable function g in class $\bar{f}$, then $\bar{f}$ is $\mathcal{B}_\mu(X)$-measurable.

To this end, let $\bar{f} \in L^1(X, \mathcal{B}(X), \mu)$, $\bar{f} \geq 0$, be μ-invariant and assume that there exists a $g \in B_b(X)$ such that g is in class $\bar{f}$.

Clearly, we may choose g such that $g \geq 0$. Therefore, we assume that $g \geq 0$.

In view of the discussion preceding Example 6.1.3, in order to prove that $\bar{f}$ is $\mathcal{B}_\mu(X)$-measurable, it is enough to prove that g is $\mathcal{B}_\mu(X)$-measurable; thus, it is enough to prove that the sets $\{x \in X \mid g(x) < \alpha\}$, $\alpha \in \mathbb{R}$, belong to $\mathcal{B}_\mu(X)$. Since $\{x \in X \mid g(x) < \alpha\}$ is the empty set (because $g \geq 0$), so it belongs to $\mathcal{B}_\mu(X)$ whenever $\alpha \leq 0$, we have to prove only that $\{x \in X \mid g(x) < \alpha\}$ belongs to $\mathcal{B}_\mu(X)$ for every $\alpha \in \mathbb{R}$, $\alpha > 0$.

To this end, let $\alpha \in \mathbb{R}$, $\alpha > 0$.

Taking into consideration that

$$(\alpha \mathbf{1}_X - (g \wedge (\alpha \mathbf{1}_X)))(x) = \begin{cases} \alpha - g(x) & \text{if } g(x) < \alpha \\ 0 & \text{if } g(x) \geq \alpha \end{cases}$$

for every $x \in X$ (that is, $\alpha \mathbf{1}_X - (g \wedge (\alpha \mathbf{1}_X)) = (\alpha \mathbf{1}_X - g)\mathbf{1}_{\{x \in X \mid g(x) < \alpha\}}$), we obtain that the sequence $((\mathbf{1}_X \wedge (n(\alpha \mathbf{1}_X - (g \wedge (\alpha \mathbf{1}_X)))))(x))_{n \in \mathbb{N}}$ converges and

$$\lim_{n\to+\infty} \mathbf{1}_X \wedge (n(\alpha\mathbf{1}_X - (g \wedge (\alpha\mathbf{1}_X))))(x) = \begin{cases} 1 & \text{if } g(x) < \alpha \\ 0 & \text{if } g(x) \geq \alpha, \end{cases}$$

or, equivalently, $\lim_{n\to+\infty}(\mathbf{1}_X \wedge (n(\alpha\mathbf{1}_X - (g \wedge (\alpha\mathbf{1}_X)))))(x) = \mathbf{1}_{\{y\in X\,|\,g(y)<\alpha\}}(x)$ for every $x \in X$.

Since the sequence $(\mathbf{1}_X \wedge (n(\alpha\mathbf{1}_X - (g \wedge (\alpha\mathbf{1}_X)))))_{n\in\mathbb{N}}$ is increasing and converges pointwise on X to $\mathbf{1}_{\{y\in X\,|\,g(y)<\alpha\}}$, it follows that $\sup_{n\in\mathbb{N}}(\mathbf{1}_X \wedge (n((\alpha\mathbf{1}_X) - (g \wedge (\alpha\mathbf{1}_X)))))$ exists and is equal to $\mathbf{1}_{\{y\in X\,|\,g(y)<\alpha\}}$.

Accordingly, $\sup_{n\in\mathbb{N}} \overline{(\mathbf{1}_X \wedge (n((\alpha\mathbf{1}_X) - (g \wedge (\alpha\mathbf{1}_X)))))}$ (or, equivalently, $\sup_{n\in\mathbb{N}}(\bar{\mathbf{1}}_X \wedge (n((\alpha\bar{\mathbf{1}}_X) - (\bar{f} \wedge (\alpha\bar{\mathbf{1}}_X)))))$) exists and is equal to $\bar{\mathbf{1}}_{\{y\in X\,|\,g(y)<\alpha\}}$.

The elements $\bar{\mathbf{1}}_X \wedge (n(\alpha\bar{\mathbf{1}}_X - (\bar{f} \wedge (\alpha\bar{\mathbf{1}}_X))))$, $n \in \mathbb{N}$, are μ-invariant because, using the first property of μ-invariant elements of $L^1(X, \mathcal{B}(X), \mu)$ discussed in *Step 1*, and the fact that $\bar{f}$ and $\bar{\mathbf{1}}_X$ are μ-invariant, we obtain that

$$V_t^{(\mu)}(\bar{\mathbf{1}}_X \wedge (n(\alpha\bar{\mathbf{1}}_X - (\bar{f} \wedge (\alpha\bar{\mathbf{1}}_X))))) = \bar{\mathbf{1}}_X \wedge V_t^{(\mu)}(n(\alpha\bar{\mathbf{1}}_X - (\bar{f} \wedge (\alpha\bar{\mathbf{1}}_X))))$$

$$= \bar{\mathbf{1}}_X \wedge (n(\alpha V_t^{(\mu)}\bar{\mathbf{1}}_X - ((V_t^{(\mu)}\bar{f}) \wedge (\alpha V_t^{(\mu)}\bar{\mathbf{1}}_X)))) = \bar{\mathbf{1}}_X \wedge (n(\alpha\bar{\mathbf{1}}_X - (\bar{f} \wedge (\alpha\bar{\mathbf{1}}_X))))$$

for every $n \in \mathbb{N}$ and $t \in [0, +\infty)$.

Therefore, using the second property of μ-invariant elements discussed in *Step 1*, we obtain that

$$V_t^{(\mu)}(\bar{\mathbf{1}}_{\{y\in X\,|\,g(y)<\alpha\}}) = V_t^{(\mu)}(\sup_{n\in\mathbb{N}}(\bar{\mathbf{1}}_X \wedge (n(\alpha\bar{\mathbf{1}}_X - (\bar{f} \wedge (\alpha\bar{\mathbf{1}}_X))))))$$

$$= \sup_{n\in\mathbb{N}}(\bar{\mathbf{1}}_X \wedge (n(\alpha\bar{\mathbf{1}}_X - (\bar{f} \wedge (\alpha\bar{\mathbf{1}}_X)))))$$

$$= \bar{\mathbf{1}}_{\{y\in X\,|\,g(y)<\alpha\}}$$

for every $t \in [0, +\infty)$.

It follows that the set $\{y \in X \mid g(y) < \alpha\}$ belongs to $\mathcal{B}_\mu(X)$ for every $\alpha \in \mathbb{R}$.

Step 3.   We will prove at this step that every μ-invariant $\bar{f} \in L^1(X, \mathcal{B}(X), \mu)$, $\bar{f} \geq 0$, is $\mathcal{B}_\mu(X)$-measurable.

To this end, let $\bar{f} \in L^1(X, \mathcal{B}(X), \mu)$, $\bar{f} \geq 0$, be μ-invariant, and set $\bar{f}_n = \bar{f} \wedge (n\bar{\mathbf{1}}_X)$ for every $n \in \mathbb{N}$.

Since $\bar{f} \geq 0$, it follows that there exists a real-valued $\mathcal{B}(X)$-measurable function g in class $\bar{f}$ such that $g \geq 0$.

We now note that $\bar{f}_n$, $n \in \mathbb{N}$, are μ-invariant elements of $L^1(X, \mathcal{B}(X), \mu)$. Indeed, using the first property of μ-invariant elements discussed in *Step 1*, and since $\bar{f}$ and $\bar{\mathbf{1}}_X$ are μ-invariant, we obtain that $V_t^{(\mu)}\bar{f}_n = (V_t^{(\mu)}\bar{f}) \wedge (V_t^{(\mu)}(n\bar{\mathbf{1}}_X)) = \bar{f} \wedge (n\bar{\mathbf{1}}_X) = \bar{f}_n$ for every $n \in \mathbb{N}$ and $t \in [0, +\infty)$.

Since the function $g_n = g \wedge (n\mathbf{1}_X)$ belongs to $B_b(X)$ and to class $\bar{f}_n$, by *Step 2*, we obtain that both g_n and $\bar{f}_n$ are $\mathcal{B}_\mu(X)$-measurable for every $n \in \mathbb{N}$.

Taking into consideration that the sequence $(g_n)_{n\in\mathbb{N}}$ converges pointwise (everywhere) on X to g, and that g_n is $\mathcal{B}_\mu(X)$-measurable for every $n \in \mathbb{N}$, we obtain that g is $\mathcal{B}_\mu(X)$-measurable, so $\bar{f}$ is $\mathcal{B}_\mu(X)$-measurable, as well.

Step 4. Finally, at this step, we prove that every μ-invariant $\bar{f} \in L^1(X, \mathcal{B}(X), \mu)$ is $\mathcal{B}_\mu(X)$-measurable.

To this end, let $\bar{f} \in L^1(X, \mathcal{B}(X), \mu)$ and assume that $\bar{f}$ is μ-invariant.

Using the first property of μ-invariant elements discussed in *Step 1* we obtain that $V_t^{(\mu)}(\bar{f} \vee \bar{0}) = V_t^{(\mu)} \bar{f} \vee \bar{0} = \bar{f} \vee \bar{0} \, (= \bar{f}^+)$ and $V_t^{(\mu)}((-\bar{f}) \vee \bar{0}) = (-V_t^{(\mu)} \bar{f}) \vee \bar{0} = (-\bar{f}) \vee \bar{0} \, (= \bar{f}^-)$ for every $t \in [0, +\infty)$. Thus, both $\bar{f}^+$ and $\bar{f}^-$ are μ-invariant, so by *Step 3* they are also $\mathcal{B}_\mu(X)$-measurable. Consequently $\bar{f}$ is $\mathcal{B}_\mu(X)$-measurable as well, because $\bar{f} = \bar{f}^+ - \bar{f}^-$. $\qquad\square$

We will now use Proposition 6.1.4 and the topics on functions constant almost everywhere discussed in Sect. 4.1 in order to present another characterization of the ergodic invariant probability measures.

Proposition 6.1.5. *The following assertions are equivalent:*

(a) The probability measure μ is ergodic.
(b) Every μ-invariant element $\bar{f}$ of $L^1(X, \mathcal{B}(X), \mu)$ is constant μ-a.e.
(c) Every μ-invariant function $g \in B_b(X)$ is constant μ-a.e.

Proof. $(a) \Rightarrow (b)$ Assume that μ is an ergodic invariant probability measure, and let $\bar{f} \in L^1(X, \mathcal{B}(X), \mu)$ be μ-invariant. Using (b) of Proposition 6.1.4, we obtain that $\bar{f}$ is $\mathcal{B}_\mu(X)$-measurable. Thus, if we let g be a real-valued measurable function in the class $\bar{f}$, then g is $\mathcal{B}_\mu(X)$-measurable.

Since μ is an ergodic probability measure, by Proposition 6.1.2, $\mathcal{B}_\mu(X)$ is trivial. Therefore, $\mu(\{x \in X \mid g(x) < \alpha\}) = 0$ or 1 for every $\alpha \in \mathbb{R}$. By Proposition 4.1.1, the function g is constant μ-a.e., so $\bar{f}$ is constant μ-a.e., as well.

$(b) \Rightarrow (c)$ Let $g \in B_b(X)$ be a μ-invariant function, and let $\bar{g}$ be the element of $L^1(X, \mathcal{B}(X), \mu)$ defined by g.

Then, using a fact pointed out in the proof of Proposition 3.2.7 and our assumption that g is μ-invariant, we obtain that $\bar{g} = \overline{S_t g} = V_t^{(\mu)} \bar{g}$ for every $t \in [0, +\infty)$, so $\bar{g}$ is a μ-invariant element of $L^1(X, \mathcal{B}(X), \mu)$. Since we assume that (b) holds true, we obtain that $\bar{g}$ is constant μ-a.e.; therefore, g is constant μ-a.e., as well.

$(c) \Rightarrow (a)$ Assume that every μ-invariant function $g \in B_b(X)$ is constant μ-a.e. In view of Proposition 6.1.2, in order to prove that μ is ergodic, it is enough to prove that $\mathcal{B}_\mu(X)$ is a trivial sub-σ-algebra of $\mathcal{B}(X)$.

To this end, let $A \in \mathcal{B}_\mu(X)$. Then $\mathbf{1}_A$ is a μ-invariant function in $B_b(X)$. Since we assume that (c) is true, we obtain that $\mathbf{1}_A$ is constant μ-a.e. Since $\mathbf{1}_A$ takes only the values 0 and 1 it follows that either $\mu(\{x \in X \mid \mathbf{1}_A(x) = 0\}) = 1$ (in which case, $\mu(X \setminus A) = 1$, so $\mu(A) = 0$), or else $\mu(\{x \in X \mid \mathbf{1}_A(x) = 0\}) = 0$ (hence, $\mu(A) = 1$).

We have therefore proved that $\mu(A) = 0$ or 1 for every $A \in \mathcal{B}_\mu(X)$; thus, $\mathcal{B}_\mu(X)$ is trivial. □

Our goal now is to discuss a version of Lemma 3.3.1 of [143]. To this end, we need some preparation.

Recall (see, for example, pp. 67–68 of Dunford and Schwartz [30]) that a subset A of $L^1(X, \widetilde{\mathcal{B}(X)}, \tilde{\mu})$ is said to be *weakly sequentially compact* if every sequence $(\tilde{\bar{f}}_n)_{n \in \mathbb{N}}$ of elements of A has a subsequence $(\widetilde{f_{n_k}})_{k \in \mathbb{N}}$ that converges weakly to some $\tilde{\bar{f}} \in L^1(X, \widetilde{\mathcal{B}(X)}, \tilde{\mu})$. Note that $\tilde{\bar{f}}$ does not necessarily belong to A. Here, as usual in this book, $(X, \widetilde{\mathcal{B}(X)}, \tilde{\mu})$ stands for the completion of the probability space $(X, \mathcal{B}(X), \mu)$.

We will use the following result:

Proposition 6.1.6. *A subset A of $L^1(X, \widetilde{\mathcal{B}(X)}, \tilde{\mu})$ is weakly sequentially compact if and only if the following two conditions are satisfied:*

(i) The set A is norm bounded (that is, $\sup_{\tilde{\bar{f}} \in A} \left\| \tilde{\bar{f}} \right\| < +\infty$).

(ii) For every monotone nonincreasing sequence $(E_n)_{n \in \mathbb{N}}$ of measurable subsets of X with empty intersection, it follows that $\lim_{n \to +\infty} \int_{E_n} \tilde{\bar{f}} \, d\mu = 0$ uniformly in $\tilde{\bar{f}} \in A$; that is, for every sequence $(E_n)_{n \in \mathbb{N}}$ of elements of $\widetilde{\mathcal{B}(X)}$ such that $E_n \supseteq E_{n+1}$ for every $n \in \mathbb{N}$ and such that $\cap_{m=1}^{\infty} E_m = \emptyset$, and for every $\varepsilon \in \mathbb{R}$, $\varepsilon > 0$, there exists an $n_\varepsilon \in \mathbb{N}$ such that $\left| \int_{E_n} \tilde{\bar{f}} \, d\mu \right| < \varepsilon$ for every $\tilde{\bar{f}} \in A$ and $n \in \mathbb{N}$, $n \geq n_\varepsilon$.

For a proof of the above proposition, see Theorem 9, Section 8, Chap. 4, pp. 292–293 of Dunford and Schwartz [30].

Lemma 6.1.7. *For every $\tilde{\bar{f}} \in L^1(X, \widetilde{\mathcal{B}(X)}, \tilde{\mu})$, the set $\{\widetilde{A_t \bar{f}} \, | t \in [0, +\infty)\}$ is weakly sequentially compact in $L^1(X, \widetilde{\mathcal{B}(X)}, \tilde{\mu})$.*

Proof. We will prove the lemma by showing that the sets $\{\tilde{A}_t \tilde{\bar{f}} \, | \, t \in [0, +\infty)\}$, $\tilde{\bar{f}} \in L^1(X, \widetilde{\mathcal{B}(X)}, \tilde{\mu})$, satisfy the conditions (i) and (ii) of Proposition 6.1.6.

Since $\widetilde{A}_t$, $t \in [0, +\infty)$, are contractions of $L^1(X, \widetilde{\mathcal{B}(X)}, \tilde{\mu})$, it follows that $\sup_{t \in [0,+\infty)} \left\| \widetilde{A}_t \tilde{\bar{f}} \right\| \leq \left\| \tilde{\bar{f}} \right\|$ for every $\tilde{\bar{f}} \in L^1(X, \widetilde{\mathcal{B}(X)}, \tilde{\mu})$, so, the sets $\{\widetilde{A}_t \tilde{\bar{f}} \, | \, t \in [0, +\infty)\}$, $\tilde{\bar{f}} \in L^1(X, \widetilde{\mathcal{B}(X)}, \tilde{\mu})$, are norm bounded; therefore, they satisfy condition (i) of Proposition 6.1.6.

We will prove that the sets $\{\widetilde{A}_t \tilde{\bar{f}} \, | \, t \in [0, +\infty)\}$, $\tilde{\bar{f}} \in L^1(X, \widetilde{\mathcal{B}(X)}, \tilde{\mu})$, satisfy condition (ii) of Proposition 6.1.6 in two steps.

Step 1. At this step we prove that $\{\widetilde{A}_t \tilde{\bar{f}} \, | \, t \in [0, +\infty)\}$ satisfies condition (ii) of Proposition 6.1.6 whenever $\tilde{\bar{f}} \in L^1(X, \widetilde{\mathcal{B}(X)}, \tilde{\mu})$ is such that $\left| \tilde{\bar{f}} \right| \leq M \tilde{\mathbf{1}}_X$ for some $M \in \mathbb{R}$, $M > 0$.

Thus, let $\bar{\tilde{f}} \in L^1(X, \widetilde{\mathcal{B}(X)}, \tilde{\mu})$, and assume that there exists an $M \in \mathbb{R}$, $M > 0$, such that $\left|\bar{\tilde{f}}\right| \le M\bar{\tilde{1}}_X$.

Now, let $(E_n)_{n\in\mathbb{N}}$ be a sequence of subsets of X such that $E_n \in \widetilde{\mathcal{B}(X)}$ and $E_n \supseteq E_{n+1}$ for every $n \in \mathbb{N}$, and such that $\cap_{n=1}^{\infty} E_n = \emptyset$. Also, let $\varepsilon \in \mathbb{R}$, $\varepsilon > 0$.

Clearly, there exists an $n_\varepsilon \in \mathbb{N}$ such that $M\tilde{\mu}(E_n) < \varepsilon$ for every $n \in \mathbb{N}$, $n \ge n_\varepsilon$. We obtain that

$$\left| \int_{E_n} \widetilde{A_t}\,\bar{\tilde{f}}\,\mathrm{d}\tilde{\mu} \right| \le \int_{E_n} \left| \widetilde{A_t}\,\bar{\tilde{f}} \right| \mathrm{d}\tilde{\mu} \le \int_{E_n} M\bar{\tilde{1}}_X\,\mathrm{d}\tilde{\mu} = M\tilde{\mu}(E_n) < \varepsilon$$

for every $n \in \mathbb{N}$, $n \ge n_\varepsilon$, and $t \in [0, +\infty)$.

We have therefore proved that $\{\widetilde{A_t}\,\bar{\tilde{f}}\mid t \in [0, +\infty)\}$ satisfies condition (ii) of Proposition 6.1.6 whenever $\bar{\tilde{f}} \in L^1(X, \widetilde{\mathcal{B}(X)}, \tilde{\mu})$ is such that $\left|\bar{\tilde{f}}\right| \le M\bar{\tilde{1}}_X$ for some $M \in \mathbb{R}$, $M > 0$.

Step 2. At this step we complete the proof of the lemma by showing that the set $\{\widetilde{A_t}\,\bar{\tilde{f}}\mid t \in [0, +\infty)\}$ satisfies condition (ii) of Proposition 6.1.6, first for $\bar{\tilde{f}} \ge 0$, and then for every $\bar{\tilde{f}} \in L^1(X, \widetilde{\mathcal{B}(X)}, \tilde{\mu})$ not necessarily positive.

Thus, let $\bar{\tilde{f}} \in L^1(X, \widetilde{\mathcal{B}(X)}, \tilde{\mu})$, $\bar{\tilde{f}} \ge 0$.

Also, let $(E_n)_{n\in\mathbb{N}}$ be a monotone nonincreasing sequence of elements of $\widetilde{\mathcal{B}(X)}$ such that $\cap_{n=1}^{\infty} E_n = \emptyset$, and let $\varepsilon \in \mathbb{R}$, $\varepsilon > 0$.

For every $k \in \mathbb{N}$, set $\widetilde{\bar{f}_k} = \bar{\tilde{f}} \wedge (k\bar{\tilde{1}}_X)$.

Clearly, $\widetilde{\bar{f}_k} \in L^1(X, \widetilde{\mathcal{B}(X)}, \tilde{\mu})$ for every $k \in \mathbb{N}$, and the sequence $(\widetilde{\bar{f}_k})_{k\in\mathbb{N}}$ converges in the L^1-norm topology of $L^1(X, \widetilde{\mathcal{B}(X)}, \tilde{\mu})$ to $\bar{\tilde{f}}$. Thus, there exists a $k_\varepsilon \in \mathbb{N}$ such that $\left\|\bar{\tilde{f}} - \widetilde{\bar{f}_k}\right\| < \frac{\varepsilon}{2}$ for every $k \in \mathbb{N}$, $k \ge k_\varepsilon$.

In view of our discussion at *Step 1*, we obtain that there exists an $n_\varepsilon \in \mathbb{N}$ such that $\left|\int_{E_n} \tilde{A}_t\,\bar{\tilde{f}}_{k_\varepsilon}\,\mathrm{d}\tilde{\mu}\right| < \frac{\varepsilon}{2}$ for every $n \in \mathbb{N}$, $n \ge n_\varepsilon$, and every $t \in [0, +\infty)$.

It follows that

$$\left| \int_{E_n} \widetilde{A_t}\,\bar{\tilde{f}}\,\mathrm{d}\tilde{\mu} \right| \le \left| \int_{E_n} \tilde{A}_t(\bar{\tilde{f}} - \bar{\tilde{f}}_{k_\varepsilon})\,\mathrm{d}\tilde{\mu} \right| + \left| \int_{E_n} \tilde{A}_t\,\bar{\tilde{f}}_{k_\varepsilon}\,\mathrm{d}\tilde{\mu} \right|$$

$$< \int_X \left|\widetilde{A_t}(f - f_{k_\varepsilon})\right|\,\mathrm{d}\tilde{\mu} + \frac{\varepsilon}{2} \le \|f - f_{k_\varepsilon}\| + \frac{\varepsilon}{2} < \frac{\varepsilon}{2} + \frac{\varepsilon}{2} = \varepsilon$$

for every $n \in \mathbb{N}$, $n \ge n_\varepsilon$, and $t \in [0, +\infty)$.

Thus, $\{\tilde{A}_t\,\bar{\tilde{f}}\mid t \in [0, +\infty)\}$ satisfies condition (ii) of Proposition 6.1.6.

Now, if $\bar{\tilde{f}} \in L^1(X, \widetilde{\mathcal{B}(X)}, \tilde{\mu})$ is not necessarily positive, then using our discussion so far we obtain that $\left\{\widetilde{A_t}\left|\bar{\tilde{f}}\right|\mid t \in [0, +\infty)\right\}$ satisfies condition (ii) of Proposition 6.1.6 because $\left|\bar{\tilde{f}}\right| \ge 0$. Since $\left|\int_A \widetilde{A_t}\,\bar{\tilde{f}}\,\mathrm{d}\tilde{\mu}\right| \le \int_A \left|\widetilde{A_t}\,\bar{\tilde{f}}\right|\,\mathrm{d}\tilde{\mu} \le$

$\int_A \widetilde{A}_t \left| \bar{\bar{f}} \right| d\tilde{\mu}$ for every $A \in \widetilde{\mathcal{B}(X)}$ and $t \in [0, +\infty)$, it follows that $\{ \widetilde{A}_t \bar{\bar{f}} \mid t \in [0, +\infty) \}$ satisfies condition (*ii*) of Proposition 6.1.6, as well. $\qquad\square$

Proposition 6.1.8. *Let* $(\widetilde{A}_t)_{t \in [0,+\infty)}$ *be the family of positive contractions of* $L^1(X, \widetilde{\mathcal{B}(X)}, \tilde{\mu})$ *defined using the one-parameter semigroup of positive contractions* $(V_t^{(\mu)})_{t \in [0,+\infty)}$ *before Lemma 3.2.10. Also, let* $\bar{\bar{f}} \in L^1(X, \widetilde{\mathcal{B}(X)}, \tilde{\mu})$, *and let* $\bar{\bar{g}} \in L^1(X, \widetilde{\mathcal{B}(X)}, \tilde{\mu})$ *be the absolute* $\tilde{\mu}$-*a.e. limit of* $(\widetilde{A}_t \bar{\bar{f}})_{t \in [0,+\infty)}$ *as* $t \to +\infty$ *(the existence of* $\bar{\bar{g}}$ *and the fact that it belongs to* $L^1(X, \widetilde{\mathcal{B}(X)}, \tilde{\mu})$ *have been proved in Theorem 3.2.11). Then* $(\widetilde{A}_t \bar{\bar{f}})_{t \in [0,+\infty)}$ *converges in the norm topology of* $L^1(X, \widetilde{\mathcal{B}(X)}, \tilde{\mu})$ *to* $\bar{\bar{g}}$ *as* $t \to +\infty$, *and* $\widetilde{V_t^{(\mu)}} \bar{\bar{g}} = \bar{\bar{g}}$ *for every* $t \in [0, +\infty)$.

Proof. We have to prove that:

(1) The family $(\widetilde{A}_t \bar{\bar{f}})_{t \in [0,+\infty)}$ converges in the norm topology of $L^1(X, \widetilde{\mathcal{B}(X)}, \tilde{\mu})$ as $t \to +\infty$.

(2) The limit of $(\widetilde{A}_t \bar{\bar{f}})_{t \in [0,+\infty)}$ in the norm topology of $L^1(X, \widetilde{\mathcal{B}(X)}, \tilde{\mu})$ as $t \to +\infty$ is equal to $\bar{\bar{g}}$.

(3) $\widetilde{V_t^{(\mu)}} \bar{\bar{g}} = \bar{\bar{g}}$ for every $t \in [0, +\infty)$.

(1) In order to prove that the family $(\widetilde{A}_t \bar{\bar{f}})_{t \in [0,+\infty)}$ converges in the norm topology of $L^1(X, \widetilde{\mathcal{B}(X)}, \tilde{\mu})$ as $t \to +\infty$, it is enough to show that the one-parameter semigroup $(V_t^{(\mu)})_{t \in [0,+\infty)}$ and the family $(\widetilde{A}_t)_{t \in [0,+\infty)}$ satisfy the conditions of Theorem 8.7.1, pp. 687–688, of Dunford and Schwartz [30].

Since (by Proposition 3.2.8) the operators $\widetilde{V_t^{(\mu)}}$, $t \in [0, +\infty)$, are (positive) contractions, it follows that, for every $\bar{\bar{f}} \in L^1(X, \widetilde{\mathcal{B}(X)}, \tilde{\mu})$, the sequence $(\frac{1}{n} \widetilde{V_n^{(\mu)}} \bar{\bar{f}})_{n \in \mathbb{N}}$ converges to zero in the norm topology of $L^1(X, \widetilde{\mathcal{B}(X)}, \tilde{\mu})$, so the first condition (condition (1)) of Theorem 8.7.1 of [30] is satisfied. Since $\widetilde{A}_t$, $t \in [0, +\infty)$, are contractions (see Lemma 3.2.10), we obtain that condition (2) of Theorem 8.7.1 of [30] is satisfied, as well. Finally, using Lemma 6.1.7, we obtain that condition (3) of Theorem 8.7.1 of [30] is also satisfied. Therefore, $(\widetilde{V_t^{(\mu)}})_{t \in [0,+\infty)}$ and $(\widetilde{A}_t)_{t \in [0,+\infty)}$ satisfy all the conditions of Theorem 8.7.1 of [30].

(2) Let $\bar{\bar{h}} \in L^1(X, \widetilde{\mathcal{B}(X)}, \tilde{\mu})$ be the limit of $(\widetilde{A}_t \bar{\bar{f}})_{t \in [0,+\infty)}$ in the norm topology of $L^1(X, \widetilde{\mathcal{B}(X)}, \tilde{\mu})$ as $t \to +\infty$.

Clearly, for every sequence $(t_k)_{k \in \mathbb{N}}$ of elements of $[0, +\infty)$ that tends to $+\infty$, it follows that the sequence $(\tilde{A}_{t_k} \bar{\bar{f}})_{k \in \mathbb{N}}$ converges to $\bar{\bar{h}}$ in the norm topology of $L^1(X, \widetilde{\mathcal{B}(X)}, \tilde{\mu})$. In particular, if we set $t_k = k$ for every $k \in \mathbb{N}$, we obtain that $(\widetilde{A_k \bar{\bar{f}}})_{k \in \mathbb{N}}$ converges to $\bar{\bar{h}}$ in the norm topology of $L^1(X, \widetilde{\mathcal{B}(X)}, \tilde{\mu})$. Using Proposition 3.1.4, p. 88 of Cohn [20] and Proposition 3.1.2, pp. 86–87 of [20], we obtain that there exists a subsequence $(\tilde{A}_{k_l} \bar{\bar{f}})_{l \in \mathbb{N}}$ that converges $\tilde{\mu}$-a.e. to $\bar{\bar{h}}$. In view of the definition of the absolute μ-a.e. limit of an $L^1(X, \widetilde{\mathcal{B}(X)}, \tilde{\mu})$-valued function defined on $[0, +\infty)$ (see the Observation at the end of Sect. 3.2.1), we obtain that the subsequence $(\tilde{A}_{k_l}\bar{\tilde{f}})_{l \in \mathbb{N}}$ converges $\tilde{\mu}$-a.e. to $\bar{\tilde{g}}$; thus, $\bar{\tilde{g}} = \bar{\tilde{h}}$, and $(\widetilde{A}_t\bar{\tilde{f}})_{t \in [0,+\infty)}$ converges in the norm topology of $L^1(X, \widetilde{\mathcal{B}(X)}, \tilde{\mu})$ to $\bar{\tilde{g}}$ as $t \to +\infty$.

(3) We first note that it is enough to prove that $V_t^{(\mu)}\bar{\tilde{g}} = \bar{\tilde{g}}$ for every $t \in (0, +\infty)$ because if $\widetilde{V_{t_0}^{(\mu)}}\bar{\tilde{g}} = \bar{\tilde{g}}$ for some $t_0 > 0$, then $\widetilde{V_0^{(\mu)}}\bar{\tilde{g}} = \widetilde{V_0^{(\mu)}}\widetilde{V_{t_0}^{(\mu)}}\bar{\tilde{g}} = \widetilde{V_{t_0}^{(\mu)}}\bar{\tilde{g}} = \bar{\tilde{g}}$.

Thus, let $t \in (0, +\infty)$.

Clearly, the assertion that we have to prove is true if $\bar{\tilde{f}} = \bar{\tilde{0}}$. Therefore, we will assume that $\bar{\tilde{f}} \neq \bar{\tilde{0}}$.

Since $\widetilde{V_t^{(\mu)}}$ is a contraction of $L^1(X, \widetilde{\mathcal{B}(X)}, \tilde{\mu})$ and since $\bar{\tilde{g}}$ is the limit of $(\widetilde{A_\alpha \tilde{f}})_{\alpha \in [0,+\infty)}$ in the norm topology of $L^1(X, \widetilde{\mathcal{B}(X)}, \tilde{\mu})$ as $\alpha \to +\infty$, it follows that $\lim_{\alpha \to +\infty}{}_{\|\cdot\|} \widetilde{V_t^{(\mu)}} \widetilde{A_\alpha \tilde{f}}$ exists and $\widetilde{V_t^{(\mu)}}\bar{\tilde{g}} = \widetilde{V_t^{(\mu)}}(\lim_{\alpha \to +\infty}{}_{\|\cdot\|}(\widetilde{A_\alpha \tilde{f}})) = \lim_{\alpha \to +\infty}{}_{\|\cdot\|} \widetilde{V_t^{(\mu)}} \widetilde{A_\alpha \tilde{f}}$, where $\lim_{\|\cdot\|}$ stands for the limit in the norm topology of $L^1(X, \widetilde{\mathcal{B}(X)}, \tilde{\mu})$.

Thus, in order to show that $\widetilde{V_t^{(\mu)}}\bar{\tilde{g}} = \bar{\tilde{g}}$, we have to prove that $\lim_{\alpha \to +\infty}{}_{\|\cdot\|} \widetilde{V_t^{(\mu)}} \widetilde{A_\alpha \tilde{f}} = \bar{\tilde{g}}$.

In order to prove the last equality, it is obviously enough to prove that $\left(\left\| \widetilde{A_\alpha \tilde{f}} - \widetilde{V_t^{(\mu)}} \widetilde{A_\alpha \tilde{f}} \right\|_1 \right)_{\alpha \in [0,+\infty)}$ tends to zero as $\alpha \to +\infty$, where $\|\cdot\|_1$ denotes the norm in $L^1(X, \widetilde{\mathcal{B}(X)}, \tilde{\mu})$. Thus, we have to prove that for every $\varepsilon \in \mathbb{R}, \varepsilon > 0$, there exists an $\alpha_\varepsilon \in [0, +\infty)$ such that $\left\| \widetilde{A_\alpha \tilde{f}} - \widetilde{V_t^{(\mu)}} \widetilde{A_\alpha \tilde{f}} \right\|_1 < \varepsilon$ for every $\alpha \geq \alpha_\varepsilon$.

To this end, let $\varepsilon \in \mathbb{R}, \varepsilon > 0$. We will show that $\alpha_\varepsilon = \frac{3t}{\varepsilon} \left\| \bar{\tilde{f}} \right\|_1$ satisfies the required conditions.

Thus, let $\alpha \in \mathbb{R}, \alpha \geq \alpha_\varepsilon$.

Using (c) of Theorem 3.2.19, pp. 113–114 of Dunford and Schwartz [30], the fact that the Lebesgue measure on $\mathbb{R}$ is translation invariant, Lemma 3.2.15, pp. 109–110 of [30], the definition of a DS-integrable function (see Sect. 3.1.3), Lemma 3.2.18, p. 113, of [30], and the fact that $\widetilde{V_s^{(\mu)}}, s \in [0, +\infty)$, are contractions, we obtain that:

$$\left\| \widetilde{A_\alpha \tilde{f}} - \widetilde{V_t^{(\mu)}} \widetilde{A_\alpha \tilde{f}} \right\|_1 = \left\| \frac{1}{\alpha} \left(\text{DS-}\int_0^\alpha \widetilde{V_s^{(\mu)}} \bar{\tilde{f}} \, ds \right) - \widetilde{V_t^{(\mu)}} \left(\frac{1}{\alpha} \left(\text{DS-}\int_0^\alpha \widetilde{V_s^{(\mu)}} \bar{\tilde{f}} \, ds \right) \right) \right\|_1$$

$$= \left\| \frac{1}{\alpha} \left(\text{DS-}\int_0^\alpha \widetilde{V_s^{(\mu)}} \bar{\tilde{f}} \, ds \right) - \frac{1}{\alpha} \left(\text{DS-}\int_0^\alpha \widetilde{V_{s+t}^{(\mu)}} \bar{\tilde{f}} \, ds \right) \right\|_1$$

$$= \left\| \frac{1}{\alpha} \left(\text{DS-}\int_0^\alpha \widetilde{V_s^{(\mu)}} \bar{\tilde{f}} \, ds \right) - \frac{1}{\alpha} \left(\text{DS-}\int_t^{\alpha+t} \widetilde{V_s^{(\mu)}} \bar{\tilde{f}} \, ds \right) \right\|_1$$

$$= \left\| \frac{1}{\alpha} \left(\mathrm{DS\text{-}} \int_0^t \widetilde{V_s^{(\mu)}} \, \bar{\bar{f}} \, \mathrm{d}s \right) + \frac{1}{\alpha} \left(\mathrm{DS\text{-}} \int_t^\alpha \widetilde{V_s^{(\mu)}} \, \bar{\bar{f}} \, \mathrm{d}s \right) \right.$$

$$\left. - \frac{1}{\alpha} \left(\mathrm{DS\text{-}} \int_t^\alpha \widetilde{V_s^{(\mu)}} \, \bar{\bar{f}} \, \mathrm{d}s \right) - \frac{1}{\alpha} \left(\mathrm{DS\text{-}} \int_\alpha^{\alpha+t} \widetilde{V_s^{(\mu)}} \, \bar{\bar{f}} \, \mathrm{d}s \right) \right\|_1$$

$$\leq \frac{1}{\alpha} \left\| \mathrm{DS\text{-}} \int_0^t \widetilde{V_s^{(\mu)}} \, \bar{\bar{f}} \, \mathrm{d}s \right\|_1 + \frac{1}{\alpha} \left\| \mathrm{DS\text{-}} \int_\alpha^{\alpha+t} \widetilde{V_s^{(\mu)}} \, \bar{\bar{f}} \, \mathrm{d}s \right\|_1$$

$$\leq \frac{1}{\alpha} \int_0^t \left\| \widetilde{V_s^{(\mu)}} \, \bar{\bar{f}} \right\|_1 \mathrm{d}s + \frac{1}{\alpha} \int_\alpha^{\alpha+t} \left\| \widetilde{V_s^{(\mu)}} \, \bar{\bar{f}} \right\|_1 \mathrm{d}s$$

$$\leq \frac{1}{\alpha} t \left\| \bar{\bar{f}} \right\|_1 + \frac{1}{\alpha} (\alpha + t - \alpha) \left\| \bar{\bar{f}} \right\|_1 = \frac{2t}{\alpha} \left\| \bar{\bar{f}} \right\|_1 < \varepsilon.$$

$$\square$$

In the next proposition, we discuss the version for transition functions of Lemma 3.3.1 of [143] that we mentioned after Proposition 6.1.5.

Proposition 6.1.9. *The following assertions are equivalent:*

(a) The probability measure μ is ergodic.

(b) There exists a Borel subset B of X such that $\mu(B) = 1$ and such that $\lim_{\alpha \to +\infty} \frac{1}{\alpha} \int_0^\alpha S_t f(x) \, \mathrm{d}t$ exists and is equal to $\langle f, \mu \rangle$ whenever $f \in C_0(X)$ and $x \in B$.

Proof. $(a) \Rightarrow (b)$ Since $C_0(X)$ is a separable Banach space, there exists a sequence $(f_n)_{n \in \mathbb{N}}$ of elements of $C_0(X)$ such that the range $\{ f_n \mid n \in \mathbb{N} \}$ of $(f_n)_{n \in \mathbb{N}}$ is dense in $C_0(X)$.

In view of the definition of the functions f_n^*, $n \in \mathbb{N}$ (the definition is given before Theorem 5.3.3), and of the fact that Γ_c is a set of maximal probability (by Corollary 5.3.2), we obtain that f_n^* is a μ-a.e. limit of $(\frac{1}{\alpha} \int_0^\alpha S_t f_n \, \mathrm{d}t)_{\alpha \in (0,+\infty)}$ as $\alpha \to +\infty$ for every $n \in \mathbb{N}$.

As usual, let $\widetilde{f_n}$ and $\widetilde{f_n^*}$ be the elements of $L^1(X, \mathcal{B}(X), \tilde{\mu})$ defined by f_n and f_n^*, respectively, for every $n \in \mathbb{N}$.

Then, the function $\eta_n : [0, +\infty) \times X \to \mathbb{R}$ defined by

$$\eta_n(\alpha, x) = \begin{cases} 0 & \text{if } \alpha = 0, \\ \frac{1}{\alpha} \int_0^\alpha S_t f_n(x) \, \mathrm{d}t & \text{if } (\alpha, x) \in (0, +\infty) \times X \end{cases}$$

is a standard measurable representation of $(\widetilde{A_\alpha \widetilde{f_n}})_{\alpha \in [0,+\infty)}$ and it converges $\tilde{\mu}$-a.e. to $\widetilde{f_n^*}$, where $\widetilde{f_n^*}$ is the function f_n^*, thought of as an element of $\mathcal{L}^1(X, \mathcal{B}(X), \tilde{\mu})$ for every $n \in \mathbb{N}$. Thus, it follows that $(\widetilde{A_\alpha \widetilde{f_n}})_{\alpha \in [0,+\infty)}$ converges absolutely $\tilde{\mu}$-almost everywhere to $\widetilde{f_n^*}$ as $\alpha \to +\infty$ for every $n \in \mathbb{N}$. Using Proposition 6.1.8, we obtain that $(\widetilde{A_\alpha \bar{\tilde{f}}_n})_{\alpha \in [0,+\infty)}$ also converges in the norm topology of $L^1(X, \widetilde{\mathcal{B}(X)}, \tilde{\mu})$ to $\bar{\tilde{f}}_n^*$ as $\alpha \to +\infty$ and that $V_t^{(\mu)} \bar{\tilde{f}}_n^* = \bar{\tilde{f}}_n^*$ for every $t \in [0, +\infty)$ and for every $n \in \mathbb{N}$.

Now, let $\bar{f}_n$ and $\bar{f}_n^*$ be the elements of $L^1(X, \mathcal{B}(X), \mu)$ defined by f_n and f_n^*, respectively, for every $n \in \mathbb{N}$ (thus, $\bar{f}_n = \varphi_1^{-1}(\bar{\tilde{f}}_n)$ and $\bar{f}_n^* = \varphi_1^{-1}(\bar{\tilde{f}}_n^*)$ for every $n \in \mathbb{N}$, where $\varphi_1 : L^1(X, \mathcal{B}(X), \mu) \to L^1(X, \widetilde{\mathcal{B}(X)}, \tilde{\mu})$ is the standard isometry defined before Proposition 3.1.4).

Since $\widetilde{V_t^{(\mu)}} = \varphi_1 V_t^{(\mu)} \varphi_1^{-1}$ for every $t \in [0, +\infty)$ (see the discussion preceding Proposition 3.2.8), we obtain that $V_t^{(\mu)} \bar{f}_n^* = \bar{f}_n^*$ (because $\widetilde{V_t^{(\mu)}} \bar{\tilde{f}}_n^* = \bar{\tilde{f}}_n^*$, so $\varphi_1^{-1}(\widetilde{V_t^{(\mu)}} \bar{\tilde{f}}_n^*) = \varphi_1^{-1}(\bar{\tilde{f}}_n^*)$ for every $t \in [0, +\infty)$ and $n \in \mathbb{N}$). Thus, $\bar{f}_n^*$, $n \in \mathbb{N}$, are μ-invariant elements of $L^1(X, \mathcal{B}(X), \mu)$.

Since we assume that μ is an ergodic probability measure, using Proposition 6.1.5, we obtain that $\bar{f}_n^*$, $n \in \mathbb{N}$, are constant μ-a.e. Accordingly, the functions f_n^*, $n \in \mathbb{N}$, are constant μ-a.e., as well.

Since Γ_c is a set of maximal probability, it follows that, for every $n \in \mathbb{N}$, there exists a Borel subset B_n of X, $B_n \subseteq \Gamma_c$, such that $\mu(B_n) = 1$ and such that f_n^* is constant on B_n. Using (a) of Corollary 3.2.12, we obtain that $f_n^*(x) = \langle f_n, \mu \rangle$ for every $n \in \mathbb{N}$ and $x \in B_n$.

Set $B = \cap_{n \in \mathbb{N}} B_n$.

It is easy to see that $\mu(B) = 1$, and that $\lim_{\alpha \to +\infty} \frac{1}{\alpha} \int_0^\alpha S_t f_n(x)\, dt$ exists and is equal to $\langle f_n, \mu \rangle$ for every $n \in \mathbb{N}$ and $x \in B$.

In order to complete the proof of the implication we will show that $\lim_{\alpha \to +\infty} \frac{1}{\alpha} \int_0^\alpha S_t f(x)\, dt$ exists and is equal to $\langle f, \mu \rangle$ for every $f \in C_0(X)$ and $x \in B$.

Thus, let $f \in C_0(X)$, $x \in B$, and $\varepsilon \in \mathbb{R}$, $\varepsilon > 0$. We will prove that there exists an $\alpha_\varepsilon \in \mathbb{R}$, $\alpha_\varepsilon > 0$, such that $\left| \frac{1}{\alpha} \int_0^\alpha S_t f(x)\, dt - \langle f, \mu \rangle \right| < \varepsilon$ for every $\alpha \in \mathbb{R}$, $\alpha \geq \alpha_\varepsilon$.

Since $\{ f_n \mid n \in \mathbb{N} \}$ is dense in $C_0(X)$, it follows that there exists an $n_\varepsilon \in \mathbb{N}$ such that $\| f - f_{n_\varepsilon} \| < \frac{\varepsilon}{3}$ (here, and throughout the proof of the implication, $\| \cdot \|$ represents the sup norm on $B_b(X)$).

Since $\lim_{\alpha \to +\infty} \frac{1}{\alpha} \int_0^\alpha S_t f_{n_\varepsilon}(x)\, dt$ exists and is equal to $\langle f_{n_\varepsilon}, \mu \rangle$, it follows that there exists an $\alpha_\varepsilon \in \mathbb{R}$, $\alpha_\varepsilon > 0$ such that $\left| \frac{1}{\alpha} \int_0^\alpha S_t f_{n_\varepsilon}(x)\, dt - \langle f_{n_\varepsilon}, \mu \rangle \right| < \frac{\varepsilon}{3}$ for every $\alpha \in \mathbb{R}$, $\alpha \geq \alpha_\varepsilon$.

We obtain that

$$\left| \frac{1}{\alpha} \int_0^\alpha S_t f(x)\, dt - \langle f, \mu \rangle \right| \leq \left| \frac{1}{\alpha} \int_0^\alpha S_t f(x)\, dt - \frac{1}{\alpha} \int_0^\alpha S_t f_{n_\varepsilon}(x)\, dt \right|$$

$$+ \left| \frac{1}{\alpha} \int_0^\alpha S_t f_{n_\varepsilon}(x)\, dt - \langle f_{n_\varepsilon}, \mu \rangle \right| + \left| \langle f_{n_\varepsilon}, \mu \rangle - \langle f, \mu \rangle \right|$$

$$< \frac{1}{\alpha} \int_0^\alpha |S_t f(x) - S_t f_{n_\varepsilon}(x)|\, dt + \frac{\varepsilon}{3} + \left| \int_X (f_{n_\varepsilon}(x) - f(x))\, d\mu(x) \right|$$

$$\leq \|S_t f - S_t f_{n_\varepsilon}\| + \frac{\varepsilon}{3} + \int_X |f_{n_\varepsilon}(x) - f(x)| \, d\mu(x)$$

$$\leq \|f - f_{n_\varepsilon}\| + \frac{\varepsilon}{3} + \|f_{n_\varepsilon} - f\| < \frac{\varepsilon}{3} + \frac{\varepsilon}{3} + \frac{\varepsilon}{3} = \varepsilon$$

for every $\alpha \in \mathbb{R}$, $\alpha \geq \alpha_\varepsilon$.

Thus, the implication holds true.

$(b) \Rightarrow (a)$ Assume that (b) holds true but that the probability measure μ is not ergodic.

Then there exist two nonzero invariant measures $\nu_1 \in \mathcal{M}(X)$ and $\nu_2 \in \mathcal{M}(X)$ such that $\mu = \nu_1 + \nu_2$, and such that ν_1 and ν_2 are concentrated on two disjoint Borel subsets A_1 and A_2 of X, respectively.

Let B be the Borel subset of X whose existence is assured by (b). Since μ is concentrated on B, it follows that we may choose A_1 and A_2 such that $A_1 \subseteq B$ and $A_2 \subseteq B$.

Set $\mu_1 = \frac{\nu_1}{\|\nu_1\|}$ and $\mu_2 = \frac{\nu_2}{\|\nu_2\|}$. Then μ_1 and μ_2 are invariant probability measures, and $\mu_1 \neq \mu_2$ because μ_1 and μ_2 are concentrated on A_1 and A_2. Thus, there exists an $f \in C_0(X)$ such that $\langle f, \mu_1 \rangle \neq \langle f, \mu_2 \rangle$.

Since we assume that (b) holds true, and since B is a subset of Γ_c, it follows that $f^*(x) = \langle f, \mu \rangle$ for every $x \in B$.

Using (a) of Corollary 3.2.12, and the fact that μ, μ_1 and μ_2 are concentrated on B, we obtain that $\langle f, \mu_1 \rangle = \int_B f^*(x) \, d\mu_1(x) = \langle f, \mu \rangle$ and $\langle f, \mu_2 \rangle = \int_B f^*(x) \, d\mu_2(x) = \langle f, \mu \rangle$. Thus, we have obtained a contradiction which stems from our assumption that μ is not ergodic. $\qquad\square$

As usual, given a real vector space V, a nonempty convex subset K of V, and $x \in K$, we say that x is an *extreme point of* K if the following condition is satisfied: if $y \in K$, $z \in K$ and $\lambda \in \mathbb{R}$, $0 < \lambda < 1$, are such that $x = \lambda y + (1 - \lambda)z$ then $x = y = z$.

Note that the set of all invariant probabilities for $(P_t)_{t \in [0,+\infty)}$ is a convex subset of $\mathcal{M}(X)$. Our goal now is to show that an invariant probability measure μ is ergodic if and only if μ is an extreme point of the set of all invariant probabilities for $(P_t)_{t \in [0,+\infty)}$. To this end, we need some preparation.

Let E be a Banach lattice.

As usual, we say that a subset A of E is *bounded from above* if there exists a $u \in E$ such that $v \leq u$ for every $v \in A$.

The Banach lattice E is said to be *order complete* (or *Dedekind complete*) if $\sup A$ exists in E for every subset A of E that is bounded from above.

Example 6.1.10. $\mathcal{M}(X)$, and, in general, every AL-space is order complete. By contrast, $C([0, 1])$, the Banach lattice of all real-valued continuous functions defined on $[0, 1]$, is not order complete. See Aliprantis and Burkinshaw [3] for details.   ∎

Proposition 6.1.11. *Assume that the Banach lattice E is order complete and let $u \in E$, $u \geq 0$, and $v \in E$, $v \geq 0$. Then:*

(i) $\sup_{n\in\mathbb{N}}(u \wedge (nv))$ *exists in* E.

(ii) If we set $u_v = \sup_{n\in\mathbb{N}}(u \wedge (nv))$, *then* $0 \le u_v \le u$ *and* $(u - u_v) \wedge u_v = 0$.

For a proof, see Aliprantis and Burkinshaw [3].

We now return to our standard setting of this section; that is, we assume given a transition function $(P_t)_{t\in\mathbb{T}}$ that has invariant probabilities, and we let μ be such an invariant probability for $(P_t)_{t\in\mathbb{T}}$.

Lemma 6.1.12. *Assume that there exist two distinct invariant probabilities* v_1 *and* v_2 *for* $(P_t)_{t\in\mathbb{T}}$, *and* $\lambda_1 \in \mathbb{R}$, $0 < \lambda_1 < 1$, *such that* $\mu = \lambda_1 v_1 + (1 - \lambda_1)v_2$. *Then there exist two invariant probabilities* μ_1 *and* μ_2 *for* $(P_t)_{t\in\mathbb{T}}$ *and* $\lambda \in \mathbb{R}$, $0 < \lambda < 1$, *such that* $\mu_1 \wedge \mu_2 = 0$ *and* $\mu = \lambda\mu_1 + (1 - \lambda)\mu_2$.

Proof. Assume that $\mu = \lambda_1 v_1 + (1 - \lambda_1)v_2$ for two distinct invariant probabilities v_1 and v_2 and for some $\lambda_1 \in \mathbb{R}, 0 < \lambda_1 < 1$.

We will prove the lemma in several steps.

Step 1.   At this step, we prove that $(v_1 - v_2)^+ \ne 0$ and $(v_1 - v_2)^- \ne 0$.

To this end, assume that $(v_1 - v_2)^+ = 0$. Then $v_1 - v_2 = -(v_1 - v_2)^- \le 0$, so $v_1 \le v_2$; that is, $v_1(A) \le v_2(A)$ for every $A \in \mathcal{B}(X)$. We have obtained a contradiction because v_1 and v_2 are distinct probability measures, so, if $v_1(B) \le v_2(B)$ and $v_1(X \setminus B) \le v_2(X \setminus B)$, then we obtain that $v_1(B) = v_2(B)$ for all $B \in \mathcal{B}(X)$.

In a similar manner it can be shown that $(v_1 - v_2)^- \ne 0$.

Step 2.   At this step we first note that since $\mathcal{M}(X)$ is an order complete Banach lattice, using (i) of Proposition 6.1.11, we obtain that $\sup_{n\in\mathbb{N}}(\mu \wedge (n(v_1 - v_2)^+))$ exists in $\mathcal{M}(X)$.

Set $\eta_1 = \sup_{n\in\mathbb{N}}(\mu \wedge (n(v_1 - v_2)^+))$ and $\eta_2 = \mu - \eta_1$.
We will now prove that $\eta_1 \ne 0$ and $\eta_2 \ne 0$.
Since $(v_1 - v_2)^+ \ne 0$ (as shown at *Step 1*), there exists an $A \in \mathcal{B}(X)$ such that $(v_1 - v_2)^+(A) > 0$.
Since $\mu \ge \lambda_1 v_1 \ge \lambda_1(v_1 - v_2)^+$, and since $n \ge \lambda_1$ for every $n \in \mathbb{N}$, it follows that

$$\eta_1(A) = (\sup_{n\in\mathbb{N}}(\mu\wedge(n(v_1-v_2)^+))) \ge \mu\wedge(\lambda_1(v_1-v_2)^+)(A) \ge \lambda_1(v_1-v_2)^+(A) > 0;$$

hence $\eta_1 \ne 0$.

In order to prove that $\eta_2 \ne 0$, we first note that $(v_1 - v_2)^- \wedge \eta_1 = 0$ because $(v_1 - v_2)^- \wedge (\vee_{n\in\mathbb{N}}(\mu \wedge (n(v_1 - v_2)^+)))$.

In order to prove that $\eta_2 \ne 0$, we first note that

$$(v_1 - v_2)^- \wedge \eta_1 = 0 \tag{6.1.2}$$

because

$$(v_1 - v_2)^- \wedge (\vee_{n \in \mathbb{N}} (\mu \wedge (n(v_1 - v_2)^+))) = \vee_{n \in \mathbb{N}} ((v_1 - v_2)^- \wedge \mu \wedge (n(v_1 - v_2)^+)).$$

Using (*ii*) of Proposition 6.1.11 and the equality (6.1.2) we obtain that

$$\begin{aligned}
(\lambda_1 (v_1 - v_2)^-) \wedge (\lambda_1 v_1) &\le (v_1 - v_2)^- \wedge \mu \\
&= (v_1 - v_2)^- \wedge (\mu - \eta_1 + \eta_1) \\
&= (v_1 - v_2)^- \wedge ((\mu - \eta_1) \vee \eta_1) \\
&= ((v_1 - v_2)^- \wedge (\mu - \eta_1)) \vee ((v_1 - v_2)^- \wedge \eta_1) \\
&= (v_1 - v_2)^- \wedge (\mu - \eta_1).
\end{aligned}$$

Since $(\lambda_1 (v_1 - v_2)^-) \wedge (\lambda_1 v_1) \ge 0$ and since $(\lambda_1 (v_1 - v_2)^-) \wedge (\lambda_1 v_1) \ne 0$ (because $\lambda_1 v_1 \ge \lambda_1 (v_1 - v_2)^- \ge 0$ and $(v_1 - v_2)^- \ne 0$ by *Step 1*), it follows that $\eta_2 = \mu - \eta_1 \ne 0$.

Step 3. At this step, we prove that η_1 and η_2 are invariant measures for $(P_t)_{t \in \mathbb{T}}$, and then we complete the proof of the lemma.

We first prove that η_1 is an invariant measure for $(P_t)_{t \in \mathbb{T}}$; that is, we prove that $T_t \eta_1 = \eta_1$ for every $t \in \mathbb{T}$.

Thus, let $t \in \mathbb{T}$.

Since T_t is a positive operator and since v_1 and v_2 are invariant probability measures, it follows that $T_t ((v_1 - v_2)^+) \ge (T_t (v_1 - v_2)) \vee 0 = (v_1 - v_2)^+$. Using the fact that T_t is a Markov operator on $\mathcal{M}(X)$, we obtain that $T_t ((v_1 - v_2)^+) = (v_1 - v_2)^+$.

Next, using the fact that μ is an invariant probability measure for $(P_s)_{s \in \mathbb{T}}$ and the fact that T_t is a positive operator, we obtain that

$$T_t (\mu \wedge (n(v_1 - v_2)^+)) \le (T_t \mu) \wedge T_t (n(v_1 - v_2)^+) = \mu \wedge (n(v_1 - v_2)^+)$$

for every $n \in \mathbb{N}$, so, since T_t is a Markov operator, it follows that $T_t (\mu \wedge (n(v_1 - v_2)^+)) = \mu \wedge (n(v_1 - v_2)^+)$ for every $n \in \mathbb{N}$.

Finally, using the positivity of T_t, we obtain that $T_t (\eta_1) \ge \sup_{n \in \mathbb{N}} T_t (\mu \wedge (n(v_1 - v_2)^+)) = \eta_1$, and since T_t is a Markov operator, it follows that $T_t \eta_1 = \eta_1$.

Clearly, in this case $T_t (\eta_2) = T_t (\mu - \eta_1) = \eta_2$.

Since $T_t \eta_i = \eta_i$ for every $t \in \mathbb{T}$ and $i = 1, 2$, it follows that η_1 and η_2 are invariant measures for $(P_s)_{s \in \mathbb{T}}$.

Since $\mu = \eta_1 + \eta_2$ and both η_1 and η_2 are nonzero measures, it follows that $\mu_1 = \frac{\eta_1}{\|\eta_1\|}$ and $\mu_2 = \frac{\eta_2}{\|\eta_2\|}$ are well-defined probability measures (in the sense that $\|\eta_i\| \ne 0$ for every $i = 1, 2$).

Since $\eta_1 \wedge \eta_2 = 0$ (by (*ii*) of Proposition 6.1.11), it follows that $\mu_1 \wedge \mu_2 = 0$.

Finally, since $\mu = \|\eta_1\| \eta_1 + \|\eta_2\| \eta_2$, if we set $\lambda = \|\eta_1\|$, we obtain that $\mu = \lambda \eta_1 + (1 - \lambda) \eta_2$. $\qquad \square$

We are now in a position to prove the fact that we mentioned before defining the order complete Banach lattices; that is, we will prove in the next proposition that the ergodic invariant probability measures for $(P_t)_{t\in\mathbb{T}}$ are precisely the extreme points of the (convex) set of all invariant probability measures for $(P_t)_{t\in\mathbb{T}}$.

Proposition 6.1.13. *The following assertions are equivalent:*

(a) The probability measure μ is ergodic.
(b) The probability measure μ is an extreme point of the set of all invariant probability measures for $(P_t)_{t\in\mathbb{T}}$.

Proof. $(a) \Rightarrow (b)$ If we assume that μ is not an extreme point of the set of all invariant probability measures for $(P_t)_{t\in\mathbb{T}}$, then by Lemma 6.1.12, there exist probability measures μ_1 and μ_2 such that $\mu_1 \wedge \mu_2 = 0$, and there exists a $\lambda \in \mathbb{R}$, $0 < \lambda < 1$, such that $\mu = \lambda\mu_1 + (1 - \lambda)\mu_2$. But the fact that we can write μ as the sum of the measures $\lambda\mu_1$ and $(1 - \lambda)\mu_2$ is in contradiction with our assumption that μ is ergodic.

$(b) \Rightarrow (a)$ If we assume that μ is not an ergodic measure, then $\mu = \nu_1 + \nu_2$ for some nonzero invariant measures ν_1 and ν_2 such that $\nu_1 \wedge \nu_2 = 0$.

But then, if we set $\mu_i = \frac{\nu_i}{\|\nu_i\|}$, $i = 1, 2$, then μ_1 and μ_2 are invariant probability measures and $\mu = \|\nu_1\|\,\mu_1 + (1 - \|\nu_1\|)\mu_2$; therefore, it follows that μ is not an extreme point of the set of all invariant probability measures. The contradiction stems, obviously, from our assumption that μ is not ergodic. $\qquad\square$

We will now summarize in the next theorem the various characterizations of the ergodic invariant probability measures obtained in this section.

Theorem 6.1.14. *The following assertions are equivalent:*

(a) μ is an ergodic invariant probability measure.
(b) $\mathcal{B}_\mu(X)$ is trivial.
(c) $\mathcal{B}_{\tilde{\mu}}(X)$ is trivial.
(d) Every μ-invariant element of $L^1(X, \mathcal{B}(X), \mu)$ is constant μ-a.e.
(e) Every μ-invariant function in $B_b(X)$ is constant μ-a.e.
(f) There exists a subset B of X, $B \in \mathcal{B}(X)$, such that $\mu(B) = 1$ and such that $\lim_{\alpha\to+\infty} \frac{1}{\alpha} \int_0^\alpha S_t f(x)\, dt$ exists and is equal to $\langle f, \mu\rangle$ whenever $f \in C_0(X)$ and $x \in B$.
(g) The probability measure μ is an extreme point of the set of all invariant probability measures for $(P_t)_{t\in\mathbb{T}}$.

Except for (f) (which is a version for transition functions of a similar statement for transition probabilities that appears in Lemma 3.3.1 of [143]), versions of all the other statements can be found as definitions of ergodic invariant probability measures for flows or semiflows in various places in the literature. A notable exception is on p. 417 in Stroock's monograph [119], where an assertion somewhat similar to (d) is used to define the ergodic invariant measures for transition functions even when these measures are σ-finite.

6.2   The Invariant Ergodic Probability Measures as Standard Elementary Measures

As pointed out in the abstract for the present chapter, in this section we study a subset of Γ_{cpi} that we denote by Γ_{cpie}, and we obtain versions for transition functions of the results obtained in Section 6 of [146] for transition probabilities. Thus, we prove that Γ_{cpie} is $\mathcal{B}(X)$-measurable and a set of maximal probability, and we study the delicate relationship between the set of all ergodic invariant probability measures for the given transition function and the structure of the set Γ_{cpie}.

In this section, we assume given a transition function $(P_t)_{t \in \mathbb{T}}$ defined on a locally compact separable metric space (X, d), we assume that $(P_t)_{t \in \mathbb{T}}$ satisfies the s.m.a. and is pointwise continuous, we let $((S_t, T_t))_{t \in \mathbb{T}}$ be the family of Markov pairs defined by $(P_t)_{t \in \mathbb{T}}$, and we use the terminology and notations introduced earlier if they make sense in this setting.

Set

$$
\Gamma_{\text{cpie}} = \left\{ x \in \Gamma_{\text{cpi}} \;\middle|\; \begin{array}{l} \int_{\Gamma_{\text{cpi}}} (f^*(y) - f^*(x))^2 \, d\varepsilon_x(y) = 0 \\ \text{for every } f \in C_0(X) \end{array} \right\}.
$$

Our goals in this section are:

(a) To prove that Γ_{cpie} is a measurable subset of X;

(b) To prove that Γ_{cpie} is a set of maximal probability;

(c) To prove that each ε_x, $x \in \Gamma_{\text{cpie}}$, is an ergodic invariant probability measure, and that conversely, each ergodic invariant probability measure μ is of the form ε_x for some $x \in \Gamma_{\text{cpie}}$;

(d) If μ is an ergodic invariant probability measure, then the set $A_\mu = \{x \in \Gamma_{\text{cpie}} \mid \mu = \varepsilon_x\}$ is $\mathcal{B}(X)$-measurable and μ is concentrated on A_μ (note that using (c), we obtain that the set A_μ is nonempty).

In order to prove that Γ_{cpie} is a $\mathcal{B}(X)$-measurable subset of X, we need the following lemma:

Lemma 6.2.1. *Let $(f_l)_{l \in \mathbb{N}}$ be a sequence of elements of $C_0(X)$ such that the range $\{f_l \mid l \in \mathbb{N}\}$ of $(f_l)_{l \in \mathbb{N}}$ is dense in $C_0(X)$ (the existence of such a sequence $(f_l)_{l \in \mathbb{N}}$ is assured by the fact that $C_0(X)$ is a separable Banach space). Also, let $x \in \Gamma_{\text{cpi}}$. Then $x \in \Gamma_{\text{cpie}}$ if and only if*

$$
\int_{\Gamma_{\text{cpi}}} (f_l^*(y) - f_l^*(x))^2 \, d\varepsilon_x(y) = 0 \tag{6.2.1}
$$

for every $l \in \mathbb{N}$.

Proof. Clearly, if $x \in \Gamma_{\text{cpie}}$, then the equality (6.2.1) holds true for every $l \in \mathbb{N}$.

Thus, in order to complete the proof of the lemma, we only have to prove that if $x \in \Gamma_{\mathrm{cpi}}$ is such that the equality (6.2.1) holds true for every $l \in \mathbb{N}$, then

$$\int_{\Gamma_{\mathrm{cpi}}} (f^*(y) - f^*(x))^2 \, \mathrm{d}\varepsilon_x(y) = 0 \tag{6.2.2}$$

for every $f \in C_0(X)$.

To this end, let $x \in \Gamma_{\mathrm{cpi}}$ and let $f \in C_0(X)$.

Note that the equality (6.2.2) is obviously satisfied if $f = 0$. Thus, assume that $f \neq 0$ and set $M = \|f\|$.

Since the left-hand side of the equality (6.2.2) is a nonnegative number, in order to show that (6.2.2) holds true for f it is enough to prove that

$$\int_{\Gamma_{\mathrm{cpi}}} (f^*(y) - f^*(x))^2 \, \mathrm{d}\varepsilon_x(y) \leq \varepsilon$$

for every $\varepsilon \in \mathbb{R}, \varepsilon > 0$.

Thus, let $\varepsilon \in \mathbb{R}, \varepsilon > 0$.

Since $\{f_l \mid l \in \mathbb{N}\}$ is dense in $C_0(X)$, it follows that there exists an $l_0 \in \mathbb{N}$ such that $\|f - f_0\| < \min\{\frac{\varepsilon}{4(2M+1)}, 1\}$.

Set $\eta = \min\{\frac{\varepsilon}{4(2M+1)}, 1\}$.

Since $S_t, t \in [0, +\infty)$, are (positive) contractions of $B_b(X)$, it follows that

$$\left| \frac{1}{\alpha} \int_0^\alpha S_t f(z) \, \mathrm{d}t - \frac{1}{\alpha} \int_0^\alpha S_t f_{l_0}(z) \, \mathrm{d}t \right| < \eta \tag{6.2.3}$$

because

$$\left| \frac{1}{\alpha} \int_0^\alpha S_t f(z) \, \mathrm{d}t - \frac{1}{\alpha} \int_0^\alpha S_t f_{l_0}(z) \, \mathrm{d}t \right| \leq \frac{1}{\alpha} \int_0^\alpha |S_t f(z) - S_t f_{l_0}(z)| \, \mathrm{d}t$$

$$\leq \frac{1}{\alpha} \alpha \sup_{t \in [0,\alpha]} \|S_t f - S_t f_{l_0}\| \leq \|f - f_{l_0}\| < \eta$$

for every $\alpha \in (0, +\infty)$ and $z \in X$.

In particular, (6.2.3) holds true for $z \in \Gamma_{\mathrm{cpi}}$. Taking into consideration that $\lim_{\alpha \to +\infty} \frac{1}{\alpha} \int_0^\alpha S_t f(z) \, \mathrm{d}t$ and $\lim_{\alpha \to +\infty} \frac{1}{\alpha} \int_0^\alpha S_t f_{l_0}(z) \, \mathrm{d}t$ exist and are equal to $f^*(z)$ and $f_{l_0}^*(z)$, respectively, whenever $z \in \Gamma_{\mathrm{cpi}}$, using (6.2.3), we obtain that

$$\left| f^*(z) - f_{l_0}^*(z) \right| \leq \eta \tag{6.2.4}$$

for every $z \in \Gamma_{\mathrm{cpi}}$.

We now note that

$$|f^*(z)| \leq M \tag{6.2.5}$$

and

$$\left| f_{l_0}^*(z) \right| \le M + 1 \qquad (6.2.6)$$

because

$$|f^*(z)| = \left| \lim_{\alpha \to +\infty} \frac{1}{\alpha} \int_0^\alpha S_t f(z) \, dt \right| \le \lim_{\alpha \to +\infty} \frac{1}{\alpha} \int_0^\alpha |S_t f(z)| \, dt \le \|f\| = M$$

and

$$\left| f_{l_0}^*(z) \right| = \left| \lim_{\alpha \to +\infty} \frac{1}{\alpha} \int_0^\alpha S_t f_{l_0}(z) \, dt \right| \le \| f_{l_0} \| \le \| f_{l_0} - f \| + \| f \| < \eta + M \le 1 + M,$$

respectively, for every $z \in \Gamma_{\mathrm{cpi}}$.

Using the inequalities (6.2.4)–(6.2.6), and the fact that (by Theorem 5.3.7) Γ_{cpi} is a set of maximal probability, we obtain that

$$\int_{\Gamma_{\mathrm{cpi}}} (f^*(y) - f^*(x))^2 \, d\varepsilon_x(y)$$

$$= \left| \int_{\Gamma_{\mathrm{cpi}}} (f^*(y) - f^*(x))^2 \, d\varepsilon_x(y) - \int_{\Gamma_{\mathrm{cpi}}} (f_{l_0}^*(y) - f_{l_0}^*(x))^2 \, d\varepsilon_x(y) \right|$$

$$\le \int_{\Gamma_{\mathrm{cpi}}} \left| (f^*(y) - f^*(x))^2 - (f_{l_0}^*(y) - f_{l_0}^*(x))^2 \right| \, d\varepsilon_x(y)$$

$$\le \int_{\Gamma_{\mathrm{cpi}}} \left| (f^*(y))^2 - (f_{l_0}^*(y))^2 \right| \, d\varepsilon_x(y) + \int_{\Gamma_{\mathrm{cpi}}} \left| (f^*(x))^2 - (f_{l_0}^*(x))^2 \right| \, d\varepsilon_x(y)$$

$$+ 2 \int_{\Gamma_{\mathrm{cpi}}} \left| f^*(y) f^*(x) - f_{l_0}^*(y) f_{l_0}^*(x) \right| \, d\varepsilon_x(y)$$

$$= \int_{\Gamma_{\mathrm{cpi}}} \left| f^*(y) - f_{l_0}^*(y) \right| \left| f^*(y) + f_{l_0}^*(y) \right| \, d\varepsilon_x(y) + \left| (f^*(x))^2 - (f_{l_0}^*(x))^2 \right| \varepsilon_x(\Gamma_{\mathrm{cpi}})$$

$$+ 2 \int_{\Gamma_{\mathrm{cpi}}} \left| f^*(y) f^*(x) - f_{l_0}^*(y) f^*(x) + f_{l_0}^*(y) f^*(x) - f_{l_0}^*(y) f_{l_0}^*(x) \right| \, d\varepsilon_x(y)$$

$$\le \eta \int_{\Gamma_{\mathrm{cpi}}} \left| f^*(y) + f_{l_0}^*(y) \right| \, d\varepsilon_x(y) + \left| f^*(x) - f_{l_0}^*(x) \right| \left| f^*(x) + f_{l_0}^*(x) \right|$$

$$+ 2 \int_{\Gamma_{\mathrm{cpi}}} \left| f^*(y) - f_{l_0}^*(y) \right| \left| f^*(x) \right| + \left| f_{l_0}^*(y) \right| \left| f^*(x) - f_{l_0}^*(x) \right| \, d\varepsilon_x(y)$$

$$\leq \eta \int_{\Gamma_{\mathrm{cpi}}} (|f^*(y)| + |f_{l_0}^*(y)|)\, d\varepsilon_x(y) + 2(\eta M + (M+1)\eta)\varepsilon_x(\Gamma_{\mathrm{cpi}})$$

$$+\eta(|f^*(x)| + |f_{l_0}^*(x)|) \leq \eta(M + M + 1)\varepsilon(\Gamma_{\mathrm{cpi}}) + 2\eta(2M+1) + \eta(2M+1)$$

$$= 4\eta(2M+1) \leq 4\frac{\varepsilon}{4(2M+1)}(2M+1) = \varepsilon.$$

We have therefore proved that $x \in \Gamma_{\mathrm{cpie}}$. $\square$

Using Lemma 6.2.1, we will now prove in the next theorem that Γ_{cpie} is a measurable subset of X.

Theorem 6.2.2. *The set* Γ_{cpie} *belong to* $\mathcal{B}(X)$.

Proof. Let $(f_l)_{l \in \mathbb{N}}$ be the sequence of elements of $C_0(X)$ defined in Lemma 6.2.1.
For every $l \in \mathbb{N}$, let $g_l : \Gamma_{\mathrm{cpi}} \to \mathbb{R}$ be defined by $g_l(x) = \int_{\Gamma_{\mathrm{cpi}}} (f_l^*(y) - f_l^*(x))^2\, d\varepsilon_x(y)$ for every $x \in \Gamma_{\mathrm{cpi}}$.
Using Lemma 6.2.1, we obtain that

$$\Gamma_{\mathrm{cpie}} = \left\{ x \in \Gamma_{\mathrm{cpi}} \,\middle|\, \begin{array}{l} \int_{\Gamma_{\mathrm{cpi}}} (f_l^*(y) - f_l^*(x))^2\, d\varepsilon_x(y) = 0 \\ \text{for every } l \in \mathbb{N} \end{array} \right\}$$

$$= \left\{ x \in \Gamma_{\mathrm{cpi}} \,\middle|\, \begin{array}{l} g_l(x) = 0 \\ \text{for every } l \in \mathbb{N} \end{array} \right\}$$

$$= \bigcap_{l \in \mathbb{N}} \{ x \in \Gamma_{\mathrm{cpi}} \mid g_l(x) = 0 \}.$$

Thus, in order to prove the theorem it is enough to show that g_l is a measurable function for every $l \in \mathbb{N}$.
Using the fact that Γ_{cpi} is a set of maximal probability, and the definition of the functions $h^\square$, $h \in B_b(X)$, (the definition appears before Lemma 5.2.5) we obtain that

$$g_l(x) = \int (f_l^*(y) - f_l^*(x))^2\, d\varepsilon_x(y)$$

$$= \int (f_l^*)^2(y)\, d\varepsilon_x(y) - 2f_l^*(x) \int f_l^*(y)\, d\varepsilon_x(y) + (f_l^*)^2(x)$$

$$= ((f_l^*)^2)^\square(x) - 2f_l^*(x)(f_l^*)^\square(x) + (f_l^*)^2(x)$$

for every $x \in \Gamma_{\mathrm{cpi}}$ and every $l \in \mathbb{N}$.
Now, using the fact that $h^* \in B_b(X)$ whenever $h \in B_b(X)$ (see the discussion preceding Theorem 5.3.3) and using Lemma 5.2.5, we obtain that the functions $((f_l^*)^2)^\square$, $f_l^*(f_l^*)^\square$, and $(f_l^*)^2$ are all measurable. Since Γ_{cpi} is a measurable subset of X, it follows that g_l is a measurable function for every $l \in \mathbb{N}$. $\square$

Our next goal is to prove that Γ_{cpie} is a set of maximal probability. To this end, we need the following lemma:

Lemma 6.2.3. *Assume that $(P_t)_{t\in\mathbb{T}}$ has invariant probability measures, and let μ be such an invariant probability. Then $\Gamma_{\mathrm{cpi}} \neq \emptyset$ and*

$$\int_{\Gamma_{\mathrm{cpi}}} \left(\int_{\Gamma_{\mathrm{cpi}}} (f^*(y) - f^*(x))^2 \, d\varepsilon_x(y) \right) d\mu(x) = 0$$

for every $f \in C_0(X)$.

Proof. Note first that, since we assume that $(P_t)_{t\in\mathbb{T}}$ has invariant probability measures, and since, by Theorem 5.3.7, Γ_{cpi} is a set of maximal probability, we obtain that Γ_{cpi} is nonempty.

Let $f \in C_0(X)$ and set

$$\mathcal{X} = \int_{\Gamma_{\mathrm{cpi}}} \left(\int_{\Gamma_{\mathrm{cpi}}} (f^*(y))^2 \, d\varepsilon_x(y) \right) d\mu(x),$$

$$\mathcal{Y} = \int_{\Gamma_{\mathrm{cpi}}} \left(\int_{\Gamma_{\mathrm{cpi}}} f^*(y) f^*(x) \, d\varepsilon_x(y) \right) d\mu(x),$$

and

$$\mathcal{Z} = \int_{\Gamma_{\mathrm{cpi}}} \left(\int_{\Gamma_{\mathrm{cpi}}} (f^*(x))^2 \, d\varepsilon_x(y) \right) d\mu(x).$$

Clearly, the proof of the lemma will be completed if we show that $\mathcal{X} - 2\mathcal{Y} + \mathcal{Z} = 0$.

To this end, we first simplify the expressions of $\mathcal{X}$, $\mathcal{Y}$ and $\mathcal{Z}$.

Using the fact that Γ_{cpi} is a set of maximal probability, the fact that ε_x, $x \in \Gamma_{\mathrm{cpi}}$ and μ are invariant probability measures, using Proposition 5.3.6, and using the fact that Γ_c is a set of maximal probability together with (a) of Corollary 3.2.12, we obtain that

$$\mathcal{X} = \int_{\Gamma_{\mathrm{cpi}}} \left(\int_X (f^*(y))^2 \, d\varepsilon_x(y) \right) d\mu(x) = \int_{\Gamma_{\mathrm{cpi}}} \langle (f^*)^2, \varepsilon_x \rangle \, d\mu(x)$$

$$= \int_{\Gamma_{\mathrm{cpi}}} ((f^*)^2)^{\square}(x) \, d\mu(x) = \int_X ((f^*)^2)^{\square}(x) \, d\mu(x)$$

$$= \int_X ((f^*)^2)^*(x) \, d\mu(x) = \int_X (f^*)^2(x) \, d\mu(x).$$

In order to simplify $\mathcal{Y}$, we use the fact that Γ_{cpi} is a set of maximal probability, the fact that ε_x, $x \in \Gamma_{\text{cpi}}$, are invariant probability measures, the definition of f^* together with the fact that Γ_c is a set of maximal probability and together with (a) of Corollary 3.2.12, the definition of $f^\square$ and Proposition 5.3.6 in order to obtain that

$$\mathcal{Y} = \int_{\Gamma_{\text{cpi}}} f^*(x) \left(\int_{\Gamma_{\text{cpi}}} f^*(y)\, \mathrm{d}\varepsilon_x(y) \right) \mathrm{d}\mu(x)$$

$$= \int_{\Gamma_{\text{cpi}}} f^*(x) \left(\int_X f^*(y)\, \mathrm{d}\varepsilon_x(y) \right) \mathrm{d}\mu(x) = \int_{\Gamma_{\text{cpi}}} f^*(x) \left(\int_X f(y)\, \mathrm{d}\varepsilon_x(y) \right) \mathrm{d}\mu(x)$$

$$= \int_{\Gamma_{\text{cpi}}} f^*(x) f^\square(x)\, \mathrm{d}\mu(x) = \int_{\Gamma_{\text{cpi}}} (f^*)^2(x)\, \mathrm{d}\mu(x) = \int_X (f^*)^2(x)\, \mathrm{d}\mu(x).$$

We now turn our attention to $\mathcal{Z}$. Using the fact that Γ_{cpi} is a set of maximal probability, and the fact that ε_x, $x \in \Gamma_{\text{cpi}}$ and μ are invariant probabilities, we obtain that

$$\mathcal{Z} = \int_{\Gamma_{\text{cpi}}} (f^*(x))^2 \left(\int_{\Gamma_{\text{cpi}}} \mathrm{d}\varepsilon_x(y) \right) \mathrm{d}\mu(x)$$

$$= \int_{\Gamma_{\text{cpi}}} (f^*(x))^2\, \mathrm{d}\mu(x) = \int_X (f^*(x))^2\, \mathrm{d}\mu(x).$$

It follows that

$$\mathcal{X} - 2\mathcal{Y} + \mathcal{Z} = \int_X (f^*(x))^2\, \mathrm{d}\mu(x) - 2\int_X (f^*(x))^2\, \mathrm{d}\mu(x) + \int_X (f^*(x))^2\, \mathrm{d}\mu(x) = 0.$$

$\square$

We now use the above lemma in order to prove that Γ_{cpie} is a set of maximal probability.

Theorem 6.2.4. *The set Γ_{cpie} is a set of maximal probability.*

Proof. In view of the definition of a set of maximal probability, we have to prove that if $(P_t)_{t \in \mathbb{T}}$ has invariant probability measures, and if μ is such an invariant probability, then $\mu(\Gamma_{\text{cpie}}) = 1$.

To this end, assume that $(P_t)_{t \in \mathbb{T}}$ has invariant probabilities and let μ be such an invariant probability.

Also, assume that $\mu(\Gamma_{\text{cpi}}) < 1$. Since Γ_{cpi} is a set of maximal probability, and since $\Gamma_{\text{cpie}} \subseteq \Gamma_{\text{cpi}}$, it follows that $\mu(\Gamma_{\text{cpi}} \setminus \Gamma_{\text{cpie}}) > 0$.

Let $(f_l)_{l \in \mathbb{N}}$ be the sequence of elements of $C_0(X)$ considered in Theorem 6.2.2, and let $(g_l)_{l \in \mathbb{N}}$ be the sequence of functions defined in the theorem, and recall that (using Lemma 6.2.1) we showed in Theorem 6.2.2 that Γ_{cpie} is the set of all $x \in \Gamma_{\mathrm{cpi}}$ such that $g_l(x) = 0$ for every $l \in \mathbb{N}$. Since $g_l(x) \geq 0$ for every $x \in \Gamma_{\mathrm{cpi}}$ and $l \in \mathbb{N}$, it follows that

$$\Gamma_{\mathrm{cpi}} \setminus \Gamma_{\mathrm{cpie}} = \bigcup_{l=1}^{\infty} \{x \in \Gamma_{\mathrm{cpi}} \mid g_l(x) > 0\}.$$

Thus, it follows that $\mu(\{x \in \Gamma_{\mathrm{cpi}} \mid g_{l_0}(x) > 0\}) > 0$ for some $l_0 \in \mathbb{N}$. But in this case

$$\int_{\Gamma_{\mathrm{cpi}}} \left(\int_{\Gamma_{\mathrm{cpi}}} (f_{l_0}^*(y) - f_{l_0}^*(x))^2 \, \mathrm{d}\varepsilon_x(y) \right) \mathrm{d}\mu(x) = \int_{\Gamma_{\mathrm{cpi}}} g_{l_0}(x) \, \mathrm{d}\mu(x)$$

$$\geq \int_A g_{l_0}(x) \, \mathrm{d}\mu(x) > 0,$$

where $A = \{x \in \Gamma_{\mathrm{cpi}} \mid g_{l_0}(x) > 0\}$.

We have obtained a contradiction because $\int_{\Gamma_{\mathrm{cpi}}} g_{l_0}(x) \, \mathrm{d}\mu(x) \ = \ 0$ by Lemma 6.2.3. $\qquad\qquad\square$

Let $\sim$ be a relation on Γ_{cpie} defined as follows: $x \sim y$ if and only if $\lim_{\alpha \to +\infty} \frac{1}{\alpha} \int_0^{\alpha} S_t f(x) \, \mathrm{d}t = \lim_{\alpha \to +\infty} \frac{1}{\alpha} \int_0^{\alpha} S_t f(y) \, \mathrm{d}t$ for every $f \in C_0(X)$ whenever $x \in \Gamma_{\mathrm{cpie}}$ and $y \in \Gamma_{\mathrm{cpie}}$. Thus, $x \sim y$ if and only if $f^*(x) = f^*(y)$ for every $f \in C_0(X)$, where $x \in \Gamma_{\mathrm{cpie}}$ and $y \in \Gamma_{\mathrm{cpie}}$.

It is easy to see that $\sim$ is an equivalence relation on Γ_{cpie}. We will denote by $[x]$ the equivalence class of $x \in \Gamma_{\mathrm{cpie}}$.

Lemma 6.2.5. *Let $(f_l)_{l \in \mathbb{N}}$ be a sequence of elements of $C_0(X)$ whose range $\{f_n \mid n \in \mathbb{N}\}$ is dense in $C_0(X)$ (the existence of the sequence is assured by the separability of $C_0(X)$). Also, let $x \in \Gamma_{\mathrm{cpi}}$ and $y \in \Gamma_{\mathrm{cpi}}$. The following assertions are equivalent:*

(a) $\lim_{\alpha \to +\infty} \frac{1}{\alpha} \int_0^{\alpha} S_t f(x) \, \mathrm{d}t = \lim_{\alpha \to +\infty} \frac{1}{\alpha} \int_0^{\alpha} S_t f(y) \, \mathrm{d}t$ *for every $f \in C_0(X)$.*
(b) $\lim_{\alpha \to +\infty} \frac{1}{\alpha} \int_0^{\alpha} S_t f_l(x) \, \mathrm{d}t = \lim_{\alpha \to +\infty} \frac{1}{\alpha} \int_0^{\alpha} S_t f_l(y) \, \mathrm{d}t$ *for every $l \in \mathbb{N}$.*

Note that in both (a) and (b) the limits exist because we assume that x and y belong to Γ_{cpi}.

Proof. Clearly, $(a) \Rightarrow (b)$, so we only have to prove that $(b) \Rightarrow (a)$.

Thus, assume that (b) holds true and let $f \in C_0(X)$. Since we have to prove that $f^*(x) = f^*(y)$, it is obviously enough to prove that $|f^*(x) - f^*(y)| < \varepsilon$ for every $\varepsilon \in \mathbb{R}$, $\varepsilon > 0$.

To this end, let $\varepsilon \in \mathbb{R}$, $\varepsilon > 0$, and let $l \in \mathbb{N}$ be such that $\| f - f_l \| < \frac{\varepsilon}{4}$.

Using the inequality (6.2.4) in the proof of Lemma 6.2.1, we obtain that $\left| f^*(x) - f_l^*(x) \right| \le \frac{\varepsilon}{4}$ and $\left| f^*(y) - f_l^*(y) \right| \le \frac{\varepsilon}{4}$.

Since we assume that (b) holds true, we obtain that

$$\left| f^*(x) - f^*(y) \right| \le \left| f^*(x) - f_l^*(x) \right| + \left| f_l^*(y) - f^*(y) \right| \le \frac{\varepsilon}{4} + \frac{\varepsilon}{4} = \frac{\varepsilon}{2} < \varepsilon.$$

Thus, $f^*(x) = f^*(y)$. $\qquad\square$

Theorem 6.2.6. *For every $x \in \Gamma_{\mathrm{cpie}}$, the set $[x]$ is $\mathcal{B}(X)$-measurable.*

Proof. Let $x \in \Gamma_{\mathrm{cpie}}$, and let $(f_l)_{l \in \mathbb{N}}$ be the sequence of elements of $C_0(X)$ that we considered in Lemma 6.2.5.

Set $A_l = \{ y \in \Gamma_{\mathrm{cpie}} \mid f_l^*(y) = \langle f, \varepsilon_x \rangle \}$ for every $l \in \mathbb{N}$.

Using Lemma 6.2.5, we obtain that

$$[x] = \left\{ y \in \Gamma_{\mathrm{cpie}} \;\middle|\; \begin{array}{l} \lim_{\alpha \to +\infty} \frac{1}{\alpha} \int_0^\alpha S_t f(y)\, \mathrm{d}t = \lim_{\alpha \to +\infty} \frac{1}{\alpha} \int_0^\alpha S_t f(x)\, \mathrm{d}t \\ \text{for every } f \in C_0(X) \end{array} \right\}$$

$$= \left\{ y \in \Gamma_{\mathrm{cpie}} \;\middle|\; \lim_{\alpha \to +\infty} \frac{1}{\alpha} \int_0^\alpha S_t f_l(y)\, \mathrm{d}t = \langle f_l, \varepsilon_x \rangle \text{ for every } l \in \mathbb{N} \right\}$$

$$= \bigcap_{l \in \mathbb{N}} A_l.$$

Since the functions f_l^*, $l \in \mathbb{N}$, are measurable, we obtain that $A_l \in \mathcal{B}(X)$ for every $l \in \mathbb{N}$; therefore, $[x]$ belongs to $\mathcal{B}(X)$, as well. $\qquad\square$

The next theorem articulates the connection between the set Γ_{cpie} and the set of all invariant ergodic probability measures of $(P_t)_{t \in \mathbb{T}}$.

Theorem 6.2.7. (a) *For every $x \in \Gamma_{\mathrm{cpie}}$, the measure ε_x is an invariant ergodic probability measure for $(P_t)_{t \in \mathbb{T}}$ and $\varepsilon_x([x]) = 1$.*

(b) *Conversely, any invariant ergodic probability measure μ for $(P_t)_{t \in \mathbb{T}}$ is of the form $\mu = \varepsilon_x$ for some $x \in \Gamma_{\mathrm{cpie}}$ (that is, for every invariant ergodic probability measure μ of $(P_t)_{t \in \mathbb{T}}$, there exists an $x \in \Gamma_{\mathrm{cpie}}$ such that $\mu = \varepsilon_x$).*

Proof. (a) Let $x \in \Gamma_{\mathrm{cpie}}$. Since $\Gamma_{\mathrm{cpie}} \subseteq \Gamma_{\mathrm{cpi}}$, it follows that the map $\varepsilon_x : C_0(X) \to \mathbb{R}$, $\varepsilon_x(f) = \lim_{\alpha \to +\infty} \frac{1}{\alpha} \int_0^\alpha S_t f(x)\, \mathrm{d}t$ for every $f \in C_0(X)$, is well defined and can be identified with an element of $\mathcal{M}(X)$ (denoted again by ε_x) which is an invariant probability measure. Thus, it only remains to prove that ε_x is ergodic and that $\varepsilon([x]) = 1$ (the last equality makes sense because, by Theorem 6.2.6, the set $[x]$ is $\mathcal{B}(X)$-measurable). Now, taking into consideration the definition of the set $[x]$ and using Proposition 6.1.9, we obtain that it is enough to prove that $\varepsilon([x]) = 1$, because, then, by Proposition 6.1.9, we can conclude that ε_x is ergodic.

Now, let $(f_l)_{l \in \mathbb{N}}$ be the sequence of elements of $C_0(X)$ that we considered in Lemma 6.2.5.

Since $x \in \Gamma_{\mathrm{cpie}}$, and since (by Theorem 6.2.4) Γ_{cpie} is a set of maximal probability, it follows that

$$0 = \int_{\Gamma_{\mathrm{cpi}}} (f_l^*(y) - f_l^*(x))^2 \, d\varepsilon_x(y) = \int_{\Gamma_{\mathrm{cpie}}} (f_l^*(y) - f_l^*(x))^2 \, d\varepsilon_x(y)$$

for every $l \in \mathbb{N}$.

Thus, for every $l \in \mathbb{N}$, there exists a $\mathcal{B}(X)$-measurable subset A_l of Γ_{cpie} such that $\varepsilon_x(A_l) = 1$ and $f_l^*(x) = f_l^*(y)$ for every $y \in A_l$.

Now set $A = \cap_{l \in \mathbb{N}} A_l$. Then A is obviously a measurable subset of Γ_{cpie} and $\varepsilon_x(A) = 1$.

Since $\{f_l \mid l \in \mathbb{N}\}$ is dense in $C_0(X)$, and since $f_l^*(y) = f_l^*(x)$ for every $y \in A$ and $l \in \mathbb{N}$, using Lemma 6.2.5, we obtain that $f^*(y) = f^*(x)$ for every $y \in A$ and $f \in C_0(X)$. Therefore, $A \subseteq [x]$. Since $\varepsilon_x(A) = 1$ and ε_x is a probability measure, it follows that $\varepsilon_x([x]) = 1$, as well.

(b) Let μ be an invariant ergodic probability measure for $(P_t)_{t \in \mathbb{T}}$.

Then, by Proposition 6.1.9, there exists a measurable subset B of X such that $\mu(B) = 1$ and such that $\lim_{\alpha \to +\infty} \frac{1}{\alpha} \int_0^\alpha S_t f(x) \, dt$ exists and is equal to $\langle f, \mu \rangle$ for every $f \in C_0(X)$ and $x \in B$.

Since μ is an invariant probability measure, it follows that $B \in \Gamma_{\mathrm{cpi}}$.

Since $\mu(B) = 1$, it follows that $B \neq \emptyset$. Thus, there exists an $x \in B$ and it is obvious that $\mu = \varepsilon_x$. Since $f^*(y) = \langle f, \varepsilon_x \rangle$ for every $y \in B$ and $f \in C_0(X)$, it follows that

$$\int_{\Gamma_{\mathrm{cpi}}} (f^*(y) - f^*(x))^2 \, d\varepsilon_x(y) = \int_B (f^*(y) - f^*(x))^2 \, d\varepsilon_x(y) = 0$$

for every $f \in C_0(X)$. Hence, $x \in \Gamma_{\mathrm{cpie}}$. $\square$

Corollary 6.2.8. *The following assertions are equivalent:*

(a) *The transition function $(P_t)_{t \in \mathbb{T}}$ has at least one invariant probability.*
(b) $\Gamma_{\mathrm{cpie}} \neq \emptyset$.

Proof. (a) $\Rightarrow$ (b) Assume that $(P_t)_{t \in \mathbb{T}}$ has invariant probabilities, and let μ be such an invariant probability measure. Since Γ_{cpie} is a set of maximal probability, it follows that $\mu(\Gamma_{\mathrm{cpie}}) = 1$; hence $\Gamma_{\mathrm{cpie}} \neq \emptyset$.

(b) $\Rightarrow$ (a) is obvious because if $\Gamma_{\mathrm{cpie}} \neq \emptyset$, then there exists an $x \in \Gamma_{\mathrm{cpie}}$, so ε_x is an invariant probability measure for $(P_t)_{t \in \mathbb{T}}$. $\square$

From our discussion so far, we see that the invariant ergodic probability measures for the transition function $(P_t)_{t \in \mathbb{T}}$ are the "building blocks" for the set of all invariant probabilities for $(P_t)_{t \in \mathbb{T}}$ in the sense that any invariant probability measure μ (for $(P_t)_{t \in \mathbb{T}}$) can be expressed (using (a) of Corollary 3.2.12, Proposition 5.3.6, and the fact that Γ_{cpie} is a set of maximal probability) as a convex combination in integral form of invariant ergodic probability measures by means of the equality

$$\int_X f(x)\,d\mu(x) = \int_{\Gamma_{\text{cpie}}} \left(\int_X f(y)\,d\varepsilon_x(y) \right) d\mu(x) \qquad (6.2.7)$$

for every $f \in C_0(X)$. Using the notation $\langle f, \varepsilon_x \rangle = \int_X f(y)\,d\varepsilon_x(y)$, the equality (6.2.7) becomes

$$\int_X f(x)\,d\mu(x) = \int_{\Gamma_{\text{cpie}}} \langle f, \varepsilon_x \rangle \,d\mu(x)$$

for every $f \in C_0(X)$.

Assume that the transition function $(P_t)_{t \in \mathbb{T}}$ has invariant probabilities. Then, by Corollary 6.2.8, $(P_t)_{t \in \mathbb{T}}$ also has invariant ergodic probability measures as $\Gamma_{\text{cpie}} \neq \emptyset$. Given an invariant probability measure μ of $(P_t)_{t \in \mathbb{T}}$, a natural question is: how do we know whether or not μ is ergodic? Of course, we could try to see if μ satisfies one of the equivalent definitions of invariant ergodic probability measures discussed in Sect. 6.1 (see Theorem 6.1.14). In particular, we could try to see if we cannot find a set $B \in \mathcal{B}(X)$ such that $\mu(B) = 1$ and such that, for every $x \in B$, the family $\left(\dfrac{1}{s} \displaystyle\int_0^s S_t f(x)\,dt \right)_{s \in (0,+\infty)}$ converges to $\langle f, \mu \rangle$ as $s \to +\infty$ for every $f \in C_0(X)$ and every $x \in B$ (that is, we could try to find out if μ satisfies (f) of Theorem 6.1.14). However, condition (f), as well as all the other equivalent definitions of an invariant ergodic probability measure stated in Theorem 6.1.14 are hard to verify. Our goal here (in the next theorem) is to obtain a sufficient condition for the ergodicity of an invariant probability measure μ of $(P_t)_{t \in \mathbb{T}}$, a condition which is significantly weaker than condition (f) of Theorem 6.1.14.

Theorem 6.2.9. *Let μ be an invariant probability of $(P_t)_{t \in \mathbb{T}}$, and set*

$$A = \left\{ x \in X \;\middle|\; \begin{array}{l} \lim\limits_{s \to +\infty} \frac{1}{s} \int_0^s S_t f(x)\,dt \text{ exists and is} \\ \text{equal to } \langle f, \mu \rangle \text{ for every } f \in C_0(X) \end{array} \right\}.$$

If $\mu(A) > 0$, then μ is an ergodic measure.

Proof. Since (by Theorem 6.2.4) Γ_{cpie} is a set of maximal probability, it follows that $\mu(A \cap \Gamma_{\text{cpie}}) > 0$. Thus, there exists an $x \in A \cap \Gamma_{\text{cpie}}$.

Then, in view of the definition of Γ_{cpie}, the measure ε_x exists and is an invariant ergodic probability; moreover, taking into consideration the manner in which ε_x and the set A are defined, we obtain that $\lim\limits_{s \to +\infty} \dfrac{1}{s} \displaystyle\int_0^s S_t f(x)\,dt$ exists and is equal to both $\langle f, \varepsilon_x \rangle$ and $\langle f, \mu \rangle$ for every $f \in C_0(X)$. Accordingly, $\mu = \varepsilon_x$; that is, μ is an invariant ergodic probability measure. $\qquad \square$

We will conclude the section with a few examples.

Example 6.2.10. Let $X = [0, 1]$ and let d be the usual metric on $[0, 1]$ defined by the absolute value. Also, let $\mathbf{w}$ be the one-parameter semigroup of elements of $B([0, 1])$ that was used to construct the transition function $(P_t^{(\mathbf{w})})_{t \in [0, +\infty)}$ in Example 2.2.3.

Then, in view of our discussion in Example 5.1.8, we obtain that $\mathcal{D} = \emptyset$, $\Gamma_c = \Gamma_{cp} = [0, 1]$, $\Gamma_{cpi} = \Gamma_{cpie} = \{0, 1\}$, the transition function is uniquely ergodic, the unique invariant (necessarily ergodic) probability measure is δ_1, and Γ_{cpie} has only one equivalence class $[0] = [1] = \{0, 1\}$. ∎

Example 6.2.11. Let Γ be a lattice in $\mathrm{SL}(2, \mathbb{R})$, let $j = 1$ or 2, let $\mathbf{w}^{(j\Gamma L)}$ be the horocycle flow on $(\mathrm{SL}(2, \mathbb{R})/\Gamma)_L$ that corresponds to j (see (b) of Example B.1.9), and let $(P_t^{(\mathbf{w}^{(j\Gamma L)})})_{t \in \mathbb{R}}$ be the transition function defined by $\mathbf{w}^{(j\Gamma L)}$ (see (c) of Example 2.2.7). Then using the Dani-Smillie theorem on the equidistribution of horocycle orbits, we obtain that for the transition function $(P_t^{(\mathbf{w}^{(j\Gamma L)})})_{t \in \mathbb{R}}$ we have $\Gamma_c = \Gamma_{cp} = \Gamma_{cpi} = \Gamma_{cpie} = (\mathrm{SL}(2, \mathbb{R})/\Gamma)_L$. If there are no periodic points for the horocycle flow, then the flow (and the transition function) are uniquely ergodic and the unique invariant ergodic probability measure is the standard $\mathrm{SL}(2, \mathbb{R})$-invariant probability measure $\nu_{(\mathrm{SL}(2,\mathbb{R})/\Gamma)_L}$.

If there are periodic points for the horocycle flow under consideration, let C_i, $i \in I$, be all the distinct orbits of periodic points of the flow; that is, C_i, $i \in I$, are orbits of periodic points such that any point in $(\mathrm{SL}(2, \mathbb{R})/\Gamma)_L \setminus (\cup_{i \in I} C_i)$ is not periodic and $C_i \cap C_j$ is the empty set whenever $i \in I$ and $j \in I$ are such that $i \neq j$. Then the invariant ergodic probability measures for the transition function and the flow are $\nu_{(\mathrm{SL}(2,\mathbb{R})/\Gamma)_L}$, which defines the equivalence class $(\mathrm{SL}(2, \mathbb{R})/\Gamma)_L \setminus (\cup_{i \in I} C_i)$ in Γ_{cpie}, and the invariant ergodic probability measures μ_i concentrated on C_i and defining the equivalence classes C_i, $i \in I$.

The above discussion can be easily adapted to the remaining three transition functions defined by the other three horocycle flows. ∎

Example 6.2.12. Let $n \in \mathbb{N}$, $n \geq 2$, let Γ be a lattice in $\mathrm{SL}(n, \mathbb{R})$, let $\mathbf{v}$ be a unipotent flow (see Sect. B.4.2 for the terminology used so far), and let $(P_t^{(\mathbf{v})})_{t \in \mathbb{R}}$ be the transition function defined by $\mathbf{v}$ (see Example 2.2.8). Using Ratner's theorem, Theorem B.4.9, we obtain that for $(P_t^{(\mathbf{v})})_{t \in \mathbb{R}}$, we have that $\Gamma_c = \Gamma_{cp} = \Gamma_{cpi} = \Gamma_{cpie} = (\mathrm{SL}(n, \mathbb{R})/\Gamma)_R$ (actually, a compact way of stating Theorem B.4.9 is: if $\mathbf{v}$ is a unipotent flow on $(\mathrm{SL}(n, \mathbb{R})/\Gamma)_R$, then $\Gamma_{cpie} = (\mathrm{SL}(n, \mathbb{R})/\Gamma)_R$ for the transition function $(P_t^{(\mathbf{v})})_{t \in \mathbb{R}}$ defined by $\mathbf{v}$). Moreover, in Ratner's Theorem B.4.10, she obtains a complete characterization of the invariant ergodic probability measures of $(P_t^{(\mathbf{v})})_{t \in \mathbb{R}}$ and their supports. ∎

6.3 More About the Set of All Invariant Ergodic Probabilities

Our goal in this section is to discuss the connection between the KBBY decomposition and various results that appear in Section 12 of Phelps' monograph [86], Rohlin [101], and Sections 8 and 9 of Oxtoby [85] on the role played by the invariant ergodic probability measures in the set of all real-valued invariant signed Borel measures for a transition function $(P_t)_{t \in \mathbb{T}}$.

Throughout the section, we assume given a locally compact separable metric space (X, d), a transition function $(P_t)_{t \in \mathbb{T}}$, and the family $((S_t, T_t))_{t \in \mathbb{T}}$ of Markov pairs defined by $(P_t)_{t \in \mathbb{T}}$. We also assume that $(P_t)_{t \in \mathbb{T}}$ satisfies the s.m.a., is pointwise continuous, and has invariant probability measures.

Let Y be the set of all equivalence classes $[x]$, $x \in \Gamma_{\mathrm{cpie}}$ (clearly, we can also think of Y as the set of all invariant ergodic probability measures of $(P_t)_{t \in \mathbb{T}}$).

We now define a collection $\mathcal{F}$ of subsets of Y as follows: a subset A of Y belongs to $\mathcal{F}$ if, by definition, the set $B_A = \{y \in \Gamma_{\mathrm{cpie}} \mid y$ belongs to $[x]$ for some $[x] \in A\}$ is $\mathcal{B}(X)$-measurable.

Lemma 6.3.1. *The collection $\mathcal{F}$ is a σ-algebra of subsets of Y.*

Proof. Clearly, the empty set $\emptyset$ belongs to $\mathcal{F}$ because $B_\emptyset = \emptyset$, and so $B_\emptyset \in \mathcal{B}(X)$. Also, Y belongs to $\mathcal{F}$ because $B_Y = \Gamma_{\mathrm{cpie}}$, and Γ_{cpie} belongs to $\mathcal{B}(X)$ by Theorem 6.2.2.

Now let $A \in \mathcal{F}$. Since $B_{Y \setminus A} = \{y \in \Gamma_{\mathrm{cpie}} \mid y \sim x$ for some $[x] \in Y \setminus A\} = \Gamma_{\mathrm{cpie}} \setminus B_A$ and since both Γ_{cpie} and B_A belong to $\mathcal{B}(X)$, we obtain that $B_{Y \setminus A} \in \mathcal{B}(X)$ so $Y \setminus A \in \mathcal{F}$.

Finally, we have to prove that $\cup_{n=1}^\infty A_n$ belongs to $\mathcal{F}$ for every sequence $(A_n)_{n \in \mathbb{N}}$ of elements of $\mathcal{F}$.

To this end, let $(A_n)_{n \in \mathbb{N}}$ be a sequence of elements of $\mathcal{F}$.

We will first prove that

$$B_{\cup_{n=1}^\infty A_n} = \cup_{n=1}^\infty B_{A_n} \qquad (6.3.1)$$

by showing that

$$B_{\cup_{n=1}^\infty A_n} \subseteq \cup_{n=1}^\infty B_{A_n} \qquad (6.3.2)$$

and that

$$B_{\cup_{n=1}^\infty A_n} \supseteq \cup_{n=1}^\infty B_{A_n}. \qquad (6.3.3)$$

If $y \in B_{\cup_{n=1}^\infty A_n}$, then $y \sim x$ for some $[x] \in \cup_{n=1}^\infty A_n$; therefore, $[x] \in A_m$ for some $m \in \mathbb{N}$; hence, $y \in B_{A_m} \subseteq \cup_{n=1}^\infty B_{A_n}$. Thus, the inclusion (6.3.2) is true.

Now let $y \in \cup_{n=1}^\infty B_{A_n}$. Then $y \in B_{A_m}$ for some $m \in \mathbb{N}$, so $y \sim x$ for some $[x] \in A_m \subseteq \cup_{n=1}^\infty A_n$; hence, $y \in B_{\cup_{n=1}^\infty A_n}$. Accordingly, the inclusion (6.3.3) holds true, as well.

Using the fact that (6.3.1) is true, we easily obtain that $\cup_{n=1}^\infty A_n \in \mathcal{F}$. Indeed, since $B_{A_n} \in \mathcal{B}(X)$ for every $n \in \mathbb{N}$, it follows that $\cup_{n=1}^\infty B_{A_n} \in \mathcal{B}(X)$; therefore, $\cup_{n=1}^\infty A_n \in \mathcal{F}$.

Thus, $\mathcal{F}$ is a σ-algebra. $\qquad\qquad\square$

We will denote by $\mathcal{M}(Y,\mathcal{F})$, or, simply, by $\mathcal{M}(Y)$ the Banach lattice of all real-valued signed measures on $(Y,\mathcal{F})$, where the norm on $\mathcal{M}(Y)$ is the total variation norm, and the order relation on $\mathcal{M}(Y)$ is the usual order relation for measures.

Let $\mathcal{M}_{\text{inv}}(X)$ be the vector space of all invariant elements for $(P_t)_{t\in\mathbb{T}}$ in $\mathcal{M}(X)$. We endow $\mathcal{M}_{\text{inv}}(X)$ with the restriction of the total variation norm of $\mathcal{M}(X)$ to $\mathcal{M}_{\text{inv}}(X)$, and with the restriction of the order relation on $\mathcal{M}(X)$ to $\mathcal{M}_{\text{inv}}(X)$. Then, as pointed out in Example 4.6.4, $\mathcal{M}_{\text{inv}}(X)$ becomes a Banach sublattice of $\mathcal{M}(X)$ and a Banach lattice in its own right.

Our goal now is to prove that $\mathcal{M}_{\text{inv}}(X)$ is isometric and Banach lattice isomorphic (in a natural way) to $\mathcal{M}(Y)$ (two Banach lattices are said to be *isomorphic* if there exists a Banach space isomorphism from one space onto the other which is also a lattice isomorphism).

To this end, we will start by associating to every probability measure in $\mathcal{M}_{\text{inv}}(X)$ a measure on $(Y,\mathcal{F})$.

Thus, let $\mu \in \mathcal{M}(X)$ be an invariant probability measure for $(P_t)_{t\in\mathbb{T}}$. We define $\mu_{\mathcal{F}} : \mathcal{F} \to \mathbb{R}$ by $\mu_{\mathcal{F}}(A) = \mu(B_A)$ for every $A \in \mathcal{F}$.

In the next lemma, we state a few useful formulas for $\mu_{\mathcal{F}}(A)$, $A \in \mathcal{F}$.

Lemma 6.3.2. *Let* $\mu \in \mathcal{M}_{inv}(X)$ *be a probability measure, let* $\mu_{\mathcal{F}} : \mathcal{F} \to \mathbb{R}$ *be defined as above, and let* $A \in \mathcal{F}$. *Then:*

(a) $\mu_{\mathcal{F}}(A) = \int \mathbf{1}^{*}_{B_A}\mathrm{d}\mu$.

(b) $\mu_{\mathcal{F}}(A) = \int_{\Gamma_{cpie}} \mathbf{1}^{\square}_{B_A}\mathrm{d}\mu$.

(c) $\mu_{\mathcal{F}}(A) = \int_{\Gamma_{cpie}} \varepsilon_x(B_A)\mathrm{d}\mu(x)$.

Proof. (a) Since $\mu_{\mathcal{F}}(A) = \int \mathbf{1}_{B_A}\mathrm{d}\mu$, and since $\mathbf{1}^{*}_{B_A}$ is a μ-a.e. limit of $(\frac{1}{\alpha}\int_0^{\alpha} S_t\mathbf{1}_{B_A}(x)\mathrm{d}t)_{\alpha\in(0,+\infty)}$ as $\alpha \to +\infty$ we can use (a) of Corollary 3.2.12 in order to conclude that $\mu_{\mathcal{F}} = \int \mathbf{1}^{*}_{B_A}\mathrm{d}\mu$.

(b) The equality holds true because Γ_{cpie} is a set of maximal probability, and because, by Proposition 5.3.6, $\mathbf{1}^{*}_{B_A} = \mathbf{1}^{\square}_{B_A}$ μ-a.e.

(c) The proof of the equality is obtained using the definition of $\mathbf{1}^{\square}_{B_A}$, the fact that both Γ_{c} and Γ_{cpie} are sets of maximal probability, and the fact that $\Gamma_{\text{cpie}} \subseteq \Gamma_{\text{c}}$.

$\square$

Lemma 6.3.3. *Let* $\mu \in \mathcal{M}_{inv}(X)$ *be a probability measure, and let* $\mu_{\mathcal{F}} : \mathcal{F} \to \mathbb{R}$ *be the map defined before Lemma 6.3.2. Then* $\mu_{\mathcal{F}}$ *is a probability measure on* $(Y,\mathcal{F})$.

Proof. We first note that $\mu_{\mathcal{F}}(\emptyset) = 0$ because $B_{\emptyset} = \emptyset$.

Next we note that $B_Y = \Gamma_{\text{cpie}}$, so using (c) of Lemma 6.3.2, we obtain that $\mu_{\mathcal{F}}(Y) = \int_{\Gamma_{\text{cpie}}} \varepsilon_x(\Gamma_{\text{cpie}})\mathrm{d}\mu(x) = \int_{\Gamma_{\text{cpie}}} \mathrm{d}\mu(x) = 1$.

We now note that given $A \in \mathcal{F}$ and $C \in \mathcal{F}$, then $A \cap C = \emptyset$ if and only if $B_A \cap B_C = \emptyset$. Indeed, $A \cap C \neq \emptyset$ if and only if there exists $[x] \in A \cap C$ if and only if the nonempty set $\{y \in \Gamma_{\text{cpie}} \mid y \sim x\}$ is included in both B_A and B_C.

Now, let $(A_n)_{n\in\mathbb{N}}$ be a sequence of mutually disjoint subsets of Y such that $A_n \in \mathcal{F}$ for every $n \in \mathbb{N}$. Then using the observation made in the previous paragraph, we obtain that $(B_{A_n})_{n\in\mathbb{N}}$ is a sequence of mutually disjoint subsets of Γ_{cpie} such that $B_{A_n} \in \mathcal{B}(X)$ for every $n \in \mathbb{N}$.

Using the equality (6.3.1) and (c) of Lemma 6.3.2, we obtain that

$$\mu_{\mathcal{F}}(\cup_{k=1}^{n} A_k) = \int_{\Gamma_{\text{cpie}}} \varepsilon_x(B_{\cup_{k=1}^{n} A_k}) d\mu(x) = \int_{\Gamma_{\text{cpie}}} \varepsilon_x(\cup_{k=1}^{n} B_{A_k}) d\mu(x)$$

$$= \int_{\Gamma_{\text{cpie}}} \sum_{k=1}^{n} \varepsilon_x(B_{A_k}) d\mu(x) = \sum_{k=1}^{n} \mu_{\mathcal{F}}(A_k)$$

for every $n \in \mathbb{N}$.

Now, let $f_n : X \to \mathbb{R}$ be defined by

$$f_n(x) = \begin{cases} \sum_{k=1}^{n} \varepsilon_x(B_{A_k}) = \varepsilon_x(\cup_{k=1}^{n} B_{A_k}) & \text{if } x \in \Gamma_{\text{c}}, \\ 0 & \text{if } x \in X \setminus \Gamma_{\text{c}} \end{cases}$$

for every $n \in \mathbb{N}$.

We now note that $f_n = \mathbf{1}_{\cup_{k=1}^{n} B_{A_k}}^{\square}$, so, by Lemma 5.2.5, f_n is a bounded measurable function for every $n \in \mathbb{N}$. Moreover, $(f_n)_{n \in \mathbb{N}}$ is a monotone nondecreasing sequence of positive functions that converges pointwise (everywhere on X) to $f = \mathbf{1}_{\cup_{n=1}^{\infty} B_{A_n}}^{\square}$. Using Lemma 5.2.5 again, we obtain that f is also a bounded measurable function. Thus, we can apply the monotone convergence theorem and we obtain that

$$\mu_{\mathcal{F}}(\cup_{n=1}^{\infty} A_n) = \int_{\Gamma_{\text{cpie}}} \mathbf{1}_{B_{\cup_{n=1}^{\infty} A_n}}^{\square} d\mu = \int_{\Gamma_{\text{cpie}}} \mathbf{1}_{\cup_{n=1}^{\infty} B_{A_n}} d\mu$$

$$= \lim_{n \to +\infty} \int_{\Gamma_{\text{cpie}}} \mathbf{1}_{\cup_{k=1}^{n} B_{A_k}}^{\square} d\mu = \lim_{n \to +\infty} \sum_{k=1}^{n} \int_{\Gamma_{\text{cpie}}} \mathbf{1}_{B_{A_k}} d\mu$$

$$= \lim_{n \to +\infty} \sum_{k=1}^{n} \mu_{\mathcal{F}}(A_k) = \sum_{n=1}^{\infty} \mu_{\mathcal{F}}(A_n).$$

We have therefore proved that $\mu_{\mathcal{F}}$ is a probability measure on $(Y, \mathcal{F})$. $\square$

Our goal now is to construct, given a probability measure ν on $(Y, \mathcal{F})$, an invariant probability measure $\nu_{\mathcal{B}}$ for $(P_t)_{t \in \mathbb{T}}$ on $(X, \mathcal{B}(X))$.

Thus, let ν be a probability measure on $(Y, \mathcal{F})$.

For every $A \in \mathcal{B}(X)$ we define a function $f_A : \Gamma_{\text{cpie}}^{\square} \to \mathbb{R}$ by $f_A(x) = \varepsilon_x(A)$ for every $x \in \Gamma_{\text{cpie}}$. Since f_A is the restriction of $\mathbf{1}_A^{\square}$ to Γ_{cpie}, and since $\mathbf{1}_A^{\square}$ is a measurable function and Γ_{cpie} is a measurable subset of X, it follows that f_A is measurable.

For every $[x] \in Y$, we denote by $\varepsilon_{[x]}$ the measure on $(X, \mathcal{B}(X))$ defined by $\varepsilon_{[x]} = \varepsilon_y$ for some y in the class $[x]$ ($y \in \Gamma_{\text{cpie}}$). Note that since $\varepsilon_y = \varepsilon_z$ whenever $y \in \Gamma_{\text{cpie}}$ and $z \in \Gamma_{\text{cpie}}$ are such that $y \sim z$, it follows that the definition of $\varepsilon_{[x]}$ is correct in the sense that it does not depend on the particular choice of the element y in the class $[x]$.

For every $A \in \mathcal{B}(X)$, we define a function $\tilde{f}_A : Y \to \mathbb{R}$ as follows: $\tilde{f}_A([x]) = \varepsilon_{[x]}(A)$ for every $[x] \in Y$.

Now, for $h \in B_b(X)$, we define a function $\tilde{f}_h : Y \to \mathbb{R}$ by $\tilde{f}_h([x]) = \langle h, \varepsilon_{[x]} \rangle = \int_X h(y) \mathrm{d}\varepsilon_{[x]}(y)$ for every $[x] \in Y$. Note that if $A \in \mathcal{B}(X)$, then $\tilde{f}_A = \tilde{f}_{\mathbf{1}_A}$ (that is, $\tilde{f}_A = \tilde{f}_h$, where $h = \mathbf{1}_A$).

Lemma 6.3.4. *The functions $\tilde{f}_h$, $h \in B_b(X)$, are $\mathcal{F}$-measurable and ν-integrable.*

Proof. We first note that taking into consideration the definition of $\tilde{f}_h$, $h \in B_b(X)$, and using the fact that $\varepsilon_{[x]}$, $[x] \in Y$, are probability measures, we obtain that $\tilde{f}_h([x]) \leq \|h\|$ for every $[x] \in Y$, where $\|h\|$ is the (sup) norm on $B_b(X)$, $h \in B_b(X)$, so $\tilde{f}_h$, $h \in B_b(X)$, are bounded functions.

Since ν is a probability measure and since $\tilde{f}_h$, $h \in B_b(X)$, are bounded functions on Y, in order to prove the lemma, it is enough to prove that the functions $\tilde{f}_h$, $h \in B_b(X)$, are $\mathcal{F}$-measurable. We will prove this fact in two steps: first for $h = \mathbf{1}_A$ for some $A \in \mathcal{B}(X)$, and then, in general, for every $h \in B_b(X)$.

Step 1.   Let $A \in \mathcal{B}(X)$. We have to prove that $\tilde{f}_A = \tilde{f}_{\mathbf{1}_A}$ is $\mathcal{F}$-measurable.

To this end, we will prove that the sets $A_r = \{[z] \in Y \mid \tilde{f}_A([z]) \leq r\}$, $r \in \mathbb{R}$, belong to $\mathcal{F}$.

Thus, let $r \in \mathbb{R}$, and note that

$$B_{A_r} = \{y \in \Gamma_{\mathrm{cpie}} \mid \tilde{f}_A([y]) \leq r\} = \{y \in \Gamma_{\mathrm{cpie}} \mid \varepsilon_{[y]}(A) \leq r\}$$
$$= \{y \in \Gamma_{\mathrm{cpie}} \mid \varepsilon_y(A) \leq r\} = \{y \in \Gamma_{\mathrm{cpie}} \mid \mathbf{1}_A^{\square}(y) \leq r\}.$$

Since the function $\mathbf{1}_A^{\square}$ is $\mathcal{B}(X)$-measurable, it follows that $B_{A_r} \in \mathcal{B}(X)$, so $A_r \in \mathcal{F}$.

Step 2.   At this step, we first note that if h is a simple $\mathcal{B}(X)$-measurable function, $h = \sum_{i=1}^n a_i \mathbf{1}_{A_i}$ for some $n \in \mathbb{N}$, n real numbers $a_1, a_2, \ldots, a_n$, and n $\mathcal{B}(X)$-measurable subsets $A_1, A_2, \ldots, A_n$ of X, then $\tilde{f}_h = \sum_{i=1}^n a_i \tilde{f}_{\mathbf{1}_{A_i}}$; so, using Step 1, we obtain that $\tilde{f}_h$ is $\mathcal{F}$-measurable.

Now, let $h \in B_b(X)$, $h \geq 0$, be a not necessarily simple function. Then there exist a monotone nondecreasing sequence $(h_n)_{n \in \mathbb{N}}$ of simple $\mathcal{B}(X)$-measurable functions such that $h_n \geq 0$ for every $n \in \mathbb{N}$, and such that $(h_n)_{n \in \mathbb{N}}$ converges uniformly to h on X.

Taking into consideration the manner in which $\tilde{f}_h$ and $\tilde{f}_{h_n}$, $n \in \mathbb{N}$, are defined, and using the monotone convergence theorem at every $[y] \in Y$, we obtain that $(\tilde{f}_{h_n})_{n \in \mathbb{N}}$ converges pointwise (everywhere) to $\tilde{f}_h$ on Y. Since $\tilde{f}_{h_n}$ is $\mathcal{F}$-measurable for every $n \in \mathbb{N}$, it follows that $\tilde{f}_h$ is $\mathcal{F}$-measurable, as well.

If $h \in B_b(X)$ is not necessarily a positive element of $B_b(X)$, then $h = h^+ - h^-$, where $h^+ = h \vee 0$ and $h^- = (-h) \vee 0$. Since $\tilde{f}_h = \tilde{f}_{h^+} - \tilde{f}_{h^-}$ and since, by our discussion so far, $\tilde{f}_{h^+}$ and $\tilde{f}_{h^-}$ are $\mathcal{F}$-measurable, it follows that $\tilde{f}_h$ is $\mathcal{F}$-measurable, as well. $\qquad\square$

We now construct (given the probability measure ν on $(Y, \mathcal{F})$) a map $\nu_{\mathcal{B}}$: $\mathcal{B}(X) \rightarrow \mathbb{R}$ as follows: $\nu_{\mathcal{B}}(A) = \int_Y \tilde{f}_A([x]) d\nu([x]) (= \int_Y \tilde{f}_{1_A}([x]) d\nu([x]))$ for every $A \in \mathcal{B}(X)$.

Clearly, in view of Lemma 6.3.4, the definitions of $\nu_{\mathcal{B}}(A)$, $A \in \mathcal{B}(X)$, are correct; that is, the integrals $\int_Y \tilde{f}_A d\nu$, $A \in \mathcal{B}(X)$, exist and are real numbers.

Lemma 6.3.5. *The map $\nu_{\mathcal{B}}$ is a probability measure on $(X, \mathcal{B}(X))$.*

Proof. Note first that $\tilde{f}_\emptyset$ is the constant zero function defined on Y (that is, $\tilde{f}_\emptyset([x]) = 0$ for every $[x] \in Y$), so it follows that $\nu_{\mathcal{B}}(\emptyset) = 0$.

Since $\tilde{f}_X([x]) = 1$ for every $[x] \in Y$, it follows that $\nu_{\mathcal{B}}(X) = 1$ because ν is a probability measure.

In order to complete the proof of the lemma we have to prove only that $\nu_{\mathcal{B}}$ is σ-additive.

To this end, let $(A_n)_{n \in \mathbb{N}}$ be a sequence of mutually disjoint measurable subsets of X.

Since $\varepsilon_{[x]}$ is a probability measure on $(X, \mathcal{B}(X))$, it follows that $\tilde{f}_{\cup_{k=1}^n A_k} = \sum_{k=1}^n \tilde{f}_{A_k}$ for every $n \in \mathbb{N}$. Moreover, using again the fact that $\varepsilon_{[x]}$, $[x] \in Y$, are measures on $(X, \mathcal{B}(X))$, we obtain that $\lim_{n \to +\infty} \tilde{f}_{\cup_{k=1}^n A_k}$ exists pointwise (for every $[x] \in Y$) and $\lim_{n \to +\infty} \tilde{f}_{\cup_{k=1}^n A_k} = \tilde{f}_{\cup_{n=1}^\infty A_n} = \sum_{n=1}^\infty \tilde{f}_{A_n}$. Thus, by applying the monotone convergence theorem to the monotone nondecreasing sequence $(\tilde{f}_{\cup_{k=1}^n A_k})_{n \in \mathbb{N}}$ of nonnegative ν-integrable functions, we obtain that

$$\nu_{\mathcal{B}}(\cup_{n=1}^\infty A_n) = \int_Y \tilde{f}_{\cup_{n=1}^\infty A_n}([x]) d\nu([x]) = \int_Y (\lim_{n \to +\infty} \tilde{f}_{\cup_{k=1}^n A_k})([x]) d\nu([x])$$

$$= \lim_{n \to +\infty} \int_Y \tilde{f}_{\cup_{k=1}^n A_k}([x]) d\nu([x]) = \lim_{n \to +\infty} \int_Y \sum_{k=1}^n \tilde{f}_{A_k}([x]) d\nu([x])$$

$$= \lim_{n \to +\infty} \sum_{k=1}^n \int_Y \tilde{f}_{A_k}([x]) d\nu([x])$$

$$= \lim_{n \to +\infty} \sum_{k=1}^n \nu_{\mathcal{B}}(A_k) = \sum_{n=1}^\infty \nu_{\mathcal{B}}(A_n).$$

We have therefore proved that $\nu_{\mathcal{B}}(\cup_{n=1}^\infty A_n) = \sum_{n=1}^\infty \nu_{\mathcal{B}}(A_n)$ for every sequence $(A_n)_{n \in \mathbb{N}}$ of mutually disjoint measurable subsets of X. $\qquad \square$

Our next goal is to prove that $\nu_{\mathcal{B}}$ is invariant for $(P_t)_{t \in \mathbb{T}}$. To this end, we need the following lemma:

Lemma 6.3.6. *The equality*

$$\langle h, \nu_{\mathcal{B}} \rangle = \int_Y \tilde{f}_h([x]) d\nu([x])$$

holds true for every $h \in B_b(X)$.

Proof. We prove the lemma in two steps: at Step 1, we prove it for $h = \mathbf{1}_A$ for some $A \in \mathcal{B}(X)$ and in the case when h is a simple measurable function; at Step 2, we prove the lemma in the general case for every $h \in B_b(X)$.

Step 1.   Let $A \in \mathcal{B}(X)$ and set $h = \mathbf{1}_A$. Then

$$\langle h, \nu_\mathcal{B} \rangle = \nu_\mathcal{B}(A) = \int_Y \tilde{f}_{\mathbf{1}_A}([x]) \mathrm{d}\nu([x]).$$

Now assume that h is a simple $\mathcal{B}(X)$-measurable function. Then there exist $n \in \mathbb{N}$, n real numbers $a_1, a_2, \ldots, a_n$ and n $\mathcal{B}(X)$-measurable subsets $A_1, A_2, \ldots, A_n$ of X such that $h = \sum_{i=1}^n a_i \mathbf{1}_{A_i}$. We obtain that

$$\langle h, \nu_\mathcal{B} \rangle = \sum_{i=1}^n a_i \langle \mathbf{1}_{A_i}, \nu_\mathcal{B} \rangle = \sum_{i=1}^n a_i \int_Y \tilde{f}_{\mathbf{1}_{A_i}}([x]) \mathrm{d}\nu([x])$$

$$= \sum_{i=1}^n a_i \int_Y \langle \mathbf{1}_{A_i}, \varepsilon_{[x]} \rangle \mathrm{d}\nu([x]) = \int_Y \left\langle \sum_{i=1}^n a_i \mathbf{1}_{A_i}, \varepsilon_{[x]} \right\rangle \mathrm{d}\nu([x])$$

$$= \int_Y \langle h, \varepsilon_{[x]} \rangle \mathrm{d}\nu([x]) = \int_Y \tilde{f}_h([x]) \mathrm{d}\nu([x]).$$

Step 2.   At this step we prove the lemma for every $h \in B_b(X)$.

To this end, note that it is enough to prove the lemma under the assumption that $h \geq 0$. Indeed, assume that the lemma is true for every $h \in B_b(X)$, $h \geq 0$, and let $g \in B_b(X)$ be a not necessarily positive function. Then $g = g^+ - g^-$ where $g^+ = g \vee 0$ and $g^- = (-g) \vee 0$. Since both g^+ and g^- are positive elements of $B_b(X)$, using our assumption, we obtain that

$$\langle g, \nu_\mathcal{B} \rangle = \langle g^+, \nu_\mathcal{B} \rangle - \langle g^-, \nu_\mathcal{B} \rangle$$

$$= \int_Y \tilde{f}_{g^+}([x]) \mathrm{d}\nu([x]) - \int_Y \tilde{f}_{g^-}([x]) \mathrm{d}\nu([x])$$

$$= \int_Y (\tilde{f}_{g^+}([x]) - \tilde{f}_{g^-}([x])) \mathrm{d}\nu([x])$$

$$= \int_Y (\langle g^+, \varepsilon_{[x]} \rangle - \langle g^-, \varepsilon_{[x]} \rangle) \mathrm{d}\nu([x])$$

$$= \int_Y \langle g, \varepsilon_{[x]} \rangle \mathrm{d}\nu([x]) = \int_Y \tilde{f}_g([x]) \mathrm{d}\nu([x]).$$

Now, let $h \in B_b(X)$, $h \geq 0$. Then there exists a monotone nondecreasing sequence $(h_n)_{n \in \mathbb{N}}$ of simple measurable functions such that $h_n \geq 0$ for every $n \in \mathbb{N}$, and such that $(h_n)_{n \in \mathbb{N}}$ converges uniformly to h on X.

Since $h_n, n \in \mathbb{N}$, and h are integrable with respect to the probability measure $\varepsilon_{[x]}$, using the monotone convergence theorem, we obtain that $(\langle h_n, \varepsilon_{[x]}\rangle)_{n \in \mathbb{N}}$ converges to $\langle h, \varepsilon_{[x]}\rangle$; that is, $(\tilde{f}_{h_n}([x]))_{n \in \mathbb{N}}$ converges to $\tilde{f}_h([x])$ for every $[x] \in Y$.

Since $(\tilde{f}_{h_n})_{n \in \mathbb{N}}$ is a monotone nondecreasing sequence of positive $\mathcal{F}$-measurable and ν-integrable functions that converges pointwise on Y to $\tilde{f}_h$, and since $\tilde{f}_h$ is ν-integrable, using one more time the monotone convergence theorem applied now to the sequence $(\tilde{f}_{h_n})_{n \in \mathbb{N}}$ of functions defined on Y and to the probability measure ν, we obtain that $(\int_Y \tilde{f}_{h_n}([x])\mathrm{d}\nu([x]))_{n \in \mathbb{N}}$ converges to $\int_Y \tilde{f}_h([x])\mathrm{d}\nu([x])$.

Using the monotone convergence theorem again and our discussion at Step 1, we obtain that

$$\langle h, \nu_\mathcal{B}\rangle = \lim_{n \to +\infty} \langle h_n, \nu_\mathcal{B}\rangle = \lim_{n \to +\infty} \int_Y \tilde{f}_{h_n}([x])\mathrm{d}\nu([x]) = \int_Y \tilde{f}_h([x])\mathrm{d}\nu([x]).$$

$\square$

As mentioned before the above lemma, we will now prove that $\nu_\mathcal{B}$ is invariant for $(P_t)_{t \in \mathbb{T}}$.

Theorem 6.3.7. $T_t \nu_\mathcal{B} = \nu_\mathcal{B}$ *for every* $t \in \mathbb{T}$; *that is, the probability measure $\nu_\mathcal{B}$ is invariant for* $(P_t)_{t \in \mathbb{T}}$.

Proof. We have to prove that $T_t \nu_\mathcal{B}(A) = \nu_\mathcal{B}(A)$ for every $t \in \mathbb{T}$ and $A \in \mathcal{B}(X)$.

Thus, let $t \in \mathbb{T}$ and $A \in \mathcal{B}(X)$.

Using the manner in which the operator T_t acts on $\nu_\mathcal{B}$, using the fact that the function $h : X \to \mathbb{R}$ defined by $h(x) = P_t(x, A)$ for every $x \in X$ belongs to $B_b(X)$, applying Lemma 6.3.6, using the definition of $\tilde{f}_h$, using the manner in which T_t acts on $\varepsilon_{[x]}$, $x \in X$, using the fact that $\varepsilon_{[x]}$ is an invariant probability measure for $(T_s)_{s \in \mathbb{T}}$, and the definitions of $\tilde{f}_A$ and $\nu_\mathcal{B}$, we obtain that

$$T_t \nu_\mathcal{B}(A) = \int_X P_t(x, A)\mathrm{d}\nu_\mathcal{B}(x) = \int_X h(x)\mathrm{d}\nu_\mathcal{B}(x)$$

$$= \int_Y \tilde{f}_h([x])\mathrm{d}\nu([x]) = \int_Y \left(\int_X h(y)\mathrm{d}\varepsilon_{[x]}(y)\right)\mathrm{d}\nu([x])$$

$$= \int_Y \left(\int_X P_t(y, A)\mathrm{d}\varepsilon_{[x]}(y)\right)\mathrm{d}\nu([x]) = \int_Y T_t\varepsilon_{[x]}(A)\mathrm{d}\nu([x])$$

$$= \int_Y \varepsilon_{[x]}(A)\mathrm{d}\nu([x]) = \int_Y \tilde{f}_A([x])\mathrm{d}\nu([x]) = \nu_\mathcal{B}(A).$$

$\square$

We now define a collection $\mathcal{F}_{\mathrm{cpie}}$ of subsets of Γ_{cpie} as follows: a subset A of Γ_{cpie} belongs to $\mathcal{F}_{\mathrm{cpie}}$ if, by definition, $A \in \mathcal{B}(X)$, and if A has the following property: $(C_{\mathcal{F}_{\mathrm{cpie}}})$. If $x \in A$ and $y \sim x$, then $y \in A$; that is, $[x] \subseteq A$ whenever $x \in A$.

Lemma 6.3.8. *The collection $\mathcal{F}_{cpie}$ of subsets of Γ_{cpie} is a σ-algebra on Γ_{cpie}.*

Proof. Clearly, the empty set belongs to $\mathcal{F}_{cpie}$ because $\emptyset \in \mathcal{B}(X)$ and condition $C_{\mathcal{F}_{cpie}}$ is satisfied since $\emptyset$ does not have elements.

Also, it is easy to see that Γ_{cpie} belongs to $\mathcal{F}_{cpie}$.

Let $A \in \mathcal{F}_{cpie}$. We will prove that $\Gamma_{cpie} \setminus A \in \mathcal{F}_{cpie}$. To this end, note that since $\Gamma_{cpie} \in \mathcal{B}(X)$ and since $A \in \mathcal{B}(X)$, as well (because we assume that $A \in \mathcal{F}_{cpie}$), it follows that $\Gamma_{cpie} \setminus A \in \mathcal{B}(X)$. Now, if we assume that $\Gamma_{cpie} \setminus A$ does not satisfy condition $(C_{\mathcal{F}_{cpie}})$, then there exists $x \in \Gamma_{cpie} \setminus A$ and $y \in A$ such that $x \sim y$. But then we have obtained a contradiction because we have assumed that $A \in \mathcal{F}_{cpie}$.

Finally, we show that $\mathcal{F}_{cpie}$ is closed under countable unions.

To this end, let $(A_n)_{n \in \mathbb{N}}$ be a sequence of elements of $\mathcal{F}_{cpie}$. Then $A_n \in \mathcal{B}(X)$ for every $n \in \mathbb{N}$, so $\cup_{n=1}^{\infty} A_n$ belongs to $\mathcal{B}(X)$, as well. Thus, in order to show that $\cup_{n=1}^{\infty} A_n \in \mathcal{F}_{cpie}$, we have to prove only that $\cup_{n=1}^{\infty} A_n$ has the property $C_{\mathcal{F}_{cpie}}$. However, this is obvious because if $x \in \cup_{n=1}^{\infty} A_n$, then $x \in A_{n_0}$ for some $n_0 \in \mathbb{N}$, so, since $A_{n_0} \in \mathcal{F}_{cpie}$, we obtain that $[x] \subseteq A_{n_0} \subseteq \cup_{n=1}^{\infty} A_n$.

Thus, $\mathcal{F}_{cpie}$ is a σ-algebra on Γ_{cpie}. $\qquad\square$

Let $B_b(\Gamma_{cpie}, \mathcal{F}_{cpie})$ be the Banach lattice of all real-valued bounded $\mathcal{F}_{cpie}$-measurable functions defined on Γ_{cpie}.

Our goal now is to prove that $B_b(\Gamma_{cpie}, \mathcal{F}_{cpie})$ and $B_b(Y, \mathcal{F})$ are Banach lattice isometric and Riesz isomorphic, where $B_b(Y, \mathcal{F})$ is the Banach lattice of all real-valued bounded $\mathcal{F}$-measurable functions defined on Y. To this end, we need some preparation.

Lemma 6.3.9. *Let $g : \Gamma_{cpie} \to \mathbb{R}$ be a function which is bounded, and measurable with respect to the restriction of $\mathcal{B}(X)$ to Γ_{cpie}. The following assertions are equivalent:*

(a) $g \in B_b(\Gamma_{cpie}, \mathcal{F}_{cpie})$.
(b) $g(x) = g(y)$ for every $x \in \Gamma_{cpie}$ and $y \in \Gamma_{cpie}$ such that $x \sim y$.

Proof. $(a) \Rightarrow (b)$ Let $g \in B_b(\Gamma_{cpie}, \mathcal{F}_{cpie})$, and assume that (b) does not hold true for g. Then, there exist $x \in \Gamma_{cpie}$ and $y \in \Gamma_{cpie}$ such that $x \sim y$ but $g(x) \neq g(y)$. Clearly, we may assume that $g(x) < g(y)$.

Let $\alpha \in \mathbb{R}$ be such that $g(x) < \alpha < g(y)$. Then the set $\{z \in \Gamma_{cpie} \mid g(z) < \alpha\}$ does not belong to $\mathcal{F}_{cpie}$ because $x \in \{z \in \Gamma_{cpie} \mid g(z) < \alpha\}$, but $[x] \not\subseteq \{z \in \Gamma_{cpie} \mid g(z) < \alpha\}$ because $y \notin \{z \in \Gamma_{cpie} \mid g(z) < \alpha\}$. Thus, g is not $\mathcal{F}_{cpie}$-measurable; that is, we have obtained a contradiction.

$(b) \Rightarrow (a)$ Now assume that (b) holds true, but g does not belong to $B_b(\Gamma_{cpie}, \mathcal{F}_{cpie})$. Since g is bounded, it follows that g is not $\mathcal{F}_{cpie}$-measurable. Thus, there exists an $\alpha \in \mathbb{R}$ such that the set $A_\alpha = \{z \in \Gamma_{cpie} \mid g(z) < \alpha\}$ does not belong to $\mathcal{F}_{cpie}$. Since g is measurable with respect to the restriction of $\mathcal{B}(X)$ to Γ_{cpie}, it follows that A_α belongs to $\mathcal{B}(X)$; therefore, the fact that A_α does not belong to $\mathcal{F}_{cpie}$ means that A_α fails to satisfy the property $(C_{\mathcal{F}_{cpie}})$ that appears in the definition of $\mathcal{F}_{cpie}$. Thus, there exist $x \in \Gamma_{cpie}$ and $y \in \Gamma_{cpie}$, $x \sim y$, such that $x \in A_\alpha$ and

$y \notin A_\alpha$; accordingly, $g(x) < \alpha$, $g(y) \geq \alpha$, and $x \sim y$; thus, we have obtained a contradiction which stems from our assumption that $(b) \Rightarrow (a)$ is false. $\square$

Now, we define two operators $\Delta : B_b(\Gamma_{cpie}, \mathcal{F}_{cpie}) \rightarrow B_b(Y, \mathcal{F})$ and $\Lambda : B_b(Y, \mathcal{F}) \rightarrow B_b(\Gamma_{cpie}, \mathcal{F}_{cpie})$ as follows: $\Delta g([x]) = g(x)$ for every $[x] \in Y$ and $g \in B_b(\Gamma_{cpie}, \mathcal{F}_{cpie})$, and $\Lambda h(x) = h([x])$ for every $x \in \Gamma_{cpie}$ and $h \in B_b(Y, \mathcal{F})$. Note that the definition of Δg is correct in the sense that, since $g \in B_b(\Gamma_{cpie}, \mathcal{F}_{cpie})$, it follows that $g(x) = g(y)$ for every $x \in \Gamma_{cpie}$ and $y \in \Gamma_{cpie}$ such that $x \sim y$ by Lemma 6.3.9, so the value $g(x)$ depends only on the class $[x]$ defined by x and not on the particular element of Γ_{cpie} that defines that class.

Lemma 6.3.10. *(a) The operator Δ is well-defined in the sense that $\Delta g \in B_b(Y, \mathcal{F})$ for every $g \in B_b(\Gamma_{cpie}, \mathcal{F}_{cpie})$.*
(b) Similarly, the operator Λ is well-defined in the sense that $\Lambda h \in B_b(\Gamma_{cpie}, \mathcal{F}_{cpie})$ for every $h \in B_b(Y, \mathcal{F})$.

Proof. (a) Let $g \in B_b(\Gamma_{cpie}, \mathcal{F}_{cpie})$.

Taking into consideration that g is a bounded function, and using the definition of Δg, we obtain that Δg is a bounded function, as well.

Thus, in order to complete the proof of (a), we only have to prove that Δg is $\mathcal{F}$-measurable.

To this end, let $\alpha \in \mathbb{R}$ and set $A_\alpha = \{[x] \in Y \mid \Delta g([x]) < \alpha\}$.

Using the definition of B_{A_α} (stated at the beginning of this section before Lemma 6.3.1) and Lemma 6.3.9, we obtain that

$$B_{A_\alpha} = \{y \in \Gamma_{cpie} \mid y \text{ belongs to } [x] \text{ for some } [x] \in A_\alpha\}$$

$$= \{y \in \Gamma_{cpie} \mid g(y) < \alpha\}.$$

Since $g \in B_b(\Gamma_{cpie}, \mathcal{F}_{cpie})$, it follows that $B_{A_\alpha} \in \mathcal{F}_{cpie} \subseteq \mathcal{B}(X)$. Taking into consideration the manner in which the σ-algebra $\mathcal{F}$ was defined, we obtain that $A_\alpha \in \mathcal{F}$.

We have therefore proved that $A_\alpha \in \mathcal{F}$ for every $\alpha \in \mathbb{R}$, so Δg is $\mathcal{F}$-measurable.

(b) We have to prove that $\Lambda h \in B_b(\Gamma_{cpie}, \mathcal{F}_{cpie})$ for every $h \in B_b(Y, \mathcal{F})$.

Therefore, let $h \in B_b(Y, \mathcal{F})$.

In a similar manner as in (a), using the definition of Λh, we obtain that Λh is a bounded function.

Using the definition of Λh again, we obtain that $\Lambda h(x) = \Lambda h(y)$ whenever $x \in \Gamma_{cpie}$, $y \in \Gamma_{cpie}$ and $x \sim y$. Thus, in view of Lemma 6.3.9, in order to complete the proof of the assertion (b) (and of the lemma), it is enough to prove that Λh is measurable with respect to the restriction of $\mathcal{B}(X)$ to Γ_{cpie}. Thus, it is enough to prove that $(\Lambda h)^{-1}(-\infty, \alpha)$ belongs to $\mathcal{B}(X)$ for every $\alpha \in \mathbb{R}$.

To this end, let $\alpha \in \mathbb{R}$ and consider the set $D_\alpha = \{[x] \in Y \mid h([x]) < \alpha\}$. Since $h \in B_b(Y, \mathcal{F})$, it follows that $D_\alpha \in \mathcal{F}$. Using one more time the definition of Λh, we obtain that $\{x \in \Gamma_{cpie} \mid \Lambda h(x) < \alpha\} = \{x \in \Gamma_{cpie} \mid h([x]) < \alpha\} =$

$\{x \in \Gamma_{\text{cpie}} \mid [x] \in D_\alpha\} = B_{D_\alpha}$. (For the definition of B_{D_α}, see the discussion preceding Lemma 6.3.1.) Since $D_\alpha \in \mathcal{F}$, it follows that $B_{D_\alpha} \in \mathcal{B}(X)$. $\square$

Theorem 6.3.11. *(a) The operators* $\Delta : B_b(\Gamma_{\text{cpie}}, \mathcal{F}_{\text{cpie}}) \to B_b(Y, \mathcal{F})$ *and* $\Lambda : B_b(Y, \mathcal{F}) \to B_b(\Gamma_{\text{cpie}}, \mathcal{F}_{\text{cpie}})$ *are linear isometries and lattice isomorphisms.*

(b) The two operators Δ *and* Λ*, discussed in (a), are inverse to each other; that is,* $\Delta\Lambda = \text{Id}_{B_b(Y,\mathcal{F})}$ *and* $\Lambda\Delta = \text{Id}_{B_b(\Gamma_{\text{cpie}}, \mathcal{F}_{\text{cpie}})}$. *Hence, both* Δ *and* Λ *are surjections, so the Banach lattices* $B_b(\Gamma_{\text{cpie}}, \mathcal{F}_{\text{cpie}})$ *and* $B_b(Y, \mathcal{F})$ *are isometric and Riesz isomorphic.*

For the definitions of and details on lattice isomorphisms and Riesz isomorphic vector lattices, see Sect. 4.6.

Proof. (a) It is easy to see that both Δ and Λ are linear positive operators.

Since $\|\Delta g\| = \sup_{[x]\in Y} |\Delta g([x])| = \sup_{x\in\Gamma_{\text{cpie}}} |g(x)| = \|g\|$ for every $g \in B_b(\Gamma_{\text{cpie}}, \mathcal{F}_{\text{cpie}})$, it follows that Δ is an isometry.

Also, in view of the fact that $\|\Lambda h\| = \sup_{x\in\Gamma_{\text{cpie}}} |\Lambda h(x)| = \sup_{[x]\in Y} |h([x])| = \|h\|$ for every $h \in B_b(Y, \mathcal{F})$, we obtain that Λ is an isometry, as well.

We now prove that both Δ and Λ are lattice isomorphisms.

Let $g_1 \in B_b(\Gamma_{\text{cpie}}, \mathcal{F}_{\text{cpie}})$ and $g_2 \in B_b(\Gamma_{\text{cpie}}, \mathcal{F}_{\text{cpie}})$.

Using the definition of Δ, and the fact that the order relation that defines the Banach lattice structure on $B_b(\Gamma_{\text{cpie}}, \mathcal{F}_{\text{cpie}})$ and on $B_b(Y, \mathcal{F})$ is the pointwise order, we obtain that

$$\Delta(g_1 \wedge g_2)([x]) = (g_1 \wedge g_2)(x) = (g_1(x)) \wedge (g_2(x))$$
$$= (\Delta g_1)([x]) \wedge (\Delta g_2)([x]) = ((\Delta g_1) \wedge (\Delta g_2))([x])$$

for every $[x] \in Y$. Using Proposition 4.6.6, we obtain that Δ is a lattice homomorphism.

Now let $h_1 \in B_b(Y, \mathcal{F})$ and $h_2 \in B_b(Y, \mathcal{F})$.

Using the definition of Λ and again using the fact that, as Banach lattices, the order on $B_b(Y, \mathcal{F})$ and on $B_b(\Gamma_{\text{cpie}}, \mathcal{F}_{\text{cpie}})$ is the pointwise order, we obtain that

$$\Lambda(h_1 \wedge h_2)(x) = (h_1 \wedge h_2)([x]) = (h_1([x])) \wedge (h_2([x]))$$
$$= (\Lambda h_1(x)) \wedge (\Lambda h_2(x)) = ((\Lambda h_1) \wedge (\Lambda h_2))(x)$$

for every $x \in \Gamma_{\text{cpie}}$. Thus, $\Lambda(h_1 \wedge h_2) = (\Lambda h_1) \wedge (\Lambda h_2)$ for every $h_1 \in B_b(Y, \mathcal{F})$ and $h_2 \in B_b(Y, \mathcal{F})$; therefore, by Proposition 4.6.6, Λ is a lattice homomorphism.

Since both Δ and Λ are one-to-one (because they are isometries), it follows that Δ and Λ are lattice isomorphisms.

(b) We first prove that $\Delta\Lambda = \text{Id}_{B_b(Y,\mathcal{F})}$. To this end, let $h \in B_b(Y, \mathcal{F})$. Then $(\Lambda h)(x) = h([x])$ and $(\Delta(\Lambda h))([x]) = (\Lambda h)(x)$ for every $[x] \in Y$, so $(\Delta\Lambda h) = h$.

We now prove that $\Lambda\Delta = \mathrm{Id}_{B_b(\Gamma_{\mathrm{cpie}},\mathcal{F}_{\mathrm{cpie}})}$. Thus, let $g \in B_b(\Gamma_{\mathrm{cpie}},\mathcal{F}_{\mathrm{cpie}})$. We obtain that $(\Delta g)([x]) = g(x)$ and $(\Lambda(\Delta g))(x) = (\Delta g)([x])$ for every $x \in \Gamma_{\mathrm{cpie}}$, so $(\Lambda\Delta g) = g$.

We have therefore proved that Λ and Δ are inverse to each other. Accordingly, both Δ and Λ are surjections, so $B_b(\Gamma_{\mathrm{cpie}},\mathcal{F}_{\mathrm{cpie}})$ and $B_b(Y,\mathcal{F})$ are isometric and Riesz isomorphic Banach lattices. $\qquad\square$

Let $\mathcal{M}(\Gamma_{\mathrm{cpie}},\mathcal{F}_{\mathrm{cpie}})$ be the Banach lattice of all signed real-valued measures defined on $(\Gamma_{\mathrm{cpie}},\mathcal{F}_{\mathrm{cpie}})$, and, as before, let $\mathcal{M}_{\mathrm{inv}}(X)$ be the Banach sublattice of all invariant elements of $\mathcal{M}(X)$ with respect to $(P_t)_{t\in\mathbb{T}}$, thought of as a Banach lattice in its own right. Our goal is to prove that $\mathcal{M}(\Gamma_{\mathrm{cpie}},\mathcal{F}_{\mathrm{cpie}})$ and $\mathcal{M}_{\mathrm{inv}}(X)$ are isometric and Riesz isomorphic.

To this end, we start by defining an operator $\mathbf{R}' : \mathcal{M}(\Gamma_{\mathrm{cpie}},\mathcal{F}_{\mathrm{cpie}}) \rightarrow \mathcal{M}(X)$ as follows: if $v \in \mathcal{M}(\Gamma_{\mathrm{cpie}},\mathcal{F}_{\mathrm{cpie}})$, then $\mathbf{R}'v : \mathcal{B}(X) \rightarrow \mathbb{R}$ is given by $\mathbf{R}'v(A) = \int_{\Gamma_{\mathrm{cpie}}} \mathbf{1}_A^{\square}(x)\mathrm{d}v(x)$ for every $A \in \mathcal{B}(X)$.

Note that the integral $\int_{\Gamma_{\mathrm{cpie}}} \mathbf{1}_A^{\square}(x)\mathrm{d}v(x)$ is well-defined in the sense that the restriction of $\mathbf{1}_A^{\square}$ to Γ_{cpie} is v-integrable. Indeed, since (by Proposition 5.2.9) the function $\mathbf{1}_A^{\square}$ is bounded and $\mathcal{B}(X)$-measurable, it follows that the restriction of $\mathbf{1}_A^{\square}$ to Γ_{cpie} is bounded and measurable with respect to the restriction of $\mathcal{B}(X)$ to Γ_{cpie}. Since $\mathbf{1}_A^{\square}(x) = \langle \mathbf{1}_A, \varepsilon_x \rangle$ for every $x \in \Gamma_{\mathrm{cpie}}$, it follows that $\mathbf{1}_A(x) = \mathbf{1}_A(y)$ for every $x \in \Gamma_{\mathrm{cpie}}$ and $y \in \Gamma_{\mathrm{cpie}}$ such that $x \sim y$. Therefore, using Lemma 6.3.9, we obtain that the restriction of $\mathbf{1}_A^{\square}$ to Γ_{cpie} belongs to $B_b(\Gamma_{\mathrm{cpie}},\mathcal{F}_{\mathrm{cpie}})$; hence, the restriction is v-integrable.

Lemma 6.3.12. *The operator $\mathbf{R}'$ is well-defined in the sense that $\mathbf{R}'v$ indeed belongs to $\mathcal{M}(X)$ for every $v \in \mathcal{M}(\Gamma_{cpie},\mathcal{F}_{cpie})$.*

Proof. We first note that it is enough to prove that $\mathbf{R}'v$ belongs to $\mathcal{M}(X)$ whenever v is a positive element of $\mathcal{M}(\Gamma_{\mathrm{cpie}},\mathcal{F}_{\mathrm{cpie}})$. Indeed, if we assume that $\mathbf{R}'v \in \mathcal{M}(X)$ whenever $v \in \mathcal{M}(\Gamma_{\mathrm{cpie}},\mathcal{F}_{\mathrm{cpie}})$, $v \geq 0$, and if $\mu \in \mathcal{M}(\Gamma_{\mathrm{cpie}},\mathcal{F}_{\mathrm{cpie}})$ is not necessarily a positive element of $\mathcal{M}(\Gamma_{\mathrm{cpie}},\mathcal{F}_{\mathrm{cpie}})$, then using our assumption, we obtain that $\mathbf{R}'\mu^+$ and $\mathbf{R}'\mu^-$ belong to $\mathcal{M}(X)$, where μ^+ and μ^- are the positive part and the negative part that appear in the Jordan decomposition of μ. Since $\mathbf{R}'\mu(A) = \mathbf{R}'\mu^+(A) - \mathbf{R}'\mu^-(A)$ for every $A \in \mathcal{B}(X)$, it follows that $\mathbf{R}'\mu$ belongs to $\mathcal{M}(X)$, as well.

Thus, let v be a positive element of $\mathcal{M}(\Gamma_{\mathrm{cpie}},\mathcal{F}_{\mathrm{cpie}})$. We have to prove that $\mathbf{R}'v$ is a finite positive measure on $(X,\mathcal{B}(X))$.

Since the empty set $\emptyset$ has the property that $\mathbf{1}_{\emptyset}^{\square}$ is the constant zero function, it follows that $\mathbf{R}'v(\emptyset) = 0$.

Accordingly, in order to complete the proof of the lemma, we have to prove that $\mathbf{R}'v$ is σ-additive and finite.

To this end, let $(A_n)_{n\in\mathbb{N}}$ be a sequence of mutually disjoint $\mathcal{B}(X)$-measurable subsets of X.

Set $f_n = \sum_{k=1}^{n} \mathbf{1}_{A_k}$ for every $n \in \mathbb{N}$.

Clearly, $(f_n)_{n\in\mathbb{N}}$ is a monotone nondecreasing sequence of positive $\mathcal{B}(X)$-measurable functions that converges everywhere to $\mathbf{1}_{\cup_{k=1}^\infty A_k}$ on X.

Since for every $x \in \Gamma_{\mathrm{cpie}}$, the functions f_n, $n \in \mathbb{N}$, and $\mathbf{1}_{\cup_{k=1}^\infty A_k}$ are integrable with respect to ε_x, using the monotone convergence theorem, we obtain that the sequence $(\langle f_n, \varepsilon_x\rangle)_{n\in\mathbb{N}}$ converges, and $\lim_{n\to+\infty} \langle f_n, \varepsilon_x\rangle = \left\langle \mathbf{1}_{\cup_{k=1}^\infty A_k}, \varepsilon_x\right\rangle$; that is, $\lim_{n\to+\infty} f_n^\square(x)$ exists and $\lim_{n\to+\infty} f_n^\square(x) = \mathbf{1}_{\cup_{k=1}^\infty A_k}^\square(x)$ for every $x \in \Gamma_{\mathrm{cpie}}$.

Using the monotone convergence theorem again but applied this time to the sequence $(g_n)_{n\in\mathbb{N}}$ and to the measure v, where $g_n = f_n^\square$ for every $n \in \mathbb{N}$, we obtain that

$$\mathbf{R}'v(\cup_{n=1}^\infty A_n) = \int_{\Gamma_{\mathrm{cpie}}} \mathbf{1}_{\cup_{k=1}^\infty A_k}^\square(x)\mathrm{d}v(x)$$

$$= \int_{\Gamma_{\mathrm{cpie}}} \lim_{n\to+\infty} \sum_{k=1}^n \mathbf{1}_{A_k}^\square(x)\mathrm{d}v(x)$$

$$= \lim_{n\to+\infty} \int_{\Gamma_{\mathrm{cpie}}} \sum_{k=1}^n \mathbf{1}_{A_k}^\square(x)\mathrm{d}v(x)$$

$$= \lim_{n\to+\infty} \sum_{k=1}^n \int_{\Gamma_{\mathrm{cpie}}} \mathbf{1}_{A_k}^\square(x)\mathrm{d}v(x)$$

$$= \sum_{n=1}^\infty \int_{\Gamma_{\mathrm{cpie}}} \mathbf{1}_{A_n}^\square(x)\mathrm{d}v(x) = \sum_{n=1}^\infty \mathbf{R}'v(A_n).$$

Since $\mathbf{R}'v(\cup_{n=1}^\infty A_n) = \sum_{n=1}^\infty \mathbf{R}'v(A_n)$ for every sequence $(A_n)_{n\in\mathbb{N}}$ of disjoint $\mathcal{B}(X)$-measurable subsets of X, it follows that $\mathbf{R}'v$ is a σ-additive measure on $(X, \mathcal{B}(X))$.

Since $\mathbf{R}'v(X) = \int_{\Gamma_{\mathrm{cpie}}} \mathbf{1}_X^\square(x)\mathrm{d}v(x) = \int_{\Gamma_{\mathrm{cpie}}} \mathrm{d}v = v(\Gamma_{\mathrm{cpie}})$ and since $v(\Gamma_{\mathrm{cpie}})$ is a real number, we obtain that $\mathbf{R}'v$ is a finite measure; thus, $\mathbf{R}'v \in \mathcal{M}(X)$. $\square$

Our next goal is to prove that the range of $\mathbf{R}'$ is equal to $\mathcal{M}_{\mathrm{inv}}(X)$. To this end, we need the following lemma:

Lemma 6.3.13. *Let $v \in \mathcal{M}(\Gamma_{cpie}, \mathcal{F}_{cpie})$. Then*

$$\int f(x)\mathrm{d}(\mathbf{R}'v)(x) = \int_{\Gamma_{cpie}} f^\square(x)\mathrm{d}v(x)$$

for every $f \in B_b(X)$.

Note that, in a similar manner as in the proof that the integral $\int_{\Gamma_{\mathrm{cpie}}} \mathbf{1}_A^\square(x)\mathrm{d}v(x)$ exists for every $A \in \mathcal{B}(X)$ and $v \in \mathcal{M}(\Gamma_{\mathrm{cpie}}, \mathcal{F}_{\mathrm{cpie}})$ (proof given before Lemma 6.3.12), using also Proposition 5.2.9 and Lemma 6.3.9, we obtain that the integral $\int_{\Gamma_{\mathrm{cpie}}} f^{\square}(x)d\nu(x)$ that appears in the above lemma exists as well, for every $f \in B_b(X)$.

Proof (of Lemma 6.3.13). We first note that taking into consideration the manner in which $\mathbf{R}'\nu$ is defined, we obtain that the lemma is true whenever $f = \mathbf{1}_A$ for some $A \in \mathcal{B}(X)$.

Using the above assertion, and Lemma 6.3.12, we obtain that the assertion of the lemma under consideration is also true if f is a simple real-valued $\mathcal{B}(X)$-measurable function because in this case there exist $n \in \mathbb{N}$, n $\mathcal{B}(X)$-measurable subsets $A_1, A_2, \cdots, A_n$ of X, and n real numbers $a_1, a_2, \cdots, a_n$ such that $f = \sum_{k=1}^{n} a_i \mathbf{1}_{A_i}$; therefore, $f^{\square} = \sum_{k=1}^{n} a_i \mathbf{1}_{A_i}^{\square}$, so

$$\langle f, \mathbf{R}'\nu \rangle = \sum_{i=1}^{n} a_i \mathbf{R}'\nu(A_i) = \sum_{i=1}^{n} a_i \int_{\Gamma_{\mathrm{cpie}}} \mathbf{1}_{A_i}^{\square}(x)d\nu(x)$$

$$= \int_{\Gamma_{\mathrm{cpie}}} (\sum_{i=1}^{n} a_i \mathbf{1}_{A_i}^{\square}(x))d\nu(x) = \int_{\Gamma_{\mathrm{cpie}}} f^{\square}(x)d\nu(x).$$

We now prove that the lemma is true in general for every $f \in B_b(X)$. To this end, we note that it is enough to prove the lemma under the assumptions that $f \geq 0$ and $\nu \geq 0$. Indeed, if we know that the lemma is true for every $f \in B_b(X)$, $f \geq 0$, and $\nu \in \mathcal{M}(\Gamma_{\mathrm{cpie}}, \mathcal{F}_{\mathrm{cpie}})$, $\nu \geq 0$, and if we let $g \in B_b(X)$ and $\mu \in \mathcal{M}(\Gamma_{\mathrm{cpie}}, \mathcal{F}_{\mathrm{cpie}})$ be not necessarily positive elements of $B_b(X)$ and $\mathcal{M}(\Gamma_{\mathrm{cpie}}, \mathcal{F}_{\mathrm{cpie}})$, respectively, then using the definition of $g^{\square}$, Lemma 6.3.12, our assumption, and the obvious fact that the operator $R' : \mathcal{M}(\Gamma_{\mathrm{cpie}}, \mathcal{F}_{\mathrm{cpie}}) \to \mathcal{M}(X)$ is linear, we obtain that

$$\int g(x)d(\mathbf{R}'\mu)(x) = \int g^{+}(x)d(\mathbf{R}'\mu^{+})(x) - \int g^{+}(x)d(\mathbf{R}'\mu^{-})(x)$$

$$- \int g^{-}(x)d(\mathbf{R}'\mu^{+})(x) + \int g^{-}(x)d(\mathbf{R}'\mu^{-})(x)$$

$$= \int_{\Gamma_{\mathrm{cpie}}} (g^{+})^{\square}(x)d\mu^{+}(x) - \int_{\Gamma_{\mathrm{cpie}}} (g^{+})^{\square}(x)d\mu^{-}(x)$$

$$- \int_{\Gamma_{\mathrm{cpie}}} (g^{-})^{\square}(x)d\mu^{+}(x) + \int_{\Gamma_{\mathrm{cpie}}} (g^{-})^{\square}(x)d\mu^{-}(x)$$

$$= \int_{\Gamma_{\mathrm{cpie}}} (g^{+})^{\square}(x)d\mu(x) - \int_{\Gamma_{\mathrm{cpie}}} (g^{-})^{\square}(x)d\mu(x)$$

$$= \int_{\Gamma_{\mathrm{cpie}}} g^{\square}(x)d\mu(x),$$

where $g^{+} = g \vee 0$, $g^{-} = (-g) \vee 0$, $\mu^{+} = \mu \vee 0$, and $\mu^{-} = (-\mu) \vee 0$.

Thus, let $f \in B_b(X)$, $f \geq 0$, and $\nu \in \mathcal{M}(\Gamma_{\text{cpie}}, \mathcal{F}_{\text{cpie}})$, $\nu \geq 0$. Also, let $(f_n)_{n \in \mathbb{N}}$ be a monotone nondecreasing sequence of positive real-valued simple $\mathcal{B}(X)$-measurable functions that converges pointwise (everywhere) to f on X (clearly, as shown in measure theory, such a sequence does exist).

By Lemma 6.3.12, $\mathbf{R}'\nu$ belongs to $\mathcal{M}(X)$, and it is easy to see that $\mathbf{R}'\nu \geq 0$ because $\nu \geq 0$. Thus, by the monotone convergence theorem applied to the sequence $(f_n)_{n \in \mathbb{N}}$, and using the fact that the lemma is true for every f_n, $n \in \mathbb{N}$, we obtain that

$$\langle f, \mathbf{R}'(\nu) \rangle = \lim_{n \to +\infty} \langle f_n, \mathbf{R}'(\nu) \rangle = \lim_{n \to +\infty} \int_{\Gamma_{\text{cpie}}} f_n^{\square}(x) d\nu(x). \tag{6.3.4}$$

We now note that using the monotone convergence theorem applied to the monotone nondecreasing sequence $(f_n)_{n \in \mathbb{N}}$ of positive measurable functions that converges everywhere to f, and to the measure ε_x, we obtain that the sequence $(f_n^{\square}(x))_{n \in \mathbb{N}}$ converges to $f^{\square}(x)$ for every $x \in \Gamma_{\text{cpie}}$. Thus, we obtain that $(\mathbf{1}_{\Gamma_{\text{cpie}}} f_n^{\square})_{n \in \mathbb{N}}$, which is a monotone nondecreasing sequence of positive functions, converges pointwise (everywhere) to $\mathbf{1}_{\Gamma_{\text{cpie}}} f^{\square}$ on X. Therefore, we can apply again the monotone convergence theorem, this time to the restrictions to $\mathbf{1}_{\Gamma_{\text{cpie}}}$ of $\mathbf{1}_{\Gamma_{\text{cpie}}} f_n^{\square}$, $n \in \mathbb{N}$, in the measure space $(\Gamma_{\text{cpie}}, \mathcal{F}_{\text{cpie}}, \nu)$ in order to obtain that

$$\lim_{n \to +\infty} \int_{\Gamma_{\text{cpie}}} f_n^{\square}(x) d\nu(x) = \int_{\Gamma_{\text{cpie}}} f^{\square}(x) d\nu(x). \tag{6.3.5}$$

Using the equalities (6.3.4) and (6.3.5), we obtain that

$$\langle f, \mathbf{R}'\nu \rangle = \int_{\Gamma_{\text{cpie}}} f^{\square}(x) d\nu(x).$$

$\square$

Our discussion so far allows us to prove that the range of $\mathbf{R}'$ is $\mathcal{M}_{\text{inv}}(X)$. We do this in the next proposition.

Proposition 6.3.14. *(a)* $\mathbf{R}'\nu \in \mathcal{M}_{inv}(X)$ *for every* $\nu \in \mathcal{M}(\Gamma_{cpie}, \mathcal{F}_{cpie})$.
(b) Conversely, if $\mu \in \mathcal{M}_{inv}(X)$*, then the restriction* ν *of* μ *to* $\mathcal{F}_{cpie}$ *(which obviously belongs to* $\mathcal{M}(\Gamma_{cpie}, \mathcal{F}_{cpie})$*) has the property that* $R'\nu = \mu$.

Proof. (a) Let $\nu \in \mathcal{M}(\Gamma_{\text{cpie}}, \mathcal{F}_{\text{cpie}})$. We have to prove that $P_t(\mathbf{R}'\nu)(A) = (\mathbf{R}'\nu)(A)$ for every $t \in \mathbb{T}$ and $A \in \mathcal{B}(X)$.

To this end, let $t \in \mathbb{T}$ and $A \in \mathcal{B}(X)$.

Since $P_t \mathbf{R}'\nu(A) = \int_X P_t(x, A) d\mathbf{R}'\nu(x)$, we have to prove that

$$\int_X P_t(x, A) d\mathbf{R}'\nu(x) = \mathbf{R}'\nu(A).$$

Using Lemma 6.3.13 applied to the function $f : X \to \mathbb{R}$ defined by $f(x) = P_t(x, A)$ for every $x \in X$ (we can apply the lemma because $f \in B_b(X)$), and the fact that $T_t \varepsilon_x(A) = \varepsilon_x(A)$ (because ε_x is an invariant measure) for every $x \in \Gamma_{\mathrm{cpie}}$, we obtain that

$$\int_X P_t(x, A)\mathrm{d}\mathbf{R}'\nu(x) = \int_X f(x)\mathrm{d}\mathbf{R}'\nu(x) = \int_{\Gamma_{\mathrm{cpie}}} f^{\square}(x)\mathrm{d}\nu(x)$$

$$= \int_{\Gamma_{\mathrm{cpie}}} \langle f, \varepsilon_x \rangle \, \mathrm{d}\nu(x) = \int_{\Gamma_{\mathrm{cpie}}} \left(\int_X P_t(y, A)\mathrm{d}\varepsilon_x(y) \right)\mathrm{d}\nu(x)$$

$$= \int_{\Gamma_{\mathrm{cpie}}} (T_t \varepsilon_x(A))\mathrm{d}\nu(x) = \int_{\Gamma_{\mathrm{cpie}}} \varepsilon_x(A)\mathrm{d}\nu(x)$$

$$= \int_{\Gamma_{\mathrm{cpie}}} \langle \mathbf{1}_A, \varepsilon_x \rangle \, \mathrm{d}\nu(x) = \int_{\Gamma_{\mathrm{cpie}}} \mathbf{1}_A^{\square}(x)\mathrm{d}\nu(x) = \mathbf{R}'\nu(A).$$

(b) We will first prove that the assertion is true whenever $\mu \in \mathcal{M}_{\mathrm{inv}}(X)$ is a probability measure, and then we will discuss the case when $\mu \in \mathcal{M}_{\mathrm{inv}}(X)$ is not necessarily a probability measure.

Thus, let $\mu \in \mathcal{M}_{\mathrm{inv}}(X)$ be a probability measure, and let ν be the restriction of μ to $\mathcal{F}_{\mathrm{cpie}}$. We have to prove that $\mathbf{R}'\nu = \mu$.

To this end, let $A \in \mathcal{B}(X)$, and note (as we did several times earlier) that using the fact that $\mathbf{1}_A^{\square}$ is $\mathcal{B}(X)$-measurable and using the Lemma 6.3.9, we obtain that $\mathbf{1}_{A|\Gamma_{\mathrm{cpie}}}^{\square} \in B_b(\Gamma_{\mathrm{cpie}}, \mathcal{F}_{\mathrm{cpie}})$, where $\mathbf{1}_{A|\Gamma_{\mathrm{cpie}}}^{\square}$ is the restriction of $\mathbf{1}_A^{\square}$ to Γ_{cpie}.

Now let $\mathcal{B}(\Gamma_{\mathrm{cpie}})$ be the restriction of $\mathcal{B}(X)$ to Γ_{cpie}. Then $\mathcal{F}_{\mathrm{cpie}} \subseteq \mathcal{B}(\Gamma_{\mathrm{cpie}})$. Since $\mathbf{1}_{A|\Gamma_{\mathrm{cpie}}}^{\square}$ is $\mathcal{F}_{\mathrm{cpie}}$-measurable, it follows that $\mathbf{1}_{A|\Gamma_{\mathrm{cpie}}}^{\square} \in E(\mathbf{1}_{A|\Gamma_{\mathrm{cpie}}}^{\square} \mid \mathcal{F}_{\mathrm{cpie}})$, where $E(\mathbf{1}_{A|\Gamma_{\mathrm{cpie}}}^{\square} \mid \mathcal{F}_{\mathrm{cpie}})$ is the conditional expectation of $\mathbf{1}_{A|\Gamma_{\mathrm{cpie}}}^{\square}$ with respect to $\mathcal{F}_{\mathrm{cpie}}$ (for the definition of conditional expectations and various properties used here, see Sect. 4.2). Thus, $\int_B \mathbf{1}_{A|\Gamma_{\mathrm{cpie}}}^{\square} \, \mathrm{d}\nu = \int_B \mathbf{1}_{A|\Gamma_{\mathrm{cpie}}}^{\square} \, \mathrm{d}\mu$ for every $B \in \mathcal{F}_{\mathrm{cpie}}$; that is, $\int_B \mathbf{1}_A^{\square}\mathrm{d}\nu = \int_B \mathbf{1}_A^{\square}\mathrm{d}\mu$ for every $B \in \mathcal{F}_{\mathrm{cpie}}$. In particular, $\int_{\Gamma_{\mathrm{cpie}}} \mathbf{1}_A^{\square}\mathrm{d}\nu = \int_{\Gamma_{\mathrm{cpie}}} \mathbf{1}_A^{\square}\mathrm{d}\mu$.

Taking into consideration that Γ_{cpie} is a set of maximal probability, and that μ is an invariant probability measure, we obtain that

$$\int_{\Gamma_{\mathrm{cpie}}} \mathbf{1}_A^{\square}\mathrm{d}\mu = \int_{\Gamma_{\mathrm{cpie}}} \mathbf{1}_A^{*} \, \mathrm{d}\mu = \mu(A).$$

Thus, $\mathbf{R}'\nu(A) = \int_{\Gamma_{\mathrm{cpie}}} \mathbf{1}_A^{\square}(x)\mathrm{d}\nu(x) = \int_{\Gamma_{\mathrm{cpie}}} \mathbf{1}_A^{\square}(x)\mathrm{d}\mu(x) = \mu(A)$.

We have therefore proved that $\mathbf{R}'\nu(A) = \mu(A)$ for every $A \in \mathcal{B}(X)$; that is, $\mathbf{R}'\nu = \mu$ whenever μ is an invariant probability measure and ν is the restriction of μ to $\mathcal{F}_{\mathrm{cpie}}$.

If $\mu \in \mathcal{M}_{\text{inv}}(X)$ is not necessarily a probability measure, then the proof of the proposition is completed if we show that assertion (b) of the proposition is true in each of the following situations:

(i) $\mu = 0$;

(ii) $\mu \geq 0$ and $\mu \neq 0$;

(iii) $\mu \leq 0$ and $\mu \neq 0$;

(iv) $\mu = \mu^+ - \mu^-$, where $\mu^+ = \mu \vee 0 \neq 0$ and $\mu^- = (-\mu) \vee 0 \neq 0$.

(i) If $\mu = 0$, then, obviously, the restriction ν of μ to $\mathcal{F}_{\text{cpie}}$ is the zero measure on $(\Gamma_{\text{cpie}}, \mathcal{F}_{\text{cpie}})$. Thus, using the definition of R', we obtain that (b) is true in this case.

(ii) Assume that $\mu \geq 0$ and $\mu \neq 0$ and set $\mu_1 = \frac{\mu}{\|\mu\|}$. Using the fact that assertion (b) of the proposition is true for probability measures (and that μ_1 is obviously an invariant probability measure for $(T_t)_{t \in \mathbb{T}}$), we obtain that the restriction ν_1 of μ_1 to $\mathcal{F}_{\text{cpie}}$ has the property that $\mathbf{R}'\nu_1 = \mu_1$.

We now note that the restriction of μ to $\mathcal{F}_{\text{cpie}}$ is $\nu = \|\mu\| \, \nu_1$ because $\mu = \|\mu\| \, \mu_1$. Since $\mathbf{R}'$ is a linear operator (the linearity follows from the definition of $\mathbf{R}'$), we obtain that $\mathbf{R}'\nu = \|\mu\| \, \mathbf{R}'\nu_1 = \|\mu\| \, \mu_1 = \mu$.

(iii) Let $\mu \leq 0$, $\mu \neq 0$, and let ν be the restriction of μ to $\mathcal{F}_{\text{cpie}}$. Using the linearity of $\mathbf{R}'$ and applying (ii) to $-\mu$ and to the restriction of $-\mu$ to $\mathcal{F}_{\text{cpie}}$, we obtain that (b) holds true in this case, as well.

(iv) Let $\mu \in \mathcal{M}_{\text{inv}}(X)$, $\mu = \mu^+ - \mu^-$, where $\mu^+ = \mu \vee 0 \neq 0$ and $\mu^- = (-\mu) \vee 0 \neq 0$. Using the linearity of $\mathbf{R}'$ and applying (ii) to μ^+ and μ^-, and to their restrictions ν_1 and ν_2 to $\mathcal{F}_{\text{cpie}}$, we obtain that (b) also holds true in this situation. $\qquad\square$

By the above proposition, the range $\mathbf{R}'$ is precisely $\mathcal{M}_{\text{inv}}(X)$. Thus, it makes sense to define $\mathbf{R} : \mathcal{M}(\Gamma_{\text{cpie}}, \mathcal{F}_{\text{cpie}}) \to \mathcal{M}_{\text{inv}}(X)$ as follows: $\mathbf{R}\nu = \mathbf{R}'\nu$ for every $\nu \in \mathcal{M}(\Gamma_{\text{cpie}}, \mathcal{F}_{\text{cpie}})$. By Proposition 6.3.14, the operator $\mathbf{R}$ is well-defined and surjective.

Note that the operator $\mathbf{R}$ has the same domain and acts exactly as $\mathbf{R}'$, but has as codomain the range of $\mathbf{R}'$.

Let $\mathbf{Q} : \mathcal{M}_{\text{inv}}(X) \to \mathcal{M}(\Gamma_{\text{cpie}}, \mathcal{F}_{\text{cpie}})$ be defined as follows: $\mathbf{Q}\mu : \mathcal{F}_{\text{cpie}} \to \mathbb{R}$ is the restriction of $\mu \in \mathcal{M}_{\text{inv}}(X)$ to $\mathcal{F}_{\text{cpie}}$. Clearly, $\mathbf{Q}$ is a positive linear contraction.

Theorem 6.3.15. (*a*) *The operators* $\mathbf{Q}$ *and* $\mathbf{R}$ *are inverse to each other; that is,*
$$\mathbf{QR} = \text{Id}_{\mathcal{M}(\Gamma_{\text{cpie}}, \mathcal{F}_{\text{cpie}})} \text{ and } \mathbf{RQ} = \text{Id}_{\mathcal{M}_{\text{inv}}(X)}.$$
(*b*) $\mathbf{Q}$ *and* $\mathbf{R}$ *are lattice isomorphisms onto and isometries. Consequently,* $\mathcal{M}(\Gamma_{\text{cpie}}, \mathcal{F}_{\text{cpie}})$ *and* $\mathcal{M}_{\text{inv}}(X)$ *are isometric and Riesz isomorphic vector lattices.*

Proof. (*a*) We first prove that $\mathbf{QR} = \text{Id}_{\mathcal{M}(\Gamma_{\text{cpie}}, \mathcal{F}_{\text{cpie}})}$. Thus, we have to prove that
$$\mathbf{QR}\nu = \nu \text{ for every } \nu \in \mathcal{M}(\Gamma_{\text{cpie}}, \mathcal{F}_{\text{cpie}}).$$
To this end, let $\nu \in \mathcal{M}(\Gamma_{\text{cpie}}, \mathcal{F}_{\text{cpie}})$.

In order to prove that $\mathbf{QR}\nu = \nu$, we have to prove that the restriction of $\mathbf{R}\nu$ to $\mathcal{F}_{\text{cpie}}$ is equal to ν; that is, we have to prove that $\mathbf{R}\nu(A) = \nu(A)$ for every $A \in \mathcal{F}_{\text{cpie}}$.

Thus, let $A \in \mathcal{F}_{\text{cpie}}$.

The restriction of $\mathbf{1}_A : X \to \mathbb{R}$ to Γ_{cpie} is equal to the restriction of $\mathbf{1}_A^{\square}$ to Γ_{cpie} because, using the definition of $\mathcal{F}_{\text{cpie}}$ and the fact that $A \in \mathcal{F}_{\text{cpie}}$, we obtain that

$$\mathbf{1}_A(x) = 1 \Leftrightarrow x \in A \Leftrightarrow [x] \subseteq A \Rightarrow \langle \mathbf{1}_A, \varepsilon_x \rangle = 1 \Rightarrow \mathbf{1}_A^{\square}(x) = 1,$$

and because, using again the fact that $A \in \mathcal{F}_{\text{cpie}}$, we obtain that

$$\mathbf{1}_A(x) = 0 \Leftrightarrow x \notin A \Leftrightarrow [x] \cap A = \emptyset \Rightarrow \langle \mathbf{1}_A, \varepsilon_x \rangle = 0 \Rightarrow \mathbf{1}_A^{\square}(x) = 0$$

for every $x \in \Gamma_{\text{cpie}}$.

Thus, using the fact that the restriction of $\mathbf{1}_A$ to Γ_{cpie} is equal to the restriction of $\mathbf{1}_A^{\square}$ to Γ_{cpie}, we obtain that

$$\mathbf{R}\nu(A) = \int_{\Gamma_{\text{cpie}}} \mathbf{1}_A^{\square}(x)\mathrm{d}\nu(x) = \int_{\Gamma_{\text{cpie}}} \mathbf{1}_A(x)\mathrm{d}\nu(x) = \nu(A).$$

Accordingly, $\mathbf{QR}\nu = \nu$ whenever $\nu \in \mathcal{M}(\Gamma_{\text{cpie}}, \mathcal{F}_{\text{cpie}})$; that is, $\mathbf{QR} = \text{Id}_{\mathcal{M}(\Gamma_{\text{cpie}}, \mathcal{F}_{\text{cpie}})}$.

We now note that the fact that $\mathbf{RQ} = \text{Id}_{\mathcal{M}_{\text{inv}}(X)}$ is a consequence of (b) of Proposition 6.3.14.

Indeed, $\mathbf{R}'\mathbf{Q}\mu = \mu$ for every $\mu \in \mathcal{M}_{\text{inv}}(X)$ by (b) of the above-mentioned proposition. Since $\mathbf{R}'\nu = \mathbf{R}\nu$ for every $\nu \in \mathcal{M}(\Gamma_{\text{cpie}}, \mathcal{F}_{\text{cpie}})$, and since $\mathbf{Q}\mu \in \mathcal{M}(\Gamma_{\text{cpie}}, \mathcal{F}_{\text{cpie}})$, it follows that $\mathbf{RQ}\mu = \mathbf{R}'\mathbf{Q}\mu = \mu$ for every $\mu \in \mathcal{M}_{\text{inv}}(X)$.

(b) Clearly, both $\mathbf{Q}$ and $\mathbf{R}$ are positive linear operators. Using (a), we obtain that $\mathbf{Q}$ and $\mathbf{R}$ are one-to-one onto, and each operator is the inverse of the other. Thus, by Proposition 4.6.7, both $\mathbf{Q}$ and $\mathbf{R}$ are lattice isomorphisms onto.

Since both $\mathbf{Q}$ and $\mathbf{R}$ are lattice homomorphisms, using Proposition 4.6.6 we obtain that:

(i) $(\mathbf{Q}\mu_1) \wedge (\mathbf{Q}\mu_2) = 0$ whenever $\mu_1 \in \mathcal{M}_{\text{inv}}(X)$ and $\mu_2 \in \mathcal{M}_{\text{inv}}(X)$ are such that $\mu_1 \wedge \mu_2 = 0$.

(ii) $(\mathbf{R}\nu_1) \wedge (\mathbf{R}\nu_2) = 0$ for every $\nu_1 \in \mathcal{M}(\Gamma_{\text{cpie}}, \mathcal{F}_{\text{cpie}})$ and $\nu_2 \in \mathcal{M}(\Gamma_{\text{cpie}}, \mathcal{F}_{\text{cpie}})$ such that $\nu_1 \wedge \nu_2 = 0$.

Since $\mathbf{Q}$ and $\mathbf{R}$ are Markov operators and satisfy (i) and (ii), respectively, it follows that both $\mathbf{Q}$ and $\mathbf{R}$ are isometries.

Thus $\mathcal{M}(\Gamma_{\text{cpie}}, \mathcal{F}_{\text{cpie}})$ and $\mathcal{M}_{\text{inv}}(X)$ are isometric and Riesz isomorphic Banach lattices. $\square$

We now prove that $\mathcal{M}(\Gamma_{\text{cpie}}, \mathcal{F}_{\text{cpie}})$ and $\mathcal{M}(Y, \mathcal{F})$ are isometric and Riesz isomorphic in a standard way in the sense that we will construct "standard" operators

$\mathbf{V} : \mathcal{M}(\Gamma_{\text{cpie}}, \mathcal{F}_{\text{cpie}}) \to \mathcal{M}(Y, \mathcal{F})$ and $\mathbf{W} : \mathcal{M}(Y, \mathcal{F}) \to \mathcal{M}(\Gamma_{\text{cpie}}, \mathcal{F}_{\text{cpie}})$ such that both $\mathbf{V}$ and $\mathbf{W}$ are isometries and lattice isomorphisms onto, and each operator is the inverse of the other.

To this end, we consider the operators $\Delta : B_b(\Gamma_{\text{cpie}}, \mathcal{F}_{\text{cpie}}) \to B_b(Y, \mathcal{F})$ and $\Lambda : B_b(Y, \mathcal{F}) \to B_b(\Gamma_{\text{cpie}}, \mathcal{F}_{\text{cpie}})$ that we defined before Lemma 6.3.10.

Using the definitions of Δ and Λ given there, we obtain that

$$\Delta \mathbf{1}_B = \mathbf{1}_{A_B} \tag{6.3.6}$$

for every $B \in \mathcal{F}_{\text{cpie}}$, where $A_B = \{[x] \in Y \mid x \in B\}$, and

$$\Lambda \mathbf{1}_A = \mathbf{1}_{B_A} \tag{6.3.7}$$

for every $A \in \mathcal{F}$, where $B_A = \{x \in \Gamma_{\text{cpie}} \mid [x] \in A\}$.

The equalities (6.3.6) and (6.3.7) suggest to us that we construct the following maps: $\boldsymbol{\Delta} : \mathcal{F}_{\text{cpie}} \to \mathcal{F}$ defined by $\boldsymbol{\Delta}(\emptyset) = \emptyset$ and $\boldsymbol{\Delta}(B) = A_B$ for every $B \in \mathcal{F}_{\text{cpie}}$, $B \neq \emptyset$ (note that, as a result of the manner in which $\mathcal{F}$ was defined, $\boldsymbol{\Delta}$ is well-defined in the sense that $\boldsymbol{\Delta}(B) \in \mathcal{F}$ for every $B \in \mathcal{F}_{\text{cpie}}$), and $\boldsymbol{\Lambda} : \mathcal{F} \to \mathcal{F}_{\text{cpie}}$ defined by $\boldsymbol{\Lambda} A = B_A$ for every $A \in \mathcal{F}$, where $B_\emptyset = \emptyset$ (as in the case of $\boldsymbol{\Delta}$, observe that the definition of $\boldsymbol{\Lambda}$ is correct in the sense that $B_A \in \mathcal{F}_{\text{cpie}}$ whenever $A \in \mathcal{F}$).

Lemma 6.3.16. (*a*) $\boldsymbol{\Lambda}\boldsymbol{\Delta} = \mathrm{Id}_{\mathcal{F}_{\text{cpie}}}$ *and* $\boldsymbol{\Delta}\boldsymbol{\Lambda} = \mathrm{Id}_{\mathcal{F}}$. *Thus,* $\boldsymbol{\Lambda}$ *and* $\boldsymbol{\Delta}$ *are one-to-one, onto, and each map is the inverse of the other.*

(*b*) *The set function* $\boldsymbol{\Delta}$ *has the following properties:*

    (*b*1) $B_1 \subseteq B_2$ *if and only if* $\boldsymbol{\Delta} B_1 \subseteq \boldsymbol{\Delta} B_2$ *for every* $B_1 \in \mathcal{F}_{\text{cpie}}$ *and* $B_2 \in \mathcal{F}_{\text{cpie}}$.
    (*b*2) $B_1 \cap B_2 = \emptyset$ *if and only if* $(\boldsymbol{\Delta} B_1) \cap (\boldsymbol{\Delta} B_2) = \emptyset$ *whenever* $B_1 \in \mathcal{F}_{\text{cpie}}$ *and* $B_2 \in \mathcal{F}_{\text{cpie}}$.

(*c*) *Similarly, the map* $\boldsymbol{\Lambda}$ *has the following properties:*

    (*c*1) $A_1 \subseteq A_2$ *if and only if* $\boldsymbol{\Lambda} A_1 \subseteq \boldsymbol{\Lambda} A_2$ *whenever* $A_1 \in \mathcal{F}$ *and* $A_2 \in \mathcal{F}$.
    (*c*2) $A_1 \cap A_2 = \emptyset$ *if and only if* $(\boldsymbol{\Lambda} A_1) \cap (\boldsymbol{\Lambda} A_2) = \emptyset$.

Proof. (a) Let $B \in \mathcal{F}_{\text{cpie}}$, and $A = \{[x] \in Y \mid x \in B\}$.

Note that B is nonempty if and only if $A \neq \emptyset$. Note also that $A = A_B$ and $B = B_A$ whenever $A \in \mathcal{F}$ or $B \in \mathcal{F}_{\text{cpie}}$.

We obtain that $\boldsymbol{\Delta}(\boldsymbol{\Lambda}(A)) = \boldsymbol{\Delta} B_A = A_{B_A} = A$, and that $\boldsymbol{\Lambda}(\boldsymbol{\Delta}(B)) = \boldsymbol{\Lambda}(A_B) = B_{A_B} = B$.

We have therefore proved that $\boldsymbol{\Delta}\boldsymbol{\Lambda} = \mathrm{Id}_{\mathcal{F}}$ and $\boldsymbol{\Lambda}\boldsymbol{\Delta} = \mathrm{Id}_{\mathcal{F}_{\text{cpie}}}$; thus, $\boldsymbol{\Delta}$ and $\boldsymbol{\Lambda}$ are also one-to-one and onto.

(b) We now prove that $\boldsymbol{\Delta}$ has the properties (b1) and (b2).

The fact that $\boldsymbol{\Delta}$ has property (b1) follows from the manner in which $\boldsymbol{\Delta}$ is defined.

We now prove that $\boldsymbol{\Delta}$ has property (b2).

To this end, assume first that $B_1 \in \mathcal{F}_{\text{cpie}}$ and $B_2 \in \mathcal{F}_{\text{cpie}}$ are such that $B_1 \cap B_2 = \emptyset$, but $(\boldsymbol{\Delta} B_1) \cap (\boldsymbol{\Delta} B_2) \neq \emptyset$. Then there exists $[x] \in (\boldsymbol{\Delta} B_1) \cap (\boldsymbol{\Delta} B_2)$;

however, since $\mathbf{\Delta}$ satisfies property (b1), it follows that $\{y \in \Gamma_{\text{cpie}} \mid y \sim x\} \subseteq B_1 \cap B_2$, so we obtain a contradiction because we assume that $B_1 \cap B_2 = \emptyset$.

Conversely, if we assume that $(\mathbf{\Delta}\, B_1) \cap (\mathbf{\Delta}\, B_2) = \emptyset$ for some $B_i \in \mathcal{F}_{\text{cpie}}$, $i = 1, 2$, but $B_1 \cap B_2 \neq \emptyset$, then there exists an $x \in B_1 \cap B_2$. Using the definition of $\mathbf{\Delta}$, we obtain that $[x] \in (\mathbf{\Delta}\, B_1) \cap (\mathbf{\Delta}\, B_2)$, so $(\mathbf{\Delta}\, B_1) \cap (\mathbf{\Delta}\, B_2) \neq \emptyset$; that is, we have obtained a contradiction.

The proof of (c) is obtained using (a) and (b). $\square$

We will now use the maps $\mathbf{\Delta}$ and $\mathbf{\Lambda}$ in order to construct the two operators V and W that we mentioned after Theorem 6.3.15.

The first operator $V : \mathcal{M}(\Gamma_{\text{cpie}}, \mathcal{F}_{\text{cpie}}) \to \mathcal{M}(Y, \mathcal{F})$ is defined in the following manner: for every $\mu \in \mathcal{M}(\Gamma_{\text{cpie}}, \mathcal{F}_{\text{cpie}})$, let $V\mu : \mathcal{F} \to \mathbb{R}$ be given by $V\mu(A) = \mu(\mathbf{\Lambda}\, A)$ for every $A \in \mathcal{F}$.

The second operator $W : \mathcal{M}(Y, \mathcal{F}) \to \mathcal{M}(\Gamma_{\text{cpie}}, \mathcal{F}_{\text{cpie}})$ is defined as follows: if $v \in \mathcal{M}(Y, \mathcal{F})$, then $Wv(B) = v(\mathbf{\Delta}\, B)$ for every $B \in \mathcal{F}_{\text{cpie}}$, where $Wv : \mathcal{F}_{\text{cpie}} \to \mathbb{R}$.

Theorem 6.3.17. *(a) V and W are linear bounded positive operators such that $\|V\mu\| = \|\mu\|$ and $\|Wv\| = \|v\|$ for every $\mu \in \mathcal{M}(\Gamma_{cpie}, \mathcal{F}_{cpie})$, $\mu \geq 0$, and every $v \in \mathcal{M}(Y, \mathcal{F})$, $v \geq 0$ (that is, both V and W are Markov operators (as defined in Sect. 4.6 before Example 4.6.5)).*

(b) $VW = \text{Id}_{\mathcal{M}(Y,\mathcal{F})}$ and $WV = \text{Id}_{\mathcal{M}(\Gamma_{cpie}, \mathcal{F}_{cpie})}$; consequently, both V and W are one-to-one, onto, and each operator is the inverse of the other.

(c) Both V and W are isometries and lattice isomorphisms onto. Therefore, the spaces $\mathcal{M}(\Gamma_{cpie}, \mathcal{F}_{cpie})$ and $\mathcal{M}(Y, \mathcal{F})$ are isometric and Riesz isomorphic.

Proof. (a) The fact V and W are Markov operators is obvious.

(b) Clearly, $VW : \mathcal{M}(Y, \mathcal{F}) \to \mathcal{M}(Y, \mathcal{F})$ and $WV : \mathcal{M}(\Gamma_{\text{cpie}}, \mathcal{F}_{\text{cpie}}) \to \mathcal{M}(\Gamma_{\text{cpie}}, \mathcal{F}_{\text{cpie}})$.

Now, let $v \in \mathcal{M}(Y, \mathcal{F})$ and $\mu \in \mathcal{M}(\Gamma_{\text{cpie}}, \mathcal{F}_{\text{cpie}})$. Using the definitions of V and W, and (a) of Lemma 6.3.16, we obtain that

$$V(Wv)(A) = Wv(\mathbf{\Lambda}\, A) = v(\mathbf{\Delta}\,\mathbf{\Lambda}\, A) = v(A)$$

for every $A \in \mathcal{F}$, and that

$$W(V\mu)(B) = V\mu(\mathbf{\Delta}\, B) = \mu(\mathbf{\Lambda}\,\mathbf{\Delta}\, B) = \mu(B)$$

for every $B \in \mathcal{F}_{\text{cpie}}$. Thus, $VWv = v$ and $WV\mu = \mu$.

(c) Since both operators V and W are positive, and each operator is the inverse of the other, by Proposition 4.6.7, we obtain that both V and W are lattice isomorphisms (onto). In particular, both operators are lattice homomorphisms. Using the fact that both V and W are Markov operators and using the implication $(a) \Rightarrow (c)$ of Proposition 4.6.6, we obtain that both V and W are isometries (onto). Thus, the spaces $\mathcal{M}(\Gamma_{\text{cpie}}, \mathcal{F}_{\text{cpie}})$ and $\mathcal{M}(Y, \mathcal{F})$ are isometric and Riesz isomorphic in a standard way (via the operators V and W). $\square$

In the next theorem we summarize the results obtained in this section.

Theorem 6.3.18. *The Banach lattice $\mathcal{M}_{inv}(X)$ is isometric and Riesz isomorphic in a standard manner to each of the spaces $\mathcal{M}(\Gamma_{cpie}, \mathcal{F}_{cpie})$ and $\mathcal{M}(Y, \mathcal{F})$.*

Proof. The proof is obvious in view of Theorems 6.3.15 and 6.3.17. $\square$

Note that Theorem 6.3.18 extends the results of Section 12 of Phelps' monograph [86] (in particular, the fact that $\mathcal{M}_{inv}(X)$ and $\mathcal{M}(Y, \mathcal{F})$ are isometric and Riesz isomorphic extends the theorem on p. 77 of Phelps [86]), certain results of Rochlin [101] and of Sections 8 and 9 of Oxtoby [85].

Chapter 7
Feller Transition Functions

As discussed at the beginning of Sect. 1.4, our goal in this chapter is to extend several results of [143] to Feller transition functions.

We obtained the results that will be discussed here during the years 2000–2007 (the main results of Sect. 7.3 have been announced in [147], see also [33] for results similar to the main results of Sect. 7.3). Later, after our papers [146] and [147] appeared in print, independently of us, Worm and Hille [133] obtained some of the results discussed in this chapter in the more general case of transition functions defined on Polish spaces.

We stress that in spite of a few overlapping results mentioned above, the paper [133] by Worm and Hille contains many topics that are new even in the setting of locally compact separable metric spaces. They consider an extension of $C_0(X)$-equicontinuity called the e-property; the e-property was introduced in Lasota and Szarek [63] and can be used whenever X is Polish and not necessarily locally compact. They study the KBBY decomposition of transition functions and transition probabilities that have the Cesàro e-property (as expected, the Cesàro e-property is an e-property defined in terms of averages); moreover, they study uniquely ergodic transition functions and uniquely ergodic transition probabilities that have the Cesàro e-property; finally, they obtain criteria for asymptotic stability for transition functions and transition probabilities.

In Szarek and Worm [122], the authors study transition functions and transition probabilities that have the Cesàro e-property or the eventual e-property (a transition function or transition probability has the eventual e-property if the condition defining the e-property is satisfied for large values of $t \in \mathbb{T}$, where $\mathbb{T} = \mathbb{N}$ or $[0, +\infty)$). They introduce various notions (like the weak concentration condition) that stem from earlier work of the first author, and study the structure of the set of ergodic measures of a given transition probability or transition function. Also, they study conditions for weakly mean ergodicity and asymptotic stability of transition probabilities or transition functions. For various other results related to the topics discussed in [122], see Kapica, Szarek and Ślęczka [51], Komorowski, Peszat and

R. Zaharopol, *Invariant Probabilities of Transition Functions*, Probability and Its Applications 44, DOI 10.1007/978-3-319-05723-1_7,
© Springer International Publishing Switzerland 2014

Szarek [52], Lasota and Szarek [63], Szarek [120] and [121], and Szarek, Ślęczka and Urbański [123].

As in the case of transition probabilities, Feller transition functions have the remarkable property that elementary measures (not necessarily standard) are invariant. We begin the first section of this chapter by discussing this fact. Then we deal with the main topic of the section, namely, the supports of the elementary measures of Feller transition functions. We also discuss here minimal and trivially minimal transition functions.

Given a transition function $(P_t)_{t \in \mathbb{T}}$, it is of interest to compare the structure of the set of invariant probabilities (existence of invariant probability measures, the supports of these invariant probabilities, and the KBBY decomposition for the transition function) of $(P_t)_{t \in \mathbb{T}}$, on one hand, with the structure of the set of invariant probabilities of the individual transition probabilities P_t, $t \in \mathbb{T}$, that form the transition function, on the other hand; we consider various results in this direction for Feller transition functions in Sect. 7.1, as well.

In Sect. 7.2, we extend the results of Chapter 3 and Section 4.1 of [143] to transition functions (these results of [143] are summarized in Sect. 1.4.1 and part of Sect. 1.4.2 in this book). We start by characterizing the support of the invariant probability of a uniquely ergodic Feller transition function, and by characterizing the closed subset F of a transition function that is not necessarily uniquely ergodic but has the property that each of its invariant probabilities have the entire F as their support. We discuss dominant generic points and their role in studying ergodic invariant probability measures of Feller transition functions. We also obtain a criterion for the unique ergodicity of a $C_0(X)$-equicontinuous Feller transition function similar to the corresponding result for transition probabilities obtained in Section 4.1 of [143].

In the last section of the chapter and of the book (Sect. 7.3), we obtain the mean ergodic theorems that we announced in [147], and which we mentioned earlier. We note that the approach to the proofs of the mean ergodic theorems for transition functions discussed in this section allows us to obtain a new proof of Proposition 4.2.2 of [143]. We also present various examples that illustrate the usefulness of the mean ergodic theorems proved in the section.

7.1   Elementary Measures and Their Supports

In this section, after discussing the fact that the elementary measures (not necessarily standard) of a Feller transition function are invariant, we obtain several criteria for the existence of invariant probabilities of Feller transition functions, we study the supports of the elementary measures of Feller transition functions, and we discuss minimal and trivially minimal transition functions.

We conclude the section with results in which we compare the structure of the set of invariant probabilities of a Feller transition function $(P_t)_{t \in \mathbb{T}}$ on one hand, with the structure of the set of invariant probabilities of the individual transition probabilities P_t, $t \in \mathbb{T}$, that form the transition function, on the other hand. Note that results of this type for not necessarily Feller transition functions appear in this book in Sect. 3.3.2 (see especially Corollaries 3.3.10–3.3.12).

Let (X, d) be a locally compact separable metric space.

Also, let $(P_t)_{t \in \mathbb{T}}$ be a Feller transition function, and assume that $(P_t)_{t \in \mathbb{T}}$ satisfies the s.m.a. and is pointwise continuous.

In the next theorem we show that since $(P_t)_{t \in \mathbb{T}}$ is a Feller transition function, each elementary measure of $(P_t)_{t \in \mathbb{T}}$ is invariant. (Recall that $\mu \in \mathcal{M}(X)$, $\mu \geq 0$, is said to be an *elementary measure of* $(P_t)_{t \in \mathbb{T}}$ if $\mu \neq 0$ and $\mu = \varepsilon_x^{(L^{(\mathrm{ct})})}$ for some $x \in X$ and some continuous-time Banach limit $L^{(\mathrm{ct})}$ (for the definition of $\varepsilon_x^{(L^{(\mathrm{ct})})}$, see Sect. 5.1)).

Theorem 7.1.1. *Let $x \in X$ and let $L^{(ct)}$ be a continuous-time Banach limit. Then the measure $\varepsilon_x^{(L^{(\mathrm{ct})})}$ is an invariant measure for $(P_t)_{t \in \mathbb{T}}$. In particular, the elementary measures are invariant for $(P_t)_{t \in \mathbb{T}}$.*

Proof. Let $x \in X$ and let $L^{(ct)}$ be a continuous-time Banach limit. We have to prove that $\varepsilon_x^{(L^{(\mathrm{ct})})}$ is an invariant measure for $(P_t)_{t \in \mathbb{T}}$.

Using Proposition 2.3.7, we obtain that in order to prove that $\varepsilon_x^{(L^{(\mathrm{ct})})}$ is an invariant measure for $(P_t)_{t \in \mathbb{R}}$, it is enough to prove that $\varepsilon_x^{(L^{(\mathrm{ct})})}$ is invariant for $(P_t)_{t \in [0,+\infty)}$.

In order to prove that $\varepsilon_x^{(L^{(\mathrm{ct})})}$ is an invariant measure for $(P_t)_{t \in [0,+\infty)}$, we will use Theorem 1.2.3 (the Lasota-Yorke lemma). To this end, let $\phi : C_b(X) \to \mathbb{R}$ be defined as follows: $\phi(f) = L^{(\mathrm{ct})}((S_t f(x))_{t \in [0,+\infty)})$ for every $f \in C_b(X)$.

Since $(P_t)_{t \in [0,+\infty)}$ satisfies the s.m.a., using Corollary 2.1.6, we obtain that the real-valued map $t \mapsto S_t f(x)$, $t \in [0, +\infty)$, is measurable with respect to the Borel σ-algebra $\mathcal{B}(\mathbb{R})$ on $\mathbb{R}$ and the σ-algebra of all Lebesgue measurable subsets of $[0, +\infty)$; also, since $|S_t f(x)| \leq \|f\|$ for every $t \in [0, +\infty)$, it follows that the map $(S_t f(x))_{t \in [0,+\infty)}$ is bounded. Therefore, $(S_t(f(x)))_{t \in [0,+\infty)}$ belongs to the Banach lattice $B_b^{(L)}([0, +\infty))$ of all real-valued bounded Lebesgue measurable functions defined on $[0, +\infty)$; accordingly, $L^{(\mathrm{ct})}((S_t f(x))_{t \in [0,+\infty)})$ is well-defined, so $\phi(f)$ is also well-defined for every $f \in C_b(X)$.

It is easy to see that ϕ is a positive linear functional of $C_b(X)$, so ϕ is continuous.

Now let $\alpha \in [0, +\infty)$. Since $L^{(ct)}$ is a continuous-time Banach limit, using condition (b) in the definition of a continuous-time Banach limit (see the beginning of Sect. 4.4), we obtain that

$$\phi(S_\alpha f) = L^{(\mathrm{ct})}((S_t S_\alpha f(x))_{t \in [0,+\infty)})$$

$$= L^{(\mathrm{ct})}((S_t f(x))_{t \in [0,+\infty)}) = \phi(f)$$

for every $f \in C_b(X)$.

Using the Lasota-Yorke lemma (see Theorem 1.2.3) and the fact that the restriction of ϕ to $C_0(X)$ is obviously $\varepsilon_x^{(L^{(\mathrm{ct})})}$, we obtain that $T_\alpha(\varepsilon_x^{(L^{(\mathrm{ct})})}) = (\varepsilon_x^{(L^{(\mathrm{ct})})})$.

Since $T_\alpha(\varepsilon_x^{(L^{(ct)})}) = \varepsilon_x^{(L^{(ct)})}$ for every $\alpha \in [0, +\infty)$, it follows that $\varepsilon_x^{(L^{(ct)})}$ is an invariant measure for $(P_t)_{t \in \mathbb{T}}$. In particular, the elementary measures are invariant for $(P_t)_{t \in \mathbb{T}}$. $\qquad\square$

While Theorem 7.1.1 is an extension of Theorem 2.1.1 of [143] to transition functions (see also Theorem 1.4.1), the next theorem is an extension of Theorem 2.1.2 of [143] to transition functions (see Theorem 1.4.2, as well).

Theorem 7.1.2. *Let $(P_t)_{t \in \mathbb{T}}$ be a Feller transition function defined on (X, d), and assume that $(P_t)_{t \in \mathbb{T}}$ satisfies the s.m.a. The following assertions are equivalent:*

(a) $(P_t)_{t \in \mathbb{T}}$ has invariant probability measures.
(b) $\Gamma \neq \emptyset$.
(c) There exists an $x_0 \in X$ and a compact subset K of X such that

$$\liminf_{t \to +\infty} \frac{1}{t} \int_0^t T_u \delta_{x_0}(K) \, du > 0.$$

Observation. (1) Recall that the set Γ was defined at the beginning of Sect. 5.1.
(2) Let $x_0 \in X$ and let K be a compact subset of X. Since $T_u \delta_{x_0}(K) = S_u 1_K(x_0)$ for every $u \in \mathbb{T}$ and since $0 \le S_u 1_K(x_0) \le 1$ for every $u \in \mathbb{T}$, it follows that the map $u \mapsto T_u \delta_{x_0}(K)$, $u \in (0, +\infty)$, is bounded; by Corollary 2.1.6, the map is also measurable with respect to the Borel σ-algebra on $\mathbb{R}$ and the σ-algebra $\mathcal{L}((0, +\infty))$ of all Lebesgue measurable subsets of $(0, +\infty)$.
    Accordingly, the integral $\int_0^t T_u \delta_{x_0}(K) \, du$ exists for every $t \in (0, +\infty)$, so $\liminf_{t \to +\infty} \frac{1}{t} \int_0^t T_u \delta_{x_0}(K) \, du$ is well-defined. $\qquad\blacktriangle$

Proof (of Theorem 7.1.2). (a) $\Rightarrow$ (c) Assume that (a) holds true, so $(P_t)_{t \in \mathbb{T}}$ has invariant probability measures. By Corollary 6.2.8, $\Gamma_{\mathrm{cpie}} \neq \emptyset$, so we may and do choose $x_0 \in \Gamma_{\mathrm{cpie}}$.
    Since ε_{x_0} is a standard elementary probability measure, there exists an $f \in C_0(X)$, $0 \le f \le 1_X$, such that $\langle f, \varepsilon_{x_0} \rangle > 0$. Since the vector space $C_c(X)$ of all real-valued continuous functions on X that have compact supports is dense in $C_0(X)$ with respect to the uniform (sup) norm on $C_0(X)$, it follows that we can choose f to have compact support.
    Let K be the support of f. Thus, K is compact and $0 \le f \le 1_K$.
    We obtain that

$$0 < \langle f, \varepsilon_{x_0} \rangle = \lim_{\substack{t \to +\infty \\ t > 0}} \frac{1}{t} \int_0^t S_u f(x_0) \, du$$

$$= \liminf_{\substack{t \to +\infty \\ t > 0}} \frac{1}{t} \int_0^t S_u f(x_0) \, du$$

$$\le \liminf_{\substack{t \to +\infty \\ t > 0}} \frac{1}{t} \int_0^t S_u 1_K(x_0) \, du$$

$$= \liminf_{\substack{t \to +\infty \\ t > 0}} \frac{1}{t} \int_0^t \langle S_u 1_K, \delta_{x_0} \rangle \, du$$

$$= \liminf_{\substack{t \to +\infty \\ t > 0}} \frac{1}{t} \int_0^t \langle 1_K, T_u \delta_{x_0} \rangle \, du$$

$$= \liminf_{\substack{t \to +\infty \\ t > 0}} \frac{1}{t} \int_0^t T_u \delta_{x_0}(K) \, du.$$

(c) $\Rightarrow$ (b) Since we assume that (c) holds true, there exist $x_0 \in X$ and a compact subset K of X such that

$$\liminf_{\substack{t \to +\infty \\ t > 0}} \frac{1}{t} \int_0^t S_u 1_K(x_0) \, du > 0.$$

Using Proposition 7.1.8, p. 199 of Cohn's book [20] we further obtain that there exists an $f \in C_0(X)$ (actually, we can choose $f \in C_0(X)$ with compact support) such that $1_K \leq f \leq 1_X$.

It follows that

$$\liminf_{\substack{t \to +\infty \\ t > 0}} \frac{1}{t} \int_0^t S_u f(x_0) \, du \geq \liminf_{\substack{t \to +\infty \\ t > 0}} \frac{1}{t} \int_0^t S_u 1_K(x_0) \, du > 0.$$

Thus, since, by Theorem 4.4.5, there exist continuous-time Banach limits, we may and do consider one such continuous-time Banach limit $L^{(\mathrm{ct})}$.

In view of the definition of $L^{(\mathrm{ct})}$ (see the beginning of Sect. 4.4), we obtain that $\varepsilon_{x_0}^{L^{(\mathrm{ct})}}$ is an elementary measure because $\varepsilon_{x_0}^{L^{(\mathrm{ct})}}(f) \neq 0$.

We have therefore proved that $\Gamma \neq \emptyset$ because $x_0 \in \Gamma$.

(b) $\Rightarrow$ (a) Assume that $\Gamma \neq \emptyset$, and let $x \in \Gamma$. In view of the definition of Γ, we obtain that there exists a continuous-time Banach limit $L^{(\mathrm{ct})}$ such that $\varepsilon_x^{(L^{(\mathrm{ct})})}$ is an elementary measure. By Theorem 7.1.1, the measure $\varepsilon_x^{(L^{(\mathrm{ct})})}$ is invariant. Since $\varepsilon_x^{(L^{(\mathrm{ct})})} \neq 0$, it follows that $(P_t)_{t \in \mathbb{T}}$ has invariant probabilities. $\quad\square$

In Sect. 5.1 we discussed supports of elementary measures of not necessarily Feller transition functions, and for standard elementary measures we actually obtained "formulas" for their supports (see Theorem 5.1.4). Naturally, when dealing with Feller transition functions, we expect significantly more relevant information concerning the supports of invariant elementary measures (recall that in the Feller case any elementary measure is invariant by Theorem 7.1.1). Thus, we will now discuss the supports of the various types of invariant (finite) measures of Feller transition functions.

Let $(P_t)_{t \in \mathbb{T}}$ be a Feller transition function defined on (X, d) and assume that $(P_t)_{t \in \mathbb{T}}$ satisfies the s.m.a. In view of Propositions 2.3.7 and 2.3.13, and of the manner in which the elementary measures are defined, we obtain that we may assume that $\mathbb{T} = [0, +\infty)$ because any result discussed here concerning invariant elements is valid for $(P_t)_{t\in\mathbb{T}}$ if and only if the result is valid for the restriction $(P_t)_{t\in[0,+\infty)}$ of $(P_t)_{t\in\mathbb{T}}$ to $[0,+\infty)$.

Thus let $(P_t)_{t\in[0,+\infty)}$ be a Feller transition function.

Recall (see the discussion preceding Proposition 2.1.7) that the orbit $\mathcal{O}(x)$ of an element x of X under the action of $(P_t)_{t\in[0,+\infty)}$ is defined by $\mathcal{O}(x) = \bigcup_{t\in[0,+\infty)} \mathrm{supp}\,(T_t\delta_x)$ and that the closure $\overline{\mathcal{O}(x)}$ of $\mathcal{O}(x)$ is called the orbit-closure of x. Note that since we assume that $\mathbb{T} = [0,+\infty)$ the orbit and the forward orbit are equal.

Theorem 7.1.3. *Assume that $\Gamma_c \neq \emptyset$ for the Feller transition function $(P_t)_{t\in[0,+\infty)}$, let $x \in \Gamma_c$, and assume that $x \in \mathrm{supp}\,\varepsilon_x$. Then $\mathrm{supp}\,\varepsilon_x = \overline{\mathcal{O}(x)}$.*

Proof. By Proposition 5.1.7, $\mathrm{supp}\,\varepsilon_x \subseteq \overline{\mathcal{O}(x)}$. Thus, in order to prove the theorem, it is enough to prove that $\overline{\mathcal{O}(x)} \subseteq \mathrm{supp}\,\varepsilon_x$.

Since $x \in \mathrm{supp}\,\varepsilon_x$, using Proposition 1.1.3 (Lasota and Myjak) that appeared in [59], we obtain that $\mathrm{supp}\,(T_t\delta_x) \subseteq \mathrm{supp}\,(T_t\varepsilon_x)$ for every $t \in [0,+\infty)$; since ε_x is an invariant measure for $(T_t)_{t\in[0,+\infty)}$, it follows that $\mathrm{supp}\,(T_t\delta_x) \subseteq \mathrm{supp}\,\varepsilon_x$ for every $t \in [0,+\infty)$; hence, $\mathcal{O}(x) \subseteq \mathrm{supp}\,\varepsilon_x$; since $\mathrm{supp}\,\varepsilon_x$ is a closed set, we therefore obtain that $\overline{\mathcal{O}(x)} \subseteq \mathrm{supp}\,\varepsilon_x$. $\square$

Note that the above theorem is an extension of Theorem 1.4.3 (see also (b) of Theorem 2.2.1 of [143]) to Feller transition functions.

In our next theorem, we extend Theorem 1.4.4 (see also Theorem 2.2.2 of [143]) to transition functions, thereby obtaining a "formula" for the supports of the invariant ergodic probability measures of Feller transition functions. In the theorem we use the fact that if a transition function has at least one invariant probability measure, then $\Gamma_{\mathrm{cpie}} \neq \emptyset$ (see Corollary 6.2.8).

Theorem 7.1.4. *Assume that the Feller transition function $(P_t)_{t\in[0,+\infty)}$ has invariant probability measures, so $\Gamma_{\mathrm{cpie}} \neq \emptyset$, and let $x \in \Gamma_{\mathrm{cpie}}$. Then*

$$\mathrm{supp}\,\varepsilon_x = \bigcap_{y\in[x]} \overline{\mathcal{O}(y)}.$$

Proof. We have to prove that

$$\mathrm{supp}\,\varepsilon_x \subseteq \overline{\mathcal{O}(y)} \tag{7.1.1}$$

for every $y \in [x]$, and that

$$\mathrm{supp}\,\varepsilon_x \supseteq \bigcap_{y\in[x]} \overline{\mathcal{O}(y)}. \tag{7.1.2}$$

Since $\varepsilon_y = \varepsilon_x$ for every $y \in [x]$, and since, by Proposition 5.1.7, $\mathrm{supp}\,\varepsilon_y \subseteq \overline{\mathcal{O}(y)}$ for every $y \in [x]$, it follows that the inclusion (7.1.1) holds true for every $y \in [x]$.

By Theorem 6.2.6, the set $[x]$ is $\mathcal{B}(X)$-measurable and by (a) of Theorem 6.2.7, $\varepsilon_x([x]) = 1$. Accordingly, $\varepsilon_x([x] \cap (\text{supp}\,\varepsilon_x)) = 1 > 0$; hence, there exists a $y \in [x] \cap (\text{supp}\,\varepsilon_x)$. Using Proposition 1.1.3 (Lasota and Myjak) (for details, see Proposition 3.1 of [59]) and the fact that ε_x is an invariant probability measure for $(T_t)_{t \in [0,+\infty)}$, we obtain that $\text{supp}\,(T_t\delta_y) \subseteq \text{supp}\,(T_t\varepsilon_x) = \text{supp}\,\varepsilon_x$ for every $t \in [0, +\infty)$. Therefore, $\mathcal{O}(y) \subseteq \text{supp}\,\varepsilon_x$, so $\overline{\mathcal{O}(y)} \subseteq \text{supp}\,\varepsilon_x$ because $\text{supp}\,\varepsilon_x$ is a closed subset of X.

It is now easy to see that the inclusion (7.1.2) holds true, as well. $\square$

Theorem 7.1.3 and the inequality (7.1.2) that appears in the proof of Theorem 7.1.4 have the following consequence.

Corollary 7.1.5. *Assume that the Feller transition function* $(P_t)_{t \in [0,+\infty)}$ *has ergodic invariant probability measures, and let* μ *be such an ergodic probability. Then there exists an* $x \in \Gamma_{\text{cpie}} \cap \text{supp}\,\mu$ *such that* $\mu = \varepsilon_x$.

Consequently, $\text{supp}\,\mu = \overline{\mathcal{O}(x)}$.

Proof. Since μ is an ergodic invariant probability measure for $(P_t)_{t \in [0,+\infty)}$, by Corollary 6.2.8, there exists an $x' \in \Gamma_{\text{cpie}}$ such that $\mu = \varepsilon_{x'}$. Using Theorem 6.2.6 and (a) of Theorem 6.2.7 as we did in the proof of the inclusion Theorem 7.1.2, we obtain that there exists an $x \in [x'] \cap (\text{supp}\,\mu) \subseteq \Gamma_{\text{cpie}} \cap (\text{supp}\,\mu)$ such that $\mu = \varepsilon_x$. Since $x \in \text{supp}\,\mu$, using Theorem 7.1.3, we also obtain that $\text{supp}\,\mu = \overline{\mathcal{O}(x)}$. $\square$

Our goal now is to discuss minimal Feller transition functions and their invariant probabilities.

As before let $(P_t)_{t \in [0,+\infty)}$ be a Feller transition function.

Recall (see Sect. 2.1) that $(P_t)_{t \in [0,+\infty)}$ is said to be minimal if the orbit $\mathcal{O}(x)$ is dense in X for every $x \in X$.

Minimal Feller transition functions have the property that, if they have invariant probabilities, then the supports of these probability measures are equal to the entire space. We discuss this fact in the next proposition.

Proposition 7.1.6. *Let* $(P_t)_{t \in [0,+\infty)}$ *be a minimal Feller transition function, and assume that* $(P_t)_{t \in [0,+\infty)}$ *has invariant probabilities. If* μ *is an invariant probability for* $(P_t)_{t \in [0,+\infty)}$, *then* $\text{supp}\,\mu = X$.

Proof. Let $(P_t)_{t \in [0,+\infty)}$ be a minimal Feller transition function, let $((S_t, T_t))_{t \in [0,+\infty)}$ be the family of Markov-Feller pairs defined by $(P_t)_{t \in [0,+\infty)}$, assume that $(P_t)_{t \in [0,+\infty)}$ has invariant probability measures, and let μ be such an invariant probability.

Now let $x \in X$. Using Proposition 1.1.3 (Lasota and Myjak), we obtain that $\text{supp}\,T_t\delta_x \subseteq \text{supp}\,T_t\mu = \text{supp}\,\mu$ for every $t \in [0, +\infty)$.

Accordingly, $\mathcal{O}(x) = \bigcup_{t \in [0,+\infty)} \text{supp}\,T_t\delta_x \subseteq \text{supp}\,\mu$; since $\text{supp}\,\mu$ is a closed set, it follows that $\overline{\mathcal{O}(x)} \subseteq \text{supp}\,\mu$. Since $(P_t)_{t \in [0,+\infty)}$ is minimal, it follows that $\overline{\mathcal{O}(x)} = X$, so $\text{supp}\,\mu = X$. $\square$

Observations.

(a) Note that in the above proposition we do not have to assume that $(P_t)_{t\in[0,+\infty)}$ satisfies the s.m.a. or that $(P_t)_{t\in[0,+\infty)}$ is pointwise continuous.
(b) Note also that the proposition is an extension of Proposition 1.4.7 (see also Proposition 2.3.2 of [143]). ▲

Our goal now is to show that under a mild additional condition, Proposition 7.1.6 has a converse.

In order to state the converse, recall that given a transition function $(P_t)_{t\in[0,+\infty)}$ on (X,d) and the family $((S_t,T_t))_{t\in[0,+\infty)}$ of Markov pairs defined by $(P_t)_{t\in[0,+\infty)}$, we denote by Γ the set of all $x\in X$ for which there exists a continuous-time Banach limit $L^{(ct)}$ and $f\in C_0(X)$ such that $L^{(ct)}(((\langle f,T_t\delta_x\rangle)_{t\in[0,+\infty)}))>0$, where $L^{(ct)}$ and f depend on x, of course (for additional details on the set X, see Sect. 5.1).

Proposition 7.1.7. *Let $(P_t)_{t\in[0,+\infty)}$ be a Feller transition function that satisfies the s.m.a., and assume that $\Gamma=X$. If every invariant probability measure μ is supported on the entire space X (that is, if $\mathrm{supp}\,\mu=X$ whenever μ is an invariant probability of $(P_t)_{t\in[0,+\infty)}$), then $(P_t)_{t\in[0,+\infty)}$ is a minimal transition function.*

Proof. Let $(P_t)_{t\in[0,+\infty)}$ be a Feller transition function as in the proposition. We have to prove that, under the conditions stated in the proposition, $\overline{\mathcal{O}(x)}=X$ for every $x\in X$.

To this end, let $x\in X$.

Since we assume that $X=\Gamma$, we obtain that there exists a continuous-time Banach limit $L^{(ct)}$ such that $\varepsilon_x^{(L^{(ct)})}$ is an elementary measure. By Theorem 7.1.1, the measure $\varepsilon_x^{(L^{(ct)})}$ is invariant for $(P_t)_{t\in[0,+\infty)}$.

Since $\varepsilon_x^{(L^{(ct)})}\neq0$, it follows that $\dfrac{\varepsilon_x^{(L^{(ct)})}}{\|\varepsilon_x^{(L^{(ct)})}\|}$ is well defined. Clearly, $\dfrac{\varepsilon_x^{(L^{(ct)})}}{\|\varepsilon_x^{(L^{(ct)})}\|}$ is an invariant probability measure for $(P_t)_{t\in[0,+\infty)}$.

Since we assume that the support of any invariant probability measure for $(P_t)_{t\in[0,+\infty)}$ is the entire space X, it follows that $\mathrm{supp}\,\left(\dfrac{\varepsilon_x^{(L^{(ct)})}}{\|\varepsilon_x^{(L^{(ct)})}\|}\right)=X$; therefore, $\mathrm{supp}\,\varepsilon_x^{(L^{(ct)})}=X$. By Proposition 5.1.7, $\mathrm{supp}\,\varepsilon_x^{(L^{(ct)})}\subseteq\overline{\mathcal{O}(x)}$. Thus, $\overline{\mathcal{O}(x)}=X$. □

Combining Propositions 7.1.6 and 7.1.7, we obtain the following theorem:

Theorem 7.1.8. *Let $(P_t)_{t\in[0,+\infty)}$ be a Feller transition function, and assume that $\Gamma=X$. Then the following assertions are equivalent:*

(a) The transition function $(P_t)_{t\in[0,+\infty)}$ is minimal.
(b) $\mathrm{supp}\,\mu=X$ whenever μ is an invariant probability for $(P_t)_{t\in[0,+\infty)}$.

Proof. The proof is obvious in view of Propositions 7.1.6 and 7.1.7. □

Note that if X is compact, then $\Gamma=X$ because $\langle\mathbf{1}_X,T_t\delta_x\rangle=1$ for every $x\in X$ and $t\in[0,+\infty)$, so taking into consideration that in this case $\mathbf{1}_X\in C_0(X)$ and using condition (d) of the definition of a continuous-time Banach limit (see

Sect. 4.4), we obtain that $L^{(\mathrm{ct})}(((\langle \mathbf{1}_X, T_t\delta_x\rangle)_{t\in[0,+\infty)}) = 1$ for every continuous-time Banach limit $L^{(\mathrm{ct})}$ and every $x \in X$.

Thus in the case of a compact space X, Theorem 7.1.8 becomes:

Corollary 7.1.9. *Assume that the metric space (X, d) is compact, and let $(P_t)_{t\in[0,+\infty)}$ be a Feller transition function on (X, d). Then the following assertions are equivalent:*

(a) The transition function $(P_t)_{t\in[0,+\infty)}$ is minimal.
(b) $\operatorname{supp}\mu = X$ whenever μ is an invariant probability for $(P_t)_{t\in[0,+\infty)}$.

Examples of minimal transition functions are easy to find. Probably the simplest such example is the restriction to $[0, +\infty)$ of the transition function defined by the flow of the rotations of the unit circle (see Example 2.2.4).

Recall that in Sect. 1.1 we have defined the trivially minimal transition probabilities (see also Section 2.3 of [143], where we have defined the trivially minimal Markov-Feller pairs). The definition of trivial minimality can easily be extended to transition functions.

Indeed, given a transition function $(P_t)_{t\in[0,+\infty)}$ defined on a locally compact separable metric space (X, d), and given the family $((S_t, T_t))_{t\in[0,+\infty)}$ of Markov pairs defined by $(P_t)_{t\in[0,+\infty)}$, we say that $(P_t)_{t\in[0,+\infty)}$ (or $((S_t, T_t))_{t\in[0,+\infty)}$, or $(T_t)_{t\in[0,+\infty)}$) is *trivially minimal* if $\operatorname{supp}(T_t\delta_x) = X$ for every $t \in (0, +\infty)$ and $x \in X$.

In general, minimal transition functions need not be trivially minimal (for example, the above mentioned restriction to $[0, +\infty)$ of the transition function defined by the flow of the rotations of the unit circle in Example 2.2.4 is minimal, but not trivially minimal). However, trivially minimal transition functions do exist, and we will now discuss an example.

Example 7.1.10. Let H be a locally compact separable metric group (for the definition of these groups, see Sect. A.2).

It is easy to see that there exist probability measures $\mu \in \mathcal{M}(H)$ such that $\operatorname{supp}\mu = H$. A simple method to construct such measures is the following: since H is separable, there exists a sequence $(h_j)_{j\in\mathbb{N}}$ of elements of H such that the range $\{h_j \mid j \in \mathbb{N}\}$ of $(h_j)_{j\in\mathbb{N}}$ is dense in H. Set $\mu = \sum_{j=1}^{\infty} \frac{1}{2^j}\delta_{h_j}$. Clearly, μ is a probability measure on $(H, \mathcal{B}(H))$. Since $\{h_j \mid j \in \mathbb{N}\}$ is dense in both H and $\operatorname{supp}\mu$, it follows that $\operatorname{supp}\mu = H$.

Now let $\mu \in M(H)$ be a probability measure such that $\operatorname{supp}\mu = H$ (for instance, we can use the probability measure constructed in the previous paragraph), let $(\mu_t)_{t\in[0,+\infty)}$ be the exponential one-parameter convolution semigroups of probability measures defined by μ described in Proposition B.3.2, and let $(P_t)_{t\in[0,+\infty)}$ and $((S_t, T_t))_{t\in[0,+\infty)}$ be the Feller transition function and the corresponding family of Markov-Feller pairs defined by $(\mu_t)_{t\in[0,+\infty)}$, respectively (see the discussion following Proposition 2.2.10).

Taking into consideration that $\mu_t = \exp_s(t\mu) = e^{-t}\sum_{k=0}^{\infty}\frac{t^k}{k!}\mu^k$ for every $t \in \mathbb{R}$, $t \geq 0$, that $\operatorname{supp}\mu = H$, that $T_t\delta_x = \mu_t * \delta_x = e^{-t}\sum_{k=0}^{\infty}\frac{t^k}{k!}\mu^k * \delta_x$ for every

$t \in \mathbb{R}, t \geq 0$, and $x \in H$ (see the definitions preceding Proposition 2.2.10), and that $Hx = H$ for every $x \in H$ (because H is a group), we obtain that supp $T_t \delta_x = H$ for every $x \in H$ and $t \in \mathbb{R}, t > 0$.

Thus $(P_t)_{t \in [0,+\infty)}$ is trivially minimal. ∎

We will now discuss results in which we compare properties of supports of invariant probability measures and the structure of the set of invariant probability measures of a transition function $(P_t)_{t \in \mathbb{T}}$ on one hand, with properties of supports of invariant probabilities and the structure of the set of invariant probabilities of the transition probabilities P_t, $t \in \mathbb{T}$, that form the transition function, on the other hand.

We will continue to assume that $\mathbb{T} = [0, +\infty)$, because using Proposition 2.3.7, we can easily extend the results that follow to transition functions where $\mathbb{T} = \mathbb{R}$.

Let $(P_t)_{t \in [0,+\infty)}$ be a Feller transition function defined on a locally compact separable metric space (X, d), and let $((S_t, T_t))_{t \in [0,+\infty)}$ be the family of Markov-Feller pairs defined by $(P_t)_{t \in [0,+\infty)}$. Also, assume that $(P_t)_{t \in [0,+\infty)}$ satisfies the s.m.a.

Let A be a nonempty subset of X.

We will use the notation $T_t(A) = \bigcup_{x \in A} \mathrm{supp}\,(T_t \delta_x)$.

Recall (see the discussion preceding Proposition 2.1.7) that the orbit-closure (which is the same as the forward orbit-closure in this case) of A under the action of $(P_t)_{t \in [0,+\infty)}$ is denoted by $\overline{\mathcal{O}(A)}$ and is defined by $\overline{\mathcal{O}(A)} = \overline{\bigcup_{x \in A} \mathcal{O}(x)}$.

Now assume that for some $\alpha \in [0, +\infty)$, $\alpha > 0$, the transition probability P_α has an invariant probability measure, say μ_α. As pointed out at the beginning of Sect. 3.3.2, the pointwise integral P-$\int_0^\alpha T_t \mu_\alpha \, dt$ is well-defined and $\nu_\alpha = \frac{1}{\alpha} \left(\text{P-}\int_0^\alpha T_t \mu_\alpha \, dt \right)$ is a probability measure. By Theorem 3.3.9, ν_α is an invariant probability measure for the transition function $(P_t)_{t \in [0,+\infty)}$.

Our goal now is to obtain characterizations of the support of ν_α in terms of supp μ_α whenever $(P_t)_{t \in [0,+\infty)}$ belongs to a fairly large class of Feller transition functions.

To this end, we need the following lemma:

Lemma 7.1.11. *As above, let* $(P_t)_{t \in [0,+\infty)}$ *be a Feller transition function defined on a locally compact separable metric space* (X, d), *let* $((S_t, T_t))_{t \in [0,+\infty)}$ *be the family of Markov-Feller pairs defined by* $(P_t)_{t \in [0,+\infty)}$, *assume that* $(P_t)_{t \in [0,+\infty)}$ *satisfies the s.m.a., and let* $\alpha \in [0, +\infty)$, $\alpha > 0$. *Then, for every* $f \in C_b(X)$, *the pointwise integral P-*$\int_0^\alpha S_t f \, dt$ *exists and is an element of* $C_b(X)$.

Proof. Let $f \in C_b(X)$.

The existence of the pointwise integral stems from Proposition 3.3.2.

Set $h = \text{P-}\int_0^\alpha S_t f \, dt$. Thus, $h : X \to \mathbb{R}$ is defined by $h(x) = \int_0^\alpha S_t f(x) \, dt$ for every $x \in X$.

Clearly, h is a bounded function because the operators S_t, $t \in [0, +\infty)$, are contractions, so $|h(x)| \leq \int_0^\alpha |S_t f(x)| \, dt \leq \alpha \|f\|$ for every $x \in X$.

We now prove that h is continuous.

Since (X, d) is a metric space, in order to prove that h is continuous, it is enough to prove that for every convergent sequence $(x_n)_{n \in \mathbb{N}}$ of elements of X, the sequence $(h(x_n))_{n \in \mathbb{N}}$ converges to $h(x)$, where $x = \lim_{n \to +\infty} x_n$.

Thus, let $(x_n)_{n \in \mathbb{N}}$ be a convergent sequence of elements of X, and set $x = \lim_{n \to +\infty} x_n$.

For every $n \in \mathbb{N}$, let $g_n : [0, \alpha] \to \mathbb{R}$ be defined by $g_n(t) = S_t f(x_n)$ for every $t \in [0, \alpha]$; also, let $g : [0, \alpha] \to \mathbb{R}$ be defined by $g(t) = S_t f(x)$ for every $t \in [0, \alpha]$.

Since $(P_t)_{t \in [0, +\infty)}$ satisfies the s.m.a., it follows that the functions g_n, $n \in \mathbb{N}$, and g are measurable.

Taking into consideration that $|g(t)| \leq \|f\|$ and that $|g_n(t)| \leq \|f\|$, $n \in \mathbb{N}$, for every $t \in [0, \alpha]$, and using the fact that the Lebesgue measure on $[0, \alpha]$ is a finite measure, we obtain that we can apply the Lebesgue dominated convergence theorem to the sequence of functions $(g_n)_{n \in \mathbb{N}}$. Accordingly, the sequence $(\int_0^\alpha g_n(t)\, dt)_{n \in \mathbb{N}}$ converges to $\int_0^\alpha g(t)\, dt$; that is, $(h(x_n))_{n \in \mathbb{N}}$ converges to $h(x)$. Thus, h is a continuous function. $\qquad\square$

Given a Feller transition function $(P_t)_{t \in \mathbb{T}}$ defined on a locally compact separable metric space (X, d), and the family $((S_t, T_t))_{t \in \mathbb{T}}$ of Markov-Feller pairs defined by $(P_t)_{t \in \mathbb{T}}$, we say that $(P_t)_{t \in \mathbb{T}}$ (or $((S_t, T_t))_{t \in \mathbb{T}}$, or $(S_t)_{t \in \mathbb{T}}$) is $C_0(X)$-*jointly continuous* if, for every $f \in C_0(X)$, the real-valued map $(t, x) \mapsto S_t f(x)$, $(t, x) \in \mathbb{T} \times X$, is continuous with respect to the standard topology on $\mathbb{R}$ and the product topology $\mathcal{T}(\mathbb{T}) \otimes \mathcal{T}_d(X)$ on $\mathbb{T} \times X$, where $\mathcal{T}(\mathbb{T})$ and $\mathcal{T}_d(X)$ are the standard topology on $\mathbb{T}$ and the metric topology on X, respectively.

Typical examples of $C_0(X)$-jointly continuous transition functions are the transition functions defined by continuous flows and continuous semiflows, and the transition functions defined by exponential one-parameter convolution semigroups of probabilities defined by equicontinuous probability measures (the transition functions of these one-parameter semigroups of probability measures are discussed in Proposition 2.2.15; it is easy to adapt the proof of the proposition in order to show that the transition functions that appear in the proposition are not only $C_0(H)$-equicontinuous, but also $C_0(H)$-jointly continuous, where $(H, \cdot, d)$ is the metric semigroup under consideration in Proposition 2.2.15).

In the next theorem we discuss the characterizations, mentioned earlier, of the supports of invariant probability measures of certain transition functions.

Theorem 7.1.12. *Let $(P_t)_{t \in [0, +\infty)}$ be a Feller transition function defined on a locally compact separable metric space (X, d), assume that $(P_t)_{t \in [0, +\infty)}$ is $C_0(X)$-jointly continuous, and let $((S_t, T_t))_{t \in [0, +\infty)}$ be the family of Markov-Feller pairs defined by $(P_t)_{t \in [0, +\infty)}$. Let $\alpha \in [0, +\infty)$, $\alpha > 0$, assume that T_α has nonzero invariant elements, let μ_α be an invariant probability measure for T_α, and let $\nu_\alpha = \frac{1}{\alpha}(P\text{-}\int_0^\alpha T_t \mu_\alpha\, dt)$ be the probability constructed at the beginning of Sect. 3.3.2. Then*

$$\operatorname{supp} \nu_\alpha = \overline{\bigcup_{0 \leq t \leq \alpha} T_t(\operatorname{supp} \mu_\alpha)} = \overline{\mathcal{O}(\operatorname{supp} \mu_\alpha)}.$$

Proof. We will prove the theorem by showing that the following inclusions hold:

$$\operatorname{supp} \nu_\alpha \subseteq \overline{\bigcup_{0 \le t \le \alpha} T_t(\operatorname{supp} \mu_\alpha)} \tag{7.1.3}$$

$$\overline{\bigcup_{0 \le t \le \alpha} T_t(\operatorname{supp} \mu_\alpha)} \subseteq \overline{\mathcal{O}(\operatorname{supp} \mu_\alpha)} \tag{7.1.4}$$

$$\overline{\mathcal{O}(\operatorname{supp} \mu_\alpha)} \subseteq \overline{\bigcup_{0 \le t \le \alpha} T_t(\operatorname{supp} \mu_\alpha)} \tag{7.1.5}$$

and

$$\overline{\bigcup_{0 \le t \le \alpha} T_t(\operatorname{supp} \mu_\alpha)} \subseteq \operatorname{supp} \nu_\alpha. \tag{7.1.6}$$

Proof of (7.1.3). We have to prove that

$$\operatorname{supp} \nu_\alpha \subseteq \overline{\bigcup_{0 \le t \le \alpha} \bigcup_{x \in \operatorname{supp}\mu_\alpha} \operatorname{supp}(T_t \delta_x)}. \tag{7.1.7}$$

To this end, let $z \in \operatorname{supp} \nu_\alpha$. In order to show that $z \in \overline{\bigcup_{0 \le t \le \alpha} \bigcup_{x \in \operatorname{supp} \nu_\alpha} \operatorname{supp}(T_t \delta_x)}$, it is enough to prove that for every $r \in \mathbb{R}$, $r > 0$, the open ball $B(z, r)$ contains at least one element of $\bigcup_{0 \le t \le \alpha} \bigcup_{x \in \operatorname{supp} \mu_\alpha} \operatorname{supp}(T_t \delta_x)$; that is, it is enough to show that there exist $t_0 \in [0, \alpha]$, $x \in \operatorname{supp} \mu_\alpha$ and $w \in \operatorname{supp} T_{t_0} \delta_x$ such that $w \in B(z, r)$.

Thus, let $r \in \mathbb{R}$, $r > 0$. Since X is locally compact, there exists a $\rho \in \mathbb{R}$, $\rho > 0$, such that $\overline{B(z, \rho)}$ is a compact subset of X.

Using Proposition 7.1.8, p. 199, of Cohn [20], we obtain that there exists an $f \in C_0(X)$ such that $\mathbf{1}_{\overline{B(z,\rho)}} \le f \le \mathbf{1}_{B(z,r)}$.

Since $z \in \operatorname{supp} \nu_\alpha$, it follows that $\nu_\alpha(\overline{B(z, \rho)}) > 0$, so $\langle f, \nu_\alpha \rangle > 0$.

Using Proposition 3.3.7, we obtain that

$$0 < \langle f, \nu_\alpha \rangle = \left\langle f, \frac{1}{\alpha}\left(\text{P-}\int_0^\alpha T_t \mu_\alpha \, dt\right)\right\rangle = \left\langle \frac{1}{\alpha}\left(\text{P-}\int_0^\alpha S_t f \, dt\right), \mu_\alpha \right\rangle.$$

Set $h = \frac{1}{\alpha}(\text{P-}\int_0^\alpha S_t f \, dt)$. By Lemma 7.1.11, the function h is continuous and bounded.

Since $\langle h, \mu_\alpha \rangle > 0$, it follows that $\{f > 0\} \cap \operatorname{supp} \mu_\alpha \ne \emptyset$; therefore, there exists an $x \in \operatorname{supp} \mu_\alpha$ such that $h(x) > 0$.

Using the definition of h and Proposition 3.3.7, we obtain that

$$0 < h(x) = \langle h, \delta_x \rangle = \left\langle \frac{1}{\alpha} \left(\text{P-} \int_0^\alpha S_t f \, dt \right), \delta_x \right\rangle$$

$$= \left\langle f, \frac{1}{\alpha} \int_0^\alpha T_t \delta_x \, dt \right\rangle.$$

In view of the fact that f is continuous, we obtain that

$$U \cap \text{supp} \left(\frac{1}{\alpha} \left(\text{P-} \int_0^\alpha T_t \delta_x \, dt \right) \right) \neq \emptyset,$$

where $U = \{f > 0\}$.

Thus, $\left(\text{P-} \int_0^\alpha T_t \delta_x \, dt \right) (U) > 0$; accordingly, there exists a $t_0 \in [0, \alpha]$ such that $(T_{t_0} \delta_x)(U) > 0$.

We obtain that $U \cap \text{supp}(T_{t_0} \delta_x) \neq \emptyset$; hence, there exists a $w \in U = \{f > 0\} \subseteq B(z, r)$ such that $w \in \text{supp}(T_{t_0} \delta_x)$.

Thus, the inclusion (7.1.3) holds true.

Proof of (7.1.4). Since

$$T_t(\text{supp } \mu_\alpha) = \bigcup_{x \in \text{supp } \mu_\alpha} (\text{supp } (T_t \delta_x))$$

$$\subseteq \bigcup_{s \in [0, +\infty)} \bigcup_{x \in \text{supp } \mu_\alpha} (\text{supp } (T_s \delta_x))$$

$$= \bigcup_{x \in (\text{supp } \mu_\alpha)} \bigcup_{s \in [0, +\infty)} (\text{supp } (T_s \delta_x)) = \bigcup_{x \in \text{supp } \mu_\alpha} \mathcal{O}(x)$$

$$= \mathcal{O}(\text{supp } \mu_\alpha)$$

for every $t \in [0, \alpha]$, it follows that $\overline{\bigcup_{0 \leq t \leq \alpha} T_t(\text{supp } \mu_\alpha)} \subseteq \overline{\mathcal{O}(\text{supp } \mu_\alpha)}$; that is, the inclusion (7.1.4) is also true.

Proof of (7.1.5). We have to prove that

$$\overline{\bigcup_{x \in (\text{supp} \mu_\alpha)} \mathcal{O}(x)} \subseteq \overline{\bigcup_{0 \leq t \leq \alpha} T_t(\text{supp } \mu_\alpha)}. \tag{7.1.8}$$

Since $\overline{T_r(\text{supp } \mu_\alpha)} \subseteq \overline{\bigcup_{0 \leq t \leq \alpha} T_t(\text{supp } \mu_\alpha)}$ for every $r \in [0, \alpha]$, it follows that, in order to prove that the inclusion (7.1.8) is true, it is enough to prove that for every $x \in \text{supp } \mu_\alpha$, for every $\gamma \in [0, +\infty)$, and for every $u \in \text{supp } T_\gamma \delta_x$, there exists an $s \in [0, \alpha]$ such that $u \in \overline{T_s(\text{supp } \mu_\alpha)}$.

To this end, let $x \in \text{supp } \mu_\alpha$, let $\gamma \in [0, +\infty)$, and let $u \in \text{supp } T_\gamma \delta_x$. Then there exist $h \in \mathbb{N} \cup \{0\}$ and $s \in [0, \alpha)$ such that $\gamma = h\alpha + s$.

Our goal is to prove that $u \in \overline{T_s(\text{supp } \mu_\alpha)}$.

Since $T_s(\text{supp } \mu_\alpha) = \bigcup_{z \in (\text{supp } \mu_\alpha)} (\text{supp}(T_s\delta_z))$, it is enough to prove that for every $r \in \mathbb{R}$, $r > 0$, we have

$$B(u, r) \cap \left(\bigcup_{z \in (\text{supp } \mu_\alpha)} (\text{supp}(T_s\delta_z)) \right) \neq \emptyset.$$

Thus, let $r \in \mathbb{R}$, $r > 0$. Clearly, we may choose $r > 0$ small enough such that $\overline{B(u, r)}$ is a compact subset of X.

Using Proposition 7.1.8, p. 199, of Cohn [20], we obtain that there exists an $f \in C_0(x)$ such that $\mathbf{1}_{\overline{B(u,r/2)}} \leq f \leq \mathbf{1}_{B(u,r)}$.

Taking into consideration that $x \in \text{supp } \mu_\alpha$, that $\gamma = h\alpha + s$, and that $T_\alpha \mu_\alpha = \mu_\alpha$, and using Proposition 1.1.3 (a result due to Lasota and Myjak [59]), we obtain that

$$\text{supp } (T_\gamma \delta_x) \subseteq \text{supp } (T_\gamma \mu_\alpha) = \text{supp } (T_{h\alpha+s} \mu_\alpha) = \text{supp } (T_s(T_\alpha^h \mu_\alpha))$$

$$= \text{supp } (T_s \mu_\alpha).$$

Since $u \in \text{supp } T_\gamma \delta_x$ and $\mathbf{1}_{\overline{B(u,r/2)}} \leq f$, it follows that $\langle f, T_\gamma \delta_x \rangle > 0$; hence,

$$0 < \langle f, T_\gamma \mu_\alpha \rangle = \langle f, T_s \mu_\alpha \rangle = \langle S_s f, \mu_\alpha \rangle.$$

Taking into consideration that $S_s f$ is a continuous function, we obtain that $\{S_s f > 0\}$ is an open set, so there exists a $z \in \text{supp } \mu_\alpha$ such that $S_s f(z) > 0$. Thus, $0 < \langle S_s f, \delta_z \rangle = \langle f, T_s \delta_z \rangle$; therefore, $\{f > 0\} \cap (\text{supp } (T_s\delta_z)) \neq \emptyset$. Since $0 \leq f \leq \mathbf{1}_{B(u,r)}$, it follows that $B(u, r) \cap \text{supp } (T_s\delta_z) \neq \emptyset$. Taking into consideration that $z \in \text{supp } \mu_\alpha$, we conclude that the inclusion (7.1.5) is true.

Proof of the Inclusion (7.1.6). We note that in order to prove that (7.1.6) holds true it is enough to prove that $\bigcup_{0 \leq t \leq \alpha} T_t(\text{supp } \mu_\alpha) \subseteq \text{supp } \nu_\alpha$; that is, it is enough to prove that

$$\bigcup_{0 \leq t \leq \alpha} \bigcup_{z \in \text{supp } \mu_\alpha} (\text{supp } (T_t\delta_z)) \subseteq \text{supp } \nu_\alpha.$$

Accordingly, let $x \in \bigcup_{0 \leq t \leq \alpha} \bigcup_{z \in \text{supp } \mu_\alpha} (\text{supp } (T_t\delta_z))$.

In order to prove that $x \in \text{supp } \nu_\alpha$, we will prove that $\nu_\alpha(B(x, r)) > 0$ for every $r \in \mathbb{R}$, $r > 0$. Clearly, it is enough to prove that $\nu_\alpha(B(x, r)) > 0$ whenever $r \in \mathbb{R}$, $r > 0$, is such that $\overline{B(x, r)}$ is a compact subset of X.

Thus, let $r \in \mathbb{R}$, $r > 0$, be such that $\overline{B(x, r)}$ is a compact subset of X. Using again Proposition 7.1.8, p. 199, of Cohn [20] (as we did in the proof of (7.1.3)), we obtain that there exists an $f \in C_0(X)$ such that $\mathbf{1}_{\overline{B(x,r/2)}} \leq f \leq \mathbf{1}_{B(x,r)}$.

Since $x \in \bigcup_{0 \leq t \leq \alpha} \bigcup_{z \in \text{supp } \mu_\alpha} (\text{supp } (T_t\delta_z))$, it follows that there exist $t_0 \in [0, \alpha]$ and $z \in \text{supp } \mu_\alpha$ such that $x \in \text{supp } (T_{t_0}\delta_z)$. Thus, $(T_{t_0}\delta_z)(\overline{B(x, r/2)}) > 0$.

We obtain that $0 < \langle \mathbf{1}_{\overline{B(x,r/2)}}, T_{t_0}\delta_z \rangle \le \langle f, T_{t_0}\delta_z \rangle = S_{t_0} f(z)$.

Since $(S_t)_{t \in [0,+\infty)}$ is $C_0(X)$-jointly continuous, it follows that there exists an $\eta \in \mathbb{R}$, $\eta > 0$, such that $S_t f(u) > 0$ for every $t \in I$ and every $u \in B(z, \eta)$, where I is a closed interval of length η such that $I \subseteq [0, \alpha]$ and one endpoint of I is t_0.

Using Proposition 3.3.7, the fact that $S_t f \ge 0$ for all t, $\mu_\alpha \ge 0$, and $\frac{1}{\alpha}(\text{P-}\int_I S_t f(u)\,dt) \ge 0$ for all $u \in B(z, \eta)$, and using the definition of η, we obtain that

$$
\begin{aligned}
\nu_\alpha((B(x,r))) &\ge \left\langle f, \frac{1}{\alpha}\Big(\text{P-}\int_0^\alpha T_t \mu_\alpha \,dt\Big) \right\rangle \\
&= \frac{1}{\alpha}\left\langle \text{P-}\int_0^\alpha S_t f \,dt, \mu_\alpha \right\rangle \ge \frac{1}{\alpha}\left\langle \text{P-}\int_I S_t f \,dt, \mu_\alpha \right\rangle \\
&= \frac{1}{\alpha}\int_X \Big(\text{P-}\int_I S_t f(u)\,dt\Big)\,d\mu_\alpha(u) \\
&\ge \frac{1}{\alpha}\int_{B(z,\eta)} \Big(\text{P-}\int_I S_t f(u)\,dt\Big)\,d\mu_\alpha(u) > 0.
\end{aligned}
$$

Thus, the inclusion (7.1.6) is also true.

$\square$

The formulas for the support of ν_α obtained in Theorem 7.1.12 are particularly useful whenever the transition function $(P_t)_{t \in [0,+\infty)}$ is defined by a continuous semiflow, or is the restriction to $[0,+\infty)$ of the transition function defined by a continuous flow. We discuss these formulas for transition functions defined by continuous flows or continuous semiflows in the next corollary.

Corollary 7.1.13. *Let $\mathbf{w} = (w_t)_{t \in \mathbb{T}}$ be a continuous flow or a continuous semiflow defined on a locally compact separable metric space (X, d), and let $(P_t^{(\mathbf{w})})_{t \in \mathbb{T}}$ be the transition function defined by $\mathbf{w}$. Also let $((S_t, T_t))_{t \in \mathbb{T}}$ be the family of Markov-Feller pairs defined by $\mathbf{w}$. Assume that for some $\alpha \in \mathbb{T}$, $\alpha > 0$, the transition probability $P_\alpha^{(\mathbf{w})}$ has invariant probabilities, let μ_α be such an invariant probability measure for $P_\alpha^{(\mathbf{w})}$, and let $\nu_\alpha = \frac{1}{\alpha}\big(\text{P-}\int_0^\alpha T_t \mu_\alpha \,dt\big)$ be the probability measure considered in Theorem 7.1.12. Then $\operatorname{supp}\nu_\alpha = \overline{\bigcup_{0 \le t \le \alpha} w_t(\operatorname{supp}\mu_\alpha)} = \overline{\mathcal{O}(\operatorname{supp}\mu_\alpha)}$.*

Proof. If $\mathbb{T} = [0, +\infty)$, then taking into consideration that

$$
w_t(A) = \bigcup_{x \in A} \{w_t(x)\} = \bigcup_{x \in A} (\operatorname{supp}(T_t \delta_x)) = T_t(A)
$$

for every nonempty subset A of X and every $t \in [0, +\infty)$, we obtain that the corollary is just a reformulation of Theorem 7.1.12.

If $\mathbb{T} = \mathbb{R}$ (that is, if $\mathbf{w}$ is a flow), then using the comments made after Proposition 2.3.7, we obtain that μ_α is an invariant probability measure for the transition probability $P_{-\alpha}$, as well.

Now, let us "split" $\mathbf{w}$ into two semiflows $\mathbf{u}$ and $\mathbf{v}$ defined as follows: $\mathbf{u} = (u_t)_{t \in [0,+\infty)}$, where $u_t = w_t$ for every $t \in [0, +\infty)$, and $\mathbf{v} = (v_t)_{t \in [0,+\infty)}$, where $v_t = w_{-t}$ for every $t \in [0, +\infty)$.

Let $((S_t^{(\mathbf{u})}, T_t^{(\mathbf{u})}))_{t \in [0,+\infty)}$ and $((S_t^{(\mathbf{v})}, T_t^{(\mathbf{v})}))_{t \in [0,+\infty)}$ be the families of Markov-Feller pairs defined by $\mathbf{u}$ and $\mathbf{v}$, respectively.

Set $v_\alpha^{(\mathbf{u})} = \frac{1}{\alpha} \mathrm{P}\!-\! \int_0^\alpha T_t^{(\mathbf{u})} \mu_\alpha \, dt$ and $v_\alpha^{(\mathbf{v})} = \frac{1}{\alpha} \mathrm{P}\!-\! \int_0^\alpha T_t^{(\mathbf{v})} \mu_\alpha \, dt$.

Clearly, $v_\alpha^{(\mathbf{u})} = v_\alpha$ (because $T_t^{(\mathbf{u})} = T_t$ for every $t \geq 0$).

Using the fact that $T_\alpha \mu_\alpha = \mu_\alpha$ and the change of variables formula (see, for instance, Theorem 6.1.6, p. 171, of Cohn [20]), we obtain that

$$
\begin{aligned}
v_\alpha^{(\mathbf{v})}(A) &= \frac{1}{\alpha} \int_0^\alpha T_t^{(\mathbf{v})} \mu_\alpha(A) \, dt \\
&= \frac{1}{\alpha} \int_0^\alpha \mu_\alpha(v_t^{-1}(A)) \, dt = \frac{1}{\alpha} \int_0^\alpha \mu_\alpha(w_{-t}^{-1}(A)) \, dt \\
&= \frac{1}{\alpha} \int_0^\alpha \mu_\alpha(w_{\alpha-t}^{-1}(A)) \, dt = \frac{1}{\alpha} \int_0^\alpha \mu_\alpha(w_s^{-1}(A)) \, ds = v(A)
\end{aligned}
$$

for every $A \in \mathcal{B}(X)$.

We have therefore proved that $v_\alpha = v_\alpha^{(\mathbf{u})} = v_\alpha^{(\mathbf{v})}$, so by applying Theorem 7.1.12 to the transition functions $(P_t^{(\mathbf{u})})_{t \in [0,+\infty)}$ and $(P_t^{(\mathbf{v})})_{t \in [0,+\infty)}$ defined by the semiflows $\mathbf{u}$ and $\mathbf{v}$, we obtain that the assertion of the corollary is true in the case in which $\mathbb{T} = \mathbb{R}$, as well. $\square$

As before, let $(P_t)_{t \in \mathbb{T}}$ be a Feller transition function defined on a locally compact separable metric space (X, d), and let $((S_t, T_t))_{t \in \mathbb{T}}$ be the family of Markov-Feller pairs defined by $(P_t)_{t \in \mathbb{T}}$.

For every $t \in \mathbb{T}$, set $\Omega_t^{(\mathrm{TP})} = \Omega^{(\mathrm{TP})}(P_t)$ and $\Gamma_t^{(\mathrm{TP})} = \Gamma^{(\mathrm{TP})}(P_t)$, where $\Omega^{(\mathrm{TP})}(P_t)$ and $\Gamma^{(\mathrm{TP})}(P_t)$ are the sets defined at the beginning of Sect. 1.3 for the transition probability P_t; thus,

$$
\Omega_t^{(\mathrm{TP})} = \left\{ x \in X \;\middle|\;
\begin{array}{l}
L^{(\mathrm{dt})}((S_t^n f(x))_{n \in \mathbb{N} \cup \{0\}}) = 0 \text{ for} \\
\text{every } f \in C_0(X) \text{ and every} \\
\text{discrete-time Banach limit } L^{(\mathrm{dt})}
\end{array} \right\}
$$

$$
= \left\{ x \in X \;\middle|\;
\begin{array}{l}
L^{(\mathrm{dt})}((S_{tn} f(x))_{n \in \mathbb{N} \cup \{0\}}) = 0 \text{ for} \\
\text{every } f \in C_0(X) \text{ and every} \\
\text{discrete-time Banach limit } L^{(\mathrm{dt})}
\end{array} \right\}
$$

and $\Gamma_t^{(\mathrm{TP})} = X \setminus \Omega_t^{(\mathrm{TP})}$.

Set $\Delta_{\mathrm{M}} = \bigcup_{\substack{t \in \mathbb{T} \\ t \neq 0}} \Gamma_t^{(\mathrm{TP})}$.

The nonemptiness of the set Δ_M is a necessary and sufficient condition for the existence of invariant probabilities of the transition function $(P_t)_{t \in \mathbb{T}}$. We discuss this fact in the next theorem.

Theorem 7.1.14. *Let $(P_t)_{t \in \mathbb{T}}$ be a Feller transition function and let $((S_t, T_t))_{t \in \mathbb{T}}$ be the family of Markov-Feller pairs defined by $(P_t)_{t \in \mathbb{T}}$. Assume that, if $\mathbb{T} = \mathbb{R}$, the transition probability P_0 is defined by $P_0(x, A) = \mathbf{1}_A(x)$ for every $x \in X$ and $A \in \mathcal{B}(X)$. The following assertions are equivalent:*

(a) $(P_t)_{t \in \mathbb{T}}$ has at least one invariant probability measure.
(b) $\Delta_M \neq \emptyset$.

Proof. (a) $\Rightarrow$ (b) Assume that $(P_t)_{t \in \mathbb{T}}$ has invariant probabilities, and let $\mu \in \mathcal{M}(X)$ be such an invariant probability. Then μ is, in particular, an invariant probability for the transition probability P_1.

Since P_1 has invariant probabilities, and since, by Proposition 1.3.9, the set

$$
\Gamma_{\mathrm{cp}}^{(\mathrm{TP})}(P_1) = \left\{ x \in X \;\middle|\; \begin{array}{l} (\frac{1}{m}\sum_{k=0}^{m-1} S_1^k f(x))_{m \in \mathbb{N}} \text{ converges} \\ \text{for every } f \in C_0(X), \text{ and the} \\ \text{resulting standard elementary measure } \varepsilon_x \text{ is} \\ \text{a probability measure} \end{array} \right\}
$$

is a set of maximal probability, it follows that $\Gamma_{\mathrm{cp}}^{(\mathrm{TP})}(P_1)$ is nonempty.

Since $\Gamma_{\mathrm{(cp)}}^{(\mathrm{TP})}(P_1) \subseteq \Gamma_1^{(\mathrm{TP})} \subseteq \Delta_M$, it follows that Δ_M is nonempty, as well.

(b) $\Rightarrow$ (a) Since we assume that $\Delta_M \neq \emptyset$, we obtain that there exists a $t_0 \in \mathbb{T}$, $t_0 \neq 0$, such that $\Gamma_{t_0}^{(\mathrm{TP})}$ is nonempty.

Let $x_0 \in \Gamma_{t_0}^{(\mathrm{TP})}$. Then there exists a discrete-time Banach limit $L^{(\mathrm{dt})}$ such that $\varepsilon_{x_0}^{L^{(\mathrm{dt})}}$ is an elementary measure.

Since $(P_t)_{t \in \mathbb{T}}$ is a Feller transition function, it follows that P_{t_0} is a Feller transition probability. Thus, using Theorem 1.4.1, we obtain that $\varepsilon_{x_0}^{(L^{(dt)})}$ is a (nonzero) invariant measure of P_{t_0}. Thus, P_{t_0} has invariant probabilities.

In view of Corollary 3.3.10, we obtain that the implication holds true. $\square$

As earlier, let $(P_t)_{t \in \mathbb{T}}$ be a Feller transition function defined on the locally compact separable metric space (X, d).

Set $\Delta_m = \bigcap_{t \in \mathbb{T}} \Gamma_t^{(\mathrm{TP})}$.

In the next theorem, we use the set Δ_m to study the minimality of the transition function $(P_t)_{t \in \mathbb{T}}$.

Theorem 7.1.15. *Assume that the transition function $(P_t)_{t \in \mathbb{T}}$ is $C_0(X)$-jointly continuous and that $\Delta_m = X$. Then the following assertions are equivalent:*

(i) The transition function $(P_t)_{t \in \mathbb{T}}$ is minimal.
(ii) Any invariant probability measure of $(P_t)_{t \in \mathbb{T}}$ is supported on the entire space X.

Proof. (i) $\Rightarrow$ (ii). Assume that $(P_t)_{t \in \mathbb{T}}$ is minimal, and let μ be an invariant probability measure of $(P_t)_{t \in \mathbb{T}}$. We have to prove that $\operatorname{supp} \mu = X$.

To this end, let $x \in \operatorname{supp} \mu$.

Since $(P_t)_{t \in \mathbb{T}}$ is a Feller transition function, it follows that P_t is a Feller transition probability for every $t \in \mathbb{T}$. Thus, by applying a result of Lasota and Myjak (Proposition 1.1.3) to the probability measures δ_x and μ we obtain that

$$\operatorname{supp}(T_t \delta_x) \subseteq \operatorname{supp}(T_t \mu) = \operatorname{supp} \mu \text{ for every } t \in \mathbb{T}.$$

Accordingly, $\mathcal{O}(x) = \bigcup_{t \in \mathbb{T}} \operatorname{supp}(T_t \delta_x) \subseteq \operatorname{supp} \mu$. Since we assume that $(P_t)_{t \in \mathbb{T}}$ is a minimal transition function, we obtain that $\operatorname{supp} \mu = X$ because $\overline{\mathcal{O}(x)} = X$.

(ii) $\Rightarrow$ (i). We have to prove that, for every $x \in X$, the orbit $\mathcal{O}(x)$ of x under the action of $(P_t)_{t \in \mathbb{T}}$ is dense in X.

To this end, let $x \in X$.

Since we assume that $\Delta_{\mathrm{m}} = X$, it follows that $\Gamma_1^{(\mathrm{TP})} = X$, so $x \in \Gamma_1^{(\mathrm{TP})}$. Accordingly, there exists a discrete-time Banach limit $L^{(\mathrm{dt})}$ such that $\varepsilon_x^{(L^{(\mathrm{dt})})}$ is an elementary measure for the transition probability P_1.

Since P_1 is a Feller transition probability, we can use (a) of Theorem 2.2.1 of [143], and we obtain that $\operatorname{supp} \varepsilon_x^{(L^{(\mathrm{dt})})} \subseteq \overline{\mathcal{O}_{P_1}^{(\mathrm{TP})}(x)}$, where $\overline{\mathcal{O}_{P_1}^{(\mathrm{TP})}(x)}$ is the orbit-closure of x under the action of the transition probability P_1.

Now let $((S_t, T_t))_{t \in \mathbb{T}}$ be the family of Markov-Feller pairs defined by $(P_t)_{t \in \mathbb{T}}$, and let $\nu_1 = \mathrm{P}\text{-}\int_0^1 T_t \mu_1 dt$ be the probability measure ν_α for $\alpha = 1$ corresponding to the probability measure $\mu_1 = \dfrac{\varepsilon_x^{(L^{(\mathrm{dt})})}}{\|\varepsilon_x^{(L^{(\mathrm{dt})})}\|}$.

Clearly, $\operatorname{supp} \mu_1 = \operatorname{supp} \varepsilon_x^{(L^{(\mathrm{dt})})} \subseteq \overline{\mathcal{O}_{P_1}^{(\mathrm{TP})}(x)} \subseteq \overline{\mathcal{O}(x)}$.

Using Theorem 7.1.12, we obtain that

$$\operatorname{supp} \nu_1 = \overline{\mathcal{O}(\operatorname{supp} \mu_1)} = \overline{\bigcup_{z \in \operatorname{supp} \mu_1} \mathcal{O}(z)} \subseteq \overline{\mathcal{O}(x)}.$$

Since ν_1 is an invariant probability for the transition function $(P_t)_{t \in \mathbb{T}}$, and we assume that (ii) holds true, it follows that $\operatorname{supp} \nu_1 = X$ so $\overline{\mathcal{O}(x)} = X$. $\qquad\square$

If $(P_t)_{t \in \mathbb{T}}$ is defined by a continuous semiflow or a continuous flow, then Theorem 7.1.15 becomes:

Corollary 7.1.16. *Let* $\mathbf{w} = (w_z)_{t \in \mathbb{T}}$ *be a continuous semiflow or flow, and let* $(P_t^{(\mathbf{w})})_{t \in \mathbb{T}}$ *be the transition function defined by* $\mathbf{w}$*. If* $\Delta_{\mathrm{m}} = X$*, then the following assertions are equivalent:*

(a) $\mathbf{w}$ *is minimal.*

(b) *Any invariant probability measure of* $\mathbf{w}$ *is supported on the entire space* X*.*

Note that since **w** is continuous, it follows that $(P_t)_{t \in \mathbb{T}}$ is $C_0(X)$-jointly continuous, so the proof of the corollary is obvious in view of Theorem 7.1.15.

For a compact metric space, Theorem 7.1.15 becomes:

Corollary 7.1.17. *Let (X, d) be a compact metric space, and let $(P_t)_{t \in \mathbb{T}}$ be a $C(X)$-jointly continuous transition function. Then $(P_t)_{t \in \mathbb{T}}$ is a minimal transition function if and only if any invariant probability measure of $(P_t)_{t \in \mathbb{T}}$ is supported on the entire space X.*

The proof of the corollary is obvious because the fact that X is compact implies that $\Delta_{\mathrm{m}} = X$.

7.2 Unique Ergodicity and Related Topics

As mentioned in the abstract for the chapter, in this section we extend the results of Chapter 3 and Section 4.1 of [143] to Feller transition functions.

The section is organized into three subsections. In the first subsection (Sect. 7.2.1) we use the sets that appear in the KBBY decomposition (see Chap. 5 and the first two sections of Chap. 6) in order to characterize the support of the invariant probability measure of a uniquely ergodic Feller transition function. Next we show that if a Feller transition function is not necessarily uniquely ergodic, but has the property that there exists a unique closed set F such that F is the support of any invariant probability measure of the transition function, then F can be characterized in a similar manner as the support of the invariant probability of a uniquely ergodic transition function.

In the second subsection, we study the transition functions that are $C_0(X)$-equicontinuous in the mean and obtain a criterion for the unique ergodicity of this type of Feller transition function.

Finally, in the last subsection (Sect. 7.2.3) we study unique ergodicity and ergodic measures using generic points.

7.2.1 Supports of Invariant Probabilities of Uniquely Ergodic Transition Functions and a Related Topic

As usual in this book, let (X, d) be a locally compact separable metric space.

Let $(P_t)_{t \in \mathbb{T}}$ be a Feller transition function defined on (X, d).

Recall (see Sect. 2.3) that $(P_t)_{t \in \mathbb{T}}$ is said to be uniquely ergodic if it has exactly one invariant probability measure.

Using Proposition 2.3.7, we obtain that, if $\mathbb{T} = \mathbb{R}$, then $(P_t)_{t \in \mathbb{R}}$ is uniquely ergodic if and only if the restriction $(P_t)_{t \in [0,+\infty)}$ of $(P_t)_{t \in \mathbb{R}}$ to $[0, +\infty)$ is uniquely ergodic. Therefore, when studying the unique ergodicity of transition functions we may and do assume that $\mathbb{T} = [0, +\infty)$.

Thus, let $(P_t)_{t \in [0, +\infty)}$ be a Feller transition function. Unless otherwise stated, we will assume that $(P_t)_{t \in [0, +\infty)}$ satisfies the s.m.a. and is pointwise continuous.

Set $\gamma = \bigcap_{x \in \Gamma} \overline{\mathcal{O}(x)}$, $\gamma_c = \bigcap_{x \in \Gamma_c} \overline{\mathcal{O}(x)}$, $\gamma_{cp} = \bigcap_{x \in \Gamma_{cp}} \overline{\mathcal{O}(x)}$, $\gamma_{cpi} = \bigcap_{x \in \Gamma_{cpi}} \overline{\mathcal{O}(x)}$, and $\gamma_{cpie} = \bigcap_{x \in \Gamma_{cpie}} \overline{\mathcal{O}(x)}$.

Note that using the fact that $(P_t)_{t \in [0, +\infty)}$ is a Feller transition function, by Theorem 7.1.1, we obtain that $\Gamma_{cp} = \Gamma_{cpi}$, so $\gamma_{cp} = \gamma_{cpi}$.

If we assume that $(P_t)_{t \in [0, +\infty)}$ is also uniquely ergodic, then using Corollary 6.2.8, we obtain that the unique invariant probability measure of $(P_t)_{t \in [0, +\infty)}$ is an ergodic measure. Thus, if μ is the unique invariant probability measure, then $\varepsilon_x = \mu$ for every $x \in \Gamma_{cpie}$ and also $\varepsilon_x = \mu$ for every $x \in \Gamma_{cpi}$. Therefore, in this case we obtain that $\Gamma_{cpi} = \Gamma_{cpie}$. In the next theorem we obtain the characterization of supp μ that we mentioned at the beginning of this section.

Note that if $(P_t)_{t \in [0, +\infty)}$ is uniquely ergodic, then Γ_{cpie} is nonempty, so Γ_{cpi}, Γ_{cp}, Γ_c and Γ are nonempty, as well.

Theorem 7.2.1. *Assume that the Feller transition function $(P_t)_{t \in [0, +\infty)}$ is uniquely ergodic, and let μ be the unique invariant probability measure of $(P_t)_{t \in [0, +\infty)}$. Then* supp $\mu = \gamma_{cpie} = \gamma_{cpi} = \gamma_{cp} = \gamma_c = \gamma$.

Proof. As pointed out before the theorem, the fact that $(P_t)_{t \in [0, +\infty)}$ is a Feller transition function implies that $\gamma_{cp} = \gamma_{cpi}$.

Also, before stating the theorem we mentioned that, since $(P_t)_{t \in [0, +\infty)}$ is uniquely ergodic, it follows that $\Gamma_{cpi} = \Gamma_{cpie}$. Thus, $\gamma_{cpi} = \gamma_{cpie}$.

Accordingly, in order to complete the proof of the theorem we only have to prove that supp $\mu = \gamma_{cp} = \gamma_c = \gamma$.

Taking into consideration that $\Gamma_{cp} \subseteq \Gamma_c \subseteq \Gamma$, we obtain that $\gamma_{cp} \supseteq \gamma_c \supseteq \gamma$. Thus, the proof of the theorem will be completed as soon as we show that:

$$(1) \qquad \text{supp } \mu = \gamma_{cp}$$

and

$$(2) \qquad \text{supp } \mu \subseteq \gamma.$$

Proof of (1). As pointed out before the theorem $\Gamma_{cpie} \neq \emptyset$ (by Corollary 6.2.8).

Since $(P_t)_{t \in [0, +\infty)}$ is uniquely ergodic, it follows that $\Gamma_{cpie} = [x]$ for every $x \in \Gamma_{cpie}$, where $[\cdot]$ is the equivalence class defined before Lemma 6.2.5.

Using (a) of Theorem 6.2.7, we obtain that $\varepsilon_x([x]) = 1$ for every $x \in \Gamma_{cpie}$. Thus, supp $\mu = $ supp $\varepsilon_x \subseteq \overline{[x]}$ for every $x \in \Gamma_{cpie}$ (where, of course, $\overline{[x]}$ is the closure of $[x]$ in the metric topology of X). Therefore, $[x] \cap$ supp$\mu \neq \emptyset$ whenever $x \in \Gamma_{cpie}$.

Now let $x \in \Gamma_{cpie}$. In view of the above discussion there exists a $y \in [x] \cap$ supp μ. Thus, $\varepsilon_y = \varepsilon_x = \mu$ and $y \in$ supp μ. Since $(P_t)_{t \in \mathbb{T}}$ is a Feller transition function, we can apply Theorem 7.1.3, and we obtain that supp $\mu = \overline{\mathcal{O}(y)}$; since $y \in [x] = \Gamma_{cpie} \subseteq \Gamma_{cp}$, it follows that supp $\mu \supseteq \gamma_{cp}$.

Since $\varepsilon_x = \mu$ for every $x \in \Gamma_{\mathrm{cp}}$ (because $(P_t)_{t \in [0,+\infty)}$ is uniquely ergodic), using Proposition 5.1.7, we obtain that $\operatorname{supp} \mu = \operatorname{supp} \varepsilon_x \subseteq \overline{\mathcal{O}(x)}$ for every $x \in \Gamma_{\mathrm{cp}}$, so $\operatorname{supp} \mu \subseteq \cap_{x \in \Gamma_{\mathrm{cp}}} \overline{\mathcal{O}(x)} = \gamma_{\mathrm{cp}}$.

We have therefore proved that $\operatorname{supp} \mu = \gamma_{\mathrm{cp}}$.

Proof of (2). In order to prove that $\operatorname{supp} \mu \subseteq \gamma$, we will prove that for every $y \in \operatorname{supp} \mu$ and every $x \in \Gamma$, it follows that $y \in \overline{\mathcal{O}(x)}$.

To this end, let $y \in \operatorname{supp} \mu$ and $x \in \Gamma$.

In order to prove that $y \in \overline{\mathcal{O}(x)}$, we will construct a sequence $(y_n)_{n \in \mathbb{N}}$ of elements of $\mathcal{O}(x)$ such that $(y_n)_{n \in \mathbb{N}}$ converges to y.

Since $x \in \Gamma$, there exists a continuous-time Banach limit $L^{(\mathrm{ct})}$ such that $\varepsilon_x^{(L^{(\mathrm{ct})})}$ is an elementary measure. Since $(P_t)_{t \in [0,+\infty)}$ is a Feller transition function, using Theorem 7.1.1, we obtain that $\varepsilon_x^{(L^{(\mathrm{ct})})}$ is also an invariant measure. Since μ is the unique invariant probability measure of $(P_t)_{t \in [0,+\infty)}$, it follows that there exists an $a \in \mathbb{R}$, $a > 0$, such that $\varepsilon_x^{(L^{(\mathrm{ct})})} = a\mu$.

Using the fact that (X, d) is locally compact we obtain that there exists an $\alpha \in \mathbb{R}$, $\alpha > 0$, such that $\overline{B(y, \alpha)}$ is a compact subset of X.

As in the proof of Theorem 3.1.1 of [143], for every $n \in \mathbb{N}$, we define a function $f_n = X \to \mathbb{R}$ as follows: $f_n(z) = d(z, X \backslash B(y, \frac{\alpha}{n}))$ for every $z \in X$.

It is easy to see (and well-known) that the functions f_n, $n \in \mathbb{N}$, are continuous. Since $f_n(z) = 0$ for every $z \in X \setminus B(y, \frac{\alpha}{n})$, it follows that $f_n \in C_0(X)$ for every $n \in \mathbb{N}$ (actually, the functions f_n, $n \in \mathbb{N}$, have compact supports).

Since $y \in \operatorname{supp} \mu$ and since $f_n(z) > 0$ for every $z \in B(y, \frac{\alpha}{n})$, it follows that
$$0 < \langle f_n, \alpha\mu \rangle = \langle f_n, \varepsilon_x^{(L^{(\mathrm{ct})})} \rangle = L^{(\mathrm{ct})}((\langle f_n, T_t \delta_x \rangle)_{t \in [0,+\infty)}) \text{ for every } n \in \mathbb{N}.$$

Thus, for every $n \in \mathbb{N}$, there exists a $t_n \in [0, +\infty)$ such that $\langle f_n, T_{t_n} \delta_x \rangle > 0$ (because, if there exists an $n \in \mathbb{N}$ such that $\langle f_n, T_t \delta_x \rangle = 0$ for all $t \in [0, +\infty)$, then $\langle f_n, \varepsilon_x^{(L^{(\mathrm{ct})})} \rangle = 0$; that is, we obtain a contradiction).

Since $\langle f_n, T_{t_n} \delta_x \rangle > 0$ and since $f_n(z) > 0$ implies that $z \in B(y, \frac{\alpha}{n})$, it follows that $B(y, \frac{\alpha}{n}) \cap (\operatorname{supp}(T_{t_n} \delta_x)) \neq \emptyset$ for every $n \in \mathbb{N}$.

Clearly, $(y_n)_{n \in \mathbb{N}}$ converges to y and $y_n \in \mathcal{O}(x)$ for every $n \in \mathbb{N}$. Thus, $y \in \overline{\mathcal{O}(x)}$.

We have therefore proved that $\operatorname{supp} \mu \subseteq \gamma$.

$\square$

Note that the above theorem is an extension of Theorem 1.4.12 (Theorem 3.1.1 of [143]) to transition functions.

Note also that, as in the case of transition probabilities, we already have a result (Theorem 7.1.4) that can be used to obtain a "formula" for the support of the invariant probability measure of a uniquely ergodic transition function (in the case of uniquely ergodic transition probabilities it is formula (3.1.1) of [143] that can be used for this purpose). However, like Theorem 3.1.1 of [143], some of the "formulas" that appear in Theorem 7.2.1 are often easier to use than the "formula" obtained from Theorem 7.1.4.

In the case in which X is compact, Theorem 7.2.1 has the following consequence:

Corollary 7.2.2. *Assume that the transition function $(P_t)_{t\in[0,+\infty)}$ is defined in a compact metric space (X,d) and is uniquely ergodic. If μ is the invariant probability measure of $(P_t)_{t\in[0,+\infty)}$, then* $\operatorname{supp}\mu = \bigcap_{x\in X}\overline{\mathcal{O}(x)}$.

Proof. Taking into consideration that X is a compact metric space, and using the definition of the set Γ (see the beginning of Sect. 5.1 for the definition of Γ), we obtain that $\Gamma = X$ in this case, because the compactness of X implies that the function $\mathbf{1}_X$ belongs to $C_0(X)$, so $L^{(ct)}(((\langle \mathbf{1}_X, T_t\delta_x\rangle)_{t\in[0,+\infty)})) = 1 > 0$ for every $x \in X$.

Thus, by Theorem 7.2.1, $\operatorname{supp}\mu = \bigcap_{x\in X}\overline{\mathcal{O}(x)}$. $\qquad\square$

Note that the above corollary is an extension of Corollary 1.4.13 (see also Corollary 3.1.2 of [143]).

We now return to the general case where (X,d) is a locally compact (not necessarily compact) separable metric space.

It is sometimes the case that $(P_t)_{t\in[0,+\infty)}$ has invariant probabilities, and even though $(P_t)_{t\in[0,+\infty)}$ is not uniquely ergodic, there exists a closed subset F of X such that for every invariant probability measure μ of $(P_t)_{t\in[0,+\infty)}$, it follows that $\operatorname{supp}\mu = F$. The questions that appear in this case are whether or not we can find "formulas" for F, and whether or not the "formulas" that appear in Theorem 7.2.1 cannot be used for F, as well. It turns out that the "formulas" that appear in Theorem 7.2.1 can also be used for F. The details are included in the next theorem.

Theorem 7.2.3. *Assume that $(P_t)_{t\in[0,+\infty)}$ has invariant probabilities, and there exists a closed subset F of X such that* $\operatorname{supp}\mu = F$ *whenever μ is an invariant probability of $(P_t)_{t\in[0,+\infty)}$. Then* $F = \gamma_{\mathrm{cpie}} = \gamma_{\mathrm{cpi}} = \gamma_{\mathrm{cp}} = \gamma_{\mathrm{c}} = \gamma$.

Proof. Since $(P_t)_{t\in[0,+\infty)}$ has invariant probabilities, using Corollary 6.2.8, we obtain that Γ_{cpie} is nonempty. Therefore, Γ is also nonempty because $\Gamma \supseteq \Gamma_{\mathrm{cpie}}$.

Taking into consideration that $\Gamma_{\mathrm{cpie}} \subseteq \Gamma_{\mathrm{cpi}} \subseteq \Gamma_{\mathrm{cp}} \subseteq \Gamma_{\mathrm{c}} \subseteq \Gamma$, we obtain that $\gamma_{\mathrm{cpie}} \supseteq \gamma_{\mathrm{cpi}} \supseteq \gamma_{\mathrm{cp}} \supseteq \gamma_{\mathrm{c}} \supseteq \gamma$; thus, in order to prove the theorem, it is enough to prove that

$$(1) \quad F \subseteq \gamma \quad \text{and} \quad (2) \quad \gamma_{\mathrm{cpie}} \subseteq F.$$

Proof of (1). The proof will be completed if we show that $F \subseteq \overline{\mathcal{O}(x)}$ for every $x \in \Gamma$.

Thus, let $x \in \Gamma$. Then there exists a continuous-time Banach limit $L^{(ct)}$ such that $\varepsilon_x^{(L^{(ct)})}$ is an elementary measure. By Proposition 5.1.7, we have that $\operatorname{supp}\varepsilon_x^{(L^{(ct)})} \subseteq \overline{\mathcal{O}(x)}$. Since $(P_t)_{t\in[0,+\infty)}$ is a Feller transition function, using Theorem 7.1.1, we obtain that $\varepsilon_x^{(L^{(ct)})}$ is also an invariant measure of $(P_t)_{t\in[0,+\infty)}$; accordingly, $\operatorname{supp}\varepsilon_x^{(L^{(ct)})} = F$.

We have therefore proved that $F \subseteq \overline{\mathcal{O}(x)}$ for every $x \in \Gamma$.

Proof of (2). In order to prove that $\gamma_{\text{cpie}} \subseteq F$, it is enough to prove that for every $x \in \Gamma_{\text{cpie}}$ there exists a $y \in \Gamma_{\text{cpie}}$ such that $\overline{\mathcal{O}(x)} \supseteq \overline{\mathcal{O}(y)} = F$ (note that we use here the fact that Γ_{cpie} is nonempty, a fact that we pointed out when we started the proof of the theorem).

So, let $x \in \Gamma_{\text{cpie}}$. Using (a) of Theorem 6.2.7, we obtain that ε_x is an ergodic invariant probability measure of $(P_t)_{t \in [0,+\infty)}$, and $\varepsilon_x([x]) = 1$. Therefore, $\varepsilon_x([x] \cap (\text{supp } \varepsilon_x)) = 1$.

Accordingly, $[x] \cap (\text{supp } \varepsilon_x) \neq \emptyset$; thus, there exists a $y \in [x] \cap (\text{supp } \varepsilon_x)$.

We obtain that $\varepsilon_x = \varepsilon_y$. Since $y \in \text{supp } \varepsilon_x = \text{supp } \varepsilon_y$, using Theorem 7.1.3, we obtain that $\text{supp } \varepsilon_y = \overline{\mathcal{O}(y)}$.

But $\text{supp } \varepsilon_y = F$ because ε_y is an invariant probability measure of $(P_t)_{t \in [0,+\infty)}$. Thus, $F = \overline{\mathcal{O}(y)}$.

By Proposition 5.1.7, $\text{supp } \varepsilon_x \subseteq \overline{\mathcal{O}(x)}$, so $\overline{\mathcal{O}(y)} = \text{supp } \varepsilon_y = F = \text{supp } \varepsilon_x \subseteq \overline{\mathcal{O}(x)}$.

We have therefore proved that for every $x \in \Gamma_{\text{cpie}}$ there exists a $y \in \Gamma_{\text{cpie}}$ such that $F = \overline{\mathcal{O}(y)} \subseteq \overline{\mathcal{O}(x)}$, so $\gamma_{\text{cpie}} \subseteq F$.

$\square$

Note that the above theorem is an extension of Theorem 1.4.12 (Theorem 3.1.1 of [143]) to transition functions. Theorem 3.1.1 of [143] answers a question that Furstenberg asked me about Feller transition probabilities in 1999. The question also makes perfect sense for Feller transition functions, so Theorem 7.2.3 can be thought of as an answer to Furstenberg's question when we are dealing with transition functions.

In a similar manner as in the case of transition probabilities, given a transition function $(P_t)_{t \in [0,+\infty)}$ and the associated family $((S_t, T_t))_{t \in [0,+\infty)}$ of Markov pairs (both defined on the locally compact separable metric space (X, d)), and given $x \in X$ and $y \in X$, we say that *x leads to y* if $y \in \overline{\mathcal{O}(x)}$.

Given a nonempty subset A of X and $y \in X$, we say that y is a *universal element with respect to A* if x leads to y for every $x \in A$ (that is, if $y \in \cap_{x \in A}\overline{\mathcal{O}(x)}$). We say that $y \in X$ is a *universal element* if y is a universal element with respect to X.

Observation. The term "universal element" was suggested to us by Furstenberg. ▲

As we did for Markov-Feller pairs at the end of Section 3.1 of [143], we can restate the results obtained so far in this section in terms of universal elements for transition functions.

Thus, Theorem 7.2.1 can be restated as follows:

Theorem 7.2.4. *Consider the Feller transition function* $(P_t)_{t \in [0,+\infty)}$, *assume that* $(P_t)_{t \in [0,+\infty)}$ *is uniquely ergodic, and let* μ *be the invariant probability measure of* $(P_t)_{t \in [0,+\infty)}$. *Then* $\text{supp } \mu$ *is equal to each of the following sets:*

- *The set of all universal elements with respect to* Γ_{cpie};
- *The set of all universal elements with respect to* Γ_{cpi};
- *The set of all universal elements with respect to* Γ_{cp};

- *The set of all universal elements with respect to Γ_c;*
- *The set of all universal elements with respect to Γ.*

Next, Corollary 7.2.2 becomes:

Corollary 7.2.5. *Assume that (X,d) is a compact metric space, and let $(P_t)_{t \in [0,+\infty)}$ be a uniquely ergodic Feller transition function. If μ is the unique invariant probability measure of $(P_t)_{t \in [0,+\infty)}$, then $\mathrm{supp}\, \mu$ is the set of all universal elements generated by $(P_t)_{t \in [0,+\infty)}$.*

Finally, Theorem 7.2.3 becomes:

Theorem 7.2.6. *Assume that the Feller transition function has at least one invariant probability, and that there exists a closed subset F of X such that F is the support of each of the invariant probabilities of $(P_t)_{t \in [0,+\infty)}$ (that is, if μ is an invariant probability for $(P_t)_{t \in [0,+\infty)}$, then $\mathrm{supp}\, \mu = F$).*
Then F is equal to each of the following sets:

- *The set of all universal elements with respect to Γ_{cpie};*
- *The set of all universal elements with respect to Γ_{cpi};*
- *The set of all universal elements with respect to Γ_{cp};*
- *The set of all universal elements with respect to Γ_c;*
- *The set of all universal elements with respect to Γ.*

7.2.2 A Criterion for Unique Ergodicity

As usual in this chapter, let $(P_t)_{t \in [0,+\infty)}$ be a Feller transition function that satisfies the s.m.a. and is pointwise continuous.

If $(P_t)_{t \in [0,+\infty)}$ is uniquely ergodic, then, using Theorem 7.2.1 we can infer that the sets γ_{cpie}, γ_{cpi}, γ_{cp}, γ_c and γ are all nonempty. Thus, it is tempting to believe that if $(P_t)_{t \in [0,+\infty)}$ satisfies the s.m.a., is pointwise continuous, and one (or several) of the sets γ_{cpie}, γ_{cpi}, γ_{cp}, γ_c or γ is (are) nonempty, then $(P_t)_{t \in [0,+\infty)}$ is uniquely ergodic. However, it can be shown that there exist transition functions $(P_t)_{t \in [0,+\infty)}$ that satisfy the s.m.a., are pointwise continuous, fail to be uniquely ergodic, and each of the sets γ_{cpie}, γ_{cpi}, γ_{cp}, γ_c and γ is equal to the entire space X (such transition functions can be obtained as follows: it can be shown that there exist minimal continuous flows $\mathbf{w} = (w_t)_{t \in \mathbb{R}}$ that fail to be uniquely ergodic; using Corollary 7.1.16, we obtain that the restriction $(P_t^{(\mathbf{w})})_{t \in [0,+\infty)}$ to $[0, +\infty)$ of the transition function $(P_t^{(\mathbf{w})})_{t \in \mathbb{R}}$ defined by $\mathbf{w}$ satisfies all the above-mentioned properties).

Thus, a natural question that appears is: can we find a large enough collection $\mathcal{C}$ of transition functions which have the property that if $(P_t)_{t \in [0,+\infty)}$ belongs to $\mathcal{C}$ and any of the sets γ_{cpie}, γ_{cpi}, γ_{cp}, γ_c or γ that correspond to $(P_t)_{t \in [0,+\infty)}$ is nonempty, then $(P_t)_{t \in [0,+\infty)}$ is uniquely ergodic.

Our goal in this subsection is to discuss such a class of transition functions.

To this end, we need some preparation.

Let Λ be a nonempty set.

As usual, let (X, d) be a locally compact separable metric space, and let $B_b(X)$ be the Banach lattice of all real-valued bounded Borel measurable functions defined on X.

A family $(Q_t)_{t\in\Lambda}$ of linear bounded operators, $Q_t = B_b(X) \to B_b(X)$ for every $t \in \Lambda$, is said to be $C_0(X)$-*equicontinuous* if the following condition is satisfied:

- for every $f \in C_0(X)$, for every convergent sequence $(x_n)_{n\in\mathbb{N}}$ of elements of X, and for every $\varepsilon \in \mathbb{R}$, $\varepsilon > 0$, there exists an $n_\varepsilon \in \mathbb{N}$ such that $|Q_t f(x_n) - Q_t f(x)| < \varepsilon$ for every $n \in \mathbb{N}$, $n \geq n_\varepsilon$, and every $t \in \Lambda$, where $x = \lim_{n\to+\infty} x_n$.

Note that for $\Lambda = \mathbb{N}\cup\{0\}$, the $C_0(X)$-equicontinuity (of sequences of operators) has been discussed in [143], where we pointed out that we can define other types of equicontinuity, that $C_0(X)$-equicontinuity is the most general type of equicontinuity among the kinds of equicontinuity that we mentioned in [143], and that, luckily, $C_0(X)$-equicontinuity was good enough for our purposes. All these observations apply also for the families of operators that we consider in this book, so we will deal with $C_0(X)$-equicontinuity only.

Now, let $(P_t)_{t\in[0,+\infty)}$ be a transition function defined on (X, d), and, as usual in this chapter, assume that $(P_t)_{t\in[0,+\infty)}$ satisfies the s.m.a. and is pointwise continuous. Also, let $((S_t, T_t))_{t\in[0,+\infty)}$ be the family of Markov pairs defined by $(P_t)_{t\in[0,+\infty)}$.

Recall (see Sect. 2.1) that we say that $(P_t)_{t\in[0,+\infty)}$ is a $C_0(X)$-equicontinuous transition function if the family of operators $(S_t)_{t\in[0,+\infty)}$ is $C_0(X)$-equicontinuous.

Since $C_0(X)$-equicontinuity is the only type of equicontinuity that will be considered in this book, we will often refer to $C_0(X)$-equicontinuity simply as equicontinuity.

Note that if $(P_t)_{t\in[0,+\infty)}$ is an equicontinuous transition function, then $(P_t)_{t\in[0,+\infty)}$ is a Feller transition function, as well.

For the transition function $(P_t)_{t\in[0,+\infty)}$ and the family $((S_t, T_t))_{t\in[0,+\infty)}$ of Markov pairs defined by $(P_t)_{t\in[0,+\infty)}$ we can use Proposition 3.3.2 in order to infer that the pointwise integral P-$\int_0^u S_t f dt$ exists for every $f \in B_b(X)$ and $u \in [0, +\infty)$. This means that given $f \in B_b(X)$ and $u \in [0, +\infty)$, the map $x \mapsto \int_0^u S_t f(x)dt$, $x \in X$, belongs to $B_b(X)$.

Thus, it makes sense to define the operators $\mathbf{A}_u = B_b(X) \to B_b(X)$, $\mathbf{A}_u f = \frac{1}{u}\left(\text{P-}\int_0^u S_t f dt\right)$ for every $f \in B_b(X)$ and $u \in (0, +\infty)$.

It is easy to see that the operators $\mathbf{A}_s$, $s \in (0, +\infty)$, are linear and positive. Therefore, the operators $\mathbf{A}_s$, $s \in (0, +\infty)$ are also bounded (continuous).

Moreover, since the function $\mathbf{1}_X$ is the largest element of the unit ball of $B_b(X)$, and since $\mathbf{A}_s \mathbf{1}_X = \mathbf{1}_X$ for every $s \in (0, +\infty)$, it follows that $\mathbf{A}_s$, $s \in (0, +\infty)$, are positive contractions of $B_b(X)$.

We say that $(P_t)_{t\in[0,+\infty)}$ (or $(S_t)_{t\in[0,+\infty)}$), is $C_0(X)$-*equicontinuous in the mean*, or, simply, *equicontinuous in the mean* if the family of operators $(\mathbf{A}_s)_{s\in(0,+\infty)}$ is equicontinuous.

Note that if $(P_t)_{t\in[0,+\infty)}$ is equicontinuous, then $(P_t)_{t\in[0,+\infty)}$ is equicontinuous in the mean, as well.

We say that $(P_t)_{t\in[0,+\infty)}$ has the *e.m.d.s. property* (has the property that its *ergodic measures have disjoint supports*) if any two distinct invariant ergodic probability measures have disjoint supports; that is, if $(\operatorname{supp}\mu) \cap (\operatorname{supp}\nu) = \emptyset$ whenever μ and ν are invariant ergodic probability measures and $\mu \neq \nu$.

Note that if $(P_t)_{t\in[0,+\infty)}$ does not have invariant probability measures (so, by Corollary 6.2.8, $(P_t)_{t\in[0,+\infty)}$ does not have invariant ergodic probability measures, either), then $(P_t)_{t\in[0,+\infty)}$ has the e.m.d.s. property. Also, if $(P_t)_{t\in[0,+\infty)}$ is uniquely ergodic (so it has exactly one ergodic invariant probability measure), then, again, $(P_t)_{t\in[0,+\infty)}$ has the e.m.d.s. property.

We need the following lemma:

Lemma 7.2.7. *Assume that the transition function $(P_t)_{t\in[0,+\infty)}$ has the property that $\Gamma_{\mathrm{cpie}} \neq \emptyset$ (that is, assume that $(P_t)_{t\in[0,+\infty)}$ has at least one invariant ergodic probability measure), let $x \in \Gamma_{\mathrm{cpie}}$, and let $y \in X$ be such that for every $f \in C_0(X)$, the limit $\lim_{u\to+\infty} \frac{1}{u}\int_0^u S_t f(y)dt$ exists and is equal to $\langle f, \varepsilon_x\rangle$. Then $y \in \Gamma_{\mathrm{cpie}}$ and $y \sim x$, where, of course, $\sim$ is the equivalence relation on Γ_{cpie} defined before Lemma 6.2.5.*

Proof. Let $x \in \Gamma_{\mathrm{cpie}}$ and let y be as in the lemma.

Taking into consideration the fact that $\lim_{u\to+\infty} \frac{1}{u}\int_0^u S_t f(y)dt$ exists and is equal to $\langle f, \varepsilon_x\rangle$ for every $f \in C_0(x)$, using the definition of Γ_{cpi} (see the end of Sect. 5.1), and using the definition of f^* (see the discussion preceding Theorem 5.3.3), we obtain that $y \in \Gamma_{\mathrm{cpi}}$, that $\varepsilon_y = \varepsilon_x$, and that $f^*(x) = f^*(y)$ for every $f \in C_0(X)$.

It follows that

$$\int_{\Gamma_{\mathrm{cpi}}} (f^*(z) - f^*(y))^2\, d\varepsilon_y(z)$$

$$= \int_{\Gamma_{\mathrm{cpi}}} (f^*(z) - f^*(x))^2\, d\varepsilon_x(z) = 0$$

for every $f \in C_0(X)$ (the last equality holds true because $x \in \Gamma_{\mathrm{cpie}}$).

In view of the definition of Γ_{cpie} (see the beginning of Sect. 6.2), we obtain that $y \in \Gamma_{\mathrm{cpie}}$.

Since $\varepsilon_x = \varepsilon_y$, it follows that $x \sim y$. $\qquad\square$

In the next theorem we discuss an important feature of the equicontinuous transition functions that have invariant probabilities; namely, that the sets $[x]$, $x \in \Gamma_{\mathrm{cpie}}$, are closed (in the metric topology of X, of course).

Theorem 7.2.8. *Assume that the transition function $(P_t)_{t\in[0,+\infty)}$ is equicontinuous in the mean and that $\Gamma_{\mathrm{cpie}} \neq \emptyset$. Then, the set $[x]$ is a closed subset of X whenever $x \in \Gamma_{\mathrm{cpie}}$.*

Proof. The theorem will be completely proved if we show that for every $x \in \Gamma_{\text{cpie}}$ and every convergent sequence $(y_n)_{n \in \mathbb{N}}$ of elements of $[x]$, the limit y of $(y_n)_{n \in \mathbb{N}}$ belongs to $[x]$, as well.

To this end, let $x \in \Gamma_{\text{cpie}}$, let $(y_n)_{n \in \mathbb{N}}$ be a convergent sequence of elements of $[x]$ and set $y = \lim_{n \to +\infty} y_n$.

In view of Lemma 7.2.7, in order to prove that $y \in [x]$, we have to prove that $\lim_{u \to +\infty} \frac{1}{u} \int_0^u S_t f(y) \, dt$ exists and is equal to $\langle f, \varepsilon_x \rangle$ for every $f \in C_0(X)$.

Thus, let $f \in C_0(X)$. We have to prove that for every $\varepsilon \in \mathbb{R}$, $\varepsilon > 0$, there exists a $u_\varepsilon \in \mathbb{R}$, $u_\varepsilon > 0$, such that $|\frac{1}{u} \int_0^u S_t f(y) \, dt - \langle f, \varepsilon_x \rangle| < \varepsilon$ for every $u \in \mathbb{R}$, $u \geq u_\varepsilon$.

So, let $\varepsilon \in \mathbb{R}$, $\varepsilon > 0$.

Since $(y_n)_{n \in \mathbb{N}}$ converges to y, and since we assume that $(P_t)_{t \in [0,+\infty)}$ is $C_0(X)$-equicontinuous in the mean, we obtain that there exists an $n_\varepsilon \in \mathbb{N}$ such that $|\frac{1}{u} \int_0^u S_t f(y_n) \, dt - \frac{1}{u} \int_0^u S_t f(y) \, dt| < \frac{\varepsilon}{2}$ for every $n \in \mathbb{N}$, $n \geq n_\varepsilon$, and every $u \in (0,+\infty)$.

Since $y_{n_\varepsilon} \in [x]$, it follows that there exists a $u_\varepsilon \in (0,+\infty)$ such that $|\frac{1}{u} \int_0^u S_t f(y_{n_\varepsilon}) \, dt| - \langle f, \varepsilon_x \rangle| < \frac{\varepsilon}{2}$ for every $u \in \mathbb{R}$, $u \geq u_\varepsilon$.

We obtain that

$$\left| \frac{1}{u} \int_0^u S_t f(y) \, dt - \langle f, \varepsilon_x \rangle \right|$$

$$\leq \left| \frac{1}{u} \int_0^u S_t f(y) \, dt - \frac{1}{u} \int_0^u S_t f(y_{n_\varepsilon}) \, dt \right|$$

$$+ \left| \frac{1}{u} \int_0^u S_t f(y_{n_\varepsilon}) \, dt - \langle f, \varepsilon_x \rangle \right| < \frac{\varepsilon}{2} + \frac{\varepsilon}{2} = \varepsilon$$

for every $u \in \mathbb{R}$, $u \geq u_\varepsilon$. $\square$

Theorem 7.2.8 has the following consequence:

Corollary 7.2.9. *If $(P_t)_{t \in [0,+\infty)}$ is equicontinuous in the mean, then $(P_t)_{t \in [0,+\infty)}$ has the e.m.d.s. property.*

Proof. As we mentioned after defining the e.m.d.s. property (before Lemma 7.2.7), if $(P_t)_{t \in [0,+\infty)}$ does not have invariant probability measures, or if $(P_t)_{t \in [0,+\infty)}$ is uniquely ergodic, then the transition function has the e.m.d.s. property.

Thus, we only have to prove the assertion of the corollary in the case when $(P_t)_{t \in [0,+\infty)}$ has at least two distinct invariant ergodic probability measures.

So, assume that $(P_t)_{t \in [0,+\infty)}$ has invariant probabilities and is not uniquely ergodic, and let μ and ν be two distinct invariant ergodic probability measures.

Using (b) of Theorem 6.2.7, we obtain that there exist $x \in \Gamma_{\text{cpie}}$ and $y \in \Gamma_{\text{cpie}}$ such that $\mu = \varepsilon_x$ and $\nu = \varepsilon_y$.

Since $\mu \neq \nu$, it follows that $[x] \cap [y] = \emptyset$ (because $[x]$ and $[y]$ are distinct equivalence classes).

Since $\varepsilon_x([x]) = 1$ and $\varepsilon_y([y]) = 1$ (by (a) of Theorem 6.2.7), and since $[x]$ and $[y]$ are closed sets (by Theorem 7.2.8), we obtain that $\operatorname{supp}\mu \subseteq [x]$ and $\operatorname{supp}\nu \subseteq [y]$; therefore, $(\operatorname{supp}\mu) \cap (\operatorname{supp}\nu) = \emptyset$.

Thus, $(P_t)_{t \in [0,+\infty)}$ has the e.m.d.s. property. $\qquad\square$

Our discussion so far allows us to prove that if $(P_t)_{t \in [0,+\infty)}$ is a transition function that is equicontinuous in the mean and has invariant probabilities, then the nonemptiness of any of the sets γ_{cpie}, γ_{cpi}, γ_{cp}, γ_{c} or γ implies the unique ergodicity of $(P_t)_{t \in [0,+\infty)}$.

To this end, we need the following proposition of independent interest:

Proposition 7.2.10. *Assume that $(P_t)_{t \in [0,+\infty)}$ is a transition function that satisfies the s.m.a. and is pointwise continuous. The following assertions are equivalent:*

(a) $(P_t)_{t \in [0,+\infty)}$ is uniquely ergodic.
(b) $(P_t)_{t \in [0,+\infty)}$ has precisely one invariant ergodic probability measure.

Proof. We will use Theorem 6.3.18 and the notations that appear there.

By definition, $(P_t)_{t \in [0,+\infty)}$ is uniquely ergodic if and only if $(P_t)_{t \in [0,+\infty)}$ has exactly one invariant probability measure.

Therefore, $(P_t)_{t \in [0,+\infty)}$ is uniquely ergodic if and only if the Banach lattice $\mathcal{M}_{\mathrm{inv}}(X)$, as a vector space, is one-dimensional, if and only if (by Theorem 6.3.18) the Banach lattice $\mathcal{M}(Y, \mathcal{F})$, thought of as a vector space, is one-dimensional. Now, $\mathcal{M}(Y, \mathcal{F})$ is one-dimensional if and only if the set Y of all equivalence classes of Γ_{cpie} (see the beginning of Sect. 6.3 for the definition of Y) is a singleton (has exactly one element). Finally, the set Y is a singleton (that is, any two elements $x \in \Gamma_{\mathrm{cpie}}$ and $y \in \Gamma_{\mathrm{cpie}}$ are equivalent) if and only if $(P_t)_{t \in [0,+\infty)}$ has exactly one invariant ergodic probability measure. $\qquad\square$

Our goal, in the next theorem, is to prove the criterion for unique ergodicity mentioned at the beginning of this subsection and before Proposition 7.2.10. Actually we will prove slightly more than what we stated before the last proposition, so, for this purpose, set $\gamma_0 = \cap_{x \in \Gamma_0} \overline{\mathcal{O}(x)}$.

Note that in Proposition 7.2.10 we do not have to assume that the transition function is Feller. However, in the next theorem we return to the usual setting in this section; that is, we assume given a Feller transition function $(P_t)_{t \in [0,+\infty)}$ (that satisfies the s.m.a., and is pointwise continuous).

Theorem 7.2.11. *Assume that the transition function $(P_t)_{t \in [0,+\infty)}$ has invariant probabilities and is equicontinuous in the mean. If any of the sets γ_{cpie}, γ_{cpi}, γ_{cp}, γ_{c}, γ or γ_0 is nonempty, then $(P_t)_{t \in [0,+\infty)}$ is uniquely ergodic.*

Proof. We first note that $\gamma_0 \subseteq \gamma_{\mathrm{cpie}}$, and that $\gamma \subseteq \gamma_{\mathrm{c}} \subseteq \gamma_{\mathrm{cp}} \subseteq \gamma_{\mathrm{cpi}} \subseteq \gamma_{\mathrm{cpie}}$. Thus, each of the subsets $\gamma_0, \gamma, \gamma_{\mathrm{c}}, \gamma_{\mathrm{cp}}$ and γ_{cpi} is a subset of γ_{cpie}. Therefore, the theorem will be completely proved if we show that if $\gamma_{\mathrm{cpie}} \neq \emptyset$, then $(P_t)_{t \in [0,+\infty)}$ is uniquely ergodic (because if any of the sets $\gamma_0, \gamma, \gamma_{\mathrm{c}}, \gamma_{\mathrm{cp}}$ or γ_{cpi} is nonempty, then γ_{cpie} is nonempty, as well).

Thus, assume that $\gamma_{\mathrm{cpie}} \neq \emptyset$.

Also assume that $(P_t)_{t \in [0,+\infty)}$ is not uniquely ergodic.

Since we also assume that $(P_t)_{t \in [0,+\infty)}$ has invariant probability measures, using Proposition 7.2.10, we obtain that $(P_t)_{t \in [0,+\infty)}$ has at least two distinct invariant ergodic probability measures, say μ and ν.

Using Theorem 7.1.4 we obtain that γ_{cpie} is included in the support of any invariant ergodic probability measure of $(P_t)_{t \in [0,+\infty)}$. In particular, $\gamma_{\mathrm{cpie}} \subseteq \mathrm{supp}\,\mu$ and $\gamma_{\mathrm{cpie}} \subseteq \mathrm{supp}\,\nu$.

Since we assume that $\gamma_{\mathrm{cpie}} \neq \emptyset$, and since, by Corollary 7.2.9, $(P_t)_{t \in [0,+\infty)}$ has the e.m.d.s. property, we obtain a contradiction, which stems from our assumption that $(P_t)_{t \in [0,+\infty)}$ is not uniquely ergodic. $\qquad\square$

Corollary 7.2.12. *Assume that $(P_t)_{t \in [0,+\infty)}$ is equicontinuous in the mean, and that at least one of the sets γ_{cpie}, γ_{cpi}, γ_{cp}, γ_{c} or γ is nonempty. Then $(P_t)_{t \in [0,+\infty)}$ is uniquely ergodic, and, if μ is the unique invariant probability of $(P_t)_{t \in [0,+\infty)}$, then*

$$\gamma = \gamma_{\mathrm{c}} = \gamma_{\mathrm{cp}} = \gamma_{\mathrm{cpi}} = \gamma_{\mathrm{cpie}} = \mathrm{supp}\,\mu. \tag{7.2.1}$$

Proof. By Theorem 7.2.11, the transition function $(P_t)_{t \in [0,+\infty)}$ is uniquely ergodic.

Let μ be the unique invariant probability measure of $(P_t)_{t \in [0,+\infty)}$.

Using Theorem 7.2.1, we obtain that the equalities (7.2.1) hold true. $\qquad\square$

We will now discuss several examples to illustrate the results studied here.

Example 7.2.13. Let $\mathbf{w} = (w_t)_{t \in \mathbb{R}}$ be the flow of the rotations of the unit circle (for the definition of $\mathbf{w}$, see Example A.3.4), and let $(P_t^{(\mathbf{w})})_{t \in \mathbb{R}}$ be the transition function defined by $\mathbf{w}$ and discussed in Example 2.2.4.

Since the unit circle $\mathbb{R}/\mathbb{Z}$ is a compact metric space, it follows that $\Gamma = \mathbb{R}/\mathbb{Z}$. Thus, we obtain that $\gamma = \mathbb{R}/\mathbb{Z}$ because the orbit $\mathcal{O}(\hat{x})$ is equal to the entire space $\mathbb{R}/\mathbb{Z}$ for every $\hat{x} \in \mathbb{R}/\mathbb{Z}$.

Since $\mathbb{R}/\mathbb{Z}$ is compact, $(P_t^{(\mathbf{w})})_{t \in \mathbb{R}}$ has invariant probabilities.

Since (as pointed out in Example 2.2.4) $(P_t^{(\mathbf{w})})_{t \in \mathbb{R}}$ is equicontinuous and γ is nonempty, using Corollary 7.2.12 (applied to the restriction $(P_t^{(\mathbf{w})})_{t \in [0,+\infty)}$ of $(P_t^{(\mathbf{w})})_{t \in \mathbb{R}}$ to $[0, +\infty)$), we obtain that $(P_t^{(\mathbf{w})})_{t \in [0,+\infty)}$ (and, hence, $(P_t^{(\mathbf{w})})_{t \in \mathbb{R}}$) is uniquely ergodic.

Since (as pointed out in Example 2.3.1) the Haar-Lebesgue measure on $\mathbb{R}/\mathbb{Z}$ is an invariant probability measure for $(P_t^{(\mathbf{w})})_{t \in \mathbb{R}}$, it follows that the Haar-Lebesgue measure on $\mathbb{R}/\mathbb{Z}$ is the unique invariant probability of $(P_t^{(\mathbf{w})})_{t \in \mathbb{R}}$. $\qquad\blacksquare$

Example 7.2.14. Let $n \in \mathbb{N}$, $n \geq 2$, let $\mathbf{v} \in \mathbb{R}^n$, $\mathbf{v} = (v_1, v_2, \ldots, v_n)$, let $\mathbf{w} = (w_t)_{t \in \mathbb{R}}$ be the rectilinear flow with velocity $\mathbf{v}$ defined on the n-dimensional torus $\mathbb{R}^n/\mathbb{Z}^n$ and discussed in Example A.3.5, and let $(P_t^{(\mathbf{w})})_{t \in \mathbb{R}}$ be the transition function defined by $\mathbf{w}$ and studied in Example 2.2.5.

We will prove that the following assertions are equivalent:

(a) The numbers $v_1, v_2, \ldots, v_n$ are rationally independent.
(b) The transition function $(P_t^{\mathbf{w}})_{t \in \mathbb{R}}$ is minimal.
(c) The transition function $(P_t^{\mathbf{w}})_{t \in \mathbb{R}}$ is strictly ergodic.

Proof of (a) $\Rightarrow$ (b). It is pointed out on p. 69 of Cornfeld, Fomin and Sinai's monograph [22] that the following assertion is true:

Assertion A. The numbers $v_1, v_2, \ldots, v_n$ are rationally independent if and only if the Haar-Lebesgue measure $v_{\mathbb{R}^n/\mathbb{Z}^n}$ on the torus is an (invariant) ergodic probability measure.

Now, assume that the numbers $v_1, v_2, \ldots, v_n$ are rationally independent. Then, using Assertion A, we obtain that $v_{\mathbb{R}^n/\mathbb{Z}^m}$ is an ergodic probability measure of $(P_t^{(\mathbf{w})})_{t \in \mathbb{R}}$.

Since $v_{\mathbb{R}^n/\mathbb{Z}^n}$ is an ergodic invariant probability measure of $(P_t^{(\mathbf{w})})_{t \in \mathbb{R}}$, using (a) of Theorem 6.2.7, we obtain that there exists an $\hat{x} \in \Gamma_{\mathrm{cpie}}$ such that $v_{\mathbb{R}^n/\mathbb{Z}^n} = \varepsilon_{\hat{x}}$.

Since (as pointed out in Example 2.2.4) the transition function $(P_t^{(\mathbf{w})})_{t \in \mathbb{R}}$ is equicontinuous (so, obviously, $(P_t^{(\mathbf{w})})_{t \in \mathbb{R}}$ is also equicontinuous in the mean), using Theorem 7.2.8, we obtain that $[\hat{x}]$ is a closed set.

Since (by (a) of Theorem 6.2.7) $\varepsilon_{\hat{x}}([\hat{x}]) = 1$, it follows that $\mathbb{R}^n/\mathbb{Z}^n = \mathrm{supp}\, v_{\mathbb{R}^n/\mathbb{Z}^n} \subseteq \mathrm{supp}\, \varepsilon_{\hat{x}}$.

It is obvious now that the transition function $(P_t^{(\mathbf{w})})_{t \in \mathbb{R}}$ has only one invariant ergodic probability measure, namely $v_{\mathbb{R}^n/\mathbb{Z}^n}$. Using Proposition 7.2.10 (applied to the restriction $(P_t^{(\mathbf{w})})_{t \in [0,+\infty)}$ of $(P_t^{(\mathbf{w})})_{t \in \mathbb{R}}$ to $[0, +\infty)$), we obtain that $(P_t^{(\mathbf{w})})_{t \in \mathbb{R}}$ is uniquely ergodic.

Since $\mathbb{R}^n/\mathbb{Z}^n$ is a compact metric space, and since $(P_t^{(\mathbf{w})})_{t \in \mathbb{R}}$ is a $C(\mathbb{R}^n/\mathbb{Z}^n)$-continuous transition function (because $\mathbf{w}$ is a continuous flow), using Corollary 7.1.17, we obtain that $(P_t^{(\mathbf{w})})_{t \in \mathbb{R}}$ is minimal.

Proof of (b) $\Rightarrow$ (c). Since $(P_t^{(\mathbf{w})})_{t \in \mathbb{R}}$ has at least one invariant probability ($v_{\mathbb{R}^n/\mathbb{Z}^n}$ is such an invariant probability), it follows that $\Gamma_{\mathrm{cpie}} \neq \emptyset$. Since we assume that $(P_t^{(\mathbf{w})})_{t \in \mathbb{R}}$ is minimal it follows that $\gamma_{\mathrm{cpie}} = \mathbb{R}^n/\mathbb{Z}^n$, so γ_{cpie} is nonempty.

Since $(P_t^{(\mathbf{w})})_{t \in \mathbb{R}}$ is equicontinuous, using Theorem 7.2.11 (or Corollary 7.2.12) applied to the restriction $(P_t^{(\mathbf{w})})_{t \in [0,+\infty)}$ of $(P_t^{(\mathbf{w})})_{t \in \mathbb{R}}$ to $[0, +\infty)$, we obtain that $(P_t^{(\mathbf{w})})_{t \in \mathbb{R}}$ is uniquely ergodic. Since the unique invariant probability measure of $(P_t^{(\mathbf{w})})_{t \in \mathbb{R}}$ is $v_{\mathbb{R}^n/\mathbb{Z}^n}$ and since $\mathrm{supp}\,(v_{\mathbb{R}^n/\mathbb{Z}^n}) = \mathbb{R}^n/\mathbb{Z}^n$, it follows that $(P_t^{(\mathbf{w})})_{t \in \mathbb{R}}$ is strictly ergodic.

Proof of (c) $\Rightarrow$ (a). Since we assume that $(P_t^{(\mathbf{w})})_{t \in \mathbb{R}}$ is strictly ergodic, and since the Haar-Lebesgue probability measure is invariant for $(P_t^{(\mathbf{w})})_{t \in \mathbb{R}}$, using Proposition 7.2.10, we obtain that the Haar-Lebesgue measure on $\mathbb{R}^n/\mathbb{Z}^n$ is actually an invariant ergodic probability. Thus, using Assertion A, we obtain that the numbers $v_1, v_2, \ldots, v_n$ are rationally independent.

Note that when proving the equivalence of the assertions (a), (b) and (c), we also showed that if any of the assertions is true, then the Haar-Lebesgue measure $\nu_{\mathbb{R}^n/\mathbb{Z}^n}$ is the only invariant probability of the transition function $(P_t^{(\mathbf{w})})_{t\in\mathbb{R}}$ and $\Gamma_{\mathrm{cpie}} = \mathbb{R}^n/\mathbb{Z}^n$.

Example 7.2.15. Let $(H,\cdot,d)$ be a locally compact separable metric semigroup with unit, and let u be the unit of H.

Let $\mu \in \mathcal{M}(H)$ be an equicontinuous probability measure on $(H,\mathcal{B}(H))$ such that $\bigcup_{n=0}^\infty(\mathrm{supp}\,(\mu^n)) = H$.

Now let $(\mu_t)_{t\in[0,+\infty)}$ be the exponential one-parameter convolution semigroup of probability measures defined by μ. Thus, $\mu_t = e^{-t}\sum_{k=0}^\infty \frac{t^k}{k!}\mu^k$ for every $t \in [0,+\infty)$. Note that $\mathrm{supp}\,\mu_t = H$ for every $t > 0$. (For additional details about these one-parameter convolution semigroups, see Sect. B.3.)

Let $(P_t)_{t\in[0,+\infty)}$ be the transition function defined by $(\mu_t)_{t\in[0,+\infty)}$ (see the beginning of Sect. 2.2.3).

Using Proposition 2.2.13, we obtain that $(P_t)_{t\in[0,+\infty)}$ satisfies the s.m.a. and is pointwise continuous. As pointed out in Sect. 2.2.3, $(P_t)_{t\in[0,+\infty)}$ is a Feller transition probability.

Since μ is an equicontinuous probability measure, using Proposition 2.2.15, we obtain that $(P_t)_{t\in[0,+\infty)}$ is $C_0(H)$-equicontinuous.

Now assume that $(P_t)_{t\in[0,+\infty)}$ has at least one invariant probability measure and that the semigroup H has left zeroids (recall, see Sect. A.1, that H has left zeroids if $\bigcap_{a\in H} Ha \neq \emptyset$; any element $z \in \bigcap_{a\in H} Ha$ is called a left zeroid). Then $(P_t)_{t\in[0,+\infty)}$ is uniquely ergodic.

Indeed, for every $x \in H$, we have

$$\mathcal{O}(x) = \bigcup_{n=0}^\infty (\mathrm{supp}\,(\mu^n * \delta_x)) = \bigcup_{n=0}^\infty \overline{(\mathrm{supp}(\mu^n))}x.$$

Therefore,

$$\bigcap_{a\in H} Ha \subseteq \overline{Hx} = \overline{(\bigcup_{n=0}^\infty (\mathrm{supp}\,(\mu^n)))}x = \overline{(\bigcup_{n=1}^\infty (\mathrm{supp}\,\mu)^n) \cup \{u\}}x$$

$= \overline{\mathcal{O}(x)}$ for every $x \in H$.

Since we assume that $(P_t)_{t\in[0,+\infty)}$ has invariant probabilities, it follows that $\Gamma_{\mathrm{cpie}} \neq \emptyset$.

Since, from the above inclusions, the left zeroids belong to γ_{cpie}, it follows that $\gamma_{\mathrm{cpie}} \neq \emptyset$. Using Corollary 7.2.12, we obtain that $(P_t)_{t\in[0,+\infty)}$ is uniquely ergodic. ∎

We conclude the subsection with a criterion for unique ergodicity that is sometimes useful.

Theorem 7.2.16. *Assume that* $(P_t)_{t\in[0,+\infty)}$ *is equicontinuous in the mean and has at least one invariant probability measure. If* $\overline{\mathcal{O}(x)} \cap \overline{\mathcal{O}(y)} \neq \emptyset$ *for every* $x \in X$ *and* $y \in X$, *then* $(P_t)_{t\in[0,+\infty)}$ *is uniquely ergodic.*

Proof. Assume that $(P_t)_{t\in[0,+\infty)}$ is not uniquely ergodic and has invariant probabilities. Then, using the implication (b) $\Rightarrow$ (a) of Proposition 7.2.10, we obtain that $(P_t)_{t\in[0,+\infty)}$ has at least two distinct invariant ergodic probability measures. Thus, there exist $x_1 \in \Gamma_{\mathrm{cpie}}$ and $y_1 \in \Gamma_{\mathrm{cpie}}$ such that $\varepsilon_{x_1} \neq \varepsilon_{y_1}$.

Using Theorem 7.2.8, we obtain that both $[x_1]$ and $[y_1]$ are closed sets.

In view of (a) of Theorem 6.2.7, we obtain that $\varepsilon_{x_1}([x_1]) = 1$ and $\varepsilon_{y_1}([y_1]) = 1$; therefore, $\operatorname{supp} \varepsilon_{x_1} \subseteq [x_1]$ and $\operatorname{supp} \varepsilon_{y_1} \subseteq [y_1]$.

Now let $x \in \operatorname{supp} \varepsilon_{x_1}$ and $y \in \operatorname{supp} \varepsilon_{y_1}$.

Since $\varepsilon_x = \varepsilon_{x_1}$ and $\varepsilon_y = \varepsilon_{y_1}$, using Theorem 7.1.3, we obtain that

$$\overline{\mathcal{O}(x)} = \operatorname{supp} \varepsilon_x = \operatorname{supp} \varepsilon_{x_1} \subseteq [x_1]$$

and

$$\overline{\mathcal{O}(y)} = \operatorname{supp} \varepsilon_y = \operatorname{supp} \varepsilon_{y_1} \subseteq [y_1].$$

Since $[x_1] \cap [y_1] = \emptyset$ (because $[x_1]$ and $[y_1]$ are distinct equivalence classes), it follows that $\overline{\mathcal{O}(x)} \cap \overline{\mathcal{O}(y)} = \emptyset$.

Thus, we have obtained a contradiction, which stems from our assumption that $(P_t)_{t\in[0,+\infty)}$ is not uniquely ergodic. $\square$

7.2.3 Generic Points

Our goal in this subsection is to extend the results of Sections 3.2 and 3.3 of [143] to transition functions.

Since in this section we deal only with invariant probability measures and invariant ergodic probability measures, using Propositions 2.3.7 and 2.3.13 we may assume that all the transition functions $(P_t)_{t\in\mathbb{T}}$ under consideration have the property that $\mathbb{T} = [0, +\infty)$.

Thus, let $(P_t)_{t\in[0,+\infty)}$ be a transition function defined on a locally compact separable metric space (X, d), and let $((S_t, T_t))_{t\in[0,+\infty)}$ be the associated family of Markov pairs.

We also assume that $(P_t)_{t\in[0,+\infty)}$ satisfies the s.m.a. and is pointwise continuous.

In a similar manner as in Section 3.2 of [143], we call $x \in X$ a *generic point* if $\lim_{u\to+\infty} \frac{1}{u} \int_0^u S_t f(x)\,dt$ exists whenever $f \in C_0(X)$. The term "generic point" is borrowed from ergodic theory (see Furstenberg's monograph [38]), where it is used for such points in the "discrete-time" case when these points are defined by a transition probability generated by a measurable transformation. Generic points are also called *quasi-regular* (see Krengel's book [53] and Oxtoby [85]).

A generic point $x \in X$ is called *nonsingular* if $\lim_{u \to +\infty} \frac{1}{u} \int_0^u S_t f(x) dt \neq 0$ for some $f \in C_0(X)$; that is, the generic point $x \in X$ is said to be nonsingular if and only if $x \in \Gamma_c$.

A point $x \in X$ is called a *dominant generic point* if x is a nonsingular generic point and if the following condition (that we call the *DGPTF condition*) is satisfied:

- if $y \in X$ and $f \in C_0(X)$, $f \geq 0$, are such that $\lim_{u \to +\infty} \frac{1}{u} \int_0^u S_t f(y) dt$ exists, then $\lim_{u \to +\infty} \frac{1}{u} \int_0^u S_t f(y) dt \leq \lim_{s \to +\infty} \frac{1}{s} \int_0^s S_t f(x) dt$.

Our goal now is to study the connection between the existence of dominant generic points on the one hand, and the existence and unicity of a standard elementary probability measure on the other hand. This study, which is of intrinsic interest in its own right, can be used to decide the unique ergodicity of a transition function. Indeed, if we know that the transition function $(P_t)_{t \in [0,+\infty)}$ has a unique standard elementary probability measure, say μ, then if we know that μ is an invariant measure for $(P_t)_{t \in [0,+\infty)}$, we obtain that $(P_t)_{t \in [0,+\infty)}$ is uniquely ergodic; for instance if $(P_t)_{t \in [0,+\infty)}$ is a Feller transition function and has a unique standard elementary probability measure, then $(P_t)_{t \in [0,+\infty)}$ is uniquely ergodic.

Our next proposition is an extension to our setting of both Propositions 3.2.1 and 3.2.2 of [143].

Proposition 7.2.17. *If $x \in X$ is a dominant generic point of $(P_t)_{t \in [0,+\infty)}$, and if $(P_t)_{t \in [0,+\infty)}$ has nonzero finite invariant measures, then $(P_t)_{t \in [0,+\infty)}$ is uniquely ergodic and the standard elementary measure ε_x is the invariant probability measure of $(P_t)_{t \in [0,+\infty)}$.*

Proof. Assume that $(P_t)_{t \in [0,+\infty)}$ has at least one dominant generic point and let $x \in X$ be such a dominant generic point.

Since we assume that $(P_t)_{t \in [0,+\infty)}$ has nonzero invariant finite measures, we obtain that $(P_t)_{t \in [0,+\infty)}$ has at least one invariant probability measure.

Since x is a dominant generic point, it is also a nonsingular generic point, so, as mentioned before the proposition, it follows that $x \in \Gamma_c$. Thus, we may and do consider the standard elementary measure ε_x defined by x.

Since (as we pointed out above) $(P_t)_{t \in [0,+\infty)}$ has at least one invariant probability, in view of our discussion so far, we obtain that in order to prove the proposition, it is enough to prove that for every invariant probability μ of $(P_t)_{t \in [0,+\infty)}$, we have that $\mu = \varepsilon_x$.

To this end, let μ be an invariant probability measure for $(P_t)_{t \in [0,+\infty)}$.

Since ε_x is a standard elementary measure, it follows that $0 < \|\varepsilon_x\| \leq 1$. Thus, in order to prove that $\mu = \varepsilon_x$, it is enough to prove that $\mu \leq \varepsilon_x$ (since, in this case, taking into consideration that μ is a probability measure, we obviously obtain that $\mu = \varepsilon_x$). Therefore, it is enough to prove that $\langle f, \mu \rangle \leq \langle f, \varepsilon_x \rangle$ for every $f \in C_0(X)$, $f \geq 0$.

Thus, let $f \in C_0(X)$, $f \geq 0$.

Also, let $f^* : X \to \mathbb{R}$ be the function generated by f and defined after Corollary 5.3.2. As pointed out after the definition of f^*, the function f^* belongs to $B_b(X)$.

Using (a) of Corollary 3.2.12, using the fact that Γ_c is a set of maximal probability (by Corollary 5.3.2), the fact that x is a dominant generic point (so $f^*(y) \le f^*(x)$ for every $y \in \Gamma_c$), and the fact that $x \in \Gamma_c$ (so $f^*(x) = \langle f, \varepsilon_x \rangle$), we obtain that

$$\langle f, \mu \rangle = \langle f^*, \mu \rangle = \int_{\Gamma_c} f^*(y) d\mu(y) \le \int_{\Gamma_c} f^*(x) d\mu(y) = f^*(x) = \langle f, \varepsilon_x \rangle.$$

Since $\langle f, \mu \rangle \le \langle f, \varepsilon_x \rangle$ for every $f \in C_0(X)$, $f \ge 0$, it follows that $\mu \le \varepsilon_x$, so $\mu = \varepsilon_x$. Thus, $(P_t)_{t \in [0,+\infty)}$ is uniquely ergodic and ε_x is the unique invariant probability of $(P_t)_{t \in [0,+\infty)}$. $\qquad\square$

Note that in the above proposition we do not have to assume that $(P_t)_{t \in [0,+\infty)}$ is a Feller transition function. However, if $(P_t)_{t \in [0,+\infty)}$ is a Feller transition function, we can also prove a converse of Proposition 7.2.17. The details are included in the next proposition.

Proposition 7.2.18. *Assume that $(P_t)_{t \in [0,+\infty)}$ is a Feller transition function. If $(P_t)_{t \in [0,+\infty)}$ is uniquely ergodic, then $(P_t)_{t \in [0,+\infty)}$ has dominant generic points.*

Proof. Since we assume that $(P_t)_{t \in [0,+\infty)}$ has an invariant probability, using Corollary 6.2.8, we obtain that $\Gamma_{\text{cpie}} \ne \emptyset$.

In order to prove the proposition we will show that every $x \in \Gamma_{\text{cpie}}$ is a dominant generic point.

To this end, let $x \in \Gamma_{\text{cpie}}$.

Clearly, x is a nonsingular generic point, so we have to prove only that x satisfies the DGPTF condition.

Thus, let $y \in X$ and $f \in C_0(X)$ be such that $\lim_{u \to +\infty} \frac{1}{u} \int_0^u S_t f(y) \, dt$ exists.

Our goal is to prove that $\lim_{u \to +\infty} \frac{1}{u} \int_0^u S_t f(y) \, dt \le \lim_{u \to +\infty} \frac{1}{u} \int_0^u S_t f(x) \, dt$ (clearly, $\lim_{u \to +\infty} \frac{1}{u} \int_0^u S_t f(x) \, dt$ exists because $x \in \Gamma_{\text{cpie}}$).

Now let $L : B_b^{(L)}([0, +\infty)) \to \mathbb{R}$ be a continuous-time Banach limit (such continuous-time Banach limits do exist – see Theorem 4.4.5).

Let $L_0 : B_b^{(L)}([0, +\infty)) \to \mathbb{R}$ be defined by $L_0(h) = L\left(\left(\frac{1}{u} \int_0^u h(t) \, dt \right)_{\overline{u \in (0,+\infty)}} \right)$ for every $h \in B_b^{(L)}([0, +\infty))$, where $\left(\frac{1}{u} \int_0^u h(t) \, dt \right)_{\overline{u \in (0,+\infty)}}$ is the real-valued function defined on $[0, +\infty)$ by

$$\left(\frac{1}{u} \int_0^u h(t) \, dt \right)_{\overline{u \in (0,+\infty)}} = \begin{cases} \frac{1}{u} \int_0^u h(t) \, dt & \text{if } u \in (0, +\infty) \\ 0 & \text{if } u = 0. \end{cases}$$

Note that $\left(\frac{1}{u} \int_0^u h(t) \, dt \right)_{\overline{u \in (0,+\infty)}}$ is well-defined for every $h \in B_b^{(L)}([0, +\infty))$ (because h is integrable on the interval $[0, v]$ for every $v \in (0, +\infty)$), and belongs to $B_b^{(L)}([0, +\infty))$ (because the map $u \mapsto \frac{1}{u} \int_0^u h(t) \, dt$, $u \in (0, +\infty)$, is continuous, so $\left(\frac{1}{u}\int_0^u h(t)\,dt\right)_{\overline{u\in(0,+\infty)}}$ is measurable, and because $\left|\frac{1}{u}\int_0^u h(t)\,dt\right| \le \|h\|$ for every $u \in (0,+\infty)$, it follows that the function $\left(\frac{1}{u}\int_0^u h(t)\,dt\right)_{\overline{u\in(0,+\infty)}}$ is also bounded).

Since $\left(\frac{1}{u}\int_0^u h(t)\,dt\right)_{\overline{u\in(0,+\infty)}}$ is well-defined and belongs to $B_b^{(L)}([0,+\infty))$, it follows that it makes sense to consider $L_0(h)$ for every $h \in B_b^{(L)}([0,+\infty))$, so the real-valued map L_0 is also well-defined. By Proposition 4.4.6, we obtain that, like L, L_0 is a continuous-time Banach limit, as well.

Now, let $\phi : C_b(X) \to \mathbb{R}$ be defined by $\phi(g) = L_0((S_t g(y))_{t\in[0,+\infty)})$ for every $g \in C_b(X)$. Note that since $(P_t)_{t\in[0,+\infty)}$ satisfies the s.m.a., using Corollary 2.1.6, we obtain that the map $t \mapsto S_t g(y)$, $t \in [0,+\infty)$ is measurable for every $g \in C_b(X)$; using the fact that S_t is a (positive) contraction of $B_b(X)$, we obtain that $|S_t g(y)| \le \|g\|$ for every $t \in [0,+\infty)$ and for every $g \in C_b(X)$, so the function $t \mapsto S_t g(y)$, $t \in [0,+\infty)$, is bounded whenever $g \in C_b(X)$. Thus, the definition of ϕ is correct.

Clearly, ϕ is a positive linear functional (ϕ is positive because S_t is a positive operator for every $t \in [0,+\infty)$).

Our goal is to apply the Lasota-Yorke lemma (Theorem 1.2.3) to ϕ and the restriction ν of ϕ to $C_0(X)$. To this end, we first have to prove that $\phi(S_r g) = \phi(g)$ for every $r \in [0,+\infty)$ and $g \in C_0(X)$.

To this end, let $r \in [0,+\infty)$. Using the fact that L_0 is a continuous-time Banach limit, so L_0 satisfies the condition (b) of the definition of a continuous-time Banach limit stated at the beginning of Sect. 4.4, we obtain that

$$\phi(S_r g) = L_0((S_t S_r g(y))_{t\in[0,+\infty)})$$
$$= L_0((S_{t+r} g(y))_{t\in[0,+\infty)}) = L_0((S_t g(y))_{t\in[0,+\infty)}) = \phi(g)$$

for every $g \in C_0(X)$. Thus, ϕ and S_r satisfy the conditions of the Lasota-Yorke lemma (Theorem 1.2.3). According to the lemma, we obtain that the restriction ν of ϕ to $C_0(X)$ satisfies the equality $T_t \nu = \nu$.

We have therefore proved that $T_t \nu = \nu$ for every $t \in [0,+\infty)$. Hence, ν is an invariant element of $(P_t)_{t\in[0,+\infty)}$. Since $\nu \ge 0$ and since $(P_t)_{t\in[0,+\infty)}$ is uniquely ergodic, we obtain that $\nu = a\mu$, where $a \in \mathbb{R}$, $a \ge 0$ and μ is the unique invariant probability of $(P_t)_{t\in[0,+\infty)}$.

Since $\phi(1_X) = L_0((S_t 1_X(y))_{t\in[0,+\infty)}) = 1$ and since ν is the restriction of ϕ to $C_0(X)$, it follows that $\nu(X) \le 1$, so $a \le 1$.

Using condition (c) in the definition of a continuous-time Banach limit, the manner in which L_0 was defined, the definition of ϕ and of ν, the fact that $x \in \Gamma_{\text{cpie}}$ (so $\varepsilon_x = \mu$), and the fact that $0 \le a \le 1$, we obtain that

$$\lim_{u\to+\infty} \frac{1}{u}\int_0^u S_t f(y)\,dt = L\left(\left(\frac{1}{u}\int_0^u S_t f(y)\,dt\right)_{\overline{u\in(0,+\infty)}}\right)$$
$$= L_0\left((S_t f(y))_{t\in[0,+\infty)}\right) = \phi(f) = \langle f, \nu\rangle = \langle f, a\mu\rangle$$

$$= a\langle f, \mu \rangle = a \lim_{u \to +\infty} \frac{1}{u} \int_0^u S_t f(x) \, dt$$

$$\leq \lim_{u \to +\infty} \frac{1}{u} \int_0^u S_t f(x) \, dt.$$

We have therefore proved that x satisfies the DGPTF condition, so x is a dominant generic point. $\qquad\qquad\square$

Since most transition functions that we use are Feller, it is a good idea to combine Propositions 7.2.17 and 7.2.18 as follows:

Theorem 7.2.19. *Let $(P_t)_{t \in [0,+\infty)}$ be a Feller transition function. The transition function is uniquely ergodic if and only if $(P_t)_{t \in [0,+\infty)}$ has at least one dominant generic point.*

The proof of the theorem is obvious in view of Propositions 7.2.17 and 7.2.18.

A transition function $(P_t)_{t \in [0,+\infty)}$ (that satisfies the s.m.a. and is pointwise continuous) is said to be *weak* mean ergodic* if every $x \in X$ is a generic point. The reason for the terminology lies in the fact that a transition function $(P_t)_{t \in [0,+\infty)}$ is weak* mean ergodic if and only if the limit in the weak* topology of $\mathcal{M}(X)$ of the family $\left(\frac{1}{u} \int_0^u T_s \mu \, ds \right)_{u \in (0,+\infty)}$ exists as u tends to $+\infty$ for every $\mu \in \mathcal{M}(X)$, where $(T_s)_{s \in [0,+\infty)}$ is the semigroup of Markov operators on $\mathcal{M}(X)$ defined by $(P_t)_{t \in [0,+\infty)}$. Clearly, $(P_t)_{t \in [0,+\infty)}$ is weak* mean ergodic if and only if $X = D \cup \Gamma_c$.

An important source of examples of weak* mean ergodic transition functions is provided by the pointwise mean ergodic theorem for equicontinuous transition functions that will be discussed in the next section. Weak* mean ergodic transition functions that are not equicontinuous are very hard to obtain; however, in spite of the tremendous difficulties encountered, Ratner was able to prove a theorem (Theorem B.4.9) from which we obtain that if $n \in \mathbb{N}$, $n \geq 2$, if Γ is a lattice in $\mathrm{SL}(n, \mathbb{R})$, if $\mathbf{v}$ is a unipotent flow on $(\mathrm{SL}(n, \mathbb{R})/\Gamma)_R$, and if $(P_t^{(\mathbf{v})})_{t \in \mathbb{R}}$ is the transition function defined by $\mathbf{v}$ (see Example 2.2.8), then the restriction $(P_t^{(\mathbf{v})})_{t \in [0,+\infty)}$ of $(P_t^{(\mathbf{v})})_{t \in \mathbb{R}}$ to $[0, +\infty)$ is weak* mean ergodic.

Clearly, if $(P_t)_{t \in [0,+\infty)}$ is weak* mean ergodic, then for every $f \in C_0(X)$, the function $f^* : X \to \mathbb{R}$ that was introduced after Corollary 5.3.2 can be defined in a compact form by $f^*(x) = \lim_{u \to +\infty} \frac{1}{u} \int_0^u S_t f(x) \, dt$ for every $x \in X$ because $X = \mathcal{D} \cup \Gamma_c$, so the limit on the right-hand side of the equality defining f^* exists for every $x \in X$ in this case.

Recall (see Sect. 1.4) that given a set $\mathcal{A}$ of real-valued functions defined on X, we say that $\mathcal{A}$ has a *common (absolute) maximum at $x_0 \in X$* if $g(x) \leq g(x_0)$ for every $x \in X$ and $g \in \mathcal{A}$ (that is, $\mathcal{A}$ has a common maximum at x_0 if every $g \in \mathcal{A}$ attains an absolute maximum value at x_0).

The notion of a common maximum of a set of functions can be used to restate Theorem 7.2.19 in the case of weak* mean ergodic transition functions as follows:

Corollary 7.2.20. *Let* $(P_t)_{t\in[0,+\infty)}$ *be a weak* mean ergodic Feller transition function. Then* $(P_t)_{t\in[0,+\infty)}$ *is uniquely ergodic if and only if there exists an* $x_0 \in X$ *such that the set* $\mathcal{A} = \{f^* \mid f \in C_0(X)\}$ *has a common maximum at* x_0.

Let $(P_t)_{t\in[0,+\infty)}$ be a transition function (that satisfies the s.m.a. and is pointwise continuous).

Now, let $\mu \in \mathcal{M}(X)$ be an invariant probability measure of $(P_t)_{t\in[0,+\infty)}$ (of course, we assume that $(P_t)_{t\in[0,+\infty)}$ has invariant probabilities). A problem that is usually highly nontrivial is to decide if μ is ergodic. That is why, in this book, we discuss various methods that might be useful to establish the ergodicity of μ (for instance, Theorem 6.2.9). However, as we alluded to earlier, no method is very general, so it is good to have as many methods as possible in our toolbox. That is why we will now discuss a few such methods based on a generalization of the notion of dominant generic points.

Let A be a nonempty subset of X, $A \in B(X)$.

A point $x_0 \in A$ is called a *dominant generic point of A defined by* $(P_t)_{t\in[0,+\infty)}$ if x_0 is a nonsingular generic point and if the following condition, called the *DGPTF-A condition*, is satisfied:

- if $y \in A$ and $f \in C_0(X)$ are such that $\lim_{u\to+\infty} \frac{1}{u} \int_0^u S_t f(y)\, dt$ exists, then

$$\lim_{u\to+\infty} \int_0^u S_t f(y)\, dt \le \lim_{v\to+\infty} \frac{1}{v} \int_0^v S_t f(x_0)\, dt.$$

Using the notion of dominant generic point of a nonempty measurable subset of X, we obtain the following strengthening of Theorem 6.2.9:

Theorem 7.2.21. *Assume that* $(P_t)_{t\in[0,+\infty)}$ *has invariant probabilities, and let* μ *be such an invariant probability. Then the following two assertions are equivalent:*

(a) The measure μ is ergodic.

(b) There exist a nonempty measurable subset A of X, and $x_0 \in A$ such that $\mu(A) > 0$, μ is a standard elementary measure with respect to x_0 (that is, $\mu = \varepsilon_{x_0}$), and x_0 is a dominant generic point of A.

Proof. (a) $\Rightarrow$ (b). Assume that μ is an invariant ergodic probability measure of $(P_t)_{t\in[0,+\infty)}$. Then, by (b) of Theorem 6.2.7, there exists an $x_0 \in \Gamma_{\mathrm{cpie}}$ such that $\mu = \varepsilon_{x_0}$ (that is, such that μ is a standard elementary probability measure with respect to x_0).

Set $A = [x_0]$. Then A is a nonempty measurable subset of X ($A \ne \emptyset$ because $x_0 \in A$). By (a) of Theorem 6.2.7, $\mu(A) = 1 > 0$. Clearly, x_0 is a dominant generic point of A (actually, any point in A is a dominant generic point of A).

(b) $\Rightarrow$ (a) Assume that the conditions at (b) are satisfied relative to μ. Since Γ_{cpie} is a set of maximal probability (by Theorem 6.2.4), it follows that $\mu(\Gamma_{\mathrm{cpie}}) = 1$; therefore, $A \cap \Gamma_{\mathrm{cpie}} \ne \emptyset$ (because if we assume that A and Γ_{cpie} are disjoint, we obtain a contradiction with the fact that μ is a probability measure, $\mu(\Gamma_{\mathrm{cpie}}) = 1$, and $\mu(A) > 0$).

Thus, there exists a $y \in A \cap \Gamma_{\text{cpie}}$. Since y is an element of Γ_{cpie}, it makes sense to consider the standard elementary measure ε_y, which is actually an invariant ergodic probability measure.

Since we assume that x_0 is a dominant generic point of A and $y \in A$, we obtain that

$$\langle f, \varepsilon_y \rangle = \lim_{u \to +\infty} \frac{1}{u} \int_0^u S_t f(y) \, dt \leq \lim_{u \to +\infty} \frac{1}{u} \int_0^u S_t f(x_0) \, dt$$
$$= \langle f, \varepsilon_{x_0} \rangle = \langle f, \mu \rangle$$

for every $f \in C_0(X)$. Thus, $\varepsilon_y \leq \mu$. Since both ε_y and μ are probability measures, it follows that $\varepsilon_y = \mu$.

Thus, μ is an invariant ergodic probability measure. $\square$

We will now conclude the subsection (and the section) with two consequences of Theorem 7.2.21. To this end, we need some preparation.

Assume that the transition function $(P_t)_{t \in [0, +\infty)}$ has invariant probabilities, and let μ be such an invariant probability.

Set

$$A_\mu = \left\{ x \in X \;\middle|\; \begin{array}{l} \text{for every } f \in C_0(X), \text{ the limit} \\ \lim_{u \to +\infty} \frac{1}{u} \int_0^u S_t f(x) \, dt \text{ exists and} \\ \text{is equal to } \langle f, \mu \rangle \end{array} \right\}.$$

Lemma 7.2.22. *Assume that $(P_t)_{t \in [0, +\infty)}$ has invariant probability measures. If μ is an invariant probability of $(P_t)_{t \in [0, +\infty)}$, then the set A_μ is Borel measurable (that is, $A_\mu \in B(X)$).*

Proof. We will use the separability of the Banach lattice $C_0(X)$ in order to prove the lemma.

To this end, we first note that, by Proposition 3.2.5, the set

$$A_{\vec{f}} = \left\{ x \in X \;\middle|\; \begin{array}{c} \lim_{\substack{u \to +\infty \\ u > 0}} \frac{1}{u} \int_0^u S_t f(x) \, dt \text{ exists and is} \\ \text{a real number} \end{array} \right\}$$

belongs to $\mathcal{B}(X)$ and the function $f^{\to} : X \to \mathbb{R}$ defined by

$$f^{\to}(x) = \begin{cases} \lim_{\substack{u \to +\infty \\ u > 0}} \frac{1}{u} \int_0^u S_t f(x) \, dt & \text{if } x \in A_{\vec{f}} \\ 0 & \text{if } x \in X \setminus A_{\vec{f}} \end{cases}$$

belongs to $B_b(X)$ for every $f \in C_0(X)$.

Accordingly, the set

$$A_\mu^{(f)} = \left\{ x \in A_{\vec{f}} \;\middle|\; \lim_{\substack{u \to +\infty \\ u > 0}} \frac{1}{u} \int_0^u S_t f(x) \, dt = \langle f, \mu \rangle \right\}$$

belongs to $\mathcal{B}(X)$.

Since $C_0(X)$ is a separable Banach lattice, we may and do pick a sequence $(g_n)_{n \in \mathbb{N}}$ of elements of $C_0(X)$ such that the range $\{g_n \mid n \in \mathbb{N}\}$ of $(g_n)_{n \in \mathbb{N}}$ is dense in $C_0(X)$.

Since $A_\mu^{(g_n)} \in \mathcal{B}(X)$ for every $n \in \mathbb{N}$, it follows that the lemma will be completely proved if we show that $A_\mu = \cap_{n \in \mathbb{N}} A_\mu^{(g_n)}$.

Clearly, $A_\mu = \cap_{f \in C_0(X)} A_\mu^{(f)} \subseteq \cap_{n \in \mathbb{N}} A_\mu^{(g_n)}$. Thus, it is enough to prove that $\cap_{n \in \mathbb{N}} A_\mu^{(g_n)} \subseteq A_\mu$.

To this end, let $x \in \cap_{n \in \mathbb{N}} A_\mu^{(g_n)}$.

We have to prove that $x \in A_\mu$; that is, we have to prove that, for every $f \in C_0(X)$, the limit $\lim_{u \to +\infty} \frac{1}{u} \int_0^u S_t f(x)\, dt$ exists and is equal to $\langle f, \mu \rangle$.

Thus, let $f \in C_0(X)$. We have to prove that for every $\varepsilon \in \mathbb{R}$, $\varepsilon > 0$, there exists a $u_\varepsilon \in \mathbb{R}$, $u_\varepsilon > 0$, such that $\left| \frac{1}{u} \int_0^u S_t f(x) dt - \langle f, \mu \rangle \right| < \varepsilon$ for every $u \in \mathbb{R}$, $u \geq u_\varepsilon$.

So, let $\varepsilon \in \mathbb{R}$, $\varepsilon > 0$. Since $\{g_n \mid n \in \mathbb{N}\}$ is dense in $C_0(X)$, it follows that there exists an $n_0 \in \mathbb{N}$ such that $\|f - g_{n_0}\| < \frac{\varepsilon}{3}$.

Since $x \in \cap_{n \in \mathbb{N}} A_\mu^{(g_n)} \subseteq A_\mu^{(g_{n_0})}$, it follows that $\lim_{u \to +\infty} \frac{1}{u} \int_0^u S_t g_{n_0}(x)\, dt$ exists and is equal to $\langle g_{n_0}, \mu \rangle$. Thus, there exists $u_\varepsilon \in \mathbb{R}$, $u_\varepsilon > 0$, such that $\left| \frac{1}{u} \int_0^u S_t g_{n_0}(x)\, dt - \langle g_{n_0}, \mu \rangle \right| < \frac{\varepsilon}{3}$ for every $u \in \mathbb{R}$, $u \geq u_\varepsilon$.

Taking into consideration the manner in which u_ε is defined, the fact that S_t, $t \in [0, +\infty)$, are positive contractions of $B_b(X)$, and the fact that μ is a probability measure, we obtain that

$$\left| \frac{1}{u} \int_0^u S_t f(x)\, dt - \langle f, \mu \rangle \right| \leq \left| \frac{1}{u} \int_0^u S_t f(x)\, dt - \frac{1}{u} \int_0^u S_t g_{n_0}(x)\, dt \right|$$

$$+ \left| \frac{1}{u} \int_0^u S_t g_{n_0}(x)\, dt - \langle g_{n_0}, \mu \rangle \right| + |\langle g_{n_0}, \mu \rangle - \langle f, \mu \rangle|$$

$$< \frac{1}{u} \left| \int_0^u (S_t f(x) - S_t g_{n_0}(x))\, dt \right| + \frac{\varepsilon}{3} + |\langle g_{n_0} - f, \mu \rangle|$$

$$\leq \frac{1}{u} \int_0^u |S_t f(x) - S_t g_{n_0}(x)|\, dt + \frac{\varepsilon}{3} + \|g_{n_0} - f\|$$

$$< \frac{1}{u} \int_0^u \|f - g_{n_0}\|\, dt + \frac{\varepsilon}{3} + \frac{\varepsilon}{3} = \|f - g_{n_0}\| + \frac{\varepsilon}{3} + \frac{\varepsilon}{3} < \frac{\varepsilon}{3} + \frac{\varepsilon}{3} + \frac{\varepsilon}{3} = \varepsilon.$$

We have therefore proved that the limit $\lim_{u \to +\infty} \frac{1}{u} \int_0^u S_t f(x) dt$ exists and is equal to $\langle f, \mu \rangle$ for every $f \in C_0(X)$.

Accordingly, $A_\mu = \cap_{n \in \mathbb{N}} A_\mu^{(g_n)}$, so A_μ is measurable because $A_\mu^{(g_n)}$ is measurable for every $n \in \mathbb{N}$. $\qquad\square$

Using the above lemma we obtain the following consequence of Theorem 7.2.21:

Corollary 7.2.23 (A "Zero-One" Law for the Invariant Probability Measures of Transition Functions). *Assume that the transition function $(P_t)_{t \in [0,+\infty)}$ has invariant probabilities, let μ be an invariant probability of $(P_t)_{t \in [0,+\infty)}$, and let*

A_μ *be the set defined before Lemma 7.2.22. Then* $\mu(A_\mu)$ *is equal to either 0, or else it is equal to 1. Moreover,* $\mu(A_\mu) = 1$ *if and only if* μ *is ergodic.*

Observation. Note that it makes sense to consider $\mu(A_\mu)$ because $A_\mu \in \mathcal{B}(X)$ by Lemma 7.2.22. ▲

Proof. Let μ be an invariant probability measure of $(P_t)_{t \in [0,+\infty)}$, and assume that $\mu(A_\mu) > 0$. Then A_μ is a nonempty set, so there exists an $x_0 \in A_\mu$. Clearly, by the definition of the set A_μ, any element $x \in A_\mu$ is a dominant generic point of A_μ; in particular, x_0 is a dominant generic point of A_μ, and $\mu = \varepsilon_{x_0}$. Thus, A_μ satisfies all the conditions of (b) of Theorem 7.2.21; therefore, using the theorem, we obtain that μ is an ergodic measure.

Since μ is an invariant ergodic probability measure, we can use (b) of Theorem 6.2.7 in order to infer that $\mu = \varepsilon_y$ for some $y \in \Gamma_{\text{cpie}}$. Using (a) of Theorem 6.2.7, we obtain that $\mu([y]) = \varepsilon_y([y]) = 1$. Using the definitions of $[y]$ and A_μ, we obtain that $[y] \subseteq A_\mu$. Since μ is a probability measure and $\mu([y]) = 1$, it follows that $\mu(A_\mu) = 1$.

We have therefore proved that if μ is an invariant probability measure for $(P_t)_{t \in [0,+\infty)}$, then $\mu(A_\mu)$ is equal to either zero, or else $\mu(A_\mu)$ is equal to 1. We also obtain from the above argument and Theorem 6.1.14 that if $\mu(A_\mu) = 1$, then the measure μ is ergodic.

In order to complete the proof of the corollary, we only have to prove that if the measure μ is ergodic, then $\mu(A_\mu) = 1$.

Thus, assume that μ is an invariant ergodic probability measure. Using (b) of Theorem 6.2.7, we obtain that $\mu = \varepsilon_y$ for some $y \in \Gamma_{\text{cpie}}$; using (a) of the same theorem, we obtain that $\mu([y]) = 1$. Since $[y] \subseteq A_\mu$, and since μ is a probability measure, it follows that $\mu(A_\mu) = 1$. □

Note that in Theorem 7.2.21, Lemma 7.2.22 and Corollary 7.2.23 we do not assume that the transition function $(P_t)_{t \in [0,+\infty)}$ is Feller.

For Feller transition functions, Theorem 7.2.21 can be restated as follows:

Corollary 7.2.24. *Let* $(P_t)_{t \in [0,+\infty)}$ *be a Feller transition function, and let* $\mu \in \mathcal{M}(X)$ *be a probability measure. The following assertions are equivalent:*

(i)　The measure μ is an invariant ergodic probability measure.

(ii)　There exist a nonempty measurable subset A of X and $x_0 \in A$ such that $\mu(A) > 0$, μ is a standard elementary measure with respect to x_0 (that is, $\mu = \varepsilon_{x_0}$), and x_0 is a dominant generic point of A.

Proof. (i) $\Rightarrow$ (ii) follows from the implication (a) $\Rightarrow$ (b) of Theorem 7.2.21 because if we assume (i), then we assume that μ is an invariant probability measure and we also assume (a) of Theorem 7.2.21; thus, by Theorem 7.2.21 the assertion (b) in the theorem holds true, as well; but (b) of Theorem 7.2.21 is precisely (ii).

(ii) $\Rightarrow$ (i) Since we assume that (ii) holds true, it follows that μ is a standard elementary probability measure. Since we also assume that $(P_t)_{t \in [0,+\infty)}$ is a Feller transition function, using Theorem 7.1.1, we obtain that μ is an invariant probability measure (note that here we also use Proposition 5.1.3). Thus, condition (b) of Theorem 7.2.21 is also satisfied; using (a) of the theorem, we obtain that (i) holds true.   □

7.3 Mean Ergodic Theorems

As pointed out in the abstract of the chapter, in this section, our goal is to extend the mean ergodic theorems obtained for equicontinuous transition probabilities in Section 4.3 of [143] to transition functions that are equicontinuous in the mean. The results discussed in this section were announced in [147] and are similar to the results obtained in [33]. In our approach here we also obtain a new proof of Proposition 4.2.2 of [143].

As usual in this chapter, let (X, d) be a locally compact separable metric space.

Let $(P_t)_{t \in \mathbb{T}}$ be a transition function. If $\mathbb{T} = \mathbb{R}$, the mean ergodic theorems involve only the restriction $(P_t)_{t \in [0,+\infty)}$ of $(P_t)_{t \in \mathbb{R}}$ to $[0, +\infty)$, so there is no loss in generality if we will consider only transition functions of the form $(P_t)_{t \in [0,+\infty)}$.

Thus, let $(P_t)_{t \in [0,+\infty)}$ be a transition function.

Unless otherwise stated, we will assume that $(P_t)_{t \in [0,+\infty)}$ satisfies the s.m.a. and is pointwise continuous.

Our goal now is to obtain an extension of Proposition 4.2.2 of [143] to transition functions.

To this end, we need the following proposition:

Proposition 7.3.1. *Let* $\overline{\mathbf{B}_{C_0(X)}(0, 1)}$ *be the closed unit ball in* $C_0(X)$; *that is,* $\overline{\mathbf{B}_{C_0(X)}(0, 1)} = \{f \in C_0(X) \mid \|f\| \leq 1\}$, *where, as usual,* $\|f\|$ *is the sup norm of* f *on* $C_0(X)$. *Set* $E = \overline{\mathbf{B}_{C_0(X)}(0, 1)} \times X$, *and let* $\rho = E \times E \to \mathbb{R}$ *be defined by* $\rho((f, x), (g, y)) = \|f - g\| + d(x, y)$ *for every* $f \in \overline{\mathbf{B}_{C_0(X)}(0, 1)}$, $g \in \overline{\mathbf{B}_{C_0(X)}(0, 1)}$, $x \in X$, *and* $y \in X$. *Then* (E, ρ) *is a separable metric space, and the topology defined by* ρ *is the product topology of* $\overline{\mathbf{B}_{C_0(X)}(0, 1)} \times X$.

Proof. Clearly, $\overline{\mathbf{B}_{C_0(X)}(0, 1)}$ is a metric space in its own right when endowed with the restriction to $\overline{\mathbf{B}_{C_0(X)}(0, 1)}$ of the metric on $C_0(X)$ defined by the sup norm.

Since $C_0(X)$ is a separable Banach space, it follows that the open unit ball $\mathbf{B}_{C_0(X)}(0, 1)$ in $C_0(X)$ is also separable because if $\mathcal{D}$ is a countable dense subset of elements of $C_0(X)$, then $\mathcal{D} \cap \mathbf{B}_{C_0(X)}(0, 1)$ is a countable dense subset of $\mathbf{B}_{C_0(X)}(0, 1)$; consequently, $\overline{\mathbf{B}_{C_0(X)}(0, 1)}$ is also separable because $\overline{\mathbf{B}_{C_0(X)}(0, 1)}$ is the closure (in the metric space $C_0(X)$) of $\mathbf{B}_{C_0(X)}(0, 1)$.

Now, given two metric spaces (Z_1, d_1) and (Z_2, d_2), it is well-known and easy to prove that if we set $Z = Z_1 \times Z_2$ and we define $d = Z \times Z \to \mathbb{R}$ by $d((a, x), (b, y)) = d_1(a, b) + d_2(x, y)$ for every $((a, x), (b, y)) \in Z \times Z$, then d is a metric on Z. Therefore, ρ is a metric on E.

Since both $\overline{\mathbf{B}_{C_0(X)}(0, 1)}$ and X are separable, it follows that there exist two countable dense subsets D_1 and D_2 of $\overline{\mathbf{B}_{C_0(X)}(0, 1)}$ and X, respectively. It is easy to see that $D_1 \times D_2$ is a countable dense subset of $\overline{\mathbf{B}_{C_0(X)}(0,1)} \times X$ in the metric topology defined by ρ on E.

It now remains to show that the topology defined by ρ is the product topology of $\overline{\mathbf{B}_{C_0(X)}(0,1)} \times X$.

To this end, let $\mathbf{B}_{C_0(X)}(g,\delta)$ be the open ball in $C_0(X)$ of center $g \in C_0(X)$ and of radius $\delta \in \mathbb{R}$, $\delta > 0$, let $\mathbf{B}_d(y,\delta)$ be the open ball in X of radius $\delta \in \mathbb{R}$, $\delta > 0$, and center $y \in X$, and let $\mathbf{B}_\rho((g,y),\delta)$ be the open ball in E of center $(g,y) \in \overline{\mathbf{B}_{C_0(X)}(0,1)} \times X$ and of radius $\delta \in \mathbb{R}$, $\delta > 0$.

Using the above notation, we obtain that

$$(\mathbf{B}_{C_0(X)}(f,\varepsilon) \cap \overline{\mathbf{B}_{C_0(X)}(0,1)}) \times \mathbf{B}_d(x,\varepsilon) \subseteq \mathbf{B}_\rho((f,x),2\varepsilon)$$

$$\subseteq (\mathbf{B}_{C_0(X)}(f,2\varepsilon) \cap \overline{\mathbf{B}_{C_0(X)}(0,1)}) \times \mathbf{B}_d(x,2\varepsilon)$$

for every $f \in \overline{\mathbf{B}_{C_0(X)}(0,1)}$, $x \in X$, and $\varepsilon \in \mathbb{R}$, $\varepsilon > 0$.

Taking into consideration the above inclusions, we obtain that the topology defined by ρ on E is the product topology of $\overline{\mathbf{B}_{C_0(X)}(0,1)} \times X$. $\qquad\square$

We will now discuss an extension of Proposition 4.2.2 of [143] (see also Proposition 1.4.36) to Feller transition functions that are equicontinuous in the mean (for the definition of an equicontinuous in the mean transition function, see the beginning of Sect. 7.2.2).

Proposition 7.3.2. *Assume that the transition function $(P_t)_{t \in [0,+\infty)}$ is Feller and equicontinuous in the mean. Also, let $((S_t, T_t))_{t \in [0,+\infty)}$ be the family of Markov-Feller pairs defined by $(P_t)_{t \in [0,+\infty)}$. If $(t_n)_{n \in \mathbb{N}}$ is a sequence of strictly positive real numbers that diverges to $+\infty$, then there exists a subsequence $(t_{n_l})_{l \in \mathbb{N}}$ of $(t_n)_{n \in \mathbb{N}}$ such that for every $f \in C_0(X)$ and every $x \in X$, the sequence $\left(\frac{1}{t_{n_l}} \int_0^{t_{n_l}} S_\alpha f(x)\, d\alpha\right)_{l \in \mathbb{N}}$ converges.*

Proof. Let $(t_n)_{n \in \mathbb{N}}$ be a sequence of real numbers that diverges to $+\infty$ such that $t_n > 0$ for every $n \in \mathbb{N}$.

Our goal is to prove that there exists a subsequence $(t_{n_l})_{l \in \mathbb{N}}$ of $(t_n)_{n \in \mathbb{N}}$ such that the sequence $\left(\frac{1}{t_{n_l}} \int_0^{t_{n_l}} S_\alpha f(x)\, d\alpha\right)_{l \in \mathbb{N}}$ converges whenever $f \in C_0(X)$ and $x \in X$.

However, note that since

$$\frac{1}{t} \int_0^t S_\alpha f(x)\, d\alpha = \|f\| \frac{1}{t} \int_0^t S_\alpha\left(\frac{f}{\|f\|}\right)(x)\, d\alpha$$

for every $t \in (0,+\infty)$, $f \in C_0(X)$, $f \neq 0$, and $x \in X$, it follows that it is enough to prove that there exists a subsequence $(t_{n_l})_{l \in \mathbb{N}}$ of $(t_n)_{n \in \mathbb{N}}$ such that the sequence $\left(\frac{1}{t_{n_l}} \int_0^{t_{n_l}} S_\alpha f(x)\, d\alpha\right)_{l \in \mathbb{N}}$ converges for every $f \in C_0(X)$, $\|f\| \leq 1$, and every $x \in X$.

To this end, let (E, ρ) be the metric space considered in Proposition 7.3.1; that is, let $E = \mathbf{B}_{C_0(X)}(0, 1) \times X$, where $\mathbf{B}_{C_0(X)}(0, 1)$ is the closed unit ball in $C_0(X)$, and let $\rho : E \times E \to \mathbb{R}$ be defined by $\rho((f, x), (g, y)) = \|f - g\| + d(x, y)$ for every $((f, x), (g, y)) \in E \times E$.

Now, for every $n \in \mathbb{N}$, let $\varphi_n : E \to \mathbb{R}$ be defined by

$$\varphi_n(f, x) = \frac{1}{t_n} \int_0^{t_n} S_\alpha f(x) \, d\alpha$$

for every $(f, x) \in E$.

Our goal is to apply Theorem 4.5.1 (Ascoli-Arzelà) to the sequence $(\varphi_n)_{n \in \mathbb{N}}$. To this end, we have to prove that $(\varphi_n)_{n \in \mathbb{N}}$ is uniformly bounded and equicontinuous.

Taking into consideration that S_α, $\alpha \in [0, +\infty)$, are positive contractions of $B_b(X)$, we obtain that

$$|\varphi_n(f, x)| = \frac{1}{t_n} \left| \int_0^{t_n} S_\alpha f(x) \, d\alpha \right| \leq \frac{1}{t_n} \int_0^{t_n} |S_\alpha f(x)| \, d\alpha$$

$$\leq \frac{1}{t_n} \int_0^{t_n} \|f\| \, d\alpha = \|f\| \leq 1$$

for every $n \in \mathbb{N}$, every $f \in C_0(X)$, $\|f\| \leq 1$, and every $x \in X$. Thus, the sequence $(\varphi_n)_{n \in \mathbb{N}}$ is uniformly bounded.

We now prove that the sequence $(\varphi_n)_{n \in \mathbb{N}}$ is equicontinuous. Thus, we have to prove that for every convergent sequence $((g_k, y_k))_{k \in \mathbb{N}}$ of elements of E ($g_k \in C_0(X)$, $\|g_k\| \leq 1$, and $y_k \in X$ for every $k \in \mathbb{N}$), and for every $\varepsilon \in \mathbb{R}$, $\varepsilon > 0$, there exists a $k_\varepsilon \in \mathbb{N}$ such that $|\varphi_n(g_k, y_k) - \varphi_n(g, y)| < \varepsilon$ for every $k \in \mathbb{N}$, $k \geq k_\varepsilon$, and every $n \in \mathbb{N}$, where $(g, y) = \lim_{k \to +\infty}(g_k, y_k)$.

To this end, let $((g_k, y_k))_{k \in \mathbb{N}}$ be a convergent sequence of elements of E, let (g, y) be the limit of $((g_k, y_k))_{k \in \mathbb{N}}$ in E and let $\varepsilon \in \mathbb{R}$, $\varepsilon > 0$.

Since $((g_k, y_k))_{k \in \mathbb{N}}$ converges to (g, y) in the metric topology of E, it follows that $(g_k)_{k \in \mathbb{N}}$ converges to g in the norm topology of $C_0(X)$ and $(y_k)_{k \in \mathbb{N}}$ converges to y in the metric topology of X.

Since $(g_k)_{k \in \mathbb{N}}$ converges to g, it follows that there exists a $k'_\varepsilon \in \mathbb{N}$ such that $\|g_k - g\| < \frac{\varepsilon}{2}$ for every $k \in \mathbb{N}$, $k \geq k'_\varepsilon$.

Now, let $(\mathbf{A}_t)_{t \in (0, +\infty)}$ be the family of operator averages generated by $(S_\alpha)_{\alpha \in [0, +\infty)}$ defined in Sect. 7.2.2 before Lemma 7.2.7; thus, $\mathbf{A}_t : B_b(X) \to B_b(X)$ is defined by $\mathbf{A}_t f = \frac{1}{t}(\text{P-}\int_0^t S_\alpha f \, d\alpha)$ for every $f \in B_b(X)$ and $t \in (0, +\infty)$.

Since we assume that $(P_t)_{t \in [0, +\infty)}$ is equicontinuous in the mean, it follows that the family $(\mathbf{A}_t)_{t \in (0, +\infty)}$ is $C_0(X)$-equicontinuous. Accordingly, there exists a $k''_\varepsilon \in \mathbb{N}$ large enough such that $|\mathbf{A}_t g(y_k) - \mathbf{A}_t g(y)| < \frac{\varepsilon}{2}$ (that is, $|\frac{1}{t} \int_0^t S_\alpha g(y_k) \, d\alpha - \frac{1}{t} \int_0^t S_\alpha g(y) \, d\alpha| < \frac{\varepsilon}{2}$) for every $k \geq k''_\varepsilon$.

Set $k_\varepsilon = \max\{k'_\varepsilon, k''_\varepsilon\}$.

Taking into consideration that S_u is a contraction for every $u \in [0, +\infty)$, we obtain that

$$|\varphi_n(g_k, y_k) - \varphi_n(g, y)| \le |\varphi_n(g_k, y_k) - \varphi_n(g, y_k)| + |\varphi_n(g, y_k) - \varphi_n(g, y)|$$

$$= \left| \frac{1}{t_n} \int_0^{t_n} S_\alpha g_k(y_k) \, d\alpha - \frac{1}{t_n} \int_0^{t_n} S_\alpha g(y_k) \, d\alpha \right|$$

$$+ \left| \frac{1}{t_n} \int_0^{t_n} S_\alpha g(y_k) \, d\alpha - \frac{1}{t_n} \int_0^{t_n} S_\alpha g(y) \, d\alpha \right|$$

$$\le \frac{1}{t_n} \int_0^{t_n} |S_\alpha g_k(y_k) - S_\alpha g(y_k)| \, d\alpha + \frac{1}{t_n} \int_0^{t_n} |S_\alpha g(y_k) - S_\alpha g(y)| \, d\alpha$$

$$< \frac{1}{t_n} \int_0^{t_n} \|g_k - g\| \, d\alpha + \frac{\varepsilon}{2} < \frac{\varepsilon}{2} + \frac{\varepsilon}{2} = \varepsilon$$

for every $n \in \mathbb{N}$, and $k \in \mathbb{N}$, $k \ge k_\varepsilon$.

Accordingly, the sequence $(\varphi_n)_{n \in \mathbb{N}}$ is equicontinuous.

By Theorem 4.5.1 (Ascoli-Arzelà), there exists a subsequence $(\varphi_{n_l})_{l \in \mathbb{N}}$ of $(\varphi_n)_{n \in \mathbb{N}}$ such that $(\varphi_{n_l}(f, x))_{l \in \mathbb{N}}$ converges for every $f \in \overline{\mathbf{B}_{C_0(X)}(0, 1)}$ and every $x \in X$; hence, $\left(\frac{1}{t_{n_l}} \int_0^{t_{n_l}} S_\alpha f(x) \, d\alpha \right)_{l \in \mathbb{N}}$ converges for every $f \in C_0(X)$ and $x \in X$. $\square$

In proving Proposition 7.3.2, we used Proposition 7.3.1 and Theorem 4.5.1 (Ascoli-Arzelà). It turns out that we can use Proposition 7.3.1 in order to give a new proof of Proposition 4.2.2 of [143], thereby obtaining a unified approach to Proposition 7.3.2 and to Proposition 4.2.2 of [143]. We discuss the new proof next.

Proposition 7.3.3 (Proposition 4.2.2 of [143]). *Let (S, T) be a $C_0(X)$-equicontinuous Markov-Feller pair defined on (X, d). Also, let $\left(\frac{1}{n_k} \sum_{i=0}^{n_k - 1} S^i \right)_{k \in \mathbb{N}}$ be a subsequence of $\left(\frac{1}{n} \sum_{i=0}^{n-1} S^i \right)_{n \in \mathbb{N}}$. Then there exists a subsequence $\left(\frac{1}{n_{k_l}} \sum_{i=0}^{n_{k_l} - 1} S^i \right)_{l \in \mathbb{N}}$ of $\left(\frac{1}{n_k} \sum_{i=0}^{n_k - 1} S^i \right)_{k \in \mathbb{N}}$ such that $\left(\frac{1}{n_{k_l}} \sum_{i=0}^{n_{k_l} - 1} S^i f(x) \right)_{l \in \mathbb{N}}$ converges for every $f \in C_0(X)$ and $x \in X$.*

Proof (using Proposition 7.3.1 and Theorem 4.5.1 (Ascoli-Arzelà)). Let $\left(\frac{1}{n_k} \sum_{i=0}^{n_k - 1} S^i \right)_{k \in \mathbb{N}}$ be a subsequence of $\left(\frac{1}{n} \sum_{i=0}^{n-1} S^i \right)_{n \in \mathbb{N}}$.

Since $\frac{1}{n} \sum_{i=0}^{n-1} S^i \frac{f}{\|f\|}(x) = \frac{1}{\|f\|} \frac{1}{n} \sum_{i=0}^{n-1} S^i f(x)$ for every $n \in \mathbb{N}$, $f \in C_0(X)$, $f \ne 0$, and $x \in X$, it follows that the proposition will be completely proved if we show that there exists a subsequence $(n_{k_l})_{l \in \mathbb{N}}$ of $(n_k)_{k \in \mathbb{N}}$ such that $\left(\frac{1}{n_{k_l}} \sum_{i=0}^{n_{k_l} - 1} S^i f(x) \right)_{l \in \mathbb{N}}$ converges for every $f \in C_0(X)$, $\|f\| \le 1$, and $x \in X$.

Now, let (E, ρ) be the separable metric space considered in Proposition 7.3.1, $E = \overline{\mathbf{B}_{C_0(X)}(0, 1)} \times X$, and $\rho((f, x), (g, y)) = \|f - g\| + d(x, y)$ for every $(f, x) \in E$ and $(g, y) \in E$.

For every $k \in \mathbb{N}$, let $\eta_k = E \to \mathbb{R}$ be defined by $\eta_k(f, x) = \frac{1}{n_k} \sum_{i=0}^{n_k-1} S^i f(x)$ for every $(f, x) \in E$.

Our goal now is to show that the sequence $(\eta_k)_{k \in \mathbb{N}}$ satisfies the conditions of Theorem 4.5.1 (Ascoli-Arzelà).

Thus, we have to prove that $(\eta_k)_{k \in \mathbb{N}}$ is uniformly bounded and equicontinuous.

Since

$$|\eta_k(f, x)| = \frac{1}{n_k} \left| \sum_{i=0}^{n_k-1} S^i f(x) \right| \le \frac{1}{n_k} \sum_{i=0}^{n_k-1} |S^i f(x)| \le \frac{1}{n_k} \sum_{i=0}^{n_k-1} \|S^i f\| \le \|f\| \le 1$$

for every $f \in \overline{\mathbf{B}_{C_0(X)}(0, 1)}$, $x \in X$, and $k \in \mathbb{N}$, it follows that $(\eta_k)_{k \in \mathbb{N}}$ is uniformly bounded.

We now prove that $(\eta_k)_{k \in \mathbb{N}}$ is equicontinuous.

To this end, let $((g_j, y_j))_{j \in \mathbb{N}}$ be a convergent sequence of elements of E, set $(g, y) = \lim_{j \to +\infty}(g_j, y_j)$, and let $\varepsilon \in \mathbb{R}, \varepsilon > 0$.

Since $(y_j)_{j \in \mathbb{N}}$ converges to y in the metric topology of X and since (S, T) is a $C_0(X)$-equicontinuous Markov-Feller pair, it follows that there exists a $j'_\varepsilon \in \mathbb{N}$ such that $|S^i g(y_j) - S^i g(y)| < \frac{\varepsilon}{2}$ for every $i \in \mathbb{N} \cup \{0\}$ and $j \in \mathbb{N}, j \ge j'_\varepsilon$.

Since $(g_j)_{j \in \mathbb{N}}$ converges to g in the norm topology of $C_0(X)$, it follows that there exists a $j''_\varepsilon \in \mathbb{N}$ such that $\|g_j - g\| < \frac{\varepsilon}{2}$ for every $j \in \mathbb{N}, j \ge j''_\varepsilon$.

Set $j_\varepsilon = \max\{j'_\varepsilon, j''_\varepsilon\}$.

It follows that

$$|\eta_k(g_j, y_j) - \eta_k(g, y)| \le |\eta_k(g_j, y_j) - \eta_k(g, y_j)| + |\eta_k(g, y_j) - \eta_k(g, y)|$$

$$= \left| \frac{1}{n_k} \sum_{i=0}^{n_k-1} S^i g_j(y_j) - \frac{1}{n_k} \sum_{i=0}^{n_k-1} S^i g(y_j) \right| + \left| \frac{1}{n_k} \sum_{i=0}^{n_k-1} S^i g(y_j) - \frac{1}{n_k} \sum_{i=0}^{n_k-1} S^i g(y) \right|$$

$$\le \frac{1}{n_k} \sum_{i=0}^{n_k-1} |S^i(g_j - g)(y_j)| + \frac{1}{n_k} \sum_{i=0}^{n_k-1} |S^i(g(y_j) - S^i g(y)|$$

$$< \frac{1}{n_k} \sum_{i=0}^{n_k-1} \|S^i(g_j - g)\| + \frac{\varepsilon}{2} < \frac{\varepsilon}{2} + \frac{\varepsilon}{2} = \varepsilon$$

for every $k \in \mathbb{N}$ and $j \in \mathbb{N}, j \ge j_\varepsilon$.

We have therefore proved that for every convergent sequence $((g_j, y_j))_{j \in \mathbb{N}}$ and for every $\varepsilon \in \mathbb{R}, \varepsilon > 0$, there exists a $j_\varepsilon \in \mathbb{N}$ such that $|\eta_k(g_j, y_j) - \eta_k(g, y)| < \varepsilon$ for every $k \in \mathbb{N}$ and $j \in \mathbb{N}, j \ge j_\varepsilon$. Thus, $(\eta_k)_{k \in \mathbb{N}}$ is an equicontinuous sequence of functions.

In view of our discussion so far, we can apply Theorem 4.5.1 (Ascoli-Arzelà) to the sequence $(\eta_k)_{k\in\mathbb{N}}$. Using the theorem we obtain that there exists a subsequence $(\eta_{k_l})_{l\in\mathbb{N}}$ of $(\eta_k)_{k\in\mathbb{N}}$ such that $(\eta_{k_l}(f,x))_{l\in\mathbb{N}}$ converges for every $(f,x)\in E$.

We have therefore proved that there exists a subsequence $\left(\frac{1}{n_{k_l}}\sum_{i=0}^{n_{k_l}-1}S^i\right)_{l\in\mathbb{N}}$ of $\left(\frac{1}{n_k}\sum_{i=0}^{n_k-1}S^i\right)_{k\in\mathbb{N}}$ such that $\left(\frac{1}{n_{k_l}}\sum_{i=0}^{n_{k_l}-1}S^i f(x)\right)_{l\in\mathbb{N}}$ converges for every $f\in C_0(X)$ and $x\in X$. $\qquad\square$

We now return to the setting of Proposition 7.3.2; that is, we assume given a Feller transition function $(P_t)_{t\in[0,+\infty)}$ (that satisfies the s.m.a. and is pointwise continuous) defined on (X,d), and we let $((S_t,T_t))_{t\in[0,+\infty)}$ be the family of Markov-Feller pairs defined by $(P_t)_{t\in[0,+\infty)}$.

We say that $(P_t)_{t\in[0,+\infty)}$ (or $(S_t)_{t\in[0,+\infty)}$, or $((S_t,T_t))_{t\in[0,+\infty)}$) is $C_0(X)$-*strongly continuous* (or *strongly continuous with respect to $C_0(X)$*) if, for every $f\in C_0(X)$, the $B_b(X)$-valued map $t\mapsto S_t f$, $t\in[0,+\infty)$, is continuous with respect to the topology of uniform convergence on $B_b(X)$ and the usual standard topology on $[0,+\infty)$. Obviously, in the above definition, we can consider a transition function $(P_t)_{t\in\mathbb{T}}$, where $\mathbb{T}=\mathbb{R}$ rather than $[0,+\infty)$.

Example 7.3.4. Let $\mathbf{w}=(w_t)_{t\in\mathbb{R}}$ be the flow of the rotations of the unit circle $\mathbb{R}/\mathbb{Z}$. Let $(P_t^{(\mathbf{w})})_{t\in\mathbb{R}}$ be the transition function defined by $\mathbf{w}$ and let $((S_t^{\mathbf{w}},T_t^{\mathbf{w}}))_{t\in\mathbb{R}}$ be the family of Markov pairs defined by $\mathbf{w}$ (for the definitions of $(P_t^{(\mathbf{w})})_{t\in\mathbb{R}}$ and $((S_t^{(\mathbf{w})},T_t^{(\mathbf{w})}))_{t\in\mathbb{R}}$, see Example 2.2.4).

We will now prove that $(P_t^{(\mathbf{w})})_{t\in\mathbb{R}}$ is $C_0(\mathbb{R}/\mathbb{Z})$-strongly continuous.

Thus, we have to prove that the $B_b(\mathbb{R}/\mathbb{Z})$-valued map $t\mapsto S_t f$, $t\in\mathbb{R}$, is continuous whenever $f\in C_0(\mathbb{R}/\mathbb{Z})$.

To this end, let $f\in C_0(\mathbb{R}/\mathbb{Z})$.

Clearly, it is enough to prove that for every convergent sequence $(t_n)_{n\in\mathbb{N}}$ of real numbers we have that $(S_{t_n}f)_{n\in\mathbb{N}}$ converges to $S_t f$ in the norm topology of $B_b(X)$, where $t=\lim_{n\to+\infty}t_n$.

Thus, let $(t_n)_{n\in\mathbb{N}}$ be a convergent sequence of real numbers, set $t=\lim_{n\to+\infty}t_n$, and let $\varepsilon\in\mathbb{R}$, $\varepsilon>0$. We want to prove that there exists an $n_\varepsilon\in\mathbb{N}$ such that $\|S_{t_n}f-S_t f\|<\varepsilon$ for every $n\in\mathbb{N}$, $n\geq n_\varepsilon$.

Since $\mathbb{R}/\mathbb{Z}$ is a compact space, it follows that f is uniformly continuous, so there exists a $\delta\in\mathbb{R}$, $\delta>0$, such that $|f(\hat{u})-f(\hat{v})|<\frac{\varepsilon}{2}$, whenever $\eta(\hat{u},\hat{v})<\delta$, where η is the metric on $\mathbb{R}/\mathbb{Z}$ defined in Example A.2.8.

Let $n_\varepsilon\in\mathbb{N}$ be such that $|t_n-t|<\delta$ for every $n\in\mathbb{N}$, $n\geq n_\varepsilon$.

Then $|S_{t_n}f(\hat{x})-S_t f(\hat{x})| = |f(\widehat{t_n+x})-f(\widehat{t+x})| < \varepsilon/2$ because $\eta(\widehat{t_n+x},\widehat{t+x})=\eta(\hat{t}_n,\hat{t})<\delta$ for every $n\in\mathbb{N}$, $n\geq n_\varepsilon$, and $\hat{x}\in\mathbb{R}/\mathbb{Z}$.

Accordingly, $\|S_{t_n}f-S_t f\|\leq\frac{\varepsilon}{2}<\varepsilon$ for every $n\geq n_\varepsilon$.

We have therefore proved that $(P_t^{\mathbf{w}})_{t\in\mathbb{R}}$ is $C_0(\mathbb{R}/\mathbb{Z})$-strongly continuous. ∎

Example 7.3.5. Let $n\in\mathbb{N}$, $n\geq 2$, let $\mathbb{R}^n/\mathbb{Z}^n$ be the n-dimensional torus defined in Example A.2.9, and let $\mathbf{v}\in\mathbb{R}^n$, $\mathbf{v}=(v_1,v_2,\ldots,v_n)$. Now let $\mathbf{w}=(w_t)_{t\in\mathbb{R}}$ be the rectilinear flow on $\mathbb{R}^n/\mathbb{Z}^n$ with velocity $\mathbf{v}$ (see Example A.3.5), let $(P_t^{(\mathbf{w})})_{t\in\mathbb{R}}$ be the transition function defined by $\mathbf{w}$, and let $((S_t, T_t))_{t \in \mathbb{R}}$ be the family of Markov-Feller pairs defined by $\mathbf{w}$.

We will now show that the transition function $(P_t^{(\mathbf{w})})_{t \in \mathbb{R}}$ is strongly continuous with respect to $C_0(\mathbb{R}^n/\mathbb{Z}^n)$.

Thus, let $f \in C_0(\mathbb{R}^n/\mathbb{Z}^n)$. Since we have to show that the $B_b(\mathbb{R}^n/\mathbb{Z}^n)$-valued map $t \mapsto S_t f$, $t \in \mathbb{R}$, is continuous, it is enough to prove that for every convergent sequence $(t_k)_{k \in \mathbb{N}}$ of real numbers, and for every $\varepsilon \in \mathbb{R}$, $\varepsilon > 0$, there exists a $k_\varepsilon \in \mathbb{N}$ such that $||S_{t_k} f - S_t f|| < \varepsilon$ for every $k \in \mathbb{N}$, $k \geq k_\varepsilon$, where $t = \lim_{k \to +\infty} t_k$.

Thus, let $(t_k)_{t \in \mathbb{N}}$ be a convergent sequence of real numbers and let $\varepsilon \in \mathbb{R}$, $\varepsilon > 0$.

Since $\mathbb{R}^n/\mathbb{Z}^n$ is a compact metric space and f is continuous on $\mathbb{R}^n/\mathbb{Z}^n$, it follows that f is uniformly continuous on $\mathbb{R}^n/\mathbb{Z}^n$. Thus, there exists a $\delta \in \mathbb{R}$, $\delta > 0$, such that $|f(\hat{\mathbf{u}}) - f(\hat{\mathbf{z}})| < \varepsilon/2$ whenever $\eta^{(n)}(\hat{\mathbf{u}}, \hat{\mathbf{z}}) < \delta$ where $\eta^{(n)}$ is the metric defined in Example A.2.9.

Now, let $t = \lim_{k \to +\infty} t_k$, and let k_ε be such that $\eta^{(n)}(t_k \hat{\mathbf{v}}, t \hat{\mathbf{v}}) < \delta$ (the existence of k_ε is obvious in view of the fact that the sequence $(t_k \hat{\mathbf{v}})_{k \in \mathbb{N}}$ converges to $t\hat{\mathbf{v}}$).

We obtain that

$$|S_{t_k} f(\hat{\mathbf{u}}) - S_t f(\hat{\mathbf{u}})| = |f(\widehat{t_k \mathbf{v} + \mathbf{u}}) - f(\widehat{t\mathbf{v} + \mathbf{u}})| < \frac{\varepsilon}{2}$$

for every $\hat{\mathbf{u}} \in \mathbb{R}^n/\mathbb{Z}^n$ because $\eta^{(n)}(\widehat{t_k \mathbf{v} + \mathbf{u}}, \widehat{t\mathbf{v} + \mathbf{u}}) = \eta^{(n)}(\widehat{t_k \mathbf{v}}, \widehat{t\mathbf{v}}) < \delta$ for every $k \geq k_\varepsilon$.

Accordingly, $||S_{t_k} f - S_t f|| \leq \frac{\varepsilon}{2} < \varepsilon$ for every $k \in \mathbb{N}$, $k \geq k_\varepsilon$.

Since, for every $f \in C_0(\mathbb{R}^n/\mathbb{Z}^n)$ and for every convergent sequence $(t_n)_{n \in \mathbb{N}}$ of real numbers we have that $(||S_{t_n} f - S_t f||)_{n \in \mathbb{N}}$ converges to zero, where $t = \lim_{n \to +\infty} t_n$, we conclude that $(P_t^{(\mathbf{w})})_{t \in \mathbb{R}}$ is $C_0(\mathbb{R}^n/\mathbb{Z}^n)$-strongly continuous. ∎

The reason for introducing the notion of $C_0(X)$-strong continuity of a transition function is, of course, that we will use $C_0(X)$-strongly continuous transition functions in this section. More precisely, we will use the fact that these transition functions have a certain nice property that we will discuss in the next lemma.

Lemma 7.3.6. *Let $(P_t)_{t \in [0,+\infty)}$ be a Feller transition function that is $C_0(X)$-strongly continuous and let $((S_t, T_t))_{t \in [0,+\infty)}$ be the family of Markov-Feller pairs defined by $(P_t)_{t \in [0,+\infty)}$.*

(a) For every $t_1 \in [0, +\infty)$ and $t_2 \in [0, +\infty)$ such that $t_1 \leq t_2$, the pointwise integral $P\text{-}\int_{t_1}^{t_2} S_t f \, dt$ exists and is an element of $C_b(X)$ whenever $f \in C_0(X)$.
(b) Let $u \in [0, +\infty)$, $t_1 \in [0, +\infty)$, and $t_2 \in [0, +\infty)$, and assume that $t_1 \leq t_2$. Then $S_u \left(P\text{-}\int_{t_1}^{t_2} S_t f \, dt \right) = P\text{-}\int_{t_1+u}^{t_2+u} S_t f \, dt$ for every $f \in C_0(X)$.

Proof. (a) Let $t_1 \in [0, +\infty)$ and $t_2 \in [0, +\infty)$ be such that $t_1 \leq t_2$ and let $f \in C_0(X)$.

The fact that the pointwise integral $P\text{-}\int_{t_1}^{t_2} S_t f \, dt$ exists is a consequence of (a) of Theorem 3.3.4.

We now prove that $P\text{-}\int_{t_1}^{t_2} S_t f \, dt$ is an element of $C_b(X)$.

Since S_t, $t \in [0, +\infty)$, are contractions of $B_b(X)$, we obtain that

$$\left| \left(\text{P-} \int_{t_1}^{t_2} S_t f \, dt \right)(x) \right| = \left| \int_{t_1}^{t_2} S_t f(x) \, dt \right| \leq \int_{t_1}^{t_2} |S_t f(x)| \, dt \leq (t_2 - t_1) \| f \|$$

for every $x \in X$. Thus, P- $\int_{t_1}^{t_2} S_t f \, dt$ is a bounded function.

We now prove that P- $\int_{t_1}^{t_2} S_t f \, dt$ is a continuous function.

To this end, let $(y_n)_{n \in \mathbb{N}}$ be a convergent sequence of elements of X, and set $y = \lim_{n \to +\infty} y_n$.

Let $\varphi = [t_1, t_2] \to \mathbb{R}$ be defined by $\varphi(t) = S_t f(y)$ for every $t \in [t_1, t_2]$, and, for every $n \in \mathbb{N}$, let $\varphi_n : [t_1, t_2] \to \mathbb{R}$ be defined by $\varphi_n(t) = S_t f(y_n)$ for every $t \in [t_1, t_2]$.

Applying the Lebesgue dominated convergence theorem to the sequence $(\varphi_n)_{n \in \mathbb{N}}$ (which converges pointwise to φ on $[t_1, t_2]$) with respect to the Lebesgue measure on $[t_1, t_2]$, we obtain that $\lim_{n \to +\infty} \int_{t_1}^{t_2} S_t f(y_n) \, dt$ exists and is equal to $\int_{t_1}^{t_2} S_t f(y) \, dt$. Consequently, $\lim_{n \to +\infty} (\text{P-} \int_{t_1}^{t_2} S_t f \, dt)(y_n)$ exists and is equal to $(\text{P-} \int_{t_1}^{t_2} S_t f \, dt)(y)$.

Thus, P- $\int_{t_1}^{t_2} S_t f \, dt$ is a continuous function.

(b) Let $t_1 \in [0, +\infty)$ and $t_2 \in [0, +\infty)$ be such that $t_1 \leq t_2$, let $u \in [0, +\infty)$, and let $f \in C_0(X)$.

Using (c) and (d) of Theorem 3.3.4, we obtain that the Dunford-Schwartz integral DS- $\int_{t_1}^{t_2} S_t f \, dt$ exists and is equal to P- $\int_{t_1}^{t_2} S_t f \, dt$.

Consequently, using (c) of Theorem 3.2.19, pp. 113–114 of Dunford and Schwartz [30] to S_u, we obtain that

$$S_u \left(\text{P-} \int_{t_1}^{t_2} S_t f \, dt \right) = S_u \left(\text{DS-} \int_{t_1}^{t_2} S_t f \, dt \right) = \text{DS-} \int_{t_1}^{t_2} S_u S_t f \, dt$$

$$= \text{DS-} \int_{t_1}^{t_2} S_t S_u f \, dt = P - \int_{t_1}^{t_2} S_t S_u f \, dt = \text{P-} \int_{t_1}^{t_2} S_{t+u} f \, dt.$$

In order to complete the proof of the lemma, it now remains to show that P- $\int_{t_1}^{t_2} S_{t+u} f \, dt = $ P- $\int_{t_1+u}^{t_2+u} S_t f \, dt$.

To this end, let $x \in X$.

Let $\gamma_x = \mathbb{R} \to \mathbb{R}$ be defined by

$$\gamma_x(t) = \begin{cases} S_t f(x) & \text{if } t \geq 0 \\ 0 & \text{if } t < 0. \end{cases}$$

Clearly γ_x is a measurable function because $(P_t)_{t \in [0, +\infty)}$ satisfies the s.m.a.

Now let $\delta_x : \mathbb{R} \to \mathbb{R}$ be defined by $\delta_x(t) = \mathbf{1}_{[u+t_1,u+t_2]}\gamma_x(t)$ for every $t \in \mathbb{R}$, and let $\rho_x : \mathbb{R} \to \mathbb{R}$ be defined by $\rho_x(t) = \delta_x(t + u)$ for every $t \in \mathbb{R}$. Note that

$$\delta_x(t + u) = \mathbf{1}_{[u+t_1,u+t_2]}(t + u)\gamma_x(t + u) = \mathbf{1}_{[t_1,t_2]}(t)\gamma_x(t + u).$$

Clearly, both δ_x and ρ_x are integrable over $\mathbb{R}$ because γ_x is a bounded measurable function.

Since the Lebesgue measure on $\mathbb{R}$ is translation invariant, we obtain that $\int_{\mathbb{R}} \delta_x(t)\,dt = \int_{\mathbb{R}} \rho_x(t)\,dt$.

The proof of the lemma ia completed by observing that

$$\int_{\mathbb{R}} \rho_x(t)\,dt = \int_{t_1}^{t_2} S_{t+u}f(x)\,dt \quad \text{and} \quad \int_{\mathbb{R}} \delta_x(t)\,dt = \int_{u+t_1}^{u+t_2} S_t f(x)\,dt.$$

$\square$

Unless otherwise stated, from now on we assume given a Feller transition function $(P_t)_{t\in[0,+\infty)}$ that is equicontinuous in the mean and $C_0(X)$-strongly continuous.

Let $(0, +\infty)^{\mathbb{N}}$ be the set of all $(0, +\infty)$-valued sequences and set

$$\mathcal{R} = \left\{ \alpha \in (0, +\infty)^{\mathbb{N}} \;\middle|\; \begin{array}{l} \alpha = (s_n)_{n\in\mathbb{N}} \text{ is a strictly increasing sequence of} \\ \text{elements of } (0, +\infty) \text{ such that } (s_n)_{n\in\mathbb{N}} \text{ diverges} \\ \text{to } +\infty, \text{ and such that the sequence} \\ (\frac{1}{s_n}\int_0^{s_n} S_t f(x)dt)_{n\in\mathbb{N}} \text{ converges for every } f \in C_0(X) \\ \text{and } x \in X \end{array} \right\}.$$

Note that in terms of $\mathcal{R}$, Proposition 7.3.2 states that for every sequence $(s_i)_{i\in\mathbb{N}}$ of strictly positive real numbers that diverges to $+\infty$, there exists a subsequence $(s_{i_k})_{k\in\mathbb{N}}$ of $(s_i)_{i\in\mathbb{N}}$ such that $(s_{i_k})_{k\in\mathbb{N}}$, as a sequence of strictly positive real numbers, belongs to $\mathcal{R}$.

Let $\alpha \in \mathcal{R}$, $\alpha = (s_n)_{n\in\mathbb{N}}$, and let $f \in C_0(X)$. In view of the definition of $\mathcal{R}$ and of the fact that $\alpha \in \mathcal{R}$, it makes sense to define a function $f_\alpha : X \to \mathbb{R}$ as follows: $f_\alpha(x) = \lim_{n\to+\infty} \frac{1}{s_n}\int_0^{s_n} S_t f(x)\,dt$ for every $x \in X$. Note that the definition of f_α is correct in the sense that the limit defining $f_\alpha(x)$ exists for every $x \in X$ because $\alpha \in \mathcal{R}$.

The next proposition is an extension of Proposition 4.2.3 of [143] to our setting.

Proposition 7.3.7. *For every $f \in C_0(X)$ and $\alpha \in \mathcal{R}$, the function f_α belongs to $C_b(X)$ and $S_t f_\alpha = f_\alpha$ for every $t \in [0, +\infty)$.*

Proof. Let $f \in C_0(X)$ and $\alpha \in \mathcal{R}$, $\alpha = (s_n)_{n\in\mathbb{N}}$.
We first prove that f_α is a bounded function.

Let $x \in X$. Since the sequence $(\frac{1}{s_n} \int_0^{s_n} S_t f(x) \, dt)_{n \in \mathbb{N}}$ converges to $f_\alpha(x)$, it follows that the sequence $(|\frac{1}{s_n} \int_0^{s_n} S_t f(x) \, dt|)_{n \in \mathbb{N}}$ converges to $|f_\alpha(x)|$ and

$$|f_\alpha(x)| = \lim_{n \to +\infty} \left| \frac{1}{s_n} \int_0^{s_n} S_t f(x) \, dt \right| \le \lim_{n \to +\infty} \frac{1}{s_n} \int_0^{s_n} |S_t f(x)| \, dt$$

$$\le \lim_{n \to +\infty} \frac{1}{s_n} \int_0^{s_n} \|f\| \, dt = \lim_{n \to +\infty} \frac{1}{s_n} s_n \|f\| = \|f\|.$$

Since $|f_\alpha(x)| \le \|f\|$ for all $x \in X$, it follows that f_α is a bounded function.

We now prove that f_α is a continuous function. Clearly, it is enough to prove that for every convergent sequence $(x_k)_{k \in \mathbb{N}}$ of elements of X, it follows that $(f_\alpha(x_k))_{k \in \mathbb{N}}$ converges to $f_\alpha(x)$, where $x = \lim_{k \to +\infty} x_k$.

To this end, let $(x_k)_{k \in \mathbb{N}}$ be a convergent sequence of elements of X and set $x = \lim_{k \to +\infty} x_k$. We have to show that for every $\varepsilon \in \mathbb{R}, \varepsilon > 0$, there exists a $k_\varepsilon \in \mathbb{N}$ such that $|f_\alpha(x_k) - f_\alpha(x)| < \varepsilon$ for every $k \ge k_\varepsilon$.

Thus, let $\varepsilon \in \mathbb{R}, \varepsilon > 0$.

Since we assume that $(S_t)_{t \in [0,+\infty)}$ is equicontinuous in the mean, we obtain that there exists a $k_\varepsilon \in \mathbb{N}$ such that

$$\left| \frac{1}{t} (\text{P-} \int_0^t S_u f \, du)(x_k) - \frac{1}{t} (\text{P-} \int_0^t S_u f \, du)(x) \right| < \frac{\varepsilon}{2},$$

that is $|\frac{1}{t} \int_0^t S_u f(x_k) \, du - \frac{1}{t} \int_0^t S_u f(x) \, du| < \frac{\varepsilon}{2}$ for every $t \in [0, +\infty)$ and every $k \in \mathbb{N}, k \ge k_\varepsilon$.

Now, let $k \in \mathbb{N}, k \ge k_\varepsilon$.

Since the sequence $(\frac{1}{s_n} \int_0^{s_n} S_t f(x_k) \, dt - \frac{1}{s_n} \int_0^{s_n} S_t f(x) \, dt)_{n \in \mathbb{N}}$ converges to $f_\alpha(x_k) - f_\alpha(x)$, it follows that the sequence

$$(|\frac{1}{s_n} \int_0^{s_n} S_t f(x_k) \, dt - \frac{1}{s_n} \int_0^{s_n} S_t f(x) \, dt|)_{n \in \mathbb{N}}$$

converges to $|f_\alpha(x_k) - f_\alpha(x)|$.

We obtain that $|f_\alpha(x_k) - f_\alpha(x)| \le \frac{\varepsilon}{2} < \varepsilon$ because

$$\left| \frac{1}{s_n} \int_0^{s_n} S_t f(x_k) \, dt - \frac{1}{s_n} \int_0^{s_n} S_t f(x) \, dt \right| < \frac{\varepsilon}{2}$$

for every $n \in \mathbb{N}$.

Finally, in order to complete the proof of the proposition, we have to show that $S_u f_\alpha = f_\alpha$ for every $u \in [0, +\infty)$.

To this end, let $u \in [0, +\infty)$. We have to show that $S_u f_\alpha(x) = f_\alpha(x)$ for every $x \in X$.

Thus, let $x \in X$.

Using the Lebesgue dominated convergence theorem and Lemma 7.3.6, we obtain that

$$|S_u f_\alpha(x) - f_\alpha(x)| = |\langle S_u f_\alpha, \delta_x\rangle - \langle f_\alpha, \delta_x\rangle|$$

$$= |\langle f_\alpha, T_u\delta_x\rangle - \langle f_\alpha, \delta_x\rangle| = \left|\int_X f_\alpha(y)\,\mathrm{d}(T_u\delta_x)(y) - \int_X f_\alpha(y)\,\mathrm{d}\delta_x(y)\right|$$

$$= \left|\int_X \left(\lim_{n\to+\infty} \frac{1}{s_n}\int_0^{s_n} S_t f(y)\,\mathrm{d}t\right)\mathrm{d}(T_u\delta_x)(y)\right.$$

$$\left. - \int_X \left(\lim_{n\to+\infty} \frac{1}{s_n}\int_0^{s_n} S_t f(y)\,\mathrm{d}t\right)\mathrm{d}\delta_x(y)\right|$$

$$= \left|\lim_{n\to+\infty}\int_X \left(\frac{1}{s_n}\int_0^{s_n} S_t f(y)\,\mathrm{d}t\right)\mathrm{d}T_u\delta_x(y)\right.$$

$$\left. - \lim_{n\to+\infty}\int_X \left(\frac{1}{s_n}\int_0^{s_n} S_t f(y)\,\mathrm{d}t\right)\mathrm{d}\delta_x(y)\right|$$

$$= \lim_{n\to+\infty}\left|\int_X \frac{1}{s_n}\int_0^{s_n} S_t f(y)\,\mathrm{d}t\,\mathrm{d}(T_u\delta_x)(y) - \int_X \left(\frac{1}{s_n}\int_0^{s_n} S_t f(y)\,\mathrm{d}t\right)\mathrm{d}\delta_x(y)\right|$$

$$= \lim_{n\to+\infty}\left|\left\langle \frac{1}{s_n}\text{P-}\int_0^{s_n} S_t f\,\mathrm{d}t, T_u\delta_x\right\rangle - \left\langle \frac{1}{s_n}\text{P-}\int_0^{s_n} S_t f\,\mathrm{d}t, \delta_x\right\rangle\right|$$

$$= \lim_{n\to+\infty}\left|\left\langle \frac{1}{s_n}S_u\left(\text{P-}\int_0^{s_n} S_t f\,\mathrm{d}t\right), \delta_x\right\rangle - \left\langle \frac{1}{s_n}\text{P-}\int_0^{s_n} S_t f\,\mathrm{d}t, \delta_x\right\rangle\right|$$

$$= \lim_{\substack{n\to+\infty\\ s_n\geq u}} \frac{1}{s_n}\left|\left\langle \text{P-}\int_u^{s_n+u} S_t f\,\mathrm{d}t, \delta_x\right\rangle - \left\langle \text{P-}\int_0^{s_n} S_t f\,\mathrm{d}t, \delta_x\right\rangle\right|$$

$$= \lim_{\substack{n\to+\infty\\ s_n\geq u}} \frac{1}{s_n}\left|\left\langle \text{P-}\int_u^{s_n} S_t f\,\mathrm{d}t, \delta_x\right\rangle + \left\langle \text{P-}\int_{s_n}^{s_n+u} S_t f\,\mathrm{d}t, \delta_x\right\rangle\right.$$

$$\left. - \left\langle \text{P-}\int_0^u S_t f\,\mathrm{d}t, \delta_x\right\rangle - \left\langle \text{P-}\int_u^{s_n} S_t f\,\mathrm{d}t, \delta_x\right\rangle\right|$$

$$= \lim_{\substack{n\to+\infty\\ s_n\geq u}} \frac{1}{s_n}\left|\int_{s_n}^{s_n+u} S_t f(x)\,\mathrm{d}t - \int_0^u S_t f(x)\,\mathrm{d}t\right|$$

$$\leq \limsup_{\substack{n\to+\infty \\ s_n \geq u}} \left(\frac{1}{s_n} \left| \int_{s_n}^{s_n+u} S_t f(x)\, dt \right| + \frac{1}{s_n} \left| \int_0^u S_t f(x)\, dt \right| \right)$$

$$\leq \limsup_{\substack{n\to+\infty \\ s_n \geq u}} \left(\frac{1}{s_n} \int_{s_n}^{s_n+u} |S_t f(x)|\, dt + \frac{1}{s_n} \int_0^u |S_t f(x)|\, dt \right)$$

$$\leq \limsup_{\substack{n\to+\infty \\ s_n \geq u}} \left(\frac{1}{s_n} \int_{s_n}^{s_n+u} \|f\|\, dt + \frac{1}{s_n} \int_0^u \|f\|\, dt \right)$$

$$= \limsup_{\substack{n\to+\infty \\ s_n \geq u}} \frac{2u}{s_n} \|f\| = 0.$$

We have therefore proved that $S_u f_\alpha(x) = f_\alpha(x)$ for every $x \in X$.

Thus, $S_u f_\alpha = f_\alpha$. $\qquad\qquad\qquad\qquad\qquad\qquad\qquad\qquad\qquad\qquad\qquad\quad$ $\square$

Let $\alpha \in \mathcal{R}$, $\alpha = (s_n)_{n\in\mathbb{N}}$, and let $\mu \in \mathcal{M}(X)$, $\mu \geq 0$.

If $f \in C_0(X)$, then the sequences $(\frac{1}{s_n} \int_0^{s_n} S_t f(x)\, dt)_{n\in\mathbb{N}}$, $x \in X$, are all convergent. Using the Lebesgue dominated convergence theorem, we obtain that the sequence $((\frac{1}{s_n}(\text{P-}\int_0^{s_n} S_t f\, dt), \mu))_{n\in\mathbb{N}}$ converges (to $\langle f_\alpha, \mu\rangle$). Using Proposition 3.3.7, we obtain that $\langle \frac{1}{s_n}(\text{P-}\int_0^{s_n} S_t f\, dt), \mu\rangle = \langle f, \frac{1}{s_n}(\text{P-}\int_0^{s_n} T_t\mu\, dt)\rangle$ for every $n \in \mathbb{N}$.

Set $\mu_n = \frac{1}{s_n}(\text{P-}\int_0^{s_n} T_t\mu\, dt)$ for every $n \in \mathbb{N}$.

From the above comments it follows that $(\langle f, \mu_n\rangle)_{n\in\mathbb{N}}$ converges for every $f \in C_0(X)$. Thus, the sequence $(\mu_n)_{n\in\mathbb{N}}$ converges in the weak* topology of $\mathcal{M}(X)$. We will denote by μ_α the weak* limit of the sequence $(\mu_n)_{n\in\mathbb{N}}$.

Now let $\mu \in \mathcal{M}(X)$ be a not necessarily positive element of $\mathcal{M}(X)$. Then $\mu = \mu^+ - \mu^-$, where $\mu^+ = \mu \vee 0$ and $\mu^- = (-\mu) \vee 0$.

As before, let $\alpha = (s_n)_{n\in\mathbb{N}}$, $\alpha \in \mathcal{R}$.

Since

$$\frac{1}{s_n} \left(\text{P-}\int_0^{s_n} T_t\mu\, dt \right) = \frac{1}{s_n} \left(\text{P-}\int_0^{s_n} T_t(\mu^+)\, dt \right) - \frac{1}{s_n} \left(\text{P-}\int_0^{s_n} T_t(\mu^-)\, dt \right)$$

for every $n \in \mathbb{N}$, and since, by our previous remarks, the sequences $(\frac{1}{s_n}\text{P-}\int_0^{s_n} T_t(\mu^+)\, dt)_{n\in\mathbb{N}}$ and $(\frac{1}{s_n}\text{P-}\int_0^{s_n} T_t(\mu^-)\, dt)_{n\in\mathbb{N}}$ converge in the weak* topology of $\mathcal{M}(X)$ (because μ^+ and μ^- are positive elements of $\mathcal{M}(X)$), it follows that the sequence $(\frac{1}{s_n}(\text{P-}\int_0^{s_n} T_t\mu\, dt))_{n\in\mathbb{N}}$ converges in the weak* topology of $\mathcal{M}(X)$ to an element that will be denoted μ_α.

We obtain that $\mu_\alpha = (\mu^+)_\alpha - (\mu^-)_\alpha$.

The next proposition is an extension of Proposition 4.2.4 of [143] to the transition functions discussed here.

Proposition 7.3.8. *Let $\alpha \in \mathcal{R}$ and let $\mu \in \mathcal{M}(X)$. Then μ_α is an invariant element of $(P_t)_{t \in [0, +\infty)}$; that is, $T_{t_0}\mu_\alpha = \mu_\alpha$ for every $t_0 \in [0, +\infty)$.*

Proof. Let $\alpha \in \mathcal{R}$, $\alpha = (s_n)_{n \in \mathbb{N}}$, let $\mu \in \mathcal{M}(X)$, and let μ_α be the limit of the sequence $(\frac{1}{s_n}(\text{P-}\int_0^{s_n} T_t\mu\, dt))_{n \in \mathbb{N}}$ in the weak* topology of $\mathcal{M}(X)$ (the limit exists as shown before the proposition).

Since $\mu_\alpha = (\mu_\alpha)^+ - (\mu_\alpha)^-$ and since $(a\mu)_\alpha = a(\mu_\alpha)$ for every $a \in \mathbb{R}$, it follows that it is enough to prove the proposition under the assumption that μ is a probability measure.

Thus, assume that $\mu \in \mathcal{M}(X)$ is such that $\mu \geq 0$ and $\|\mu\| = 1$.

It is easy to see that in order to prove that $T_{t_0}\mu_\alpha = \mu_\alpha$ for every $t_0 \in [0, +\infty)$, it is enough to consider the case when $t_0 > 0$.

Thus, let $t_0 \in (0, +\infty)$. In order to prove that $T_{t_0}\mu_\alpha = \mu_\alpha$ we will use Theorem 1.2.3 (The Lasota-Yorke lemma; see also Theorem 1.2.4 of [143]).

Let L be a usual Banach limit (defined on sequences), and let $\phi : C_b(X) \to \mathbb{R}$ be defined by $\phi(f) = L(((\langle f, \frac{1}{s_n}(\text{P-}\int_0^{s_n} T_t\mu\, dt)\rangle))_{n \in \mathbb{N}})$ for every $f \in C_b(X)$.

We now note that ϕ is well-defined. Indeed, since $\mu \geq 0$, $\|\mu\| = 1$, it follows that P-$\int_0^{s_n} T_t\mu\, dt \geq 0$ for every $n \in \mathbb{N}$ and $T_t\mu(X) = 1$ for every $t \in [0, +\infty)$, so $\|\frac{1}{s_n}\text{P-}\int_0^{s_n} T_t\mu\, dt\| = \frac{1}{s_n}\int_0^{s_n} T_t\mu(X)\, dt = 1$ for every $n \in \mathbb{N}$, so the sequence $((\langle f, \frac{1}{s_n}(\text{P-}\int_0^{s_n} T_t\mu\, dt)\rangle))_{t \in \mathbb{N}}$ is bounded whenever $f \in C_b(X)$, so $L(((\langle f, \frac{1}{s_n}(\text{P-}\int_0^{s_n} T_t\mu\, dt)\rangle))_{n \in \mathbb{N}})$ exists; therefore, $\phi(f)$ is well-defined for every $f \in C_b(X)$.

Clearly, ϕ is a positive linear functional of $C_b(X)$.

Observe that the restriction of ϕ to $C_0(X)$ is μ_α because taking into consideration that $\alpha \in \mathcal{R}$ and using our discussion preceding this proposition, we obtain that the sequence $((\langle f, \frac{1}{s_n}(\text{P-}\int_0^{s_n} T_t\mu\, dt)\rangle))_{n \in \mathbb{N}}$ converges to $\langle f, \mu_\alpha\rangle$ for every $f \in C_0(X)$. Thus, $\phi(f) = \langle f, \mu_\alpha\rangle$ for every $f \in C_0(X)$.

In order to be able to use the Lasota-Yorke lemma to conclude that $T_{t_0}\mu_\alpha = \mu_\alpha$, it only remains to prove that $\phi(S_{t_0}f) = \phi(f)$ for every $f \in C_0(X)$.

We now note that it is enough to prove that $\phi(S_{t_0}f) = \phi(f)$ whenever $f \in C_0(X)$, $f \geq 0$. Indeed, if $g \in C_0(X)$, if g^+ and g^- are the positive and the negative parts of g, respectively, and if $\phi(S_{t_0}g^+) = \phi(g^+)$ and $\phi(S_{t_0}g^-) = \phi(g^-)$, then $\phi(S_0g) = \phi(g)$.

We also point out that if $f = 0$, then the equality $\phi(S_{t_0}f) = \phi(f)$ is obvious, so we may assume that $f \neq 0$.

Finally, note that we may assume that $\|f\| = 1$, because if $g \in C_0(X)$, $g \neq 0$, and if $\phi(S_{t_0}\frac{g}{\|g\|}) = \phi(\frac{g}{\|g\|})$, then $\phi(S_{t_0}g) = \phi(g)$.

Thus, let $f \in C_0(X)$ be such that $f \geq 0$ and $\|f\| = 1$.

We have to prove that $\phi(S_{t_0}f) = \phi(f)$.

Since

$$\phi(S_{t_0}f) - \phi(f) = \phi(S_{t_0}f - f) = L\left(\left(\left\langle S_{t_0}f - f, \frac{1}{s_n}\left(\text{P-}\int_0^{s_n} T_t\mu\, dt\right)\right\rangle\right)_{n \in \mathbb{N}}\right),$$

it follows that it is enough to prove that the sequence

$$\left(\left\langle S_{t_0} f - f, \frac{1}{s_n}\left(\text{P-}\int_0^{s_n} T_t \mu\, dt\right)\right\rangle\right)_{n\in\mathbb{N}}$$

converges to zero; that is, it is enough to prove that for every $\varepsilon \in \mathbb{R}$, $\varepsilon > 0$, there exists an $n_\varepsilon \in \mathbb{N}$ such that $\left|\left\langle S_{t_0} f - f, \frac{1}{s_n}(\text{P-}\int_0^{s_n} T_t \mu\, dt)\right\rangle\right| < \varepsilon$ for every $n \in \mathbb{N}$, $n \geq n_\varepsilon$.

To this end, let $\varepsilon \in \mathbb{R}$, $\varepsilon > 0$.

Since $(s_n)_{n\in\mathbb{N}} \in \mathcal{R}$, it follows that $(s_n)_{n\in\mathbb{N}}$ is a strictly increasing sequence of elements of $(0, +\infty)$ such that $(s_n)_{n\in\mathbb{N}}$ diverges to $+\infty$. Therefore, there exists an $n_\varepsilon \in \mathbb{N}$ such that $t_0 < s_{n_\varepsilon}$ and $\frac{2t_0}{s_{n_\varepsilon}} < \varepsilon$.

Using Proposition 3.3.7 and (b) of Lemma 7.3.6, we obtain that

$$\left|\left\langle S_{t_0} f - f, \frac{1}{s_n}\left(\text{P-}\int_0^{s_n} T_t \mu\, dt\right)\right\rangle\right|$$

$$= \left|\left\langle \frac{1}{s_n}\left(\text{P-}\int_0^{s_n} S_t(S_{t_0} f - f)\, dt\right), \mu\right\rangle\right|$$

$$= \left|\left\langle \frac{1}{s_n}\left(\text{P-}\int_0^{s_n} S_t(S_{t_0} f)\, dt\right) - \frac{1}{s_n}\left(\text{P-}\int_0^{s_n} S_t f\, dt\right), \mu\right\rangle\right|$$

$$= \left|\left\langle \frac{1}{s_n}\left(\text{P-}\int_0^{s_n} S_{t+t_0} f\, dt\right) - \frac{1}{s_n}\left(\text{P-}\int_0^{s_n} S_t f\, dt\right), \mu\right\rangle\right|$$

$$= \left|\left\langle \frac{1}{s_n}\left(\text{P-}\int_{t_0}^{t_0+s_n} S_t f\, dt\right) - \frac{1}{s_n}\left(\text{P-}\int_0^{s_n} S_t f\, dt\right), \mu\right\rangle\right|$$

$$= \left|\left\langle \frac{1}{s_n}\left(\left(\text{P-}\int_{t_0}^{s_n} S_t f\, dt\right) + \left(\text{P-}\int_{s_n}^{s_n+t_0} S_t f\, dt\right)\right.\right.\right.$$

$$\left.\left.\left. - \left(\text{P-}\int_0^{t_0} S_t f\, dt\right) - \left(\text{P-}\int_{t_0}^{s_n} S_t f\, dt\right)\right), \mu\right\rangle\right|$$

$$\leq \left|\left\langle \frac{1}{s_n}\left(\text{P-}\int_{s_n}^{s_n+t_0} S_t f\, dt\right), \mu\right\rangle\right| + \left|\left\langle \frac{1}{s_n}\left(\text{P-}\int_0^{t_0} S_t f\, dt\right), \mu\right\rangle\right|$$

$$= \left\langle \frac{1}{s_n}\left(\text{P-}\int_{s_n}^{s_n+t_0} S_t f\, dt\right), \mu\right\rangle + \left\langle \frac{1}{s_n}\left(\text{P-}\int_0^{t_0} S_t f\, dt\right), \mu\right\rangle$$

$$\leq \frac{1}{s_n}\int_X t_0\, d\mu(x) + \frac{1}{s_n}\int_X t_0\, d\mu(x) = \frac{2t_0}{s_n} < \varepsilon$$

for every $n \in \mathbb{N}$, $n \geq n_\varepsilon$.

Thus, $(\langle S_{t_0} f - f, \frac{1}{s_n}(\text{P-}\int_0^{s_n} T_t \mu\, dt)\rangle)_{n\in\mathbb{N}}$ converges to zero.

It follows that $\phi(S_{t_0} f) = \phi(f)$ for every $f \in C_0(X)$, so by using the Lasota-Yorke lemma, we obtain that $T_{t_0}\mu_\alpha = \mu_\alpha$ for every $t_0 \in [0, +\infty)$. $\square$

We are now in a position to state and prove a weak* mean ergodic theorem for transition functions, a theorem that extends Theorem 4.3.1 of [143] to transition functions.

Theorem 7.3.9 (Weak* Mean Ergodic Theorem for Transition Functions). *For every $\mu \in \mathcal{M}(X)$, the weak* limit $\lim_{s\to+\infty} w^* \frac{1}{s}(\text{P-}\int_0^s T_t\mu\, dt)$ exists and is an invariant element for $(T_t)_{t\in[0,+\infty)}$. If $\mu \geq 0$, then $\lim_{s\to+\infty} w^* \frac{1}{s}(\text{P-}\int_0^s T_t\mu\, dt)$ is also a positive element of $\mathcal{M}(X)$.*

Proof. Clearly, the weak* convergence and the other assertions in the theorem are true if $\mu = 0$. Thus, we may assume that $\mu \neq 0$.

If $\mu_1 \in \mathcal{M}(X)$ and $\mu_2 \in \mathcal{M}(X)$ are such that $\lim_{s\to+\infty} w^* \frac{1}{s}(\text{P-}\int_0^s T_t\mu_i\, dt)$ exist for every $i = 1$ or 2, and if $a_1 \in \mathbb{R}$ and $a_2 \in \mathbb{R}$, then, obviously, $\lim_{s\to+\infty} w^* \frac{1}{s}(\text{P-}\int_0^s T_t(a_1\mu_1 + a_2\mu_2)\, dt)$ exists and is equal to

$$a_1 \lim_{\substack{s\to+\infty \\ w^*}} \frac{1}{s}\left(\text{P-}\int_0^s T_t\mu_1\, dt\right) + a_2 \lim_{\substack{s\to+\infty \\ w^*}} \frac{1}{s}\left(\text{P-}\int_0^s T_t\mu_2\, dt\right).$$

In view of the above two observations, we obtain that it is enough to prove the theorem for $\mu \in \mathcal{M}(X)$, $\mu \geq 0$, $\|\mu\| = 1$.

Thus, let $\mu \in \mathcal{M}(X)$, $\mu \geq 0$, $\|\mu\| = 1$, and let $\mathcal{R}$ be the subset of $(0, +\infty)^{\mathbb{N}}$ defined before Proposition 7.3.7.

Using the discussion preceding Proposition 7.3.8, we obtain that for every $\alpha \in \mathcal{R}$, $\alpha = (s_n)_{n\in\mathbb{N}}$, the sequence $(\frac{1}{s_n}(\text{P-}\int_0^{s_n} T_t\mu\, dt))_{n\in\mathbb{N}}$ converges in the weak* topology of $\mathcal{M}(X)$ to an element $\mu_\alpha \in \mathcal{M}(X)$ (note that since $\mu \geq 0$ and in view of the definition of the pointwise integrals $\text{P-}\int_0^{s_n} T_t\mu\, dt$, $n \in \mathbb{N}$, we obtain that $\mu_\alpha \geq 0$).

By Proposition 7.3.2, for every sequence $(s_n)_{n\in\mathbb{N}}$ of elements of $[0, +\infty)$ that diverges to $+\infty$, there exists a subsequence $(s_{n_k})_{k\in\mathbb{N}}$ that belongs to $\mathcal{R}$. Thus, in view of Proposition 4.3.1 applied to the map $\varphi : [0, +\infty) \to \mathcal{M}(X)$, $\varphi(s) = \text{P-}\int_0^s T_t\mu\, dt$ for every $s \in [0, +\infty)$, in order to prove that $\lim_{s\to+\infty} w^* \frac{1}{s}(\text{P-}\int_0^s T_t\mu\, dt)$ exists, it is enough to prove that $\mu_\alpha = \mu_\beta$ for every $\alpha \in \mathcal{R}$ and $\beta \in \mathcal{R}$.

Thus, let $\alpha \in \mathcal{R}$, $\alpha = (s_n)_{n\in\mathbb{N}}$, and $\beta \in \mathcal{R}$, $\beta = (s'_n)_{n\in\mathbb{N}}$. We have to prove that $\mu_\alpha \leq \mu_\beta$ and that $\mu_\beta \leq \mu_\alpha$; however, it is enough to prove only that $\mu_\alpha \leq \mu_\beta$ because the proof of the fact that $\mu_\beta \leq \mu_\alpha$ is obtained by switching the roles of α and β in the proof of $\mu_\alpha \leq \mu_\beta$.

We now prove that $\mu_\alpha \leq \mu_\beta$.

To this end, let $\tilde{\mu}_\alpha : C_b(X) \to \mathbb{R}$ be defined by $\tilde{\mu}_\alpha(g) = \int g(x)\, d\mu_\alpha(x)$ for every $g \in C_b(X)$. Note that $\tilde{\mu}_\alpha$ is the standard extension of μ_α that was defined on p. 18 of [143] when discussing the Lasota-Yorke lemma.

Now let L be a usual Banach limit (defined on sequences) and let μ_α^* : $C_b(X) \to \mathbb{R}$ be defined by $\mu_\alpha^*(g) = L((\langle g, \frac{1}{s_n}(\text{P-}\int_0^{s_n} T_t\mu\, dt)\rangle)_{n\in\mathbb{N}})$ for every $g \in C_b(X)$. Note that since $\mu \in \mathcal{M}(X)$ is a probability measure, it follows that for every $n \in \mathbb{N}$, the pointwise integral $\text{P-}\int_0^{s_n} T_t\mu\, dt$ is a positive measure, so $||\text{P-}\int_0^{s_n} T_t\mu\, dt|| = (\text{P-}\int_0^{s_n} T_t\mu\, dt)(X) = \int_0^{s_n} T_t\mu(X)\, dt = \int_0^{s_n} dt = s_n$. Therefore, $\frac{1}{s_n}\int_0^{s_n} T_t\mu\, dt$ is a probability measure for every $n \in \mathbb{N}$; hence, the sequence $((\langle g, \frac{1}{s_n}(\text{P-}\int_0^{s_n} T_t\mu\, dt)\rangle)_{n\in\mathbb{N}}$ is a bounded sequence of real numbers, so $L(((\langle g, \frac{1}{s_n}(\text{P-}\int_0^{s_n} T_t\mu\, dt)\rangle)_{n\in\mathbb{N}})$ is well-defined for every $g \in C_b(X)$. Thus, μ_α^* is also well-defined.

It is easy to see that μ_α^* is a positive (hence bounded) linear functional.

In view of the definition of μ_α, we obtain that the restriction of μ_α^* to $C_0(X)$ is equal to μ_α. Therefore, using an argument that appears in the proof of Theorem 1.2.4 (The Lasota-Yorke lemma) of [143], we further obtain that $\tilde{\mu}_\alpha(g) \leq \mu_\alpha^*(g)$ for every $g \in C_b(X)$, $g \geq 0$.

Clearly, the proof of the inequality $\mu_\alpha \leq \mu_\beta$ will be completed if we show that $\langle f, \mu_\alpha \rangle \leq \langle f, \mu_\beta \rangle$ for every $f \in C_0(X)$, $f \geq 0$.

Thus, let $f \in C_0(X)$, $f \geq 0$.

By Proposition 7.3.8, we have $T_t\mu_\alpha = \mu_\alpha$ for every $t \in [0, +\infty)$, so $\text{P-}\int_0^{s_n'} T_t\mu_\alpha\, dt = s_n'\mu_\alpha$; hence, $\frac{1}{s_n'}(\text{P-}\int_0^{s_n'} T_t\mu_\alpha\, dt) = \mu_\alpha$ for every $n \in \mathbb{N}$.

Using Proposition 3.3.7, we obtain that

$$\langle f, \mu_\alpha \rangle = \langle f, \frac{1}{s_n'}(\text{P-}\int_0^{s_n'} T_t\mu_\alpha\, dt)\rangle = \langle \frac{1}{s_n'}(\text{P-}\int_0^{s_n'} S_t f\, dt), \mu_\alpha \rangle$$

for every $n \in \mathbb{N}$.

Since $\beta \in \mathcal{R}$ and since $f \in C_0(X)$, it follows that the sequence $(\frac{1}{s_n'}(\text{P-}\int_0^{s_n'} S_t f\, dt))_{n\in\mathbb{N}}$ converges pointwise (on X). Let $f_\beta = X \to \mathbb{R}$ be the pointwise limit of $(\frac{1}{s_n'}(\text{P-}\int_0^{s_n'} S_t f\, dt))_{n\in\mathbb{N}}$.

By Proposition 7.3.7, the function f_β belongs to $C_b(X)$.

Now note that

$$\left| \frac{1}{s_n'}\int_0^{s_n'} S_t f(x)\, dt \right| \leq \frac{1}{s_n'}\int_0^{s_n'} |S_t f(x)|\, dt \leq \frac{1}{s_n'}\int_0^{s_n'} ||f||\, dt = ||f||$$

for every $n \in \mathbb{N}$ and $x \in X$.

Since $|\frac{1}{s_n'}(\text{P-}\int_0^{s_n'} S_t f\, dt)| \leq ||f|| \mathbf{1}_X$ for every $n \in \mathbb{N}$, it follows that we can apply the Lebesgue dominated convergence theorem to the sequence $(\frac{1}{s_n'}(\text{P-}\int_0^{s_n'} S_t f\, dt))_{n\in\mathbb{N}}$ and to the function $||f|| \mathbf{1}_X$ (which is integrable with respect to μ_α) in order to obtain that the sequence $((\langle \frac{1}{s_n'}(\text{P-}\int_0^{s_n'} S_t f\, dt), \mu_\alpha \rangle)_{n\in\mathbb{N}}$ converges to $\langle f_\beta, \mu_\alpha \rangle = \tilde{\mu}_\alpha(f_\beta)$.

Using Proposition 3.3.7, we obtain that

$$\langle f, \mu_\alpha \rangle = \lim_{n \to +\infty} \left\langle \frac{1}{s'_n} \left(\text{P-} \int_0^{s'_n} S_t f \, dt \right), \mu_\alpha \right\rangle = \tilde{\mu}_\alpha(f_\beta)$$

$$\leq \mu_\alpha^*(f_\beta) = L\left(\left(\left\langle f_\beta, \frac{1}{s_n} \left(\text{P-} \int_0^{s_n} T_t \mu \, dt \right) \right\rangle \right)_{n \in \mathbb{N}} \right)$$

$$= L\left(\left(\left\langle \frac{1}{s_n} \left(\text{P-} \int_0^{s_n} S_t f_\beta \, dt \right), \mu \right\rangle \right)_{n \in \mathbb{N}} \right).$$

Now, since $S_t f_\beta = f_\beta$ for every $t \in [0, +\infty)$, we obtain that

$$L\left(\left(\left\langle \frac{1}{s_n} \left(\text{P-} \int_0^{s_n} S_t f_\beta \, dt \right), \mu \right\rangle \right)_{m \in \mathbb{N}} \right) = \langle f_\beta, \mu \rangle. \tag{7.3.1}$$

Since f_β is the pointwise limit of the sequence $(\frac{1}{s'_n}(\text{P-} \int_0^{s'_n} S_t f \, dt))_{n \in \mathbb{N}}$ and since we have that

$$\left| \frac{1}{s'_n} \int_0^{s'_n} S_t f(x) \, dt \right| = \frac{1}{s'_n} \int_0^{s'_n} S_t f(x) \, dt \leq \frac{1}{s'_n} \int_0^{s'_n} \|f\| \, dt = \|f\|$$

for every $n \in \mathbb{N}$ and $x \in X$, we obtain that $|\frac{1}{s'_n}(\text{P-} \int_0^{s'_n} S_t f \, dt)| \leq \|f\| \mathbf{1}_X$ for every $n \in \mathbb{N}$; therefore, $|f_\beta| \leq \|f\| \mathbf{1}_X$, as well. Thus, we can apply the Lebesgue dominated convergence theorem to the sequence $(\frac{1}{s'_n}(\text{P-} \int_0^{s'_n} S_t f \, dt))_{n \in \mathbb{N}}$, which converges to f_β and is uniformly bounded by the μ-integrable function $\|f\| \mathbf{1}_X$. We conclude that $(\langle \frac{1}{s'_n}(\text{P-} \int_0^{s'_n} S_t f \, dt), \mu \rangle)_{n \in \mathbb{N}}$ converges, and

$$\lim_{n \to +\infty} \left\langle \frac{1}{s'_n} \left(\text{P-} \int_0^{s'_n} S_t f \, dt \right), \mu \right\rangle = \langle f_\beta, \mu \rangle. \tag{7.3.2}$$

Using Proposition 3.3.7 in (7.3.2), we obtain that

$$\lim_{n \to +\infty} \left\langle f, \frac{1}{s'_n} \left(\text{P-} \int_0^{s'_n} T_t \mu \, dt \right) \right\rangle = \langle f_\beta, \mu \rangle. \tag{7.3.3}$$

Since $\beta \in \mathcal{R}$, we obtain that the sequence $(\frac{1}{s'_n}(\text{P-} \int_0^{s'_n} T_t \mu \, dt))_{n \in \mathbb{N}}$ converges in the weak* topology of $\mathcal{M}(X)$ to μ_β, so since $f \in C_0(X)$, it follows that

$$\lim_{n \to +\infty} \left\langle f, \frac{1}{s'_n} \left(\text{P-} \int_0^{s'_n} T_t \mu \, dt \right) \right\rangle = \langle f, \mu_\beta \rangle. \tag{7.3.4}$$

Using the fact that

$$\langle f, \mu_\alpha \rangle \le L\left(\left(\left|\frac{1}{s_n}\left(\text{P-}\int_0^{s_n} S_t f_\beta \, dt\right), \mu\right\rangle\right)_{m\in\mathbb{N}}\right)$$

(proved before the equality (7.3.1)) and using the equalities (7.3.1)–(7.3.4), we obtain that $\mu_\alpha \le \mu_\beta$ for every $\alpha \in \mathcal{R}$ and $\beta \in \mathcal{R}$. Thus, $\mu_\alpha = \mu_\beta$ for every $\alpha \in \mathcal{R}$ and $\beta \in \mathcal{R}$.

We obtain that $\lim_{s\to+\infty} w^* \frac{1}{s}(\text{P-}\int_0^s T_t \mu \, dt)$ exists for every $\mu \in \mathcal{M}(X)$.

Note that using Proposition 4.3.1 and the discussion preceding Proposition 7.3.8, we obtain that the set $\mathcal{R}$ is nonempty. We also obtain that under the conditions of Theorem 7.3.9 (the weak* mean ergodic theorem for transition functions) the set $\mathcal{R}$ is the collection of all strictly increasing sequences that tend to $+\infty$.

Let us fix such a sequence α in $\mathcal{R}$ and let $\mu \in \mathcal{M}(X)$. Then, from the proof of Proposition 4.3.1, we obtain that $\mu_\alpha = \mu^*$. By Proposition 7.3.8, μ^* is an invariant element for $(T_t)_{t\in[0,+\infty)}$.

Finally, if $\mu \in \mathcal{M}(X)$ is such that $\mu \ge 0$, then it is obvious from the definition of the weak* limit $\lim_{s\to+\infty} w^* \frac{1}{s}(\text{P-}\int_0^s T_t \mu \, dt)$ that

$$\mu^* = \lim_{\substack{s\to+\infty \\ w^*}} \frac{1}{s}\left(\text{P-}\int_0^s T_t \mu \, dt\right)$$

is also a positive element of $\mathcal{M}(X)$. $\qquad\qquad\square$

Theorem 7.3.9 has the following consequence:

Corollary 7.3.10 (Pointwise Mean Ergodic Theorem for Transition Functions).
For every $f \in C_0(X)$ and every $x \in X$, the limit

$$\lim_{s\to+\infty} \frac{1}{s}\int_0^s S_t f(x) \, dt$$

exists and is a real number.

Proof. By Theorem 7.3.9, for every $x \in X$, the weak* limit $\lim_{s\to+\infty} w^* \frac{1}{s}\text{P-}\int_0^s T_t \delta_x \, dt$ exists and is an element of $\mathcal{M}(X)$; that is, for every $f \in C_0(X)$ and every $x \in X$, the limit $\lim_{s\to+\infty}\langle f, \frac{1}{s}(\text{P-}\int_0^s T_t \delta_x \, dt)\rangle$ exists and is a real number.

Since (by Proposition 3.3.7) $\langle f, \frac{1}{s}(\text{P-}\int_0^s T_t \delta_x \, dt)\rangle = \langle \frac{1}{s}(\text{P-}\int_0^s S_t f \, dt), \delta_x \rangle$ for every $x \in X$, every $s \in (0, +\infty)$, and every $f \in C_0(X)$, it follows that $\lim_{s\to+\infty} \frac{1}{s}\int_0^s S_t f(x) \, dt$ exists and is a real number for every $x \in X$ and $f \in C_0(X)$; that is, the family of functions $(\frac{1}{s}(\text{P-}\int_0^s S_t f \, dt))_{s\in(0,+\infty)}$ converges pointwise (everywhere) on X as $s \to +\infty$. $\qquad\square$

We will conclude the section with several examples.

Example 7.3.11 (The Mean Ergodic Theorems for the Transition Function of the Rotations of the Unit Circle). Let, as usual, $\mathbb{R}/\mathbb{Z}$ be the unit circle, let $\mathbf{w} = (w_t)_{t\in\mathbb{R}}$

be the flow of the rotations of the unit circle, let $(P_t^{(\mathbf{w})})_{t \in \mathbb{R}}$ be the transition function defined by $\mathbf{w}$, and let $(P_t^{(\mathbf{w})})_{t \in [0,+\infty)}$ be the restriction of the transition function $(P_t^{(\mathbf{w})})_{t \in \mathbb{R}}$ to $[0, +\infty)$.

Using the discussion of $(P_t^{(\mathbf{w})})_{t \in \mathbb{R}}$ in Example 2.2.4, we obtain that $(P_t^{(\mathbf{w})})_{t \in [0,+\infty)}$ is an equicontinuous Feller transition function that satisfies the s.m.a. and is pointwise continuous. Using the discussion of $(P_t^{(\mathbf{w})})_{t \in \mathbb{R}}$ in Example 7.3.4, we obtain that $(P_t^{(\mathbf{w})})_{t \in [0,+\infty)}$ is also $C_0(\mathbb{R}/\mathbb{Z})$-strongly continuous. Thus, both Theorem 7.3.9 (The Weak* Mean Ergodic Theorem for Transition Functions) and Corollary 7.3.10 (The Pointwise Mean Ergodic Theorem for Transition Functions) hold true for $(P_t^{\mathbf{w}})_{t \in [0,+\infty)}$. ∎

Example 7.3.12 (The Mean Ergodic Theorems for the Transition Function of the Rectilinear Flow on the Torus). Let $n \in \mathbb{N}, n \geq 2$, let $\mathbf{v} \in \mathbb{R}^n$, $\mathbf{v} = (v_1, v_2, \ldots, v_n)$, let $\mathbf{w} = (w_t)_{t \in \mathbb{R}}$ be the rectilinear flow on the n-dimensional torus $\mathbb{R}^n/\mathbb{Z}^n$, let $(P_t^{(\mathbf{w})})_{t \in \mathbb{R}}$ be the transition function defined by $\mathbf{w}$, and let $(P_t^{(\mathbf{w})})_{t \in [0,+\infty)}$ be the restriction to $[0, +\infty)$ of $(P_t^{(\mathbf{w})})_{t \in \mathbb{R}}$.

Using the study of $(P_t^{(\mathbf{w})})_{t \in \mathbb{R}}$ in Example 2.2.5, we obtain that the transition function $(P_t^{(\mathbf{w})})_{t \in [0,+\infty)}$ is Feller, is equicontinuous, satisfies the s.m.a., and is pointwise continuous. Using Example 7.3.5, we further obtain that $(P_t^{(\mathbf{w})})_{t \in [0,+\infty)}$ is $C_0(\mathbb{R}^n/\mathbb{Z}^n)$-strongly continuous.

Thus, $(P_t^{(\mathbf{w})})_{t \in [0,+\infty)}$ satisfies the weak* and pointwise mean ergodic theorems discussed in Theorem 7.3.9 and Corollary 7.3.10, respectively. ∎

Example 7.3.13 (The Mean Ergodic Theorems for the Transition Functions Defined by the Semiflows $\pi^{(A)}$). Let $n \in \mathbb{N}, n \geq 2$, and let $\mathbb{S}_n$ be the compact metrizable semigroup of all $n \times n$ stochastic matrices (see Proposition B.4.1 for details).

Let $A \in \mathbb{S}_n$, and let $\boldsymbol{\pi}^{(A)} = (\pi_t^{(A)})_{t \in [0,+\infty)}$ be the semiflow defined before Proposition B.4.6. Also, let $(P_t^{(\pi^{(A)})})_{t \in [0,+\infty)}$ be the transition function defined by $\boldsymbol{\pi}^{(A)}$ (see (a) of Example 2.2.9). It is shown in (a) of Example 2.2.9 that $(P_t^{\pi^{(A)}})_{t \in [0,+\infty)}$ is a Feller transition function that satisfies the s.m.a., and is pointwise continuous and equicontinuous.

Since $\pi^{(A)}$ is continuous with respect to $t \in [0, +\infty)$, uniformly in $X \in \mathbb{S}_n$ (by Proposition B.4.6), and since $\mathbb{S}_n$ is compact, it follows that $(P_t^{(\pi^{(A)})})_{t \in [0,+\infty)}$ is also $C_0(\mathbb{S}_n)$-strongly continuous.

Therefore, $(P_t^{(\pi^{(A)})})_{t \in [0,+\infty)}$ satisfies the mean ergodic theorems stated in Theorem 7.3.9 and Corollary 7.3.10. ∎

Example 7.3.14 (The Mean Ergodic Theorems for the Transition Functions Defined by the Semiflows $\varphi^{(A)}$). Let $n \in \mathbb{N}, n \geq 2$, let $\mathbb{S}_n$ be the compact semigroup of all $n \times n$ stochastic matrices and let $\mathbb{P}_n$ be the set of all column stochastic vectors with n entries. Let $A \in \mathbb{S}_n$ and let $\varphi^{(A)} = (\varphi_t^{(A)})_{t \in [0,+\infty)}$ be the semiflow on $\mathbb{P}^n$ defined before Proposition B.4.7.

Now let $(P_t^{(\varphi^{(A)})})_{t \in [0,+\infty)}$ be the transition function defined by $\varphi^{(A)}$ (see (b) of Example 2.2.9).

It is shown at (b) of Example 2.2.9 that $(P_t^{(\varphi^{(A)})})_{t\in[0,+\infty)}$ is an equicontinuous transition function that satisfies the s.m.a., is pointwise continuous, and is also Feller.

Since (by Proposition B.4.7) the semiflow $\varphi^{(A)}$ is continuous with respect to $t \in [0, +\infty)$ uniformly in $\mathbf{x} \in \mathbb{P}_n$ and since $\mathbb{P}_n$ is a compact space, it follows that $(P_t^{(\varphi^{(A)})})_{t\in[0,+\infty)}$ is $C_0(\mathbb{P}_n)$-strongly continuous.

Therefore, $(P_t^{(\varphi^{(A)})})_{t\in[0,+\infty)}$ satisfies Theorem 7.3.9 and Corollary 7.3.10.   ∎

Example 7.3.15 (The Mean Ergodic Theorems for the Transition Functions Defined by Exponential One-Parameter Convolution Semigroups of Probability Measures). Let $(H, \cdot, d)$ be a locally compact separable metric semigroup, and assume that H has a neutral element.

Let $\mu \in \mathcal{M}(H)$ be an equicontinuous probability measure, let $(\mu_t)_{t\in[0,+\infty)}$ be the exponential one-parameter convolution semigroup of probability measures defined by μ, and let $(P_t)_{t\in[0,+\infty)}$ be the transition function defined by $(\mu_t)_{t\in[0,+\infty)}$. As pointed out in Sect. 2.2.3, $(P_t)_{t\in[0,+\infty)}$ is a Feller transition function. By Proposition 2.2.13, $(P_t)_{t\in[0,+\infty)}$ satisfies the s.m.a. and is $C_0(H)$-pointwise continuous. By Proposition 2.2.15, $(P_t)_{t\in[0,+\infty)}$ is $C_0(H)$-equicontinuous.

We now show that $(P_t)_{t\in[0,+\infty)}$ is $C_0(H)$-strongly continuous.

Thus, we have to show that for every convergent sequence $(t_n)_{n\in\mathbb{N}}$ of elements of $[0, +\infty)$ we have that $(\|S_{t_n} f - S_t f\|)_{n\in\mathbb{N}}$ converges to zero, whenever $f \in C_0(H)$, where $t = \lim_{n\to+\infty} t_n$.

To this end, let $f \in C_0(H)$, let $(t_n)_{n\in\mathbb{N}}$ be a convergent sequence of elements of $[0, +\infty)$, and set $t = \lim_{n\to+\infty} t_n$. We have to show that for every $\varepsilon \in \mathbb{R}, \varepsilon > 0$, there exists an $n_\varepsilon \in \mathbb{N}$ such that $\|S_{t_n} f - S_t f\| < \varepsilon$ for every $n \in \mathbb{N}, n \geq n_\varepsilon$.

Clearly, we may assume that $f \neq 0$ because if $f = 0$, then $S_u f = 0$ for every $u \in [0, +\infty)$.

Thus, assume that $f \neq 0$ and let $\varepsilon \in \mathbb{R}, \varepsilon > 0$.

By Proposition B.2.10, the map $u \mapsto \mu_u, u \in [0, +\infty)$, is continuous with respect to the norm topology of $M(H)$ and the standard topology of $[0, +\infty)$. Therefore, there exists an $n_\varepsilon \in \mathbb{N}$ such that $\|\mu_{t_n} - \mu_t\| < \frac{\varepsilon}{2\|f\|}$ for every $n \in \mathbb{N}, n \geq n_\varepsilon$.

It follows that

$$\begin{aligned}
|S_{t_n} f(x) - S_t f(x)| &= |\langle S_{t_n} f, \delta_x\rangle - \langle S_t f, \delta_x\rangle| \\
&= |\langle f, T_{t_n}\delta_x\rangle - \langle f, T_t\delta_x\rangle| = |\langle f, (T_{t_n} - T_t)\delta_x\rangle| \\
&= |\langle f, (\mu_{t_n} - \mu_t) * \delta_x\rangle| \leq \|f\|\|(\mu_{t_n} - \mu_t) * \delta_x\| \\
&\leq \|f\|\|\mu_{t_n} - \mu_t\|\|\delta_x\| = \|f\|\|\mu_{t_n} - \mu_t\| < \|f\|\frac{\varepsilon}{2\|f\|} \\
&= \frac{\varepsilon}{2}
\end{aligned}$$

for every $x \in H$, and every $n \in \mathbb{N}, n \geq n_\varepsilon$.

Accordingly, $\|S_{t_n} f - S_t f\| \leq \frac{\varepsilon}{2} < \varepsilon$ for every $n \in \mathbb{N}, n \geq n_\varepsilon$.

We have therefore proved that $(P_t)_{t\in[0,+\infty)}$ is $C_0(H)$-strongly continuous.

Accordingly, $(P_t)_{t\in[0,+\infty)}$ satisfies Theorem 7.3.9 and Corollary 7.3.10.   ∎

Appendix A
Semiflows and Flows: The Algebraic and Topological Setting, and First Examples

Generally, most of (but not all) the examples that we use in order to illustrate the results discussed in this book are transition functions defined by semiflows, flows, or one-parameter convolution semigroups of probability measures. In this appendix (Appendix A) we start by briefly discussing several facts about semigroups, groups and coset spaces in Sect. A.1. Next, in Sect. A.2, we outline natural topologies that can be defined on these algebraic structures. Finally, in Sect. A.3, we discuss actions, semiflows and flows: their definitions, some of their basic properties, and a few examples.

In the appendix we will use various general notations established elsewhere in the work. For instance, if (X, d) is a locally compact separable metric space (which may or may not be a semigroup, a group or a coset space), $\mathcal{B}(X)$ will denote the σ-algebra of all the Borel subsets of X.

Given a nonempty set A, the identity map from A onto A will be denoted by Id_A; that is, $\mathrm{Id}_A : A \to A$ is the function defined by $\mathrm{Id}_A(x) = x$ for every $x \in A$.

If A and B are nonempty sets, and $f : A \to B$ is a function, we will use the subscript notation f_x for $f(x)$, $x \in A$, whenever convenient. It will also be convenient from time to time to denote the function f by $(f_x)_{x \in A}$.

A.1 Semigroups, Groups and Coset Spaces

Let H be a nonempty set, and let $\cdot$ be a binary operation on H. If the operation $\cdot$ is associative, we call the ordered pair $(H, \cdot)$ a *semigroup*. The algebraic operation $\cdot$ is called *multiplication*.

A semigroup $(H, \cdot)$ is called *commutative* or *abelian* if $a \cdot b = b \cdot a$ for every $a \in H$ and $b \in H$.

Often, if $(H, \cdot)$ is a commutative semigroup, the algebraic operation $\cdot$ is denoted by $+$ and is called *addition*.

R. Zaharopol, *Invariant Probabilities of Transition Functions*, Probability and Its Applications 44, DOI 10.1007/978-3-319-05723-1,
© Springer International Publishing Switzerland 2014

If, in multiplicative or additive notation, an algebraic operation defines a semigroup structure on a nonempty set H, and if the operation is well-understood from the context (and usually this is indeed so), we will simply refer to H as the semigroup $(H, \cdot)$ or $(H, +)$, respectively.

When dealing with a semigroup $(H, \cdot)$, we will use the following notation:

$$ab = a \cdot b, \quad a \in H, \quad b \in H;$$

$$a^n = \underbrace{aaa \cdots a}_{n \text{ times}}, \quad a \in H, \quad n \in \mathbb{N}; \tag{A.1.1}$$

$$aA = \{y \in H \mid y = ax \text{ for some } x \in A\}, \quad a \in H, \quad A \subseteq H; \tag{A.1.2}$$

$$Aa = \{y \in H \mid y = xa \text{ for some } x \in A\}, \quad a \in H, \quad A \subseteq H; \tag{A.1.3}$$

$$a^{-1}A = \{x \in H \mid ax \in A\}, \ Aa^{-1} = \{x \in H \mid xa \in A\}, \ a \in H, \ A \subseteq H; \tag{A.1.4}$$

$$AB = \{z \in H \mid z = xy \text{ for some } x \in A \text{ and } y \in B\}, \ A \subseteq H, \ B \subseteq H; \tag{A.1.5}$$

$$A^n = \underbrace{AAA \cdots A}_{n \text{ times}}, \quad A \subseteq H. \tag{A.1.6}$$

If the algebraic operation on a (commutative) semigroup H is given in additive notation, then the above expressions (A.1.1)–(A.1.6) become, respectively,

$$na = \underbrace{a + a + a + \cdots + a}_{n \text{ times}}, \quad a \in H, \quad n \in \mathbb{N};$$

$$a + A = \{y \in H \mid y = a + x \text{ for some } x \in A\}, \quad a \in H, \quad A \subseteq H;$$

$$A + a = \{y \in H \mid y = x + a \text{ for some } x \in A\}, \quad a \in H, \quad A \subseteq H;$$

$$-a + A = \{x \in H \mid a + x \in A\}, \quad A - a = \{x \in H \mid x + a \in A\}, \quad a \in H, \quad A \subseteq H;$$

$$A + B = \{z \in H \mid z = x + y \text{ for some } x \in A \text{ and } y \in B\}, \quad A \subseteq H, \quad B \subseteq H;$$

$$nA = \underbrace{A + A + A + \cdots + A}_{n \text{ times}}, \quad A \subseteq H.$$

Note that since we assume that H is a commutative semigroup, it follows that $a + A = A + a$ and $-a + A = A - a$ for every $a \in H$ and $A \subseteq H$.

Let H' and H'' be two semigroups (not necessarily distinct).

A function $f : H' \to H''$ is called a *semigroup homomorphism* if

$$f(xy) = f(x)f(y) \tag{A.1.7}$$

for every $x \in H'$ and $y \in H'$.

Let $f : H' \to H''$ be a semigroup homomorphism. It will be convenient, at times, to use the subscript notation f_x for $f(x)$, $x \in H'$, and if we do so, the above equality (A.1.7) becomes

$$f_{xy} = f_x f_y$$

for every $x \in H'$ and $y \in H'$.

If H' is the additive semigroup $[0, +\infty)$, then f is often called a *one-parameter semigroup* of elements of H''; in this case, we use the notation $(f_t)_{t \in [0,+\infty)}$ for f, and the subscript notation f_t for $f(t)$, $t \in [0, +\infty)$, is preferred.

If $f : H' \to H''$ is a one-to-one semigroup homomorphism onto H'', then we call f a *semigroup isomorphism*. Note that if f is a semigroup isomorphism, then the inverse function f^{-1} of f (which exists because we assume that f is a bijection) is again a semigroup isomorphism. The semigroups H' and H'' are said to be *semigroup isomorphic* if there exists a semigroup isomorphism $g : H' \to H''$.

Let $(H, \cdot)$ be a semigroup.

An element $e \in H$ is called a *neutral element* of H if $ea = ae = a$ for every $a \in H$. If H has a neutral element, then the neutral element is unique, and we will often (but not always) denote this element by e or 1 (if the algebraic operation on H is given in additive notation, the neutral element will often (but, again, not always) be denoted by 0).

It is important to stress that if H' and H'' are two semigroups (not necessarily distinct) that have neutral elements e' and e'', respectively, and if $f : H' \to H''$ is a semigroup homomorphism, we cannot infer that $f(e') = e''$; that is, it may well happen that $f(e')$ and e'' are distinct. In particular, if H' is the additive semigroup $[0, +\infty)$ (so, $e' = 0$ in this case), and if $(f_t)_{t \in [0,+\infty)}$ is a one-parameter semigroup of elements of H'', it can happen that $f_0 \neq e''$. An instance of such a one-parameter semigroup $(f_t)_{t \in [0,+\infty)}$ of elements of H'' which has the property that $f_0 \neq e''$ is discussed in Sect. 2.2.

An element $z \in H$ is called a *left (right) zero element* if $za = z$ $(az = z)$ for every $a \in H$. A *two-sided zero element*, or, simply, a *zero* of H is an element $z \in H$ which is both a left zero and a right zero of H. If H has a zero, then the zero is unique. Even though a (two-sided) zero will sometimes be denoted by 0, there will be no danger of confusion because the context will clearly indicate if 0 stands for a zero or for a neutral element of a commutative semigroup whose algebraic operation is given in additive notation.

An element $z \in H$ is called a *left (right) zeroid* if for every $a \in H$ there exists an $x \in H$ such that $xa = z$ $(ax = z)$. Thus, $z \in H$ is a left (right) zeroid if and only if $z \in \bigcap_{a \in H} Ha$ $(z \in \bigcap_{a \in H} aH)$. An element $z \in H$ which is both a left and a right zeroid of H is called a *two-sided zeroid*, or, simply, a *zeroid* of H. The semigroup H is called *left (right) simple* if all its elements are left (right) zeroids. For additional details on zeroids, see Clifford and Miller [19].

Let H be a semigroup that has a neutral element, let $e \in H$ be its neutral element, and let $a \in H$. We say that a has an *inverse* if there exists a $b \in H$ such that

$ab = ba = e$. If a has an inverse, then the inverse is unique. If the algebraic operation on H is given in multiplicative notation, then the inverse of an element $a \in H$ is denoted by a^{-1}, and we denote by $-a$ the inverse of $a \in H$ whenever the algebraic operation on H is given in additive notation (provided, in both cases, that the inverse exists, of course).

Let G be a nonempty set, and let $\cdot$ be a binary operation on G. As usual, we say that $(G, \cdot)$ is a *group* if $(G, \cdot)$ is a semigroup that has a neutral element, and if every $a \in G$ has an inverse.

A group $(G, \cdot)$ is said to be *commutative* or *abelian* if $(G, \cdot)$, thought of as a semigroup, is commutative. As in the case of semigroups, if a group is commutative, we will often use the additive rather than the multiplicative notation for the algebraic operation that defines the group structure. Also, if $(G, \cdot)$ (or $(G, +)$) is a group and the algebraic operation is clearly understood from the context, we will refer to G as the group $(G, \cdot)$ (or $(G, +)$).

It can be shown (see Proposition 1.1, p. 3 of Högnäs and Mukherjea's monograph [48]) that a semigroup H is a group if and only if all its elements are (two-sided) zeroids (that is, if and only if H is (left and right) simple).

Let G_1 and G_2 be two groups.

A function $f : G_1 \rightarrow G_2$ is called a *group homomorphism* if

$$f(xy) = f(x)f(y) \tag{A.1.8}$$

for every $x \in G_1$ and $y \in G_1$.

If $f : G_1 \rightarrow G_2$ is a group homomorphism, then

$$f(e_1) = e_2, \tag{A.1.9}$$

where e_1 and e_2 are the neutral elements of G_1 and G_2, respectively, and

$$f(x^{-1}) = (f(x))^{-1} \tag{A.1.10}$$

whenever $x \in G_1$.

In subscript notation (that is, if we use the notation f_z for $f(z)$, $z \in G_1$), the above equalities (A.1.8)–(A.1.10) become $f_{xy} = f_x f_y$, $f_{e_1} = e_2$, and $f_{x^{-1}} = (f_x)^{-1}$, respectively.

If G_1 is the additive group $\mathbb{R}$, and $f : \mathbb{R} \rightarrow G_2$ is a group homomorphism, then we refer to f as a *one-parameter group* of elements of G_2; in this case, we prefer the notations $(f_t)_{t \in \mathbb{R}}$ and f_s rather than f and $f(s)$, respectively, $s \in \mathbb{R}$. We will use this terminology and notation even when G_2 is only a semigroup rather than a group and, of course, $f(s + t) = f(s)f(t)$ for every $s \in \mathbb{R}$ and $t \in \mathbb{R}$ (the reason for using the same terminology and notation is that, in this case, the restriction of the algebraic operation of G_2 to $f(\mathbb{R}) \times f(\mathbb{R})$ is a well-defined algebraic operation on $f(\mathbb{R})$ that generates a group structure on $f(\mathbb{R})$ with neutral element $f(0)$, and the function f is a group homomorphism if we think of it as a map from $\mathbb{R}$ onto $f(\mathbb{R})$).

A group homomorphism which is also a bijective function is called a *group isomorphism*. The groups G_1 and G_2 are said to be *group isomorphic* if there exists a group isomorphism $g : G_1 \to G_2$.

Let $(H, \cdot)$ be a semigroup. A nonempty subset H_1 of H is called a *subsemigroup* of H if $H_1 H_1 \subseteq H_1$; that is, H_1 is a subsemigroup of H if and only if $ab \in H_1$ whenever $a \in H_1$ and $b \in H_1$. Thus, if H_1 is a subsemigroup of H, the range of the restriction of the binary operation $\cdot$ to $H_1 \times H_1$ is included in H_1, so we may and do think of $\cdot$ as a binary operation on H_1, and it is obvious that H_1 becomes a semigroup in its own right when endowed with $\cdot$, thought of as an operation on H_1; the restriction of $\cdot$ to $H_1 \times H_1$ is referred to as the binary operation on H_1 *induced* by (or *inherited* from) the binary operation on H. Note that H could be a group; that is, we may (and sometimes we do) consider subsemigroups of groups.

A nonempty subset M of H is called a *subgroup* of H if M is a subsemigroup of H, and if the binary operation on M inherited from H defines a group structure on M. Although we will deal mostly with the case when H itself is a group (that is, we will consider subgroups of groups most of the time), there are instances when subgroups of semigroups that are not necessarily groups are of interest. For example, it can be shown (see Clifford and Miller [19]) that the set of all the zeroids of a semigroup H is either empty, or else, it is a subgroup of H.

Let $(G, \cdot)$ be a group, and let M be a subgroup of G.

A *left coset* of M in G is a set of the form xM for some $x \in G$. An element of G that belongs to a left coset xM for some $x \in G$ is called a *left coset representative* of xM. It is easy to see that if $x \in G$ and $y \in G$, then either $xM = yM$, or else $(xM) \cap (yM) = \emptyset$. Since $G = \bigcup_{x \in G} xM$, it follows that the collection of all left cosets of M in G is a partition of G. The set of all (distinct) left cosets of M in G is denoted by $(G/M)_{\mathrm{L}}$ and is called the *left coset space* of G defined by M.

If we define the relation $\sim_{\mathrm{L}} \subseteq G \times G$ by $(x, y) \in \sim_{\mathrm{L}}$ (or, using the standard notation for relations, $x \sim_{\mathrm{L}} y$) if and only if $x = ya$ for some $a \in M$ whenever $x \in G$ and $y \in G$, then it is easy to see that $\sim_{\mathrm{L}}$ is an equivalence relation on G, and that the set of all equivalence classes defined by $\sim_{\mathrm{L}}$ is precisely the left coset space $(G/M)_{\mathrm{L}}$. If $x \in G$, we will often denote by $\hat{x}$ the equivalence class xM. This alternative "hat" notation, which is quite often used when dealing with equivalence classes, will allow us to simplify the writing of many expressions.

A *right coset* of M in G is a set of the form Mx for some $x \in G$, and an element of G that belongs to Mx is called a *right coset representative* of Mx. In a similar way as in the case of left cosets, it can be shown that the collection of all the right cosets of M in G is a partition of G. The set of all right cosets of M in G is denoted by $(G/M)_{\mathrm{R}}$ and is called the *right coset space* of G defined by M. The relation $\sim_{\mathrm{R}} \subseteq G \times G$ defined by $x \sim_{\mathrm{R}} y$ (that is, $(x, y) \in \sim_{\mathrm{R}}$) if and only if $x = ay$ for some $a \in M$ is an equivalence relation, and a subset A of G is an equivalence class for $\sim_{\mathrm{R}}$ if and only if A is a right coset of M in G (that is, if and only if $A = Mx$ for some $x \in G$). As in the case of left cosets, we will often denote by $\hat{x}$ the right coset Mx. Note that in order to distinguish a left coset from a right coset in "hat" notation, it would be better to use $\hat{x}^{\mathrm{L}}$ and $\hat{x}^{\mathrm{R}}$ for xM and Mx, respectively.

However, in order to make various expressions easier to grasp throughout the book, we will use the notation $\hat{x}$ for both xM and Mx since it will be clear from the context whether we are considering a left coset or a right coset.

For subsets of the coset spaces $(G/M)_\mathrm{L}$ and $(G/M)_\mathrm{R}$, the following notation will be used throughout the book:

$$xA = \{xyM \mid yM \in A\} \quad \text{whenever} \quad A \subseteq (G/M)_\mathrm{L} \quad \text{and} \quad x \in G,$$

and

$$Ax = \{Myx \mid My \in A\} \quad \text{whenever} \quad A \subseteq (G/M)_\mathrm{R} \quad \text{and} \quad x \in G.$$

When dealing with the coset spaces $(G/M)_\mathrm{L}$ and $(G/M)_\mathrm{R}$, the maps

$$\varphi_\mathrm{L} : G \to (G/M)_\mathrm{L}, \quad \varphi_\mathrm{L}(x) = xM \text{ for every } x \in G$$

and

$$\varphi_\mathrm{R} : G \to (G/M)_\mathrm{R}, \quad \varphi_\mathrm{R}(x) = Mx \text{ for every } x \in G$$

turn out to be quite useful for many purposes. It is easy to see that φ_L and φ_R are onto $(G/M)_\mathrm{L}$ and $(G/M)_\mathrm{R}$, and, therefore, they are called the *standard surjections* on $(G/M)_\mathrm{L}$ and $(G/M)_\mathrm{R}$, respectively.

As usual, we say that M is a *normal subgroup* of G if $xM = Mx$ (or, equivalently, if $xMx^{-1} = M$) for every $x \in G$. Note that if G is a commutative group, any subgroup of G is normal.

Let N be a normal subgroup of a (not necessarily commutative) group G. Then each left coset is a right coset and vice versa; therefore, the left coset space $(G/N)_\mathrm{L}$ is equal to the right coset space $(G/N)_\mathrm{R}$. In this case, we simply call $(G/N)_\mathrm{L}$ or $(G/N)_\mathrm{R}$ the *coset space* of G defined by N, and we denote it by G/N. One of the main features of the coset space G/N is that the group structure of G induces in a natural manner a group structure on G/N. More precisely, using the binary operation that defines a group structure on G, we can define a binary operation $\odot$ on G/N as follows: $(xN) \odot (yN) = xyN$ (or in the "hat" notation $\hat{x} \odot \hat{y} = \widehat{xy}$) for all cosets xN (or $\hat{x}$) and yN (or $\hat{y}$); it is easy to see that $\odot$ is well-defined in the sense that if $xN = x'N$ and $yN = y'N$ for some elements x, x', y and y' of G, then $(xN) \odot (yN) = (x'N) \odot (y'N)$ (or, in the "hat" notation, if $\hat{x} = \widehat{x'}$ and $\hat{y} = \widehat{y'}$, then $\hat{x} \odot \hat{y} = \widehat{x'} \odot \widehat{y'}$, $x \in G$, $x' \in G$, $y \in G$, and $y' \in G$), so $\odot$ is indeed a binary operation on G/N. It is well known (see, for example, p. 15 of Lang [55]) and easy to prove that G/N, when endowed with the binary operation $\odot$, becomes a group with neutral element $N = \hat{e}$, where e is the neutral element of G.

A.2 Topologies on Semigroups, Groups and Coset Spaces

Let $(H, \cdot)$ be a semigroup and let τ be a topology on H.

As usual, we say that the ordered triple $(H, \cdot, \tau)$ is a *topological semigroup* if the map $\varphi : H \times H \to H$, $\varphi(x, y) = xy$ for every $x \in H$ and $y \in H$, is continuous with respect to the topology τ on H and the product topology on $H \times H$.

Naturally, if $(H_1, \cdot, \tau_1)$ and $(H_2, \cdot, \tau_2)$ are two topological semigroups, and if $f : H_1 \to H_2$ is a continuous function and a semigroup homomorphism, then f is called a *continuous (semigroup) homomorphism.*

An ordered triple $(G, \cdot, \tau)$ is called a *topological group* if the binary operation $\cdot$ defines a group structure on G (that is, if $(G, \cdot)$ is a group), if $(G, \cdot, \tau)$ is a topological semigroup, and if, in addition, the map $\psi : G \to G$, $\psi(x) = x^{-1}$ for every $x \in G$, is continuous (with respect to the topology τ on G, of course). *Continuous group homomorphisms* are defined in a similar manner as the continuous semigroup homomorphisms.

Let $(H, \cdot, \tau)$ be a topological semigroup (or a topological group). If the topology τ is Hausdorff, locally compact, compact, or separable, then we say that $(H, \cdot, \tau)$ is a *Hausdorff, locally compact, compact,* or *separable semigroup* (or *group*), respectively.

If $(H, \cdot, \tau)$ is a topological semigroup (or a topological group), and if the algebraic operation $\cdot$ and the topology τ are well-understood from the context (in general, this is often the case), we will simply refer to H as the topological semigroup (or the topological group) $(H, \cdot, \tau)$.

Let H be a topological semigroup and let H_1 be a subsemigroup of H. Since H_1, as a subsemigroup, is also a subset of H, it follows that the topology of H induces a topology on H_1. It is not difficult to see that H_1 is a topological semigroup in its own right when endowed with the topology induced by the topology of H on H_1.

Similarly, if G is a topological group and if G_1 is a subgroup of G, then the topology of G induces a topology on G_1, and G_1 becomes a topological group in its own right when endowed with the induced topology.

Let G be a topological group and let G_1 be a subgroup of G. The subgroup G_1 is said to be *discrete* if the topology of G induces the discrete topology on G_1. We say that G_1 is a *closed subgroup* if G_1 is closed as a subset of G.

Let H be a topological semigroup (or a topological group), and assume that the topology on H that generates the topological semigroup (or topological group) structure is defined by a metric d. Then the pair (H, d) is called a *metric semigroup* (or a *metric group*). If d is well-understood from the context, we will refer to H as the metric semigroup (or metric group) (H, d); on the other hand, if neither the algebraic operation $\cdot$ that defines the semigroup (or group) structure on H, nor the metric d that defines the structure of metric semigroup (or group) on H are clear in a certain context, then the metric semigroup (or the metric group) will be denoted by $(H, \cdot, d)$.

Let H be a metric semigroup whose topology is generated by a metric d, and let H_1 be a subsemigroup of H. Then the restriction d_1 of d to $H_1 \times H_1$ is a metric on

H_1. The topology generated by d_1 on H_1 is equal to the topology induced on H_1 by the topology of H, and (H_1, d_1) is a metric semigroup in its own right. Similarly, if (G, ρ) is a metric group, and if G_1 is a subgroup of G, then the restriction ρ_1 of ρ to $G_1 \times G_1$ is a metric on G_1 that defines the same topology as the topology induced by the topology of G on G_1; hence, (G_1, ρ_1) is a metric group in its own right.

A discrete subgroup of a metric group is closed. For future reference, we state the assertion as a proposition; for a more general type of topological group in which discrete subgroups are closed, see Theorem 5.10, p. 35 of Hewitt and Ross [44].

Proposition A.2.1. *Let G be a metric group, and let G_1 be a discrete subgroup of G. Then G_1 is closed.*

Let $(G, \cdot)$ be a group, and let d be a metric on G. The metric d is said to be *left invariant* if $d(ax, ay) = d(x, y)$ for every $x \in G$, $y \in G$, and $a \in G$. Similarly, we say that d is a *right invariant metric* if $d(xa, ya) = d(x, y)$ for every $x \in G$, $y \in G$, and $a \in G$.

A natural question at this point is the following: given a metric group G, can we construct two new metrics on G, one left invariant and another one right invariant, such that all three metrics (the given one and the two newly constructed ones) generate the same topology? The answer (which, of course, is of interest in itself) will be used in order to deal with metrics on spaces of cosets, and is discussed in the next proposition.

Proposition A.2.2. *Let $(G, \cdot, \tau)$ be a topological group, and assume that the topology τ is generated by a metric. Then there exist a left invariant metric d_L and a right invariant metric d_R such that each of the two metrics d_L and d_R generate τ.*

A proof of the existence of a left invariant metric appears in Theorem 8.3, Chapter 2, p. 70 of Hewitt and Ross [44]; for a proof of the existence of a right invariant metric, see Section 1.22 of Montgomery and Zippin [78].

In general, the metrics d_L and d_R whose existence is stated in Proposition A.2.2 are not equal.

Let $(G, \cdot, \tau)$ be a topological group, and let M be a subgroup of G.

Recall that $(G/M)_L$ and φ_L denote the left coset space and the standard surjection on $(G/M)_L$, respectively.

Let $\mathcal{T}((G/M)_L)$ be the collection of all the subsets U of $(G/M)_L$ such that $\varphi_L^{-1}(U)$ is an open set in G.

It can be shown (see, for instance, Theorem 5.16, Chapter 2, p. 36, of Hewitt and Ross [44]) that $\mathcal{T}((G/M)_L)$ is a topology on $(G/M)_L$, and that $\mathcal{T}((G/M)_L)$ is the strongest (that is, largest) topology on $(G/M)_L$ for which φ_L is continuous. We call $\mathcal{T}((G/M)_L)$ the *standard topology* on $(G/M)_L$ defined by the topology τ on G and we call the ordered pair $((G/M)_L, \mathcal{T}((G/M)_L))$ the *topological left coset space* of G defined by M.

Now consider the right coset space $(G/M)_R$ and the standard surjection φ_R on $(G/M)_R$. In a similar manner as in the case of $\mathcal{T}((G/M)_L)$ we define $\mathcal{T}((G/M)_R)$ as the collection of all the subsets U of $(G/M)_R$ such that $\varphi_R^{-1}(U)$ is an open subset of G. It is easy to modify the arguments that appear in the proof of the above-mentioned Theorem 5.16 on p. 36 of Hewitt and Ross [44] in order to show that $\mathcal{T}((G/M)_R)$ is the strongest topology on $(G/M)_R$ for which φ_R is continuous. We call $\mathcal{T}((G/M)_R)$ the *standard topology* on $(G/M)_R$ defined by the topology τ on G and we call the pair $((G/M)_R, \mathcal{T}((G/M)_R))$ the *topological right coset space* of G defined by M.

Since φ_L and φ_R are continuous surjective functions, it follows that if G is a separable topological group, then the topological spaces $((G/M)_L, \mathcal{T}((G/M)_L))$ and $((G/M)_R, \mathcal{T}((G/M)_R))$ are separable.

It can be shown that both φ_L and φ_R are open maps (see Theorem 5.17, Chapter 2, p. 37 of Hewitt and Ross [44]). Since φ_L and φ_R are also continuous and surjective, it follows that if G is a locally compact topological group, then $((G/M)_L, \mathcal{T}((G/M)_L))$ and $((G/M)_R, \mathcal{T}((G/M)_R))$ are locally compact topological spaces.

In the next proposition we discuss sufficient conditions for the metrizability of the topological spaces $((G/M)_L, \mathcal{T}((G/M)_L))$ and $((G/M)_R, \mathcal{T}((G/M)_R))$.

Proposition A.2.3. *Let G be a metric group and let M be a closed subgroup of G. Then $((G/M)_L, \mathcal{T}((G/M)_L))$ and $((G/M)_R, \mathcal{T}((G/M)_R))$ are metrizable topological spaces. Moreover, one can explicitly define two metrics η_L and η_R on $(G/M)_L$ and on $(G/M)_R$ that generate the topologies $\mathcal{T}((G/M)_L)$ and $\mathcal{T}((G/M)_R)$, respectively, as follows: let d_L and d_R be left invariant and right invariant metrics on G, respectively, whose existence is assured by Proposition A.2.2, and set*

$$\eta_L(xM, yM) = \inf\{d_R(a,b) \mid a \in xM, \, b \in yM\}$$

for every $xM \in (G/M)_L$ and $yM \in (G/M)_L$, and

$$\eta_R(Mx, My) = \inf\{d_L(a,b) \mid a \in Mx, \, b \in My\}$$

for every $Mx \in (G/M)_R$ and $My \in (G/M)_R$.

Note that in the above proposition d_R is used to define η_L and d_L is used to define η_R.

Obviously, in order to prove Proposition A.2.3 it is enough to prove that η_L and η_R are metrics on $(G/M)_L$ and $(G/M)_R$, and that they generate the topologies $\mathcal{T}((G/M)_L)$ and $\mathcal{T}((G/M)_R)$, respectively. For the proof that η_L is a metric that defines $\mathcal{T}((G/M)_L)$, see Section 1.23, pp. 36–37 of Montgomery and Zippin [78], or (8.14)-(a) and (8.14)-(b), pp. 76–77 of Hewitt and Ross [44]. The proof that η_R is a metric that generates $\mathcal{T}((G/M)_R)$ is obtained by straightforward modifications of the arguments used to prove the corresponding assertion for η_L.

For future reference, in the next proposition, we restate Proposition A.2.3 for the case in which G is a locally compact separable metric group.

Proposition A.2.4. *Let G be a locally compact separable metric group, and let M be a closed subgroup of G. Then $((G/M)_L, \eta_L)$ and $((G/M)_R, \eta_R)$ are locally* *compact separable metric spaces, where η_L and η_R are the metrics defined in Proposition A.2.3.*

Combining Propositions A.2.1 and A.2.4, we obtain the following proposition:

Proposition A.2.5. *If G is a locally compact separable metric group and M is a discrete subgroup, then the maps η_L and η_R defined in Proposition A.2.3 are metrics on $(G/M)_L$ and $(G/M)_R$ which define the topologies $T((G/M)_L)$ and $T((G/M)_R)$, respectively, and $((G/M)_L, \eta_L)$ and $((G/M)_R, \eta_R)$ are locally compact separable metric spaces.*

If G is a topological group and N is a normal subgroup of G, then, as pointed out earlier, the left coset space of G defined by N is equal to the corresponding right coset space and we use the notation G/N for any of these coset spaces. It is easy to see that if we think of G/N as a left coset space and as a right coset space, then the corresponding standard topologies are equal; therefore, we will use the notation $T(G/N)$ for any of these standard topologies and we call $T(G/N)$ the *standard topology* of G/N. Since G/N has a group structure defined in terms of the group structure of G, a natural question is whether or not G/N is a topological group when endowed with its standard topology. The answer is provided by the next proposition.

Proposition A.2.6. *If G is a topological group and N is a normal subgroup of G, then G/N is a topological group when endowed with its standard topology.*

For a proof of the proposition and additional information about G/N, see Theorem 5.26, p. 40 of Hewitt and Ross [44].

Let G be a locally compact separable metric group, and let N be a closed normal subgroup of G. Taking into consideration that $xN = Nx$ for every $x \in G$, we obtain that $\eta_L = \eta_R$ where η_L and η_R are the metrics defined in Proposition A.2.3. Let η be any of the metrics η_L or η_R. Then using Propositions A.2.3, A.2.4 and A.2.6 we obtain the following proposition:

Proposition A.2.7. *If G is a locally compact separable metric group, if N is a closed normal subgroup of G, and if η is the metric on G/N defined above, then $(G/N, \eta)$ is a locally compact separable metric group.*

In particular, $(G/N, \eta)$ is a locally compact separable metric group whenever N is a discrete normal subgroup of G, and G and η are as in Proposition A.2.7. This is so because, in this case, N is a closed subgroup of G by Proposition A.2.1.

Let G be a commutative metric group. Then any left invariant metric on G is also right invariant, and vice versa; therefore, if d is a left or right invariant metric on G, we simply say that d is *invariant*. Note that, by Proposition A.2.2, such invariant metrics do exist. Now let M be a closed subgroup of G. Since any subgroup of G is normal, it follows that the coset space G/M exists and is a group. Since G is commutative, G/M is also commutative. Using Propositions A.2.3 and A.2.6, we obtain that G/M is a metrizable topological group. By Proposition A.2.7, if G is locally compact and separable, then G/M is locally compact and separable, as well.

Note also that if η is the metric on G/M defined before Proposition A.2.7, then η is an invariant metric.

We will conclude this section with several examples that will be used throughout the book.

Example A.2.8 (The Unit Circle). Let $\mathbb{R}$ be the additive group of all real numbers endowed with the usual metric d defined in terms of the absolute value ($d(x, y) = |x - y|$ for every $x \in \mathbb{R}$ and $y \in \mathbb{R}$) and endowed with the topology defined by d. Then $\mathbb{R}$ is a commutative locally compact separable metric group, and d is both left and right invariant. Let $\mathbb{Z}$ be the additive group of all integers. Then $\mathbb{Z}$, thought of as a subgroup of $\mathbb{R}$, is discrete (hence closed). Since $\mathbb{R}$ is a commutative group, it follows that $\mathbb{Z}$ is a normal subgroup, so the coset space $\mathbb{R}/\mathbb{Z}$ is well-defined. Using the discussion preceding this example, we obtain that $\mathbb{R}/\mathbb{Z}$ is a locally compact separable metric commutative group when endowed with the metric η defined as follows:

$$\eta(x + \mathbb{Z}, y + \mathbb{Z}) = \min\{|a - b|, 1 - |a - b|\},$$

where a and b are the unique coset representatives of $x + \mathbb{Z}$ and $y + \mathbb{Z}$, respectively, that belong to the interval $[0, 1)$ (note that η is precisely the metric that appears in the more general setting of Proposition A.2.7).

Now let $\mathbf{C}_1$ be the unit circle in the complex plane $\mathbb{C}$; that is, let

$$\mathbf{C}_1 = \{z \in \mathbb{C} \mid |z| = 1\} = \{z \in \mathbb{C} \mid z = e^{2\pi t i} \text{ for some } t \in [0, 1)\}.$$

It is easy to see that the restriction of the multiplication of complex numbers to $\mathbf{C}_1 \times \mathbf{C}_1$ is a well-defined binary operation on $\mathbf{C}_1$ in the sense that if $z_1 \in \mathbf{C}_1$ and $z_2 \in \mathbf{C}_1$, then $z_1 z_2$ belongs to $\mathbf{C}_1$, as well. We will denote this binary operation on $\mathbf{C}_1$ by $\odot$. Clearly, $\mathbf{C}_1$ becomes a commutative group when endowed with $\odot$.

Let $\zeta : \mathbf{C}_1 \times \mathbf{C}_1 \to \mathbb{R}$ be a map defined as follows:

$$\zeta(z_1, z_2) = \frac{1}{2\pi} \text{ arclength } (z_1, z_2)$$

for every $z_1 \in \mathbf{C}_1$ and $z_2 \in \mathbf{C}_1$, where arclength (z_1, z_2) is the length of the shorter arc joining the points z_1 and z_2 on the unit circle. Then ζ is a metric on $\mathbf{C}_1$, and the topology defined by ζ on $\mathbf{C}_1$ is the topology induced by the standard topology of $\mathbb{C}$ on $\mathbf{C}_1$. Note that the topology defined by ζ on $\mathbf{C}_1$ is also the topology defined by the restriction d_1 of the standard metric on $\mathbb{C}$ to $\mathbf{C}_1 \times \mathbf{C}_1$ (the metric $d_1 : \mathbf{C}_1 \times \mathbf{C}_1 \to \mathbb{R}$ is defined as follows:

$$d_1(z_1, z_2) = |z_1 - z_2| = \quad \text{the length of the chord joining the points } z_1 \text{ and } z_2$$
$$\text{on the unit circle}$$

for every $z_1 \in \mathbf{C}_1$ and $z_2 \in \mathbf{C}_1$). However, for reasons that will become apparent very soon, we prefer to work with the metric ζ rather than d_1.

Clearly, $\mathbf{C}_1$ is a compact metric commutative group when endowed with the algebraic operation $\odot$ and the metric ζ.

The map $z \mapsto t + \mathbb{Z}$, $z \in \mathbf{C}_1$, $z = e^{2\pi t i}$, $t \in [0, 1)$, from $(\mathbf{C}_1, \odot, \zeta)$ onto $(\mathbb{R}/\mathbb{Z}, \oplus, \eta)$, where $\oplus$ is the algebraic operation that defines the group structure on $\mathbb{R}/\mathbb{Z}$, is an isometry and a group isomorphism. Therefore, we can use the map in order to identify $(\mathbf{C}_1, \odot, \zeta)$ with $(\mathbb{R}/\mathbb{Z}, \oplus, \eta)$ and to identify the points of $\mathbf{C}_1$ with the corresponding elements of $\mathbb{R}/\mathbb{Z}$ and vice versa. Note that since $(\mathbf{C}_1, \odot, \zeta)$ is a compact metric commutative group, the existence of an isometric group isomorphism between $(\mathbf{C}_1, \odot, \zeta)$ and $(\mathbb{R}/\mathbb{Z}, \oplus, \eta)$ implies that $(\mathbb{R}/\mathbb{Z}, \oplus, \eta)$ is actually a compact metric group, rather than just a locally compact separable metric group.

Consider the interval $[0, 1)$ in $\mathbb{R}$, and let $\boxplus$ be addition modulo 1 defined on $[0, 1)$; that is, let $\boxplus$ be the map on $[0, 1) \times [0, 1)$ defined as follows:

$$a \boxplus b = \begin{cases} a + b & \text{if } a + b < 1 \\ a + b - 1 & \text{if } a + b \geq 1 \end{cases}$$

for every $a \in [0, 1)$ and $b \in [0, 1)$. Note that $\boxplus$ is a well-defined algebraic operation on $[0, 1)$ in the sense that $a \boxplus b$ is indeed an element of $[0, 1)$ whenever $a \in [0, 1)$ and $b \in [0, 1)$. Note also that $\boxplus$ defines a group structure on $[0, 1)$.

Let $\rho : [0, 1) \times [0, 1) \to \mathbb{R}$ be a map defined as follows:

$$\rho(a, b) = \min\{|a - b|, 1 - |a - b|\}$$

for every $a \in [0, 1)$ and $b \in [0, 1)$. It is easy to see that ρ is a metric on $[0, 1)$ and that $([0, 1), \boxplus, \rho)$ is a metric commutative group.

Since the map $t \mapsto e^{2\pi t i}$, $t \in [0, 1)$ from $([0, 1), \boxplus, \rho)$ onto $(\mathbf{C}_1, \odot, \zeta)$ is a group isomorphism and an isometry, it follows that we can use the map in order to identify $(\mathbf{C}_1, \odot, \zeta)$ with $([0, 1), \boxplus, \rho)$, and the points on the unit circle $\mathbf{C}_1$ with the corresponding points in the interval $[0, 1)$ and vice versa. Note that the existence of an isometric group isomorphism from $([0, 1), \boxplus, \rho)$ onto $(\mathbf{C}_1, \odot, \zeta)$ implies that $([0, 1), \boxplus, \rho)$ is in fact a compact metric commutative group, rather than just a metric commutative group.

In view of our discussion so far, we note that we can (and often do) identify $(\mathbb{R}/\mathbb{Z}, \oplus, \eta)$ with the unit circle $(\mathbf{C}_1, \odot, \zeta)$ or with $([0, 1), \boxplus, \rho)$; also, we can (and, whenever convenient, do) identify the elements of $\mathbb{R}/\mathbb{Z}$ with the corresponding points in $\mathbf{C}_1$ or in $[0, 1)$ via the isometric group isomorphisms discussed above. The fact that the coset space $\mathbb{R}/\mathbb{Z}$ can be identified with the unit circle $\mathbf{C}_1$ is the reason why people often refer to $\mathbb{R}/\mathbb{Z}$ as the *unit circle*.   ∎

The next example is an extension of Example A.2.8.

Example A.2.9 (The n-Dimensional Torus). Let $n \in \mathbb{N}$, $n \geq 2$, and consider $\mathbb{R}^n$ as a locally compact separable metric commutative group, where the algebraic operation that defines the group structure on $\mathbb{R}^n$ is addition. The metric defined by the Euclidean norm (or by any other norm) on $\mathbb{R}^n$ is (both left and right) invariant. Consider also $\mathbb{Z}^n$, thought of as a subgroup of $\mathbb{R}^n$. Since $\mathbb{R}^n$ is commutative, it follows that $\mathbb{Z}^n$ is a normal subgroup, and $\mathbb{R}^n/\mathbb{Z}^n$ is itself a commutative group.

Let $\mathbf{d}_n$ be the Euclidean metric on $\mathbb{R}^n$, and let $\eta^{(n)}$ be the invariant metric on $\mathbb{R}^n/\mathbb{Z}^n$ defined using $\mathbf{d}_n$ by the procedure outlined in the discussion preceding Example A.2.8. Also, let $\oplus_n$ be the algebraic operation that defines the group structure of $\mathbb{R}^n/\mathbb{Z}^n$. Then $(\mathbb{R}^n/\mathbb{Z}^n, \oplus_n, \eta^{(n)})$ is a locally compact separable metric group. Note that $\eta^{(n)}(\mathbf{x} + \mathbb{Z}^n, \mathbf{y} + \mathbb{Z}^n) = \left(\sum_{i=1}^{n} \eta^2(x_i + \mathbb{Z}, y_i + \mathbb{Z}) \right)^{\frac{1}{2}}$ for every $\mathbf{x} \in \mathbb{R}^n$, $\mathbf{x} = (x_1, x_2, \ldots, x_n)$ and $\mathbf{y} \in \mathbb{R}^n$, $\mathbf{y} = (y_1, y_2, \ldots, y_n)$, where η is the metric on $\mathbb{R}/\mathbb{Z}$ defined in Example A.2.8.

Set

$$\mathbf{C}_1^n = \underbrace{\mathbf{C}_1 \times \mathbf{C}_1 \times \cdots \times \mathbf{C}_1}_{n \text{ times}} = \left\{ \mathbf{z} \in \mathbb{C}^n \;\middle|\; \begin{array}{l} \mathbf{z} = (e^{2\pi t_1 i}, e^{2\pi t_2 i}, \ldots, e^{2\pi t_n i}) \\ \text{for some } t_i \in [0, 1), i = 1, 2, \ldots, n \end{array} \right\}$$

where $\mathbf{C}_1$ is the unit circle in the complex plane $\mathbb{C}$ discussed in Example A.2.8.

Let $\odot_n$ be the restriction to $\mathbf{C}_1^n$ of the algebraic operation on $\mathbb{C}^n$ that consists of componentwise multiplication of the corresponding complex numbers; thus, $\odot_n$ is defined as follows:

$$\mathbf{z}^{(1)} \odot_n \mathbf{z}^{(2)} = \left(z_1^{(1)} \odot z_1^{(2)}, z_2^{(1)} \odot z_2^{(2)}, z_3^{(1)} \odot z_3^{(2)}, \ldots, z_n^{(1)} \odot z_n^{(2)} \right)$$

for every $\mathbf{z}^{(1)} \in \mathbf{C}_1^n$, $\mathbf{z}^{(1)} = \left(z_1^{(1)}, z_2^{(1)}, z_3^{(1)}, \ldots, z_n^{(1)} \right)$ and $\mathbf{z}^{(2)} \in \mathbf{C}_1^n$, $\mathbf{z}^{(2)} = \left(z_1^{(2)}, z_2^{(2)}, z_3^{(2)}, \ldots, z_n^{(2)} \right)$, where $\odot$ is the algebraic operation on $\mathbf{C}_1$ defined in Example A.2.8 (that is, $\odot$ is the restriction to $\mathbf{C}_1 \times \mathbf{C}_1$ of the multiplication of complex numbers). Clearly, $\odot_n$ is a well-defined algebraic operation on $\mathbf{C}_1^n$ (in the sense that $\mathbf{z}^{(1)} \odot_n \mathbf{z}^{(2)}$ belongs to $\mathbf{C}_1^n$ for every $\mathbf{z}^{(1)} \in \mathbf{C}_1^n$ and $\mathbf{z}^{(2)} \in \mathbf{C}_1^n$) and defines a group structure on $\mathbf{C}_1^n$ (note that $(\mathbf{C}_1^n, \odot_n)$ is the direct sum of n copies of $(\mathbf{C}_1, \odot)$) (for the definition of direct sums of groups, see, for instance, Section 13, Chap. 2 of Warner's book [127])).

Let $\zeta^{(n)} : \mathbf{C}_1^n \times \mathbf{C}_1^n \to \mathbb{R}$ be a map defined as follows:

$$\zeta^{(n)}(\mathbf{z}^{(1)}, \mathbf{z}^{(2)}) = \left(\sum_{i=1}^{n} \zeta^2(z_i^{(1)}, z_i^{(2)}) \right)^{\frac{1}{2}}$$

for every $\mathbf{z}^{(1)} \in \mathbf{C}_1^n$, $\mathbf{z}^{(1)} = \left(z_1^{(1)}, z_2^{(1)}, z_3^{(1)}, \ldots, z_n^{(1)}\right)$ and $\mathbf{z}^{(2)} \in \mathbf{C}_1^n$, $\mathbf{z}^{(2)} = \left(z_1^{(2)}, z_2^{(2)}, z_3^{(2)}, \ldots, z_n^{(2)}\right)$, where ζ is the metric on $\mathbf{C}_1$ defined in Example A.2.8. Then $\zeta^{(n)}$ is a metric on $\mathbf{C}_1^n$, and the topology defined by $\zeta^{(n)}$ on $\mathbf{C}_1^n$ is the topology induced by the standard topology of $\mathbb{C}^n$ on $\mathbf{C}_1^n$.

We obtain that $\left(\mathbf{C}_1^n, \odot_n, \zeta^{(n)}\right)$ is a compact metric commutative group.

The map $\mathbf{z} \mapsto \mathbf{t} + \mathbb{Z}^n$, $\mathbf{z} \in \mathbf{C}_1^n$, $\mathbf{z} = (e^{2\pi t_1 i}, e^{2\pi t_2 i}, \ldots, e^{2\pi t_n i})$, $t_i \in [0, 1)$, $i = 1, 2, \ldots, n$, $\mathbf{t} = (t_1, t_2, \ldots, t_n)$, from $\left(\mathbf{C}_1^n, \odot_n, \zeta^{(n)}\right)$ onto $\left(\mathbb{R}^n / \mathbb{Z}^n, \oplus_n, \eta^{(n)}\right)$ is a group isomorphism and an isometry. Thus, the map can be used to identify $\left(\mathbf{C}_1^n, \odot_n, \zeta^{(n)}\right)$ with $\left(\mathbb{R}^n / \mathbb{Z}^n, \oplus_n, \eta^{(n)}\right)$, and to identify the points of $\mathbf{C}_1^n$ with the corresponding elements of $\mathbb{R}^n / \mathbb{Z}^n$ and vice versa. Since $\left(\mathbf{C}_1^n, \odot_n, \zeta^{(n)}\right)$ is a compact metric commutative group, and since $\left(\mathbb{R}^n / \mathbb{Z}^n, \oplus_n, \eta^{(n)}\right)$ is isometric and group isomorphic to $\left(\mathbf{C}_1^n, \odot_n, \zeta^{(n)}\right)$, it follows that $\left(\mathbb{R}^n / \mathbb{Z}^n, \oplus_n, \eta^{(n)}\right)$ is also a compact metric commutative group, rather than just a locally compact separable metric commutative group. The topological group $\left(\mathbf{C}_1^n, \odot_n, \zeta^{(n)}\right)$ is known as the *n-dimensional torus*, and since the map described in this paragraph can be thought of as a standard isometry and group isomorphism from $\left(\mathbf{C}_1^n, \odot_n, \zeta^{(n)}\right)$ onto $\left(\mathbb{R}^n / \mathbb{Z}^n, \oplus_n, \eta^{(n)}\right)$, people often refer to $\left(\mathbb{R}^n / \mathbb{Z}^n, \oplus_n, \eta^{(n)}\right)$ as the *n-dimensional torus*, as well.

Let $([0, 1)^n, \boxplus_n)$ be the direct sum of n copies of the group $([0, 1), \boxplus)$ that was defined in Example A.2.8. Also, let $\rho_n : [0, 1)^n \times [0, 1)^n \to \mathbb{R}$ be defined as follows: $\rho_n(\mathbf{a}, \mathbf{b}) = \left(\sum_{i=1}^n \rho^2(a_i, b_i)\right)^{\frac{1}{2}}$ for every $\mathbf{a} \in [0, 1)^n$, $\mathbf{a} = (a_1, a_2, \ldots, a_n)$, $\mathbf{b} \in [0, 1)^n$, $\mathbf{b} = (b_1, b_2, \ldots, b_n)$, where ρ is the metric on $[0, 1)$ that was defined in Example A.2.8. Then, it is easy to see that ρ_n is a metric on $[0, 1)^n$, and that $([0, 1)^n, \boxplus_n, \rho_n)$ is a locally compact separable metric commutative group.

Since the map $\mathbf{t} \mapsto (e^{2\pi t_1 i}, e^{2\pi t_2 i}, \ldots, e^{2\pi t_n i})$, $\mathbf{t} = (t_1, t_2, \ldots, t_n)$, $t_i \in [0, 1)$, $i = 1, 2, \ldots, n$, from $([0, 1)^n, \boxplus_n, \rho_n)$ onto $\left(\mathbf{C}_1^n, \odot_n, \zeta^{(n)}\right)$ is an isometry and a group isomorphism, it follows that we can use this map to identify $([0, 1)^n, \boxplus_n, \rho_n)$ with $\left(\mathbf{C}_1^n, \odot_n, \zeta^{(n)}\right)$, and the points of $[0, 1)^n$ with the corresponding points of $\mathbf{C}_1^n$ and vice versa. The existence of the map allows us to infer that $([0, 1)^n, \boxplus_n, \rho_n)$ is actually a compact metric group rather than a locally compact separable metric group.

Taking into consideration the isometric group isomorphism between $\left(\mathbf{C}_1^n, \odot_n, \zeta^{(n)}\right)$ and $\left(\mathbb{R}^n / \mathbb{Z}^n, \oplus_n, \eta^{(n)}\right)$ that we described above, we obtain that we may (and whenever convenient we do) identify $\left(\mathbb{R}^n / \mathbb{Z}^n, \oplus_n, \eta^{(n)}\right)$ with $\left(\mathbf{C}_1^n, \odot_n, \zeta^{(n)}\right)$ or with $([0, 1)^n, \boxplus_n, \rho_n)$, and the elements of $\mathbb{R}^n / \mathbb{Z}^n$ with the corresponding points of $\mathbf{C}_1^n$ or $[0, 1)^n$.   ∎

We will denote by $\mathbf{M}(n, A)$ the set of all $n \times n$ matrices with A-valued entries, where A is a nonempty set. In this book A stands for $\mathbb{R}$ or $\mathbb{Z}$

Our last example in this section deals with certain noncommutative groups and some of their subgroups.

Example A.2.10. Let $n \in \mathbb{N}$, $n \geq 2$. As usual (see, for instance, the monographs of Bekka and Mayer [10] or Starkov [113]), we denote by $SL(n, \mathbb{R})$ the set of all $n \times n$ matrices with real entries and determinant equal to 1.

Clearly, matrix multiplication is a well-defined algebraic operation on $SL(n, \mathbb{R})$ in the sense that if $A \in SL(n, \mathbb{R})$ and $B \in SL(n, \mathbb{R})$, then AB is an element of $SL(n, \mathbb{R})$, as well. It is also easy to see that matrix multiplication defines a group structure on $SL(n, \mathbb{R})$. When endowed with matrix multiplication, $SL(n, \mathbb{R})$ is known as the *special linear group* of $n \times n$ matrices with real-valued entries.

Since $\mathbf{M}(n, \mathbb{R})$ can be identified with $\mathbb{R}^{n^2}$, the standard topology of $\mathbb{R}^{n^2}$ as an n^2-dimensional Euclidean space defines a topology on $\mathbf{M}(n, \mathbb{R})$ via the identification, a topology which is also called *standard*. Note that the standard topology on $\mathbf{M}(n, \mathbb{R})$ is the same as the Banach space topology of $\mathbf{M}(n, \mathbb{R})$, where $\mathbf{M}(n, \mathbb{R})$ is thought of as the Banach space of all linear bounded operators from $\mathbb{R}^n$ to $\mathbb{R}^n$, where $\mathbb{R}^n$ is endowed with the Euclidean norm (or any other norm). Since $SL(n, \mathbb{R})$ is a subset of $\mathbf{M}(n, \mathbb{R})$, we may and do consider on $SL(n, \mathbb{R})$ the topology induced by $\mathbf{M}(n, \mathbb{R})$. We denote the induced topology by $\mathcal{T}$ and call it the *standard topology of* $SL(n, \mathbb{R})$. Since $SL(n, \mathbb{R})$, as a subset of $\mathbf{M}(n, \mathbb{R})$ is closed, and since the restriction to $SL(n, \mathbb{R}) \times SL(n, \mathbb{R})$ of any metric that generates the standard topology of $\mathbf{M}(n, \mathbb{R})$ is a metric on $SL(n, \mathbb{R})$ that generates the topology $\mathcal{T}$, it follows that $(SL(n, \mathbb{R}), \mathcal{T})$ is a locally compact separable metrizable topological space. Since the maps $(A, B) \mapsto AB$, $(A, B) \in SL(n, \mathbb{R}) \times SL(n, \mathbb{R})$, and $A \mapsto A^{-1}$, $A \in SL(n, \mathbb{R})$, are continuous with respect to the topology $\mathcal{T}$ on $SL(n, \mathbb{R})$ and the product topology on $SL(n, \mathbb{R}) \times SL(n, \mathbb{R})$, it follows that $SL(n, \mathbb{R})$ is a locally compact separable metrizable topological group. Note that, when restricted to $SL(n, \mathbb{R}) \times SL(n, \mathbb{R})$, neither the metrics defined by norms on $\mathbb{R}^{n^2}$, nor the metrics defined by the operator norms on $\mathbf{M}(n, \mathbb{R})$, where $\mathbf{M}(n, \mathbb{R})$ is thought of as the Banach space of all linear (bounded) operators on $\mathbb{R}^n$ (each operator norm being defined by the norm under consideration on $\mathbb{R}^n$), are left or right invariant. However, by Proposition A.2.2, there exist a left invariant metric d_L and a right invariant metric d_R such that each of these two metrics generates the standard topology of $SL(n, \mathbb{R})$.

Now let $L = \{\mathbf{I}_2, -\mathbf{I}_2\}$, where $\mathbf{I}_2$ is the 2×2 identity matrix. Clearly, L is a closed normal subgroup of $SL(2, \mathbb{R})$; therefore, it makes sense to consider $SL(2, \mathbb{R})/L$. The coset space $SL(2, \mathbb{R})/L$ has a group structure when endowed with the natural algebraic operation for coset spaces defined at the end of Sect. A.1. By Proposition A.2.6, $SL(2, \mathbb{R})/L$ is a topological group when endowed with the standard topology $\mathcal{T}(SL(2, \mathbb{R})/L)$. By Proposition A.2.7, the topological group $SL(2, \mathbb{R})/L$ is locally compact, separable and metrizable. As usual (see, for instance, Bekka and Mayer [10]), we denote $SL(2, \mathbb{R})/L$ by $PSL(2, \mathbb{R})$.

If M is a discrete subgroup of $SL(n, \mathbb{R})$, $n \in \mathbb{N}$, $n \geq 2$, then using Proposition A.2.5, we obtain that the topological spaces $((SL(n, \mathbb{R})/M)_L, \mathcal{T}((SL(n, \mathbb{R})/M)_L))$ and $((SL(n, \mathbb{R})/M)_R, \mathcal{T}((SL(n, \mathbb{R})/M)_R))$ are locally compact, separable and metrizable.

An example of a discrete subgroup of $SL(n, \mathbb{R})$ of special interest is obtained as follows: note that matrix multiplication is a well-defined algebraic operation on $SL(n, \mathbb{Z}) = $ the set of all $n \times n$ matrices with determinant equal to 1 and with entries in $\mathbb{Z}$; moreover, matrix multiplication defines a group structure on $SL(n, \mathbb{Z})$, and therefore, $SL(n, \mathbb{Z})$ can be thought of as a subgroup of $SL(n, \mathbb{R})$; it is easy to see that, as a subgroup of $SL(n, \mathbb{R})$, $SL(n, \mathbb{Z})$ is discrete. It follows that $((SL(n, \mathbb{R})/SL(n, \mathbb{Z}))_{\mathrm{L}}, \mathcal{T}((SL(n, \mathbb{R})/SL(n, \mathbb{Z}))_{\mathrm{L}}))$ and $((SL(n, \mathbb{R})/SL(n, \mathbb{Z}))_{\mathrm{R}}, \mathcal{T}((SL(n, \mathbb{R})/SL(n, \mathbb{Z}))_{\mathrm{R}}))$ are locally compact separable metrizable topological spaces.

Similar observations are valid for discrete subgroups of $PSL(2, \mathbb{R})$. If M is a discrete subgroup of $PSL(2, \mathbb{R})$, then $((PSL(2, \mathbb{R})/M)_{\mathrm{L}}, \mathcal{T}((PSL(2, \mathbb{R})/M)_{\mathrm{L}}))$ and $((PSL(2, \mathbb{R})/M)_{\mathrm{R}}, \mathcal{T}((PSL(2, \mathbb{R})/M)_{\mathrm{R}}))$ are locally compact separable metrizable topological spaces. As in the case of $PSL(2, \mathbb{R})$, it is easy to see that $L = \{\mathbf{I}_2, -\mathbf{I}_2\}$ is a normal subgroup of $SL(2, \mathbb{Z})$, so it makes sense to consider the group $SL(2, \mathbb{Z})/L$, which is denoted by $PSL(2, \mathbb{Z})$. If we think of $PSL(2, \mathbb{Z})$ as a subgroup of $PSL(2, \mathbb{R})$, then it is not difficult to see that $PSL(2, \mathbb{Z})$ is discrete. Thus, the topological spaces

$$((PSL(2, \mathbb{R})/PSL(2, \mathbb{Z}))_{\mathrm{L}}, \mathcal{T}((PSL(2, \mathbb{R})/PSL(2, \mathbb{Z}))_{\mathrm{L}}))$$

and

$$((PSL(2, \mathbb{R})/PSL(2, \mathbb{Z}))_{\mathrm{R}}, \mathcal{T}((PSL(2, \mathbb{R})/PSL(2, \mathbb{Z}))_{\mathrm{R}}))$$

are locally compact, separable and metrizable.

We conclude this example with a characterization of the elements of $SL(2, \mathbb{R})$ which is often useful.

Set

$$\mathbf{K} = \left\{ \begin{pmatrix} \cos\theta & -\sin\theta \\ \sin\theta & \cos\theta \end{pmatrix} \mid 0 \leq \theta \leq 2\pi \right\},$$

$$\mathbf{A} = \left\{ \begin{pmatrix} e^t & 0 \\ 0 & e^{-t} \end{pmatrix} \mid t \in \mathbb{R} \right\},$$

$$\mathbf{N} = \left\{ \begin{pmatrix} 1 & x \\ 0 & 1 \end{pmatrix} \mid x \in \mathbb{R} \right\},$$

and

$$\mathbf{P} = \left\{ \begin{pmatrix} e^t & x \\ 0 & e^{-t} \end{pmatrix} \mid t \in \mathbb{R}, x \in \mathbb{R} \right\}.$$

Note that $\mathbf{K}$, $\mathbf{A}$, $\mathbf{N}$ and $\mathbf{P}$ are subgroups of $SL(2, \mathbb{R})$. Note also that $\mathbf{P} = \mathbf{AN}$.

It can be shown that $SL(2, \mathbb{R}) = \mathbf{KAN} = \mathbf{KP}$.

The above equality is known as the *Iwasawa decomposition* (see the introduction to Chapter 4, pp. 110–111 of Bekka and Mayer [10] and Section 3.1 in Lang's monograph [56]).   ∎

A.3 Actions, Semiflows and Flows

Our goal in this section is to discuss a few basic facts about actions, semiflows and flows, and to describe several flows that will be used throughout the book to construct examples illustrating various notions and results.

Let $(H, \cdot)$ be a semigroup which has a neutral element, and let e be the neutral element of H. Also, let X be a nonempty set.

A map $w^{(L)} : H \times X \to X$ is called a *left semigroup action* (of H on X) if $w^{(L)}$ satisfies the following two conditions:

$$w^{(L)}(gh, x) = w^{(L)}(g, w^{(L)}(h, x)) \qquad (A.3.1)$$

for every $g \in H$, $h \in H$, and $x \in X$;

$$w^{(L)}(e, x) = x \qquad (A.3.2)$$

for every $x \in X$.

Given a left semigroup action $w^{(L)}$ of H on X, it is the custom to define the functions $w_h^{(L)} : X \to X$, $w_h^{(L)}(x) = w^{(L)}(h, x)$ for every $x \in X$ and $h \in H$, and to use the notation $\left(w_h^{(L)} \right)_{h \in H}$ for $w^{(L)}$. In terms of the functions $w_h^{(L)}$, $h \in H$, the above conditions (A.3.1) and (A.3.2) become

$$w_{gh}^{(L)} = w_g^{(L)} w_h^{(L)} \qquad (A.3.3)$$

for every $g \in H$ and $h \in H$, and

$$w_e^{(L)} = \mathrm{Id}_X, \qquad (A.3.4)$$

respectively.

Let $\left(w_h^{(L)} \right)_{h \in H}$ be a left semigroup action of H on X, and let $x \in X$. The subset $\left\{ w_h^{(L)}(x) \mid h \in H \right\}$ of X is called the *orbit of x under the action of* $\left(w_h^{(L)} \right)_{h \in H}$ *in X* or, simply, the *orbit of x* if there is no danger of confusion. Since, in general, the action and the set X will be well-understood from the context, we will denote the orbit of x by $\mathcal{O}(x)$; a more complete notation that will sometimes be used is $\mathcal{O}_{(w_h^{(L)})_{h \in H}}(x)$.

A map $w^{(R)} : H \times X \to X$ is called a *right semigroup action (of H on X)* if the following two conditions are satisfied:

$$w^{(R)}(gh, x) = w^{(R)}(h, w^{(R)}(g, x)) \qquad (A.3.5)$$

for every $g \in H$, $h \in H$, and $x \in X$;

$$w^{(R)}(e, x) = x \qquad (A.3.6)$$

for every $x \in X$.

As in the case of left semigroup actions, given a right semigroup action $w^{(R)}$ of H on X, it is often more convenient to define the functions $w_h^{(R)} : X \to X$, $w_h^{(R)}(x) = w^{(R)}(h, x)$ for every $x \in X$ and $h \in H$, and to denote the action $w^{(R)}$ by $\left(w_h^{(R)} \right)_{h \in H}$. Using the functions $w_h^{(R)}$, $h \in H$, we can write the above conditions (A.3.5) and (A.3.6) as

$$w_{gh}^{(R)} = w_h^{(R)} w_g^{(R)} \tag{A.3.7}$$

for every $g \in H$ and $h \in H$, and

$$w_e^{(R)} = \mathrm{Id}_X, \tag{A.3.8}$$

respectively.

For right semigroup actions, we can define the orbits under the actions as in the case of left semigroup actions; that is, if $\left(w_h^{(R)} \right)_{h \in H}$ is a right semigroup action of H on X and if $x \in X$, then the *orbit of x* under the action is the subset $\left\{ w_h^{(R)}(x) \mid h \in H \right\}$ of X. The orbit will be denoted by $\mathcal{O}(x)$, or $\mathcal{O}_{(w_h^{(R)})_{h \in H}}(x)$ if we have to emphasize that the orbit is under the action $\left(w_h^{(R)} \right)_{h \in H}$.

If H is a group and $w^{(L)}$ is a left semigroup action of H on X (that is, $w^{(L)} : H \times X \to X$ satisfies the above conditions (A.3.1) and (A.3.2) or, equivalently, the conditions (A.3.3) and (A.3.4) in this section), then, naturally, we call $w^{(L)}$ a *left group action*. In this case, using the above-mentioned equalities (A.3.3) and (A.3.4), we obtain that

$$\mathrm{Id}_X = w_{hh^{-1}}^{(L)} = w_h^{(L)} w_{h^{-1}}^{(L)}$$

and

$$\mathrm{Id}_X = w_{h^{-1}h}^{(L)} = w_{h^{-1}}^{(L)} w_h^{(L)}$$

for every $h \in H$. Thus, if H is a group, the functions $w_h^{(L)}$, $h \in H$, are bijections.

Similarly, if H is a group and $w^{(R)}$ is a right semigroup action, then we call $w^{(R)}$ a *right group action*. As in the case of left group actions, but using the equalities (A.3.7) and (A.3.8) in this section instead of (A.3.3) and (A.3.4), we obtain that

$$\mathrm{Id}_X = w_{hh^{-1}}^{(R)} = w_{h^{-1}}^{(R)} w_h^{(R)}$$

and

$$\mathrm{Id}_X = w_{h^{-1}h}^{(R)} = w_h^{(R)} w_{h^{-1}}^{(R)}$$

for every $h \in H$; hence, $w_h^{(R)}$, $h \in H$, are bijections.

Naturally, since group actions are particular cases of semigroup actions, the *orbits under group actions* are defined as in the case of semigroup actions.

Left and right actions appear quite naturally in various situations as the next examples illustrate.

Example A.3.1. Let, as before, H be a semigroup which has a neutral element e, let X be a semigroup which has a neutral element, as well, and let e_X be the neutral element of X.

Let $g : H \to X$ be a semigroup homomorphism such that

$$g(e) = e_X. \tag{A.3.9}$$

We will use the subscript notation g_h for $g(h)$, $h \in H$. Thus, in subscript notation, the above equality (A.3.9) becomes $g_e = e_X$.

Using g, we can define two maps $w^{(L)} : H \times X \to X$ and $w^{(R)} : H \times X \to X$ as follows: $w^{(L)}(h, x) = g_h x$ and $w^{(R)}(h, x) = x g_h$ for every $(h, x) \in H \times X$. It is easy to see that $w^{(L)}$ and $w^{(R)}$ are left and right semigroup actions of H on X, respectively.

A particular family of such actions can be obtained whenever H is a subsemigroup of X such that $e_X \in H$, and $g : H \to X$ is the *standard injection* (that is, $g_h = h$ for every $h \in H$). In this case, $w^{(L)}(h, x) = hx$ and $w^{(R)}(h, x) = xh$ for every $(h, x) \in H \times X$.

If H and X are groups and g is a group homomorphism (in particular, if X is a group, H is a subgroup of X, and g is the standard injection of H into X), then $w^{(L)}$ and $w^{(R)}$ become left and right group actions, respectively. ∎

Example A.3.2. Let G and L be two groups and let $g : L \to G$ be a group homomorphism. Also, let M be a subgroup of G, and consider the left and right coset spaces $(G/M)_L$ and $(G/M)_R$ of G defined by M, respectively.

In this setting, let $w^{(L)} : L \times (G/M)_L \to (G/M)_L$ and $w^{(R)} : L \times (G/M)_R \to (G/M)_R$ be defined as follows:

$$w^{(L)}(h, xM) = g_h x M \tag{A.3.10}$$

for every $(h, xM) \in L \times (G/M)_L$ and

$$w^{(R)}(h, Mx) = M x g_h \tag{A.3.11}$$

for every $(h, Mx) \in L \times (G/M)_R$. Note that in the above equalities (A.3.10) and (A.3.11) we used the subscript notation g_h for $g(h)$; we will continue to use the subscript notation for g throughout this example. If we use the "hat" notation for cosets, then the above equalities (A.3.10) and (A.3.11) defining $w^{(L)}$ and $w^{(R)}$ become

$$w^{(L)}(h, \hat{x}) = \widehat{g_h x} \tag{A.3.12}$$

for every $(h, \hat{x}) \in L \times (G/M)_\mathrm{L}$ and

$$w^{(\mathrm{R})}(h, \hat{x}) = \widehat{x g_h} \tag{A.3.13}$$

for every $(h, \hat{x}) \in L \times (G/M)_\mathrm{R}$, respectively.

It is easy to see that $w^{(\mathrm{L})}$ and $w^{(\mathrm{R})}$ are left and right group actions of L on $(G/M)_\mathrm{L}$ and on $(G/M)_\mathrm{R}$, respectively.

If L is just a semigroup, rather than a group, and has a neutral element, say e_L, and if g is a semigroup homomorphism such that $g_{e_L} = e_G$, where e_G is the neutral element of G, then we still can define the maps $w^{(\mathrm{L})}$ and $w^{(\mathrm{R})}$ using the above formulas (A.3.10) and (A.3.11), or (A.3.12) and (A.3.13) in "hat" notation, respectively; in this case $w^{(\mathrm{L})}$ and $w^{(\mathrm{R})}$ are left and right semigroup actions of L on $(G/M)_\mathrm{L}$ and $(G/M)_\mathrm{R}$, respectively.

If L is a subgroup of G, or a subsemigroup of G such that the neutral element e_G of G belongs to L, and if $g : L \to G$ is the standard injection (see Example A.3.1 for the definition of the standard injection), then $w^{(\mathrm{L})}$ and $w^{(\mathrm{R})}$ are defined by $w^{(\mathrm{L})}(h, xM) = hxM$ for every $(h, xM) \in L \times (G/M)_\mathrm{L}$ and $w^{(\mathrm{R})}(h, Mx) = Mxh$ for every $(h, Mx) \in L \times (G/M)_\mathrm{R}$, or, in "hat" notation $w^{(\mathrm{L})}(h, \hat{x}) = \widehat{hx}$ for every $(h, \hat{x}) \in L \times (G/M)_\mathrm{L}$ and $w^{(\mathrm{R})}(h, \hat{x}) = \widehat{xh}$ for every $(h, \hat{x}) \in L \times (G/M)_\mathrm{R}$. ∎

If H is a commutative semigroup (or group) and, as before, X is a nonempty set, then any left semigroup (or group) action of H on X is also a right semigroup (or group) action and vice versa. Naturally, in this case, we call a left or right action, simply, an *action* (of H on X) and we do not use the superscripts $^{(\mathrm{L})}$ or $^{(\mathrm{R})}$ in the notation of such an action.

Now let $(H, \cdot, \tau)$ be a topological semigroup or group, and let $(X, \mathcal{T})$ be a topological space. A left action $w^{(\mathrm{L})} : H \times X \to X$ (or a right action $w^{(\mathrm{R})} : H \times X \to X$) of H on X is said to be *continuous* if $w^{(\mathrm{L})}$ (or $w^{(\mathrm{R})}$) is continuous with respect to the topology $\mathcal{T}$ on X and the product topology of τ and $\mathcal{T}$ on $H \times X$.

Our goal in the next proposition is to show that, under rather natural and general conditions, the actions described in Examples A.3.1 and A.3.2 are continuous.

Proposition A.3.3. *(a) Let $(H, \cdot, \tau)$ be a topological semigroup that has a neutral element e, let $(X, \cdot, \mathcal{T})$ be a topological semigroup that has a neutral element e_X, and let $g : H \to X$ be a continuous semigroup homomorphism such that $g_e = e_X$ (note that, as in Example A.3.1, we use the subscript notation g_h for $g(h)$, $h \in H$). Then the actions $w^{(\mathrm{L})} : H \times X \to X$, $w^{(\mathrm{L})}(h, x) = g_h x$ for every $(h, x) \in H \times X$, and $w^{(\mathrm{R})} : H \times X \to X$, $w^{(\mathrm{R})}(h, x) = x g_h$ for every $(h, x) \in H \times X$, are continuous. In particular, if H is a subsemigroup of X and is endowed with the topology induced by the topology of X on H, if $e_X \in H$, and if g is the standard injection of H into X, then $w^{(\mathrm{L})}$ and $w^{(\mathrm{R})}$ are continuous.*

If H and X are topological groups and g is a continuous group homomorphism (in particular, if H is a subgroup of X and is endowed with the topology induced by the topology of X on H, and if g is the standard injection of H into X), then $w^{(\mathrm{L})}$ and $w^{(\mathrm{R})}$ are continuous group actions.

(b) Let $(L, \cdot, \tau)$ and $(G, \cdot, \mathcal{T})$ be two topological groups, let $g : L \to G$ be a continuous group homomorphism, and let M be a subgroup of G. Then the left action $w^{(L)}$ and the right action $w^{(R)}$ of the topological group L on the topological spaces $((G/M)_L, \mathcal{T}((G/M)_L))$ and $((G/M)_R, \mathcal{T}((G/M)_R))$ defined by the equalities (A.3.10) and (A.3.11) in Example A.3.2, respectively, are continuous, where $\mathcal{T}((G/M)_L)$ and $\mathcal{T}((G/M)_R)$ are the standard topologies on $(G/M)_L$ and $(G/M)_R$ defined by the topology $\mathcal{T}$ on G, respectively.

If L is a topological semigroup with a neutral element and g is a continuous semigroup homomorphism such that $g_{e_L} = e_G$, where e_L and e_G are the neutral elements of L and G, respectively, then the actions $w^{(L)}$ and $w^{(R)}$ are still continuous.

Finally, if L is a subgroup of G, or a subsemigroup of G such that the neutral element e_G of G belongs to L, if $g : L \to G$ is the standard injection, and if L is endowed with the topology induced by $\mathcal{T}$ on L, then $w^{(L)}$ and $w^{(R)}$ are again continuous.

Proof. (a) We will prove only the continuity of $w^{(L)}$ in the case in which H and X are topological semigroups because the proofs of all the other assertions are similar.

To this end, let $u : H \times X \to X \times X$ be defined by $u(h, x) = (g_h, x)$ for every $(h, x) \in H \times X$, and $v : X \times X \to X$ be defined by $v(x, y) = xy$ for every $(x, y) \in X \times X$.

Clearly, v is a continuous function because X is a topological semigroup. Since g is a continuous function and taking into consideration that $u^{-1}(U \times V) = g^{-1}(U) \times V$ for all (open) subsets U and V of X, we obtain that $w^{(L)}$ is continuous because $w^{(L)} = v \circ u$.

(b) Let $(G, \cdot, \mathcal{T})$ be a topological group, and let M be a subgroup of G.

We will only prove that $w^{(L)}$ is continuous in the case in which L is a topological group and $g : L \to G$ is a continuous group homomorphism since the proofs of all the remaining assertions are similar.

In order to prove that $w^{(L)}$ is continuous, it is enough to prove that for every $(h, xM) \in L \times (G/M)_L$ and for every open neighborhood W of $g_h xM$ in $(G/M)_L$ it follows that $\left(w^{(L)}\right)^{-1}(W)$ is a neighborhood of (h, xM) in $L \times (G/M)_L$.

To this end, let $(h, xM) \in L \times (G/M)_L$, let w be an open neighborhood of $g_h xM$ in $(G/M)_L$, and recall (see Sect. A.1 of this appendix) that $\varphi_L : G \to (G/M)_L$ denotes the standard surjection of G on $(G/M)_L$.

Since φ_L is continuous and since $g_h x$ is a coset representative of $g_h xM$ ($g_h x$ is a coset representative of $g_h xM$ because $g_h x = g_h x e_G$, where e_G is the neutral element of G), it follows that $\varphi_L^{-1}(W)$ is an open neighborhood of $g_h x$ in G. Since the algebraic operation that defines the group structure of G is continuous, it follows that there exist open neighborhoods U and V of g_h and x in G, respectively, such that $UV \subseteq \varphi_L^{-1}(W)$; since g is continuous, it follows that $g^{-1}(U)$ is an open neighborhood of h in L. Taking into consideration that φ_L is an open map, we obtain that $g^{-1}(U) \times \varphi_L(V)$ is an open neighborhood of (h, xM) in $L \times (G/M)_L$.

Since for every $h' \in g^{-1}(U)$ and $y \in V$, it follows that $w^{(L)}(h', yM) = g_{h'}yM \in W$ ($g_{h'}yM$ belongs to W because $g_{h'}y \in UV$ and UV is a subset of $\varphi_L^{-1}(W)$), we obtain that $g^{-1}(U) \times \varphi_L(V)$ is a subset of $(w^{(L)})^{-1}(W)$; hence, $(w^{(L)})^{-1}(W)$ is a neighborhood of (h, xM) in $L \times (G/M)_L$. $\square$

In general, when dealing with continuous actions on a topological space $(X, \mathcal{T})$, we will often be interested in studying topics involving the closures of orbits under actions with respect to the topology $\mathcal{T}$ in X. We will call these closed subsets of X *orbit-closures*. Naturally, if $x \in X$, we will denote such an orbit-closure by $\overline{\mathcal{O}(x)}$. If we have to emphasize that the orbit-closure is under a certain action, the action will appear as a subscript; for example, the orbit-closure of $x \in X$ under a left semigroup action $\left(w_h^{(L)}\right)_{h \in H}$ will be denoted $\overline{\mathcal{O}_{\left(w_h^{(L)}\right)_{h \in H}}(x)}$.

A continuous action on $(X, \mathcal{T})$ is said to be *minimal* if all its orbits are dense in X

If H is the additive semigroup $[0, +\infty)$ and X is a nonempty set, any semigroup action of H on X is called a *semiflow*. When dealing with a semiflow $w : [0, +\infty) \times X \to X$, we will, generally, prefer to use the notation $(w_h)_{h \in [0, +\infty)}$ for w, and, subsequently, we will use the notation $w_h(x)$ for $w(h, x)$, $h \in [0, +\infty)$, $x \in X$.

Similarly, if H is the additive group $\mathbb{R}$ of all real numbers, a group action of H on X is called a *flow*. As in the case of semiflows, a flow w will be denoted by $(w_h)_{h \in \mathbb{R}}$, and we will prefer to use $w_h(x)$ instead of $w(h, x)$, $h \in \mathbb{R}$, $x \in X$.

As usual in this book, let $\mathbb{T}$ stand for the interval $[0, +\infty)$ or the set $\mathbb{R}$ of all real numbers, and let (X, d) be a locally compact separable metric space. A semiflow or flow $(w_h)_{h \in \mathbb{T}}$ is said to be *continuous* if $(w_h)_{h \in \mathbb{T}}$ is a continuous action; here, of course, we think of $\mathbb{T}$ as the topological additive semigroup $[0, +\infty)$ or the additive topological group $\mathbb{R}$, where the topologies on $[0, +\infty)$ and $\mathbb{R}$ are the standard topologies generated by the usual metric $\mathbf{d}$ defined by $\mathbf{d}(x, y) = |x - y|$ for every $x \in \mathbb{T}$ and $y \in \mathbb{T}$, and we think of (X, d) as the topological space whose topology is defined by the metric d. The semiflow or flow $(w_h)_{h \in \mathbb{T}}$ is said to be *measurable* if the map $(h, x) \mapsto w_h(x)$ from $\mathbb{T} \times X$ to X is measurable with respect to $\mathcal{B}(X)$ and with respect to the product σ-algebra $\mathcal{L} \otimes \mathcal{B}(X)$, where $\mathcal{L}(\mathbb{T})$ is the σ-algebra of all Lebesgue measurable subsets of $\mathbb{T}$.

When dealing with semiflows and flows, besides the usual orbits and orbit-closures, we can also define forward orbits and forward orbit-closures. To be precise, let $(w_t)_{t \in \mathbb{T}}$ be a semiflow or a flow defined on a nonempty set X. Given $x \in X$, the *forward orbit $\mathcal{O}^{(F)}(x)$ of x under the action of $(w_t)_{t \in \mathbb{T}}$* is the set $\{w_t(x) \mid t \in \mathbb{T}, t \geq 0\}$. If X is a topological space and $(w_t)_{t \in \mathbb{T}}$ is a continuous semiflow or flow, we are often interested in the closure $\overline{\mathcal{O}^{(F)}(x)}$ of $\mathcal{O}^{(F)}(x)$ in the topology of X. The set $\overline{\mathcal{O}^{(F)}(x)}$ is called the *forward orbit-closure of x under the action of $(w_t)_{t \in \mathbb{T}}$*. Note that if $(w_t)_{t \in \mathbb{T}}$ is a semiflow (that is, if $\mathbb{T} = [0, +\infty)$), then every forward orbit is a usual orbit and vice versa. Thus, the forward orbits and the forward orbit-closures are of interest only when $\mathbb{T} = \mathbb{R}$. The semiflow or flow $(w_t)_{t \in \mathbb{T}}$ is said to be *forward minimal* if all the forward orbits are dense in X. Note that if a semiflow or flow is forward minimal, then the semiflow or flow is minimal. Obviously, a minimal semiflow is forward minimal. However, a minimal flow is not necessarily forward minimal. For instance, if we think of $\mathbb{R}$ as being endowed with its standard topology defined by the distance $d : \mathbb{R} \times \mathbb{R} \to \mathbb{R}$, $d(x, y) = |x - y|$ for every $(x, y) \in \mathbb{R} \times \mathbb{R}$, and if $(w_t)_{t \in \mathbb{R}}$ is the (continuous) flow defined by $w_t(x) = x + t$ for every $x \in \mathbb{R}$ and $t \in \mathbb{R}$, then $(w_t)_{t \in \mathbb{R}}$ is minimal but not forward minimal.

Note that if $(w_h)_{h \in \mathbb{R}}$ is a flow, then the restriction $(w_h)_{h \in [0,+\infty)}$ of the flow to $[0, +\infty) \times X$ is a semiflow, which will be called the *semiflow associated to* $(w_h)_{h \in \mathbb{R}}$.

We will now discuss several examples of flows.

Example A.3.4 (Rotations of the Unit Circle). Let $\mathbb{R}/\mathbb{Z}$ be the commutative compact metric group defined in Example A.2.8. For every $t \in \mathbb{R}$, let $w_t : \mathbb{R}/\mathbb{Z} \to \mathbb{R}/\mathbb{Z}$ be defined by $w_t(x + \mathbb{Z}) = t + x + \mathbb{Z}$ for every $x + \mathbb{Z} \in \mathbb{R}/\mathbb{Z}$ (or, in the "hat" notation, $w_t(\hat{x}) = \widehat{t + x}$ for every $\hat{x} \in \mathbb{R}/\mathbb{Z}$). It is easy to see that $(w_t)_{t \in \mathbb{R}}$ is a continuous flow on $\mathbb{R}/\mathbb{Z}$; the flow is known as the *flow of the rotations of the unit circle*. Of course, if we think of $(w_t)_{t \in \mathbb{R}}$ as a flow on $\mathbf{C}_1$, then $w_t(z) = e^{2\pi(t+s)i}$ for every $z \in \mathbf{C}_1$, $z = e^{2\pi s i}$ for some $s \in [0, 1)$, and for every $t \in \mathbb{R}$. In $([0, 1), \boxplus, \rho)$, the flow $(w_t)_{t \in \mathbb{R}}$ is defined as follows: $w_t(s) = \bar{t} \boxplus s$ for every $s \in [0, 1)$ and every $t \in \mathbb{R}$, where $\bar{t}$ is the unique coset representative of $t + \mathbb{Z}$ in $[0, 1)$.

Note that for every $\hat{x} \in \mathbb{R}/\mathbb{Z}$, both the orbit and the forward orbit of $\hat{x}$ under the action of $(w_t)_{t \in \mathbb{R}}$ are the entire space $\mathbb{R}/\mathbb{Z}$. ∎

In the next example we will use the notion of rational independence. Let $n \in \mathbb{N}$ and let $\alpha_1, \alpha_2, \ldots, \alpha_n$ be n real numbers; we say that $\alpha_1, \alpha_2, \ldots, \alpha_n$ are *rationally independent* if whenever $q_1 \in \mathbb{Q}$, $q_2 \in \mathbb{Q}$, $\ldots$, $q_n \in \mathbb{Q}$ are such that $q_1\alpha_1 + q_2\alpha_2 + \cdots + q_n\alpha_n = 0$, then it follows that $q_1 = q_2 = \cdots = q_n = 0$; naturally, if $\alpha_1, \alpha_2, \ldots, \alpha_n$ are not rationally independent, we say that the n numbers are *rationally dependent*.

Example A.3.5 (The Rectilinear Flow on the Torus). Let $n \in \mathbb{N}$, $n \geq 2$, and let $\mathbf{v} \in \mathbb{R}^n$, $\mathbf{v} = (v_1, v_2, \ldots, v_n)$. Also, let $\mathbb{R}^n/\mathbb{Z}^n$ be the commutative compact metric group discussed in Example A.2.9. For every $t \in \mathbb{R}$, let $w_t : \mathbb{R}^n/\mathbb{Z}^n \to \mathbb{R}^n/\mathbb{Z}^n$ be defined by $w_t(\mathbf{x} + \mathbb{Z}^n) = t\mathbf{v} + \mathbf{x} + \mathbb{Z}^n$ for every $\mathbf{x} + \mathbb{Z}^n \in \mathbb{R}^n/\mathbb{Z}^n$ (or, in the "hat" notation, $w_t(\hat{\mathbf{x}}) = \widehat{t\mathbf{v} + \mathbf{x}}$ for every $\hat{\mathbf{x}} \in \mathbb{R}^n/\mathbb{Z}^n$). Clearly, $(w_t)_{t \in \mathbb{R}}$ is a continuous flow; the flow is called the *rectilinear flow on* $\mathbb{R}^n/\mathbb{Z}^n$ *with velocity* $\mathbf{v}$ (or *defined by* $\mathbf{v}$). If we consider $(w_t)_{t \in \mathbb{R}}$ as a flow on $\mathbf{C}_1^n$, then

$$w_t(\mathbf{z}) = \left(e^{2\pi(tv_1+s_1)i}, e^{2\pi(tv_2+s_2)i}, \ldots, e^{2\pi(tv_n+s_n)i}\right)$$

for every $\mathbf{z} \in \mathbf{C}_1^n$, $\mathbf{z} = \left(e^{2\pi s_1 i}, e^{2\pi s_2 i}, \ldots, e^{2\pi s_n i}\right)$ for some $s_i \in [0, 1)$, $i = 1, 2, \ldots, n$, and for every $t \in \mathbb{R}$.

If we think of $(w_t)_{t \in \mathbb{R}}$ as a flow on $([0, 1)^n, \boxplus_n, \rho_n)$, then

$$w_t(\mathbf{z}) = (t\mathbf{v}) \boxplus_n \mathbf{z} = ((tv_1) \boxplus z_1, (tv_2) \boxplus z_2, \ldots, (tv_n) \boxplus z_n)$$

for every $\mathbf{z} \in [0, 1)^n$, $\mathbf{z} = (z_1, z_2, \ldots, z_n)$, and every $t \in \mathbb{R}$.

In contrast with the flow of the rotations of the unit circle discussed in Example A.3.4, the orbits of the rectilinear flow on the torus are not even dense in the torus in general. However, if the entries $v_1, v_2, \ldots, v_n$ of $\mathbf{v}$ are rationally independent, then the orbit of every $\hat{x} \in \mathbb{R}^n/\mathbb{Z}^n$ is dense under the action of the flow. Indeed, assume that $v_1, v_2, \ldots, v_n$ are rationally independent; then there exists an $\alpha \in \mathbb{R}$ such that $v_1, v_2, \ldots, v_n, \alpha^{-1}$ are rationally independent; consequently, $\alpha v_1, \alpha v_2, \ldots, \alpha v_n, 1$ are rationally independent; therefore, we may apply Lemma 1 of Section 3.1 on p. 66 of Cornfeld, Fomin and Sinai's monograph [22] to the map w_α; according to the above-mentioned lemma, the set $\mathcal{O}_\alpha(\hat{\mathbf{x}}) = \left\{\widehat{m\alpha\mathbf{v} + \mathbf{x}} \mid m \in \mathbb{Z}\right\}$ is dense in $\mathbb{R}^n/\mathbb{Z}^n$ for every $\hat{\mathbf{x}} \in \mathbb{R}^n/\mathbb{Z}^n$; since $\mathcal{O}_\alpha(\hat{\mathbf{x}})$ is included in the orbit $\mathcal{O}(\hat{\mathbf{x}})$ of $\hat{\mathbf{x}}$ under the action of the flow $(w_t)_{t \in \mathbb{R}}$ for every $\hat{\mathbf{x}} \in \mathbb{R}^n/\mathbb{Z}^n$, it follows that the orbit of every $\hat{\mathbf{x}} \in \mathbb{R}^n/\mathbb{Z}^n$ under the action of $(w_t)_{t \in \mathbb{R}}$ is dense in the torus.

Like the flow of the rotations of the unit circle (Example A.3.4), the rectilinear flow on the torus is discussed in almost every book that, even remotely, deals with flows; see, for instance, the monographs by Bekka and Mayer [10], Cornfeld, Fomin and Sinai [22], Lasota and Mackey [57], and Starkov [113]. ∎

Example A.3.6 (The Geodesic Flow on $\mathrm{PSL}(2, \mathbb{R})$*).* There are several geodesic flows. Their definition depends on the space (or the type of spaces) on which the flow acts. For the reason for calling these flows geodesic and for additional details on these flows, see Chapter 2 of Bekka and Mayer's book [10] and Section 14 of Starkov's monograph [113].

For every $t \in \mathbb{R}$, set $g_t = \begin{pmatrix} e^{\frac{t}{2}} & 0 \\ 0 & e^{-\frac{t}{2}} \end{pmatrix}$, and let $w_t : \mathrm{PSL}(2, \mathbb{R}) \to \mathrm{PSL}(2, \mathbb{R})$ be defined by $w_t(\hat{g}) = \hat{g}\hat{g}_t$ for every $\hat{g} \in \mathrm{PSL}(2, \mathbb{R})$, where $\mathrm{PSL}(2, \mathbb{R})$ is the locally compact separable metrizable group defined in Example A.2.10, and $\hat{g}_t$ is, of course, the equivalence class $g_t L = \{g_t, -g_t\}$ (where $L = \{\mathbf{I}_2, -\mathbf{I}_2\}$), $t \in \mathbb{R}$. Clearly, $(w_t)_{t \in \mathbb{R}}$ is a flow. Since the standard surjection of $\mathrm{SL}(2, \mathbb{R})$ onto $\mathrm{PSL}(2, \mathbb{R})$ is continuous and the map $t \mapsto g_t$, $t \in \mathbb{R}$, from $\mathbb{R}$ to $\mathrm{SL}(2, \mathbb{R})$ is a continuous group homomorphism, it follows that the map $t \mapsto \hat{g}_t$, $t \in \mathbb{R}$, from $\mathbb{R}$ to $\mathrm{PSL}(2, \mathbb{R})$ is also a continuous group homomorphism; thus, by (*a*) of Proposition A.3.3, the flow $(w_t)_{t \in \mathbb{R}}$ is continuous. The flow is known as the *geodesic flow on* $\mathrm{PSL}(2, \mathbb{R})$, and, as we mentioned above, $(w_t)_{t \in \mathbb{R}}$ is not the only flow that is called geodesic. Other geodesic flows defined on certain topological coset spaces of $\mathrm{PSL}(2, \mathbb{R})$ will be discussed later in Appendix B. ∎

Example A.3.7 (The Horocycle Flows on $\mathrm{PSL}(2, \mathbb{R})$*).* As in the case of the geodesic flows, there are various horocycle flows depending on the space or type of spaces on which the flows can be constructed, and for each such space there are two horocycle flows that can be defined. In this example, we will discuss the two horocycle flows on $\mathrm{PSL}(2, \mathbb{R})$. Later, in Appendix B, we will present other horocycle flows defined on certain topological coset spaces. Even though the geodesic and the horocycle flows are strikingly different, they are connected. For details on this connection, for the reason behind the use of the term horocycle flows, and for additional information on these flows, see Chapter 4 of Bekka and Mayer's monograph [10] and Section 14 of Starkov's book [113].

For every $t \in \mathbb{R}$, set $h_t^{(1)} = \begin{pmatrix} 1 & t \\ 0 & 1 \end{pmatrix}$ and $h_t^{(2)} = \begin{pmatrix} 1 & 0 \\ t & 1 \end{pmatrix}$, and let $v_t^{(i)} : \mathrm{PSL}(2, \mathbb{R}) \to$ $\mathrm{PSL}(2, \mathbb{R})$ be defined by $v_t^{(i)}(\hat{h}) = \hat{h}\hat{h}_t^{(i)}$ for every $\hat{h} \in \mathrm{PSL}(2, \mathbb{R})$, where $\hat{h}_t^{(i)}$ is the element of $\mathrm{PSL}(2, \mathbb{R})$ defined as the equivalence class of $h_t^{(i)}$ in $\mathrm{PSL}(2, \mathbb{R})$ (that is, $\hat{h}_t^{(i)} = h_t^{(i)} L = \{h_t^{(i)}, -h_t^{(i)}\}$), $t \in \mathbb{R}$, $i = 1, 2$. Since for every $i = 1, 2$, the maps $t \mapsto h_t^{(i)}$, $t \in \mathbb{R}$, are group homomorphisms from $\mathbb{R}$ to $\mathrm{SL}(2, \mathbb{R})$, it follows that $\left(v_t^{(i)}\right)_{t \in \mathbb{R}}$, $i = 1, 2$, are flows. Arguments similar to those used in Example A.3.6 allow us to conclude that, for $i = 1, 2$, the group homomorphisms $t \mapsto \hat{h}_t^{(i)}$, $t \in \mathbb{R}$, are continuous; therefore, using (a) of Proposition A.3.3 as in Example A.3.6, we obtain that the flows $\left(v_t^{(i)}\right)_{t \in \mathbb{R}}$, $i = 1, 2$, are continuous, as well. The two flows $\left(v_t^{(1)}\right)_{t \in \mathbb{R}}$ and $\left(v_t^{(2)}\right)_{t \in \mathbb{R}}$ are called the *horocycle flows on* $\mathrm{PSL}(2, \mathbb{R})$. ∎

Let $(w_t)_{t \in \mathbb{T}}$ be a semiflow or a flow defined on a nonempty set X.

An element x of X is called a *periodic point* for $(w_t)_{t \in \mathbb{T}}$ if there exists a $t_1 \in \mathbb{T}$, $t_1 > 0$, such that $w_{t_1}(x) = x$ and there exists a $t \in (0, t_1)$ such that $w_t(x) \neq x$.

Proposition A.3.8. *Let $(w_t)_{t \in \mathbb{T}}$ be a continuous semiflow or flow defined on a Hausdorff topological space $(X, \mathcal{T})$, and let $x_0 \in X$. If x_0 is a periodic point for $(w_t)_{t \in \mathbb{T}}$, then there exists a $t_0 \in \mathbb{T}$, $t_0 > 0$, such that $w_{t_0}(x_0) = x_0$ and $w_t(x_0) \neq x_0$ for every t in the open interval $(0, t_0)$.*

Proof. Let x_0 be a periodic point for $(w_t)_{t \in \mathbb{T}}$, and consider the function $f : \mathbb{T} \to X$ defined by $f(t) = w_t(x_0)$ for every $t \in \mathbb{T}$. Since $(w_t)_{t \in \mathbb{T}}$ is a continuous semiflow or flow, it follows that f is a continuous function.

Set $A = \{t \in \mathbb{T} \mid t > 0, f(t) = x_0\}$. Then A is a nonempty set because x_0 is a periodic point, and $A \subseteq [0, +\infty)$.

Let $t_0 = \inf A$, and note that the proof of the proposition will be completed if we show that $t_0 > 0$. Indeed, $t_0 \in A$ because f is continuous and there exists a sequence $(t_n)_{n \in \mathbb{N}}$ of elements of A that converges to t_0, so, $f(t_0) = x_0$; thus, taking into consideration the definition of A, we obtain that if $t_0 > 0$, then the proposition is true.

In order to prove that $t_0 > 0$, assume that $t_0 = 0$. Then, for every $\varepsilon \in \mathbb{R}$, $\varepsilon > 0$, there exists an $s_\varepsilon \in (0, \varepsilon)$ such that $s_\varepsilon \in A$; therefore, the set $\{ns_\varepsilon \mid n \in \mathbb{N}, \varepsilon \in \mathbb{R}, \varepsilon > 0\}$ is dense in $[0, +\infty)$ and is a subset of A. Since f is continuous, it follows that $A = [0, +\infty)$. We have obtained a contradiction since $0 \notin A$. □

The number t_0 whose existence is discussed in Proposition A.3.8 is called the *minimal period* of x_0 (under the action of $(w_t)_{t \in \mathbb{T}}$).

If $(w_t)_{t \in \mathbb{T}}$ is a continuous semiflow or flow as in Proposition A.3.8, if x_0 is a periodic point, and if t_0 is the minimal period of x_0, then, clearly, $\mathcal{O}(x_0) = \overline{\mathcal{O}(x_0)} = \{w_t(x_0) \mid t \in [0, t_0]\}$, so the orbit $\mathcal{O}(x_0)$ of x_0 is a compact subset of X.

Note that if $X = \mathbb{R}/\mathbb{Z}$, $\mathbb{T} = \mathbb{R}$, and $(w_t)_{t \in \mathbb{R}}$ is the flow of the rotations of the unit circle discussed in Example A.3.4, then all the elements of $\mathbb{R}/\mathbb{Z}$ are periodic points of minimal period 1.

If $(w_t)_{t \in \mathbb{R}}$ is a rectilinear flow on a torus $\mathbb{R}^n/\mathbb{Z}^n$ for some $n \in \mathbb{N}$, $n \geq 2$, and if the flow has velocity $\mathbf{v} \in \mathbb{R}^n$, $\mathbf{v} = (v_1, v_2, \ldots, v_n)$, then $(w_t)_{t \in \mathbb{R}}$ has periodic points if and only if the following condition is satisfied: there exist $\alpha \in \mathbb{R}$, $\alpha \neq 0$, and n rational numbers $q_1, q_2, \ldots, q_n$ not all equal to zero such that $v_i = q_i \alpha$ for every $i = 1, 2, \ldots, n$. If $(w_t)_{t \in \mathbb{R}}$ has periodic points, then all the elements of $\mathbb{R}^n/\mathbb{Z}^n$ are periodic points and they have the same minimal period. The common minimal period is equal to $\dfrac{m}{\alpha}$ if $\alpha \notin \mathbb{Q}$ where m is the smallest natural number such that $mq_i \in \mathbb{Z}$ for every $i = 1, 2, \ldots, n$; if $\alpha \in \mathbb{Q}$, then the common minimal period is the smallest natural number m such that $mq_i \alpha \in \mathbb{Z}$ for every $i = 1, 2, \ldots, n$.

In the above examples, if a flow has periodic points, then all the elements of the space X on which the flow is defined are periodic points and they have a common minimal period. Of course, there exist flows that have periodic as well as nonperiodic points, and the minimal period of a periodic point may differ from point to point. We will discuss such situations later in Appendix B.

Appendix B
Invariant Measures, One-Parameter Convolution Semigroups, and Additional Examples of Semiflows and Flows

In this appendix we continue the discussion started in Appendix A. Thus, we will use the terminology and notation introduced there.

We start (in Sect. B.1) by briefly reviewing various types of invariant measures, including the Haar measure. When dealing with invariant measures on coset spaces, we define the finite volume coset spaces, and then we conclude Sect. B.1 by discussing the geodesic and the horocycle flows on finite volume spaces. Next (in Sect. B.2) we define the convolution of measures and the exponential function in Banach algebras with unit. This allows us to discuss one-parameter convolution semigroups of probability measures in Sect. B.3. In Sect. B.4 (the last section) we discuss exponential semiflows on the space of $n \times n$ column stochastic matrices, $n \geq 2$, and on the space of n-dimensional probability vectors, $n \geq 2$. Also in this section, we discuss exponential flows on coset spaces of $SL(n, \mathbb{R})$, $n \geq 2$, including the unipotent flows.

B.1 Invariant Measures

Our goal in this section is to briefly review several facts about invariant Borel measures on certain topological semigroups, topological groups, and coset spaces endowed with standard topologies. The discussion about invariant measures on coset spaces will allow us to recall from the literature the geodesic and the horocycle flows on certain coset spaces.

Let (H, d) be a locally compact separable metric semigroup, and, as usual when dealing with locally compact separable metric spaces, let $\mathcal{B}(H)$ be the Borel σ-algebra on H.

Let $\mu : \mathcal{B}(H) \to [0, +\infty]$ be a measure (not necessarily finite); here, of course, $[0, +\infty]$ stands for the extended half-line $[0, +\infty) \cup \{+\infty\}$.

R. Zaharopol, *Invariant Probabilities of Transition Functions*, Probability and Its Applications 44, DOI 10.1007/978-3-319-05723-1,
© Springer International Publishing Switzerland 2014

We say that μ is *left translation invariant* (or *l.t. invariant*) if $\mu(x^{-1}A) = \mu(A)$ for every $A \in \mathcal{B}(H)$ and $x \in H$; similarly, we call μ *right translation invariant* (or *r.t. invariant*) if $\mu(Ax^{-1}) = \mu(A)$ for every $A \in \mathcal{B}(H)$ and $x \in H$.

We say that μ is *finite on the compact subsets* of H if, as expected, $\mu(K) < +\infty$ for every compact subset K of H.

We say that μ is a *left* or *right Haar measure* on H if μ is a nonzero measure which is finite on the compact subsets of H, and if μ is l.t. or r.t. invariant, respectively. Using Proposition 7.2.3 of Cohn [20], pp. 206–207, and the fact that a metric space is second countable if and only if the space is separable (see, for instance, D10 and D32 of Appendix D in Cohn's book [20]), we obtain that in our setting the left and the right Haar measures on semigroups are regular measures. Therefore, in our setting, the left and right Haar measures are precisely the nonzero l^*-invariant and r^*-invariant measures defined on p. 77 in Högnäs and Mukherjea's monograph [48], respectively. Note also that if H is a locally compact separable metric group, then the left and right Haar measures defined here are the usual such measures that appear in the literature (see, for instance, Section 2 of Chapter 9 in Cohn's book [20]).

A natural question concerning left and right Haar measures is that of their existence. Another natural question is whether or not such measures are unique for a given locally compact separable metric semigroup or group. Finally, a third question is whether or not there exists a natural meaningful connection between left and right Haar measures.

Note that the second question (the unicity of the left or right Haar measures) is trivial as stated because such measures are never unique; indeed, if μ is a left or right Haar measure on H, then, for every $c \in (0, +\infty)$, the measure $c\mu : \mathcal{B}(H) \to \mathbb{R}$ defined by $(c\mu)(A) = c\mu(A)$ for every $A \in \mathcal{B}(H)$ is also a left or right Haar measure on H, respectively. Thus, we are led to the following question, which is indeed significant: for what locally compact separable metric semigroups (H, d) is the following assertion true?

(UMC)   If μ and ν are two left or two right Haar measures on (H, d), then there exists a $c \in (0, +\infty)$ such that $\mu = c\nu$.

If (H, d) has left or right Haar measures and if (UMC) holds true for the left or right Haar measures of (H, d), then we say that the left or right Haar measures of (H, d) exist and are *unique up to a multiplicative constant*, respectively. In general, if $(Y, \mathcal{Y})$ is a measurable space and $\mathcal{P}$ is a property that a (positive) not necessarily finite measure on $(Y, \mathcal{Y})$ may or may not possess, then we say that the measures on $(Y, \mathcal{Y})$ that have property $\mathcal{P}$ are *unique up to a multiplicative constant* if whenever a measure μ on $(Y, \mathcal{Y})$ has property $\mathcal{P}$, it follows that all the measures $c\mu$, $c \in (0, +\infty)$, have property $\mathcal{P}$, and if, whenever μ and ν are two measures on $(Y, \mathcal{Y})$ that have property $\mathcal{P}$, it follows that $\nu = c\mu$ for some $c \in (0, +\infty)$.

It turns out that if (H, d) is a locally compact separable metric group, then (H, d) has left and right Haar measures, and these measures are unique up to a multiplicative constant; moreover, there is a nice connection between the right and the left Haar measures of (H, d). For future reference, we state these results in the next theorem. The results are basic facts about Haar measures, and their proofs can be found in many textbooks and monographs; see, for instance, Sections 2 and 3 of Chapter 9 in Cohn [20], or Sections 4 and 6 of Chapter 14 in Royden [103].

Theorem B.1.1. *Let (G, d) be a locally compact separable metric group, and let $\mathcal{B}(G)$ be the Borel σ-algebra on G.*

(a) There exist left and right Haar measures on (G, d). In general, these measures are distinct; that is, a left Haar measure may or may not be a right Haar measure and vice versa.

(b) The left and right Haar measures on (G, d) are unique up to a multiplicative constant.

(c) There exists a continuous group homomorphism $\Delta : G \to (0, +\infty)$, where $(0, +\infty)$ is thought of as a topological group endowed with the usual multiplication of real numbers and the usual topology generated by the distance defined by the absolute value in $\mathbb{R}$, such that for every left Haar measure h_{L} and every right Haar measure h_{R} on (G, d) there exists a strictly positive real number c such that $h_{\mathrm{R}} = c\Delta h_{\mathrm{L}}$ (in the sense that $h_{\mathrm{R}}(A) = c \int_A \Delta(x)\, \mathrm{d}h_{\mathrm{L}}(x)$ for every $A \in \mathcal{B}(G)$).

The homomorphism Δ that appears in (c) of Theorem B.1.1 is called the *modular function* of G. Note that given a group G as in the theorem, the modular function of G is unique; that is, there exists exactly one continuous homomorphism $\Delta : G \to (0, +\infty)$ for which assertion (c) of Theorem B.1.1 holds true.

A locally compact separable metric group G is said to be *unimodular* if its modular function Δ is the constant function 1 (that is, if $\Delta(x) = 1$ for every $x \in G$).

Observations. (1) Let G be a locally compact separable metric group. The following assertions are equivalent:

    (a) G is unimodular.

    (b) Every left Haar measure on G is also a right Haar measure.

    (c) Each right Haar measure on G is a left Haar measure, as well.

(2) Clearly, any commutative locally compact separable metric group is unimodular. It can be shown that every compact metric group is unimodular (see, for instance, Proposition 9.3.5, p. 314, of Cohn [20]). Since there exist noncommutative compact metric groups, it follows that there exist noncommutative compact metric groups that are unimodular. Moreover, there exist locally compact separable metric groups that are unimodular even though they are neither compact nor commutative; an example of such a group is $\mathrm{SL}(2, \mathbb{R})$ (the fact that $\mathrm{SL}(2, \mathbb{R})$ is unimodular can be deduced from Example 2.12, p. 49, of Bekka and Mayer's monograph [10] and the comments that appear after Definition 2.1, p. 42, in [10]; we will return to this topic in Example B.1.5). Finally, we note that, as expected, not every locally compact separable metric

group is unimodular (for an example see, for instance, Exercise 3, Section 9.3, p. 316, of Cohn [20]).￼▲

Let $(H, \cdot)$ be a semigroup that has a neutral element (note that the case in which $(H, \cdot)$ is a group is under consideration, as well).

Also, let (X, d) be a locally compact separable metric space.

Let μ be a (positive) not necessarily finite Borel measure. If $\left(w_h^{(L)}\right)_{h \in H}$ is a left semigroup or group action of H on X, we say that μ is an *invariant measure* for (or of) $\left(w_h^{(L)}\right)_{h \in H}$ if $\mu\left(\left(w_h^{(L)}\right)^{-1}(A)\right) = \mu(A)$ for every $h \in H$ and $A \in \mathcal{B}(X)$. Similarly, if $\left(w_h^{(R)}\right)_{h \in H}$ is a right semigroup or group action, we say that μ is an *invariant measure* for (of) $\left(w_h^{(R)}\right)_{h \in H}$ if $\mu\left(\left(w_h^{(R)}\right)^{-1}(A)\right) = \mu(A)$ for every $h \in H$ and $A \in \mathcal{B}(X)$.

An element μ of $\mathcal{M}(X)$ is said to be an *invariant element* for (of) a left semigroup or group action $\left(w_h^{(L)}\right)_{h \in H}$ if the positive part μ^+ and the negative part μ^- of μ are invariant measures for $\left(w_h^{(L)}\right)_{h \in H}$. In a similar manner, an element μ of $\mathcal{M}(X)$ is said to be an *invariant element* for (of) a right semigroup or group action $\left(w_h^{(R)}\right)_{h \in H}$ if both μ^+ and μ^- are invariant measures for $\left(w_h^{(R)}\right)_{h \in H}$.

In general, in this book, whenever we deal with infinite (positive) Borel measures on (X, d) that are invariant for some action, we assume that these measures are regular (note that in our setting a (positive) measure on X is regular if and only if the measure is finite on the compact subsets of X).

Example B.1.2. Let (G, d) be a locally compact separable metric group, let M be a closed subgroup of G, and let $(G/M)_L$ and $(G/M)_R$ be the left and right coset spaces of G defined by M, respectively.

We endow $(G/M)_L$ and $(G/M)_R$ with their standard topologies defined by the metric topology on G, and with the metrics η_L and η_R, respectively, defined in Proposition A.2.3. Thus, using Proposition A.2.4, we obtain that $((G/M)_L, \eta_L)$ and $((G/M)_R, \eta_R)$ are locally compact separable metric spaces.

The left action $w^{(L)} : G \times (G/M)_L \to (G/M)_L$ defined by $w^{(L)}(y, xM) = yxM$ for every $(y, xM) \in G \times (G/M)_L$ is called the *standard action of G on $(G/M)_L$* or, simply, the *action of G on $(G/M)_L$* if there is no danger of confusion. Note that in the "hat" notation, the action of G on $(G/M)_L$ is $w^{(L)}(y, \hat{x}) = \widehat{yx}$ for every $(y, \hat{x}) \in G \times (G/M)_L$.

Similarly, the right action $w^{(R)} : G \times (G/M)_R \to (G/M)_R$ defined by $w^{(R)}(y, Mx) = Mxy$ for every $(y, Mx) \in G \times (G/M)_R$ is called the *standard action of G on $(G/M)_R$*, or the *action of G on $(G/M)_R$*. Even though the "hat" notation for the action of G on $(G/M)_R$ is somewhat similar to that for the action of G on $(G/M)_L$ ($w^{(R)}(y, \hat{x}) = \widehat{xy}$ for every $(y, \hat{x}) \in G \times (G/M)_R$), it will be clear from the context whether we are dealing with the action of G on $(G/M)_L$ or on $(G/M)_R$.

An invariant Borel measure for $w^{(L)}$ or for $w^{(R)}$ is called a *G-invariant measure* on $(G/M)_L$ or on $(G/M)_R$, respectively. Similarly, an element of $\mathcal{M}((G/M)_L)$ or of $\mathcal{M}((G/M)_R)$ that is invariant for $w^{(L)}$ or for $w^{(R)}$ is called a *G-invariant element* of $\mathcal{M}((G/M)_L)$ or of $\mathcal{M}((G/M)_R)$, respectively.

One of the main features of the Borel G-invariant real-valued signed or positive infinite measures that we have just defined, is the following: if μ is such a Borel G-invariant measure on $(G/M)_L$, then μ will be an invariant measure for any of the left actions defined in Example A.3.2, and, similarly, if μ is a Borel G-invariant measure on $(G/M)_R$, then μ is an invariant measure for all the right actions defined in Example A.3.2. ∎

Let (G, d) be a locally compact separable metric group, and let M be a closed subgroup of G. Thus, M is a locally compact separable metric group in its own right when endowed with the metric d_M and the topology defined by d_M on M, where d_M is the restriction of d to $M \times M$. Since, by Theorem B.1.1, both G and M have left and right Haar measures, it is intuitively plausible that, perhaps under some additional conditions, these Haar measures could be used to define in a natural manner G-invariant measures on $(G/M)_L$ and on $(G/M)_R$. It turns out that our intuition is correct in this case, and we will now briefly review the results that we need concerning the existence and unicity of these G-invariant measures.

To this end, let (G, d) and M be as above, and let $h_M^{(L)}$ be a left Haar measure on M.

If $f : G \to \mathbb{R}$ is a continuous function with compact support (that is, if $f \in C_c(G)$), then, clearly, for every $x \in G$, the map $\xi \mapsto f(x\xi)$, $\xi \in M$, is a real-valued function on M, which is continuous and with compact support, so the map is integrable with respect to $h_M^{(L)}$. Therefore, it makes sense to define the function $w_f^{(L)} : G \to \mathbb{R}$, $w_f^{(L)}(x) = \int_M f(x\xi)\, dh_M^{(L)}(\xi)$ for every $x \in G$.

Now, $w_f^{(L)}$ is constant on the left cosets of M in G in the sense that $w_f^{(L)}(x) = w_f^{(L)}(y)$ whenever $xM = yM$; indeed, if $y \in xM$, then $y = xa$ for some $a \in M$; therefore, using the fact that $h_M^{(L)}$ is left translation invariant, we obtain that

$$w_f^{(L)}(y) = \int_M f(xa\xi)\, dh_M^{(L)}(\xi) = \int_M f(x\xi)\, dh_M^{(L)}(\xi) = w_f^{(L)}(x) \text{ for every } x \in G.$$

Accordingly, it makes sense to define the function $g_f^{(L)} : (G/M)_L \to \mathbb{R}$, $g_f^{(L)}(\hat{x}) = w_f^{(L)}(x)$ for every $\hat{x} \in (G/M)_L$ because, for every $\hat{x} \in (G/M)_L$, $\hat{x} = xM$, the definition of $g_f^{(L)}(\hat{x})$ does not depend on the particular coset representative of xM used.

It can be shown (see Subsection 3.2 of Chapter 3 on pp. 58–59 of Reiter's monograph [96]) that, for every $f \in C_c(G)$, the function $g_f^{(L)}$ is continuous and has compact support (that is, $g_f^{(L)} \in C_c((G/M)_L)$). Thus, it makes sense to define the operator $\mathbf{Q}_{h_M^{(L)}} : C_c(G) \to C_c((G/M)_L)$, $\mathbf{Q}_{h_M^{(L)}} f = g_f^{(L)}$ for every $f \in C_c(G)$.

It is easy to see that $\mathbf{Q}_{h_M^{(L)}}$ is a linear operator, and it can be shown that $\mathbf{Q}_{h_M^{(L)}}$ is also a surjection (see Subsection 4.2 of Chapter 3 of Reiter [96], pp. 68–69, and the footnote on p. 68 in [96]).

Similarly, if $h_M^{(R)}$ is a right Haar measure on M, then for every $f \in C_c(G)$, the function $w_f^{(R)} : G \to \mathbb{R}$, $w_f^{(R)}(x) = \int_M f(\xi x)\, dh_M^{(R)}(\xi)$ for every $x \in G$, is well-defined (in the sense that, for every $x \in G$, the map $\xi \mapsto f(\xi x)$, $\xi \in M$, is integrable; the integrability of the maps $\xi \mapsto f(\xi x)$, $\xi \in M$, $x \in G$, is a consequence of the fact that these maps are continuous and have compact supports) and constant on the right cosets of M in G; thus, for every $f \in C_c(G)$, the function $g_f^{(R)} : (G/M)_R \to \mathbb{R}$, $g_f^{(R)}(\hat{x}) = w_f^{(R)}(x)$ for every $\hat{x} \in (G/M)_R$, is well-defined and, by making straightforward modifications to the arguments given in Subsection 3.2 of Chapter 3 on pp. 58–59 of Reiter [96], we obtain that $g_f^{(R)} \in C_c((G/M)_R)$; subsequently, the map $\mathbf{Q}_{h_M^{(R)}} : C_c(G) \to C_c((G/M)_R)$, $\mathbf{Q}_{h_M^{(R)}} f = g_f^{(R)}$ for every $f \in C_c(G)$, is well-defined. Clearly, $\mathbf{Q}_{h_M^{(R)}}$ is a linear operator, and, by modifying the arguments given in Subsection 4.2 of Chapter 3 on pp. 68–69 of Reiter [96], we obtain that $\mathbf{Q}_{h_M^{(R)}}$ is a surjective operator.

Using the operators $\mathbf{Q}_{h_M^{(L)}}$ and $\mathbf{Q}_{h_M^{(R)}}$, we can now state the results concerning the existence and unicity of G-invariant measures on $(G/M)_L$ and $(G/M)_R$.

Theorem B.1.3. *Let, as above, (G, d) be a locally compact separable metric group, and let M be a closed subgroup of G. Also, let Δ_G and Δ_M be the modular functions of G and M, respectively, and assume that $\Delta_{G|M} = \Delta_M$; that is, assume that the restriction of Δ_G to M is equal to Δ_M.*

(a) If $h_G^{(L)}$ and $h_M^{(L)}$ are left Haar measures on G and M, respectively, then there exists a unique regular G-invariant measure $v_{\left(h_G^{(L)}, h_M^{(L)}\right)}$ on $((G/M)_L, \mathcal{B}((G/M)_L))$ such that

$$\int_{(G/M)_L} \mathbf{Q}_{h_M^{(L)}} f(\hat{x})\, dv_{\left(h_G^{(L)}, h_M^{(L)}\right)}(\hat{x}) = \int_G f(x)\, dh_G^{(L)}(x) \tag{B.1.1}$$

for every $f \in C_c(G)$. Moreover, the measures $v_{\left(h_G^{(L)}, h_M^{(L)}\right)}$ that we obtain using the above equality (B.1.1) for various left Haar measures $h_G^{(L)}$ and $h_M^{(L)}$ on G and M, respectively, are unique up to a multiplicative constant, in the sense that, if $h_G^{(L)'}$ and $h_M^{(L)'}$ are two other left Haar measures (not necessarily distinct from $h_G^{(L)}$ and $h_M^{(L)}$) on G and M, respectively, and if $v_{\left(h_G^{(L)'}, h_M^{(L)'}\right)}$ is the resulting regular G-invariant measure on $((G/M)_L, \mathcal{B}((G/M)_L))$ obtained using $h_G^{(L)'}$ and $h_M^{(L)'}$ in the above equality (B.1.1) instead of $h_G^{(L)}$ and $h_M^{(L)}$, respectively, then $v_{\left(h_G^{(L)'}, h_M^{(L)'}\right)} = c\, v_{\left(h_G^{(L)}, h_M^{(L)}\right)}$ for some $c \in (0, +\infty)$.

*(b) Similar assertions as in (a) hold true for the measurable space $((G/M)_\mathrm{R},$
$\mathcal{B}((G/M)_\mathrm{R}))$. That is, if $h_G^{(\mathrm{R})}$ and $h_M^{(\mathrm{R})}$ are right Haar measures on G and M,
respectively, then there exists a unique regular G-invariant measure $v_{\left(h_G^{(\mathrm{R})},h_M^{(\mathrm{R})}\right)}$
on $((G/M)_\mathrm{R}, \mathcal{B}((G/M)_\mathrm{R}))$ such that*

$$\int\limits_{(G/M)_\mathrm{R}} \mathbf{Q}_{h_M^{(\mathrm{R})}} f(\hat{x})\, \mathrm{d}v_{\left(h_G^{(\mathrm{R})},h_M^{(\mathrm{R})}\right)}(\hat{x}) = \int\limits_G f(x)\, \mathrm{d}h_G^{(\mathrm{R})}(x) \qquad \text{(B.1.2)}$$

*for every $f \in C_c(G)$. The measures $v_{\left(h_G^{(\mathrm{R})},h_M^{(\mathrm{R})}\right)}$ obtained using the above
equality (B.1.2) for various right Haar measures $h_G^{(\mathrm{R})}$ and $h_M^{(\mathrm{R})}$ on G and M,
respectively, are unique up to a multiplicative constant.*

It is the custom to use the less rigorous, but intuitively more appealing notation
$\int\limits_M f(x\xi)\, \mathrm{d}h_M^{(\mathrm{L})}(\xi)$ and $\int\limits_M f(\xi x)\, \mathrm{d}h_M^{(\mathrm{R})}(\xi)$ for $\mathbf{Q}_{h_M^{(\mathrm{L})}} f(\hat{x})$ and $\mathbf{Q}_{h_M^{(\mathrm{R})}} f(\hat{x})$, respectively
(see p. 58 of Reiter [96]; the notation was introduced by A. Weil in his monograph
[129], p. 42). Using Weil's notation, we can rewrite the two equalities (B.1.1)
and (B.1.2) of Theorem B.1.3 as follows:

$$\int\limits_{(G/M)_\mathrm{L}} \left(\int\limits_M f(x\xi)\, \mathrm{d}h_M^{(\mathrm{L})}(\xi) \right) \mathrm{d}v_{\left(h_G^{(\mathrm{L})},h_M^{(\mathrm{L})}\right)}(\hat{x}) = \int\limits_G f(x)\, \mathrm{d}h_G^{(\mathrm{L})}(x) \qquad \text{(B.1.3)}$$

for every $f \in C_c(G)$, where $\hat{x} = xM$, and

$$\int\limits_{(G/M)_\mathrm{R}} \left(\int\limits_M f(\xi x)\, \mathrm{d}h_M^{(\mathrm{R})}(\xi) \right) \mathrm{d}v_{\left(h_G^{(\mathrm{R})},h_M^{(\mathrm{R})}\right)}(\hat{x}) = \int\limits_G f(x)\, \mathrm{d}h_G^{(\mathrm{R})}(x) \qquad \text{(B.1.4)}$$

for every $f \in C_c(G)$, where $\hat{x} = Mx$, respectively.

The above formulas (B.1.3) and (B.1.4) are called *Weil's formulas*, especially
if M is a normal closed subgroup of G, because the results of Theorem B.1.3 in
a much more general form were obtained by Weil in his pioneering monograph
[129]. We have followed here Reiter [96], and the results stated in Theorem B.1.3
are proved, in a significantly more general form, in Subsection 1.1 of Chapter 8
of Reiter [96], pp. 157–159 (the statements of Theorem B.1.3 are good enough for
our purposes, so, in order to simplify the exposition we have avoided presenting the
results in full generality).

We will call the measures on $(G/M)_\mathrm{L}$ and on $(G/M)_\mathrm{R}$ whose existence is
asserted at (a) and (b) of Theorem B.1.3 *standard G-invariant measures on $(G/M)_\mathrm{L}$
and on $(G/M)_\mathrm{R}$*, respectively.

Since the standard G-invariant measures on $(G/M)_\mathrm{L}$ are unique up to a
multiplicative constant, it follows that these measures can either be all infinite

measures, or else, all these measures have to be finite; if the standard G-invariant measures are finite, then $(G/M)_L$ is said to be a *finite volume (left cosets) space*. If $(G/M)_L$ is a finite volume space, then there exists exactly one *standard G-invariant probability measure on* $(G/M)_L$; this probability measure will be denoted by $v_{(G/M)_L}$.

In a similar manner, the standard G-invariant measures on $(G/M)_R$ can either be all infinite, or else these measures have to be all finite. If all the standard G-invariant measures on $(G/M)_R$ are finite, then $(G/M)_R$ is said to be a *finite volume (right cosets) space*, and the unique *standard G-invariant probability measure on* $(G/M)_R$ is denoted by $v_{(G/M)_R}$.

Assume that G and M satisfy the conditions of Theorem B.1.3. In applications, it is often of interest to know conditions under which we can infer that either both spaces $(G/M)_L$ and $(G/M)_R$ are of finite volume, or else, any standard G-invariant measure on $(G/M)_L$ or on $(G/M)_R$ is infinite. In the next proposition we discuss such conditions.

Proposition B.1.4. *Let (G, d) be a locally compact separable metric group, let M be a closed subgroup of G, and assume that both G and M are unimodular. Then G and M satisfy the conditions of Theorem B.1.3, and either both $(G/M)_L$ and $(G/M)_R$ are finite volume spaces, or else, any standard G-invariant measure on $(G/M)_L$ or on $(G/M)_R$ is infinite.*

Proof. The fact that G and M satisfy the conditions of Theorem B.1.3 is obviously true.

Now, assume that the other conclusion of the proposition is false; that is, assume that one of the spaces $(G/M)_L$ and $(G/M)_R$ is of finite volume, while the other is not. For instance, assume that $(G/M)_R$ is a finite volume space, while any standard G-invariant measure on $(G/M)_L$ is infinite (the case in which $(G/M)_L$ is of finite volume, while $(G/M)_R$ is not, can be dealt with in a similar manner).

Let h_G and h_M be Haar measures on G and M, respectively (note that we do not use the subscripts $^{(L)}$ or $^{(R)}$ for the Haar measures, because G and M are unimodular, so h_G and h_M are both left and right translation invariant).

Let $v_{(h_G,h_M)}^{(L)}$ and $v_{(h_G,h_M)}^{(R)}$ be the standard G-invariant measures on $(G/M)_L$ and on $(G/M)_R$ obtained using (a) and (b) of Theorem B.1.3, respectively.

Then $v_{(h_G,h_M)}^{(R)}$ is a finite measure, while $v_{(h_G,h_M)}^{(L)}$ is not. Set $\alpha = v_{(h_G,h_M)}^{(R)}((G/M)_R)$. Then $\alpha \in \mathbb{R}$, $\alpha > 0$.

Since $v_{(h_G,h_M)}^{(L)}$ is an infinite measure, there exists a $u \in C_c((G/M)_L)$ such that $0 \le u \le 1$ and $\int u(xM)\, dv_{(h_G,h_M)}^{(L)}(xM) \ge 2\alpha$.

Since the operator $\mathbf{Q}_{h_M^{(L)}}$ that appears in formula (B.1.1) of Theorem B.1.3 is surjective (we mentioned this fact before stating the theorem), it follows that there exists an $f \in C_c(G)$ such that $u = \mathbf{Q}_{h_M^{(L)}} f$.

Let $g : G \to \mathbb{R}$ be defined by $g(x) = f(x^{-1})$ for every $x \in G$.

Then $g \in C_c(G)$ because the map $x \mapsto x^{-1}$, $x \in G$, is a homeomorphism. Thus, it makes sense to apply the operator $\mathbf{Q}_{h_M^{(\mathrm{R})}}$ that appears in the equality (B.1.2) of Theorem B.1.3 to g. Set $v = \mathbf{Q}_{h_M^{(\mathrm{R})}} g$.

Using the fact that h_M is both left and right invariant, and using Proposition 9.3.6, pp. 314–316 of Cohn [20], we obtain that

$$v(Mx) = \int_M g(\xi x)\,dh_M(\xi) = \int_M g(\xi^{-1}x)\,dh_M(\xi)$$

$$= \int_M f(x^{-1}\xi)\,dh_M(\xi) = u(x^{-1}M)$$

for every $x \in G$. Thus, $0 \le v \le 1$.

Using formula (B.1.1) of Theorem B.1.3, we obtain that

$$2\alpha \ge \int_{(G/M)_{\mathrm{L}}} u(xM)\,dv^{(\mathrm{L})}_{(h_G,h_M)}(xM) = \int_G f(x)\,dh_G(x),$$

while using the equality (B.1.2) of the same theorem, we obtain that

$$\alpha \le \int_{(G/M)_{\mathrm{R}}} v(Mx)\,dv^{(\mathrm{R})}_{(h_G,h_M)}(Mx) = \int_G g(x)\,dh_G(x) = \int_G f(x^{-1})\,dh_G(x).$$

We have obtained a contradiction because h_G is both left and right invariant, so, by Proposition 9.3.6, pp. 314–316 of Cohn [20],

$$\int_G f(x)\,dh_G(x) = \int_G f(x^{-1})\,dh_G(x).$$

$\square$

If M is a discrete subgroup of a locally compact separable metric group G, then, as pointed out in Proposition A.2.1, M is closed; therefore, if, in addition, $\Delta_{G|M} = \Delta_M$, then we can use Theorem B.1.3 to study the G-invariant measures of $(G/M)_{\mathrm{L}}$ and $(G/M)_{\mathrm{R}}$. A discrete subgroup M of G such that $\Delta_{G|M} = \Delta_M$ is called a *lattice* in G if $(G/M)_{\mathrm{L}}$ or $(G/M)_{\mathrm{R}}$ is a finite volume space.

It can be shown that if a locally compact separable metric group G contains a lattice, then G is unimodular (see p. 42 of Bekka and Mayer [10] or Remark 1.9, p. 21 of Raghunathan's monograph [87]). Thus, at times one can prove the unimodularity of a locally compact separable metric group G by exhibiting a lattice in G. This is illustrated in the next example:

Example B.1.5. For every $n \in \mathbb{N}$, $n \ge 2$, the group $\mathrm{SL}(n,\mathbb{R})$ is unimodular because it can be shown that $\mathrm{SL}(n,\mathbb{Z})$ is a lattice in $\mathrm{SL}(n,\mathbb{R})$ (see Example 2.12 on p. 49

of Bekka and Mayer [10] for the case $n = 2$; for $n > 2$, see Theorem 2.7, p. 146 of Bekka and Mayer [10]). For the definitions of $SL(n, \mathbb{R})$ and $SL(n, \mathbb{Z})$, see Example A.2.10.   ∎

Note that, in view of Proposition B.1.4, if G and M are as in Theorem B.1.3, and M is a lattice in G, then, necessarily, both coset spaces $(G/M)_L$ and $(G/M)_R$ are finite volume spaces.

Let G and M be as in Theorem B.1.3. Clearly, if any of the coset spaces $(G/M)_L$ or $(G/M)_R$ is compact, then the space is of finite volume. However, the converse is not true; for instance, $(SL(n, \mathbb{R})/SL(n, \mathbb{Z}))_L$ and $(SL(n, \mathbb{R})/SL(n, \mathbb{Z}))_R$, $n \in \mathbb{N}$, $n \geq 2$, are noncompact, but are finite volume spaces.

Again, let G and M be as in Theorem B.1.3, and assume that M is a normal subgroup of G. Then G/M is a locally compact separable metric group (see Proposition A.2.7); therefore, by Theorem B.1.1, G/M has left and right Haar measures. It can be shown (see p. 59 of Reiter [96], or pp. 42–45 of Weil [129]) that, in this case, the standard G-invariant measures whose existence is stated in (a) and (b) of Theorem B.1.3 are precisely the left and right Haar measures on G/M, respectively. That is why the standard G-invariant measures on coset spaces are sometimes called Haar measures even when the coset spaces are defined by nonnormal subgroups (see the footnote on p. 16 of Starkov [113]). Note that if G is a commutative group, then any subgroup M of G is normal and the standard G-invariant measures on the commutative group G/M whose existence is stated in (a) and (b) of Theorem B.1.3 are the same; in this case, and if, in addition, the standard G-invariant measures on G/M are finite, we will denote by $\nu_{G/M}$ the unique G-invariant probability measure on G/M (which is also the unique left and right Haar probability measure on G/M).

We will conclude this section with several examples dealing with invariant probability measures of flows. To this end, we need two notions that we now introduce.

Let $(w_t)_{t \in \mathbb{T}}$ be a semiflow or flow defined on a locally compact separable metric space (X, d).

We say that $(w_t)_{t \in \mathbb{T}}$ is *uniquely ergodic* if $(w_t)_{t \in \mathbb{T}}$ has exactly one invariant probability measure; that is, $(w_t)_{t \in \mathbb{T}}$ is uniquely ergodic if it has nonzero finite invariant measures, and the nonzero finite invariant measures of $(w_t)_{t \in \mathbb{T}}$ are unique up to a multiplicative constant.

Assume that $(w_t)_{t \in \mathbb{T}}$ has finite nonzero invariant (positive) measures, and let μ be such a measure. We say that μ is an *ergodic measure* if μ *cannot* be written as a sum of two nonzero mutually singular invariant measures. Thus, the measure μ is ergodic if it *does not* satisfy the following condition:

(E)   There exists a Borel measurable subset A of X such that $\mu(A) > 0$, $\mu(X \setminus A) > 0$, and such that the measures $\mu_1 : \mathcal{B}(X) \to \mathbb{R}$ and $\mu_2 : \mathcal{B}(X) \to \mathbb{R}$ defined by $\mu_1(B) = \mu(B \cap A)$ and $\mu_2(B) = \mu(B \cap (X \setminus A))$ for every $B \in \mathcal{B}(X)$ are both invariant for $(w_t)_{t \in \mathbb{T}}$.

It is easy to see that if $(w_t)_{t \in \mathbb{T}}$ is uniquely ergodic, then all the invariant nonzero finite (positive) measures of $(w_t)_{t \in \mathbb{T}}$ are ergodic; in particular, the unique invariant probability measure of $(w_t)_{t \in \mathbb{T}}$ is ergodic.

Example B.1.6 (The Haar-Lebesgue Measure on $\mathbb{R}/\mathbb{Z}$ and the Rotations of the Unit Circle). Let $\mathbb{R}/\mathbb{Z}$ be the group defined in Example A.2.8. Since $\mathbb{R}/\mathbb{Z}$ is a commutative compact metric group, by Theorem B.1.1 and the observations following the theorem, $\mathbb{R}/\mathbb{Z}$ has finite Haar measures which are both left and right Haar measures; therefore, $\mathbb{R}/\mathbb{Z}$ has precisely one two-sided Haar probability measure. We call the Haar probability measure on $\mathbb{R}/\mathbb{Z}$ the *Haar-Lebesgue (probability) measure on $\mathbb{R}/\mathbb{Z}$*, because, as pointed out in Example A.2.8, we can identify $\mathbb{R}/\mathbb{Z}$ with the interval $[0, 1)$, and, via this identification, the Haar probability measure on $\mathbb{R}/\mathbb{Z}$ is the Lebesgue measure on $[0, 1)$. Since $\mathbb{R}$ is a commutative group and $\mathbb{Z}$ is a discrete subgroup of $\mathbb{R}$, using the comments on Theorem B.1.3, we obtain that the standard $\mathbb{R}$-invariant measures on $\mathbb{R}/\mathbb{Z}$ whose existence are stated at (a) and (b) of Theorem B.1.3 are precisely the Haar measures on $\mathbb{R}/\mathbb{Z}$. In particular, the unique $\mathbb{R}$-invariant standard probability measure on $\mathbb{R}/\mathbb{Z}$ is the Haar-Lebesgue probability measure.

Now, let $(w_t)_{t \in \mathbb{R}}$ be the flow of the rotations of the unit circle defined in Example A.3.4, $w_t : \mathbb{R}/\mathbb{Z} \to \mathbb{R}/\mathbb{Z}$, $w_t(\hat{x}) = \widehat{t + x}$ for every $\hat{x} \in \mathbb{R}/\mathbb{Z}$ and $t \in \mathbb{R}$. Since a Borel measure μ on $\mathbb{R}/\mathbb{Z}$ is an invariant measure for $(w_t)_{t \in \mathbb{R}}$ if and only if μ is an $\mathbb{R}$-invariant measure in the sense of Theorem B.1.3, it follows that $(w_t)_{t \in \mathbb{R}}$ is uniquely ergodic, and the (unique) invariant probability measure of $(w_t)_{t \in \mathbb{R}}$ is the Haar-Lebesgue measure on $\mathbb{R}/\mathbb{Z}$, which is necessarily ergodic.   ∎

Example B.1.7 (The Haar-Lebesgue Measure on the n-Dimensional Torus and the Rectilinear Flow). Let $n \in \mathbb{N}$, $n \geq 2$, and consider the n-dimensional torus $\mathbb{R}^n/\mathbb{Z}^n$ defined in Example A.2.9. As in the case of $\mathbb{R}/\mathbb{Z}$ in Example B.1.6, using Theorem B.1.1 and the observations that follow the theorem, we obtain that there exists exactly one Haar probability measure on $\mathbb{R}^n/\mathbb{Z}^n$, which is both left and right invariant. Since, as described in Example A.2.9, we can identify $\left(\mathbb{R}^n/\mathbb{Z}^n, \oplus_n, \eta^{(n)}\right)$ with $([0, 1)^n, \boxplus_n, \rho_n)$, and since, in terms of this identification, the Haar measure on $\mathbb{R}^n/\mathbb{Z}^n$ corresponds to the Lebesgue measure on $[0, 1)^n$, we call the Haar probability measure on $\mathbb{R}^n/\mathbb{Z}^n$ the *Haar-Lebesgue probability measure*. Again as in Example B.1.6, using the comments on Theorem B.1.3, we obtain that there is a unique $\mathbb{R}^n$-invariant standard probability measure on $\mathbb{R}^n/\mathbb{Z}^n$, and that this probability measure is the Haar-Lebesgue probability measure on $\mathbb{R}^n/\mathbb{Z}^n$.

Let $\mathbf{v} \in \mathbb{R}^n$, $\mathbf{v} = (v_1, v_2, \ldots, v_n)$, and let $(w_t)_{t \in \mathbb{R}}$ be the rectilinear flow on $\mathbb{R}^n/\mathbb{Z}^n$ with velocity $\mathbf{v}$ defined in Example A.3.5, $w_t : \mathbb{R}^n/\mathbb{Z}^n \to \mathbb{R}^n/\mathbb{Z}^n$, $w_t(\hat{\mathbf{x}}) = \widehat{t\mathbf{v} + \mathbf{x}}$ for every $\hat{\mathbf{x}} \in \mathbb{R}^n/\mathbb{Z}^n$, $t \in \mathbb{R}$. It is easy to see that the Haar-Lebesgue probability measure $\nu_{\mathbb{R}^n/\mathbb{Z}^n}$ on $\mathbb{R}^n/\mathbb{Z}^n$ (which is also the unique $\mathbb{R}^n$-invariant probability measure on $\mathbb{R}^n/\mathbb{Z}^n$) is an invariant probability measure for $(w_t)_{t \in \mathbb{R}}$. However, in general, $\nu_{\mathbb{R}^n/\mathbb{Z}^n}$ is not an ergodic measure for $(w_t)_{t \in \mathbb{R}}$. It can be shown (see Section 3.1, pp. 64–69, of Cornfeld, Fomin and Sinai's monograph [22] that $\nu_{\mathbb{R}^n/\mathbb{Z}^n}$ is an ergodic measure for $(w_t)_{t \in \mathbb{R}}$ if and only if the numbers $v_1, v_2, \ldots, v_n$

are rationally independent, and that in this case, and only in this case, the flow is uniquely ergodic.   ∎

The theoretical results summarized in this section allow us to discuss the geodesic and the horocycle flows in certain spaces of cosets as promised in Examples A.3.6 and A.3.7, respectively.

Example B.1.8 (The Geodesic Flow on Certain Spaces of Cosets of $PSL(2, \mathbb{R})$*).* It is well-known that the locally compact separable metric group $PSL(2, \mathbb{R})$ defined in Example A.2.10 has lattices (for a specific example of a lattice in $PSL(2, \mathbb{R})$, see the beginning of the second subsection of Section 3, Chapter 2 of Bekka and Mayer [10], pp. 58–59). Let Γ be a lattice in $PSL(2, \mathbb{R})$.

As in Example A.3.6, set $g_t = \begin{pmatrix} e^{\frac{t}{2}} & 0 \\ 0 & e^{-\frac{t}{2}} \end{pmatrix}$ for every $t \in \mathbb{R}$, consider the right coset space $(PSL(2, \mathbb{R})/\Gamma)_{\mathrm{R}}$, and let $w_t^{(\Gamma)} : (PSL(2, \mathbb{R})/\Gamma)_{\mathrm{R}} \to (PSL(2, \mathbb{R})/\Gamma)_{\mathrm{R}}$ be defined by $w_t^{(\Gamma)}(\Gamma \hat{x}) = \Gamma \hat{x} \hat{g}_t$ for every $\Gamma \hat{x} \in (PSL(2, \mathbb{R})/\Gamma)_{\mathrm{R}}$ and every $t \in \mathbb{R}$; the hats over x and g_t indicate that, as elements of $PSL(2, \mathbb{R})$, $\hat{x}$ and $\hat{g}_t$ are equivalence classes since $PSL(2, \mathbb{R}) = SL(2, \mathbb{R})/L$, where $L = \{\mathbf{I}_2, -\mathbf{I}_2\}$ (see Example A.2.10). It is easy to see that $\left(w_t^{(\Gamma)} \right)_{t \in \mathbb{R}}$ is a continuous flow (the continuity of $\left(w_t^{(\Gamma)} \right)_{t \in \mathbb{R}}$ is obtained by a straightforward application of Proposition A.3.3-(b)). The flow $\left(w_t^{(\Gamma)} \right)_{t \in \mathbb{R}}$ is called the *geodesic flow on* $(PSL(2, \mathbb{R})/\Gamma)_{\mathrm{R}}$.

Since Γ is a lattice in $PSL(2, \mathbb{R})$, it follows that $(PSL(2, \mathbb{R})/\Gamma)_{\mathrm{R}}$ is a finite volume space; so, there exists a (unique) $PSL(2, \mathbb{R})$-invariant probability measure on $(PSL(2, \mathbb{R})/\Gamma)_{\mathrm{R}}$, the standard $PSL(2, \mathbb{R})$-invariant probability measure $\nu_{(PSL(2,\mathbb{R})/\Gamma)_{\mathrm{R}}}$. Since $\left(w_t^{(\Gamma)} \right)_{t \in \mathbb{R}}$ is a particular case of the more general actions discussed in Example A.3.2, it follows that, as emphasized in Example B.1.2, any $PSL(2, \mathbb{R})$-invariant measure on $(PSL(2, \mathbb{R})/\Gamma)_{\mathrm{R}}$ is also an invariant measure for $\left(w_t^{(\Gamma)} \right)_{t \in \mathbb{R}}$; in particular, $\nu_{(PSL(2,\mathbb{R})/\Gamma)_{\mathrm{R}}}$ is an invariant probability measure for $\left(w_t^{(\Gamma)} \right)_{t \in \mathbb{R}}$. The natural question now is whether $\nu_{(PSL(2,\mathbb{R})/\Gamma)_{\mathrm{R}}}$ is an ergodic measure for $\left(w_t^{(\Gamma)} \right)_{t \in \mathbb{R}}$, or not. It is a quite significant result of Hedlund [41] that $\nu_{(PSL(2,\mathbb{R})/\Gamma)_{\mathrm{R}}}$ is ergodic (for a proof, see Bekka and Mayer [10], Corollary 2.3.8 on p. 61).

The next question that comes to mind is whether or not $\nu_{(PSL(2,\mathbb{R})/\Gamma)_{\mathrm{R}}}$ is the only invariant probability of $\left(w_t^{(\Gamma)} \right)_{t \in \mathbb{R}}$. It can be shown that the geodesic flow discussed in this example has uncountably many ergodic invariant probability measures (see Sinai [107]). For additional information about the geodesic flows, see Chapter 2 of Bekka and Mayer [10], Ratner [89], and Section 14, Chapter 2 of Starkov [113]. ∎

Let (G, d) be a locally compact separable metric group, and let M be a lattice in G. Then, it is easy to see that $(G/M)_{\mathrm{L}}$ is a compact space if and only if $(G/M)_{\mathrm{R}}$ is compact; in this case, it is the custom to say that M is a *cocompact lattice*.

Example B.1.9 (The Horocycle Flows on Certain Spaces of Cosets of $\mathrm{PSL}(2,\mathbb{R})$ *and* $\mathrm{SL}(2,\mathbb{R})$). As in Example A.3.7, we will use the notation $h_t^{(1)} = \begin{pmatrix} 1 & t \\ 0 & 1 \end{pmatrix}$ and $h_t^{(2)} = \begin{pmatrix} 1 & 0 \\ t & 1 \end{pmatrix}$ for every $t \in \mathbb{R}$.

(a) *The Horocycle Flows on Spaces of Cosets of* $\mathrm{PSL}(2,\mathbb{R})$. Let Γ be a lattice in $\mathrm{PSL}(2,\mathbb{R})$.

   Let $\bar{v}_t^{(j\Gamma)} : (\mathrm{PSL}(2,\mathbb{R})/\Gamma)_{\mathrm{R}} \to (\mathrm{PSL}(2,\mathbb{R})/\Gamma)_{\mathrm{R}}$ be defined by $\bar{v}_t^{(j\Gamma)}(\Gamma\hat{x}) = \Gamma\hat{x}\hat{h}_t^{(j)}$ for every $\Gamma\hat{x} \in (\mathrm{PSL}(2,\mathbb{R})/\Gamma)_{\mathrm{R}}$, $t \in \mathbb{R}$, and $j = 1,2$ (as in Example A.3.7, the hats over x and $h_t^{(j)}$ indicate that $\hat{x}$ and $\hat{h}_t^{(j)}$ are the equivalence classes of $x \in \mathrm{SL}(2,\mathbb{R})$ and $h_t^{(j)} \in \mathrm{SL}(2,\mathbb{R})$ in $\mathrm{PSL}(2,\mathbb{R})$).

   In a similar manner as in the previous example (Example B.1.8) we obtain that $\left(\bar{v}_t^{(1\Gamma)}\right)_{t\in\mathbb{R}}$ and $\left(\bar{v}_t^{(2\Gamma)}\right)_{t\in\mathbb{R}}$ are continuous flows (the continuity of the flows is proved using Proposition A.3.3-(b)). The flows $\left(\bar{v}_t^{(j\Gamma)}\right)_{t\in\mathbb{R}}$, $j = 1,2$, are called the *horocycle flows on* $(\mathrm{PSL}(2,\mathbb{R})/\Gamma)_{\mathrm{R}}$.

   A remarkable result of Hedlund (see Theorem 1.9 in Chapter 4 of Bekka and Mayer [10], p. 116, asserts that for each of the horocycle flows $\left(\bar{v}_t^{(1\Gamma)}\right)_{t\in\mathbb{R}}$ and $\left(\bar{v}_t^{(2\Gamma)}\right)_{t\in\mathbb{R}}$, and for every element $\Gamma\hat{x}$ of $(\mathrm{PSL}(2,\mathbb{R})/\Gamma)_{\mathrm{R}}$, either the orbit of $\Gamma\hat{x}$ under the action of the flow is dense in $(\mathrm{PSL}(2,\mathbb{R})/\Gamma)_{\mathrm{R}}$, or else $\Gamma\hat{x}$ is a periodic point for that flow. If Γ is cocompact, then the two flows are minimal.

(b) *The Horocycle Flows on Spaces of Cosets of* $\mathrm{SL}(2,\mathbb{R})$. Let Γ be a lattice in $\mathrm{SL}(2,\mathbb{R})$. In a similar manner as in (a), one can define two *horocycle flows* $\left(v_t^{(1\Gamma L)}\right)_{t\in\mathbb{R}}$ and $\left(v_t^{(2\Gamma L)}\right)_{t\in\mathbb{R}}$ on $(\mathrm{SL}(2,\mathbb{R})/\Gamma)_{\mathrm{L}}$ and two *horocycle flows* $\left(v_t^{(1\Gamma R)}\right)_{t\in\mathbb{R}}$ and $\left(v_t^{(2\Gamma R)}\right)_{t\in\mathbb{R}}$ on $(\mathrm{SL}(2,\mathbb{R})/\Gamma)_{\mathrm{R}}$; that is, for every $j = 1,2$ and every $t \in \mathbb{R}$, let $v_t^{(j\Gamma L)} : (\mathrm{SL}(2,\mathbb{R})/\Gamma)_{\mathrm{L}} \to (\mathrm{SL}(2,\mathbb{R})/\Gamma)_{\mathrm{L}}$ be defined by $v_t^{(j\Gamma L)}(x\Gamma) = h_t^{(j)}x\Gamma$ for every $x\Gamma \in (\mathrm{SL}(2,\mathbb{R})/\Gamma)_{\mathrm{L}}$, and let $v_t^{(j\Gamma R)} : (\mathrm{SL}(2,\mathbb{R})/\Gamma)_{\mathrm{R}} \to (\mathrm{SL}(2,\mathbb{R})/\Gamma)_{\mathrm{R}}$ be defined by $v_t^{(j\Gamma R)}(\Gamma x) = \Gamma x h_t^{(j)}$ for every $\Gamma x \in (\mathrm{SL}(2,\mathbb{R})/\Gamma)_{\mathrm{R}}$.

   Similar arguments as in (a) can be used to show that $\left(v_t^{(j\Gamma L)}\right)_{t\in\mathbb{R}}$ and $\left(v_t^{(j\Gamma R)}\right)_{t\in\mathbb{R}}$, $j = 1,2$, are continuous flows.

   Since Γ is a lattice in $\mathrm{SL}(2,\mathbb{R})$, it makes sense to consider the standard $\mathrm{SL}(2,\mathbb{R})$-invariant probability measures $v_{(\mathrm{SL}(2,\mathbb{R})/\Gamma)_{\mathrm{L}}}$ and $v_{(\mathrm{SL}(2,\mathbb{R})/\Gamma)_{\mathrm{R}}}$ on $(\mathrm{SL}(2,\mathbb{R})/\Gamma)_{\mathrm{L}}$ and on $(\mathrm{SL}(2,\mathbb{R})/\Gamma)_{\mathrm{R}}$, respectively. Of course, $v_{(\mathrm{SL}(2,\mathbb{R})/\Gamma)_{\mathrm{L}}}$ is an invariant probability measure for the flows $\left(v_t^{(j\Gamma L)}\right)_{t\in\mathbb{R}}$, $j = 1,2$, and $v_{(\mathrm{SL}(2,\mathbb{R})/\Gamma)_{\mathrm{R}}}$ is an invariant probability measure for $\left(v_t^{(j\Gamma R)}\right)_{t\in\mathbb{R}}$, $j = 1,2$. Thus, a natural question is: are the two probability measures $v_{(\mathrm{SL}(2,\mathbb{R})/\Gamma)_{\mathrm{L}}}$ and $v_{(\mathrm{SL}(2,\mathbb{R})/\Gamma)_{\mathrm{R}}}$ ergodic for the corresponding horocycle flows? If the two measures are ergodic for the corresponding flows, then the next obvious question is: are the four horocycle flows uniquely ergodic?

It is quite remarkable that, for the four flows that we are currently discussing, these questions have been completely answered in a series of outstandingly nice results that we will review very briefly here. These results offer a fairly complete picture of the structure of the set of invariant probabilities of these horocycle flows.

The first significant result in this direction is a theorem of Furstenberg [37]: if Γ is cocompact, then the four horocycle flows are uniquely ergodic. Thus, if Γ is cocompact, then $\nu_{(\mathrm{SL}(2,\mathbb{R})/\Gamma)_{\mathrm{L}}}$ is the unique invariant probability measure of the flows $\left(v_t^{(j\Gamma\mathrm{L})}\right)_{t\in\mathbb{R}}$, $j = 1, 2$, and $\nu_{(\mathrm{SL}(2,\mathbb{R})/\Gamma)_{\mathrm{R}}}$ is the unique invariant probability of $\left(v_t^{(j\Gamma\mathrm{R})}\right)_{t\in\mathbb{R}}$, $j = 1, 2$.

Now, in order to get an idea of what happens if Γ is not cocompact, let us consider the case when $\Gamma = \mathrm{SL}(2,\mathbb{Z})$ (note that as pointed out in Example B.1.5, $\mathrm{SL}(2,\mathbb{Z})$ is a lattice in $\mathrm{SL}(2,\mathbb{R})$ (see Example 2.12 on p. 49 of Bekka and Mayer [10])). Then, cosets of the form $\mathrm{SL}(2,\mathbb{Z})p$, $p \in \mathbf{P}$, thought of as elements of $(\mathrm{SL}(2,\mathbb{R})/\mathrm{SL}(2,\mathbb{Z}))_{\mathrm{R}}$, are periodic points for the horocycle flow $\left(v_t^{(1\,\mathrm{SL}(2,\mathbb{Z})\,\mathrm{R})}\right)_{t\in\mathbb{R}}$, where $\mathbf{P}$ is the subgroup of $\mathrm{SL}(2,\mathbb{R})$ defined at the end of Example A.2.10 in connection with the Iwasawa decomposition. It can be shown that given a continuous flow (or semiflow) on a locally compact separable metric space (X, d) and a periodic point $x_0 \in X$ for the flow (or semiflow), then there exists an invariant ergodic probability measure for the flow (or semiflow) whose support is precisely the orbit (which in this case is a closed set) of x_0 under the action of the flow (or semiflow), and this measure is the only invariant ergodic probability measure whose support is the orbit. This means that the horocycle flow $\left(v_t^{(1\,\mathrm{SL}(2,\mathbb{Z})\,\mathrm{R})}\right)_{t\in\mathbb{R}}$ cannot be uniquely ergodic (incidentally, we obtain from the above remarks and Furstenberg's theorem [37] mentioned earlier that $\mathrm{SL}(2,\mathbb{Z})$ cannot be a cocompact lattice in $\mathrm{SL}(2,\mathbb{R})$).

In view of the discussion so far, a natural question that comes to mind is: given a lattice Γ of $\mathrm{SL}(2,\mathbb{R})$, could it be possible that the only invariant ergodic probability measures of the four horocycle flows are the corresponding standard $\mathrm{SL}(2,\mathbb{R})$-invariant probability measures $\nu_{(\mathrm{SL}(2,\mathbb{R})/\Gamma)_{\mathrm{L}}}$ (for $\left(v_t^{(j\Gamma\mathrm{L})}\right)_{t\in\mathbb{R}}$, $j = 1, 2$) and $\nu_{(\mathrm{SL}(2,\mathbb{R})/\Gamma)_{\mathrm{R}}}$ (for $\left(v_t^{(j\Gamma\mathrm{R})}\right)_{t\in\mathbb{R}}$, $j = 1, 2$), and the invariant ergodic probability measures whose supports are orbits of periodic points for the corresponding flows? Surprisingly enough, the answer is yes. The result stating this fact was obtained by Dani in [23]. For various proofs of Dani's theorem, see Ratner [95], Section 4.3 of Bekka and Mayer [10], and, of course, [23].

We conclude our discussion of horocycle flows with a result obtained by Dani and Smillie in [28] and known as the Dani-Smillie theorem on the equidistribution of horocycle orbits.

Let Γ be a lattice of $\mathrm{SL}(2,\mathbb{R})$, and let $j = 1$ or 2. The Dani-Smillie theorem on the equidistribution of horocycle orbits states that the following two assertions hold true:

(a) Let $x\Gamma$, $x \in SL(2,\mathbb{R})$, be an element of $(SL(2,\mathbb{R})/\Gamma)_L$ that is not periodic for the horocycle flow $\left(v_t^{(j\Gamma L)}\right)_{t\in\mathbb{R}}$. Then for every real-valued continuous bounded function f on $(SL(2,\mathbb{R})/\Gamma)_L$, the limit $\lim\limits_{t\to+\infty} \dfrac{1}{t}\displaystyle\int_0^t f\left(v_s^{(j\Gamma L)}(x\Gamma)\right)\,\mathrm{d}s$ exists and is equal to $\displaystyle\int_{(SL(2,\mathbb{R})/\Gamma)_L} f\,\mathrm{d}\nu_{(SL(2,\mathbb{R})/\Gamma)_L}.$

(b) Let Γx, $x \in SL(2,\mathbb{R})$, be an element of $(SL(2,\mathbb{R})/\Gamma)_R$ that is not periodic for the horocycle flow $\left(v_t^{(j\Gamma R)}\right)_{t\in\mathbb{R}}$. Then for every real-valued continuous bounded function f on $(SL(2,\mathbb{R})/\Gamma)_R$, the limit $\lim\limits_{t\to+\infty} \dfrac{1}{t}\displaystyle\int_0^t f\left(v_s^{(j\Gamma R)}(\Gamma x)\right)\,\mathrm{d}s$ exists and is equal to $\displaystyle\int_{(SL(2,\mathbb{R})/\Gamma)_R} f\,\mathrm{d}\nu_{(SL(2,\mathbb{R})/\Gamma)_R}.$

Besides the proof given by Dani and Smillie in [28], other proofs of the result can be found in Ratner [95] and in Section 4.4 of Bekka and Mayer [10]. ∎

B.2 Banach Algebras, Convolutions of Measures, and the Exponential Function

Our goal in this section is to briefly discuss Banach algebras with unit and the exponential function. Since the spaces of real-valued signed Borel measures defined on locally compact separable metric semigroups with neutral elements are Banach algebras with units when endowed with the operation of convolution, and since these spaces are used in some examples in the book, we will also present the operation of convolution of measures.

Let $\mathbf{A}$ be a real vector space. We say that $\mathbf{A}$ is an *algebra* if, in addition to its vector space structure, $\mathbf{A}$ is also endowed with an algebraic operation $\cdot$ defined on $\mathbf{A} \times \mathbf{A}$ with values in $\mathbf{A}$ such that $(\mathbf{A}, \cdot)$ is a semigroup, and such that the following conditions are satisfied:

$$u \cdot (v + w) = u \cdot v + u \cdot w$$

$$(u + v) \cdot w = u \cdot w + v \cdot w$$

$$a(u \cdot v) = (au) \cdot v = u \cdot (av)$$

for all $u \in \mathbf{A}$, $v \in \mathbf{A}$, $w \in \mathbf{A}$ and $a \in \mathbb{R}$. The algebraic operation is called multiplication. Generally, when using $\cdot$ we will follow the same notational conventions as in the case of the multiplication in any semigroup; for instance, uv stands for $u \cdot v$ whenever $u \in \mathbf{A}$ and $v \in \mathbf{A}$.

A *Banach algebra* is an algebra $\mathbf{E}$ endowed with a norm $\|\cdot\|$ that defines a Banach space structure on $\mathbf{E}$ such that $\|uv\| \leq \|u\|\|v\|$ for every $u \in \mathbf{E}$ and $v \in \mathbf{E}$.

A Banach algebra $\mathbf{E}$ is said to be a *Banach algebra with unit* if the multiplication defined on $\mathbf{E} \times \mathbf{E}$ has a neutral element $\mathbf{e}$ and if $\|\mathbf{e}\| = 1$. In this case, the neutral element $\mathbf{e}$ is called the *unit* of $\mathbf{E}$. Although we will be concerned only with Banach algebras with unit, there exist Banach algebras that do not have a unit. For additional details on Banach algebras with or without units, see Appendix C of Hewitt and Ross [44].

As expected, there are many examples of Banach algebras. Probably, the simplest Banach algebras are obtained as follows: let $n \in \mathbb{N}$, and let $\mathbf{M}(n, \mathbb{R})$ be the Banach space of all $n \times n$ matrices with real entries, thought of as the Banach space of all linear bounded operators on $\mathbb{R}^n$, where $\mathbb{R}^n$ is thought of as endowed with some norm, say, the Euclidean norm. Then $\mathbf{M}(n, \mathbb{R})$ becomes a Banach algebra when endowed with the usual matrix multiplication (which corresponds to the operation of operator composition, of course). Note that $\mathbf{M}(n, \mathbb{R})$ is a Banach algebra with unit because the $n \times n$ identity matrix $\mathbf{I}_n$ is a neutral element for matrix multiplication in $\mathbf{M}(n, \mathbb{R})$ and $\|\mathbf{I}_n\| = 1$ no matter what norm we consider on $\mathbb{R}^n$.

Our goal now is to discuss a significantly more sophisticated example of Banach algebra. The basic idea is as follows: if (X, d) is a locally compact separable metric space, then $\mathcal{M}(X)$ is a Banach space; however, if X has also a semigroup structure with respect to which (X, d) is a locally compact separable metric semigroup, then one can define an algebraic operation on $\mathcal{M}(X)$ called *convolution*, such that, when endowed with the operation of convolution, $\mathcal{M}(X)$ becomes a Banach algebra. When dealing with a locally compact separable metric semigroup, we will use the notation H rather than X.

Thus, let $(H, \cdot, d)$ be a locally compact separable metric semigroup.

We will need the following technical result:

Proposition B.2.1. *Let $f \in B_b(H)$, and let $\mu \in \mathcal{M}(H)$. Then the function $g_f^{(\mu)}$:* $H \to \mathbb{R}$ *defined by* $g_f^{(\mu)}(x) = \int_H f(ux)\,d\mu(u)$ *for every $x \in H$ belongs to $B_b(H)$.*

If $f \in C_b(H)$, then $g_f^{(\mu)}$ belongs to $C_b(H)$, as well.

Proof. If $\mu \in \mathcal{M}(H)$ and if $\mu = \mu^+ - \mu^-$ is the Jordan decomposition of μ (for the definition of the Jordan decomposition, see, for instance, pp. 125–126 of Cohn's book [20]), then $g_f^{(\mu)} = g_f^{(\mu^+)} - g_f^{(\mu^-)}$; therefore, it is enough to prove the proposition under the assumption that $\mu \geq 0$.

To this end, assume that $\mu \geq 0$ and let $f \in B_b(H)$.

Since the algebraic operation that defines the semigroup structure on H is continuous with respect to the metric topology on H and the product topology on $H \times H$, it follows that the real-valued map $(u, v) \mapsto f(uv)$, $(u, v) \in H \times H$, is measurable with respect to the Borel σ-algebra on $\mathbb{R}$ and the Borel σ-algebra

$\mathcal{B}(H \times H)$ defined by the product topology on $H \times H$. Since H is a locally compact separable metric space, using D32, p. 348, of Cohn [20], we obtain that H has a countable basis for its topology, so by Proposition 7.6.2, p. 242, of Cohn [20], $\mathcal{B}(H \times H)$ is equal to the σ-algebra product generated by $\mathcal{B}(H) \times \mathcal{B}(H)$ (that is, the σ-algebra generated by all the sets of the form $A \times B$, where $A \in \mathcal{B}(H)$ and $B \in \mathcal{B}(H)$). Using (a) of Proposition 5.2.1, p. 159, of Cohn [20], we obtain that $g_f^{(\mu)}$ is measurable (note that even though Proposition 5.2.1 is stated for two measure spaces, we can also state (a) of Proposition 5.2.1 in the case in which we deal with a measure space and a measurable space, rather than two measure spaces). Since $g_f^{(\mu)}$ is bounded, it follows that $g_f^{(\mu)} \in B_b(H)$.

If $f \in C_b(H)$, then using the fact that μ is a finite regular measure, we obtain that $g_f^{(\mu)}$ is continuous, so $g_f^{(\mu)} \in C_b(H)$. $\square$

For every $\mu \in \mathcal{M}(H)$ and $\nu \in \mathcal{M}(H)$, let $I_{\mu,\nu} : C_0(H) \to \mathbb{R}$ be defined by

$$I_{\mu,\nu}(f) = \int_H \int_H f(xy) \, d\mu(x) \, d\nu(y) \text{ for every } f \in C_0(H). \text{ By Proposition B.2.1,}$$

the double iterated integrals defining $I_{\mu,\nu}(f)$, $f \in C_0(H)$, exist, so $I_{\mu,\nu}(f)$ is well-defined for every $f \in C_0(H)$.

Clearly, $I_{\mu,\nu}$ is a linear functional for every $\mu \in \mathcal{M}(H)$ and $\nu \in \mathcal{M}(H)$. If, in addition, $\mu \geq 0$ and $\nu \geq 0$, then $I_{\mu,\nu}$ is also positive in the sense that $I_{\mu,\nu}(f) \geq 0$ whenever $f \in C_0(H)$, $f \geq 0$; hence, $I_{\mu,\nu}$ is a continuous linear functional in this case.

Now, in general, let $\mu \in \mathcal{M}(H)$ and $\nu \in \mathcal{M}(H)$ be not necessarily positive. Set $\mu^+ = \mu \vee 0$, $\mu^- = (-\mu) \vee 0$, $\nu^+ = \nu \vee 0$, and $\nu^- = (-\nu) \vee 0$. Since $\mu = \mu^+ - \mu^-$ and $\nu = \nu^+ - \nu^-$, it follows that

$$I_{\mu,\nu} = I_{\mu^+,\nu} - I_{\mu^-,\nu} = I_{\mu^+,\nu^+} - I_{\mu^+,\nu^-} - I_{\mu^-,\nu^+} + I_{\mu^-,\nu^-}.$$

Taking into consideration that I_{μ^+,ν^+}, I_{μ^+,ν^-}, I_{μ^-,ν^+} and I_{μ^-,ν^-} are all continuous linear functionals, we obtain that $I_{\mu,\nu}$ is a continuous linear functional, as well. Thus, we may and do think of $I_{\mu,\nu}$ as an element of $\mathcal{M}(H)$.

For every $\mu \in \mathcal{M}(H)$ and $\nu \in \mathcal{M}(H)$, we denote $I_{\mu,\nu}$ by $\mu * \nu$ and we call it the *convolution of μ and ν* (in this order).

Lemma B.2.2. (a) *The convolution operation $*$ is associative; that is, $(\mu * \nu) * \lambda = \mu * (\nu * \lambda)$ for every $\mu \in \mathcal{M}(H)$, $\nu \in \mathcal{M}(H)$, and $\lambda \in \mathcal{M}(H)$.*
(b) *If $n \in \mathbb{N}$, $n \geq 2$, and if $\mu_1, \mu_2, \ldots, \mu_n$ are n probability measures in $\mathcal{M}(H)$, then $\mathrm{supp}(\mu_1 * \mu_2 * \cdots * \mu_n) = (\mathrm{supp}\, \mu_1)(\mathrm{supp}\, \mu_2) \cdots (\mathrm{supp}\, \mu_n)$.*

Proof. We will prove only (a) because (b) is easy to verify directly.

To this end, let $f \in C_0(H)$.

Using the associativity of the algebraic operation that defines the semigroup structure on H, Fubini's theorem for functions of three or more variables (see pp. 160–161 of Cohn's book [20]), the usual Fubini theorem (for functions of two variables), and noting that Fubini's theorem (for any number of variables) is also valid for real-valued signed measures, we obtain that

$$\langle f, (\mu * \nu) * \lambda \rangle = \int_H \left(\int_H f(yz)\, d(\mu * \nu)(y) \right) d\lambda(z)$$

$$= \int_H \left(\int_H \left(\int_H f((x_1 x_2)z)\, d\mu(x_1) \right) d\nu(x_2) \right) d\lambda(z)$$

$$= \int_H \left(\int_H \left(\int_H f(x_1(x_2 z))\, d\mu(x_1) \right) d\nu(x_2) \right) d\lambda(z)$$

$$= \int_H \left(\int_H \left(\int_H f(x_1(x_2 z))\, d\nu(x_2) \right) d\lambda(z) \right) d\mu(x_1)$$

$$= \int_H \left(\int_H f(x_1 u)\, d(\nu * \lambda)(u) \right) d\mu(x_1)$$

$$= \int_H \left(\int_H f(x_1 u)\, d\mu(x_1) \right) d(\nu * \lambda)(u) = \langle f, \mu * (\nu * \lambda) \rangle.$$

Since the above equalities hold true for every $f \in C_0(H)$, it follows that the operation of convolution is associative. $\qquad\square$

Lemma B.2.3. *(a) $\mathcal{M}(H)$ is an algebra when endowed with the operation of convolution.*

(b) $\left| \int_H f\, d(\mu * \nu) \right| \le \int_H |f|\, d(|\mu| * |\nu|)$ *for every* $f \in B_b(X)$, *and every* $\mu \in \mathcal{M}(H)$ *and* $\nu \in \mathcal{M}(H)$, *where, as usual,* $|\mu| = \mu^+ + \mu^-$ *and* $|\nu| = \nu^+ + \nu^-$ *are the variations of* μ *and* ν, *respectively, and* $|f| = f^+ + f^-$.

(c) $\|\mu * \nu\| \le \|\mu\|\|\nu\|$ *for every* $\mu \in \mathcal{M}(H)$ *and* $\nu \in \mathcal{M}(H)$.

Proof. *(a)* It is easy to see that $\mu*(\nu+\lambda) = \mu*\nu+\mu*\lambda$, $(\mu+\nu)*\lambda = \mu*\lambda+\nu*\lambda$, and $a(\mu * \nu) = (a\mu) * \nu = \mu * (a\nu)$ for every $\mu \in \mathcal{M}(H)$, $\nu \in \mathcal{M}(H)$, $\lambda \in \mathcal{M}(H)$, and $a \in \mathbb{R}$. Thus, using Lemma B.2.2, we obtain that $\mathcal{M}(H)$ is an algebra.

(*b*) Let $\mu \in \mathcal{M}(H)$ and $v \in \mathcal{M}(H)$.

  Since $\mathcal{M}(H)$ is an algebra, it follows that

$$\mu * v = \mu * v^+ - \mu * v^- = \mu^+ * v^+ - \mu^- * v^+ - \mu^+ * v^- + \mu^- * v^-.$$

  Thus, it is easy to see that the required inequality holds true for every $f \in B_b(H)$ by showing first that the inequality is true for $f \in B_b(H)$, $f \geq 0$, and using the fact that $f = f^+ - f^-$ whenever $f \in B_b(H)$ is not necessarily a positive function.

(*c*) Let $\mu \in \mathcal{M}(H)$ and $v \in \mathcal{M}(H)$.

  Using (*b*), we obtain that

$$\left| \int_H f\, d(\mu * v) \right| \leq \int_H |f|\, d(|\mu| * |v|)$$

$$\leq \|f\| \int_H \left(\int_H \mathbf{1}_H(xy)\, d|\mu|(x) \right) d|v|(y) = \|f\|\|\mu\|\|v\|$$

for every $f \in C_0(H)$.

  Since the above inequalities are true, in particular, for every $f \in C_0(H)$, $\|f\| \leq 1$, it follows that $\|\mu * v\| \leq \|\mu\|\|v\|$. $\square$

Using (*a*) and (*c*) of Lemma B.2.3 (and, of course, the fact that $\mathcal{M}(H)$ is a Banach space), we obtain:

Proposition B.2.4. $\mathcal{M}(H)$ *is a Banach algebra when endowed with the operation of convolution.*

Note that if H has a neutral element, say e, then $\mathcal{M}(H)$ is a Banach algebra with unit because the Dirac measure δ_e concentrated at e is a neutral element for the operation of convolution. If H does not have a neutral element, then $\mathcal{M}(H)$ is a Banach algebra that fails to be a Banach algebra with unit; thus, for instance, if H is the interval $(0, +\infty)$ in $\mathbb{R}$, thought of as a locally compact separable metric additive semigroup, (endowed with the usual topology inherited from $\mathbb{R}$), then $\mathcal{M}((0, +\infty))$ is a Banach algebra without a unit.

In order to continue our discussion of Banach algebras, we need the following lemma:

Lemma B.2.5. *Let (Y, d_1) and (Z, d_2) be two metric spaces, and let $T(Y)$ and $T(Z)$ be the metric topologies on Y and Z defined by the metrics d_1 and d_2, respectively. Then the product topology $T(Y) \otimes T(Z)$ on $Y \times Z$ is metrizable.*

Proof. Let $\rho : (Y \times Z) \times (Y \times Z) \to \mathbb{R}$ be defined by

$$\rho((y_1, z_1), (y_2, z_2)) = d_1(y_1, y_2) + d_2(z_1, z_2)$$

for every $(y_1, z_1) \in Y \times Z$ and $(y_2, z_2) \in Y \times Z$. Then, it is easy to see that ρ is a metric on $Y \times Z$.

Since

$$B_{d_1}\left(y, \frac{1}{n}\right) \times B_{d_2}\left(z, \frac{1}{n}\right) \subseteq B_{\rho}\left((y, z), \frac{2}{n}\right) \subseteq B_{d_1}\left(y, \frac{2}{n}\right) \times B_{d_2}\left(z, \frac{2}{n}\right)$$

for every $n \in \mathbb{N}$, $y \in Y$, and $z \in Z$ (here $B_{d_1}(y, a)$, $B_{d_2}(z, a)$ and $B_{\rho}((y, z), a)$ are the open balls centered at y, z and (y, z) of radius $a \in \mathbb{R}$, $a > 0$, in the metric spaces (Y, d_1), (Z, d_2) and $(Y \times Z, \rho)$, respectively), it follows that the topology defined by ρ is the product topology $\mathcal{T}(Y) \otimes \mathcal{T}(Z)$. $\qquad\square$

We will now use Lemma B.2.5 to discuss a useful property of Banach algebras.

Proposition B.2.6. *Let $\mathbf{E}$ be a Banach algebra. Then the multiplication* $\cdot : \mathbf{E} \times \mathbf{E} \to \mathbf{E}$ *which defines the algebra structure on $\mathbf{E}$ is continuous.*

Proof. Since, by Lemma B.2.5, the product topology on $\mathbf{E} \times \mathbf{E}$ is metrizable, it follows that in order to prove the proposition, it is enough to prove that if $((u_n, v_n))_{n \in \mathbb{N}}$ is a sequence of elements of $\mathbf{E} \times \mathbf{E}$ that converges with respect to the product topology of $\mathbf{E} \times \mathbf{E}$, and if (u, v) is the limit of $((u_n, v_n))_{n \in \mathbb{N}}$, then the sequence $(u_n v_n)_{n \in \mathbb{N}}$ converges to uv in the norm topology of $\mathbf{E}$.

To this end, let $((u_n, v_n))_{n \in \mathbb{N}}$ be a convergent sequence in the product topology of $\mathbf{E} \times \mathbf{E}$, and let (u, v) be the limit of $((u_n, v_n))_{n \in \mathbb{N}}$. Then $(u_n)_{n \in \mathbb{N}}$ and $(v_n)_{n \in \mathbb{N}}$ converge to u and v, respectively, in the norm topology of $\mathbf{E}$.

Since $(u_n)_{n \in \mathbb{N}}$ and $(v_n)_{n \in \mathbb{N}}$ are convergent sequences in the norm topology of $\mathbf{E}$, it follows that $\sup_{n \in \mathbb{N}} \|u_n\| < +\infty$ and $\sup_{n \in \mathbb{N}} \|v_n\| < +\infty$. Thus, there exists an $M \in \mathbb{R}$, $M > 0$, such that $\|u_n\| \leq M$ and $\|v_n\| \leq M$ for every $n \in \mathbb{N}$.

Now let $\varepsilon \in \mathbb{R}$, $\varepsilon > 0$. Then there exists an $n_\varepsilon \in \mathbb{N}$ such that $\|u_n - u\| < \dfrac{\varepsilon}{2M}$ and $\|v_n - v\| < \dfrac{\varepsilon}{2M}$ for every $n \in \mathbb{N}$, $n \geq n_\varepsilon$.

It follows that

$$\|u_n v_n - uv\| \leq \|u_n v_n - u_n v\| + \|u_n v - uv\|$$

$$\leq \|u_n\| \|v_n - v\| + \|u_n - u\| \|v\| < M \frac{\varepsilon}{2M} + M \frac{\varepsilon}{2M} = \varepsilon$$

for every $n \in \mathbb{N}$, $n \geq n_\varepsilon$.

Since for every $\varepsilon \in \mathbb{R}$, $\varepsilon > 0$, there exists an $n_\varepsilon \in \mathbb{N}$ such that $\|u_n v_n - uv\| < \varepsilon$ for every $n \in \mathbb{N}$, $n \geq n_\varepsilon$, it follows that $(u_n v_n)_{n \in \mathbb{N}}$ converges to uv. $\qquad\square$

Let $\mathbf{E}$ be a Banach algebra.

As usual, given a sequence $(u_k)_{k \in \mathbb{N}}$ of elements of $\mathbf{E}$ and $n \in \mathbb{N}$, we call $\displaystyle\sum_{k=1}^{n} u_k$ a *partial sum* (or, if we have to be more precise, the *nth partial sum*) of $(u_k)_{k \in \mathbb{N}}$. The sequence of partial sums $\left(\displaystyle\sum_{k=1}^{n} u_k \right)_{n \in \mathbb{N}}$ is called a *series* and is denoted by $\displaystyle\sum_{k=1}^{\infty} u_k$; when dealing with series, the terms u_l, $l \in \mathbb{N}$, of the sequence $(u_k)_{k \in \mathbb{N}}$ are called *terms* of the series. If the sequence of partial sums converges in the norm topology of **E**, then we say that the series *converges conditionally* (or, simply, that the series *converges*). If the series converges conditionally, and if $s \in$ **E** is the limit of the sequence $\left(\displaystyle\sum_{k=1}^{n} u_k \right)_{n \in \mathbb{N}}$ of partial sums of $(u_k)_{k \in \mathbb{N}}$, then s is called the *sum* of the series, and it is the custom to denote the sum s of the series by $\displaystyle\sum_{k=1}^{\infty} u_k$. Note that $\displaystyle\sum_{k=1}^{\infty} u_k$ stands for both the sequence of partial sums of $(u_k)_{k \in \mathbb{N}}$, as well as for the limit of the sequence of partial sums whenever it exists; this widely used abuse of notation does not lead to confusion since the meaning of $\displaystyle\sum_{k=1}^{\infty} u_k$ is always quite clear from the context, so we will also indulge in this notation.

A series $\displaystyle\sum_{k=1}^{\infty} u_k$ whose terms are elements of **E** is said to be *absolutely convergent* if the series of real numbers $\displaystyle\sum_{k=1}^{\infty} \|u_k\|$ converges.

Let $\displaystyle\sum_{k=1}^{\infty} u_k$ be a series whose terms are elements of **E**. Since every Cauchy sequence of elements of **E** is convergent, and since $\left\| \displaystyle\sum_{k=m}^{n} u_k \right\| \leq \displaystyle\sum_{k=m}^{n} \|u_k\|$ for every $m \in \mathbb{N}$ and $n \in \mathbb{N}$, $m \leq n$, it follows that if the series converges absolutely, then the series converges conditionally, as well.

Let $\displaystyle\sum_{k=0}^{\infty} u_k$ and $\displaystyle\sum_{k=0}^{\infty} v_k$ be two series whose terms are elements of **E**. Using the two series, we can construct a series $\displaystyle\sum_{k=0}^{\infty} w_k$ as follows: $w_0 = u_0 v_0$, $w_1 = u_0 v_1 + u_1 v_0$, $w_2 = u_0 v_2 + u_1 v_1 + u_2 v_0$, and, in general,

$$w_k = \sum_{j=0}^{k} u_j v_{k-j} \tag{B.2.1}$$

for every $k \in \mathbb{N} \cup \{0\}$. The series $\sum_{k=0}^{\infty} w_k$ is called the *product* of the series $\sum_{k=0}^{\infty} u_k$ and $\sum_{k=0}^{\infty} v_k$ (in this order). The reason for calling $\sum_{k=0}^{\infty} w_k$ the product of $\sum_{k=0}^{\infty} u_k$ and $\sum_{k=0}^{\infty} v_k$ stems from the validity of the following theorem:

Theorem B.2.7. *Let* $\sum_{k=0}^{\infty} u_k$ *and* $\sum_{k=0}^{\infty} v_k$ *be two conditionally convergent series whose terms are elements of* **E**, *and assume that at least one of the series is absolutely convergent. Then the series* $\sum_{k=0}^{\infty} w_k$ *converges conditionally to* $\left(\sum_{k=0}^{\infty} u_k \right) \left(\sum_{k=0}^{\infty} v_k \right)$, *where* w_k, $k \in \mathbb{N} \cup \{0\}$, *are given by the equality (B.2.1) of this section.*

Proof. Assume that $\sum_{k=0}^{\infty} v_k$ is absolutely convergent (the proof in the case when $\sum_{k=0}^{\infty} u_k$ converges absolutely is similar, and the changes that have to be made are obvious; therefore, we will not discuss this case). We will also assume that $v_k \neq 0$ for some $k \in \mathbb{N} \cup \{0\}$; we can make this assumption because if $v_k = 0$ for every $k \in \mathbb{N} \cup \{0\}$, then the assertion of the theorem is obviously true.

In order to keep the notation simple, let u and v be the sums of the series $\sum_{k=0}^{\infty} u_k$ and $\sum_{k=0}^{\infty} v_k$, respectively; also, set $q_n = \sum_{k=0}^{n} u_k$, $r_n = \sum_{k=0}^{n} v_k$, and $s_n = \sum_{k=0}^{n} w_k$ for every $n \in \mathbb{N} \cup \{0\}$.

Using the equality (B.2.1) of this section, and by changing the order of summation, we obtain that

$$s_n = \sum_{k=0}^{n} \sum_{j=0}^{k} u_j v_{k-j} = \sum_{k=0}^{n} \sum_{j=0}^{k} u_{k-j} v_j = \sum_{j=0}^{n} \left(\sum_{k=j}^{n} u_{k-j} \right) v_j$$

$$= \sum_{j=0}^{n} q_{n-j} v_j - u r_n + u r_n = \sum_{j=0}^{n} (q_{n-j} - u) v_j + u r_n$$

for every $n \in \mathbb{N} \cup \{0\}$.

Since the multiplication that defines the algebraic structure on $\mathbf{E}$ is continuous (see Proposition B.2.6), it follows that $(ur_n)_{n\in\mathbb{N}\cup\{0\}}$ converges to uv. Therefore, using the above equalities, we obtain that the theorem will be completely proved

if we show that the sequence $\left(\displaystyle\sum_{j=0}^{n}(q_{n-j}-u)v_j\right)_{n\in\mathbb{N}\cup\{0\}}$ converges to zero in the

norm topology of $\mathbf{E}$.

To this end, let $\varepsilon \in \mathbb{R}$, $\varepsilon > 0$.

Since the sequence of partial sums $(q_n)_{n\in\mathbb{N}\cup\{0\}}$ converges in the norm topology of $\mathbf{E}$ to u, it follows that the sequence $(\|q_n-u\|)_{n\in\mathbb{N}\cup\{0\}}$ converges to zero; therefore, $(\|q_n - u\|)_{n\in\mathbb{N}\cup\{0\}}$ is a bounded sequence, so, there exists an $M \in \mathbb{R}$, $M > 0$, such that $\|q_n - u\| \leq M$ for every $n \in \mathbb{N} \cup \{0\}$.

Since we assume that $\displaystyle\sum_{k=0}^{\infty} v_k$ converges absolutely, and that there exists a $k \in$

$\mathbb{N} \cup \{0\}$ such that $v_k \neq 0$, we obtain that the series $\displaystyle\sum_{k=0}^{\infty} \|v_k\|$ converges to a strictly

positive real number. Let c be the sum of the series $\displaystyle\sum_{k=0}^{\infty} \|v_k\|$.

Since $(q_n)_{n\in\mathbb{N}\cup\{0\}}$ converges to u, and since the series $\displaystyle\sum_{k=0}^{\infty} \|v_k\|$ converges to c, it

follows that there exists an $n'_\varepsilon \in \mathbb{N}$ large enough such that $\|q_n - u\| < \dfrac{\varepsilon}{2c}$ for every

$n \in \mathbb{N}$, $n \geq n'_\varepsilon$, and such that $\displaystyle\sum_{j=n'_\varepsilon+1}^{\infty} \|v_j\| < \dfrac{\varepsilon}{2M}$.

Set $n_\varepsilon = 2n'_\varepsilon$. Then

$$\left\|\sum_{j=0}^{n}(q_{n-j}-u)v_j\right\| \leq \sum_{j=0}^{n'_\varepsilon} \|q_{n-j}-u\|\|v_j\| + \sum_{j=n'_\varepsilon+1}^{n} \|q_{n-j}-u\|\|v_j\|$$

$$\leq \frac{\varepsilon}{2c}\sum_{j=0}^{n'_\varepsilon} \|v_j\| + M\sum_{j=n'_\varepsilon+1}^{n} \|v_j\| < \frac{\varepsilon}{2c}\cdot c + M\frac{\varepsilon}{2M} = \frac{\varepsilon}{2} + \frac{\varepsilon}{2} = \varepsilon$$

for every $n \in \mathbb{N}$, $n \geq n_\varepsilon$.

We have therefore proved that for every $\varepsilon \in \mathbb{R}$, $\varepsilon > 0$, there exists an

$n_\varepsilon \in \mathbb{N}$ such that $\left\|\displaystyle\sum_{j=0}^{n}(q_{n-j}-u)v_j\right\| < \varepsilon$ for every $n \in \mathbb{N}$, $n \geq n_\varepsilon$. Thus,

$\left(\displaystyle\sum_{j=0}^{n}(q_{n-j}-u)v_j\right)_{n\in\mathbb{N}\cup\{0\}}$ converges to zero. $\qquad\square$

An interesting feature of Banach algebras with unit is the fact that in such a Banach algebra we can define an exponential function. We will conclude the section with a brief discussion of this function.

Let $\mathbf{E}$ be a Banach algebra with unit, and let $\mathbf{e}$ be the unit of $\mathbf{E}$. As usual, we make the convention that $u^0 = \mathbf{e}$ for every $u \in \mathbf{E}$.

Note that, for every $u \in \mathbf{E}$, the series $\sum_{k=0}^{\infty} \dfrac{u^k}{k!}$ converges absolutely because

$$\left\| \frac{u^k}{k!} \right\| \leq \frac{\|u\|^k}{k!} \quad \text{for every } k \in \mathbb{N} \cup \{0\} \text{ and the series of real numbers } \sum_{k=0}^{\infty} \frac{\|u\|^k}{k!}$$

converges (to $e^{\|u\|}$, where $x \mapsto e^x$, $x \in \mathbb{R}$, is the usual real-valued exponential function). Thus, if we let $\exp : \mathbf{E} \to \mathbf{E}$ be the function defined by $\exp(u) = \sum_{k=0}^{\infty} \dfrac{u^k}{k!}$ for every $u \in \mathbf{E}$, then the definition of $\exp$ is correct in the sense that the series $\sum_{k=0}^{\infty} \dfrac{u^k}{k!}$ converges (even absolutely) for every $u \in \mathbf{E}$. The function $\exp$ is called the *exponential function on* $\mathbf{E}$. If $\mathbf{E} = \mathbb{R}$ or $\mathbf{M}(n, \mathbb{R})$ for some $n \in \mathbb{N}$, $n \geq 2$, we will also use the notation e^u for $u \in \mathbf{E}$.

Using the exponential function, we can define another function $\exp_s : \mathbf{E} \to \mathbf{E}$ as follows: $\exp_s(u) = e^{-\|u\|} \exp(u)$ for every $u \in \mathbf{E}$. Note that the range of $\exp_s$ is included in the closed unit ball $\{u \in \mathbf{E} \mid \|u\| \leq 1\}$ of $\mathbf{E}$. We call $\exp_s$ the *scaled exponential function* on $\mathbf{E}$.

A basic property of the standard real-valued exponential function on $\mathbb{R}$ is that $e^{s+t} = e^s e^t$ for every $s \in \mathbb{R}$ and $t \in \mathbb{R}$. As shown in the next proposition, the exponential function, even when extended to Banach algebras with unit, still has this property.

Proposition B.2.8. $\exp(x + y) = \exp(x) \exp(y)$ *for every* $x \in \mathbf{E}$ *and* $y \in \mathbf{E}$.

Proof. Let $x \in \mathbf{E}$ and $y \in \mathbf{E}$, and set $u_k = \dfrac{x^k}{k!}$ and $v_k = \dfrac{y^k}{k!}$ for every $k \in \mathbb{N} \cup \{0\}$. Then $\exp(x) = \sum_{k=0}^{\infty} u_k$ and $\exp(y) = \sum_{k=0}^{\infty} v_k$.

Let $\sum_{k=0}^{\infty} w_k$ be the product of the series $\sum_{k=0}^{\infty} u_k$ and $\sum_{k=0}^{\infty} v_k$. By Theorem B.2.7, the series $\sum_{k=0}^{\infty} w_k$ is convergent, and its sum is $\left(\sum_{k=0}^{\infty} u_k \right) \left(\sum_{k=0}^{\infty} v_k \right)$.

Taking into consideration that

$$\frac{(x + y)^k}{k!} = \frac{\sum_{j=0}^{k} \frac{k!}{j!(k-j)!} x^j y^{k-j}}{k!} = \sum_{j=0}^{k} u_j v_{k-j} = w_k$$

for every $k \in \mathbb{N} \cup \{0\}$, we obtain that

$$\exp(x+y) = \sum_{k=0}^{\infty} \frac{(x+y)^k}{k!} = \sum_{k=0}^{\infty} w_k = \left(\sum_{k=0}^{\infty} u_k\right)\left(\sum_{k=0}^{\infty} v_k\right) = \exp(x)\exp(y).$$

$\square$

Note that we can use Proposition B.2.8 to construct one-parameter semigroups and one-parameter groups of elements of $\mathbf{E}$, where $\mathbf{E}$ is thought of as a multiplicative semigroup with neutral element. Indeed, if $x \in \mathbf{E}$, then $(\exp(tx))_{t\in[0,+\infty)}$ and $(\exp_s(tx))_{t\in[0,+\infty)}$ are one-parameter semigroups, and $(\exp(tx))_{t\in\mathbb{R}}$ is a one-parameter group of elements of $\mathbf{E}$.

Our goal now is to prove that all these one-parameter semigroups and groups are continuous; that is, we will prove that, for every $x \in \mathbf{E}$, the maps $t \mapsto \exp(tx)$, $t \in [0,+\infty)$, $t \mapsto \exp_s(tx)$, $t \in [0,+\infty)$, and $t \mapsto \exp(tx)$, $t \in \mathbb{R}$, are continuous with respect to the norm topology on $\mathbf{E}$, and the standard topologies of $[0,+\infty)$ and of $\mathbb{R}$. When dealing with the continuity of one-parameter semigroups and groups, we will also consider other types of continuity with respect to other topologies on $\mathbf{E}$, so we will also use the term *norm continuity* for continuity with respect to the norm topology of $\mathbf{E}$ in order to emphasize the type of continuity under consideration.

We will need the following lemma:

Lemma B.2.9. *Let $\beta \in \mathbb{R}$, $\beta \geq 0$. Then for every $\varepsilon \in \mathbb{R}$, $\varepsilon > 0$, there exists an $n_\varepsilon \in \mathbb{N} \cup \{0\}$ such that $\displaystyle\sum_{n=n_\varepsilon+1}^{\infty} \frac{\alpha^n}{n!} < \varepsilon$ for every $\alpha \in [0,\beta]$.*

Proof. Let $\varepsilon \in \mathbb{R}$, $\varepsilon > 0$. Since the series $\displaystyle\sum_{n=0}^{\infty} \frac{\beta^n}{n!}$ converges (to e^β), it follows that there exists an $n_\varepsilon \in \mathbb{N}$ such that $\displaystyle\sum_{n=n_\varepsilon+1}^{\infty} \frac{\beta^n}{n!} < \varepsilon$. If $\alpha \in [0,\beta]$, then $\dfrac{\alpha^n}{n!} \leq \dfrac{\beta^n}{n!}$ for every $n \in \mathbb{N} \cup \{0\}$; hence, $\displaystyle\sum_{n=n_\varepsilon+1}^{\infty} \frac{\alpha^n}{n!} < \varepsilon$. $\square$

Proposition B.2.10. *Let $x \in \mathbf{E}$. The one-parameter semigroups $(\exp(tx))_{t\in[0,+\infty)}$ and $(\exp_s(tx))_{t\in[0,+\infty)}$, and the one-parameter group $(\exp(tx))_{t\in\mathbb{R}}$ are all continuous.*

Proof. Note that it is enough to prove the continuity of $(\exp(tx))_{t\in\mathbb{R}}$ and $(\exp_s(tx))_{t\in[0,+\infty)}$ because $(\exp(tx))_{t\in[0,+\infty)}$ is the restriction of $(\exp(tx))_{t\in\mathbb{R}}$ to $[0,+\infty)$.

Furthermore, the proposition will be completely proved if we show that $(\exp(tx))_{t\in\mathbb{R}}$ is continuous because $(\exp_s(tx))_{t\in[0,+\infty)}$ is the product of the real-valued continuous function $(e^{-t\|x\|})_{t\in[0,+\infty)}$ and the $\mathbf{E}$-valued function $(\exp(tx))_{t\in[0,+\infty)}$.

Thus, the proof of the proposition will be completed if we show that for every convergent sequence $(s_l)_{l \in \mathbb{N}}$ of real numbers it follows that $(\exp(s_l x))_{l \in \mathbb{N}}$ converges in the norm topology of $\mathbf{E}$ to $\exp(sx)$, where $s = \lim_{l \to +\infty} s_l$.

To this end, let $(s_l)_{l \in \mathbb{N}}$ be a convergent sequence of real numbers and let $s = \lim_{l \to +\infty} s_l$.

Since $(s_l)_{l \in \mathbb{N}}$ is convergent, it follows that $(s_l)_{l \in \mathbb{N}}$ is also a bounded sequence. Therefore, there exists an $M \in \mathbb{R}$, $M \geq 0$, such that $|s_l| \leq M$ for every $l \in \mathbb{N}$ and such that $|s| \leq M$.

Set $\beta = M \|x\|$.

Let $\varepsilon \in \mathbb{R}$, $\varepsilon > 0$.

By Lemma B.2.9, there exists an $n_\varepsilon \in \mathbb{N} \cup 0$ such that $\displaystyle\sum_{n=n_\varepsilon+1}^{\infty} \frac{t^n}{n!} < \frac{\varepsilon}{3}$ for

every $t \in [0, \beta]$; in particular, $\displaystyle\sum_{n=n_\varepsilon+1}^{\infty} \frac{|s_l|^n \|x\|^n}{n!} < \frac{\varepsilon}{3}$ for every $l \in \mathbb{N}$, and

$$\sum_{n=n_\varepsilon+1}^{\infty} \frac{|s|^n \|x\|^n}{n!} < \frac{\varepsilon}{3}.$$

Since $(s_l)_{l \in \mathbb{N}}$ converges to s, it follows that $\left(s_l^k \|x\|^k\right)_{l \in \mathbb{N}}$ converges to $s^k \|x\|^k$ for every $k \in \mathbb{N} \cup \{0\}$. Thus, there exists an $l_\varepsilon \in \mathbb{N}$ such that

$$\left|s_l^k - s^k\right| \|x\|^k < \frac{\varepsilon}{3(n_\varepsilon + 1)}$$

for every $l \geq l_\varepsilon$ and $k = 0, 1, 2, \ldots, n_\varepsilon$.

We obtain that

$$\|\exp(s_l x) - \exp(sx)\| \leq \left\| \sum_{n=0}^{n_\varepsilon} \left(\frac{s_l^n x^n}{n!} - \frac{s^n x^n}{n!} \right) \right\|$$

$$+ \left\| \left(\sum_{n=n_\varepsilon+1}^{\infty} \frac{s_l^n x^n}{n!} \right) - \left(\sum_{n=n_\varepsilon+1}^{\infty} \frac{s^n x^n}{n!} \right) \right\| \leq \sum_{n=0}^{n_\varepsilon} \frac{|s_l^n - s^n| \|x\|^n}{n!}$$

$$+ \sum_{n=n_\varepsilon+1}^{\infty} \frac{|s_l|^n \|x\|^n}{n!} + \sum_{n=n_\varepsilon+1}^{\infty} \frac{|s|^n \|x\|^n}{n!} < (n_\varepsilon + 1) \frac{\varepsilon}{3(n_\varepsilon + 1)} + \frac{\varepsilon}{3} + \frac{\varepsilon}{3} = \varepsilon$$

for every $l \geq l_\varepsilon$.

We have therefore proved that for every $\varepsilon \in \mathbb{R}$, $\varepsilon > 0$, there exists an $l_\varepsilon \in \mathbb{N}$ such that $\|\exp(s_l x) - \exp(sx)\| < \varepsilon$ for every $l \geq l_\varepsilon$; therefore, the sequence $(\exp(s_l x))_{l \in \mathbb{N}}$ converges to $\exp(sx)$. $\qquad\qquad\square$

B.3 One-Parameter Convolution Semigroups of Probability Measures

Our goal in this section is to use the operation of convolution of measures and the exponential function in order to briefly present certain facts about one-parameter convolution semigroups of probability measures. These one-parameter semigroups are used in the book in order to illustrate various concepts and results.

Let $(H, \cdot, d)$ be a locally compact separable metric semigroup, assume that H has a neutral element, and let e be the neutral element of H.

Using Proposition B.2.4 and the comment made in the paragraph following the proposition, we obtain that $\mathcal{M}(H)$ is a Banach algebra with unit δ_e. In view of the role played by the operation of convolution in defining the Banach algebra structure of $\mathcal{M}(H)$, when dealing with one-parameter semigroups and one-parameter groups of elements of $\mathcal{M}(H)$, we will call them *one-parameter convolution semigroups* and *one-parameter convolution groups*, respectively, in order to stress the fact that the Banach algebra under consideration is $\mathcal{M}(H)$. Even though one can define one-parameter convolution groups of elements of $\mathcal{M}(H)$ (for instance, if H is a locally compact separable metric group and $\phi : \mathbb{R} \to H$ is a group homomorphism, then $(\delta_{\phi(t)})_{t \in \mathbb{R}}$ is such a one-parameter convolution group), in this book we will mostly be concerned with one-parameter convolution semigroups $(\mu_t)_{t \in [0,+\infty)}$ of elements of $\mathcal{M}(H)$ that are probability measures (on $(H, \mathcal{B}(H))$, of course).

Let $(\mu_t)_{t \in [0,+\infty)}$ be a one-parameter convolution semigroup of elements of $\mathcal{M}(H)$.

We say that $(\mu_t)_{t \in [0,+\infty)}$ is *weak* continuous* if the map $t \mapsto \mu_t, t \in [0, +\infty)$, is continuous with respect to the weak* topology of $\mathcal{M}(H)$ and the standard topology of $[0, +\infty)$. Thus, $(\mu_t)_{t \in [0,+\infty)}$ is weak* continuous if and only if for every convergent sequence $(s_n)_{n \in \mathbb{N}}$ of elements of $[0, +\infty)$ and for every $f \in C_0(H)$, the sequence $(\langle f, \mu_{s_n} \rangle)_{n \in \mathbb{N}}$ converges to $\langle f, \mu_s \rangle$, where $s = \lim_{n \to +\infty} s_n$.

Clearly, if $(\mu_t)_{t \in [0,+\infty)}$ is norm continuous, then $(\mu_t)_{t \in [0,+\infty)}$ is weak* continuous, as well; also, it is easy to construct examples in order to show that the weak* continuity of a one-parameter convolution semigroup does not imply its norm continuity in general.

The convolution semigroups that are of interest to us in this book are *one-parameter convolution semigroups of probability measures* (that is, convolution semigroups $(\mu_t)_{t \in [0,+\infty)}$, where $\mu_t, t \in [0, +\infty)$, are probability measures; for other types of convolution semigroups, for a more general treatment of, and for additional details on these semigroups, see Heyer's monographs [45] and [46]).

Many one-parameter convolution semigroups of probability measures can be obtained using the scaled exponential function as follows: let $\mu \in \mathcal{M}(H)$ be a probability measure and set $\mu_t = \exp_s(t\mu)$ for every $t \in \mathbb{R}, t \geq 0$. It is easy to see that $\mu_t = e^{-t} \exp(t\mu) = e^{-t} \sum_{k=0}^{\infty} \frac{t^k}{k!} \mu^k$ for every $t \in \mathbb{R}, t \geq 0$ (here, and throughout the book when dealing with convolutions, $\mu^k = \underbrace{\mu * \mu * \cdots * \mu}_{k \text{ times}}, k \in \mathbb{N}$,

and $\mu^0 = \delta_e = $ the unit of the Banach algebra $\mathcal{M}(H)$). Using Proposition B.2.8 and the fact that $e^{s+t} = e^s e^t$ for every $s \in \mathbb{R}$ and $t \in \mathbb{R}$, we obtain that $\mu_{s+t} = \mu_s * \mu_t$ for every $s \in \mathbb{R}$, $s \geq 0$, and $t \in \mathbb{R}$, $t \geq 0$. Thus, $(\mu_t)_{t \in [0,+\infty)}$ is a one-parameter convolution semigroup. In order to show that μ_t, $t \in [0, +\infty)$, are probability measures, we need the following proposition:

Proposition B.3.1. *If $v \in \mathcal{M}(H)$ and $\lambda \in \mathcal{M}(H)$ are probability measures, then $v * \lambda$ is a probability measure, as well.*

Proof. Let v and λ be probability measures in $\mathcal{M}(H)$. Taking into consideration the manner in which the convolution operation is defined, we obtain that $v * \lambda$ is a positive element of $\mathcal{M}(H)$ (that is, $v * \lambda$ is a measure on $(H, \mathcal{B}(H))$), and since $\mathcal{M}(H)$ is a Banach algebra, it follows that $\|v * \lambda\| \leq \|v\|\|\lambda\| = 1$. In view of the identification of $\mathcal{M}(H)$ with the dual of $C_0(H)$ and since $v * \lambda$ is a positive element of $\mathcal{M}(H)$, it follows that, in order to prove that $v * \lambda$ is a probability measure, it is enough to show that, for every $\varepsilon \in \mathbb{R}$, $\varepsilon > 0$, there exists an $f \in C_0(H), 0 \leq f \leq 1$, such that $\int f(x) \, d(v * \lambda)(x) > 1 - \varepsilon$.

To this end, let $\varepsilon \in \mathbb{R}$, $\varepsilon > 0$. Since v and λ are regular probability measures (because H is a locally compact separable metric semigroup, so all the positive elements of $\mathcal{M}(H)$ are regular measures on $(H, \mathcal{B}(H))$), it follows that there exists a compact subset K of H such that $v(K) > 1 - \dfrac{\varepsilon}{2}$ and $\lambda(K) > 1 - \dfrac{\varepsilon}{2}$. Since the algebraic operation that defines the semigroup structure on H is continuous, it follows that K^2 is a compact subset of H. By Proposition 7.1.8, p. 199 of Cohn's book [20], there exists an $f \in C_0(H), 0 \leq f \leq 1$, such that $f(x) = 1$ for every $x \in K^2$.

Using the fact that $\mathbf{1}_{K^2}(xy) \geq \mathbf{1}_K(x)\mathbf{1}_K(y)$ for every $x \in H$ and $y \in H$, we obtain that

$$\int f(z) \, d(v * \lambda)(z) = \int \int f(xy) \, dv(x) \, d\lambda(y) \geq \int \int \mathbf{1}_{K^2}(xy) \, dv(x) \, d\lambda(y)$$

$$\geq \int \int \mathbf{1}_K(x)\mathbf{1}_K(y) \, dv(x) \, d\lambda(y) = v(K)\lambda(K) > \left(1 - \frac{\varepsilon}{2}\right)^2 > 1 - \varepsilon.$$

$$\square$$

Now, let us return to the one-parameter convolution semigroup $(\mu_t)_{t \in [0,+\infty)}$, $\mu_t = \exp_s(t\mu)$ for every $t \in [0, +\infty)$, defined before Proposition B.3.1.

Let $t \in [0, +\infty)$, and note that the sequence of partial sums $\left(\displaystyle\sum_{k=0}^{n} \frac{t^k}{k!} \mu^k\right)_{n \in \mathbb{N} \cup \{0\}}$, which converges to $\exp(t\mu)$ in the norm topology of $\mathcal{M}(H)$, has the property that each term in the sequence is a sum of positive elements of $\mathcal{M}(H)$. Since, by Proposition B.3.1 $1 = \|\mu^k\| \ (= \mu^k(H))$ for every $k \in \mathbb{N} \cup \{0\}$, it follows that

$$\left\| \sum_{k=0}^{n} \frac{t^k}{k!} \mu^k \right\| = \sum_{k=0}^{n} \frac{t^k}{k!} \|\mu^k\| = \sum_{k=0}^{n} \frac{t^k}{k!}$$

for every $n \in \mathbb{N} \cup \{0\}$. Taking into consideration that the sequence $\left(\left\| \sum_{k=0}^{n} \frac{t^k}{k!} \mu^k \right\| \right)_{n \in \mathbb{N} \cup \{0\}}$ converges to $\|\exp(t\mu)\|$, we obtain that

$$\|\exp(t\mu)\| = \lim_{n \to +\infty} \left\| \sum_{k=0}^{n} \frac{t^k}{k!} \mu^k \right\| = \lim_{n \to +\infty} \sum_{k=0}^{n} \frac{t^k}{k!} = e^t.$$

Thus, $\|\mu_t\| = 1$; since $\mu_t \geq 0$, it follows that μ_t is a probability measure.

For future reference, we summarize our discussion of μ_t, $t \in [0, +\infty)$ in the next proposition.

Proposition B.3.2. *If $\mu \in \mathcal{M}(H)$ is a probability measure, and $\mu_t = \exp_s(t\mu)$ for every $t \in [0, +\infty)$, then $(\mu_t)_{t \in [0,+\infty)}$ is a one-parameter convolution semigroup of probability measures.*

We will often refer to the one-parameter semigroup $(\mu_t)_{t \in [0,+\infty)}$ described in Proposition B.3.2 as an *exponential one-parameter convolution semigroup of probability measures defined by a probability measure.*

Using Proposition B.2.10 we obtain that the exponential one-parameter convolution semigroups of probability measures are norm continuous, so they are weak* continuous, as well. For future reference, we state these facts in the next proposition.

Proposition B.3.3. *Let $\mu \in \mathcal{M}(H)$ be a probability measure, and let $\mu_t = \exp_s(t\mu)$ for every $t \in \mathbb{R}$, $t \geq 0$. Then $(\mu_t)_{t \in [0,+\infty)}$ is a norm continuous one-parameter convolution semigroup. Consequently, $(\mu_t)_{t \in [0,+\infty)}$ is also weak* continuous.*

B.4 Exponential Semiflows and Flows

By exponential semiflows and flows we mean semiflows and flows that are defined using the exponential function. For instance, the geodesic flows defined in Examples A.3.6 and B.1.8 are exponential flows, while the horocycle flows defined in Examples A.3.7 and B.1.9 are not.

In this section of Appendix B we will discuss two families of exponential semiflows and a rather large family of quite sophisticated exponential flows defined on spaces of cosets of $\mathrm{SL}(n, \mathbb{R})$, $n \geq 2$. One of the families of semiflows consists of semiflows defined on spaces of (column) stochastic $n \times n$ matrices, $n \geq 2$, while the semiflows that belong to the other family are defined on the space of n-dimensional probability vectors, $n \geq 2$. We start by discussing the two families of semiflows.

B.4.1   Stochastic Matrices and Semiflows

Let $n \in \mathbb{N}, n \geq 2$.

An $n \times n$ matrix $A = \begin{bmatrix} a_{11} & a_{12} & a_{13} & \dots & a_{1n} \\ a_{21} & a_{22} & a_{23} & \dots & a_{2n} \\ \multicolumn{5}{c}{\dotfill} \\ a_{n1} & a_{n2} & a_{n3} & \dots & a_{nn} \end{bmatrix}$ is said to be *column stochastic*

if $a_{ij} \geq 0$ for every $i = 1, 2, \dots, n$ and $j = 1, 2, \dots, n$, and if $\sum_{i=1}^{n} a_{ij} = 1$ for
every $j = 1, 2, \dots, n$. Of course, we can define *row stochastic matrices* in a similar
manner; however, in this book we will only deal with column stochastic matrices,
and we will often call these matrices simply *stochastic matrices*.

Let $\mathbb{S}_n$ be the set of all $n \times n$ stochastic matrices. On $\mathbb{S}_n$ we will consider the
topology $\mathcal{T}_{\mathbb{S}_n}$ induced by the standard topology of $\mathbf{M}(n, \mathbb{R})$ on $\mathbb{S}_n$, and we will call
$\mathcal{T}_{\mathbb{S}_n}$ the *standard topology of* $\mathbb{S}_n$.

Proposition B.4.1. *(a) The usual matrix multiplication $\cdot$ is a well-defined algebraic operation on $\mathbb{S}_n$.*
*(b) $(\mathbb{S}_n, \cdot, \mathcal{T}_{\mathbb{S}_n})$ is a compact metrizable semigroup with neutral element, and the
neutral element is the $n \times n$ identity matrix $\mathbf{I}_n$. Also, $\mathbb{S}_n$ is convex as a subset of
$\mathbf{M}(n, \mathbb{R})$.*

Proof. (a) We have to prove that if $A = [a_{ij}]_{\substack{i=1,\dots,n \\ j=1,\dots,n}}$ and $B = [b_{ij}]_{\substack{i=1,\dots,n \\ j=1,\dots,n}}$ are

two column stochastic matrices, then their product AB is a column stochastic
matrix, as well.

To this end, let $AB = [c_{ij}]_{\substack{i=1,\dots,n \\ j=1,\dots,n}}$. Then

$$\sum_{i=1}^{n} c_{ij} = \sum_{i=1}^{n} \left(\sum_{k=1}^{n} a_{ik} b_{kj} \right) = \sum_{k=1}^{n} \left(\sum_{i=1}^{n} a_{ik} \right) b_{kj} = \sum_{k=1}^{n} b_{kj} = 1$$

for every $j = 1, \dots, n$.
(b) Since $\mathbb{S}_n$ is a closed bounded subset of $\mathbf{M}(n, \mathbb{R})$, and since matrix multiplication
and the standard topology of $\mathbf{M}(n, \mathbb{R})$ define the structure of a topological
semigroup on $\mathbf{M}(n, \mathbb{R})$, using (a) and well-known facts about $\mathbf{M}(n, \mathbb{R})$, we
obtain that all the assertions of (b) hold true (the fact that $\mathbb{S}_n$ is a bounded subset
of $\mathbf{M}(n, \mathbb{R})$ means, of course, that $\mathbb{S}_n$ is bounded with respect to any norm on
$\mathbf{M}(n, \mathbb{R})$). $\qquad\square$

Our goal is to use the scaled exponential function defined on $\mathbf{M}(n, \mathbb{R})$ in order to
construct a family of semiflows on $\mathbb{S}_n$.

To this end, we need the following proposition:

Proposition B.4.2. *Let $(A_l)_{l\in\mathbb{N}}$ be a sequence of elements of $\mathbb{S}_n$, and let $(p_l)_{l\in\mathbb{N}}$ be a sequence of real numbers such that $p_l \geq 0$ for every $l \in \mathbb{N}$, and such that $\sum_{l=1}^{\infty} p_l = 1$. Then the series $\sum_{l=1}^{\infty} p_l A_l$ converges absolutely in $\mathbf{M}(n,\mathbb{R})$ with respect to any norm on $\mathbf{M}(n,\mathbb{R})$ to an element of $\mathbb{S}_n$.*

Proof. Let $\|\cdot\|$ be a norm on $\mathbf{M}(n,\mathbb{R})$. Since $\mathbb{S}_n$ is a bounded subset of $\mathbf{M}(n,\mathbb{R})$ with respect to $\|\cdot\|$, it follows that there exists an $M \in \mathbb{R}$, $M > 0$, such that $\|A\| \leq M$ for every $A \in \mathbb{S}_n$. We obtain that

$$\sum_{l=1}^{\infty} \|p_l A_l\| = \sum_{l=1}^{\infty} p_l \|A_l\| \leq M < +\infty,$$

so the series $\sum_{l=1}^{\infty} p_l A_l$ converges absolutely.

Let A be the sum of $\sum_{l=1}^{\infty} p_l A_l$. Also, set $B_k = \dfrac{\sum_{l=1}^{k} p_l A_l}{\sum_{i=1}^{k} p_i}$ for every $k \in \mathbb{N}$.

Since $\mathbb{S}_n$ is a convex subset of $\mathbf{M}(n,\mathbb{R})$, and since B_k is a convex combination of $A_1, A_2, \ldots, A_k$, it follows that $B_k \in \mathbb{S}_n$ for every $k \in \mathbb{N}$. Taking into consideration that $\left(\sum_{l=1}^{k} p_l\right)_{k\in\mathbb{N}}$ converges to 1 and that $\left(\sum_{l=1}^{k} p_l A_l\right)_{k\in\mathbb{N}}$ converges to A, we obtain that $(B_k)_{k\in\mathbb{N}}$ converges to A in the standard topology of $\mathbf{M}(n,\mathbb{R})$. Finally, since $\mathbb{S}_n$ is a closed subset of $\mathbf{M}(n,\mathbb{R})$, it follows that $A \in \mathbb{S}_n$. □

Recall that, as pointed out at the beginning of Sect. B.2 after the definition of Banach algebras with unit, $\mathbf{M}(n,\mathbb{R})$ is such a Banach algebra when we consider a norm on $\mathbb{R}^n$, we think of $\mathbf{M}(n,\mathbb{R})$ as the space of all linear bounded operators on $\mathbb{R}^n$ endowed with that norm, and we consider on $\mathbf{M}(n,\mathbb{R})$ the resulting operator norm. Thus, it makes sense to define the exponential function exp and the scaled exponential function $\exp_s$ on $\mathbf{M}(n,\mathbb{R})$. Since the series $\sum_{k=0}^{\infty} \dfrac{A^k}{k!}$ converges absolutely with respect to the operator norm mentioned above for every $A \in \mathbf{M}(n,\mathbb{R})$, we obtain that, actually, the series converges with respect to any norm on $\mathbf{M}(n,\mathbb{R})$ because any two norms on $\mathbf{M}(n,\mathbb{R})$ are equivalent. Note however that the scaled exponential function is defined only with respect to a norm that generates a Banach algebra with unit structure on $\mathbf{M}(n,\mathbb{R})$.

In order to define the semiflows that we have in mind, we need a norm on $\mathbf{M}(n,\mathbb{R})$ which defines a Banach algebra with unit structure on $\mathbf{M}(n,\mathbb{R})$ (because we have to be able to use the scaled exponential function), and which also has the property that the norm of any column stochastic matrix in $\mathbf{M}(n,\mathbb{R})$ is equal to one. We will now discuss such a norm.

Let us consider $\mathbb{R}^n$ as a vector space of column vectors, and let $\|\cdot\|_1$ be the norm on $\mathbb{R}^n$ defined by $\|\mathbf{x}\|_1 = \sum_{i=1}^{n} |x_i|$ for every $\mathbf{x} \in \mathbb{R}^n$, $\mathbf{x} = \begin{bmatrix} x_1 \\ x_2 \\ \cdot \\ \cdot \\ \cdot \\ x_n \end{bmatrix}$. Now, let $\|\cdot\|_1$ be the norm on $\mathbf{M}(n, \mathbb{R})$ defined as follows: for every $A \in \mathbf{M}(n, \mathbb{R})$, $\|A\|_1$ is the operator norm of A, where A is thought of as a linear bounded operator on the Banach space $\mathbb{R}^n$, the Banach space structure of $\mathbb{R}^n$ being defined by $\|\cdot\|_1$; thus,

$$\|A\|_1 = \sup_{\substack{\mathbf{x} \in \mathbb{R}^n \\ \|\mathbf{x}\|_1 \leq 1}} \|A\mathbf{x}\|_1 .$$

In the next proposition and throughout the book, the superscript $^\mathrm{T}$ is used to denote the transpose of a vector or a matrix.

Proposition B.4.3. $\|A\|_1 = 1$ for every $A \in \mathbb{S}_n$.

Proof. Let $A \in \mathbb{S}_n$, $A = [a_{ij}]_{\substack{i=1,2,\dots,n \\ j=1,2,\dots,n}}$. Taking into consideration that, as a linear operator, A is positive with respect to the coordinatewise order (in the sense that $A\mathbf{x} \geq 0$ whenever $\mathbf{x} \geq 0$), we obtain that

$$\|A\|_1 = \sup_{\substack{\mathbf{x} \in \mathbb{R}^n \\ \|\mathbf{x}\|_1 \leq 1}} \|A\mathbf{x}\|_1 = \sup_{\substack{\mathbf{x} \in \mathbb{R}^n \\ \mathbf{x}=(x_1,x_2,\dots,x_n)^\mathrm{T} \\ x_i \geq 0 \text{ for every } i}} \sum_{i=1}^{n}\sum_{j=1}^{n} a_{ij} x_j$$

$$= \sup_{\substack{\mathbf{x} \in \mathbb{R}^n \\ \mathbf{x}=(x_1,x_2,\dots,x_n)^\mathrm{T} \\ x_i \geq 0 \text{ for every } i}} \sum_{j=1}^{n}\left(\sum_{i=1}^{n} a_{ij}\right) x_j = 1 .$$

$\square$

The first family of semiflows that we are going to define and study here are semiflows of the type described in (a) of Proposition A.3.3. But first, we need some preparation.

Unless otherwise stated, from now on in this subsection, the scaled exponential function $\exp_s$ is always defined with respect to the norm $\|\cdot\|_1$ on $\mathbf{M}(n, \mathbb{R})$; that is, $\exp_s(A) = e^{-\|A\|_1} \exp(A)$ for every $A \in \mathbf{M}(n, \mathbb{R})$.

Proposition B.4.4. *If $A \in \mathbb{S}_n$, then $\exp_s(\alpha A)$ belongs to $\mathbb{S}_n$, as well, for every $\alpha \in \mathbb{R}$, $\alpha \geq 0$.*

Proof. Let $A \in \mathbb{S}_n$ and $\alpha \in \mathbb{R}$, $\alpha \geq 0$.

Using Proposition B.4.3, we obtain that

$$\exp_s(\alpha A) = e^{-\alpha} \exp(\alpha A) = e^{-\alpha} \sum_{k=0}^{\infty} \frac{\alpha^k}{k!} A^k.$$

Set $A_l = A^{l-1}$ and $p_l = e^{-\alpha} \frac{\alpha^{l-1}}{(l-1)!}$ for every $l \in \mathbb{N}$.

Then $A_l, l \in \mathbb{N}$, are column stochastic matrices by (a) of Proposition B.4.1, and it is easy to see that $p_l \geq 0$ for every $l \in \mathbb{N}$ and that $\sum_{l=1}^{\infty} p_l = 1$. By Proposition B.4.2, $\sum_{l=1}^{\infty} p_l A_l$ converges absolutely and its sum is an element of $\mathbb{S}_n$. This completes the proof because $\exp_s(\alpha A) = \sum_{l=1}^{\infty} p_l A_l$. $\square$

Proposition B.4.5. *Let $A \in \mathbb{S}_n$. Then the function $g^{(A)} : [0, +\infty) \to \mathbb{S}_n$, $g_t^{(A)} = \exp_s(tA)$ for every $t \in [0, +\infty)$, is a well-defined continuous semigroup homomorphism.*

Proof. By Proposition B.4.4, the function $g^{(A)}$ is well-defined in the sense that $g_t^{(A)}$ belongs to $\mathbb{S}_n$ for every $t \in [0, +\infty)$.

Using Proposition B.2.8, we obtain that $g^{(A)}$ is a semigroup homomorphism.

Finally, using Proposition B.2.10, we obtain that $g^{(A)}$ is continuous. $\square$

Let $\mathbf{A} \in \mathbb{S}_n$.

Let $\pi^{(\mathbf{A})} : [0, +\infty) \times \mathbb{S}_n \to \mathbb{S}_n$ be defined by $\pi^{(\mathbf{A})}(t, \mathbf{X}) = \exp_s(t\mathbf{A}) \cdot \mathbf{X}$ (or, in subscript notation, $\pi_t^{(\mathbf{A})}(\mathbf{X}) = \exp_s(t\mathbf{A})\mathbf{X}$) for every $(t, \mathbf{X}) \in [0, +\infty) \times \mathbb{S}_n$. It is easy to see that $\pi^{(\mathbf{A})}$ is a semiflow on the compact metric multiplicative semigroup $\mathbb{S}_n$. Note that $\pi^{(\mathbf{A})}$ is a semiflow of the type described in Example A.3.1. Using Proposition B.4.5 and (a) of Proposition A.3.3, we obtain that $\pi^{(\mathbf{A})}$ is a continuous semiflow.

Let (X, d) be a locally compact separable metric space, let $\mathbb{T}$ stand for $\mathbb{R}$ or the interval $[0, +\infty)$ in $\mathbb{R}$, and let $(w_t)_{t \in \mathbb{T}}$ be a flow or a semiflow on X.

We say that $(w_t)_{t \in \mathbb{T}}$ is *continuous with respect to $t \in \mathbb{T}$ uniformly in $x \in X$* if the following condition is satisfied:

- For every convergent sequence $(s_n)_{n \in \mathbb{N}}$ of elements of $\mathbb{T}$, and for every $\varepsilon \in \mathbb{R}$, $\varepsilon > 0$, there exists an $n_\varepsilon \in \mathbb{N}$ such that $d\left(w_{s_n}(x), w_s(x)\right) < \varepsilon$ for every $n \geq n_\varepsilon$ and $x \in X$, where $s = \lim_{n \to +\infty} s_n$.

We say that $(w_t)_{t \in \mathbb{T}}$ is *equicontinuous with respect to $t \in \mathbb{T}$* if $(w_t)_{t \in \mathbb{T}}$ has the following property:

- For every convergent sequence $(x_n)_{n \in \mathbb{N}}$ of elements of X and for every $\varepsilon \in \mathbb{R}$, $\varepsilon > 0$, there exists an $n_\varepsilon \in \mathbb{N}$ such that $d(w_t(x_n), w_t(x)) < \varepsilon$ for every $n \geq n_\varepsilon$ and every $t \in \mathbb{T}$, where $x = \lim_{n \to +\infty} x_n$.

Flows and semiflows that have the above two properties have a certain nice behavior that is discussed in detail in the book. In the next proposition we show that the semiflows $\pi^{(\mathbf{A})}$, $\mathbf{A} \in \mathbb{S}_n$, have these two properties.

Proposition B.4.6. *Let* $\mathbf{A} \in \mathbb{S}_n$. *The semiflow* $\pi^{(\mathbf{A})}$ *is continuous with respect to* $t \in [0, +\infty)$ *uniformly in* $\mathbf{X} \in \mathbb{S}_n$, *and is equicontinuous with respect to* $t \in [0, +\infty)$.

Proof. Taking into consideration that $\|\mathbf{X}\|_1 = 1$ for every $\mathbf{X} \in \mathbb{S}_n$ by Proposition B.4.3, and the fact that the norm $\|\cdot\|_1$ defines a Banach algebra structure on $\mathbf{M}(n, \mathbb{R})$, we obtain that

$$\|\pi^{(\mathbf{A})}_{s_k}(\mathbf{X}) - \pi^{(\mathbf{A})}_{s}(\mathbf{X})\|_1 = \|g^{(\mathbf{A})}_{s_k}\mathbf{X} - g^{(\mathbf{A})}_{s}\mathbf{X}\|_1$$

$$\leq \|g^{(\mathbf{A})}_{s_k} - g^{(\mathbf{A})}_{s}\|_1\|\mathbf{X}\|_1 = \|g^{(\mathbf{A})}_{s_k} - g^{(\mathbf{A})}_{s}\|_1 \quad \text{(B.4.1)}$$

for every convergent sequence $(s_m)_{m \in \mathbb{N}}$ of elements of $[0, +\infty)$ and every $k \in \mathbb{N}$, where $s = \lim_{m \to +\infty} s_m$ and $g^{(\mathbf{A})}$ is the function defined in Proposition B.4.5. Using the above relations (B.4.1), and the fact that, by Proposition B.4.5, the function $g^{(\mathbf{A})}$ is continuous, we obtain that $\pi^{(\mathbf{A})}$ is continuous with respect to $t \in [0, +\infty)$ uniformly in $\mathbf{X} \in \mathbb{S}_n$.

Using again the fact that the norm $\|\cdot\|_1$ defines a Banach algebra structure on $\mathbf{M}(n, \mathbb{R})$ and since, by Proposition B.4.4, the matrices $g^{(\mathbf{A})}_{s}$, $s \in [0, +\infty)$, belong to $\mathbb{S}_n$, where $g^{(\mathbf{A})}$ is the semigroup homomorphism defined in Proposition B.4.5, we obtain that

$$\|\pi^{(\mathbf{A})}_{t}(\mathbf{X}_k) - \pi^{(\mathbf{A})}_{t}(\mathbf{X})\|_1 = \|g^{(\mathbf{A})}_{t}(\mathbf{X}_k - \mathbf{X})\|_1$$

$$\leq \|g^{(\mathbf{A})}_{t}\|_1\|\mathbf{X}_k - \mathbf{X}\|_1 = \|\mathbf{X}_k - \mathbf{X}\|_1 \quad \text{(B.4.2)}$$

for every $t \in [0, +\infty)$, every convergent sequence $(\mathbf{X}_m)_{m \in \mathbb{N}}$ of elements of $\mathbb{S}_n$, and every $k \in \mathbb{N}$, where $\mathbf{X} = \lim_{m \to +\infty} \mathbf{X}_m$. Taking into consideration the above relations (B.4.2), we easily obtain that $\pi^{(\mathbf{A})}$ is equicontinuous with respect to $t \in [0, +\infty)$. $\square$

We will now start to discuss a second family of semiflows closely related to the semiflows $\pi^{(\mathbf{A})}$, $\mathbf{A} \in \mathbb{S}_n$. As pointed out at the beginning of this section, the semiflows that we are going to discuss now are defined on the set of all n-dimensional probability vectors. By an *n-dimensional probability vector*, or (an *n-dimensional*) *column stochastic vector*, we mean a column vector $\mathbf{x} \in \mathbb{R}^n$, $\mathbf{x} = (x_1, x_2, \ldots, x_n)^{\mathrm{T}}$ such that $x_i \geq 0$ for every $i = 1, 2, \ldots, n$, and such that $\sum_{i=1}^{n} x_i = 1$. The set of all n-dimensional column stochastic vectors is denoted by $\mathbb{P}_n$.

Note that $\mathbf{A}\mathbf{x} \in \mathbb{P}_n$ whenever $\mathbf{A} \in \mathbb{S}_n$ and $\mathbf{x} \in \mathbb{P}_n$.

Now let $\mathbf{A} \in \mathbb{S}_n$, and note that using Proposition B.4.4 and the above observation, we obtain that $\exp_s(t\mathbf{A})\mathbf{x}$ is an element of $\mathbb{P}_n$. Thus, the function $\varphi^{(\mathbf{A})} : [0, +\infty) \times \mathbb{P}_n \to \mathbb{P}_n$ defined by $\varphi^{(\mathbf{A})}(t, \mathbf{x}) = g_t^{(\mathbf{A})}\mathbf{x}$ (or, in subscript notation, $\varphi_t^{(\mathbf{A})}(\mathbf{x}) = g_t^{(\mathbf{A})}\mathbf{x}$) for every $(t, \mathbf{x}) \in [0, +\infty) \times \mathbb{P}_n$ is well-defined, where $g^{(\mathbf{A})}$ is the function defined in Proposition B.4.5.

Using the associativity of matrix multiplication and the fact that, by Proposition B.4.5, $g^{(\mathbf{A})}$ is a semigroup homomorphism, we obtain that $\left(\varphi_t^{(\mathbf{A})}\right)_{t \in [0,+\infty)}$ is a semiflow.

Since

$$\|\varphi_{s_m}^{(\mathbf{A})}(\mathbf{x}_m) - \varphi_s^{(\mathbf{A})}(\mathbf{x})\|_1 \leq \|g_{s_m}^{(\mathbf{A})}\mathbf{x}_m - g_s^{(\mathbf{A})}\mathbf{x}_m\|_1 + \|g_s^{(\mathbf{A})}\mathbf{x}_m - g_s^{(\mathbf{A})}\mathbf{x}\|_1$$

$$\leq \|g_{s_m}^{(\mathbf{A})} - g_s^{(\mathbf{A})}\|_1 \|\mathbf{x}_m\|_1 + \|g_s^{(\mathbf{A})}\|_1 \|\mathbf{x}_m - \mathbf{x}\|_1 = \|g_{s_m}^{(\mathbf{A})} - g_s^{(\mathbf{A})}\|_1 + \|\mathbf{x}_m - \mathbf{x}\|_1$$

for all convergent sequences $(s_m)_{m\in\mathbb{N}}$ and $(\mathbf{x}_m)_{m\in\mathbb{N}}$ of elements of $[0, +\infty)$ and $\mathbb{P}_n$, respectively, where $s = \lim_{m\to+\infty} s_m$ and $\mathbf{x} = \lim_{m\to+\infty} \mathbf{x}_m$, and since, by Proposition B.4.5, the semigroup homomorphism $g^{(\mathbf{A})}$ is continuous, it follows that $\left(\varphi_t^{(\mathbf{A})}\right)_{t \in [0,+\infty)}$ is a continuous semiflow.

Taking into consideration that

$$\|\varphi_{s_m}^{(\mathbf{A})}(\mathbf{x}) - \varphi_s^{(\mathbf{A})}(\mathbf{x})\|_1 = \|g_{s_m}^{(\mathbf{A})}\mathbf{x} - g_s^{(\mathbf{A})}\mathbf{x}\|_1 \leq \|g_{s_m}^{(\mathbf{A})} - g_s^{(\mathbf{A})}\|_1 \|\mathbf{x}\|_1 = \|g_{s_m}^{(\mathbf{A})} - g_s^{(\mathbf{A})}\|_1$$

for every $\mathbf{x} \in \mathbb{P}_n$ and every convergent sequence $(s_m)_{m\in\mathbb{N}}$ of elements of $[0, +\infty)$, where $s = \lim_{m\to+\infty} s_m$, and using again the fact that $g^{(\mathbf{A})}$ is continuous, we obtain that the semiflow $\left(\varphi_t^{(\mathbf{A})}\right)_{t \in [0,+\infty)}$ is continuous with respect to $t \in [0, +\infty)$ uniformly in $\mathbf{x} \in \mathbb{P}_n$.

Finally, since

$$\|\varphi_t^{(\mathbf{A})}(\mathbf{x}_m) - \varphi_t^{(\mathbf{A})}(\mathbf{x})\|_1 = \|g_t^{(\mathbf{A})}(\mathbf{x}_m - \mathbf{x})\|_1 \leq \|g_t^{(\mathbf{A})}\|_1 \|\mathbf{x}_m - \mathbf{x}\|_1 = \|\mathbf{x}_m - \mathbf{x}\|_1$$

for every $t \in [0, +\infty)$ and every convergent sequence $(x_m)_{m\in\mathbb{N}}$ of elements of $\mathbb{P}_n$, where $\mathbf{x} = \lim_{m\to+\infty} \mathbf{x}_m$, it follows that $\left(\varphi_t^{(\mathbf{A})}\right)_{t \in [0,+\infty)}$ is equicontinuous with respect to $t \in [0, +\infty)$.

For future reference, we summarize the information about $\varphi^{(\mathbf{A})}$ that we have obtained here in the following proposition:

Proposition B.4.7. *Let $\mathbf{A} \in \mathbb{S}_n$ and let $\varphi^{(\mathbf{A})} : [0, +\infty) \times \mathbb{P}_n \to \mathbb{P}_n$ be defined by $\varphi^{(\mathbf{A})}(t, \mathbf{x}) = g_t^{(\mathbf{A})}\mathbf{x}$ (or $\varphi_t^{(\mathbf{A})}(\mathbf{x}) = g_t^{(\mathbf{A})}\mathbf{x}$ in subscript notation) for every $(t, \mathbf{x}) \in [0, +\infty) \times \mathbb{P}_n$. Then $\left(\varphi_t^{(\mathbf{A})}\right)_{t \in [0,+\infty)}$ is a well-defined (in the sense that $\varphi_t^{(\mathbf{A})}(\mathbf{x}) \in \mathbb{P}_n$ for every $\mathbf{x} \in \mathbb{P}_n$) continuous semiflow. Moreover, $\left(\varphi_t^{(\mathbf{A})}\right)_{t \in [0,+\infty)}$ is continuous with* *respect to* $t \in [0, +\infty)$ *uniformly in* $\mathbf{x} \in \mathbb{P}_n$ *and is equicontinuous with respect to* $t \in [0, +\infty)$.

B.4.2   Exponential Flows on Spaces of Cosets of $\mathrm{SL}(n, \mathbb{R})$

In order to define exponential flows on spaces of cosets of $\mathrm{SL}(n, \mathbb{R})$ we need a result (Proposition B.4.8 below) on the trace of $n \times n$ matrices, $n \in \mathbb{N}$, $n \geq 2$.

Let $n \in \mathbb{N}$, $n \geq 2$, and let $\mathbf{A} \in \mathbf{M}(n, \mathbb{R})$, $\mathbf{A} = [a_{ij}]_{\substack{i=1,2,\ldots,n \\ j=1,2,\ldots,n}}$. The *trace* of $\mathbf{A}$ is the sum of the entries on the main diagonal of $\mathbf{A}$ and is denoted $\mathrm{trace}(\mathbf{A})$. Thus, $\mathrm{trace}(\mathbf{A}) = \sum_{i=1}^{n} a_{ii}$. The following proposition is well-known (for a proof see, for instance, the proof of formula (8), p. 106, of F. W. Warner's book [128]).

Proposition B.4.8. $\det(\exp(\mathbf{A})) = e^{\mathrm{trace}(\mathbf{A})}$ *for every* $n \times n$ *matrix* $\mathbf{A}$, $n \in \mathbb{N}$, $n \geq 2$. *In particular, if* $\mathrm{trace}(\mathbf{A}) = 0$, *then* $\exp(\mathbf{A}) \in \mathrm{SL}(n, \mathbb{R})$.

The matrices $\mathbf{A} \in \mathbf{M}(n, \mathbb{R})$ which have the property that $\mathrm{trace}(\mathbf{A}) = 0$ will be called *trace zero matrices*.

Now, let $\mathbf{A} \in \mathbf{M}(n, \mathbb{R})$ be a trace zero matrix. Then, obviously, $t\mathbf{A}$ is a trace zero matrix for every $t \in \mathbb{R}$; hence, the map $\mathbf{g}^{(\mathbf{A})} : \mathbb{R} \to \mathrm{SL}(n, \mathbb{R})$, $\mathbf{g}_t^{(\mathbf{A})} = \exp(t\mathbf{A})$ for every $t \in \mathbb{R}$, is well-defined (in the sense that $\mathbf{g}_t^{(\mathbf{A})} \in \mathrm{SL}(n, \mathbb{R})$ for every $t \in \mathbb{R}$). Note that $\mathbf{g}^{(\mathbf{A})}$ is a group homomorphism.

Next, let M be a closed subgroup of $\mathrm{SL}(n, \mathbb{R})$. Then the maps $u : \mathbb{R} \times (\mathrm{SL}(n, \mathbb{R})/M)_{\mathrm{L}} \to (\mathrm{SL}(n, \mathbb{R})/M)_{\mathrm{L}}$ defined by $u(t, \mathbf{x}M) = \mathbf{g}_t^{(\mathbf{A})}\mathbf{x}M$ (or $u_t(\mathbf{x}M) = \mathbf{g}_t^{(\mathbf{A})}\mathbf{x}M$) for every $(t, \mathbf{x}M) \in \mathbb{R} \times (\mathrm{SL}(n, \mathbb{R})/M)_{\mathrm{L}}$, and $v : \mathbb{R} \times (\mathrm{SL}(n, \mathbb{R})/M)_{\mathrm{R}} \to (\mathrm{SL}(n, \mathbb{R})/M)_{\mathrm{R}}$ defined by $v(t, M\mathbf{x}) = M\mathbf{x}\mathbf{g}_t^{(\mathbf{A})}$ (or $v_t(M\mathbf{x}) = M\mathbf{x}\mathbf{g}_t^{(\mathbf{A})}$) for every $(t, M\mathbf{x}) \in \mathbb{R} \times (\mathrm{SL}(n, \mathbb{R})/M)_{\mathrm{R}}$ are flows of the type described in Example A.3.2. Since, by Proposition A.2.4, $(\mathrm{SL}(n, \mathbb{R})/M)_{\mathrm{L}}$ and $(\mathrm{SL}(n, \mathbb{R})/M)_{\mathrm{R}}$ are (locally compact separable metrizable) topological spaces, and since the group homomorphism $\mathbf{g}^{(\mathbf{A})}$ is continuous (the continuity of $\mathbf{g}^{(\mathbf{A})}$ is a consequence of the fact that, by Proposition B.2.10, $\mathbf{g}^{(\mathbf{A})}$ is continuous when thought of as a one-parameter group with values in the Banach algebra $\mathbf{M}(n, \mathbb{R})$), it follows that we can apply (*b*) of Proposition A.3.3 in order to infer that the flows $(u_t)_{t \in \mathbb{R}}$ and $(v_t)_{t \in \mathbb{R}}$ are continuous. We call $(u_t)_{t \in \mathbb{R}}$ and $(v_t)_{t \in \mathbb{R}}$ the *exponential flows* on $(\mathrm{SL}(n, \mathbb{R})/M)_{\mathrm{L}}$ and $(\mathrm{SL}(n, \mathbb{R})/M)_{\mathrm{R}}$, respectively, defined by the matrix $\mathbf{A}$.

The study of these exponential flows is quite challenging. Usually, they are studied in a significantly more general setting, even though, often, the subgroup M is assumed to be a lattice. One of the reasons for the interest in the case when M is a lattice is the fact that, if M is a lattice, the coset spaces are finite volume spaces, so the exponential flows have invariant probability measures.

Of particular interest among the exponential flows are the unipotent ones. In our setting, an exponential flow defined on a coset space of $\mathrm{SL}(n, \mathbb{R})$ is *unipotent* if the matrix $\mathbf{A}$ that defines the flow is *nilpotent* (that is, if there exists a $k \in \mathbb{N}$ such that $\mathbf{A}^k$ is the zero matrix). Various results about unipotent flows have applications in number theory.

The reader interested in learning more about exponential flows should consult Chapters 5 and 6 of Bekka and Mayer's monograph [10] and Starkov's book [113]. The papers by Brezin and Moore [16], Margulis [75], Moore [79], and Starkov [109–112] deal with flows that are not necessarily unipotent. Unipotent flows are discussed in the papers by Dani [24–27], Margulis [72–75], Ratner [90–95], and Shah [106]. Naturally, the reader should also consult the references given in the above-mentioned works.

We will now briefly describe several quite impressive results of Ratner [92, 93] that we will use in the book in order to illustrate our results. It is important to stress that Ratner obtained her results in a much more general setting than our setting here, and that our setting is by no means the only interesting setting for the results to follow.

Theorem B.4.9 (Ratner [92]). *Let $n \in \mathbb{N}$, $n \geq 2$, let Γ be a lattice in $\mathrm{SL}(n, \mathbb{R})$, and let $\mathbf{v} = (v_t)_{t \in \mathbb{R}}$ be a unipotent flow on $(\mathrm{SL}(n, \mathbb{R})/\Gamma)_{\mathrm{R}}$. Then for every $\Gamma\mathbf{X} \in (\mathrm{SL}(n, \mathbb{R})/\Gamma)_{\mathrm{R}}$ there exists an ergodic invariant probability measure $\varepsilon_{\Gamma\mathbf{X}}$ for $(v_t)_{t \in \mathbb{R}}$ such that the following assertion holds true: for every real-valued continuous bounded function f on $(\mathrm{SL}(n, \mathbb{R})/\Gamma)_{\mathrm{R}}$ that vanishes at infinity, the limit*

$$\lim_{t \to +\infty} \frac{1}{t} \int_0^t f(v_t(\Gamma\mathbf{X})) \, \mathrm{d}t \ \textit{exists and is equal to} \ \int_{(\mathrm{SL}(n,\mathbb{R})/\Gamma)_{\mathrm{R}}} f \, \mathrm{d}\varepsilon_{\Gamma\mathbf{X}}.$$

It can be shown that, under the conditions of Theorem B.4.9, any invariant ergodic probability measure for the flow $(v_t)_{t \in \mathbb{R}}$ is of the form $\varepsilon_{\Gamma\mathbf{X}}$ for some $\Gamma\mathbf{X} \in (\mathrm{SL}(n, \mathbb{R})/\Gamma)_{\mathrm{R}}$. Thus, the set $\{\varepsilon_{\Gamma\mathbf{X}} \mid \Gamma\mathbf{X} \in (\mathrm{SL}(n, \mathbb{R})/\Gamma)_{\mathrm{R}}\}$ is the set of all invariant ergodic probability measures of $(v_t)_{t \in \mathbb{R}}$.

Moreover, in addition to Theorem B.4.9 and various other results, Ratner obtained in [92] a remarkably nice and complete characterization of the invariant ergodic probability measures of $(v_t)_{t \in \mathbb{R}}$ in terms of the algebraic, topological, and measure theoretical setting in which $(v_t)_{t \in \mathbb{R}}$ is defined. Let us briefly outline this characterization.

To this end, let L be a topological commutative group, let $n \in \mathbb{N}$, $n \geq 2$, let $h : L \to \mathrm{SL}(n, \mathbb{R})$, $h = (h_t)_{t \in L}$ be a continuous group homomorphism, and let $\mathbf{w} = (w_t)_{t \in L}$ be the action of L on $(\mathrm{SL}(n, \mathbb{R})/\Gamma)_{\mathrm{R}}$ defined by $w_t(\Gamma\mathbf{X}) = \Gamma\mathbf{X}h_t$ for every $t \in L$ and $\Gamma\mathbf{X} \in (\mathrm{SL}(n, \mathbb{R})/\Gamma)_{\mathrm{R}}$, where, as before, Γ is a lattice in $\mathrm{SL}(n, \mathbb{R})$.

Let $\Gamma\mathbf{X} \in (\mathrm{SL}(n, \mathbb{R})/\Gamma)_{\mathrm{R}}$ and let $\mathbf{S}$ be a closed subgroup of $\mathrm{SL}(n, \mathbb{R})$. We say that the orbit-closure $\overline{\theta_{\mathbf{w}}(\Gamma\mathbf{X})}$ of $\Gamma\mathbf{X}$ under the action of $\mathbf{w}$ is *homogeneous* with respect to $\mathbf{S}$ if $h(L) \subseteq \mathbf{S}$, $\overline{\theta_{\mathbf{w}}(\Gamma\mathbf{X})} = \Gamma\mathbf{X}\mathbf{S}$, and if $(\mathbf{X}\mathbf{S}\mathbf{X}^{-1}) \cap \Gamma$ is a lattice in $\mathbf{X}\mathbf{S}\mathbf{X}^{-1}$. Note that $\mathbf{X}\mathbf{S}\mathbf{X}^{-1}$ is a closed subgroup of $\mathrm{SL}(n, \mathbb{R})$, so $\mathbf{X}\mathbf{S}\mathbf{X}^{-1}$ is a locally compact separable metric group in its own right when endowed with the topology induced by the topology of $\mathrm{SL}(n, \mathbb{R})$ on $\mathbf{X}\mathbf{S}\mathbf{X}^{-1}$, and $(\mathbf{X}\mathbf{S}\mathbf{X}^{-1}) \cap \Gamma$ is a discrete subgroup of $\mathbf{X}\mathbf{S}\mathbf{X}^{-1}$, so, the assumption that $(\mathbf{X}\mathbf{S}\mathbf{X}^{-1}) \cap \Gamma$ is a lattice in $\mathbf{X}\mathbf{S}\mathbf{X}^{-1}$ makes sense.

In general, given $\Gamma\mathbf{X} \in (\mathrm{SL}(n,\mathbb{R})/\Gamma)_\mathrm{R}$, we say that $\overline{\mathcal{O}_\mathbf{w}(\Gamma\mathbf{X})}$ is *homogeneous* if there exists a closed subgroup $\mathbf{S}$ of $\mathrm{SL}(n,\mathbb{R})$ such that $\overline{\mathcal{O}_\mathbf{w}(\Gamma\mathbf{X})}$ is homogeneous with respect to $\mathbf{S}$.

Now let $\Gamma\mathbf{X} \in (\mathrm{SL}(n,\mathbb{R})/\Gamma)_\mathrm{R}$, let $\mathbf{S}$ be a closed subgroup of $\mathrm{SL}(n,\mathbb{R})$, assume that $\overline{\mathcal{O}_\mathbf{w}(\Gamma\mathbf{X})}$ is homogeneous with respect to $\mathbf{S}$, and set $\Gamma_1 = (\mathbf{XSX}^{-1}) \cap \Gamma$. Since Γ_1 is a lattice in $\mathbf{XSX}^{-1}$, using the discussion preceding Proposition B.1.4, we obtain that there exists a unique $(\mathbf{XSX}^{-1})$-invariant probability measure $\nu_{((\mathbf{XSX}^{-1})/\Gamma_1)_\mathrm{R}}$ on $((\mathbf{XSX}^{-1})/\Gamma_1)_\mathrm{R}$. Let $\varphi : ((\mathbf{XSX}^{-1})/\Gamma_1)_\mathrm{R} \to \Gamma\mathbf{XS}$ be defined by $\varphi(\Gamma_1\mathbf{XYX}^{-1}) = \Gamma\mathbf{XY}$ for every $\mathbf{Y} \in \mathbf{S}$ (that is, for every $\Gamma_1\mathbf{XYX}^{-1} \in ((\mathbf{XSX}^{-1})/\Gamma_1)_\mathrm{R}$). It can be shown that φ is a homeomorphism onto $\Gamma\mathbf{XS}$, where on $\Gamma\mathbf{XS}$ we consider the topology induced by the topology of $(\mathrm{SL}(n,\mathbb{R})/\Gamma)_\mathrm{R}$, and $((\mathbf{XSX}^{-1})/\Gamma_1)_\mathrm{R}$ is endowed with its standard topology as a coset space; therefore, on the Borel σ-algebra of $\Gamma\mathbf{XS}$ we can define a probability measure $\mu^{(1)}_{\overline{\mathcal{O}_\mathbf{w}(\Gamma\mathbf{X})}}$ as follows:

$$\mu^{(1)}_{\overline{\mathcal{O}_\mathbf{w}(\Gamma\mathbf{X})}}(A) = \nu_{((\mathbf{XSX}^{-1})/\Gamma_1)_\mathrm{R}}(\varphi^{-1}(A)) \text{ for every Borel subset } A \text{ of } \Gamma\mathbf{XS} (= \overline{\mathcal{O}_\mathbf{w}(\Gamma\mathbf{X})}).$$

Finally, the measure $\mu^{(1)}_{\overline{\mathcal{O}_\mathbf{w}(\Gamma\mathbf{X})}}$ can be extended in a standard manner to a probability measure $\mu_{\overline{\mathcal{O}_\mathbf{w}(\Gamma\mathbf{X})}}$ on the Borel σ-algebra of $(\mathrm{SL}(n,\mathbb{R})/\Gamma)_\mathrm{R}$ by defining $\mu_{\overline{\mathcal{O}_\mathbf{w}(\Gamma\mathbf{X})}}(A) = \mu^{(1)}_{\overline{\mathcal{O}_\mathbf{w}(\Gamma\mathbf{X})}}(A \cap \overline{\mathcal{O}_\mathbf{w}(\Gamma\mathbf{X})})$ for every Borel subset A of $(\mathrm{SL}(n,\mathbb{R})/\Gamma)_\mathrm{R}$. Note that the support of $\mu_{\overline{\mathcal{O}_\mathbf{w}(\Gamma\mathbf{X})}}$ is the orbit-closure $\overline{\mathcal{O}_\mathbf{w}(\Gamma\mathbf{X})}$ of $\Gamma\mathbf{X}$. It can be shown that the measure $\mu_{\overline{\mathcal{O}_\mathbf{w}(\Gamma\mathbf{X})}}$ is independent of the closed subgroup $\mathbf{S}$ with respect to which the orbit-closure is homogeneous, and of the choice of coset representative $\mathbf{X}$ in $\Gamma\mathbf{X}$. We call $\mu_{\overline{\mathcal{O}_\mathbf{w}(\Gamma\mathbf{X})}}$ the *probability measure defined by the homogeneous orbit-closure* $\overline{\mathcal{O}_\mathbf{w}(\Gamma\mathbf{X})}$.

A natural question at this point is whether or not the orbit-closures of a unipotent flow as in Theorem B.4.9 are homogeneous. Actually, Raghunathan conjectured (in a significantly more general setting and for significantly more general actions) that all orbit-closures are homogeneous; the conjecture is known as Raghunathan's topological conjecture (see [92]). In her paper [92], Ratner proved that Raghunathan's topological conjecture is true in its full generality. In the setting of Theorem B.4.9, this means that all the orbit-closures of a unipotent flow as in the theorem are homogeneous. Moreover, she also proved that the invariant ergodic probability measures that appear in the theorem are the probability measures defined by the corresponding (homogeneous) orbit-closures. For future reference, we state this result in the following theorem:

Theorem B.4.10 (Ratner [92]). *Let* $\mathbf{v} = (v_t)_{t \in \mathbb{R}}$ *be a unipotent flow on* $(\mathrm{SL}(n,\mathbb{R})/\Gamma)_\mathrm{R}$ *as in Theorem B.4.9. Then, for every* $\Gamma\mathbf{X} \in (\mathrm{SL}(n,\mathbb{R})/\Gamma)_\mathrm{R}$, *the orbit-closure* $\overline{\mathcal{O}_\mathbf{v}(\Gamma\mathbf{X})}$ *of* $\Gamma\mathbf{X}$ *under the action of* $\mathbf{v}$ *is homogeneous and* $\varepsilon_{\Gamma\mathbf{X}} = \mu_{\overline{\mathcal{O}_\mathbf{v}(\Gamma\mathbf{X})}}$, *where* $\varepsilon_{\Gamma\mathbf{X}}$ *is the invariant ergodic probability measure discussed in Theorem B.4.9, and* $\mu_{\overline{\mathcal{O}_\mathbf{v}(\Gamma\mathbf{X})}}$ *is the probability measure defined by* $\overline{\mathcal{O}_\mathbf{v}(\Gamma\mathbf{X})}$.

In [92], Ratner also obtained "discrete-time" versions of Theorems B.4.9 and B.4.10 (again, in a significantly more general setting). In order to state these "discrete-time" results, we need some preparation.

An $n \times n$ matrix $\mathbf{Y} \in \mathrm{SL}(n, \mathbb{R})$ is said to be *unipotent* if $(\mathbf{Y} - \mathbf{I}_n)^l = \mathbf{0}$ for some $l \in \mathbb{N}$. Typical examples of such elements are $\exp(\mathbf{A})$, where $\mathbf{A}$ is a trace zero nilpotent $n \times n$ matrix.

Now let $\mathbf{Y} \in \mathrm{SL}(n, \mathbb{R})$ and let $\mathbf{h}^{(\mathbf{Y})} : \mathbb{Z} \to \mathrm{SL}(n, \mathbb{R})$ be defined by $\mathbf{h}_k^{(\mathbf{Y})} = \mathbf{Y}^k$ for every $k \in \mathbb{Z}$. Clearly, $\mathbf{h}^{(\mathbf{Y})} = \left(\mathbf{h}_k^{(\mathbf{Y})} \right)_{k \in \mathbb{Z}}$ is a group homomorphism, and we call it the *standard group homomorphism defined by* $\mathbf{Y}$. Given a lattice Γ in $\mathrm{SL}(n, \mathbb{R})$, we can then define an action $w^{(\mathbf{Y})} : \mathbb{Z} \times (\mathrm{SL}(n, \mathbb{R})/\Gamma)_{\mathrm{R}} \to (\mathrm{SL}(n, \mathbb{R})/\Gamma)_{\mathrm{R}}$ as follows: $w_k^{(\mathbf{Y})}(\Gamma\mathbf{X}) = \Gamma\mathbf{X}\mathbf{Y}^k$ for every $k \in \mathbb{Z}$ and $\Gamma\mathbf{X} \in (\mathrm{SL}(n, \mathbb{R})/\Gamma)_{\mathrm{R}}$. We call $w^{(\mathbf{Y})} = \left(w_k^{(\mathbf{Y})} \right)_{k \in \mathbb{Z}}$ the *standard action of* $\mathbb{Z}$ *on* $(\mathrm{SL}(n, \mathbb{R})/\Gamma)_{\mathrm{R}}$ *defined by* $\mathbf{Y}$, or the *standard* $\mathbb{Z}$*-action on* $(\mathrm{SL}(n, \mathbb{R})/\Gamma)_{\mathrm{R}}$ *defined by* $\mathbf{Y}$.

Let $\mathbf{Y} \in \mathrm{SL}(n, \mathbb{R})$, let Γ be a lattice in $\mathrm{SL}(n, \mathbb{R})$, and let $w^{(\mathbf{Y})}$ be the standard $\mathbb{Z}$-action on $(\mathrm{SL}(n, \mathbb{R})/\Gamma)_{\mathrm{R}}$ defined by $\mathbf{Y}$.

Since $w^{(\mathbf{Y})}$ is an action, it makes sense to consider the invariant Borel measures of $w^{(\mathbf{Y})}$ (we defined these measures before Example B.1.2). Also, it makes sense to consider finite invariant ergodic measures of the $\mathbb{Z}$-action $w^{(\mathbf{Y})}$; these measures are defined in exactly the same manner as the finite invariant ergodic measures of semiflows and flows, which were defined before Example B.1.6, provided that we assume that $\mathbb{T} = \mathbb{Z}$.

The "discrete-time" versions of Theorems B.4.9 and B.4.10 are, respectively, Theorems B.4.11 and B.4.12 below.

Theorem B.4.11 (Ratner [92]). *Let $n \in \mathbb{N}$, $n \geq 2$, let Γ be a lattice in $\mathrm{SL}(n, \mathbb{R})$, let $\mathbf{Y}$ be a unipotent element of $\mathrm{SL}(n, \mathbb{R})$, and let $w^{(\mathbf{Y})} = \left(w_k^{(\mathbf{Y})} \right)_{k \in \mathbb{Z}}$ be the standard $\mathbb{Z}$-action on $(\mathrm{SL}(n, \mathbb{R})/\Gamma)_{\mathrm{R}}$ defined by $\mathbf{Y}$. Then for every $\Gamma\mathbf{X} \in (\mathrm{SL}(n, \mathbb{R})/\Gamma)_{\mathrm{R}}$ there exists an invariant ergodic probability measure $\varepsilon_{\Gamma\mathbf{X}}$ for the action $w^{(\mathbf{Y})}$ such that the sequence $\left(\dfrac{1}{m} \displaystyle\sum_{k=0}^{m-1} f\left(w_k^{(\mathbf{Y})}(\Gamma\mathbf{X}) \right) \right)_{m \in \mathbb{N}}$ converges to $\displaystyle\int_{(\mathrm{SL}(n,\mathbb{R})/\Gamma)_{\mathrm{R}}} f \, \mathrm{d}\varepsilon_{\Gamma\mathbf{X}}$ for every continuous bounded real-valued function f on $(\mathrm{SL}(n, \mathbb{R})/\Gamma)_{\mathrm{R}}$ that vanishes at infinity.*

As in the case of the unipotent flows discussed in Theorem B.4.9, it can be shown that any invariant ergodic probability measure for $w^{(\mathbf{Y})}$ is of the form $\varepsilon_{\Gamma\mathbf{X}}$ for some $\Gamma\mathbf{X} \in (\mathrm{SL}(n, \mathbb{R})/\Gamma)_{\mathrm{R}}$.

Theorem B.4.12 (Ratner [92]). *As in Theorem B.4.11, let $\mathbf{Y}$ be a unipotent element of $\mathrm{SL}(n, \mathbb{R})$, and let $w^{(\mathbf{Y})}$ be the standard $\mathbb{Z}$-action on $(\mathrm{SL}(n, \mathbb{R})/\Gamma)_{\mathrm{R}}$ defined by $\mathbf{Y}$. Then, for every $\Gamma\mathbf{X} \in (\mathrm{SL}(n, \mathbb{R})/\Gamma)_{\mathrm{R}}$, the orbit-closure $\overline{\mathcal{O}_{w^{(\mathbf{Y})}}(\Gamma\mathbf{X})}$ of $\Gamma\mathbf{X}$ under the action of $w^{(\mathbf{Y})}$ is homogeneous and $\varepsilon_{\Gamma\mathbf{X}} = \mu_{\overline{\mathcal{O}_{w^{(\mathbf{Y})}}(\Gamma\mathbf{X})}}$, where $\varepsilon_{\Gamma\mathbf{X}}$ is the invariant ergodic probability measure discussed in Theorem B.4.11 and $\mu_{\overline{\mathcal{O}_{w^{(\mathbf{Y})}}(\Gamma\mathbf{X})}}$ is the probability measure defined by $\overline{\mathcal{O}_{w^{(\mathbf{Y})}}(\Gamma\mathbf{X})}$.*

Bibliography

1. Y.A. Abramovich, C.D. Aliprantis, *An Invitation to Operator Theory*. Graduate Studies in Mathematics, vol. 50 (American Mathematical Society, Providence, 2002)
2. Y.A. Abramovich, C.D. Aliprantis, *Problems in Operator Theory*. Graduate Studies in Mathematics, vol. 51 (American Mathematical Society, Providence, 2002)
3. C.D. Aliprantis, O. Burkinshaw, *Positive Operators* (Academic, Orlando, 1985)
4. T. Alkurdi, S.C. Hille, O. van Gaans, Ergodicity and stability of a dynamical system perturbed by impulsive random interventions. J. Math. Anal. Appl. **407**, 480–494 (2013)
5. M.F. Barnsley, Iterated function systems for lossless data compression, in *Fractals in Multimedia*, ed. by M.F. Barnsley, D. Saupe, E.R. Vrskay (Springer, New York, 2002), pp. 33–63
6. M.F. Barnsley, S. Demko, Iterated function systems and the global construction of fractals. Proc. R. Soc. Lond. A **399**, 243–275 (1985)
7. M.F. Barnsley, S.G. Demko, J.H. Elton, J.S. Geronimo, Invariant measures for Markov processes arising from iterated function systems with place-dependent probabilities. Ann. Inst. H. Poincaré, Probab. et Statist. **24**, 367–394 (1988); **25**, 589–590 (1989)
8. H. Bauer, *Probability Theory* (de Gruyter, Berlin/New York, 1996)
9. M. Beboutoff, Markoff chains with a compact state space. Rec. Math. (Mat. Sbornik) N. S. **10**(52), 213–238 (1942)
10. M.B. Bekka, M. Mayer, *Ergodic Theory and Topological Dynamics of Group Actions on Homogeneous Spaces* (Cambridge University Press, Cambridge, 2000)
11. L. Beznea, N. Boboc, *Potential Theory and Right Processes* (Kluwer, Dordrecht, 2004)
12. P. Billingsley, *Probability and Measure*, 2nd edn. (Wiley, New York/Chichester/Brisbane/Toronto/Singapore, 1986)
13. P. Billingsley, *Convergence of Probability Measures*, 2nd edn. (Wiley, New York/Chichester/Weinheim/Brisbane/Singapore/Toronto, 1999)
14. R.M. Blumenthal, R.K. Getoor, *Markov Processes and Potential Theory* (Academic, New York, 1968)
15. A.A. Borovkov, *Ergodicity and Stability of Stochastic Processes* (Wiley, Chichester, 1998)
16. J. Brezin, C.C. Moore, Flows on homogeneous spaces: a new look. Am. J. Math. **103**, 571–613 (1981)
17. P.M. Centore, E.R. Vrscay, Continuity of attractors and invariant measures for iterated function systems. Can. Math. Bull. **37**, 315–329 (1994)
18. K.L. Chung, J.B. Walsh, *Markov Processes, Brownian Motion, and Time Symmetry*, 2nd edn. (Springer, New York, 2005)
19. A.H. Clifford, D.D. Miller, Semigroups having zeroid elements. Am. J. Math. **70**, 117–125 (1948)

R. Zaharopol, *Invariant Probabilities of Transition Functions*, Probability and Its Applications 44, DOI 10.1007/978-3-319-05723-1,
© Springer International Publishing Switzerland 2014

20. D.L. Cohn, *Measure Theory* (Birkhäuser, Boston, 1980)
21. J.B. Conway, *A Course in Functional Analysis*. Graduate Texts in Mathematics, vol. 96 (Springer, New York, 1985)
22. I.P. Cornfeld, S.V. Fomin, Ya.G. Sinai, *Ergodic Theory* (Springer, Berlin/New York, 1982)
23. S.G. Dani, Invariant measures of horospherical flows on noncompact homogeneous spaces. Invent. Math. **47**, 101–138 (1978)
24. S.G. Dani, On invariant measures, minimal sets and a lemma of Margulis. Invent. Math. **51**, 239–260 (1979)
25. S.G. Dani, On orbits of unipotent flows on homogeneous spaces. Ergod. Theory Dyn. Syst. **4**, 25–34 (1984)
26. S.G. Dani, On orbits of unipotent flows on homogeneous spaces, II. Ergod. Theory Dyn. Syst. **6**, 167–182 (1986)
27. S.G. Dani, Flows on homogeneous spaces: a review, in *Ergodic Theory of $\mathbb{Z}^d$-Actions, Warwick, 1993–1994*, ed. by M. Pollicott, K. Schmidt. London Mathematical Society Lecture Note Series, vol. 228 (Cambridge University Press, Cambridge, 1996), pp. 63–112
28. S.G. Dani, J. Smillie, Uniform distribution of horocycle orbits for Fuchsian groups. Duke Math. J. **51**, 185–194 (1984)
29. J.-D. Deuschel, D.W. Stroock, *Large Deviations*. Pure and Applied Mathematics, vol. 137 (Academic, San Diego, 1989)
30. N. Dunford, J.T. Schwartz, *Linear Operators, Part I: General Theory* (Wiley, New York, 1988)
31. E.B. Dynkin, *Markov Processes*, vol. 2 (Springer, Berlin/Göttingen/Heidelberg, 1965)
32. A. Edalat, Power domains and iterated function systems. Inf. Comput. **124**, 182–197 (1996)
33. E.Yu. Emel'yanov, R. Zaharopol, Convergence of Lotz-Räbiger nets of operators on spaces of continuous functions. Rev. Roum. Math. Pures Appl. **55**, 1–26 (2010)
34. K.-J. Engel, R. Nagel, *One-Parameter Semigroups for Linear Evolution Equations* (Springer, New York, 2000)
35. S.N. Ethier, T.G. Kurtz, *Markov Processes: Characterization and Convergence* (Wiley, Hoboken, 2005)
36. M. Fukushima, Y. Oshima, M. Takeda, *Dirichlet Forms and Symmetric Markov Processes* (de Gruyter, Berlin/New York, 1994)
37. H. Furstenberg, The unique ergodicity of the horocycle flow, in *Recent Advances in Topological Dynamics*, ed. by A. Beck. Lecture Notes in Mathematics, vol. 318 (Springer, Berlin/Heidelberg/New York, 1973), pp. 95–115
38. H. Furstenberg, *Recurrence in Ergodic Theory and Combinatorial Number Theory* (Princeton University Press, Princeton, 1981)
39. I.I. Gihman, A.V. Skorohod, *The Theory of Stochastic Processes*, vol. 2 (Springer, Berlin/Heidelberg, 1975)
40. W.H. Gottschalk, G.A. Hedlund, *Topological Dynamics*. Colloquium Publications, vol. 36 (American Mathematical Society, Providence, 1955)
41. G. Hedlund, Dynamics of geodesic flows. Bull. Am. Math. Soc. **45**, 241–260 (1939)
42. O. Hernández-Lerma, J.B. Lasserre, Ergodic theorems and ergodic decompositions for Markov chains. Acta Appl. Math. **54**, 99–119 (1988)
43. O. Hernández-Lerma, J.B. Lasserre, *Markov Chains and Invariant Probabilities* (Birkhäuser, Basel, 2003)
44. E. Hewitt, K.A. Ross, *Abstract Harmonic Analysis*, vol. 1, 2nd edn. (Springer, Berlin/Heidelberg, 1979)
45. H. Heyer, *Probability Measures on Locally Compact Groups* (Springer, Berlin/Heidelberg, 1977)
46. H. Heyer, *Structural Aspects in the Theory of Probability: A Primer in Probabilities on Algebraic-Topological Structures* (World Scientific, Singapore/River Edge/London, 2004)
47. T. Hida, *Brownian Motion*. Applications of Mathematics, vol. 11 (Springer, New York, 1980)
48. G. Högnäs, A. Mukherjea, *Probability Measures on Semigroups: Convolution Products, Random Walks, and Random Matrices* (Plenum Press, New York, 1995)

49. M. Iosifescu, S. Grigorescu, *Dependence with Complete Connections and Its Applications* (Cambridge University Press, Cambridge, 1990)
50. M. Iosifescu, R. Theodorescu, *Random Processes and Learning* (Springer, New York, 1969)
51. R. Kapica, T. Szarek, M. Ślęczka, On a unique ergodicity of some Markov processes. Potential Anal. **36**, 589–606 (2012)
52. T. Komorowski, S. Peszat, T. Szarek, On ergodicity of some Markov processes. Ann. Probab. **38**, 1401–1443 (2010)
53. U. Krengel, *Ergodic Theorems* (de Gruyter, Berlin/New York, 1985)
54. N. Kryloff, N. Bogoliouboff, La théorie générale de la mesure dans son application à l'étude des systèmes de la mécanique nonlinéaires. Ann. Math. **38**, 65–113 (1937)
55. S. Lang, *Algebra* (Addison-Wesley, Reading/Menlo Park/London/Don Mills, 1971)
56. S. Lang, $SL_2(\mathbf{R})$ (Springer, New York, 1985)
57. A. Lasota, M.C. Mackey, *Chaos, Fractals, and Noise: Stochastic Aspects of Dynamics* (Springer, New York, 1994)
58. A. Lasota, J. Myjak, Semifractals. Bull. Pol. Acad. Sci. Math. **44**, 5–21 (1996)
59. A. Lasota, J. Myjak, Markov operators and fractals. Bull. Pol. Acad. Sci. Math. **45**, 197–210 (1997)
60. A. Lasota, J. Myjak, Semifractals on Polish spaces. Bull. Pol. Acad. Sci. Math. **46**, 179–196 (1998)
61. A. Lasota, J. Myjak, Fractals, semifractals and Markov operators. Int. J. Bifurc. Chaos **9**, 307–325 (1999)
62. A. Lasota, J. Myjak, Attractors of multifunctions. Bull. Pol. Acad. Sci. Math. **48**, 319–334 (2000)
63. A. Lasota, T. Szarek, Lower bound technique in the theory of a stochastical differential equation. J. Differ. Equ. **231**, 513–533 (2006)
64. A. Lasota, J.A. Yorke, Lower bound technique for Markov operators and iterated function systems. Random Comput. Dyn. **2**, 41–77 (1994)
65. T.M. Liggett, *Interacting Particle Systems* (Springer, New York, 1985)
66. T.M. Liggett, *Stochastic Interacting Systems: Contact, Voter and Exclusion Processes* (Springer, Berlin, 1999)
67. M. Lin, Conservative Markov processes on a topological space. Isr. J. Math. **8**, 165–186 (1970)
68. W.A.J. Luxemburg, A.C. Zaanen, *Riesz Spaces I* (North-Holland, Amsterdam, 1971)
69. P. Mandl, *Analytical Treatment of One-Dimensional Markov Processes* (Academia, Prague/Springer, Berlin/Heidelberg/New York, 1968)
70. R. Mañé, *Ergodic Theory and Differentiable Dynamics* (Springer, Berlin/Heidelberg, 1987)
71. M.B. Marcus, J. Rosen, *Markov Processes, Gaussian Processes, and Local Times*. Cambridge Studies in Advanced Mathematics, vol. 100 (Cambridge University Press, New York, 2006)
72. G.A. Margulis, On the action of unipotent groups in the space of lattices, in *Lie Groups and Their Representations*, Budapest, 1971, ed. by I.M. Gelfand (Akadémiai Kiado, Budapest/Wiley, New York/Toronto, 1975), pp. 365–370
73. G.A. Margulis, Lie groups and ergodic theory, in *Algebra – Some Current Trends*, Varna, 1986, ed. by L.L. Avramov, K.B. Tchakerian. Lecture Notes in Mathematics, vol. 1352 (Springer, Berlin/Heidelberg, 1988), pp. 130–146
74. G.A. Margulis, Discrete subgroups and ergodic theory, in *Number Theory, Trace Formulas and Discrete Groups*, ed. by K.E. Aubert, E. Bombieri, D. Goldfeld. Proceedings of the Conference in Honor of A. Selberg, Oslo, 1987 (Academic, San Diego/London, 1989), pp. 377–398
75. G.A. Margulis, Dynamical and ergodic properties of subgroup actions on homogeneous spaces with applications to number theory, in *Proceedings of the International Congress of Mathematicians*, Kyoto, 1990, vol. 1 (Mathematical Society of Japan/Springer, Kyoto, 1991), pp. 193–215
76. P.-A. Meyer, *Probabilités et Potentiel* (Hermann, Paris, 1966)
77. S.P. Meyn, R.L. Tweedie, *Markov Chains and Stochastic Stability* (Springer, London, 1993)

78. D. Montgomery, L. Zippin, *Topological Transformation Groups* (Interscience, New York/London, 1955)
79. C.C. Moore, Ergodicity of flows on homogeneous spaces. Am. J. Math. **88**, 154–178 (1966)
80. J. Myjak, T. Szarek, On Hausdorff dimension of invariant measures arising from non-contractive iterated function systems. Annali di Matematica **181**, 223–237 (2002)
81. J. Neveu, *Bases Mathématiques du Calcul des Probabilités* (Masson, Paris, 1964)
82. M. Nicol, N. Sidorov, D. Broomhead, On the fine structure of stationary measures in systems which contract-on-average. J. Theor. Probab. **15**, 715–730 (2002)
83. F. Norman, *Markov Processes and Learning Models* (Academic, New York/London, 1972)
84. O. Onicescu, G. Mihoc, Sur les chaines de variables statistiques. Bull. Sci. Math. **59**, 174–192 (1935)
85. J.C. Oxtoby, Ergodic sets. Bull. Am. Math. Soc. **58**, 116–136 (1952)
86. R.R. Phelps, *Lectures on Choquet's Theorem*, 2nd edn. Lecture Notes in Mathematics, vol. 1757 (Springer, Berlin/Heidelberg, 2001)
87. M.S. Raghunathan, *Discrete Subgroups of Lie Groups* (Springer, Berlin/Heidelberg, 1972)
88. M.M. Rao, *Stochastic Processes: General Theory* (Kluwer, Dordrecht, 1995)
89. M. Ratner, Ergodic theory in hyperbolic space. Contemp. Math. **26**, 309–334 (1984)
90. M. Ratner, Strict measure rigidity for unipotent subgroups of solvable groups. Invent. Math. **101**, 449–482 (1990)
91. M. Ratner, On measure rigidity of unipotent subgroups of semisimple groups. Acta Math. **165**, 229–309 (1990)
92. M. Ratner, Raghunathan's topological conjecture and distributions of unipotent flows. Duke Math. J. **63**, 235–280 (1991)
93. M. Ratner, Distribution rigidity for unipotent actions on homogeneous spaces. Bull. Am. Math. Soc. **24**, 321–325 (1991)
94. M. Ratner, On Raghunathan's measure conjecture. Ann. Math. **134**, 545–607 (1991)
95. M. Ratner, Raghunathan's conjectures for SL(2, $\mathbb{R}$). Isr. J. Math. **80**, 1–31 (1992)
96. H. Reiter, *Classical Harmonic Analysis and Locally Compact Groups* (Oxford University Press, Oxford, 1968)
97. D. Revuz, *Markov Chains* (North-Holland, Amsterdam/American Elsevier, New York, 1975)
98. D. Revuz, M. Yor, *Continuous Martingales and Brownian Motion*, 2nd edn. (Springer, Berlin/Heidelberg, 1994)
99. C. Robinson, *Dynamical Systems: Stability, Symbolic Dynamics, and Chaos* (CRC Press, Boca Raton/Ann Arbor/London/Tokyo, 1995)
100. L.C.G. Rogers, D. Williams, *Diffusions, Markov Processes, and Martingales, Volume 1: Foundations*, 2nd edn. (Wiley, Chichester, 1994)
101. V.A. Rohlin, On the fundamental ideas of measure theory. (Russian) Mat. Sbornik N. S. **25**(67), 107–150 (1949) (English Translation in: Am. Math. Soc. Transl. **71**, 1–55 (1952))
102. M. Rosenblatt, *Markov Processes. Structure and Asymptotic Behavior* (Springer, Berlin/Heidelberg, 1971)
103. H.L. Royden, *Real Analysis*, 3rd edn. (Macmillan, New York, 1988)
104. D.J. Rudolph, *Fundamentals of Measurable Dynamics: Ergodic Theory on Lebesgue Spaces* (Clarendon, Oxford, 1990)
105. H.H. Schaefer, *Banach Lattices and Positive Operators* (Springer, Berlin/New York, 1974)
106. N.A. Shah, Uniformly distributed orbits of certain flows on homogeneous spaces. Math. Ann. **289**, 315–334 (1991)
107. Ya.G. Sinai, Gibbs measures in ergodic theory. Russ. Math. Surv. **27**, 21–69 (1972)
108. A.V. Skorokhod, Topologically recurrent Markov chains: ergodic properties. Theor. Probab. Appl. **31**, 563–571 (1986)
109. A.N. Starkov, The ergodic behavior of flows on homogeneous spaces. Sov. Math. Dokl. **28**, 675–676 (1983)
110. A.N. Starkov, Nonergodic homogeneous flows. Sov. Math. Dokl. **33**, 729–731 (1986)
111. A.N. Starkov, Ergodic decomposition of flows on homogeneous spaces of finite volume. Math. USSR Sb. **68**, 483–502 (1991)

112. A.N. Starkov, Minimal sets of homogeneous flows. Ergod. Theory Dyn. Syst. **15**, 361–377 (1995)
113. A.N. Starkov, *Dynamical Systems on Homogeneous Spaces* (AMS, Providence, 2000)
114. Ö. Stenflo, Ergodic theorems for time-dependent random iterations of functions, in *Fractals and Beyond*, Valletta, 1998 (World Scientific Publishing, River Edge, 1998), pp. 129–136
115. Ö. Stenflo, Markov chains in random environments and random iterated function systems. Trans. Am. Math. Soc. **353**, 3547–3562 (2001)
116. Ö. Stenflo, A note on a theorem of Karlin. Stat. Probab. Lett. **54**, 183–187 (2001)
117. Ö. Stenflo, Uniqueness of invariant measures for place-dependent random iterations of functions, in *Fractals in Multimedia*, ed. by M.F. Barnsley, D. Saupe, E.R. Vrscay (Springer, New York, 2002), pp. 13–32
118. Ö. Stenflo, Uniqueness in g-measures. Nonlinearity **16**, 403–410 (2003)
119. D.W. Stroock, *Probability Theory, An Analytic View* (Cambridge University Press, New York, 1993)
120. T. Szarek, Invariant measures for Markov operators with application to function systems. Studia Math. **154**, 207–222 (2003)
121. T. Szarek, The uniqueness of invariant measures for Markov operators. Studia Math. **189**, 225–233 (2008)
122. T. Szarek, D.T.H. Worm, Ergodic measures of Markov semigroups with the e-property. Ergodic Theory Dyn. Syst. **32**, 1117–1135 (2012)
123. T. Szarek, M. Ślęczka, M. Urbański, On stability of velocity vectors for some passive tracer models. Bull. Lond. Math. Soc. **42**, 923–936 (2010)
124. K. Taira, *Semigroups, Boundary Value Problems and Markov Processes* (Springer, Berlin/Heidelberg, 2004)
125. E.R. Vrscay, From fractal image compression to fractal-based methods in mathematics, in *Fractals in Multimedia*, ed. by M.F. Barnsley, D. Saupe, E.R. Vrscay (Springer, New York, 2002), pp. 65–106
126. P. Walters, *An Introduction to Ergodic Theory* (Springer, New York, 1982)
127. S. Warner, *Modern Algebra*, vol. 1 (Prentice Hall, Englewood Cliffs, 1965)
128. F.W. Warner, *Foundations of Differentiable Manifolds and Lie Groups* (Scott, Foresman&Co., Glenview, 1971)
129. A. Weil, *L'Intégration dans les Groupes Topologiques et ses Applications*, 2nd edn. (Hermann, Paris, 1951)
130. D.T.H. Worm, *Semigroups on Spaces of Measures*. Ph.D. Thesis, Thomas Stieltjes Institute for Mathematics, Leiden (2010)
131. D.T.H. Worm, S.C. Hille, Ergodic decompositions associated to regular Markov operators on Polish spaces. Ergod. Theory Dyn. Syst. **31**, 571–597 (2011)
132. D.T.H. Worm, S.C. Hille, An ergodic decomposition defined by regular jointly measurable Markov semigroups on Polish spaces. Acta Appl. Math. **116**, 27–53 (2011)
133. D.T.H. Worm, S.C. Hille, Equicontinuous families of Markov operators on complete separable metric spaces with applications to ergodic decompositions and existence, uniqueness, and stability of invariant measures (preprint)
134. D. Worm, R. Zaharopol, On an ergodic decomposition defined in terms of certain generators: the case when the generator is defined on the entire space $C_b(X)$, to appear in Rev. Roum. Math. Pures Appl. 1, 2015
135. K. Yosida, Markoff processes with a stable distribution. Proc. Imp. Acad. Tokyo **16**, 43–48 (1940)
136. K. Yosida, Simple Markoff process with a locally compact phase space. Math. Jpn. **1**, 99–103 (1948)
137. K. Yosida, *Lectures on Differential and Integral Equations* (Interscience, New York, 1960)
138. K. Yosida, *Functional Analysis*, 3rd edn. (Springer, New York, 1971)
139. A.C. Zaanen, *Introduction to Operator Theory in Riesz Spaces* (Springer, Berlin/Heidelberg, 1997)

140. R. Zaharopol, Iterated function systems generated by strict contractions and place-dependent probabilities. Bull. Pol. Acad. Sci. Math. **48**, 429–438 (2000)
141. R. Zaharopol, Fortet-Mourier norms associated with some iterated function systems. Stat. Probab. Lett. **50**, 149–154 (2000)
142. R. Zaharopol, Attractive probability measures and their supports. Rev. Roum. Math. Pures Appl. **49**, 397–418 (2004)
143. R. Zaharopol, *Invariant Probabilities of Markov-Feller Operators and Their Supports* (Birkhäuser, Basel/Boston/Berlin, 2005)
144. R. Zaharopol, Invariant probabilities of convolution operators. Rev. Roum. Math. Pures Appl. **50**, 387–405 (2005)
145. R. Zaharopol, Equicontinuity and the existence of attractive probability measures for some iterated function systems. Rev. Roum. Math. Pures Appl. **52**, 259–286 (2007); Erratum in Vol. **52**(6) (2007)
146. R. Zaharopol, An ergodic decomposition defined by transition probabilities. Acta Appl. Math. **104**, 47–81 (2008)
147. R. Zaharopol, Vector integrals and a pointwise mean ergodic theorem for transition functions. Rev. Roum. Math. Pures Appl. **53**, 63–78 (2008)
148. R. Zaharopol, Transition probabilities, transition functions, and an ergodic decomposition. Bull. Transilv. Univ. Braş. **2**(51), 149–170 (2009)

Index

R. Zaharopol, *Invariant Probabilities of Transition Functions*, Probability and Its
Applications 44, DOI 10.1007/978-3-319-05723-1,
© Springer International Publishing Switzerland 2014

GPSR Compliance
The European Union's (EU) General Product Safety Regulation (GPSR) is a set of rules that requires consumer products to be safe and our obligations to ensure this.

If you have any concerns about our products, you can contact us on

ProductSafety@springernature.com

In case Publisher is established outside the EU, the EU authorized representative is:

Springer Nature Customer Service Center GmbH
Europaplatz 3
69115 Heidelberg, Germany